T0180313

IFIP Advances in Information and Communication Technology 452

Editor-in-Chief

Kai Rannenberg, Goethe University Frankfurt, Germany

Editorial Board

Foundation of Computer Science
Jacques Sakarovitch, Télécom ParisTech, France
Software: Theory and Practice
Michael Goedicke, University of Duisburg-Essen, Germany
Education
Arthur Tatnall, Victoria University, Melbourne, Australia
Information Technology Applications
Erich J. Neuhold, University of Vienna, Austria
Communication Systems
Aiko Pras, University of Twente, Enschede, The Netherlands
System Modeling and Optimization
Fredi Tröltzsch, TU Berlin, Germany
Information Systems
Jan Pries-Heje, Roskilde University, Denmark
ICT and Society
Diane Whitehouse, The Castlegate Consultancy, Malton, UK
Computer Systems Technology
Ricardo Reis, Federal University of Rio Grande do Sul, Porto Alegre, Brazil
Security and Privacy Protection in Information Processing Systems
Yuko Murayama, Iwate Prefectural University, Japan
Artificial Intelligence
Tharam Dillon, La Trobe University, Melbourne, Australia
Human-Computer Interaction
Jan Gulliksen, KTH Royal Institute of Technology, Stockholm, Sweden
Entertainment Computing
Matthias Rauterberg, Eindhoven University of Technology, The Netherlands

IFIP – The International Federation for Information Processing

IFIP was founded in 1960 under the auspices of UNESCO, following the First World Computer Congress held in Paris the previous year. An umbrella organization for societies working in information processing, IFIP's aim is two-fold: to support information processing within its member countries and to encourage technology transfer to developing nations. As its mission statement clearly states,

> *IFIP's mission is to be the leading, truly international, apolitical organization which encourages and assists in the development, exploitation and application of information technology for the benefit of all people.*

IFIP is a non-profitmaking organization, run almost solely by 2500 volunteers. It operates through a number of technical committees, which organize events and publications. IFIP's events range from an international congress to local seminars, but the most important are:

- The IFIP World Computer Congress, held every second year;
- Open conferences;
- Working conferences.

The flagship event is the IFIP World Computer Congress, at which both invited and contributed papers are presented. Contributed papers are rigorously refereed and the rejection rate is high.

As with the Congress, participation in the open conferences is open to all and papers may be invited or submitted. Again, submitted papers are stringently refereed.

The working conferences are structured differently. They are usually run by a working group and attendance is small and by invitation only. Their purpose is to create an atmosphere conducive to innovation and development. Refereeing is also rigorous and papers are subjected to extensive group discussion.

Publications arising from IFIP events vary. The papers presented at the IFIP World Computer Congress and at open conferences are published as conference proceedings, while the results of the working conferences are often published as collections of selected and edited papers.

Any national society whose primary activity is about information processing may apply to become a full member of IFIP, although full membership is restricted to one society per country. Full members are entitled to vote at the annual General Assembly, National societies preferring a less committed involvement may apply for associate or corresponding membership. Associate members enjoy the same benefits as full members, but without voting rights. Corresponding members are not represented in IFIP bodies. Affiliated membership is open to non-national societies, and individual and honorary membership schemes are also offered.

More information about this series at http://www.springer.com/series/6102

Daoliang Li · Yingyi Chen (Eds.)

Computer and Computing Technologies in Agriculture VIII

8th IFIP WG 5.14 International Conference, CCTA 2014 Beijing, China, September 16–19, 2014 Revised Selected Papers

Editors
Daoliang Li
China Agricultural University
Beijing
China

Yingyi Chen
China Agricultural University
Beijing
China

ISSN 1868-4238 ISSN 1868-422X (electronic)
IFIP Advances in Information and Communication Technology
ISBN 978-3-319-79282-8 ISBN 978-3-319-19620-6 (eBook)
DOI 10.1007/978-3-319-19620-6

Springer Cham Heidelberg New York Dordrecht London
© IFIP International Federation for Information Processing 2015
Softcover re-print of the Hardcover 1st edition 2015
This work is subject to copyright. All rights are reserved by the Publisher, whether the whole or part of the material is concerned, specifically the rights of translation, reprinting, reuse of illustrations, recitation, broadcasting, reproduction on microfilms or in any other physical way, and transmission or information storage and retrieval, electronic adaptation, computer software, or by similar or dissimilar methodology now known or hereafter developed.
The use of general descriptive names, registered names, trademarks, service marks, etc. in this publication does not imply, even in the absence of a specific statement, that such names are exempt from the relevant protective laws and regulations and therefore free for general use.
The publisher, the authors and the editors are safe to assume that the advice and information in this book are believed to be true and accurate at the date of publication. Neither the publisher nor the authors or the editors give a warranty, express or implied, with respect to the material contained herein or for any errors or omissions that may have been made.

Printed on acid-free paper

Springer International Publishing AG Switzerland is part of Springer Science+Business Media
(www.springer.com)

Preface

The 8th International Conference on Computer and Computing Technologies in Agriculture (CCTA 2014) was held in Beijing, China, during September 16-19, 2014.

This conference was hosted by the China Agricultural University; Chinese Academy of Agricultural Mechanization Sciences; Chinese Academy of Agricultural Engineering; Beijing Academy of Agriculture and Forestry Science; and East China Jiaotong University. It was sponsored by the International Commission of Agricultural and Biosystems Engineering, Session VII; Chinese Society for Agricultural Machinery (CSAM); Chinese Society of Agricultural Engineering (CSAE); WG 5.14, International Federation for Information Processing; National Natural Science Foundation of China; Da Bei Nong Agricultural Education Fund; and the Wang Kuancheng Education Foundation.

In recent years, modern information technology and intelligent equipment techniques have spread into all the pre-/inter-/post-production processes in agriculture and are becoming an important means of transforming traditional agriculture and developing modern agriculture. This international academic exchange plays an important role in improving the innovation and development of intelligent agricultural information technology and in promoting the wide application of the Internet of Things and computing technology. CCTA aims to provide an academic platform for the integration of information on agricultural modernization, to share new research theories, methods, and achievements of sci-tech innovation and industrial technology progress, and to promote the understanding and cooperation among international communities and scientists. Eight International Conferences on Computer and Computing Technologies in Agriculture have been held since 2007.

The topics of CCTA 2014 cover the interesting theory and applications of all kinds of technology in agriculture, including intelligent sensing, monitoring, and automatic control technology models; the key technology and model of the Internet of Things; agricultural intelligent equipment technology; computer vision; computer graphics and virtual reality; computer simulation, optimization, and modeling; cloud computing and agricultural applications; agricultural, big data; decision support systems and expert system; technology and precision agriculture; the quality and safety of agricultural products; detection and tracing technology; and agricultural electronic commerce technology.

We selected the 81 best papers among the 216 papers submitted to CCTA 2014 for these proceedings. All papers underwent two reviews from the Special Interest Group on Advanced Information Processing in Agriculture (AIPA), IFIP. In these proceedings, creative thoughts and inspirations can be discovered, discussed, and disseminated. It is always exciting to have experts, professionals, and scholars with creative contributions getting together to share inspiring ideas and accomplish great developments in the field.

I would like to express my sincere thanks to all authors who submitted research papers to the conference. Finally, I would also like to express my sincere thanks to all speakers, session chairs, and attendees, both national and international, for their active participation and support of this conference.

April 2015 Daoliang Li

Organization

The 8th International Conference on Computer & Computing Technologies in Agriculture (CCTA 2014) was held during September 16-19, 2014, in Beijing, China.

Symposium Topics

Agricultural Information Sensing and Intelligent Control
Precision Agriculture Technology and Equipment
Agricultural Remote Sensing and Agricultural Aviation Application
Agricultural Intelligent Decision-making and Information Service Technology
Agricultural Product Safety Control and Traceability

Organizers

China Agricultural University
Chinese Academy of Agricultural Mechanization Sciences
Chinese Academy of Agricultural Engineering
Beijing Academy of Agriculture and Forestry Science
East China Jiaotong University

Sponsors

International Commission of Agricultural and Biosystems Engineering, Session VII
Chinese Society for Agricultural Machinery (CSAM)
Chinese Society of Agricultural Engineering (CSAE)
WG 5.14, International Federation for Information Processing
National Natural Science Foundation of China
Da Bei Nong Agricultural Education Fund
Wang Kuancheng Education Foundation

Organizing Committee

Chairs

Daoliang Li	China Agricultural University, China
Yande Liu	East China Jiaotong University

Invited Speaker

Daoliang Li	China Agricultural University, China
John Victor Stafford	8th Silsoe Solutions, International Society of Precision Agriculture
Nick Sigrimis	Agricultural University of Athens, Greece
Georg Staaks	Leibniz Institute of Freshwater Ecology and Inland Fisheries (IGB), Germany
Changying Li	University of Georgia
Lehmann Alexandra	EU Delegation in Beijing, China
Csukás Béla	8th Kaposvar University, Hungary
Divas Karimanzira	Fraunhofer Application Center System Technology, Germany
Arnfinn Morvik	Havforskningsinstituttet/Institute of Marine Research Norsk Marint Datasenter, Norway
Hongxin Cao	Jiangsu Academy of Agricultural Sciences, China
Shuangyin Liu	Guangdong Ocean University, China
Yang Chen	China Agricultural University
Yu Zhang	Wuhan University, China
Yuan Yuan	Chinese Academy of Sciences, China
Wenzhu Yang	Hebei University, China
Miao Zhang	China Agricultural University
Chaofan Wu	Zhejiang University, China
Yunlong Kong	Chinese Academy of Sciences, China
Wei Yang	China Agricultural University
Linjun Yu	Chinese Academy of Sciences, China
Shuangxi Liu	Shandong Agricultural University, China
Xiaochen Kang	Chinese Academy of Surveying and Mapping, China
Shahbaz Gul Hassan	China Agricultural University
Changyi Xiao	China Agricultural University

Secretary General

Lihong Shen	China Agricultural University
Dongbin Chen	East China Jiaotong University, China

Contents

High-Throughput Estimation of Yield for Individual Rice Plant Using Multi-angle RGB Imaging

Lingfeng Duan[1,2], Chenglong Huang[1,2], Guoxing Chen[3], Lizhong Xiong[4], Qian Liu[5], and Wanneng Yang[1,2,4,*]

[1] Agricultural Bioinformatics Key Laboratory of Hubei Province, Huazhong Agricultural University, Wuhan 430070, P.R. China
ywn@mail.hzau.edu.cn
[2] College of Engineering, Huazhong Agricultural University, Wuhan 430070, P.R. China
[3] MOA Key Laboratory of Crop Ecophysiology and Farming System in the Middle Reaches of the Yangtze River, Huazhong Agricultural University, Wuhan 430070, China
[4] National Key Laboratory of Crop Genetic Improvement and National Center of Plant Gene Research, Huazhong Agricultural University, Wuhan 430070, P.R. China
[5] Britton Chance Center for Biomedical Photonics, Wuhan National Laboratory for Optoelectronics-Huazhong University of Science and Technology, Wuhan 430074, P.R. China

Abstract. Modern breeding technologies are capable of producing hundreds of new varieties daily, so fast, simple and effective methods for screening valuable candidate plant materials are urgently needed. Final yield is a significant agricultural trait in rice breeding. In the screening and evaluation of the rice varieties, measuring and evaluating rice yield is essential. Conventional means of measuring rice yield mainly depend on manual determination, which is tedious, labor-intensive, subjective and error-prone, especially when large-scale plants were to be investigated. This paper presented an in vivo, automatic and high-throughput method to estimate the yield of individual pot-grown rice plant using multi-angle RGB imaging and image analysis. In this work, we demonstrated a new idea of estimating rice yield from projected panicle area, projected area of leaf and stem and fractal dimension. 5-fold cross validation showed that the predictive error was 7.45%. The constructed model achieved promising results on rice plants grown both in-door and out-door. The presented work has the potential of accelerating yield estimation and would be a promising impetus for plant phenomics.

Keywords: yield estimation, multi-angle imaging, individual rice plant, plant phenotyping, high-throughput.

1 Introduction

The growing population and the impact of changing climate on agriculture have put major crisis on world's food supply[1]. 70% more food is needed by 2050 according to

* Corresponding author.

© IFIP International Federation for Information Processing 2015
D. Li and Y. Chen (Eds.): CCTA 2014, IFIP AICT 452, pp. 1–12, 2015.
DOI: 10.1007/978-3-319-19620-6_1

the recent Declaration of World Summit on Food Security (www.fao.org/wsfs/world-summit/en/). As the staple food for over half of the world's population[2], rice production is significant for the food security of the world. Modern breeding technologies are capable of producing hundreds of new varieties daily, so fast, simple and effective methods for screening valuable candidate plant materials are urgently needed[3]. Phenotyping the interested populations is widely deemed as the most laborious, costly and technically challenging part in plant breeding. Plant phenotyping has become one of the new bottlenecks in plant science and plant breeding[4, 5]. Plant phenomics enables the acceleration of progress in linking gene function, phenotype, and environmental responses, which will eventually accelerate plant breeding to develop new germplasm to meet the future food demanding.

Digital image analysis has been proved an effective tool in plant phenomics. There has been research on the application of image analysis for phenotyping of an individual plant. Yang *et. al* adopted x-ray computed tomography to measure rice tiller number. Utilization of digital imaging and image analysis, Granier *et al.* estimate growth rates from projected leaf area for Arabidopsis plants [6]. Golzarian *et. al* [7]used the projected shoot area of the cereal plants on images at three orthogonal views and plant age in days after planting as predictors to infer the cereal biomass. Jones *et al.* quantified plant responses to water stress from thermal infrared images of crop canopies[8].

Final yield is a significant agricultural trait in rice breeding. In the screening and evaluation of the rice varieties, measuring and evaluating rice yield is essential. Conventional means of measuring rice yield mainly depend on manual determination. In this method, spikelets were threshed from the panicles and filled spikelets were weighed. Generally, manual measurement is tedious, labor-intensive, subjective and error-prone, especially when large-scale plants were to be investigated.

There has been work on yield-related trait measurement[9, 10]. In these works, rice panicles were harvested from the plant and spikelets were threshed manually or by the threshing machine. The major limit is that, the threshing machine had a threshing error of 3.33% for filled spikelets and 2.27% for total spikelets, and was prone to break the spikelets. Precision agriculture using remote sensing technology, in which the yield of the above-ground canopy for a large area is estimated from satellite and airborne images, is a wide-spread method for yield prediction [11-14]. However, this technology is not capable of estimating yield of an individual plant. For yield estimation of individual plant in vivo, academic publication is unavailable.

The objective of the present study is to develop an in vivo, automatic and high-throughput method to estimate the yield of individual pot-grown rice plant using multi-angle RGB imaging and image analysis. In this work, we demonstrated a model that used mixed variables of projected panicle area, projected area of leaf and stem, and fractal dimension achieved good performance in predicting rice yield. In order to improve the generalization ability of our method, we tested our method on rice plants grown both in-door and out-door and achieved promising results.

2 Experiments and Methods

2.1 Rice Sample Preparation

390 pot-grown rice plants, including 302 rice plants growing in the greenhouse (greenhouse rice) and 88 rice plants growing outdoor (outdoor rice), were used in this study. The samples were harvested at maturing stage and rice yield were measured manually. Yield varied from 7.04 *g*-43.10 *g* for the greenhouse rice and 23.68 *g*-58.43 *g* for the outdoor rice. The average yield for the greenhouse rice and outdoor rice were 27.55 g and 41.20 g, respectively. The samples were randomly split into two datasets at ratio of 2:1: a training dataset for model construction (260 plants) and a validation dataset for model validation (130 plants).

2.2 Hardware Setup and Image Acquisition

Previously, our group developed a high-throughput facility dubbed H-SMART for measuring rice tiller number[3]. The H-SMART facility used an industrial conveyor to transfer pot-grown rice plant to the imaging area for image acquisition. A rotation platform enabled the rice plants to be lifted and turned. A barcode scanner read barcode of each pot for indexing purpose. To allow measurement of rice yield, we incorporated visible light imaging into the H-SMART facility. As occlusion becomes problematic in rice plants at maturing stage, multi-angle imaging was adopted in the research. Plants were illustrated by fluorescent light tubes both from side and top. Images were taken at every 30°intervals by a Charge Coupled Device (CCD) camera (Stingray F-504C, Applied Vision Technologies, Germany) as the plant was rotated. For each rice plant, 12 images (2452×2056) at different angles were taken. Lighting conditions were constant throughout the research. Image acquisition was performed by NI-IMAQ Virtual Instruments (VI) Library for LabVIEW (National Instruments Corporation, USA). More details concerning the H-SMART system can be found in Yang et al. 2011[3].

2.3 Image Analysis and Feature Extraction

Glozarian et al. proved projected shoot area on two dimensional images was a good predictor of shoot biomass of the cereal plants[7]. And rice yield was the product of the harvest index and the biomass (more precisely, the total dry weight)[15]. It can be inferred that projected shoot area should have a good correlation with yield. Panicle was the organ where grains grow on. So projected panicle area may correlate well with yield. For this reason, we raised a hypothesis that separating the projected shoot area into two parts: projected panicle area and projected area of leaves and stems would reduce the bias in yield estimation. Gong demonstrated the feasibility of predicting yield per m2 for field rice from fractal dimension and texture features of the rice image[16]. In this study, we developed a model using projected panicle area, projected area of leaves and stems, fractal dimension and texture features.

The image analysis and feature extraction was performed in LabVIEW(National Instruments Corporation, USA). Fig. 1 shows the image analysis pipeline for this study. Firstly, the background was removed from the image to get the binary image of rice plant, from which projected area of rice plant (A) and information dimension (IFD) were calculated. Intensity image of the RGB image was extracted according to Eq. 1. Using the binary image as the mask, intensity component of rice plant, from which texture features including differential box counting dimension (DBC) [17] and 5 histogram features were calculated, can be obtained. 5 histogram features include the mean value (M), the standard deviation (S), the third moment (mu3), the uniformity (U) and the entropy (E). After that, pixel discrimination and region recognition were performed to obtain the binary image of panicle, from which projected panicle area (PA) was calculated.

$$I=(R+G+B)/3 \tag{1}$$

where R, G, B was the R, G, B component of the RGB image.

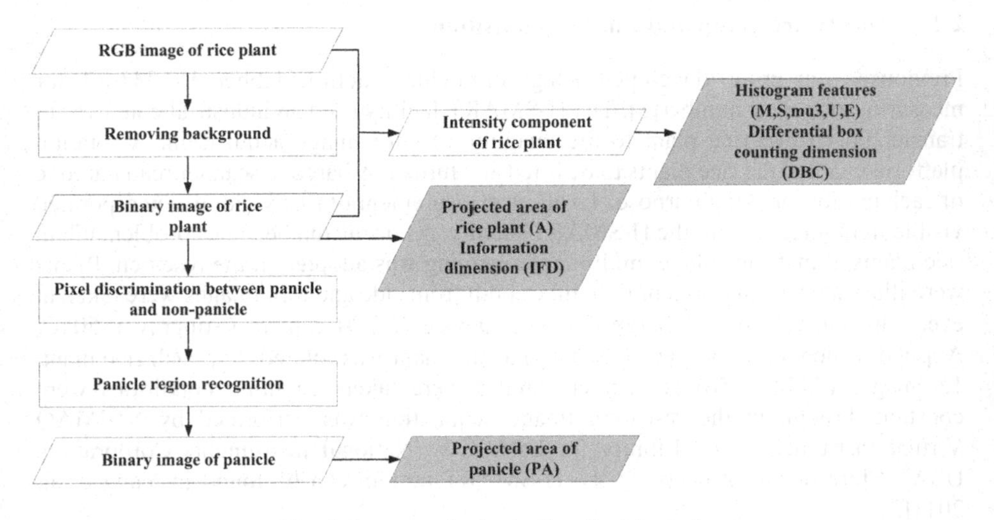

Fig. 1. Image analysis pipeline for this study

Fig. 2 illustrates the image processing diagram for removing background. First, the ExG component and i2 component were extracted according to Eqns. 2-3. The ExG component was used to segment the green parts of the rice plant, and the i2 component was used to segment the yellow parts. The images in this study were captured in a constant illustration environment, so both the ExG component and i2 component were binarized using fixed thresholds. Then the union of the two binary images was calculated. In the end, regions whose area was less than the predefined threshold were removed. Projected area of rice plant (A) was computed by counting the number of foreground pixels in the binary image of rice plant.

$$ExG = 2Ng-Nr-Nb \tag{2}$$

$$i2= (R-B)/2 \tag{3}$$

where Ng, Nr, Nb was the normalized r,g,b component, defined by Eqns. 4-6.

$$Nr=R/(R+G+B) \tag{4}$$

$$Ng=G/(R+G+B) \tag{5}$$

$$Nb=B/(R+G+B) \tag{6}$$

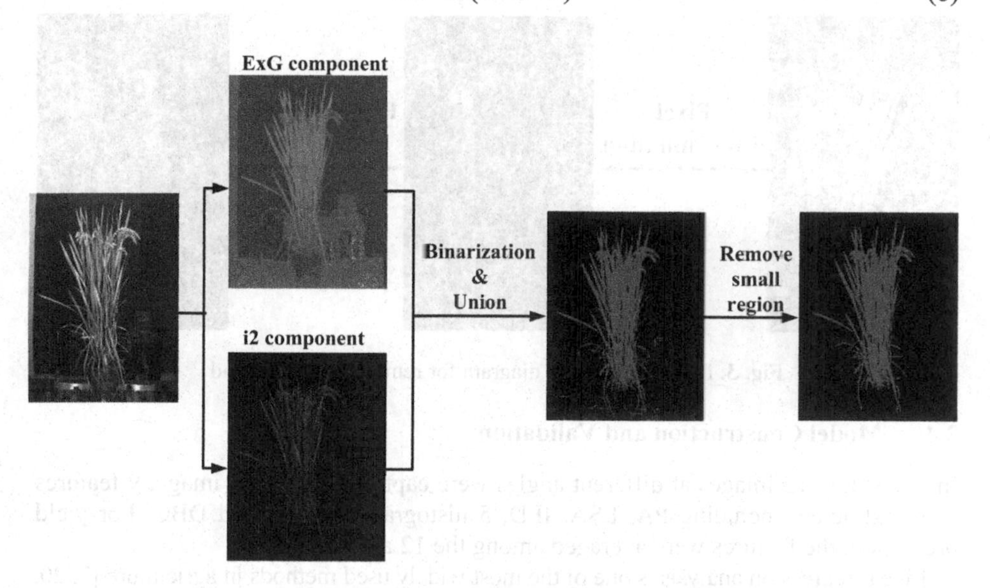

Fig. 2. Image processing diagram for removing background

Fractal dimension can be used to describe shape complexity[18]. The most commonly used method for calculating fractal dimension is box-counting method as it is simple to compute. In this research, the authors adopted information dimension, as it provides a more precise estimate of the fractal dimension than the box-counting method and meanwhile is easy to compute. IFD is calculated by plotting information I(ε) (defined as Eq. 7) against the natural logarithm of box size ε and IFD is computed as the slope of the regression line.

$$I_{\varepsilon} = \sum_{i=1}^{N_{\varepsilon}} p_i \ln \frac{1}{p_i} \tag{7}$$

Where N_{ε} is the number of boxes, ε is the box size, p_i is the probability of foreground pixels falling into the ith box.

Fig. 3 indicates the image processing diagram for generating binary image of panicle. Pixel discrimination between panicle and stem/leaf (non-panicle) based on discriminant analysis was performed to segment the panicle region[19]. There may be non-panicle region in the extracted panicle regions after pixel discrimination. So, region recognition was used to remove non-panicle regions. Rice yield is mainly contributed by effective panicles, which generally locates in the uppermost one third region of the rice plant. Therefore, in this study, a simple location discrimination method was used to recognize the panicle region. More specifically, the image was equally divided into three sub-regions according to the plant height. Regions that were located in the uppermost region were regarded as panicle region and other regions were treated as non-panicle regions. Projected panicle area (PA) was computed by counting the number of foreground pixels in the binary image of panicle. Projected area of leaf and stem (LSA) was computed by A minus PA.

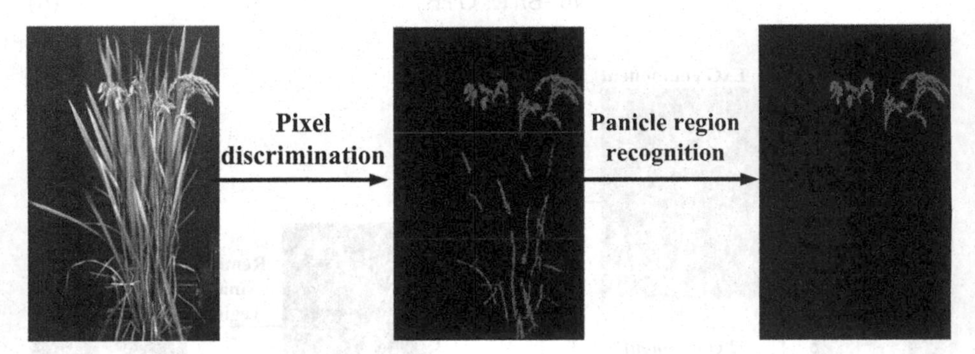

Fig. 3. Image processing diagram for removing background

2.4 Model Construction and Validation

In this study, 12 images at different angles were captured. For each image, 9 features were extracted, including PA, LSA, IFD, 5 histogram features and DBC. For yield prediction, the features were averaged among the 12 angles.

Linear regression analysis is one of the most widely used methods in agriculture [7, 20, 21]. In this research, linear regression was used to predict rice yield. Before modeling, all the data were log-transformed in order to meet normality assumptions and increase model performance[22]. For variable selection and optimal model identification, all-possible regression method was performed, using criteria of Akaike's information (AIC)[23], adjusted coefficient of determination (adjusted R^2), and prediction error sum of squares (PRESS statistic)[24]. Adjusted R^2 indicates the proportion of variation explained by the predictors and increases on adding a variable only when the variable decreases the residual mean square. Taking the complexity of the model into account, AIC is generally regarded as a measure of the goodness of fit of the model. PRESS statistic indicates the prediction ability of the model. A model with a highest adjusted R^2 and lowest AIC and PRESS statistic was selected as the final model. Other assumptions requisite to meet the specification of the multiple regression analysis, such as significance of the regression, significance of individual regression coefficients and multi-collinearity, were also checked. Model construction was performed in SAS (version 9.2, SAS Institute Inc., USA).

For model validation, the final model was tested on the validation dataset. In addition, cross validation technique was used to evaluate the prediction error of the model[7]. In the cross validation method, the samples are randomly partitioned into K approximately equal-sized K groups. In each iteration, the i-th (i=1,2…K) group is treated as validation set, and the remaining K-1 groups are used to fit the model. This process are repeated for K iterations, and the prediction errors, such as root mean square error (RMSE) and mean absolute percentage error (MAPE), are averaged over the iterations. Model validation was performed in Matlab ((Mathworks Inc., USA).

3 Results and Discussion

3.1 Image Analysis

Rice yield is mainly contributed by effective panicles, which generally locates in the uppermost one third region of the rice plant. In this study, the candidate panicle regions were firstly segmented using discriminant analysis. Then, the panicle regions were discriminated from non-panicle regions according to their location. After that, projected panicle area (PA) was computed. This method would be fit for most rice varieties. However, some rice varieties may have effective panicle that locates in the middle of the plant. In this occasion, some panicle regions may be mistaken treated as non-panicle regions and consequently the computed projected panicle area may be biased.

This paper mainly focuses on demonstrating the feasibility of estimating rice yield from projected panicle area, projected area of leaf and stem and fractal dimension. Improvement of projected panicle area extraction would be our future orientation.

3.2 Model Development and Validation

In variable selection and optimal model identification, we noticed that when the number of variables in the model was fixed, the model with the highest R^2 generally had the lowest AIC and PRESS statistic. Therefore, among models with the same number of variables, the model with the highest R^2 was selected for further consideration to accelerate the optimal model identification. This process reduced the number of candidate models to nine. Table 1 shows the performance of the nine candidate models.

When using alone, IFD was the most powerful variable for predicting yield, explaining 62.8% of the variation in the 260 training samples. Combination of PA and LSA outperformed other models with two variables, increasing adjusted R^2 to 0.87, and dramatically reducing AIC and PRESS. When the number of variables in the model was more than 4 (excluding constant), adding more variables to the model did not produce less AIC and PRESS. The model with four variables IFD, E, PA and LSA had the highest Adjusted R^2 and lowest AIC and PRESS. However, multi-collinearity was noticed in this model (VIF>5).Therefore, the model with three variables (Eq.8 and Table 2) was selected as the final model. Model diagnostics suggested that underlying assumptions of the regression model were met.

$$\text{Model 1: } \ln(Y) = a_0 + a_1 \cdot \ln(PA) + a_2 \cdot \ln(LSA) + a_3 \cdot \ln(IFD) \quad (8)$$

Table 1. Performance of the nine candidate model based on training samples

Number of variables	Adjusted R^2	AIC	PRESS	Variables in the model
1	0.6284	-940.80	7.01	IFD
2	0.8725	-1218.04	2.40	PA, LSA
3	0.8904	-1256.28	2.08	IFD, PA, LSA
4	0.8917	-1258.37	2.06	IFD, E, PA, LSA
5	0.8916	-1257.08	2.07	DBC, IFD, E, PA, LSA
6	0.8917	-1256.44	2.08	IFD, M, S, mu3, PA, LSA
7	0.8915	-1255.06	2.10	IFD, E, M, S, mu3, PA, LSA
8	0.8915	-1253.94	2.11	IFD, E, M, S, U, mu3, PA, LSA
9	0.8910	-1251.98	2.13	DBC, IFD, E, M, S, U, mu3, PA, LSA

Table 2 shows the summary of the final model based on the training samples. F-test value and its significance (P>F) showed that the overall fit of the model was significant. The t-test values and its significance (P>|t|) indicated that the individual independent variables were statistically significant. Variance inflation factor (VIF) for each individual independent variable showed that there was no multi-collinearity between the variables.

Table 2. Summary of the final model based on training samples

	Coefficients		t	P>|t|	VIF	F	P>F	R^2	RMSE
	Value	Standard error							
a_0	-10.613	0.39223	-27.06	<.0001	0	702.32	<.0001	0.89	2.71 g
a_1	0.4181	0.01892	22.1	<.0001	1.36293				
a_2	0.64369	0.0439	14.66	<.0001	2.09423				
a_3	3.67453	0.56133	6.55	<.0001	2.59348				

Fig. 4 shows the plot of residuals versus fitted values $\ln(\hat{Y})$ of the final model. It indicated that the residuals were evenly distributed at the two sides of y=0 and can be contained in a horizontal band, suggesting no obvious model defects were existed. In addition, no significant differences were presented among outdoor rice and greenhouse rice. This meant that the model performed well on both outdoor rice and greenhouse rice.

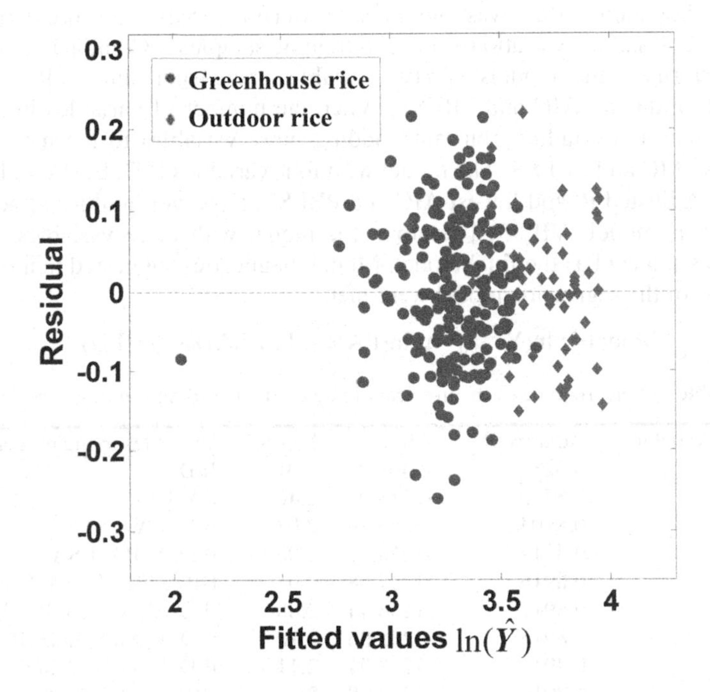

Fig. 4. Plot of residuals against the fitted values $\ln(\hat{Y})$ of the final model

The final model with three variables was then applied to the validation dataset and performed well (Fig. 5). The RMSE and MAPE for the validation dataset were 3.04 g and 7.74%.

5-fold cross validation showed that the RMSE and MAPE of the final model were 2.83 g and 7.45%.

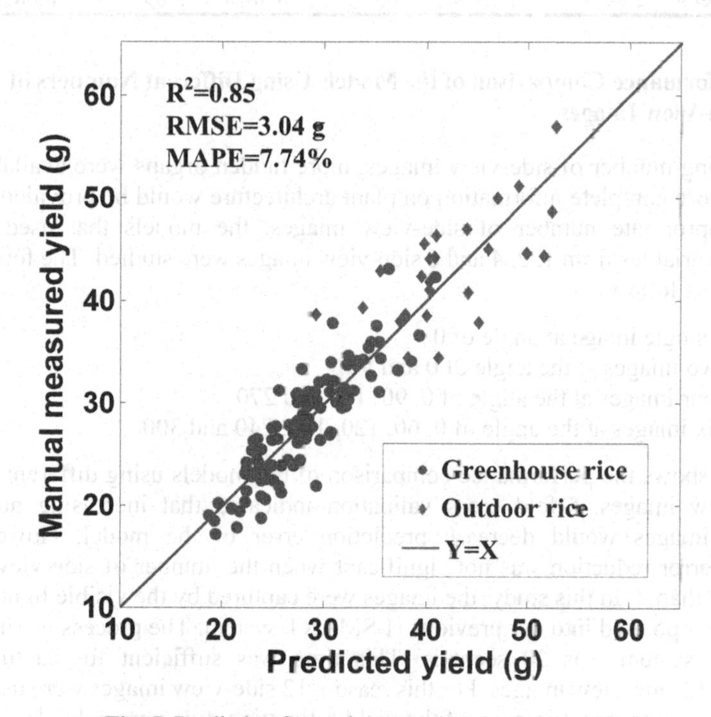

Fig. 5. Prediction for testing set using the final model

3.3 Performance Comparison of the Different Models

For comparison, we also tested the model using projected shoot area (Eq.9) and the model using projected panicle area and projected area of leaf and stem (Eq.10). And compared the two models with the model used in this study (Table 3). Separating A into PA and LSA significantly reduces the estimation error in yield prediction, which proved our hypothesis. And Adding fractal dimension into the model had a slight promotion on model performance. This was because that fractal dimension describes shape complexity of rice and can compensate the influence of organ occlusion problem at a certain degree.

$$\text{Model 2: } \ln(Y)=a_0+a_1\cdot\ln(A) \tag{9}$$

$$\text{Model 3: } \ln(Y)=a_0+a_1\cdot\ln(PA)+a_2\cdot\ln(LSA) \tag{10}$$

Table 3. Performance comparison of the three models

	Training set		Validation set		Cross validation	
	RMSE (*g*)	MAPE	RMSE (*g*)	MAPE	RMSE (*g*)	MAPE
Model 1	2.71	7.16%	3.04	7.74%	2.83	7.45%
Model 2	4.79	12.03%	5.48	13.82%	5.04	12.74%
Model 3	2.94	7.76%	3.35	8.48%	3.09	8.08%

3.4 Performance Comparison of the Models Using Different Numbers of Side-View Images

By increasing number of side-view images, more hidden organs were available to be seen and more complete information on plant architecture would be provided. To find out the appropriate number of side-view images, the models that used average prediction variables from 1, 2, 4 and 6 side-view images were studied. The four models were listed as follows.

Model 4: a single image at angle of 0
Model 5: two images at the angle of 0 and 90
Model 6: four images at the angle of 0, 90, 180 and 270
Model 7: six images at the angle of 0, 60, 120, 180, 240 and 300

Table 4 shows the performance comparison of the models using different numbers of side-view images. 5-fold cross validation indicated that increasing number of side-view images would decrease prediction error of the model. However, the prediction error reduction was not significant when the number of side-view images were larger than 4. In this study, the images were captured by the visible light imaging unit that incorporated into our previous H-SMART system. The processing time of the H-SMART system was 20 seconds. This time was sufficient for capturing and processing 12 side-view images. For this reason, 12 side-view images were used in this study. To promote the efficiency of the yield estimation, two perpendicular side-view images would generate acceptable results.

Table 4. Performance comparison of the models using different numbers of side-view images

Model	Number of side-view images	5-fold Cross validation	
		RMSE (*g*)	MAPE
Model 4	1	3.80	10.08%
Model 5	2	3.13	8.15%
Model 6	4	3.01	7.87 %
Model 7	6	2.92	7.77%
Model 1	12	2.83	7.45%

4 Conclusions

This paper presented a non-destructive and high-throughput method to estimate the yield of individual pot-grown rice plant using photonics-based technology. And we demonstrated a new idea of estimating rice yield from projected panicle area, projected area of leaf and stem and fractal dimension. 5-fold cross validation showed that the predictive error was 7.45% with testing rice plants grown both in-door and out-door. The presented work has the potential of accelerating yield estimation and would be a promising impetus for plant phenomics and high-throughput phenotyping.

Acknowledgment. This work was supported by grants from the Program for the National Program on High Technology Development (2013AA102403), the National Natural Science Foundation of China (30921091, 31200274), and the Fundamental Research Funds for the Central Universities (2013PY034), New Century Excellent Talents in University (No. NCET-10-0386).

Additional equations: Definition of the five histogram features.

$$M = \sum_{i=0}^{L-1} G_i p(G_i) \tag{1}$$

$$S = \sqrt{\sum_{i=0}^{L-1} (G_i - m)^2 p(G_i)} \tag{2}$$

$$mu3 = \sum_{i=0}^{L-1} (G_i - m)^3 p(G_i) \tag{3}$$

$$U = \sum_{i=0}^{L-1} p^2(G_i) \tag{4}$$

$$E = \sum_{i=0}^{L-1} p(G_i) \log_2 p(G_i) \tag{5}$$

Where G_i was the i-th graylevel, and $p(G_i)$ was the probability of G_i.

References

[1] Furbank, R.T.: Plant phenomics: from gene to form and function. Functional Plant Biology 36(10), 5–6 (2009)

[2] Zhou, Y., Zhu, J., Li, Z., et al.: Deletion in a quantitative trait gene qPE9-1 associated with panicle erectness improves plant architecture during rice domestication. Genetics 183(1), 315–324 (2009)

[3] Yang, W., Xu, X., Duan, L., et al.: High-throughput measurement of rice tillers using a conveyor equipped with x-ray computed tomography. Review of Scientific Instruments 82(2), 025102–025107 (2011)

[4] Cabrera-Bosquet, L., Crossa, J., von Zitzewitz, J., et al.: High-throughput phenotyping and genomic selection: The frontiers of crop breeding converge. Journal of Integrative Plant Biology 54(5), 312–320 (2012)

[5] Furbank, R.T., Tester, M.: Phenomics–technologies to relieve the phenotyping bottleneck. Trends in Plant Science 16(12), 635–644 (2011)

[6] Christine, G., Luis, A., Karine, C., et al.: PHENOPSIS, an automated platform for reproducible phenotyping of plant responses to soil water deficit in Arabidopsis thaliana permitted the identification of an accession with low sensitivity to soil water deficit. New Phytologist 169(3), 623–635 (2005)
[7] Golzarian, M.R., Frick, R.A., Rajendran, K., et al.: Accurate inference of shoot biomass from high-throughput images of cereal plants. Plant Methods 7(1), 2 (2011)
[8] Jones, H.G., Serraj, R., Loveys, B.R., et al.: Thermal infrared imaging of crop canopies for the remote diagnosis and quantification of plant responses to water stress in the field. Functional Plant Biology 36(11), 978–989 (2009)
[9] Duan, L., Yang, W., Bi, K., et al.: Fast discrimination and counting of filled/unfilled rice spikelets based on bi-modal imaging. Computers and Electronics in Agriculture 75(1), 196–203 (2011)
[10] Duan, L., Yang, W., Huang, C., et al.: A novel machine-vision-based facility for the automatic evaluation of yield-related traits in rice. Plant Methods 7(1), 44 (2011)
[11] Wang, Y.-P., Chang, K.-W., Chen, R.-K., et al.: Large-area rice yield forecasting using satellite imageries. International Journal of Applied Earth Observation and Geoinformation 12(1), 27–35 (2010)
[12] Wang, P., Sun, R., Zhang, J., et al.: Yield estimation of winter wheat in the North China Plain using the remote-sensing–photosynthesis–yield estimation for crops (RS–P–YEC) model. International Journal of Remote Sensing 32(21), 6335–6348 (2011)
[13] Jégo, G., Pattey, E., Liu, J.: Using Leaf Area Index, retrieved from optical imagery, in the STICS crop model for predicting yield and biomass of field crops. Field Crops Research 131, 63–74 (2012)
[14] Gitelson, A.A., Kaufman, Y.J., Merzlyak, M.N.: Use of a green channel in remote sensing of global vegetation from EOS-MODIS. Remote Sensing of Environment 58(3), 289–298 (1996)
[15] Shouichi, Y.: Fundamentals of rice crop science. International Rice Research Institute (1981)
[16] Hongju, G.: Estimating paddy yield based on fractal and image texture analysis. Nanjing Agricultural University (2008)
[17] Sarkar, N., Chaudhuri, B.B.: An efficient differential box-counting approach to compute fractal dimension of image. IEEE Transactions on Systems, Man and Cybernetics 24(1), 115–120 (1994)
[18] Bruno, O.M., De Oliveira Plotze, P., Falvo, M., et al.: Fractal dimension applied to plant identification. Information Sciences 178(12), 2722–2733 (2008)
[19] Okamoto, H., Lee, W.S.: Green citrus detection using hyperspectral imaging. Computers and Electronics in Agriculture 66(2), 201–208 (2009)
[20] Overman, J.P.M., Witte, H.J.L., Saldarriaga, J.G.: Evaluation of regression models for above-ground biomass determination in Amazon rainforest. Journal of Tropical Ecology 10(2), 207–218 (1994)
[21] Pompelli, M.F., Antunes, W.C., Ferreira, D.T.R.G., et al.: Allometric models for non-destructive leaf area estimation of Jatropha curcas. Biomass and Bioenergy 36, 77–85 (2012)
[22] Cristian, G., David, B., Kristin, S., et al.: Temperature-controlled organic carbon mineralization in lake sediments. Nature 466(7305), 478–481 (2010)
[23] Hirotugu, A.: A new look at the statistical model identification. IEEE Transactions on Automatic Control 19(6), 716–723 (1974)
[24] Montgomery, D.C., Peck, E.A., Vining, G.G.: Introduction to linear regression analysis. Wiley (2012)

A Method to Determine the Maximum Side Perspective of Satellite with the Constraints of Mapping Accuracy

Jihong Yang[1], Haiwei Li[2,4], Yin Zhan[1], Liangshu Shi[1], Jinqiang Wang[3], and Zhengchao Chen[4,*]

[1] Chinese Land Surveying and Planning Institute, Beijing 100035, China
[2] School of Geosciences and Info-Physics, Central South University, Changsha 410083, China
[3] Surveying and Mapping Bureau of Yunnan Province, Kunming 650034, Yunnan, China
[4] Institute of Remote Sensing and Digital Earth, Chinese Academy of Sciences, Beijing 100094, China
zcchen@ceode.ac.cn

Abstract. In order to shorten the revisit period, and improve the efficiency of imaging, currently, all the in orbit high resolution remote sensing satellites adopt side-view imaging technology. When satellites imaging with a side perspective, it would inevitably cause the degradation of image quality, like spatial resolution reduction, image deformation incensement, and the reduction of positioning accuracy. In addition, with the increase of the side perspective, the error of DEM data will be amplified in the process of ortho-rectification, which, thus, will bring bigger error for image mapping accuracy. To address this problem, embarked from the impact of side view and DEM precision on image point error, this paper used Quick bird image data of the same area but different side perspectives, and simulate the impacts of different DEM elevation error on different side view image. At last we put forward a method to determine the maximum side perspective with map precision constraints, and set up a conversion relationship among satellite side view, DEM error and image point error. Finally, the paper combines the errors of 1:50000 DEM, ASTERGDEM and the requirement of ortho-image drawing, discusses and gives the values of maximum side perspective in the mountains and plains.

Keywords: DEM accuracy, satellite side perspective, ortho-rectification, geolocation accuracy.

1 Introduction

Along with the continuous improvement of remote sensing satellites' spatial resolution, it also brought adverse effects of the image width and image acquisition efficiency reduction. In order to shorten the return cycle and to improve the efficiency of imaging, currently, on-orbit high resolution remote sensing satellites use the side-view imaging technology to improve the flexibility towards observation objects and

* Corresponding author.

© IFIP International Federation for Information Processing 2015
D. Li and Y. Chen (Eds.): CCTA 2014, IFIP AICT 452, pp. 13–22, 2015.
DOI: 10.1007/978-3-319-19620-6_2

image acquisition efficiency. As is well-known, when the satellites side-view imaging, the observation path is increased, the relative geometric relationship between satellite and ground objects is changed, which caused the degeneration of image quality [1] (such as the spatial resolution reduced, image deformation increase, etc.) According to the principle of projection, from a point in space (viewpoint) observing the surface, each point on the surface along with topographic relief will generate geometric distortion on the image. The specific performance on the optical remote sensing imageries is as follows: the ground point, which is away from the viewpoint and with higher elevation, will fall to the opposite side of the viewpoint[2]. Ortho-rectification is to make use of terrain elevation model (DEM) to perform terrain distortion correction for each pixel in image, and make the image meet the orthographic projection, thus eliminating this distortion.

According to the error analysis of satellite imagery ortho-rectification model, the main factors affect the accuracy of ortho-rectification are as follows: ortho-rectification model error, satellite orbital attitude error, control point accuracy, DEM error, etc[3]. In which the ortho-rectification model commonly used strict physical model and rational polynomial model (RPC). Strict physical model are considered to be theoretically rigorous, thus, It can be considered without error;The RPC model is a simulation of strict physical model, the model conversion error is generally less than 1m (Liu Shijie, 2008); The error of satellite orbit and attitude is decided by the control accuracy of the satellite itself and the in-orbit state, which cannot be changed by the average user. The precision of control points can be controlled through field measurement, currently using GPS can get higher precision. DEM data are input parameters of ortho-rectification, which generally directly adopt national fundamental geographic information data, the precision of is slightly different in the mountains and plains. In addition, when the satellites side-view imaging, their orbit and attitude control precision and measurement accuracy will decrease because of satellites or remote sensors' side swing[4]; Furthermore, according to the theory of projection, the increase of side perspective will definitely magnify the effects caused by DEM error. Therefore, the satellite side perspective is also one of the main error sources of satellite images ortho-rectification.

It can be seen from the above error analysis of ortho-rectification, in general case, the factors to be considered most by users are side perspective and DEM errors. Many scholars have carried out some research work about the effects on image quality degradation and ortho-rectification precision caused by side perspective, and the effects on ortho-rectification precision caused by DEM errors. Such as Han Wenli, Yuan Xiuxiao, and Wang Xuejun, etc. Han Wenli(2010) quantitatively analyzed the impact on ortho-rectification precision caused by side perspective in theory, and proved the validity of its conclusions through experiments. The conclusions indicate that with increment of the side perspective, the image ortho-rectification accuracy will gradually decline, and the decrease amplitude is quite obvious[3]. Taking into account changes of the side perspective, Yuan Xiuxiao et al., 2009, established the geometry model of high-resolution satellite remote sensing image processing, which improved the image positioning accuracy on the ground targets. It concluded that the change of the side perspective is a factor that cannot be neglected in the high-resolution satellite remote sensing image geometry processing model[5]. Wang Xuejun et al., 2008, selecting different DEM scales, conducted ortho-rectification to the data with different side perspectives acquired by SPOT5. It certified that when SPOT5 imaging, the side perspective is the key factor leading layer-over phenomenon of SPOT5 orthophoto products, with the premise of ensuring the accuracy of ground control points. It also certified that in the calibration process the improvement of DEM accuracy cannot effectively eliminate layer-over phenomenon of the mountains with complex terrain and large elevation drop. About the impact on image correcting accuracy caused by DEM data, Shi Yuhua (2007), using two different DEM data, calibrated the same scene image. The results show that the ortho-rectification accuracy is higher with small basic grid and high-precision DEM data[6].

However, these studies only consider the impact of unilateral side perspective or DEM error on ortho-rectification precision, and failed to combine both. Actually it is a coupling effect between side perspective and DEM error: DEM error will be amplified with the increase of side perspective; Errors caused by side perspective will change along with the change DEM data. As the input conditions of ortho-rectification the accuracy of DEM data are often unable to change. Therefore, in practice, it is more significant to study how to determine the maximum satellite observation side perspective in the premise that there is some error in DEM data, as well as guarantee certain accuracy of ortho-rectification and mapping. Aiming at this problem, this paper adopted Quickbird image data with different side perspectives in Beijing, proposed a method to determine the maximum satellite imaging side perspective with certain constraints of mapping accuracy by simulating different influence law on the accuracy of image ortho-rectification caused by DEM elevation error, and established a quantitative conversion relationship among satellite side perspective, DEM error, and the error of ortho-rectification image points.

2 Experiments and Methods

2.1 Experimental DATA

The experimental area is the northern part of Beijing, including plain and mountain. Experimental data are multispectral image collected by using high-resolution Quickbird satellite in 2002 to 2012. With 2.44 m resolution, side perspective spanning from 5.6°to 43.7°ten different angles in total. Image side perspective spanned broadly, evenly distributed, specifically details can be seen in the table below.

Table 1. Test data tables side perspective Statistics

Satellite Name	Side perspective (°)									
QuickBird	5.6	10.4	14.4	16.3	22.1	28.2	31.8	34.3	38.9	43.7

2.2 Experimental Methods

According to the principle of photogrammetry, the comparation between image point error caused by the side perspective and imaginary horizontal photograph as Figure 1(a) shows.

P is the tilted photograph and *P0* is the imaginary horizontal photograph with the same observation conditions. They intersect at isocon line *hc*. Imagine that ground point M is (m) on tilted photograph P and (m_0) on horizontal photograph P_0, respectively. After two photograph overlapping along isocon line, the position of (m) in the horizontal photograph is m, obviously mm_0 is the tilt error. With the increase of side perspective, the tilt error will become bigger. Image point error caused by the relief is shown as follows(right). To facilitate discussion, assume that photograph is horizontal, and the height difference between ground point A and the datum is h. The image point of ground piont A is point a on the photograph; the projection of A on the datum is A_0 , whose image point is a_0 . aa_0 is the displacement of image points caused by the relief. As can be seen, the displacement of image points will increase with the increasement of height difference h[7-9].

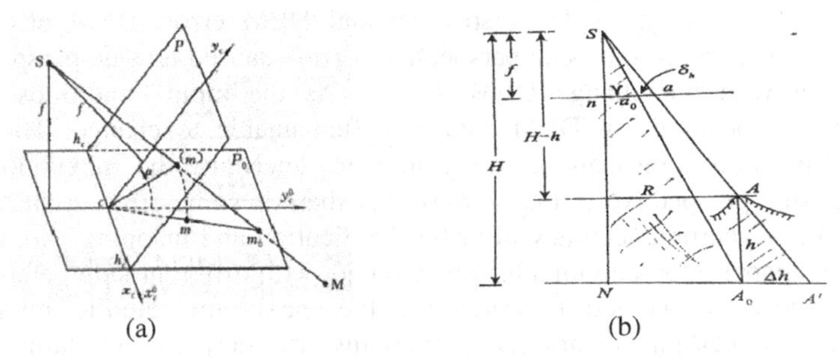

Fig. 1. The error of displacement of image points caused by side perspective and relief ((a): side perspective; (b): relief)

During the process of ortho-rectification, image points error will be affected by side perspective and relief at the same time. In the existing research results, different influence factors are often analyzed independently, which ignores the mutual influence and internal relations of different factors and is not real and comprehensive enough to reflect the objective situation of data. Because the strict physical model to calibrate the satellite image contains quantities of trigonometric functions and complex variables. If directly use strict physical model to deduce the error model of ortho-rectification, it will be very complicated. It is also why many scholars assume that the variables are independent in the derivation of the law of ortho-rectification error[10][11]. In addition, RPC model is an approximate mathematical simulation of the physical model. RPC model without satellite orbit and attitude parameters, there is no way to establish the error model of ortho-rectification, DEM and orbit attitude parameters. Based on these considerations, this paper used the real images with different side perspectives and corresponding parameters, analyzed the relationship among side perspective, DEM accuracy, and ortho-rectification precision, when inputting different DEM elevation data, and derived the function among variables based on experimental data statistics. The specific process is shown in figure 2.

According to the figure 2, the concrete steps and methods are as follows:

Input the satellite images and their corresponding parameters, original DEM data into RPC model, when the side perspective is given. Using RPC model, conduct ortho-rectification against specified pixel. Record the calibrated coordinates as the initial value of the corrected geographical coordinates.

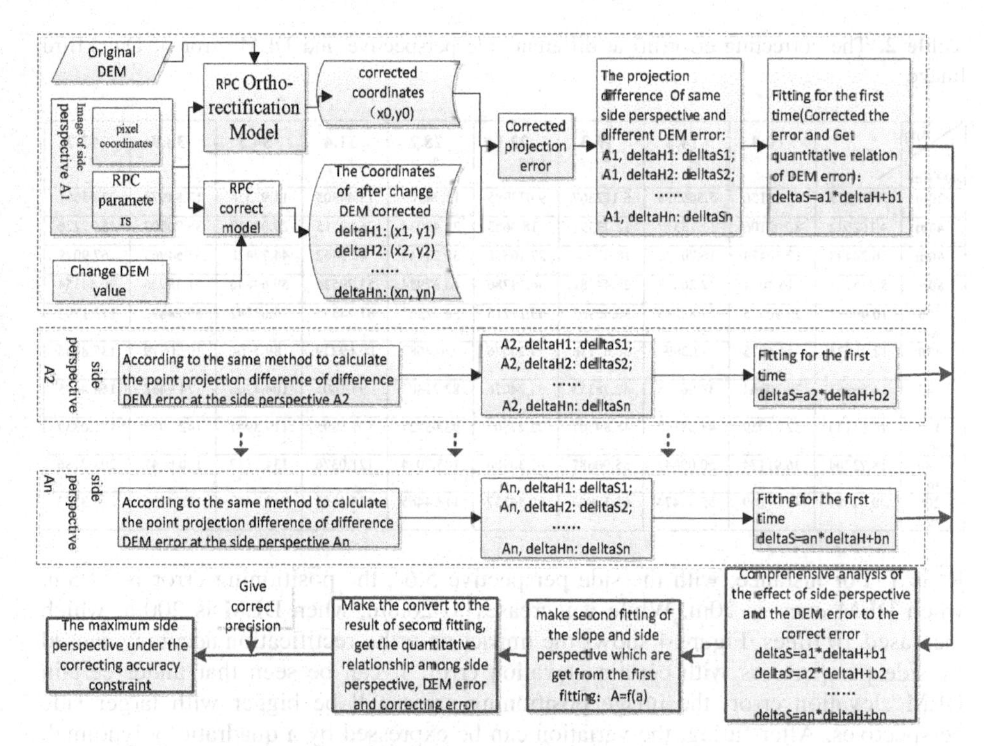

Fig. 2. The analysis process of the effects on ortho-rectification precision caused by side perspective and DEM error

Change the DEM elevation value without other parameters changed. Conduct ortho-rectification against specified pixel by using RPC ortho-rectification model, then Obtain the corrected value of geographical coordinates which is derived from formula (1). The displacement of the coordinates in step ① is also calculated by formula (1). Given the different DEM elevation values, obtain the ortho-rectification error(one column in Table 2)with the same side perspective and different DEM elevation error.

$$\delta S = \sqrt{(\Delta x)^2 + (\Delta y)^2} \tag{1}$$

3 Results and Discussion

Figure 3 shows that using experimental data with different side perspectives, the paper calculated the influence law distribution on ortho-rectification accuracy caused by DEM error. From the numerical distribution and the fitting results in figure 3, it can be seen that under the same side perspective, pixel positioning errors increase with the increasement of DEM elevation values; the larger the side perspective is, the greater the pixel positioning is. And the pixel positioning is more obviously affected by DEM values. The change is completely linear relationship, and fitting precision

Table 2. The correcting error(m) at different side perspective and DEM error of QuickBird image

Angle / DEM error	5.6°	10.4°	14.4°	16.3°	22.1°	28.2°	31.4°	34.3°	38.9°	43.7°
20m	2.080815	4.101156	5.566553	6.112367	9.033995	11.74509	13.44905	14.92516	17.84973	22.63661
40m	4.162622	8.202106	11.13377	12.22391	18.0683	23.49034	26.89715	29.84912	35.70007	45.27226
60m	6.24443	12.30424	16.70032	18.33628	27.10355	35.23445	40.3462	44.77428	53.54963	67.9079
80m	8.325118	16.4054	22.26658	24.44782	36.13786	46.97954	53.79524	59.69846	71.39936	90.54354
100 m	10.40693	20.50635	27.83285	30.55962	45.17216	58.72381	67.24335	74.62242	89.24892	113.1792
120 m	12.48873	24.6075	33.3994	36.67116	54.20646	70.4689	80.69239	89.5466	107.0979	135.8148
140 m	14.56955	28.70964	38.96566	42.78353	63.24076	82.21301	94.1405	104.4708	124.9468	158.4495
160 m	16.65135	32.81059	44.53192	48.89507	72.27507	93.95727	107.5895	119.3947	142.7957	181.0839
180 m	18.73204	36.91174	50.09847	55.00687	81.30906	105.7014	121.0376	134.3189	160.6447	203.7186
200 m	20.81385	41.01269	55.66474	61.11842	90.34337	117.4465	134.4854	149.2429	178.4936	226.3533

R^2 is 1. For instance, with the side perspective 5.6°, the positioning error is 2.08m, when DEM error is 20m. While it increases to 20.8m, when DEM is 200m, which increased 10 times. Figure 4 shows the impact on ortho-rectification accuracy caused by side perspectives with certain elevation error. It can be seen that under certain DEM elevation error, the image positioning error will be bigger with larger side perspectives. After fitting, the variation can be expressed by a quadratic polynomial. Fitting accuracy R^2 is almost equal to 1, like, when the elevation error is 20m, $R^2 = 0.9996$.

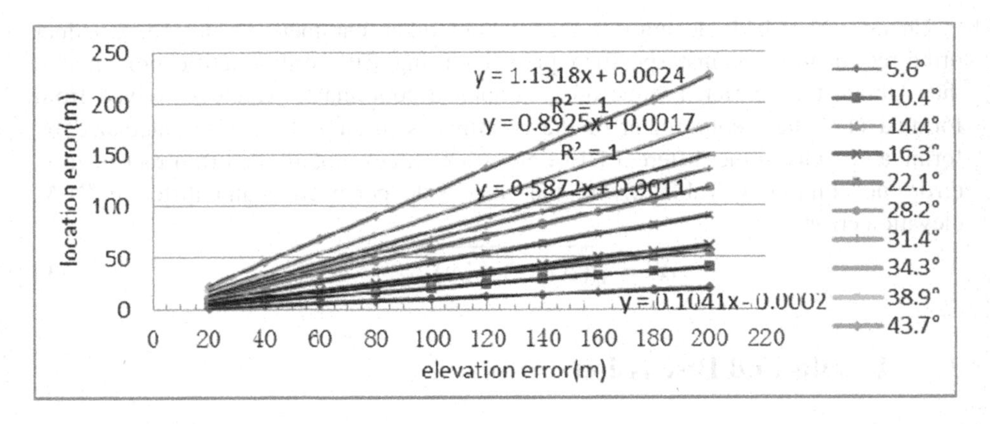

Fig. 3. The effect of elevation error to the position error at different side perspective

Comprehensively analyze the impact on ortho-rectification accuracy caused by side perspective and DEM error. Table 2 and Figure 3, Figure 4, various variations combine the impact on image points displacement caused by DEM elevation values and side perspective. They reflect the variations and quantitative relationship of the three variables.

According to figure 3, it can be seen that with certain side perspective, the impact relationship on ortho-rectification accuracy caused by DEM error is completely linear, and the slope of the fitting line increases with the increasement of side perspectives.

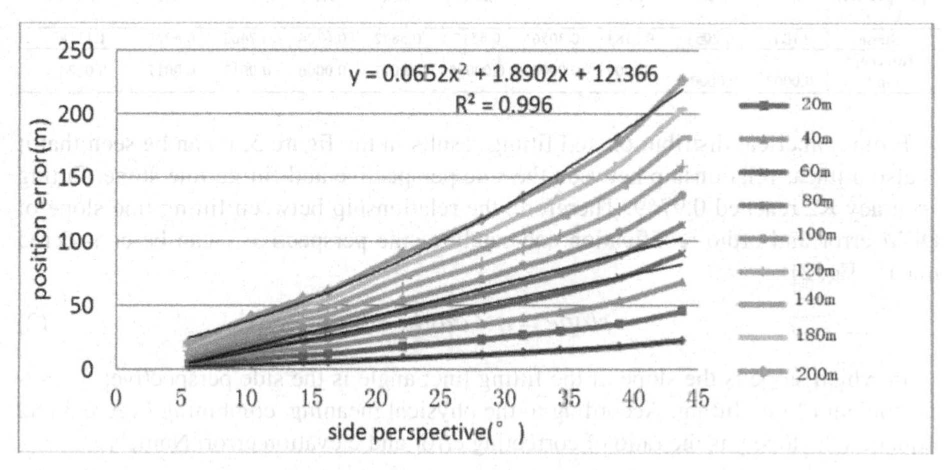

Fig. 4. The effect of side perspective to the position error at different elevation error

In order to investigate the relationship between the fitting line slope and side perspective, the slope and intercept of all the test data are presented in table 3. As can be seen from the table, the slope of the fitted line increases with the increasement of side perspectives, but the intercepts are relatively small. The Maximum of intercept is no more than 0.0024 cm, which can be ignored as infinitesimal in practical applications. In order to further give the quantitative relationship of fitting line slope and side perspective, Figure 5 shows the slope distribution curve of the fitting line along with the change of side perspective and fitting results of experimental data in this paper.

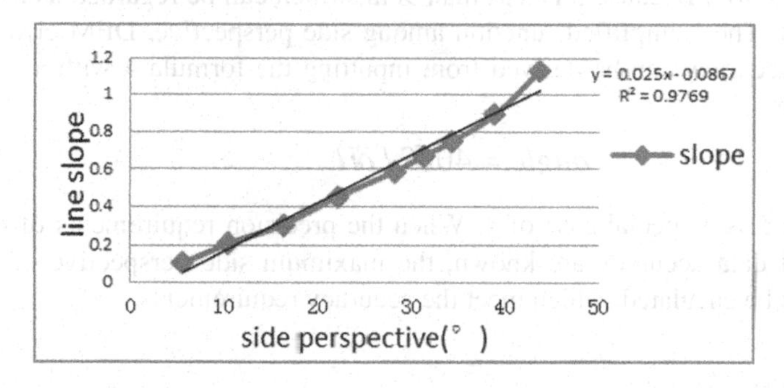

Fig. 5. The relationship of side perspective and line slope

Table 3. The slope and intercept of fitting straight line between DEM error and correcting precision with different side perspectives

Side perspective (°)	5.6	10.4	14.4	16.3	22.1	28.2	31.4	34.3	38.9	43.7
Slope	0.1041	0.2051	0.2783	0.3056	0.4517	0.5872	0.6724	0.7462	0.8925	1.1318
Intercept (m)	-0.0002	-0.00009	0.0007	0.0006	0.00005	0.0011	0.0006	0.0013	0.0017	0.0024

From numerical distribution and fitting results in the figure 5, it can be seen that it is also a linear relationship between the side perspective and fitting line slope. Fitting accuracy R2 reached 0.9769. Therefore, the relationship between fitting line slope of DEM error and ortho-rectification and satellite side perspective can be considered linear. (Eq. 2):

$$slope = a \times angle + b \tag{2}$$

In which slope is the slope of the fitting line; angle is the side perspective; a, b is coefficient of line fitting. According to the physical meaning, combining Figure 3 and Equation 2, slope y is the ratio of correcting error and elevation error. Namely:

$$slope = \delta S / \delta H \tag{3}$$

Where δS is correcting error; δH is DEM error.

Equation 4 can be derived from combining Equations 2 and 3:

$$\delta S / \delta H = a \times angle + b \tag{4}$$

Equation 4 describes the approximate relationship among side perspective, DEM error and error of pixel correcting.

Using the experimental data in this paper, it can derive that a equals to 0.025, and b equals to -0.0867. Because b is less than 9cm, which can be regarded as infinitesimal and ignored. The simplified function among side perspective, DEM error and pixel correcting accuracy can be derived from inputting the formula 4 with a = 0.025 and discarding b.

$$angle = 40 \delta S / \delta H \tag{5}$$

Equation 5 is a special case of 4. When the precision requirements of positioning error, DEM data accuracy are known, the maximum side perspective of QuickBird satellite can be caculated, which meet the accuracy requirements.

4 Application

This paper, by using QuickBird satellite images in Beijing, deduced the effects on the precision of image ortho-rectification caused by side perspective and DEM error, and provided the quantitative functional relationship among the three. With this function, if given the actuary request of image ortho-rectification and DEM accuracy, it is possible to determine the largest satellite imaging side perspective. Thus, before collecting data, it can help avoid the potential trouble of ortho-rectification accuracy overrunning in later phase.

Based on the Quickbird test data statistics model, this paper, using the 1:50 000 DEM of China and the American ASTER GDEM to perform ortho-rectification, respectively calculated the maximum allowable side perspective on the mountains and plains.

According to the requirements of GBT 13977-1992 "1:5000,1:10000 topographic maps and aerial photogrammetry field work norm", plane position mean error of feature points on the map is less than 0.5mm on plains and hilly lands, 0.75mm on mountainous and alpine areas.

Corresponding 1:10000 topographic mapping requirements, the error of image calibration is no more than 5m on plains and hilly lands,7.5m on mountainous and alpine areas.

In addition, the grid size of our nationwide 1:50000DEM 25×25 m is known. The elevation mean error of grid points comparing to the near field elevation control point is not more than 4m on plains, not more than 11m on mountains[1]. The elevation accuracy of American ASTER GDEM is 20m(ASTER Global DEM Validation Summary Report. 2009).

1) According to formula (5), When the national 1:50000 DEM is used of δH plains is 4m, and δS is 5m. Then in order to meet the mapping requirements of 1:10000, the side perspective on the plain areas of Quickbird image must be less than or equal to 50 °. δH of mountains is 11m, and δS is 7.5m, the corresponding side perspective must be less than or equal to 27.3°.

2)when American ASTER GDEM is used, δH of plains and mountains is 20m, in order to meet the mapping requirements of 1:10000, the error of image calibration is no more than 5m on plains and hilly lands and 7.5m on mountainous and alpine areas. Taking elevation value 5m into computation, the imaging side perspective of Quickbird should be less than or equal to 10 °。

Note that the formula (6) is deduced based on the results of data regression statistics in Beijing experimental zone. Due to the limitations of statistical samples and terrain differences, these results are only for reference.

In addition, given the current main on-orbit high-resolution commercial satellites, including QuickBird, WordView, IKONOS, etc. Due to the similarity of Ortho-rectification error and solutions, the method should also be applicable to other similar type of satellite platforms.

5 Conclusions

This paper established the quantitative conversion relationship among satellite side perspective, DEM error, ortho-rectification, and image rectification, proposed a method to determine the maximum imaging side perspective with map precision constraints, and provided guidance on data selection and processing. When known DEM data accuracy and the accuracy that ortho-rectification expected to achieve, the maximum side perspective of the satellite can be calculated to meet those conditions.

On the other hand, when the existing satellite data side perspective is known, the minimum accuracy of DEM data can be calculated to achieve the expected ortho-rectification precision. Because the model in this article is obtained by the analysis of real data, thus, the authenticity of the results is guaranteed, and the conclusion is also in line with the actual situation. In addition, the ideas and methods adopted by this paper are also applicable to the analysis and study of other sensors with the same kind.

Acknowledgment. In this paper, the test data are provided by Beijing Aerospace Shi Jing Information Technology Co, Ltd.

References

[1] GBT_13977-1992. 1:5000, 1:10000 topographic maps and aerial photogrammetry field work norms
[2] Yong, W., Xue, H.: Remote Sensing Refined Explanation. Surveying and Mapping Press (1993)
[3] Han, W.L.: Research on the Influence of Correcting Precision Caused by Satellite Incidence Angle. Beijing Surveying and Mapping 04, 20–22 (2010)
[4] He, H.Y., Wu, C.D., Wang, X.: Study of Influence of Swinging on the Systemic Parameters of the Satellite and CCD Camera. Spacecraft Recovery and Remote Sensing 04, 14–18 (2003)
[5] Yuan, X.X., Cao, J.S., Yao, N.: A Rigorous Geometric Model Considering the Variety of Side Watch Angle for High-resolution Satellite Imagery. Acta Geodaeticaet Carto graphica Sinica 02, 120–124 (2009)
[6] Shi, Y.H.: Research of Quickbird Image on Orthocorrection and Updating of Relief Map. Geospatial Information 06, 45–47 (2007)
[7] Zhang, J.: Photogrammetry. Wuhan University Press, Wuhan (1995)
[8] Okamoto, A., Akamatu, S., Hasegawa, H.: Orientation Theory for Satellite CCD Line-Scanner Imageries of Mountainous Terrains. International Archives of Photogrammetry and Remote Sensing 29, 205–209 (1993)
[9] Okamoto, A.: Orientation theory of CCD line-scanner images. International Archives of Photogrammetry and Remote Sensing 27(B3), 609–617 (1988)
[10] Yan-Wei①②, R., Xin-Ming②, T., Hua-Bin②, W., et al.: The SPOT HRG images ortho-rectification experiment and precision analysis without GCPs. Science of Surveying and Mapping 3, 029 (2008)
[11] Zhang, L.-P., Cui, Y.-L., Wang, B.-S.: 1.The Third Heilongjiang Surveying and Mapping Engineering Institute,Harbin 150086,China;2.The Second Heilongjiang Surveying and Mapping Engineering Institute,Harbin 150086,China); Analysis on the Accuracy of Quick Bird Image Ortho-Rectification under Different Terrain Condition. Geomatics & Spatial Information Technology, March 2009

Using Hyperspectral Remote Sensing Identification of Wheat Take-All Based on SVM

Hongbo Qiao, Hongtao Jiao, Yue Shi, Lei Shi, Wei Guo, and Xinming Ma

College of Information and Management Science,
Henan Agriculture University, Zhengzhou, 450046, China
{qiaohb,sleicn,guoweiworkhome,xinmingma}@126.com,
python.jiao@gmail.com, 353303543@qq.com

Abstract. Wheat take-all is quarantine diseaseand took place more and more severer in recent years, It is important to monitor it effectively. This article using hyperspectral remote sensing, through the different levels of the incidence of wheat take-all canopy spectral reflectance data collection analysis and processing, using support vector machine(SVM) classification method to build Wheat Take-all disease level prediction model for the prediction and prevention for wheat take-all to provide technical support. Results shows that the wheat canopy spectral reflectance change significantly under the influence of the disease; through data analysis, choose 700~900nm wavelength band training as sensitive to model the performance of the best results; Upon examination, constructed the forecasting model based on this band to predict when the predicted value and the actual value of the correlation coefficient up to 0.9434. The results of this study will not only provide theoretical and technical support for wheat no-destructive detection and safety production, but also shed light on the development of novel strategy to detect and control crop pest and disease, which has great significance to the food safety.

Keywords: Wheat, Wheat Take-all, hyperspectral, support vector machine, forecasting model.

1 Introduction

Wheat is China's major grain crops, Invasion by the diseases of wheat in the growth process, resulting in lower yields, even crops. Wheat Take-all is quarantine disease by soil-borne. In recent years, with the transport seeds and combined harvest, there is increasing trend occurred in our country. Therefore, analysis and research on wheat canopy spectral reflectance analysis of the differences in different Wheat Take-all disease infestation level in different canopy reflectance, Establishment of wheat take-all prediction models for large-scale Wheat Take-all prediction and prevention is important. Crops affected by different levels of disease stress because of their cell structure, pigment, water, nitrogen content and external shape changes result in changes in canopy spectra, for use fast hyperspectral remote sensing technology, predicted a large area of wheat take-all disease possible. Currently, some domestic

© IFIP International Federation for Information Processing 2015
D. Li and Y. Chen (Eds.): CCTA 2014, IFIP AICT 452, pp. 23–30, 2015.
DOI: 10.1007/978-3-319-19620-6_3

scholars use hyperspectral remote sensing technology for species identification and recognition of crop disease research; *Liu et.al* using hyperspectral data for fir and Masson pine identify research, by extracting the important feature of the band to complete the fir and Masson pine classification[1]. Xiu *et.al* using hyperspectral technology to identify rice cadmium pollution, establish a model to predict the cadmium content in rice[2]. Wang *et.al* using hyperspectral wheat stripe rust infestation level classification study to conduct of the study laid the theoretical foundation and reference[3-10]. Summary, we can build wheat canopy spectral reflectance and diseases rank correlation model, previous studies of the spectral data processing methods are classified by traditional statistical or neural networks algorithm, etc, which are not able to better handle large, complex multi-dimensional data. In this study, by using support vector machine technology for classification, it can better solve these problems. This study selected a different wheat take-all levels wheat experiment, Observed under different disease severity, wheat canopy spectral characteristics, By selecting the optimal parameters and kernel function to complete the training classification of spectral data and build predictive models, and provide a theoretical basis and technical support for further use of hyperspectral remote sensing to predict wheat take-all levels.

2 Materials and Methods

2.1 Data Collection

We collect data by using ASD FieldSpec FR, this device can continuous measurement wavelength in the range of 325 ~ 1075 nm, and select the measurement time in fine weather, no wind, no clouds 10:00 - 14:00 noon (solar elevation angle greater than 45 degrees). Before measurement, preheat the instrument should be set under natural light for 20 minutes, then the instrument calibration standard white, ground spectral measurements in natural light conditions, fiber optic probe vertically downward, with the canopy to keep the measured distance of 50cm and probe beneath the area to ensure that no shadows. Each sample collecting 10 spectral curve, after taking an average of these spectral curve as a representative curve, the instrument must be done before each measurement to optimize and make a standard white calibration. In this study, we collected 25 samples, and each sample collecting 10 spectral curve to researching. In 2012, April 21, 2013 and May 16 on wheat canopy spectra were measured.

2.2 Data Analysis and Processing

In order to improve the stability and adaptability experiment, pretreatment raw spectral data by using data processing software ASD ViewSpec Pro 6.0.1. Taking an average of each sample spectral curve as a representative curve.

In this study, support vector machine approach to training data classification, support vector machine (SVM) to deal successfully with a small sample of training data set and produce higher classification accuracy, which has been widely used in pattern recognition[10-12]. The key is to choose SVM classification kernel function and parameters C and Gama (parameter g). Finally we using Matlab 7.14 and Taiwan Professor Zhiren Lin Libsvm development kit combines training data to build predictive models.

3 Results and Analysis

3.1 Wheat Take-All Disease Canopy Spectral Reflectance Characteristics

Wheat canopy spectral relatively easy to measure, and less affected by atmospheric effects and directly reflect the spectral characteristics of the victim status of wheat. Figure 1 shows, in the infrared spectral reflectance of 400~700nm trend is not obvious, due to physiological changes in the structure of the incidence of wheat caused plants containing chlorophyll, water, nitrogen and the cell structure is changed, the light absorption and reflection with the normal difference, in particular in the range of 700 ~ 900nm (Figure 2), increasing the spectral reflectance with disease severity decreased.

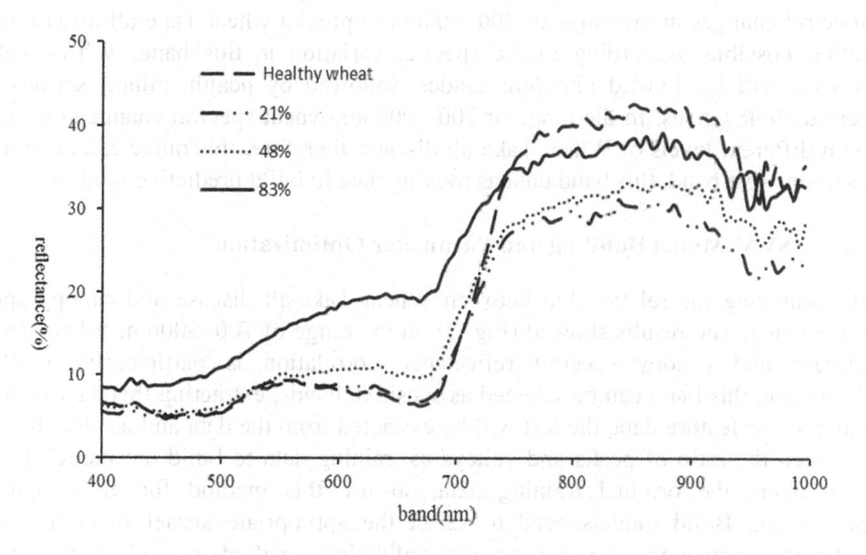

Fig. 1. Normal wheat canopy spectral curve comparison with diseases of wheat

3.2 Canopy Spectral Reflectance Characteristics of Different Degrees of Wheat Take-All Disease

Figures 1 and 2 show that in the infrared range of 400~700nm, the victim canopy reflectance and canopy reflectance curve of normal growth over the same period the basic agreement, and in the near infrared wavelength range 700~900nm, Wheat Take-all disease of spectral reflectance changed significantly, with the aggravation of the

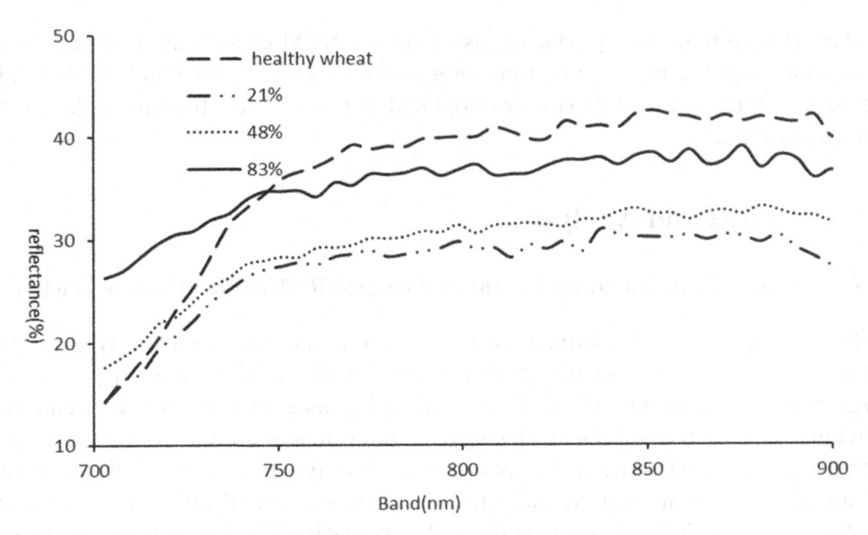

Fig. 2. Comparison reflectance spectrum at 700～900nm

disease, canopy reflectance decreased significantly. Therefore, by wheat canopy spectral changes in the range of 700~900nm to predict wheat Take-all disease hazard rating possible. According to the spectral variation in this band, Wheat Take-all disease will be divided into four grades, followed by health, minor, serious, very serious four grades. In the range of 700 ~900nm, wheat spectral change significantly with different levels of Wheat Take-all disease, therefore, this range can be extracted as a sensitive band, this band data as training data to build predictive models.

3.3 SVM Model Building and Parameter Optimization

By analyzing the relationship between Wheat Take-all disease and canopy spectral reflectance. The results showed (Fig. 2), in the range of 700~900nm, Wheat Take-all disease and canopy spectral reflectance correlation is particularly significant. Therefore, this band can be selected as sensitive bands, extracting this data within the band as the feature data, the test will be extracted from the data and use the difference between the ratio of peaks and valleys as training data to build the model, is better than using the original training data, so use this method for all sample data processing. Build models need to select the appropriate kernel function and the optimal parameters, we using cross-validation method to select the optimal parameters, and select the application more widely RBF kernel function[13-15]. The May 2013 data as training data for all samples, randomly selected in the same period in 2012 as a validation of the five groups of data, using Matlab and Libsvm toolbox training the best parameters C and Gama (parameter g) as shown in Figure 3, and using the best parameter and radial basis function to create predictive models [16-18].

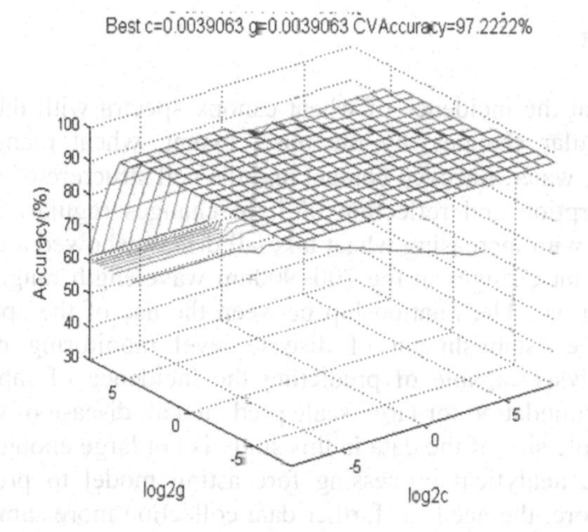

Fig. 3. By Matlab and Libsvm to find the optimal parameters C and g

3.4 The Accuracy of Prediction Model

Using Libsvm toolbox svm-predic forecasting tool to predict the five sets of unknown data, and analyzing the correlation between predicted and actual values to verify the accuracy of the model. The May 2013 data for all samples as training data for training, and all of the 2012 data in May as the validation sample data, different bands were selected as the training data as sensitive bands, different optimal parameters obtained through training use these parameters to establish predictive models, by randomly selected group to verify the predicted effects of different bands on the model accuracy through three validation to evaluate the robustness of this modeling approach. Use correlation analysis to verify the actual value of the sample and comparing the predicted value derived model predictive accuracy, validated (Table 1) shows that, by choosing different canopy spectral bands and select different parameters to establish the accuracy of the prediction model is not same[19-22], table 1 shows when the selected band is 700~900nm when the predicted value and the actual value of the correlation coefficient of 0.9434, and the root mean square (RMSE) reached 0.623, a good prediction.

Table 1. Verify the accuracy of the model under different bands and different parameters

Selected wave	C&Gama(g)	RMES	r
700~900nm	C=0.0039 g=0.0039	0.623	0.9434
325~1075nm	C=0.0039 g=3.0314	0.583	0.6377
325~1075nm	C=0.0125 g=0.0078125	0.512	0.833

4 Discussion

The study found that the incidence of wheat canopy spectra with the severity of the disease exhibit regular changes, by the onset extent, wheat plant canopy leaves contain chlorophyll, water, nitrogen content and the cell structure of wheat leaves are changing, the absorption and reflection of light changes regularly. Wheat canopy spectral reflectance with increasing wheat take-all disease showed a downward trend, particularly significant changes in the 700~900nm wavelength range, can be clearly observed spectral curve. The relationship between the use of the spectral curve and disease severity, the establishment of disease level monitoring model based on spectral image analysis capable of predicting the incidence of rapid and efficient wheat , can lay the foundation for large-scale predicted all disease of wheat.

Because the sample size of the data in this study is not large enough to take, so this sample through the analytical processing forecasting model to predict, or have a certain bias. Therefore, the need for further data collection more samples for training model to improve prediction accuracy. Meanwhile, the data collected from the hand-held spectrometer derived little effect other surface features of the spectrum, but hyperspectral remote sensing image contains a wealth of information, so that in the future this method for remote sensing image analysis and processing , the need for further spectral image noise reduction, and the SVM algorithm optimization, select the optimal parameters in order to reduce the impact of other surface features of the spectrum, to improve prediction accuracy.

5 Conclusions

Showing between Wheat Take-all disease and wheat canopy spectral reflectance changes regularly, canopy spectral reflectance with disease exacerbation showed a decreasing trend, Particularly in the near infrared (700~900nm) bands, the spectral reflectivity is particularly evident. So you can observe the wheat canopy spectral reflectance curves situation to determine the extent of the incidence of all disease of wheat. A prediction model based on these studies corresponding diseases, including the highest accuracy using the model 700~900nm wavelength band established as sensitive and can identify the model as the best of Wheat Take-all disease.

In this study, we using support vector machine approach to establish a model to identify Wheat Take-all disease. Because the support vector machine technology with high efficiency in a small sample of the training and multi-dimensional data processing, forecasting precision characteristics, and for this method on Wheat Take-all pest identification lay some foundation.

Acknowledgment. Funds for this research was provided by the National Natural Science foundation of China (31301604), Henan Science and technology plan (122102110045).

References

[1] Liu, X.-Y., Zang, Z., Sun, H., Lin, H.: Discrimination of Cunninghamia lanceolata and Pinus massoniana based on hyperspectral data. Journal of Central South University of Forestry & Technology (11), 30–33 (2011)
[2] Xiu, L.-N., Liu, X.-N., Liu, M.-L.: Analysis and Modeling of Hyperspectral Singularity in Rice under Cd Pollution. Spectroscopy and Spectral Analysis (01), 0192–0196 (2011)
[3] Wang, H.-G., Ma, Z.-H., Wang, T., Cai, C.-J., An, H., Zhang, L.-D.: Application of Hyperspectral Data to the Classification and Identification of Severity of Wheat Stripe Rust (09), 1811–1844 (2007)
[4] Liu, B., Fang, J.Y., Liu, X., Zhang, L.F., Zhang, B., Tong, Q.X.: Research on Crop-Weed Discrimination Using a Field Imaging Spectrometer. Spectroscopy and Spectral Analysis (07), 1830–1833 (2010)
[5] Li, W.-G., Zhao, C.-J., Wang, J.-H., Liu, L.-Y., Song, X.-Y.: Monitoring the Growth Condition of Winter Wheat in Jointing Stage Based on Land Sat TM Image. Journal of Triticeae Crops 27(3), 523–527 (2007)
[6] Zhang, D.-Y., Zhang, J.-C., Zhu, D.-Z., Wang, J.-H., Luo, J.-H., Zhao, J.-L., Huang, W.-J.: Investigation of the Hyperspectral Image Characteristics of Wheat Leaves under Different Stress. Spectroscopy and Spectral Analysis (04), 1101–1105 (2011)
[7] Huang, M.-Y., Huang, W.-J., Liu, L.-Y., Huang, Y.-D., Wang, J.-H., Zhao, C.-J., Wan, A.-M.: Spectral reflectance feature of winter wheat single leaf infected with stripe rust and severity level inversion. Transactions of the CSAE 20(1), 176–180 (2004)
[8] Chen, W., Yu, X.-C., Zhang, P.-Q., Wang, Z.-C., Wang, H.: Object recognition based on one-class support vector machine in hyperspectral image. Journal of Computer Applications 2011(08), 02092–02096 (2096)
[9] Qiao, H.-B., Cheng, D.-F., Sun, J.-R., Tian, Z., Chen, L., Lin, F.-R.: Effects of wheat aphid on spectrum reflectance of the wheat canopy. Plant Protection 31(2), 21–26 (2005)
[10] Wang, Z.-H., Ding, L.-X.: Tree Species Discrimination Based on Leaf-Level Hyperspectral Characteristic Analysis. Spectroscopy and Spectral Analysis 30(7), 1825–1829 (2010)
[11] Tan, K., Du, P.-J.: hyperspectral remote sensing image classification based on support vector machine. J. Infrared Millim. Waves (2), 0123–0128 (2008)
[12] Deng, R., Ma, Y.-J., Liu, Y.-M.: Support Vector Machine Multi-class Classification Based on an Improved Cross Validation Algorithm. Journal of Tianjin University of Science & Technology (06), 0058–0061 (2007)
[13] Feng, G.-H.: Parameter optimizing for Support Vector Machines classification. Computer Engineering and Applications 47(3), 123–124 (2011)
[14] Zhang, X.-G.: Introduction statistical learning theory and support vector machines. Act Automatica Sinica 1(26), 32–42 (2000)
[15] Li, X.-Y., Zhang, X.-F., Shen, L.-S.: A Selection Means on the Parameter of Radius Basis Function. Acta Electronica Sinica (12A), 2459–2463 (2005)
[16] Du, P.-J., Lin, H., Sun, D.-X.: On Progress of Support Vector Machine Based Hyperspectral RS Classification. Acta Geodaetica Et Cartographica Sinica (12), 0037–0040 (2006)
[17] Deng, W., Zhang, L.-D., He, X.-K., Mueller, J., Zeng, A.-J.: SVM-Based Spectral Recognition of Corn and Weeds at Seedling Stage in Fields. Spectroscopy and Spectral Analysis (7), 1906–1910 (2009)
[18] Ma, X.-L., Ren, Z.-Y., Wang, Y.-L.: Research on Hyperspectral Remote Sensing Image Classification Based on SVM. System Sciences and Comprehensive Studies in Agriculture (2), 0204–0207 (2009)

[19] Jie, L.V., Xiang-nan, L.I.U.: Hyperspectral Remote Sensing Estimation Model for Cd Concentration in Rice Using Support Vector Machines. Journal Of Applied Sciences-Electronics Information Engineering (01), 0105-0111 (2012)

[20] Mahlein, A.-K., Rumpf, T., Welke, P., Dehne, H.-W., Plumer, L., Steiner, U., Oerke, E.-C.: Development of spectral indices for detecting and identifying plan diseases. Remote Sensing of Environment (128), 21–30 (2013)

[21] Zhang, M., Qin, Z., Liu, X., Ustin, S.L.: Detection of stress in tomatoes induced by late blight disease in California, USA, using hyperspectral remote sensing. International Journal of Applied Earth Observation and Geoinformation (4), 295–310 (2003)

[22] Ryu, C., Suguri, M., Umeda, M.: Multivariate analysis of nitrogen content for rice at the heading stage using reflectance of airborne hyperspectral remote sensing. Field Crops Research (122), 214–224 (2011)

Evaluation Research of the Influence of Small Hydropower Station for Fuel Project on Social Development Impact

——taking Majiang of Guizhou Province as an example

Zhengqi He[1,2], Dechun Huang[1,2], Changzheng Zhang[1,2], Bo Wang[1,2], and Zhijie Ma[3]

[1] Business School of Hohai University, Nanjing, China, 211100
qi7761769@163.com, huangdechun@hhu.edu.cn,
{hhu2007,sensitivebaby}@126.com
[2] Institute of Industrial Economics of Hohai University, Nanjing, China, 211100
[3] China Institute of Water Resource and Hydropower Research (IWHR), Beijing, China, 100038
mazj@jwhr.com

Abstract. To change energy construction in the backward areas and protect the ecological construction achievements of conversion of cropland to forest, China starts the small hydropower station for fuel project, which has promoted the local social development. To evaluate the influence of the small hydropower station for fuel project on social development in the backward areas, this paper has a comprehensive consideration of the social environment and ecological environment, establishes the relative evaluation index system about influence, introduces the normal cloud model, focusing on the association of fuzzy and random problems in the artificial intelligence, into the evaluation model, and establishes a comprehensive evaluation model based on the normal cloud model and composite entropy weight. Taking the pilot Majiang County of Guizhou Province as an example, this paper evaluates the influence of the small hydropower station for fuel project on social development in the backward areas, and the results are made that the small hydropower station for fuel project has played a role in promoting the social development, and with the implement of the project, the effects become more and more obvious.

Keywords: small hydropower station for fuel project, social development, cloud model, composite entropy weight method.

1 Introduction

Water resources and forest resources are abundant in Southwest China, but for the long-term construction lag of the substitute energy in the rural areas, firewood is still the energy commonly used, which causes soil erosion, low forest coverage rate, deterioration of the ecological environment, and some important achievements of ecological construction like conversion of cropland to forest and natural forest protection cannot be consolidated. To solve the problem, the government has invested

© IFIP International Federation for Information Processing 2015
D. Li and Y. Chen (Eds.): CCTA 2014, IFIP AICT 452, pp. 31–39, 2015.
DOI: 10.1007/978-3-319-19620-6_4

lots of money to start the small hydropower station for fuel project. China is rich in the small hydropower resources, and the exploitable capacity is 1.28 billion kilowatts with great exploitation potential, though the exploitation rate is only 36.6% at present. After years of development, the small hydropower has become an important basis and strong driving force for China's economical and social development in the rural areas. Developing the small hydropower station for fuel project reasonably and efficiently can provide farmers with living fuel and rural energy steadily for a long period, protect the ecological construction achievements of conversion of cropland to forest, reduce farmers' burden of cutting firewood and make great contribution to social development. But a quantitative research is urgently needed to evaluate the project impact on social development and how it can change people's living standard and ecological environment in the backward areas.

2 Literature Review

The small hydropower station for fuel project, in a certain sense, belongs to poverty alleviation project. The organizations like World Bank and Asian Development Bank has a deep research on the poverty alleviation project that they participated in. Dani.A[1] considers how to realize the development target for the better mainly through the characteristic analysis of local social culture and organizational structure, and studies the influence of poverty alleviation project on changing demographic, social and cultural characteristics, production activities and residents' living conditions in the backward project areas. Another group of scholars mainly study the patterns for poverty alleviation, taking the pro-Poor Tourism pattern as an example, Caroline Ashley[2] pointed out that tourism development for poverty alleviation can create employment and income opportunities, promote small business management and improve the living standard of the poor. The domestic scholars mainly focus on the patterns for poverty alleviation and poverty alleviation benefits. Wang Sangui[3], who began to study the patterns for poverty alleviation in the early ninety's, pointed out that the economical development patterns can be divided into relying on resources type, asset accumulation type and technology driven type; starting from the benefit evaluation of the poverty alleviation project, Lin Boqian[4] emphasized that poverty alleviation is the most difficult challenge in the process of social development, and he believed that poverty alleviation benefits should be evaluated by both of number measuring method and monetary measuring method, and he proposed the benefit distribution analysis (BDA); taking conversion of cropland to forest project in the West China as an example, Kong Fanbin[5] studied its social impact and had a comprehensive exposition under the background of national microeconomic policy, which offered a way for the scholars to study the social impact evaluation of some important projects.

The current domestic research on the small hydropower station for fuel project mainly concentrates on feasibility evaluation and social development effect evaluation. Liu Tao[6] believes that the project is a public welfare project which focuses on the ecological benefits, and the normal small hydropower evaluation indexes and evaluation methods are unsuitable for the feasibility evaluation of the small hydropower station for fuel project. Shi Canxi[7] constructed three aspects of evaluation indexes during the research on the social impact evaluation, including political benefits, economic benefits and other social benefits (including environmental benefits and spiritual benefits), but using forest coverage rate as the only index evaluation can't truly reflects the environmental benefits; Zou Tifeng[8] noticed the influence of the small hydropower on the ecological environment, and according to the situations like the disordered exploitation caused by the small hydropower development fever, he pointed out various problems that affects the ecological environment during the construction process and proposed reason analysis, and offered a series of policy recommendations that coordinate the small hydropower development and the ecological protection, but he just did the research on the ecological environment without considering other indexes.

All the researches fail to build a perfect evaluation system and comprehensively evaluate the influence of the small hydropower station for fuel project on the social environment and ecological environment. Thus, this paper builds the evaluation index system of the project effect on social development from the two aspects of social environment and ecological environment, proposes the comprehensive method which combines the normal cloud model with composite entropy method, and finally has an empirical analysis through the example of Majiang County of Guizhou Province.

3 The Evaluation Index System Construction of Social Development Influence of the Small Hydropower Station for Fuel Project on Backward Areas

The small hydropower station for fuel project aims to improve rural energy structure, consolidate ecological construction achievements, improve rural living conditions, and improve farmers' living standard and habitat environment. Some changes have been taken place in the project areas since the project was implemented, and many aspects should be considered to evaluate the project effect on social development. According to the evaluation index system about social development built by Pan Ane[9], other aspects of indexes as well as the single aspect of economic development should be considered to evaluate the social development. Therefore, apart from the economic development, this paper has introduced living quality index and population quality index and reduces the three indexes to the social environment index, together with the ecological environment, building the comprehensive evaluation index system about social development. In the social environment, the economic level is measured by per capita GDP, per capita disposable income, unemployment rate and the third industry proportion in GDP; the living standard is measured by the Engel coefficient, consumer price index, per capita daily leisure time and small hydropower proportion in energy utilization; population quality is measured by natural growth rate of population, per capita schooling years and average span. In the ecological environment, water environment, atmospheric environment, solid waste and environment protection are measured by the eight indexes: per capita annual wastewater emission, sewage treatment rate, per capita annual toxic gas emission, number of polluted days in a year, per capita annual solid waste production, harmless treatment rate of daily garbage, forest coverage rate and environment investment proportion in GDP. Table 1 is the evaluation index system.

Table 1. The evaluation index system of the social development influence of the small hydropower station for fuel project on backward areas

the social development impact evaluation of the small hydropower station for fuel project on backward areas	social environment A	economic level A_1	per capita GDP A_{11} (RMB)
			per capita disposable income A_{12} (RMB)
			unemployment rate A_{13} (%)
			the third industry proportion in GDP A_{14} (%)
		living standard A_2	Engel coefficient A_{21}
			consumer price index A_{22}
			per capita daily leisure time A_{23} (hour)
			small hydropower proportion in energy utilization A_{24} (%)
		population quality A_3	natural growth rate of population A_{31} (%)
			per capita schooling years A_{32}
			Average life span A_{33} (year)
	ecological environment B	water environment B_1	per capita annual wastewater emission B_{11} (tons/ person)
			sewage treatment rate B_{12} (%)
		atmospheric environment B_2	per capita annual toxic gas emission B_{21} (tons/ person)
			number of polluted days in a year B_{22} (days)
		solid waste B_3	per capita annual solid waste production B_{31}(tons/ person)
			harmless treatment rate of daily garbage B_{32} (%)
		environment protection B_4	forest coverage rate B_{41} (%)
			environment investment proportion in GDP B_{42} (%)

4 The Comprehensive Evaluation Model Based on the Normal Cloud Model and Composite Entropy Method

4.1 Weight Determination Based on the Composite Entropy Method

To have a scientific determination of evaluation index weight, this paper combines the subjective weight with the objective weight according to the tendency coefficient method[10] of Yan Yi, If the subjective weight of the evaluation index j is ${W^{(1)}}_j$ and the objective weight of which is $W_j^{(2)}$, then the combined composite entropy coefficient is:

$$W_j = \gamma {W^{(1)}}_j + (1-\gamma){W^{(2)}}_j$$

γ is the tendency coefficient, which expresses the important and being tended degree of the subjective weight and the objective weight and to make a better use of the advantages, this paper takes γ=0.5, that's to say the composite weight is the arithmetic average value.

Entropy method[11] is a kind of objective weight method, and when using the entropy method, this paper uses 0-1 standardization to have a normalized processing of the data in the judgment matrix R, and haves the $r_{\max}$ and $r_{\min}$ adjustments of about 20% to avoid the zero in the standard matrix. The subjective entropy takes the expert scoring method, that's to say by anonymous advice from the relative experts, we can collect, analyze and induce the experts' opinion and determine each index weight according to most experts' experience and subjective judge.

4.2 The Comprehensive Evaluation Model

The cloud model is proposed by the Chinese Engineering academician Li Deyi, and it's a conversion model used to deal with the randomness and fuzziness which widely exist in the qualitative concept and it connects the qualitative concept with its qualitative numerical value representation. The digital characteristics in the normal cloud reflects the quantitative concept and quantitative characteristics, which is represented by expectation Ex, entropy En and excess entropy He[12]. According to the digital characteristics (Ex, En, He) in the normal cloud, we use the normal cloud generator to have the cloud droplets, and then finish the mapping from qualitative analysis to quantitative analysis.[13]

In the comprehensive evaluation, weight is the key evaluation content, and this paper determines the size of index weight by the composite entropy method. The comprehensive evaluation model and its steps are as follows:

(1) Establish the factor domain of the evaluation object $U = \{u_1, u_2, \cdots, u_n\}$, establish the comment domain $V = \{v_1, v_2, \cdots, v_m\}$;

(2) Calculate index weight $W = \{w_1, w_2, \cdots, w_n\}$ by the composite entropy method;

(3) Have a single factor evaluation between factor domain U and comment domain V, and establish fuzzy relation matrix R. This paper calculates the membership degree of the evaluation factor by the normal cloud model. Let the upper boundary value and the lower boundary value of grade j a and b, which corresponds to factor i, then the quantitative concept grade j can be represented by the normal cloud model, and among them

$$Ex = \frac{a+b}{2} \quad ①$$

For the boundary value is a transition value from one grade to another grade, and it's a fuzzy boundary which belongs to both of the two corresponding grades, that's to say the membership degree of the two grades is equivalent, therefore it has:

$$\exp\left\{-\frac{(a-b)^2}{8(En)^2}\right\} = 0.5 \quad ②$$

It gets $$En = \frac{a-b}{2.355} \quad ③$$

The excess entropy He expresses the uncertain measure of entropy, which reflects the condensation degree of cloud droplets, the larger the excess entropy value is, the thicker the cloud is, and vice versa. This paper takes the value of excess entropy He according to experience.

(4) Determine the membership degree matrix $Z = \{z_{ij}\}_{n\times m}$ of cloud model of every grade corresponded to each index. What needs to be noticed is that in order to enhance the evaluation credibility, we need to have a repeated function of normal cloud

generator N times, and calculate the average comprehensive evaluation value under different membership degree situations: $z_{ij} = \sum_{k=1}^{N} \frac{z_{ij}^{k}}{N}$;

(5) get the fuzzy subset B by the fuzzy conversion of weight set W and membership degree matrix Z: $B = W \bullet Z = (b_1, b_2, \ldots\ldots, b_m)$

In the equation: $b_j = \sum_{i=1}^{n} w_i z_{ij}$, b_j expresses the membership degree of the objects to be evaluated to the jth comment. According to the maximum membership degree principle, the ith grade which corresponds to the maximum membership degree is chosen as the comprehensive evaluation result.

5 Social Development Influence Evaluation of the Small Hydropower Station for Fuel Project in Backward Areas

5.1 Date Source

Majiang county of Guizhou Province is one of the pilot of China's small hydropower station for fuel project, and the project was officially started in 2005, therefore all the data are derived from the statistical yearbook and local survey research in Majiang county of Guizhou Province from 2006 to 2010.

5.2 The Empirical Results and Analysis

On the basis of the evaluation index system of the social development influence on the small hydropower station for fuel project, the evaluation criteria of the project needs to been determined to get the evaluation results conveniently. According to evaluation index system and evaluation index criteria, this paper expresses the corresponding grades of each index with the relative normal cloud model by the formula ①—③(see appendix 1). Assume that N=100, we can use the positive generator algorithm to generate the membership degree matrix, substitute the data of years in the positive generator of the grade cloud model described above, and calculate the average comprehensive evaluation value under the different membership degree situations by a repeated calculation of 100 times. According to the evaluation factor quantitative data of years, get each index weight by composite entropy method:

W={0.0465,0.0476,0.0392,0.0392,0.0451,0.0453,0.0565,0.0554,0.0549,0.0640,0.0550,0.0481,0.0558,0.0519,0.0519,0.0483,0.0563,0.0615,0.0776}.

Finally, we can have a fuzzy conversion of weight set W and membership degree matrix Z according to step (5) and get the membership degree of the object to be evaluated to the jth comment and according to the maximum membership degree principle, the ith evaluation grade, which is corresponded by the maximum membership degree , was chosen as the comprehensive evaluation result(table2).

According to the evaluation result table and run chart, the small hydropower station for fuel project plays a significant role in promoting the social development in Majiang County. (take the year 2012 as an example, the huge promotion: 0.6101, the greater

Table 2. The evaluation results of the influence of the small hydropower station for fuel of different years on social development

year	Membership degree of the normal cloud				evaluation results of The normal cloud model
	Huge promotion	Greater promotion	General promotion	Smaller promotion	
2006	0.1831	0.0932	0.2046	0.5482	Smaller
2007	0.1903	0.1844	0.4641	0.3149	General
2008	0.2117	0.4842	0.2107	0.2014	Greater
2009	0.4209	0.5119	0.1015	0.1004	Greater
2010	0.6101	0.3517	0.0189	0.0338	Huge

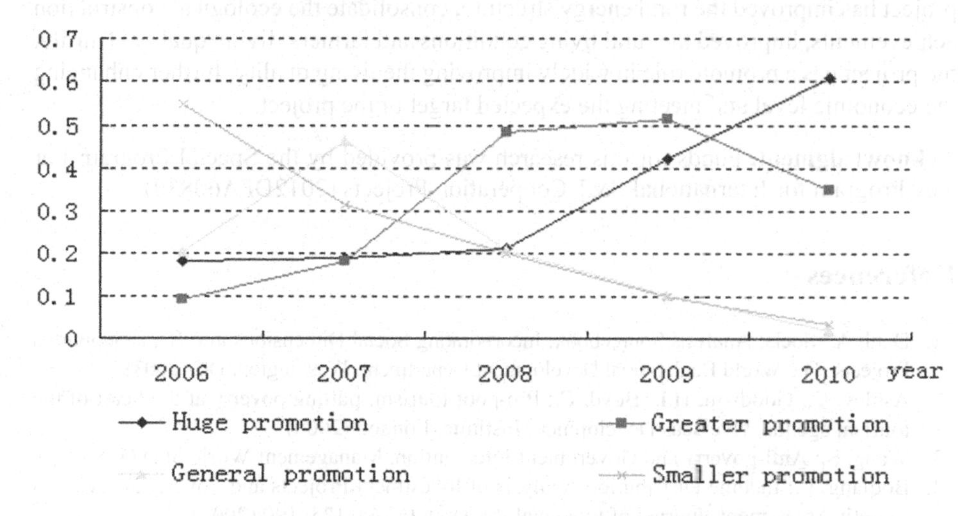

Fig. 1. Run chart of the influence of the small hydropower station for fuel project on social development effect

promotion: 0.3517), from the implement of the project in 2006, the huge promotion effect has increased to 0.6101, comparing to 0.1831 in 2006. From the data analysis, it can been seen that with the implement of the small hydropower station for fuel, the local living quality index A_2, the atmospheric environment index B_2 and solid waste index B_3 has changed obviously, and because of the usage of the small hydropower, the living quality has been improved, farmers have more leisure time with enhanced subjective happiness, and for the decrease of the firewood combustion, the atmospheric environment and solid waste pollution is improved, which just fits the purpose of the small hydropower station for fuel project implement, improving the living quality of people in backward areas and improving the ecological environment.

6 Conclusion

This paper applies the normal cloud model, which studies the association problems of fuzziness and randomness, into the social development influence evaluation to realize the uncertain mapping of evaluation factor value to comment, randomness as well as fuzziness can be considered in the social development influence evaluation and qualitative analysis and quantitative analysis can be combined. Build the evaluation index system with the starting point of social environment and ecological environment, and determine every index weight by the combination of the subjective and objective composite entropy method, we can have a more scientific reflection of every index weight , and have an evaluation of the influence of the small hydropower station for fuel project in Majiang County of Guizhou Province on local social development.

The results show that: firstly, under the common effects of many factors, the small hydropower station for fuel has improved the local ecological environment, controlled the air pollution and reduced the solid waste pollution. Secondly, the implement of the project has improved the rural energy structure, consolidate the ecological construction achievements, improved the rural living conditions and farmers' living quality. Thirdly, the project has a promote role in widely improving the living quality, further enhancing the economic level and meeting the expected target of the project.

Acknowledgment. Funds for this research was provided by the Special Program For Key Program for International S&T Cooperation Projects (2012DFA60830).

References

1. Dani, A.: Social Analysis Sourcebook: Incorporating Social Dimensions into Bank-Supported Projects. The World Bank Social Development Department. Washington, DC (2003)
2. Ashley, C., Goodwin, H.I., Boyd, C.: Pro-poor tourism: putting poverty at the heart of the tourism agenda. Overseas Development Institute, London (2000)
3. Wang, S.: Anti-poverty and Government Intervention. Management World 3(1) (1994)
4. Boqiang, L.: Income Distribution Analysis of Investment Projects and Poverty Alleviation Benefit Assessment. Journal of Financial Research (03A), 175–190 (2007)
5. Fanbin, K.: Social Influence Research on Conversion of Cropland to Forest in West China. Journal of Jiangxi University of Finance and Economics (1), 50–52 (2006)
6. Tao, L.: Feasibilty Evaluation of the Small Hydropower Station for Fuel Ecological Protection Project Based on Grey AHP. Journal of Water Resources and Water Engineering 17(3), 46–49 (2006)
7. Canxi, S.: Comprehensive Evaluation of the Social Benefits of the Small Hydropower. Small Hydropower (3), 1–4 (1998)
8. Zou, T., Wang, Y., Fei, L.: Analysis of the Ecological Environment Protection Problems of the Small Hydropower Development. China Rural Water and Hydropower (3), 97-98 (2008)
9. Ane, P.: Comprehensive Evaluation of Economic and Social Development of Wuhan City Based on Principal Component Analysis. China Soft Science(7), 118–121 (2005)
10. Yi, Y.: A study on Supply Chain Coordination of Engineering Project Manufacturing System with Large Logistics its Synergetic Evaluation. Tientsin University (2007)
11. Sui, Z., Mei, Z.: Science and Technology Evaluation Model and Empirical Study Based on Entropy Method. Chinese Journal of Management 7(1), 34–42 (2010)
12. Deyi, L., Changyu, L.: The Universality of the Normal Cloud Model. Engineering Sciences 6(8), 28–34 (2004)
13. Wang, S., Zhang, L., Li, H.: A Subjective Trust Evaluation Method Based on Cloud Model. Journal of Software 21(6), 1341–1352 (2010)

Appendix 1. The normal cloud standard of the evaluation index of the social development influence on the small hydropower station for fuel project

Index	The social development influence on small hydropower station for fuel project			
	Huge promotion	Greater promotion	General promotion	Smaller promotion
A_{11}	(7750,636.94,80)	(6250,636.94,80)	(4750,636.94,80)	(3250,636.94,80)
A_{12}	(3400,169.85,60)	(3000,169.85,60)	(2600,169.85,60)	(2200,169.85,60)
A_{13}	(3.5,0.085,0.02)	(3.7,0.085,0.02)	(3.9,0.085,0.02)	(4.1,0.085,0.02)
A_{14}	(48,0.85,0.2)	(46,0.85,0.2)	(44,0.85,0.2)	(42,0.85,0.2)
A_{21}	(58,1.69,0.4)	(62,1.69,0.4)	(66,1.69,0.4)	(70,1.69,0.4)
A_{22}	(97.5,1.27,0.3)	(100.5,1.27,0.3)	(103.5,1.27,0.3)	(106.5,1.27,0.3)
A_{23}	(2.8,0.17,0.04)	(2.4,0.17,0.04)	(2.0,0.17,0.04)	(1.6,0.17,0.04)
A_{24}	(21,2.55,0.6)	(15,2.55,0.6)	(9,2.55,0.6)	(3,2.55,0.6)
A_{31}	(7.35,0.13,0.03)	(7.05,0.13,0.03)	(6.75,0.13,0.03)	(6.45,0.13,0.03)
A_{32}	(9.3,0.085,0.02)	(9.1,0.085,0.02)	(8.9,0.085,0.02)	(8.7,0.085,0.02)
A_{33}	(72,0.85,0.2)	(70,0.85,0.2)	(68,0.85,0.2)	(66,0.85,0.2)
B_{11}	(14.5,1.27,0.3)	(17.5,1.27,0.3)	(20.5,1.27,0.3)	(23.5,1.27,0.3)
B_{12}	(51.5,3.82,0.9)	(42.5,3.82,0.9)	(33.5,3.82,0.9)	(24.5,3.82,0.9)
B_{21}	(0.067,0.002,0.001)	(0.071,0.002,0.001)	(0.075,0.002,0.001)	(0.079,0.002,0.001)
B_{22}	(38.5,1.27,0.3)	(41.5,1.27,0.3)	(44.5,1.27,0.3)	(47.5,1.27,0.3)
B_{31}	(1.7,0.085,0.2)	(1.9,0.085,0.2)	(2.1,0.085,0.2)	(2.3,0.085,0.2)
B_{32}	(44,1.69,0.4)	(40,1.69,0.4)	(36,1.69,0.4)	(32,1.69,0.4)
B_{41}	(73,3.39,0.8)	(65,3.39,0.8)	(57,3.39,0.8)	(49,3.39,0.8)
B_{42}	(1.9,0.085,0.2)	(1.7,0.085,0.2)	(1.5,0.085,0.2)	(1.3,0.085,0.2)

Study on Survey Methods for Crop Area Change Reasons at National Scale

Quan Wu[1], Hualang Hu[1], Yanxia Liu[2], Danqiong Wang[1], Xinyu Duan[2], Lijuan Jia[1], and Yajuan He[1]

[1] Remote Sensing Application Centre,
Chinese Academy of Agricultural Engineering, Beijing 100125, China
wuquan95@tom.com, landson_hu@sohu.com, 1226773217@.qq.com
[2] Heilongjiang Academy of Agricultural Sciences, Harbin 150086, China
liuyanxia_2001@163.com

Abstract. Remote Sensing Application Centre (RSAC) has been working on monitoring the main crop planting areas as an operational task and a research project based on its system for several years. A problem has been proposed from the RSAC'S technology system, which is that the monitoring result is only an estimate to main crop planting areas without quantitative explanation to crop area change reasons. The study as a project approved by RSAC attempts to solve the problem. A survey index system about crop area change reasons has been established by questionnaire and expert consultation. Through further research, 9 driving factors of crop area change have been set up by assembling 5 survey indexes. Based on triangular whitenization weight function of grey system theory, the crop Area Change Driving Model (ACDM) was designed. The factor weights and factor grey classes were produced through expert consultation. Heilongjiang province as the experiment region and single cropping rice as the experiment crop were selected in this study. The experiment result produced by ACDM is consistent with the estimate produced in monitoring crop area by RSAC in 2013.

Keywords: Crop area, Crop area change, Survey index, Driving factor, Grey system, Whitenization weight function, Estimate, RS.

1 Introduction

Crop area change is an important project paid much attention by governments in the world because it is closely related with food security. Sampling methods are widely used to estimate main crop area change by many countries and regions, such as US, China, EU, etc.[1][2][3]. Monitoring Agriculture with Remote Sensing (MARS) is a project facing Europe in order to obtain crop yield information constituted by European Union Committee[4]. It is a kind of three-stage sampling based on unsupervised classification with multitemporal RS data. In America, the prediction of total crop yield is acquired from crop acreage and crop yield per unit. The crop acreage data is mainly gotten by June Agricultural Survey(JAS). Two different

© IFIP International Federation for Information Processing 2015
D. Li and Y. Chen (Eds.): CCTA 2014, IFIP AICT 452, pp. 40–47, 2015.
DOI: 10.1007/978-3-319-19620-6_5

sampling units used by JAS are area frame covering America and name list frame consisted of the names of registered farmers[4]. In China, the operating prediction of crop area change is mainly provided by RSAC[5]. RSAC adopts two methods to obtain the acreage of main crops such as wheat, corn, cotton, soybean, rice, etc.. One method is stratified sampling with RS and the other is ground random sampling using GPS[1][2]. RSAC has been working on monitoring the main crop area change as an operational task and a research project based on the complete organization system and the maturity of a technical system for several years. RSAC submits monitoring results to Ministry of Agriculture at prescribed time according to the crop monitoring calendar every year[6]. The monitoring results are only estimates for interannual variation of crop areas while the explanation for the reasons of crop area change is qualitative not quantitative[7][8]. At the same time, at the beginning of the year planting intent survey is done by relevant departments of the government in China, which is a kind of sampling survey to peasant households through the interview. Although the interannual variation of crop areas can be calculated based on data from investigated peasant households, there is not quantitative interpretation for the reasons of crop area change.

2 Objectives

It is the first objective of the research to theoretically establish a set of survey index system for crop area change reasons. Going a step further, the ultimate goal is to find driving factors of crop area change and set up the Area Change Driving Model (ACDM) for estimating the trend of crop area change based on the driving factors. So, the reasons of crop area change will be explained quantitatively.

3 Materials and Methods

The study was acted as a research project authorized by CAAE (Chinese Academy of Agricultural Engineering), launched by RSAC at the beginning of 2013. In order to achieve the research goal RSAC form a group to carry out the work through selecting a study region and a crop for experiment. The members of the group come from the team specialized in monitoring Agricultural condition in RSAC, which have rich experience in Agricultural survey.

3.1 The Study Region and Crop for Experiment

Taking into account the combination of research and operational task, Heilongjiang province acted as study region was selected and single cropping rice acted as experimental crop was chosen from several crops. RSAC monitors single cropping rice area of Heilongjiang every year. The monitoring result can be used to test the study result. The following figure shows the spatial distribution of single cropping rice of Heilongjiang province in 2010.

Fig. 1. The spatial distribution of single cropping rice of Heilongjiang province

3.2 Candidate Indexes for Estimating the Reasons of Crop Area Change

Because RSAC's operational system of crop area change monitoring has been working well for many years while the members of the study group are also main participants for monitoring task[5], the candidate indexes can be produced by the researchers based on rich experience to Agricultural survey. The indexes are the purchase price and sales revenue of main crops, production cost, natural disaster, field management, occupation of cultivated land, labor transfer, irrigation and water conservancy, Agricultural mechanization, tax and subsidy, planting structure adjustment, crop rotation, land transfer, product sales, etc..

3.3 Survey Indexes for Estimating the Reasons of Crop Area Change

When the candidate indexes were built, the next step is to extract survey indexes from them. The extraction rules are some characteristics about the indexes, which are integrity, importance, quantifiability and availability. How the job for extracting indexes was done? Based on analysis about the extraction rules to every index, the study group designed survey forms to investigate peasant household in the study region. There are 100 peasant households, distributed on 57 counties, which were surveyed. After this step, three survey indexes were picked up from 14 candidate indexes, which are production cost, yield, purchasing price. The indexes forming an index system can be classified into three categories, which are investment, production and yield. The process of extraction survey indexes from candidate indexes is shown below by figure 2.

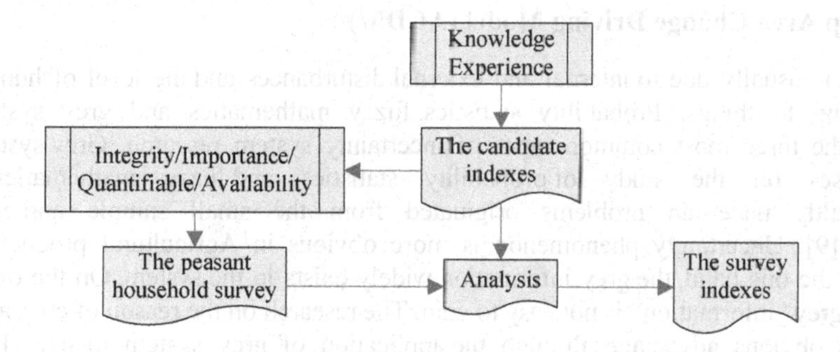

Fig. 2. The extraction flow of survey indexes from candidate indexes

3.4 Driving Factors for Estimating Crop Area Change

The driving factors are the main contradiction of the existence and development of things. It determines the direction of the development of things. They are the index which have greater influence to interannual variation of crop area, or directly influence planting intentions of peasants. According to the survey indexes system mentioned above, the driving factors were produced based on further research. They are price difference, cost difference, and per unit yield difference, which are divided into 7 factors. The table 1 shows the situaton.

Table 1. Driving indexes for paddy area change in Heilongjiang province

The driving factors	The meanings of driving factors	Marks
Sale	The difference of purchasing price of rice between the current year with last year	X_1
	The difference of purchasing price of rice between the last year with the year before last	X_2
	The difference between the purchasing price of three year average to rice with the average purchasing price of the main crops	X_3
Investment	The difference of production cost of rice between the current year with last year	X_4
	The difference of production cost of rice between the last year with the year before last	X_5
	The difference between the three year average production cost of rice with the average production cost of the main crops	X_6
Yield	The difference of rice yield between the current year with last year	X_7

3.5 Crop Area Change Driving Model (ACDM)

Uncertainty is usually due to internal and external disturbances and the level of human understanding to things. Probability statistics, fuzzy mathematics and grey system theory are the three most common types of uncertainty system research. Grey system theory focuses on the study of probability statistics and fuzzy mathematics to solve difficultly uncertain problems originated from the small sample and poor information[9]. Uncertainty phenomenon is more obvious in Agricultural production system. On the one hand, the grey information widely exists in the system. On the other hand, this "grey" information" is not easy to gain. The research on the reason of crop area change has obvious advantages through the application of grey system theory. This research adopts the method of grey clustering evaluation to establish a driving model of crop area change. Grey clustering includes grey relational clustering and grey whitenization weight function clustering. The grey whitenization weight function clustering includes grey fixed weight clustering, grey regulable weight clustering and whitenization weight function clustering based on endpoint or centre of triangle. The study adopts whitenization weight function clustering, based on triangular centre, acted as ACDM prototype. The general form of based on centre triangular whitenization weight function clustering is shown below.

$$f_j^k(\bullet) = \begin{cases} 0, x \notin \left[\lambda_j^{k-1}, \lambda_j^{k+1}\right] \\ \dfrac{x-\lambda_j^{k-1}}{\lambda_j^k - \lambda_j^{k-1}}, x \in \left(\lambda_j^{k-1}, \lambda_j^k\right] \\ \dfrac{\lambda_j^{k+1} - x}{\lambda_j^{k+1} - \lambda_J^k}, x \in \left(\lambda_j^k, \lambda_j^{k+1}\right) \end{cases} \tag{1}$$

Where $j(j = 1,2,\cdots,m)$ is survey index, $k(k = 1,2,\cdots,s)$ is grey class, $\lambda(\lambda = 1,2,\cdots,s)$ is the centre of grey class, $f_j^k(x)$ is the membership function belonging to grey class k of observed value x of index j. The Comprehensive clustering coefficient of object $i(i = 1,2,\cdots,n)$ belonging to grey class k is σ_i^k.

$$\sigma_i^k = \sum_{j=1}^{m} f_j^k(x_{ij}) \bullet \eta_j \tag{2}$$

Where $f_j^k(x_{ij})$ is whitenization weight function of index j belonging to k subclass. η_j is weight of j index in comprehensive clustering. The formula of object i belonging to k grey class is shown below.

$$\max_{1 \le k \le s}\left\{\sigma_i^k\right\} = \sigma_i^k \tag{3}$$

Through expert consultation with experience, the weight of every driving factor and the grey class centre of every factor can be determined.

3.6 Experiment Result

The Values of Survey Indexes. The surveyed crops are not only rice but other main crops which are soybean and corn acted as main crops in the region. Based on peasant household survey the average values of 3 survey indexes were gained. The data came from 3 years when there were not serious drought and snowstorm in this region from 2011 to 2013. This table shown below is presented the survey result to rice production region in study region.

Table 2. The survey result to main crops in Heilongjiang province

Crops	Years	Yield(Kg/mu)	Cost(Yuan/mu)	Purchase price(Yuan/50 Kg)
	2011	466.7	742.0	128.0
Rice	2012	471.5	874.8	140.0
	2013	483.0	1031.4	150.0
	2011	112.7	271.0	200.0
Soybean	2012	116.0	306.5	225.0
	2013	110.0	346.7	231.0
	2011	505.6	398.2	86.3
Corn	2012	482.2	451.6	89.3
	2013	494.8	512.1	96.5

The Observed Values and Weights of Driving Factors. Using the data in the table 2, it is easy to calculate the values of the driving factors for estimating rice area change. The seven driving factors were used to estimate rice area change with ACDM in 2013. The values of the seven driving factors are shown in table 3. The weights of driving factors are obtained by expert investigation, which is also listed in table 3.

Table 3. The survey values and weights of driving factors

Driving factors	X_1	X_2	X_3	X_4	X_5	X_6	X_7
The observed values	10	12	−15	157	133	502	12
The weights	0. 20	0. 15	0. 10	0. 20	0. 15	0. 10	0. 10

The Grey Class Centres of 7 Driving Factors. In order to establish ACDM with the driving factors mentioned above based on its prototype, the first step is to determine the grey class centers of 7 driving factors. The result was gained through study and expert investigation, which is shown in table 4 below. Three grey classes on behalf of decreasing, unchanged and increasing situation of rice area from 2012 to 2013 were designed, whose centers are respectively marked by λ_1, λ_2 and λ_3. The λ_0 and λ_4 are centers of grey class continuation.

Table 4. The grey class centers of driving factors

Grey class centers	X_1	X_2	X_3	X_4	X_5	X_6	X_7
λ_0	-50	-50	-30	300	300	700	-20
λ_1	-20	-20	-20	200	200	600	-10
λ_2	0	0	-10	150	150	500	0
λ_3	20	20	0	100	100	400	10
λ_4	50	50	10	50	50	300	20

Setting Up ACDM. Using the values of grey class centers of every driving factor in table 4, the ACDM can be easily established based on its prototype presented by formula (1). Due to space limit, only the ACDM of the first and second driving factors is listed below.

$$f_{1,2}^{1}(x)=\begin{cases}0, x\notin[-50,0]\\ \frac{x+50}{30}, x\in(-50,-20]\\ \frac{-x}{20}, x\in(-20,0)\end{cases} \quad (4) \qquad f_{1,2}^{2}(x)=\begin{cases}0, x\notin[-20,20]\\ \frac{x+20}{20}, x\in(-20,0]\\ \frac{20-x}{20}, x\in(0,20)\end{cases} \quad (5) \qquad f_{1,2}^{3}(x)=\begin{cases}0, x\notin[0,50]\\ \frac{x}{20}, x\in(0,20]\\ \frac{50-x}{30}, x\in(20,50)\end{cases} \quad (6)$$

The Grey Clustering Based on ACDM. When all ACDMs are finished, based on the models the next step is calculate the value of whitenization weight function of every driving factor with survey values in table 3, which are 3 figures presented 3 grey classes. When values of whitenization weight function of 7 driving factors, listed in table 5, are worked out, based on formula (2) the final step is calculate comprehensive coefficient with the weights of driving factors in table 3. The result is shown in the last column in table 5.

Table 5. The comprehensive coefficient of each grey class

Grey classes	X_1	X_2	X_3	X_4	X_5	X_6	X_7	X
1	0. 0000	0. 0000	0. 0500	0. 0280	0. 0000	0. 0020	0. 0000	0. 0800
2	0. 1000	0. 0600	0. 0500	0. 1720	0. 0990	0. 0980	0. 0000	0. 5790
3	0. 1000	0. 0900	0. 0000	0. 0000	0. 0510	0. 0000	0. 0800	0. 3210

3.7 Analysis for the Experiment Result

Using the formula (3), the result of the experiment is $\max_{1\le k\le 3}\{\sigma_1^k\}=\sigma_1^2=0.5790$. This suggests that the rice area of study region from 2012 to 2013 is unchangeable. Because $\sigma_1^2=0.5790$ is close to $\sigma_1^3=0.3210$, it demonstrates the rice area has increasing trend. In 2013, the RSAC's monitoring result is that the rate of area change is 2.1%, which confidence interval is from -7.24% to 12.36% based on 95% degree of confidence. This shows the result of the experiment is consistent with the monitoring result produced by RSAC.

4 Conclusion and Problems

Based on triangular whitenization weight function grey clustering can be used to establish ACDM. The model accurately reflects change situation of rice area from 2012 to 2013 in Heilongjiang province, Referencing RSAC' monitoring result. The survey indexes and driving factors founded in the study better reflect objective condition of reasons to rice area change, which better meet demands of the model.

Including survey indexes and driving factors, the ACDM prototype suits not only for rice but also for other main crops, such as winter wheat, soybean, corn, etc.. At the same time, the ACDM prototype suits not only for Heilongjiang province but also for other regions, such as northeast China, southwest China, etc.. Obviously, the parameters of the model designed in this study need to be recalculated because of different crops and different regions.

Acknowledgement. This paper is supported by the project named Study on survey methods for crop area change reasons at national scale, authorized by CAAE in 2012.

References

[1] Wu, Q., Sun, L.: Sampling methods using RS and GPS in crops acreage monitoring at a national scale in China. Remote Sensing and Spatial Information Sciences XXXVII (Part B7) [WG VII/7], 1337–1342 (2008)
[2] Wu, Q., Sun, L., Wang, F., Jia, S.: Theory of double sampling applied to main crops acreage monitoring at national scale based on 3S in china—CT316. In: Li, D., Liu, Y., Chen, Y. (eds.) CCTA 2010, Part III. IFIP AICT, vol. 346, pp. 198–211. Springer, Heidelberg (2011)
[3] Wu, Q., Pei, Z., Wang, F., Zhao, H., Guo, L., Sun, J., Jia, L.: A Sampling Design for Monitoring of the Cultivated Areas of Main Crops at National Scale Based 3S Technologies in China. In: Li, D., Chen, Y. (eds.) Computer and Computing Technologies in Agriculture VI, Part II. IFIP AICT, vol. 393, pp. 10–19. Springer, Heidelberg (2013)
[4] Yang, B.: Monitoring the agricultural condition using RS, pp. 19–29. China Agriculture Press, Beijing (2005) (in Chinese)
[5] Wu, Q., Sun, L., Wang, F.: The applications of 3S in operational monitoring system of main crops acreage in China. In: Computer and Computing Technologies in Agriculture-II, vol. 24, pp. 319–324. TSI Press, USA (2010)
[6] Wu, Q., Pei, Z., Zhang, S., Wang, F., Wang, Q.: The methods for monitoring land-use change with RS at non-large scale, pp. 33–40. China Agriculture Press, Beijing (2010) (in Chinese)
[7] Wu, Q., Sun, L., Wang, F., Jia, S.: The Quantificational Evaluation of a Sampling Unit Error Derived from Main Crop Area Monitorings at National Scale Based 3S in China. Sensor Lett. 10, 213–220 (2012)
[8] Wu, Q., Pei, Z., Guo, L., Liu, Y., Zhao, Z.: A study of two methods for accuracy assessment to RS classification. In: 2012 First International Conference on Agro-Geoinformatics, pp. 1–5 (2012)
[9] Liu, S., Dang, Y., Fang, Z., Xie, N.: Grey system theory and application, pp. 122–130. Science Press, Beijing (2010) (in Chinese)

Path Analysis on Effects of Main Economic Traits on the Yield of YU6, A Japonica x Indica Hybrid Rice Line

Weiming Liu

Taizhou Vocational College of Science & Technology, Zhejiang, Taizhou 318020, China

15157622288@139.com

Abstract. To further explore the effect of major factors on the yield of a line of japonica x indica hybrid rice, to clarify technical approaches to high efficient high yield production, the author analyzed 237 groups of data from 9 years in 6 counties of Taizhou. With correlation, regression, and path analyses, effects of major economic traits on the yield were determined. An effective approach to high efficient high yield production is to increase a seed setting rate and to produce large panicles while ensuring the number of productive panicles.

Keywords: japonica x indica hybrid rice, economic trait, yield, path analysis, technical approach.

1 Introduction

Hybridization of japonica rice with indica rice can potentially produce superior hybrids. YU6 is a hybrid that was created by mating between Yg2a and K6001. This line has been evaluated by Zhejiang Science and Technology Department in June 2005. YU6 is the first japonica x indica hybrid rice variety that passed the provincial evaluation (Zhe Ke Jian Zi [2005]#235). It has also passed the examination of experts organized by the Department of Agriculture of China in August, 2006, which confirmed that YU6 is the first recommended japonica x indica hybrid line (Nong Ke Ban [2006] #36). YU6 participated in the single cropping rice regional test of Zhejiang Province in 2002 and 2003. Its average yield was 563.3 kg/mu, which was 11.4% greater than that of the control, Xiushui 63. It takes 156.4 days to mature, which is 4.7 days longer than that of the control. YU6 has 13.4 thousand productive panicles per 667 m2 and 210.1 spikes per panicle on average. Its seed setting rate is 72.9% and the 1000-seed weight is 24.7 g. It is medium in resistance to rice blast disease and bacterial leaf blight, but susceptible to brown planthopper. Its grain quality is very good, with a head rice rate 66.9%, kernel chalkiness rate 16.4%, chalky 1.9%, and transparent 2.5, gel consistency 69.5 mm, the length to width ratio 2.3, and amylose 14.0%. YU6 inherited both the loose aromatic quality of indica rice and the elastic, sweet and glutinous character of japonica rice. YU6 has high hills, strong stems, straight leaves, large panicles, well developed primary branches, thick leaf sheath and excellent maturity looking. It is resistant to cold and lodging.

In recent years, along with the successful creation and application of YU6 hybrid rice, studies on the cultivation of this type of rice started to erupt [1-9]. In order to

© IFIP International Federation for Information Processing 2015
D. Li and Y. Chen (Eds.): CCTA 2014, IFIP AICT 452, pp. 48–53, 2015.
DOI: 10.1007/978-3-319-19620-6_6

explore the effect of its main economic traits on the yield, clarify the technical approach for high efficiency, high yield production, the author has analyzed data from the past on economic trait and yield, conducted regression, and path analysis before [10]. Now, the author has collected more data from additional years and sites, conducted further analysis on the influences of major economic traits on the yield of japonica x indica hybrids, using YU6 as a representative.

2 Materials and Methods

2.1 Data for Panicle, Grain Weights

Table 1 shows the data for panicles and grains of YU6 cultivated in 6 counties, districts or cities.

Table 1. Panicle and grain characteristics of YU6 based on the yield category

Yield level/ (kg/hm^2)	Paddy	Average productive panicles/ (million/ hm^2)	Average grains/ panicle	Average filled grains/ panicle	Average seed setting /%	1000-seed weigh t/g	Average yield/ (kg/ hm^2)
>11250	16	1.7795	322.48	290.98	89.75	23.28	11616.13
≤11250—>10500	28	1.7758	310.28	275.75	88.75	23.50	10782.43
≤10500—>9750	55	1.6255	321.31	284.67	88.36	23.66	10064.32
≤9750—>9000	48	1.5552	324.97	285.31	86.70	23.70	9385.09
≤9000—>8250	50	1.6347	304.51	251.45	82.52	23.71	8657.76
≤8250—>7500	30	1.5678	299.65	238.75	79.58	23.60	7958.60
≤7500—>6495	10	1.6979	288.30	219.21	76.64	23.20	7093.35

2.2 Analysis Method

Using YU6 as a representative, the author performed correlation and regression analyses, and finally, path analysis to determine the relationship between each of the major economic traits and yield, the relationship among these traits in the japonica x indica rice line, to clarify the approaches for high yield and high efficiency.

3 Results and Analysis

3.1 Correlation Analysis

As shown in Table 2, among the 5 traits, 4 traits showed a very significant positive correlation with the yield, except that the 1000-seed weight displayed a slight negative correlation. Further analysis on relationships among the five traits indicates that there were very significant negative relationship between the number of productive panicles and the number of gains per panicle as well as the number of filled grains per panicle. There was a significant negative relationship between the number of productive panicles and the seed setting rate. The number of grains per panicle was negatively correlated with the number of filled grains per panicle. The number of filled grains per panicle was very significantly correlated with the seed setting rate. The 1000-seed weight was weakly negatively correlated with the other four traits. Thus, among the five main economic traits, although there were very significant correlations among four traits, the 1000-seed weight was weakly correlated with the yield. The relationships among the five traits were also complex. To explore and formulate a standard for high efficient cultivation of the hybrid rice, it is therefore important to pay attention to the compatibility among these economic traits, so that the positive contribution of each trait can be maximized.

Table 2. Correlation coefficients among traits

	grains per panicle	filled grains per panicle	seed setting rate	1000-seed weight	actual yield
productive panicles	-0.6852**	-0.6268**	-0.1355*	-0.1254	0.1948**
grains per panicle		0.8146**	0.0620	-0.0183	0.2295**
filled grains per panicle			0.5985**	-0.0760	0.4584**
seed setting rate				-0.0749	0.4973**
1000-seed weight					-0.0482

Note: * P < 0.05, ** P < 0.01

3.2 Stepwise Regression Analysis

A regression equation was obtained from a stepwise regression analysis on 237 groups of data, based on the principle of maximizing the correlation coefficient: $\hat{y}$ =-13011.07+44.65 x_1 +7.70 x_2 +18.37 x_3 +35.47 x_4 +201.54 x_5 ±721.04

where, x_1 is the number of productive panicles, x_2 the number of grains per panicle, x_3 the number of filled grains, x_4 seed setting rate, x_5 the 1000-seed weight. Their ranges were, respectively, 1108.5~2161.5 thousand/hm^2, 214.10~453.18 grains/panicle, 168.70~407.00 grains/panicle, 65.82%~96.60% and 21.42~25.2 g. The range of y (yield) was 6495.00~12540.00 kg/hm^2。

The F value of the above equation was 73.52, which is very significant. An analysis of the partial regression coefficients indicates that the number of productive panicles per hm had the greatest effect ($r(y, x_1)$ = 0.6768), followed by sequentially the number of grains per panicle ($r(y, x_2) = 0.0937$), the number of filled grains per panicle ($r(y, x_3) = 0.1871$) and the seed setting rate ($r(y, x_4) = 0.1101$). However, these traits were interdependent to a certain extent. Because multivariate regression analysis has multiple co-linearity, it is not easy to judge contribution of each trait. The path analysis has effectively determined the direct effect of each variable on the result, allowed estimating the indirect effect of an independent variable, thus enabled the direct comparison of the importance of each trait on the yield [8, 9, 10]. Therefore, the author also performed a path analysis to determine effects of the major economic traits on yield in japonica x indica rice.

3.3 Path Analysis

A path analysis on five major economic traits selected from the stepwise regression analysis was conducted. Results indicates that the number of productive panicles had the greatest contribution, followed by the number of filled grains per panicle, the number of grains per panicle, the seed setting rate and the 1000-seed weight (Table 3).

Table 3. Path analysis

Factor	Direct effect	via x_1	via x_2	via x_3	via x_4	via x_5
x_1	0.8171		-0.1734	-0.4041	-0.0292	-0.0155
x_2	0.2531	-0.5598		0.5252	0.0134	-0.0023
x_3	0.6447	-0.5122	0.2062		0.1291	-0.0094
x_4	0.2157	-0.1107	0.0157	0.3859		-0.0093
x_5	0.1240	-0.1024	-0.0046	-0.0490	-0.0161	

The number of productive panicles had the greatest direct effect, with a direct path coefficient 0.8171. But it had negative effects through the other four traits, especially through the number of filled grains per panicle, with the path coefficient -0.4041. Thus, an increase in the number of productive panicles affected not only the number of grains per year, seed setting rate and the 1000-seed weight, but largely the number of filled grains. The overall effect of the number of productive panicles on the yield was r = 0.1948.

The number of grains per panicle had little effect on the yield, with a direct path coefficient 0.2531. It had a large indirect effect through the number of productive panicles negatively and through the number of filled grains per panicle positively, with coefficients -0.5598 and 0.5252. Its positive indirect effect through seed setting rate and negative indirect effect through the weight of ones thousand grains were small, being 0.0134 and -0.0023. The overall effect on yield from the number of grains per panicle was r = 0.2295.

The number of filled grains had a large direct effect on the yield, with a direct path coefficient 0.6447. Its negative indirect effect was large through the number of productive panicles, with a coefficient -0.5122. Its indirect positive effects through the number of grains per panicle and through the seed setting rate were smaller, with coefficients 0.2062 and 0.1291. Its negative effect through the 1000-seed weight was weak. The number of filled grains per panicle had not only a large direct effect, but also a large overall effect on the yield, being 0.4584. Therefore, while considering other traits especially the number of productive panicles, attention should be paid to increase the number of grains per panicle, especially the number of filled grains per panicle.

The seed setting rate had a direct path coefficient 0.2157, a large positive indirect effect coefficient 0.3859 through the number of filled grains per panicle. Its negative indirect effects through the number productive panicles and the 1000-seed weight, positive indirect effect through the number of grains per panicle were small. The overall effect of the seed setting rate on the yield was 0.4973.

The 1000-seed weight had the smallest direct effect on the yield, with a direct path coefficient 0.1240. It had not only small negative values for indirect effects through other traits, but also a negative overall effect on the yield. Thus, it is not advisable to increase the 1000-seed weight towards the goal of high efficient production.

4 Discussion and Conclusion

Economic traits affect not only directly the yield, but also indirectly through their influence on other traits. Through the above analysis on YU6, a representative of japonica x indica hybrid rice lines, it can be seen that the technical road to high efficient high yield production is to increase seed setting rate and produce large panicles while ensuring the number of productive panicles.

The stepwise regression analysis retained 5 factors that affect the yield. Path analysis results indicate that the number of productive panicles has the greatest influence on the yield, followed by the number of filled grains per panicle, the number of grains per panicle, seed setting rate and the 1000-seed weight. The previous stepwise regression analysis [7] removed the number of grains per panicle due to its insignificant effect. It indicates that the number of filled grains per panicle contributed the greatest influence, followed by the number of productive panicles, the 1000-seed weight and the seed setting rate. Although the first two most import factors are the same between the current analysis and the previous analysis, the order of these two factors has changed. This is due to the retention of the number of grains per panicle in the stepwise analysis in the current analysis. It is clear that the number of filled grains per panicle is highly dependent on the number of grains per panicle, and they were collinear. The current identified technical approach to high efficient, high yield production is the same as previously reported. The current study is more reliable since it had more data and these data covered a longer period.

References

[1] Zhang, Z., Yang, J., Zhou, C.: Characteristics of matter production of intersubspecific hybrid rice combinations. Jiangsu Agricultural Research 20(1), 1–8 (1999)

[2] Ye, S.: Performance of New Japonica Hybrid Rice Combination Yongyou 6 and Its High-yielding Cultural Techniques at Wenzhou, Zhejiang. Hybrid Rice 21(3), 58–59 (2006)

[3] Zheng, J.-C.: Main characters and cultural techniques of the new hybrid rice combination "Yongyou No.6". 2006(4), 409–410 (2006)
[4] Li, G.-F., Song, P., Cao, X.-Z.: Studies on the relationship between grain sink activity of Japonica/Indica hybrid rice and its grain filling. Acta Botanica Boreali-Occidentalia Sinica 20(2), 179–186 (2000)
[5] Shen, B., Wang, X.: Changes of Root Exudate of Indica-Japonica Hybrid Rice and Its Relation to Leaf Physiological Traits. Chinese Journal of Rice Science 14(2), 122–124 (2000)
[6] Liu, W.-M., Zhao, Y.-F., Yan, B.-L.: Study on Transplanting Technology and N-application Technology of Yongyou 6 as Single Cropping Late Hybrid Rice. SEED 26(12), 112–114 (2007)
[7] Gong, L.-Y., Du, X.-B., Liu, G.-L.: Research on seed production of water-saving and drought-resistant japonica hybrid rice Hanyou 8 'by machine' transplanting. Acta Agriculturae Shanghai 29(4), 18–22
[8] Bai, P.: High-efficient Ecological Cultivation Techniques of New Indica-japonica Hybrid Rice Yongyou 9. Hybrid Rice 27(2), 41–43 (2012)
[9] Liu, W.-M.: Study on the High Yield and High Efficiency Cultivation Technique of Indica-japonica Intersubspecific Hybrid Rice as Single-cropping Late Rice. Hubei Agricultural Sciences 50(18), 3684–3686 (2011)
[10] Liu, W.: Correlation, Multiple Regression and Path Analysis between Yield Traits and Yield on Intersubspecific Hybrid Rice. Chinese Agricultural Science Bulletin 25(1), 232–235 (2009)

Nitrogen Revising of Rapeseed (*Brassica napus* L.) Phenology and Leaf Number Models

Hongxin Cao[1], Yan Liu[1], Wenyu Zhang[1], Yeping Zhu[2], Daokuo Ge[1], Yanbin Yue[3], Yongxia Liu[4], Jinying Sun[5], Zhiyou Zhang[6], Yuli Chen[1], Weixin Zhang[1], Kunya Fu[1], Na Liu[7], Chunhuan Feng[1], and Taiming Yang[8]

[1] Institute of Agricultural Economics and Information, Engineering Research Center for Digital Agriculture, Jiangsu Academy of Agricultural Sciences, Nanjing 210014, Jiangsu Province, P.R. China
caohongxin@hotmail.com, liuyan0203@yahoo.com.cn, gedk@sina.com, nkyzwx@126.com

[2] Institute of Agricultural Information, China Academy of Agricultural Sciences, Beijing 100081, P.R. China
zhuyeping@caas.cn

[3] Institute of Agricultural Sci-tech Information, Guizhou Academy of Agricultural Sciences, Guiyang 550000, Guizhou Province, P.R. China
yanbin1220@163.com

[4] Institute of Banana and Plantain/Haikou Experimental Station, Chinese Academy of Tropical Agricultural Sciences, Haikou 570102, Hainan Province, P.R. China
liuyongxia0926@163.com

[5] Agricultural Technological Extensive Station of Luntai County in Xinjiang, Luntai 841600, Xinjiang, P.R. China
sunjinying240@sina.com

[6] Institute of Agricultural Sci-tech Information, Hunan Academy of Agricultural Sciences, Changsha 410000, Hunan Province, P.R. China
zhiyouzhang@sina.com

[7] Center for China Meteorological Information, China Meteorological Bureau, Beijing 100000, P.R. China

[8] Institute of Agricultural Meteorology, Anhui Provincial Meteorological Bureau, Hefei 230000, P.R. China
aunote@163.com, ytm0305@126.com, {921186907,1286234727}@qq.com, cdc@cma.gov.cn

Abstract. The Decision-making System for Rapeseed Optimization-Digital Cultivation Based on Simulation Models, DSRODCBSM, is a dynamic model that describes the growth and development of winter rapeseed. In order to perfect rapeseed growth models, Ningyou16 (NY16), Ningyou 18 (NY18), and Ningza 19 (NZ19) were adopted as materials, and the field experiments with 2 cultivars and 2 nitrogen levels, and pot experiment with 3 cultivars and 2 nitrogen levels were conducted during 2007-2008, 2008-2009, and 2011-2012 in Nanjing, respectively. The experimental results showed that the phenology and leaf number in rapeseed models had obvious difference for the same

© IFIP International Federation for Information Processing 2015
D. Li and Y. Chen (Eds.): CCTA 2014, IFIP AICT 452, pp. 54–66, 2015.
DOI: 10.1007/978-3-319-19620-6_7

cultivars under different nitrogen levels. Thus, the nitrogen effect factor, F (N), was put forward, used in the phenology sub-model in rapeseed growth models, and the verification of the leaf number sub-model can be done through model parameter adjusting. The simulated values before and after using F (N) and the observed values were compared, and the precision for the phenology sub-models in rapeseed growth models were raised further.

Keywords: nitrogen impact, rapeseed (*Brassica napus* L.), phenology models, leaf number models, revising.

1 Introduction

Rapeseed is one of very important oilseed crops in the world, and its plant area in normal year is about 18-30 million ha. The plant area of rapeseed in China is about 6-7 million ha, and its total yields is about 10-13 million tons, which ranks the fifth place in crop production in China [1]. It plays a very significant role in ensuring cooking oil and plant protein supply, and promoting farmer income increase that makes rapeseed production stable sustainable growth. However, the good cultivars and the relevant advanced management techniques are very important to promote rapeseed production with high yield, good quality, high benefit, ecology, and safety. In that the rapeseed growth models is an important basis of rapeseed precision management techniques.

In recent years, studies on rapeseed crop models have been made rapid progress. Notably, some rapeseed growth and development models, and ecological system models, e.g. EPR95 (erosion-productivity influence calculator, EPIC-Rape) [2], DAR95 (differential algebra for identifiability of systems, DAISY- Rape) [3], LINTUL-BRASNAP (light interception and utilization simulator) [4], CERES-rape (crop environment resource synthesis) [5], APSIM-Canola (agricultural production systems simulator) [6], and CECOL [7], etc. had been developed which can simulate rapeseed growth and development in real time. In China, the research on rapeseed growth model was not more. Liu and Jin [8], and Liu et al. [9] set up rapeseed phenology model etc. Zhang et al. [10], Cao et al. [11-14], and Tang et al. [15,16] studied the rapeseed growth and development simulation models, optimization models for rapeseed cultivation, and soil moisture and nitrogen dynamic models during rapeseed growth season, and the Decision-making System for Rapeseed Optimization-Digital Cultivation Based on Simulation Models (DSRODCBSM) were developed combining the rapeseed growth models (including phenology, leaf number, biomass, leaf area index (LAI), and shoot number dynamic models, etc.), the rapeseed optimization models (including the optimum season, the optimum LAI, the optimum shoot numbers, the optimum sowing rate, the optimum fertilization rate, and the optimum soil moisture, etc.), and expert knowledge of rapeseed plant diseases and insect pests, based on field experiments in Yangtz river middle valley of China [11], employing ideas of Rice or Wheat Cultivation-Simulation-Optimization-Decision making System (R/WCSODS) [17,18]. However, the rapeseed phenological models, and the leaf number models in DSRODCBSM were established under the optimum soil nitrogen, and water conditions, etc., if they were used in different soil nitrogen, and water conditions, there must be some differences in their results.

The objectives of this study were to introduce the effect factor of nitrogen in the phenology and leaf number sub-model (APPENDIX A, and B) in rapeseed growth models based on the field and pot experiments during 2007-2008, 2008-2009, and 2011-2012 in Nanjing, test, and perfect rapeseed growth models.

2 Materials and Methods

2.1 Materials

"Ningyou16" (NY16, conventional), "Ningyou18" (NY18, conventional), and "Ningza 19" (NZ19, hybrid) (breed by Institute of Economic Crops Research, Jiangsu Academy of Agricultural Sciences) were used in the experiments.

2.2 Methods

Experiment 1: The rapeseed cultivars, "NY16", and "NY18", were grown in the field from 2007 to 2008 on Yellow umber soil with higher fertility in pre-planting in soil in Nanjing (32°03'N), Jiangsu Province. The experiment included 2 cultivars and 2 nitrogen levels (Fertilizer: 0.018kg $N{\cdot}m^{-2}$; 0.012kg $P_2O_5{\cdot}m^{-2}$; 0.018kg $K_2O{\cdot}m^{-2}$; and 0.0015kg $borax{\cdot}m^{-2}$; CK: no fertilizer), 4 treatments, 3 replications, 12 subplots arranged random with 40.0-cm row spacing, 17-20 cm plant spacing in 7.00- by 4.30-m area, and the sowing date was on 26 SEP 2007. Fertilizing and other field managements in plots were the same.

Experiment 2: The rapeseed cultivars, "NY16", "NY18", and "NZ 19", were grown in the pot from 2008 to 2009 on Yellow umber soil with higher fertility in pre-planting in soil in Nanjing (32°03'N), Jiangsu Province. The experiment included 3 cultivars and 2 nitrogen levels (Fertilizer: 0.018kg $N{\cdot}m^{-2}$; 0.012kg $P_2O_5{\cdot}m^{-2}$; 0.018kg $K_2O{\cdot}m^{-2}$; and 0.0015kg $borax{\cdot}m^{-2}$; CK: no fertilizer), 4 treatments, 5 replications, and 20 pots, and the sowing date was on 28 SEP 2008. Fertilizing and other field managements in plots were the same.

Experiment 3: The rapeseed cultivars, "NY16", "NY18", and "NZ 19", were grown in the field from 2011 to 2012 on Yellow umber soil with higher fertility in pre-planting in soil in Nanjing (32°03′ N), Jiangsu Province. The experiment included 3 cultivars and 2 nitrogen levels (Fertilizer: 0.018kg $N{\cdot}m^{-2}$; 0.012kg $P_2O_5{\cdot}m^{-2}$; 0.018kg $K_2O{\cdot}m^{-2}$; and 0.0015kg $borax{\cdot}m^{-2}$; CK: no fertilizer), 6 treatments, 3 replications, and 18 subplots arranged random with 40.0-cm row spacing, 17-20 cm plant spacing in 7.00- by 4.30-m area, and the sowing date was on 15 OCT 2011. Fertilizing and other field managements in plots were the same.

The soil type of the experimental area is a hydragric anthrosol. Soil test results indicated the following: organic carbon, 13.7 g kg^{-1}; total nitrogen, 54.95 g kg^{-1}; available phosphorus, 24.25 g kg^{-1}; available potassium, 105.03 g kg^{-1}; and pH, 7.84.

2.2.1 Data Collection

The phenophase, LAI, the total shoot numbers, dry matter, leaf number, leaf photosynthesis, plant characters, and soil data, etc. were observed during rapeseed growth or after harvest.

The meteorological data during the experiments were down from Center for China Meteorological Information of China Meteorological Bureau.

2.2.2 Data Process

In this study, Excel.2007 and SPSS V 16.0 were used to analysis experimental data. The experiment data in 2008-2009 were applied to model establishment and parameter determination, and the experiment data in 2007-2008, and 2011-2012 were applied to model verification.

2.2.3 Model Verification

Simulation values were calculated in DSRODCBSM, and model precision was verified using root mean squared error (*RMSE*), mean absolute error (d_a), the ratio of d_a to the mean observation (d_{ap}) [19], the determined coefficient (R^2), and 1:1 plotting between measured values and simulated values. If *da* and *RMSE* were smaller and R^2 was larger, the simulated values were better agree with measured values, i.e. the deviation between simulated values and measured values was smaller, and simulation results of model were more accurate and reliable. The calculation formula of *RMSE* and *da* can be expressed as follows:

$$\mathrm{RMSE} = \sqrt{\frac{1}{n}\sum_{i=1}^{n}(X_{Oi} - X_{Si})^2}$$

$$d_a = \frac{1}{n}\sum_{i=1}^{n}(X_{Oi} - X_{Si})$$

$$d_{ap}(\%) = |d_a|/\overline{X}_O \times 100$$

$$\overline{X}_O = \frac{1}{n}\sum_{i=1}^{n} X_{Oi}$$

where X_{Oi} is observed values, X_{Si} is simulated values, d_a is absolute error, $|d_a|$ is a absolute value of d_a, d_{ap} is the ratio of d_a to the mean observation, and n is sample numbers.

3 Results

3.1 The Phenology and Leaf Number under the Different Nitrogen Rate

3.1.1 Phenology

Under the local normal sowing date in 2007-2008, the phenology of different nitrogen levels for same cultivars had obvious difference at enlongation, and the enlongation date under N application conditions were later than that of CK. But the mature dates were not difference (Table 1).

Table 1. The phenology under different cultivars and nitrogen rate in 2007-2008

Cultivars	Fertilizer	Sowing date (M-D)	Emergence date (M-D)	Enlongation date (M-D)	Early anthesis date (M-D)	Mature date (M-D)
NY16	N	09-26	09-29	**03-11**	03-25	05-16
	CK	09-26	09-29	**03-08**	03-25	05-16
NY18	N	09-26	09-29	**03-11**	03-25	05-16
	CK	09-26	09-29	**03-08**	03-25	05-16

Note: N represents Fertilizer: 0.018kg N·m^{-2}; 0.012kg P_2O_5·m^{-2}; 0.018kg K_2O·m^{-2}; and 0.0015kg borax·m^{-2}; and CK represents no fertilizer. The same as Table 2, 3, and 4.

Under the local late sowing date in 2011-2012, the phenology of different nitrogen levels for same cultivars had obvious difference at mature date, and the mature date under N application conditions were later than that of CK (Table 2).

Table 2. The phenology under different cultivars and nitrogen rate in 2011-2012

Cultivars	Fertilizer	Sowing date (M-D)	Emergence date (M-D)	Enlongation date (M-D)	Early anthesis date (M-D)	Mature date(M-D)
NY16	N	10-15	10-20	03-17	04-05	**05-22**
	CK	10-15	10-20	03-17	04-05	**05-20**
NY18	N	10-15	10-22	03-17	04-04	**05-21**
	CK	10-15	10-22	03-17	04-03	**05-18**
NZ19	N	10-15	10-20	03-17	04-04	**05-21**
	CK	10-15	10-20	03-17	04-03	**05-18**

3.1.2 Leaf Number

Under the local normal sowing date in 2007-2008, and in 2008-2009, the leaf number in main stem of different nitrogen levels had obvious difference only for NY16 (Table 3, and 4), and the leaf number in main stem of different nitrogen levels for NY18, and NZ19 had no obvious difference (Table 3, and 4).

Table 3. The leaf number in main stem under different cultivars and nitrogen rate in 2007-2008

Cultivar	Fertilizer	Date (M-D)							
		10-07	11-03	11-25	01-10	02-20	03-04	03-12	03-27
NY16	N	1.5	3.1	8.9	11.2	14.2	19.1	22.2	25.8
	CK	1.5	3.1	8.8	10.4	12.5	17.5	20.9	23.3
	±(N-CK)	**0**	**0**	**0.1**	**0.8**	**1.7**	**1.6**	**1.3**	**2.5**
NY18	N	1.5	4.1	8.9	10.7	12.8	18.1	22.6	27
	CK	1.5	4.1	9.6	11.1	13.9	18.8	22.8	27.1
	±(N-CK)	**0**	**0**	**-0.7**	**-0.4**	**-1.1**	**-0.7**	**-0.2**	**-0.1**

Table 4. The leaf number in main stem under different cultivars and nitrogen rate in 2008-2009

Cultivar	Fertilizer	Date (M-D)				
		11-25	12-16	02-24	03-03	03-07
NY16	N	10.2	13.9	28.1	28.7	35.3
	CK	10.6	13.7	26.1	27.5	33.7
	±(N-CK)	**-0.3**	**0.2**	**2.1**	**1.2**	**1.7**
NY18	N	8.2	11.1	31.6	32.0	30.3
	CK	8.2	11.1	30.9	32.5	31.0
	±(N-CK)	**-0.1**	**0.0**	**0.7**	**-0.5**	**-0.7**
NZ19	N	10.3	13.2	27.2	28.0	33.3
	CK	10.2	12.9	27.1	28.0	34.0
	±(N-CK)	**0.1**	**0.3**	**0.1**	**0.0**	**-0.7**

3.2 The Effect Factors of Nitrogen in the Phenology and Leaf Number Sub-model

3.2.1 Nitrogen Content in Leaf and Silique

Nitrogen content in leaf and silique of various cultivars and nitrogen rate in 2007-2008 shown in Fig. 1, and Fig. 2, and the results showed that the nitrogen content in leaf had a peak value at pre-over-wintering (8 JAN 2008) under nitrogen application conditions, in contrast, had a vale value at the same time under CK conditions (Fig. 1); the nitrogen content in silique had a peak value around end anthesis under nitrogen application conditions, in contrast, had a vale value at the same time under CK conditions (Fig. 2). It set a basis for developing the effect factor of nitrogen in the next step.

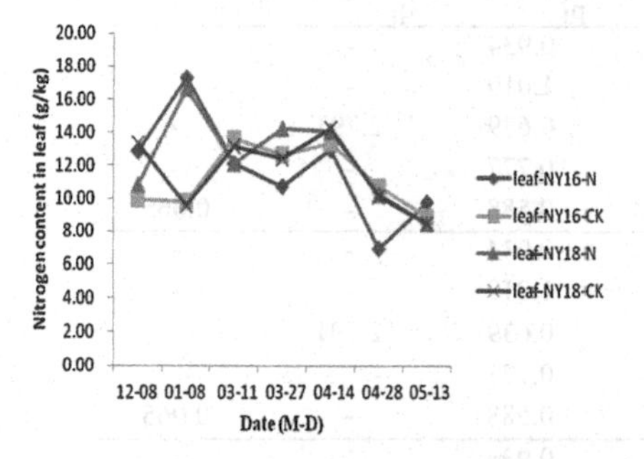

Fig. 1. The nitrogen content in leaf of various cultivars and nitrogen rate in 2007-2008

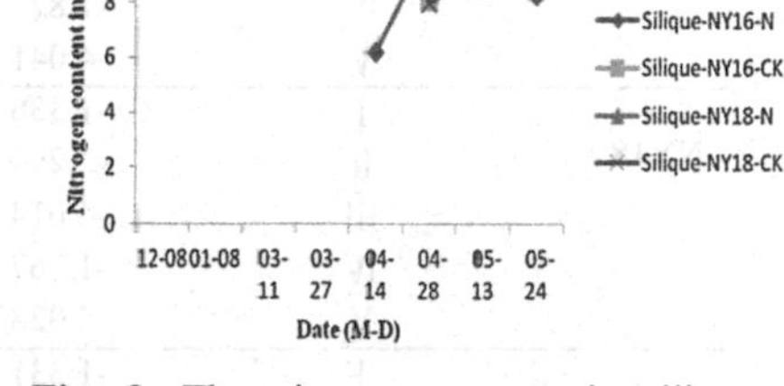

Fig. 2. The nitrogen content in silique of various cultivars and nitrogen rate in 2007-2008

3.2.2 Effect Factor of Nitrogen

According to experiment data in 2007-2008, the enlongation date was postponed with nitrogen application, and the mature date was postponed under late sowing condition in 2011-2012. Therefore, the effect factor of nitrogen, *F (N)*, in the phenology sub-model can be expressed as follows:

$$F(N) = \begin{cases} 1 & TRN < TCN \\ \frac{TRN - TCN}{TRN - TLN} & TRN \geq TCN \end{cases} \quad (1)$$

where *TRN* is the actual leaf nitrogen content (g kg^{-1}) around 10 d after fertilizing at pre-over-wintering, *TLN* is the lowest leaf nitrogen content (g kg^{-1}) in the same time for CK, and *TCN* is the critical leaf nitrogen content (g kg^{-1}). In that *TLN* and *TCN* can be obtained using the experiment data in 2007-2008, taking *TLN*=9.58 g kg^{-1} (Fig. 1) for CK at pre-over-wintering, and *TCN*=9.88 g kg^{-1}.

Due to the effects of nitrogen application on leaf numbers in main stem were different with cultivars, and years, the leaf numbers in main stem in sub-model can be verified through adjusting cultivar parameters in leaf number sub-model.

3.3 The Validation of the Phenology Sub-model After Revising

3.3.1 Parameters of the Phenology Sub-model

The various parameters of the phenology sub-model were determined using the experiment data in 2008 to 2009 (Table 5). We can see that *kj*, basic development coefficient which was determined by cultivar heredity, was different for various cultivars in the same development stages apart from stage II (emergence to vernalization), and *pj* (the genotypic coefficient of temperature effects for increasing), *qj* (the genotypic coefficient of temperature effects for decreasing), and *Gj* (the genotypic coefficient of photoperiod effects) were the same for various cultivars in the same development stages.

Table 5. Parameters of models in various development stages for winter rapeseed

Cultivar	Development Stage	Parameter of model			
		kj	pj	qj	Gj
NY16	I	-1.365	0.934	-	-
	II	-2.294	1.019	-	-
	III	-3.964	0.639	2.791	-
	IV	-1.827	0.777	-	-
	V	-4.041	0.588	-	0.065
NY18	I	-1.336	0.934	-	-
	II	-2.294	1.019	-	-
	III	-4.014	0.639	2.791	-
	IV	-1.767	0.777	-	-
	V	-4.028	0.588	-	0.065
NZ19	I	-1.331	0.934	-	-
	II	-2.294	1.019	-	-
	III	-3.778	0.639	2.791	-
	IV	-1.539	0.777	-	-
	V	-3.905	0.588	-	0.065

Note: j represents development stage I (planting to emergence), II (emergence to vernalization), III (vernalization to enlongation), IV (enlongation to early anthesis), and V (early anthesis to mature).

3.3.2 The Validation of the Phenology Sub-model After Revising

The comparison between phenology with F (N) and no F (N) were shown in Table 6, and Table 7, Fig. 3, and Fig. 4. The results showed that the phenology with F (N) were more close to the observed values of nitrogen treatments (Table 1, and Table 2), and the *RMSE*, R^2, d_a, and d_{ap} between observed and simulated values with for the same cultivars in 2007-2008, and in 2011-2012 were 0.77 d, 0.9998, -0.20 d, and 0.1715%; 1.45 d, 0.9998, -0.90 d, and 0.8295%. We can see that precision of the phenology sub-model was raised further.

Table 6. The comparison between phenology with F (N) and no F (N) in 2007-2008

Cultivars	F(N)	Sowing date (M-D)	Emergence (M-D)	Enlongation (M-D)	Early anthesis (M-D)	Mature (M-D)
NY16	—	09-26	09-29	03-08	03-25	05-16
	+	09-26	09-29	03-10	03-26	05-17
NY18	—	09-26	09-29	03-08	03-25	05-16
	+	09-26	09-29	03-10	03-26	05-17

Note: + and — denoted the phenology with F (N) and no F (N). The same as Table 7.

Table 7. The comparison between phenology with F (N) and no F (N) in 2011-2012

Cultivars	F(N)	Sowing date (M-D)	Emergence (M-D)	Enlongation (M-D)	Early anthesis (M-D)	Mature (M-D)
NY16	—	10-15	10-19	03-18	03-31	05-19
	+	10-15	10-19	03-18	04-03	05-21
NY18	—	10-15	10-19	03-20	04-01	05-19
	+	10-15	10-19	03-20	04-03	05-22

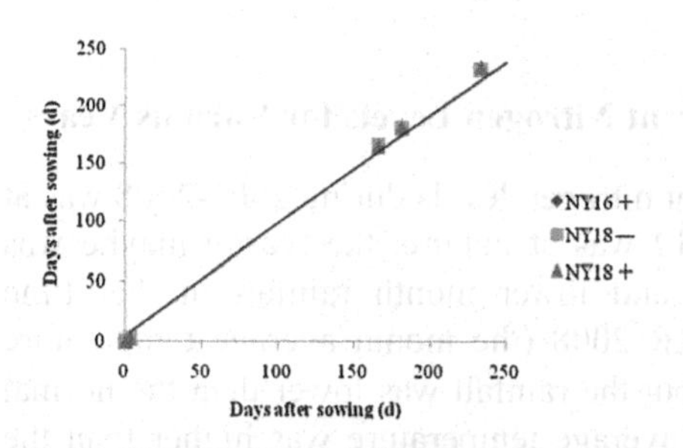

Fig. 3. The 1:1 diagram between phenology with F (N) and no F (N) in 2007-2008

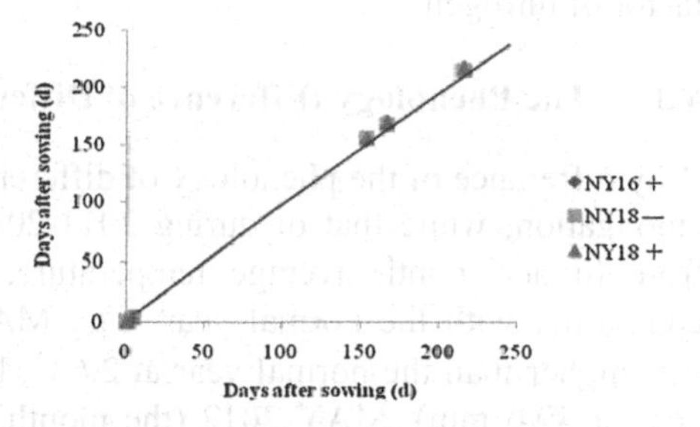

Fig. 4. The 1:1 diagram between phenology with F (N) and no F (N) in 2011-2012

4 Discussion

4.1 The Application of Phenology in Rapeseed Growth Models in Region and Site Scales

The rapeseed growth models need to be verified using the nitrogen impact factor when they will be applied in region and site scales. The effect factor of nitrogen, *F(N),* was gained according to different yield level types in region scale, and according to fertilizer level in site scale. Due to the calculation of *F(N)* need to obtain the *TRN* (actual leaf nitrogen content), *TCN* (critical leaf nitrogen content), and *TLN* (lowest leaf nitrogen content (g kg^{-1}) around 10 d after fertilizing at pre-over-wintering for CK) values under the local condition, and it should be tested in different sites. CERES-Rape [5] also had modules for crop phenology, net photosynthesis, leaf area development, and grain filling, as influenced by crop N status. Chen *et al.* [20] studied fruit-per-plant model for protected tomato using fertilizer factor, describing the effects of nitrogen on fruit-per-plant for protected tomato.

The phenology and leaf number in rapeseed were affected by multi-factors, which were decided by genotypes and environmental factors, and temperature and light in environmental factors were main factors of them. In addition, the phenology was also affected by fertilizer, water, and so on. However, the nitrogen impact factor, *F(N),* was only introduced into the phenology model in this paper. The water impact factor should be considered in future studies. As to the relationship between leaf number in main stem in rapeseed and nitrogen application, it was different under various cultivars, and needed to be studied further.

4.2 The Effect Factor of Nitrogen

It was determined according to the changes in nitrogen content in leaf and silique of various cultivars and nitrogen rates (Fig. 1, and Fig. 2), and because leaf nitrogen content, *TRN,* can be acquired easily comparing with silique, the actual leaf nitrogen content around 10 d after fertilizing at pre-over-wintering was introduced to the effect factor of nitrogen.

4.3 The Phenology Difference of Different Nitrogen Levels for Various Years

Why difference of the phenology of different nitrogen levels during 2007-2008 was at enlongation, while that of during 2011-2012 was at mature, the reason maybe was from higher month average temperature, and lower month rainfall in that time comparing with the normal year, e.g., MAR 2008 (the month average temperature was higher than the normal year at 2.6℃, but the rainfall was lower than the normal year at 49.6 mm), MAY 2012 (the month average temperature was higher than the normal year at 1.4℃, but the rainfall was lower than the normal year at 39.7 mm) (Table 8 and Table 9).

Table 8. The comparison of month value of meteorological conditions during 2007 to 2008 and that of the normal year in Nanjing (data from Center for China Meteorological Information of China Meteorological Bureau)

Year	Month	Average temperature (℃)	±(AT-NY) (℃)	Average min temperature (℃)	±(ANT-NY) (℃)	Average max temperature (℃)	±(AXT-NY) (℃)	Sun times (hr.)	±(ST-NY) (hr.)	Rainfall (mm)	±(R-NY) (mm)
2007	9	24.1	1.3	20.5	1.3	28.2	0.9	135.3	-31.9	127.4	55.3
	NY	22.8		19.2		27.3		167.2		72.1	
	10	18.5	1.4	15	2.1	23.1	0.9	153.4	-15.7	39.7	-25.4
	NY	17.1		12.9		22.2		169.1		65.1	
	11	11.2	0.8	6.7	0.6	16.8	0.9	161.3	7.8	23.7	-27.1
	NY	10.4		6.1		15.9		153.5		50.8	
	12	6.7	2.2	4.1	3.7	10.2	0.2	49.6	-100.6	40.7	16.2
	NY	4.5		0.4		10		150.2		24.5	
2008	1	1.5	-0.9	-1	0.1	4.9	-2.1	56.1	-73	110.1	72.7
	NY	2.4		-1.1		7		129.1		37.4	
	2	2.5	-1.7	-1.7	-2.3	7.6	-1.2	194.1	70.8	18.9	-28.2
	NY	4.2		0.6		8.8		123.3		47.1	
	3	11.3	2.6	6.7	1.9	16.9	3.5	185.8	49.7	32.2	-49.6
	NY	8.7		4.8		13.4		136.1		81.8	
	4	15.6	0.4	11.7	1.1	20.4	0.1	146	-22.1	90	16.6
	NY	15.2		10.6		20.3		168.1		73.4	
	5	22.2	1.7	17.3	1.4	28.1	2.5	242.4	48.4	81.4	-20.7
	NY	20.5		15.9		25.6		194		102.1	
	6	23.5	-0.9	20.6	-0.1	27.2	-1.6	71.1	-100.8	131.7	-61.7
	NY	24.4		20.7		28.8		171.9		193.4	

Note: NY, AT, ANT, AXT, ST, and R denoted the normal year, average temperature, average min temperature, average max temperature, sun times, and rainfall, respectively. The same as Table 8.

Table 9. The comparison of month value of meteorological conditions during 2011- 2012 and that of the normal year in Nanjing (data from Center for China Meteorological Information of China Meteorological Bureau)

Year	Month	Average temperature (℃)	±(AT-NY) (℃)	Average min temperature (℃)	±(ANT-NY) (℃)	Average max temperature (℃)	±(AXT-NY) (℃)	Sun times (hr.)	±(ST-NY) (hr.)	Rainfall (mm)	±(R-NY) (mm)
2011	9	23.2	0.4	20.1	0.9	27.3	0	157.2	-10	12.6	-59.5
	NY	22.8		19.2		27.3		167.2		72.1	
	10	17.6	0.5	14.3	1.4	21.9	-0.3	155.3	-13.8	28.7	-36.4
	NY	17.1		12.9		22.2		169.1		65.1	
	11	14.7	4.3	11.6	5.5	19.3	3.4	133.4	-20.1	21.3	-29.5
	NY	10.4		6.1		15.9		153.5		50.8	
	12	4.2	-0.3	0.9	0.5	8.6	-1.4	161.2	11	15.8	-8.7
	NY	4.5		0.4		10		150.2		24.5	
2012	1	2.9	0.5	0.2	1.3	6.4	-0.6	100.5	-28.6	21	-16.4
	NY	2.4		-1.1		7		129.1		37.4	
	2	3	-1.2	0.3	-0.3	6.3	-2.5	87.9	-35.4	73.3	26.2
	NY	4.2		0.6		8.8		123.3		47.1	
	3	9	0.3	5.4	0.6	13.4	0	128.8	-7.3	79.3	-2.5
	NY	8.7		4.8		13.4		136.1		81.8	
	4	17.9	2.7	13.3	2.7	23.4	3.1	189.2	21.1	56.2	-17.2
	NY	15.2		10.6		20.3		168.1		73.4	
	5	21.9	1.4	17.5	1.6	26.9	1.3	198	4	62.4	-39.7
	NY	20.5		15.9		25.6		194		102.1	
	6	25.5	1.1	22.1	1.4	29.6	0.8	141.6	-30.3	17.8	-175.6
		24.4		20.7		28.8		171.9		193.4	

5 Conclusions

This paper presented an attempt at validating and perfecting of phenology, and leaf number sub-model in rapeseed growth models. Through the 3 year field experiment data analysis, we can conclude that the phenology in rapeseed models had obvious difference for the same cultivar under different nitrogen levels. Thus, the nitrogen effect factors were put forward and used in the phenology sub-model in rapeseed growth models, and the verification of the leaf number sub-model can be done through model parameter adjusting. The simulated values before and after using nitrogen effect factors and the observed values were compared, and the precision for the phenology sub-models with nitrogen effect factors in rapeseed growth models were raised further.

Appendix

A Phenology

The basic models of rapeseed phenology were developed in the thesis through employing ideal of "Rice Clock Models" [11-14][17-18].

$$dP_j/dt = 1/D_{Sj} = e^{kj} \cdot (T_{ebj})^{pj} \cdot (T_{euj})^{qj} \cdot (P_{ej})^{Gj} \cdot f(E_{Ci})$$

$$T_{ebj} = (T_i - T_{bj}) / (T_{oj} - T_{bj}), \text{ when } T_i < T_{bj}, T_i = T_{bj}; \text{ when } T_i > T_{oj}, T_i = T_{oj}$$

$$T_{euj} = (T_{uj} - T_i) / (T_{uj} - T_{oj}), \text{ when } T_i > T_{uj}, T_i = T_{uj}$$

$$P_{ej} = (P_i - P_{bj}) / (P_{oj} - P_{bj}), \text{ when } P_i < P_{bj}, P_i = P_{bj}; \text{ when } P_i > P_{oj}, P_i = P_{oj}$$

where dP_j/dt is the development rate at the jth stages, D_{Sj} is the days at the jth stages, T_{ebj} and T_{euj} are the effective factors for temperature, respectively, kj is basic development parameter which is determined by cultivar heredity, pj and qj are the genotypic coefficient of temperature effects, P_{ej} is the effective factor of photoperiod, Gj is the genotypic coefficient of photoperiod effects, and $f(E_{Ci})$ is the effective function of agronomic practice factors for rapeseed, T_i is the daily mean temperature (℃) in the jth stage, T_{bj}, T_{oj} and T_{uj} are lower, optimum, and upper limit temperature (℃) demanded in the jth stage for rapeseed, respectively, and P_{bj}, P_{oj} are the critical and optimum day length (h) demanded in j^{th} stage for rapeseed, respectively.

Vernalization models can be described as following through employing ideals of "wheat clock models":

$$dV/dt = 1 / D_{s2} = e^{k2} \cdot (V_e)^C$$

If a cultivar was winter or semi-winter rapeseed, the expression of V_e was:

$$V_e = \begin{cases} \frac{V_{ti}+4}{9}, & -4 < V_{ti} \le 5°C \\ 1.0, & 5 < V_{ti} \le 10°C \\ \frac{20-V_{ti}}{10}, & 10 < V_{ti} \le 20°C \\ 0, & V_{ti} \le -4\,°C \text{ or } V_{ti} > 20°C \end{cases}$$

However, if it was spring rapeseed, the expression of V_e was:

$$V_e = \begin{cases} \frac{V_{ti}}{5}, & 0 < V_{ti} \le 5°C \\ 1.0, & 5 < V_{ti} \le 20°C \\ \frac{30-V_{ti}}{10}, & 20 < V_{ti} \le 30°C \\ 0, & V_{ti} \le 0\,°C \text{ or } V_{ti} > 30°C \end{cases}$$

where $K2$ and C are the parameters of vernalization, V_e is the factor of rapeseed vernalization effect, V_{ti} is the daily mean temperature in vernalization phase. It will finish vernalization phase when V_e equal to some extent accumulation days; the vernalization days of the winter rapeseed were 30 to 40 days, the semi-winter rapeseed with 20 to 30 days, and the spring rapeseed with 15 to 20 days.

B Leaf Number

The growth rate of rapeseed leaf were different in different varieties, development stages, temperature, and nutrition conditions etc., when nutrition condition was optimum, the models of rapeseed leaf number were [11-14][17-18] :

$$dL_j/dt = f(L_j) = 1/D_{Lj} = D_{Loj} \cdot (T_t/T_o)^{La/Lb}$$

$$T_t = \begin{cases} 0, & \text{when } T_t < T_{bj} \\ T_o, & \text{when } T_t > T_o \end{cases}$$

$$D_{Loj} = e^{LK}$$

where dL_j/dt is the development rate of the jth leaf , $f(L_j)$ is the basic development function, D_{Lj} is the development days demanded from emergence to the jth leaf number, D_{Loj} is the development days demanded from emergence to the jth leaf number under the optimum conditions, T_t and T_o are the daily mean temperature (℃) of the tth day, and the optimum temperature for rapeseed leaf number development, respectively, and *La*, *Lb*, and *LK* are the parameters of leaf models, respectively.

Acknowledgement. The authors would like to express their appreciation to Prof. Cunkou Qi, Huiming Pu, and Xinjun Chen of the Institute of Economic Crop Research, Jiangsu Academy of Agricultural Sciences, China, for providing trial materials. This work was supported by the National Natural Science Foundation of China (31171455; 31201127; 31471415), the National High-tech R&D Project (2013AA102305), the Jiangsu Province Agricultural Scientific Technology Innovation Fund, China (CX(14) 2114), the Agricultural Scientific Technology Support Program, Jiangsu Province, China (BE2011342; BE2012386), the No-Profit Industry (Meteorology) Research Program, China (GYHY201106027), the Jiangsu Academy of Agricultural Sciences Basic Scientific Reserch Work Special Fund [ZX(15)2008], China, and the Jiangsu Government Scholarship for Overseas Studies, China.

References

1. National Bureau of Statistics of China (2009), `http://www.stats.gov.cn/tjsj/qtsj/gjsj/` (September 26, 2013)
2. Kiniry, J.R., Major, D.J., Izaurralde, R.C., et al.: EPIC model parameters for cereal, oilseed, and forage crop in the north Great Plain region. Canadian Journal of Plant Science 63, 1063–1065 (1983)
3. Petersen, C.T., Svendsen, H., Hansen, S., et al.: Parameter assessment for simulation of biomass production and nitrogen uptake in winter rape. Europe Journal of Agronomy 4, 77–89 (1995)
4. Habekotté, B.: A model of the penological development of winter oilseed rape (Brassica napus L.). Field Crop Research 54, 137–153 (1997)
5. Gabrielle, B., Denoroy, P., Gosse, G., et al.: Development and evaluation of a CERES-type model for winter oilseed rape. Field Crop Research 57, 95–111 (1998)

6. Farré, M.J., Robertson, G.H., Walton, S.A.: Simulating response of canola to sowing data in Western Australia. In: Proceedings of the 10th Australia Agronomy Conference, Hobart, pp. 36–40 (2001)
7. Husson, F., Wallach, D., Vandeputte, A.: Evaluation of CECOL, a model of winter rape (Brassica napus L.). Europe Journal of Agronomy 8, 205–214 (1998)
8. Liu, H., Jin, Z.: The simulation models of rape development dynamic. Journal of Application Meteorology 14(5), 634–640 (2003) (in Chinese with English Abstract)
9. Liu, T., Hu, L., Zhao, Z., et al.: A mechanistic of phasic and phenological development in rape. I. Description of the model. Chinese Journal of Oil Crop Sciences 26(1), 27–30 (2004)
10. Zhang, C., Li, G., Cao, H., et al.: Simulating growth and development of winter rape in Yangtze river valley. In: Proceedings of 11th International Rapeseed Congress, Copenhagen, Denmark, July 6-10, p. 835 (2003)
11. Cao, H., Zhang, C., Li, G., et al.: Researches of Decision-making System for Rape Optimization-Digital Cultivation Based on Simulation Models. In: The Third International Conference on Intelligent Agricultural Information Technology, pp. 285–292. China Agricultural Sciences and Technology Press, Beijing (2005)
12. Cao, H., Zhang, C., Li, G., et al.: Researches of Simulation Models of Rape (Brassica napus L.) Growth and Development. Acta Agronomica Sinica 32(10), 1530–1536 (2006) (in Chinese with English Abstract)
13. Cao, H., Zhang, C., Li, G., et al.: Researches of Optimum Leaf Area Index Dynamic Models for Rape (Brassica napus L.). In: Zhao, C., Li, D. (eds.) Computer and Computing Technologies in Agriculture II, Volume 3. IFIP AICT, vol. 295, pp. 1585–1594. Springer, Boston (2009)
14. Cao, H., Zhang, C., Li, G., et al.: Researches of Optimum Shoot and Ramification Number Dynamic Models for Rapeseed (Brassica napus L.). In: World Automation Congress (WAC), pp. 129–135 (2010)
15. Tang, L.: Rapeseed growth simulation and decision-making support system. Dissertation for Ph.D. of Nanjing Agricultural University (2006)
16. Tang, L., Zhu, Y., Liu, T., et al.: A Process-Based Model for Simulating Phenological Development in Rapeseed. Scientia Agricultura Sinica 41(8), 2493–2498 (2008) (in Chinese with English Abstract)
17. Gao, L., Jin, Z., Huang, Y., et al.: Rice cultivational simulation-optimization-decision making system. Chinese Agricultural Sciences and Technology Publication House, Beijing (1992) (in Chinese with English Abstract)
18. Gao, L., Jin, Z., Zheng, G., et al.: Wheat cultivational simulation-optimization-decision making system. Journal of Jiangsu Agriculture 16, 65–72 (2000) (in Chinese with English Abstract)
19. Cao, H., Hanan, J.S., Liu, Y., et al.: Comparison of crop model validation methods. Journal of Integrative Agriculture (Formerly Agricultural Sciences in China) 11(8), 1274–1285 (2012)
20. Chen, Y.: Modeling of nitrogen accumulation and partitioning in plant and yield formation for protected cultivated tomato (Lycopersicon esculentum Mill.). M.D Paper of Nanjing Agricultural University (2012)

Research on the Principles of User Behavior in Building Information Resource Sharing System

Qi Wang and Wenyong Chen

Library of JiLin Agriculture University, ChangChun 130118, China
wangqibios@163.com, hongsepiaochong@126.com

Abstract. The information resource sharing system as a information movement, will guide users to access and use knowledge or information as the information commons. Many barriers existed in information sharing systems due to the less consideration of the users' behavior in information utilization by designers of the information sharing systems. Therefore, the design of information sharing systems should be fully aware of and respect for user information behaviors, and on this basis, to adhere to the design concept of the user information behavior laws in order to truly complete the mission of the information sharing systems. Otherwise, the principle of user information behaviors will most likely constitute the information sharing "regularity disorder".

Keywords: information sharing, information behaviors, information behavior laws.

1 Introduction

Information sharing system as a knowledge or information commons, will be a main physical location of users to complete their information behaviors. To build such system, users' needs should be the driving force, users' information behaviors should be respected, users' information use behaviors should be coincided. This is the only way to get users' support and extensive use and complete the constructed mission and achieve its ultimate goal.

2 Behavioral Obstacles of Users to Use Information Resources in Information Commons

At present, domestic and foreign scholars have conducted extensive research on behavioral obstacles of users using information under network environment. It is significant to summarizes those behavioral obstacles in establishment of information commons. Research by Zhang Yiyan et al. shows [1] that: behavioral obstacles of the users using information are as follow: (1) information identification obstacles, including: literature demand barriers, available information obstacles, information sources integration obstacles, information sources classification barriers, information language barriers; (2) information access barriers, including: information availability

© IFIP International Federation for Information Processing 2015
D. Li and Y. Chen (Eds.): CCTA 2014, IFIP AICT 452, pp. 67–72, 2015.
DOI: 10.1007/978-3-319-19620-6_8

barriers, information sources response speed, information sources cost barriers, the carrier type and literature format of information sources obstacles; (3) information use obstacles, including: information quality barriers, information reveal accuracy barriers, information not comparable obstacles. Zhao Yalan et al. have pointed out that [2], users information obstacles under the conditions of modern technology literature information sharing, including: (1) information identify obstacles; (2) information access obstacles; (3) reading habit obstacles; (4) information screening obstacles, and so on. According to JiLin Agricultural University library's actual experience to participate in the CALIS and CASHL as well as other types of literature information resources sharing systems, users meet the following obstacles in all kinds of literature at various levels of information sharing system, including: (1) Demand obstacles, for which performance are that: resources time span covered in the system are limited, especially in the seventies and eighties and the previous literature of last century only covering relatively small amount, access to the full text of such documents is relatively difficult, update of these data is much slower, with a single document type, document delivery request fill rate is low and sometimes there is deviation in billing system(CASHL). (2) Availability obstacles, e.g. searching platform overmuch and easy to confuse, searching interface not so concise or clear, and search fields limited(CASHL). (3) Tool obstacles, such as system itself not so stable, especially the CALIS.

Users information behaviors are less considered by the designer of information commons which lead the above obstacles[3]. Therefore, to realize information sharing, users information behaviors should be fully understood and respected through information resources sharing space platform design and the principle should be insisted for meeting the mission of information sharing platform.

3 User Information Behavior Principle in Building Information Resource Sharing System

3.1 Moores Law

Calvin Mooers-one of the pioneers of the information retrieval research, after a number of studies he found out that users will tend to not use information retrieval system if it brings more trouble to use. The law reveals the rule of users using information retrieval system. For users, the most accessible information is not the most useful information, which means the availability and validity of information is not consistent. The users select a retrieval system putting the information availability as the primary basis.

At present, the more popular understanding of the Moores law focus on the efforts of users obtain the required information in library and information science. Thus, the law can be expressed as: the more difficult using, the more time spending on an information retrieval system, the less using the information retrieval system by users. Roger K Summit, Dialog founder, express it by another way: “Moores law tells us that information using and easiness to access information are proportional.” [7]

According to the law, information availability is the primary basis for designing of information retrieval and sharing systems. Therefore, the Moores law is not only the rule of designing information retrieval system, but also the guide of designing collection and information system.

3.2 Zipf Law

In 1949, linguistics professor George Kingsye Zipf, Harvard University, pointed out "minimum power-saving principle" in his book called "Human behavior and Minimum Power-saving Principle --Human Ecology Introduction" [8]. It put forward that human social activities are affected by this principle, people always hope to use the minimum cost to obtain the maximum benefit. In other words, human behavior is based on the minimum power-saving principle. In fact, this principle is using throughout the whole users information retrieval behavior. Therefore, this law should be followed in the information sharing system, the users will give up obtaining sharing information if his work intensity is too large.

It reveals two characteristics of users psychology in information sources selection: (1) The most readily available source of information is the first choice when user selects information source, and then its reliability and quality. (2)In the process of obtaining information, the user is more likely to minimize the cost of obtaining information at the expense of the information quality. Hence, there are two matters should be considered for designing information sharing system: first, those systems with cumbersome operation, too complicated path in building information commons are not viable; Second, the database of information commons must be context unified, standard terminology, semantics accurate, and expression concise, minimizing the work of understanding the information content. It should build and provide service from user perspective, then research the information commons accessibility how to influence the user requirements from the user psychology, ensure the minimum power-saving principle implementation to meet user needs.

Wang Xiaona put forward seven measures to improve the accessibility of building information commons[9]: (1) physical environment availability: the using interface should be friendly except the geographical position, and can use multimedia to guide users; (2) retrieval function and access points availability: the retrieval ways and types should be varied to fit users' different retrieval habit; (3) database availability: increase the content of the database, such as subject terms, keywords and citation, etc.; (4) system openness and interconnectedness: the system can be connected with others; (5) remote communication: exchange information to the distance; (6) users assistance: with a complete operating instructions; (7)retrieval system structure availability: promote system efficiency.

In addition, the law is also applicable to the information retrieval list compilement and information document structure designment of information resources sharing space.

3.3 Robin Hood Effect

Developing with the science technology and informatization, the social members information need level has improved. Majority users' need of information tends to average, thus the information resources using also presents a trend of average. The users' information demand will be affected when the information resource is not sufficient as the Robin Hood effect. At present, the trend will be more obvious due to the network equality.

Robin Hood effect reveals two characteristics of the information resources use from the perspective of users behavior[10]: first, the users' information need must be affected if the information resources is not sufficient; Second, most users' information need level tends to average. Understanding the Robin Hood effect helps to reduce the "block" grade, break mind-set grade, and determine the priority according to the information value maximization standard in setting up information sharing space. The key users' information need should be considered priority, taking into account all users' information resources sharing power to satisfy all users' need.

3.4 Information Absorption Limit Regulation

Information absorption process is the process of cognitive activities between users and information. Users' information absorption includes information accepting, processing, understanding and utilizing etc. Users reaction and absorption to the information is faster following the informatization acceleration. But the users' information absorption is limited, in a certain range, when information input and absorption rate exceeds the critical value, the users' information response and absorption rate will be slower, then it will appear the information overload phenomenon, which we called information absorption limitation[11]. Generally speaking, under the certain period of time, large amount of outside information is provided to users which is far larger than the amount of information the user can absorb, users have to choose and filter those information,and to keep the most important and valuable information.

According to this regulation, designers of information commons, especially in the design of retrieval interfaces, should strictly follow users' information absorption regulation to design multi-types retrieval paths and provide related methods to avoid narrow or expanded searching result. In addition, to avoid reducing the users' distress of information accession, the design of retrieval platform classification system can be neither too professional nor too simplified [12].

3.5 Information Foraging Regulation

In the early 1990s, Peter Pirolli and Stuart Card,from California Palo Alto Research Centre (PARC) of United States, found that people's information search pattern under the network environment is similar to animal foraging strategies. They studied users' information behavior and behavior navigation of information landscape (links, description, and other data)under the network environment cooperating with psychologists, then they found that the time, money and energy of users in obtaining information and the needed information obtained needs to reach an optimal balance, like the animal foraging strategies. Thus, they put forward the information foraging theory first time, and pointed out that in the process of information searching and absorption, people should adjust their own information foraging strategy constantly to achieve maximization of information profit according to the information environment[13].

The information foraging theory has three main elements: information patches, information scent and information diet[14 to 15]. Information patches is the abundant information resource in information environment, it can be a website, a paper, a book, a web page, a document through that users search information. Information scent is the extension of information concept clues, it is a subjective assessment for contained information and required information correlation in the information resources during the process of searching information, and it plays a navigated role in information searching through some certain value or utility menu labeling, which decides the path information searching along. Users find the optimal information "patches" with the correlated information clues during searching information on the Internet.Information diet is the kind of information which users choose when they forage information . Due to the regularity of network information resources distribution it consumes scarce resource of time,money and energy when user foraging information, so user needs to select information during searching information. If the searching scope is too narrow, user may need to spend more time to build retrieval model, which makes the searching results partial; If the searching range is too expanded, user may be submerged by information retrieved again. So, in constructing information commons, we should design the information diet to conform user information behavior .

The designers have been paying attention to interface design in the information searching system design. They think the interface must be powerful, too simple interface can not customize questions according to users' need and obtain ideal results. However, actual survey showed that users prefer simple interface, people hope use patterns instead of words to increase the interface usability[9]. Henk J.Voorbij through the network information searching survey pointed: "The advanced retrieval utilization rate is very low"[16]. Bernard j. Jansen also found in the experiment research: "Retrieval results are not obvious affected by complex questions" [17-18]. These research results show that the retrieval system designers should not have to spend energy on the theory of design advanced retrieval functions, they should focus on how to design better and related online help, how to insert the efficient automatic correction method, how to filter the search results and carry effective sorting, how to provide users with retrieval vocabulary control, etc. Jason Vaughan also pointed out: "Retrieval system designers must consider problems from users perspective.The retrieval system designers are engineers and computer scientists, they attach more importance to comprehensiveness and flexibility of functions, but common users have no experience on retrieval system or database, they just require the design of retrieval tools are easy to use, so there should be a bridge between the thinking ways of these two types people ." [19]

4 Conclusions

Information commons will be a knowledge commons that is a special designed one-stop service center and collaborative information sharing environment. The design of it can be based on two ideas: first is open access, second is service. Information commons needs to correspond with hommization design requirements, user information behavior principles and user information behavior regulations.

Acknowledgment. The corresponding author of this paper is Wenyong Chen. This paper is supported by the Youth Foundation of Jilin Agricultural University (Grant NO. 201337), CALIS Programs of the National Agronomy Literature Information Center(Grant NO. 2014026) .

References

1. Zhang, Y., Yang, Y.: The Study of Information Process Gaps. Document, Information and Knowledge (5), 78–81 (2008)
2. Zhao, Y., Li, P.: Analysis of the Document Information Sharing Barriers in the Conditions of Modern Technology. Modern Information (2), 9–11 (2004)
3. Cooper, A., Liu, S. (trans.): About Face 3 Interactive Essence of the Design, pp. 38–49. Publishing House of Electronics Industry, Beijing (2008) (in Chinese)
4. Seal, R.A.: The Information Commons: New Pathways to Digital Resources and Knowledge Management. Reprint for the 3rd China/U.S Conference on Libraries, Shanghai (March 2005)
5. Sun, Y.: Visualization Analysis on Knowledge Foundations, Research Hotpots and Frontiers of Information Behavior. Document, Information and Knowledge (1), 108–116 (2012)
6. Zhang, Z., Zhang, H.: Avoiding Information Sharing "Regulation Obstacles". Libration Army Daily (7) (May 24, 2012)
7. Summit, R.K., Roger, K.: Summit. Online Information Review 15(3/4), 123–128 (1991)
8. Zipf, G.K.: Human Behaviour and the Principle of Least Effort: An Introduction to Human Ecology, 1st edn., pp. 19–55. Hafner, New York (1972)
9. Wang, X.: Accessibility of Minimum Power- saving Principle to Information Retrieval System. Information Science 18(2), 135–136 (2000) (in Chinese)
10. Zhou, W.: Regularity Study of Network Information users' information needs. Information Research (3), 12–14 (2008)
11. Yang, C.C., Chen, H., Honga, K.: Visualization of large category map for Internet browsing. Decision Support Systems 35(1), 89–102 (2003)
12. Wu, P., Zhang, P., Gan, L.: Research on the Mental Model for Website Users in the Process of Information Access. Journal of the China Society for Scientific and Technical Information 30(9), 935–945 (2011)
13. Pirolli, P., Card, S.K.: Information foraging in information access environments[C/OL]. In: Proceedings of the Conference on Human Factors in Computing Systems, CHI 1995, pp. 51–58. Association for Computing Machinery, New York (1995), `http://www.sigchi.org/chi95/proceedings/papers/ppp_bdy.htm` (July 8, 2010)
14. Yang, Y., Zhang, X.: Advance in Information Foraging Theory. New Technology of Library and Information Service (1), 73–79 (2009)
15. Yang, Y., Zhang, X.: Empirical Research of Network Information Environment Based on Foraging Theory. Journal of the China Society for Scientific and Technical Information 29(1), 169–176 (2010)
16. Voorbij, H.J.: Searching scientific information on the Internet: A Dutch academic user survey. Journal of the American Society for Information Science 50(7), 598–615 (1999)
17. Jansen, B.J., Spink, A., Saracevic, T.: Real life, real users, and real needs: a study and analysis of user queries on the web. Information Processing and Management 36(2), 207–227 (2000)
18. Jansen, B.J.: The effect of query complexity on Web searching results. Information Research 6(1) (October 2000), `http://informationr.net/ir/6-1/paper87.html`
19. Vaughan, J.: Considerations in the choice of an Internet search tool. Library Hi Tech 17(1), 89–106 (1999)

Research on Three Dimensional Reconstruction of the Ancient Building Based on Images

Yingfeng Hu

School of Railway Tracks and Transportation,
East China Jiaotong University, Nanchang, China, 330013
huyingfeng86@sina.com

Abstract. Three dimensional reconstruction of ancient building using the theory of computer vision is one of the main ways to protect, promote and disseminate the value of ancient building. Due to there exist some shortcomings in the current digitization process of ancient building, such as large-scale scene reconstruction algorithm is complex and the accuracy is low, we present a three dimensional reconstruction approach to point cloud of ancient building based on multiple view images. The proposed approach uses an ordinary digital camera to capture multiple images of the ancient building to be reconstructed, takes advantage of affine decomposition principle to calculate the depth information of SIFT feature point on the ancient building with no damage and convenient operation. Through three dimensional reconstructing of the shape of the Tengwang Pavilion lying in Nanchang city, one of the three famous pavilions in the south of Yangtze River in China, to verify the validity of the proposed approach.

Keywords: ancient building, three dimensional reconstruction, multiple view images, point cloud data, affine decomposition.

1 Introduction

China is an ancient country with long history of civilization, and has vast territory and abundant resources. Our forefathers carried down a rich heritage of ancient buildings to us, the ancient buildings have a long history of tradition and great achievement. However, different form the ancient stone buildings of the European and Arabia countries, the ancient buildings of China are mostly wooden structure, that supplemented with painted and sculpture, and usually are very complex. Wood characteristics of China's ancient buildings make it faces huge challenges on its protection. In history, many ancient buildings in China were destroyed in the war and natural disaster. For example, the Old Summer Place, resulting in long time labor and wisdom by ancient people and managing carefully by 6 dynasties emperors for more than 150 years in China, was destroyed seriously by the allied forces in 1860, and the Tengwang Pavilion that lies in Nanchang City Jiangxi Province, ranking first among the three famous pavilions in the south of Yangtze River, was burned down in the war in 1926, and so on. In recent years, people pay attention to protect the ancient buildings by utilizing the

© IFIP International Federation for Information Processing 2015
D. Li and Y. Chen (Eds.): CCTA 2014, IFIP AICT 452, pp. 73–79, 2015.
DOI: 10.1007/978-3-319-19620-6_9

theory of machine vision to digitize the ancient buildings with the development of economy, and the auxiliary technologies related to the protection of the ancient buildings have become a hot issue. Digitizing the ancient building of China is a powerful method in the protection, promotion, and dissemination the value and the design concept of the ancient buildings. Thus, three dimensional digitization preservation of ancient buildings of China will have an important impact in terms of design and innovation.

2 Related Work

As a kind of special buildings, the approach of three dimensional reconstruction of the ancient buildings can learn from the approach of three dimensional reconstruction of buildings in general. Zuxun Zhang etc. introduces several representative three dimensional reconstruction approaches, as well as the basic principle of the approaches, for ordinary buildings from the aspects of geometry modeling and texture mapping [1]. And Jia Hu etc. proposed three dimensional reconstruction approach for ordinary building based on graph understanding [2].

Different from buildings in general, the ancient buildings usually have complex structure and rich texture, which contains a lot of sculpture and painting. As far as the ancient building is concerned, Fei Deng etc. take advantage of the auxiliary equipment, the large three dimensional laser scanner, to reconstruct the ancient building [3]. But the large three dimensional laser scanner is not easy to install and not easy to calibrate the cameras, its operation is limited for huge ancient buildings. According to the basic elements of the ancient buildings, Ru Wang presents the geometric approach that studied the basic elements from different viewpoints of a single image [4]. Zhang Feng etc. present the automatic three dimensional reconstruction approach to distant scenes from wide baseline images [5].

The three dimensional reconstruction of the ancient buildings is also a kind of the reconstruction of big scene, the approach to three dimensional reconstruction of big scene have important reference meaning for it. As far as the three dimensional reconstruction of big scene is concerned, Carlo Tomasi and Takeo Kanade develop a factorization approach to reconstruct scene from a stream of images based on orthography projection camera model [6]; Noah Snavely etc. represent structure-from-motion and image-based rendering algorithms for image sequence [7]; Pollefeys M etc. present a system for automatic three dimensional reconstruction from video of scenes captured by handheld video camera; and Changchang Wu introduces a fast three-dimensional reconstruction approach from motion images that required only O(n) time complexity [9].

Inspired by the approaches of three dimensional reconstruction of big scene, we propose a three dimensional reconstruction approach to point cloud of the ancient building based on multiple view images. Firstly, many images are captured from multiple views by using an ordinary Single Lens Reflex (SLR) camera. Secondary, the common SIFT features on the images of the ancient buildings are searched and matched. Finally, the depth information of SIFT feature points on the ancient building are calculated that used the affine factorization principle, by those step we can obtain the point cloud data of the ancient building.

3 Approach to Three Dimensional Reconstruction of the Ancient Building

3.1 The SIFT Feature Points Matching

When using a regular SLR camera to take pictures of the ancient building, it is vulnerable to affect by some negative factors that lead to match the feature pixel points on multiple images difficultly, such as external ambient light and lens distortion etc. We use the SIFT algorithm to match the feature pixel points.

SIFT is a similarity invariant, it stays the same when images scale and rotate. When the two corresponding digital images of the ancient buildings are transforming, rotating, or affine transforming etc., SIFT feature matching algorithm can match the strong texture feature points accurately. Even for the images that captured under random angles, it can work stably. In order to distinguish with the general feature points, it is noted that David G. Lowe called the feature points as key points, which maintain the scale and orientation constant. The detailed steps of SIFT algorithm are as follows [10, 11]:

(1) Build scale-space and detect extrema: Through exploiting a difference-of-Gaussian function, the potential key points are effectively identified which are invariant to scale and orientation. This is the first stage of computation searches over all scales and image locations.
(2) Locate key points: For each candidate key points, their location and scale are determined by simulating 3D quadratic function. At the same time, the nonstable bordering key points are deleted.
(3) Set the direction vector of key points: Based on local image properties, each key point is assigned a consistent orientation, then the key point descriptor can be represented relative to this orientation and therefore achieve invariance to image rotation.
(4) Generate the local key point descriptor: By computing the gradient magnitude and orientation at each image sample point in a region, such as 4x4 or 2x2 window, around the key point location, the key point descriptor is created, namely the feature vector of SIFT algorithm.

The Euclidean distances between the feature vectors are exploited to measure the similarity of key points in different digital images after the SIFT feature vectors of the key points are created. A feature point on one image is randomly chosen, and then another two feature points are found, having the shortest and sub shortest Euclidean distances, by traversing all the feature points in another corresponding images. Among these two feature points, if the divisor between the shortest and sub shortest Euclidean distances is less than a threshold value, then they are judged to be paired homologous feature points.

3.2 The Point Cloud Reconstruction Based on Affine Factorization

According to the theory of computer vision, the general affine camera model is used when mapping the three dimensional space point P with coordinate values (x, y, z) to two dimensional pixel point p with coordinate values (u, v). The relationship between the space point P (x, y, z) and the pixel point p (u, v) is as follows:

$$\begin{bmatrix} su \\ sv \\ s \end{bmatrix} = \begin{bmatrix} m_{11} & m_{12} & m_{13} & m_{14} \\ m_{21} & m_{22} & m_{23} & m_{24} \\ m_{31} & m_{32} & m_{33} & m_{34} \end{bmatrix} \begin{bmatrix} x \\ y \\ z \\ 1 \end{bmatrix} \tag{1}$$

In equation (1), the parameter s is scale factor, and the matrix at right is projection matrix, denoted as M, the elements m_{ij} (i = 1, 2, 3; j=1, 2, 3, 4) in the projection matrix M is the inside and outside parameters of the camera. If the image coordinate and the spatial coordinate are transformed to generalized coordinates, equation (1) can be converted to equation (2):

$$s\,p' = M\,P' \tag{2}$$

We get m vision images from different perspective for the ancient building, each image has n SIFT feature points, and those feature points have been matched through the SIFT algorithm, namely give a three dimensional space point P with unknown coordinates values through different projection matrices project to m pixel points on m images. Fig.1 gives a toy example that shows the space points P projects to three matched pixel points p_1, p_2, and p_3 on three images.

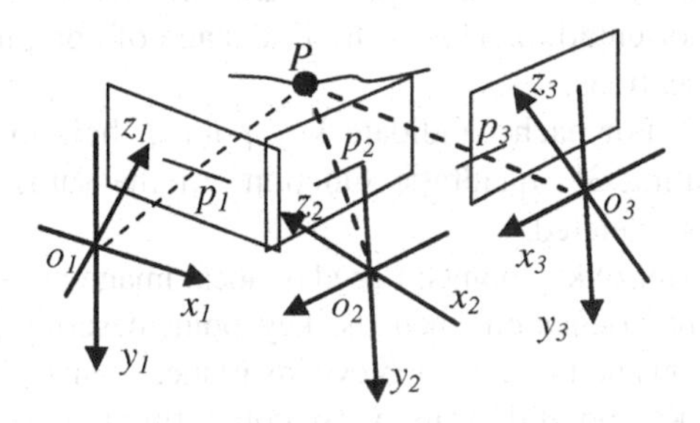

Fig. 1. A toy example of the space point projects to three images

The n space points with unknown coordinates values are denoted as P_i (i = 1, 2, …, n), the j-th pixel point mapped to P_i is denoted as p_{ij} (j = 1, 2, …, m), then we can have

$$s_{ij}\,p_{ij}' = M_j P_i' \quad (i = 1, 2, \ldots, n; \;\; j = 1, 2, \ldots, m) \tag{3}$$

For all the space point P_i (i = 1, 2, …, n) and their corresponding pixel points p_{ij} (j = 1, 2, …, m), the following simultaneous equations is set up:

$$\begin{bmatrix} s_{11}p_{11}' & s_{12}p_{12}' & \cdots & s_{1n}p_{1n}' \\ s_{21}p_{21}' & s_{22}p_{22}' & \cdots & s_{2n}p_{2n}' \\ \cdots & \cdots & \cdots & \cdots \\ s_{m1}p_{m1}' & s_{m2}p_{m2}' & \cdots & s_{mn}p_{mn}' \end{bmatrix} = \begin{bmatrix} M_1 \\ M_2 \\ \cdots \\ M_m \end{bmatrix} \begin{bmatrix} P_1' & P_2' & \cdots & P_n' \end{bmatrix} \tag{4}$$

The left side of the equations (4) is a $3m \times n$ order matrix, called measurement matrix, and it can be broken down into $3m \times 4$ order movement of the camera matrix and $4 \times n$ order generalized coordinate matrix on the left side of the equation. Due to the rank of measurement matrix is equal to 4, we can calculate the unknown coordinates values of the space points by singular value decomposition and others matrix operations.

The point cloud obtained by the above approach may be a local view of the ancient building, therefore, in order to get the whole outline point cloud of the ancient building, we must repeat the above operations to get the point cloud of the other local view, and merge them together.

4 Experimental Results

To test the validity of the proposed approach, we reconstruct the shape of the Tengwang Pavilion lying in Nanchang city Jiangxi Province in China. The Tengwang Pavilion is one of the three famous pavilions in the south of Yangtze River, it is made of wood and is like the building style of the Song Dynasty. In the experiment, the front shape of the building has been reconstructed.

We use the ordinary SLR camera, Canon EOS 70D, with standard zoom lens Canon EF 24-105mm f/4L IS USM, and collect 291 images of the Tengwang Pavilion from different perspectives. In order to reduce noise that has nothing to do with the topic, we try to avoid the surrounding trees and tourists. Before processing, the original images scale are reduced to 1800×1200 pixels, Fig.2 gives a brief view of all the collecting images.

Fig.3 gives the point cloud of the Tengwang Pavilion after three dimensional reconstruction. And Fig.4 gives the rendered model of the point cloud used the open source package, MeshLab[1]. Because there are many tourists wandered around at the bottom of the Tengwang Pavilion during taking pictures, the point cloud seems to be incomplete at this part. However, in addition to this part, the data remains intact.

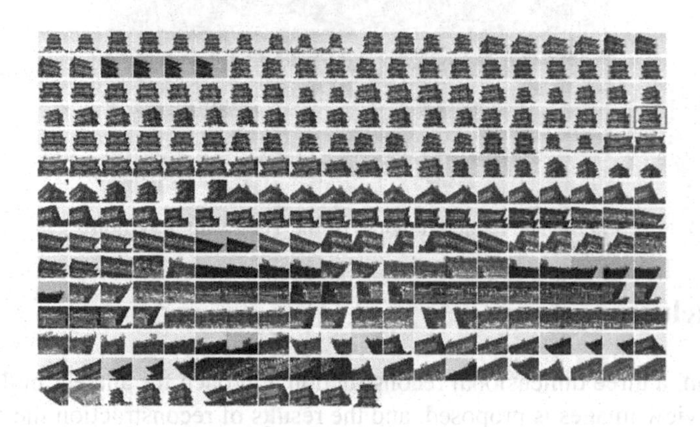

Fig. 2. The 291 images captured from different perspectives of the Tengwang Pavilion

[1] http://meshlab.sourceforge.net/

Fig. 3. The point cloud of the Tengwang Pavilion

Fig. 4. The rendered model of the Tengwang Pavilion

5 Conclusion

In conclusion, a three dimensional reconstruction approach for ancient building based on multiple view images is proposed, and the results of reconstruction the front shape of the Tengwang Pavilion prove that the approach can be used to digitalize the ancient building, which will provide the technical support for the digital preservation of the ancient buildings.

Acknowledgment. This research has been funded by the Natural Science Foundation of Jiangxi Educational Committee of China under Grant No. GJJ13316.

References

[1] Zhang, Z., Zhang, J.: 3D Reconstruction and Visualization of House in Creating Digital City. In: Procedings of ASIA GIS 2001- Collaboration Through GIS in the Internet Era, pp. 20–22 (2001)

[2] Hu, J., Yang, R.-Y., Cao, Y., et al.: 3D Reconstruction Technology for Architectural Structure Based on Graphics Understanding. Journal of Software 13(9), 1873–1880 (2002) (in Chinese)

[3] Deng, F., Zhang, Z., Zhang, J.: 3D reconstruction of old architecture by laser scanner and digital camera. Science of Surveying and Mapping 32(2), 29–30 (2007)

[4] Wang, R.: Research on the key technique of digitalization of ancient building and 3D modeling. Thesis of Northwestern University (2010)

[5] Zhang, F., Xu, Z., Shi, L., et al.: Automatic Reconstruction of Distant Scenes from Wide Baseline Images. Journal of Computer-Aided Design & Computer Graphics 22(2), 256–263 (2010) (in Chinese)

[6] Tomasi, C., Kanade, T.: Shape and motion from image streams under orthography: a factorization method. International Journal of Computer Vision 9(2), 137–154 (1992)

[7] Snavely, N., Seitz, S.M., Szeliski, R.: Modeling the World from Internet Photo Collections. International Journal of Computer Vision (2007)

[8] Pollefeys, M., Nistér, D., Frahm, J.M.: Detailed real-time urban 3D reconstruction from video. International Journal of Computer Vision 78(23), 143–167 (2008)

[9] Wu, C.: Towards Linear-time Incremental Structure from Motion. In: 3DV (2013)

[10] Lowe, D.G.: Distinctive image features from scale-invariant keypoints. International Journal of Computer Vision 60(2), 91–110 (2004)

[11] Hu, Y.: Research on a three-dimensional reconstruction method based on the feature matching algorithm of a scale-invariant feature transform. Mathematical and Computer Modelling 54, 919–923 (2011)

Development and Current Situation of Agricultural Scientific Data Sharing in China

Hua Zhao and Jian Wang

Agricultural Information Institute of CAAS, Beijing 100081

Abstract. Since agricultural scientific data are valuable resources in the field of agricultural science and technology, it is important to share scientific data in agricultural science. The paper analyses current situation and existing problems on data sharing in our country, based on expounding the connotation of agricultural scientific data. Then some countermeasures and suggestions are put forward to promote the level of the agricultural scientific data sharing.

Keywords: Scientific Data, Agriculture, Data Sharing.

Introduction

In recent years, with the rapid development of science and technology, the amount of scientific data accumulated by human society is more than the sum of the past five thousand years , the number of scientific data collection ,storage ,processing and dissemination has got a sharp increase. The scientific community has unanimously recognized that scientific data have a close relationship with knowledge innovation, which are the important resources for the sustainable development of modern science. Scientific data management and sharing have acquired unprecedented attention in various research fields. Many International organizations, such as the World Data System, Committee on Data for Science and Technology, Research Data Alliance, have been established to promote scientific data management and sharing on a global scale. Similarly, China has always paid great importance to agricultural science research. As important resources in agricultural science and technology activities, scientific data need to be shared in the fields of science, technology and education, such as government departments, researchers of agricultural scientific research institutes, persons of management-decisions and other research groups to promote agricultural science and technology innovation. In this paper, we analyze the current situation and the existing problems of agricultural scientific data sharing in China, and put forward improvement measures and suggestions for promoting the continuous development of and maximize value of agriculture scientific data.

1 The Connotation of Agricultural Scientific Data

According to national scientific data sharing project, scientific data refer to the primitive, basic data generated by the human from activities of science and technology

© IFIP International Federation for Information Processing 2015
D. Li and Y. Chen (Eds.): CCTA 2014, IFIP AICT 452, pp. 80–86, 2015.
DOI: 10.1007/978-3-319-19620-6_10

which they are engaged in during the period of perceiving and changing the world. They also contain processing data products and relevant information according to the different requirements of system [1]. Scientific data include both massive data from large-scale observation, exploration, investigation, experiment and comprehensive analysis of the long-term accumulation and integration carried out by the social public welfare sector, and a large amount of data generated by scientists through their practice of long-term taking part in the national science and technology planning projects.

Agricultural scientific data refers to basic data resulting from agricultural science and technology activities, as well as systematically processing data products and related information according to the different requirements, which include agricultural scientific experiment data, essential data and achievements transformation data of agricultural science and technology [2]. Agricultural scientific data is closely connected with agricultural scientific research, for the reason of agriculture science which belongs to a comprehensive discipline, involves many different fields such as biology, technology, environment and economy. Agricultural scientific data has three obvious characteristics [3]: (1) Complexity, dispersion and intersectionality. Scientific data on agricultural biodiversity, agro-ecological environment, soil and fertilizer research field have a clear overlap with research data from other disciplines; (2) large amounts and diverse types. Agricultural scientific research data contain both large-scale observation, detection and investigation data, and experimental data, which have extensive sources , many types, complex structure, wide spatial and temporal distribution; (3) diverse data service levels. Agricultural research data service objects include policy makers, scientific researchers, technology extension workers, agricultural business owners, farmers and so on.

As special agricultural information resource, agricultural scientific data have three aspects of important significance to be shared[4]: (1) For researchers, agricultural scientific data is the basis of their engaging in agricultural scientific research; (2)For the governmental functional departments, agricultural scientific data provides reference to the decision-making for agricultural science and technology management; (3) For the whole society, agricultural scientific data support the realization of agriculture informatization, as precious resources of agriculture and rural economic development. Realization of agricultural scientific data sharing is not only the foundation but also driving force to develop agricultural science and technology.

2 Current Situation and Existing Problems of Agricultural Scientific Data Sharing

2.1 Current Situation of Sharing

Chinese government always attaches great importance to the development of agriculture, increases investment in agricultural science and technology year by year, and organizes to carry out a large number of agricultural data collections, initially shaping scientific data work pattern to department as the main body and the scientific research institutes and universities as complementation. China has rich agricultural scientific data resources. However, the current generation and distribution of scientific data are particular. Collection, management and maintenance of massive scientific data are done by government investment. However, data resources are dispersed in different departments, institutions or individuals [5]. The national agricultural research system is composed of agricultural research institutions at all levels, the agricultural colleges and universities and agricultural enterprises, and accumulated data are also scattered in the three subsystems. Although agricultural scientific data resources are abundant, each agency goes its own way. Most of the existing data resources are only limited to construct the database in institution inside or among cooperative organizations to carry out data exchange and provide limited data service, which restricts the agricultural scientific data value and results in the waste of resources of science and technology , as well as financial and human resources.

In 2003, Ministry of Science and Technology started to launch national scientific data sharing project, which tried out developing scientific data sharing centers in the field of nine disciplines, including the agricultural scientific data sharing center. After 5 to 6 years of efforts, the agricultural data sharing center had been accomplished, which based on the agricultural sector and integrated many agricultural scientific data resources by using modern information technology. Meanwhile the service system had realized standardization management and high efficient utilization of agricultural scientific data resources, strongly supporting our country's agricultural science and technology innovation. At present, the agricultural data sharing center has got some achievements in many aspects, such as integrating scientific data resources, formulating sharing standards and scheme, developing sharing system, providing sharing service, constructing talent team and so on. The platform is playing an important role to promote scientific data management and sharing in our country. By the end of 2013, the platform had finished integrating 60 agricultural core databases from 12 categories of agricultural disciplines, which included a total of more than 700 data sets. The amount of data was up to 3217 GB, accounting for 80% of the total stock of agricultural scientific data resources in China. In addition to the national agricultural scientific data center, there are many other information sharing platforms, which are established by local agricultural scientific research units or other research institution with the overlapping of agriculture. All these platforms have provided good conditions for agriculture scientific data sharing in our country. Agricultural scientific data sharing of our country is steadily moving forward.

2.2 Existing Problems

In the field of agriculture, scientific data sharing and management has been paid great attention by the government departments and scientific research institutions at different levels, and the various resources sharing platforms are being built and perfected constantly for promoting the level of data sharing and improving management efficiency. However, some problems still exist in the implementation of data sharing which are mainly reflected as follows.

2.2.1 Lack of Effective Sharing Mechanism

Compared with the developed countries, the practice of the scientific data sharing in China is still in the exploratory stage, accompanied by lacking macro and micro level of data sharing policies and laws and regulations, and also fails to establish an effective mechanism of sharing data. It's mainly embodied that numerous scientific data are produced in implementing agricultural scientific research projects invested by government, while the data become data producer own groups' private property, only used in a small scale, most data are difficult to be touched by other persons or groups outside producers. Proper value of data has not been fully exploited, which results in low using rate of agricultural scientific data. Against this kind of strange phenomenon, there is no appropriate restraint mechanism and guiding mechanism to deal with it.

2.2.2 Lack of Strong Data Sharing Consciousness and Effective Data Processing

Scientific data in the field of agriculture has more types of users which are not limited to researchers than other areas and their levels of understanding data sharing are different. Agricultural researchers are not only the main users, but also data producers. However data sharing consciousness among agricultural researchers is not strong. It is prevalent that most researchers are willing to use data which are owned by someone else and unwilling to provide their own data to be shared by others. During their engaging in research activities, the majorities of researchers focus on the archiving of experimental data and observational data, and regard the scientific data only as a segment of scientific research, while they don't thoroughly take into account how to process the obtained data and increase the value of data.

2.2.3 Data Processing Technology Is Not Yet Mature

Modern scientific research needs to extract information from data, and then extract knowledge from information, and achieve the goals of forming scientific conclusions as well. New techniques and tools, including data acquisition, data integration, data analysis, data visualization, data distribution, data interoperability and the technology of refining information and knowledge based on the data are required during current scientific research [6]. In the field of agricultural sciences in China, some technologies related to the agricultural data management, online application service system and the distributed database management are not mature yet, which need to be further strengthened.

2.2.4 Deficiency of Data Resources Integration and Low Levels of Sharing Service

Due to wide range of agricultural science research, the data resources may be scattered in various research fields, which expands the breadth and depth of resources integration. Standards of data classification, collection, organization and quality specifications in various areas are different, which increase the difficulty of data resources integration. Existing data sharing service ability has limitations and service pattern is relatively simple. There are a lot of works to do in order to increase the value of the scientific data and satisfy the innovation needs of agricultural science and technology [7].

2.2.5 Lack of Enough Talents in Data Sharing and Management

Compared with other areas, it is relatively lack of compound talents who are proficient in expertise and information management in agricultural field. Most of the existing data sharing talents only have computer professional background, which are not familiar with areas related to agriculture. So there are many problems for data collecting, processing, preserving and sharing. It is difficult to adequately reflect the unique nature of agricultural scientific data and meet the needs of data sharing, the development of the talent team of scientific data sharing and management is hysteretic.

3 Countermeasures and Suggestions

In view of the above analysis on existing problems about agricultural scientific data sharing, the author proposes the following suggestions for expecting to be helpful to the development of agricultural scientific data sharing and management in our country.

3.1 Strengthening to Make All Aspects of Policies and Regulations and Establishing Open Sharing Mechanism of Agricultural Scientific Data

Large amounts of agricultural scientific data are generated through national investment, and have the strong public welfare, which belong to the state-owned assets. Much data are accumulated relying on the implementation of scientific research project and operation of agriculture-related departments. Therefore, it is necessary to overall plan and coordinate production, management and use of the scientific data between the various departments or institutions at national level, for making each research institution clearly recognize that management and sharing of scientific data are their obligation and responsibility. It is necessary to make multiple levels of relevant policies and regulations, including national level, the public sector level, as well as the specific research institutions so as to establish a good mechanism of scientific data sharing [8]. Many developed countries such as America and Britain, have ever established some policies and regulations to promote scientific data sharing. For example, the U.S. National Science Foundation (NSF) has required that the project applicant must simultaneously submit the "Data Management Plan" (DMP) at the time of submitting the project application since 2011. The U.S. National Institutes of Health (NIH) data sharing plan has required that since 2003, Project application for funding in more than 500 thousand dollars must be accompanied by explanations about data sharing or not, including sharing protocols ,methods, and prospective institution [9]. Therefore, our management departments and institutions of science and technology related to agriculture can learn from foreign practices to make some policies and regulations suitable for the situation of our country for establishing and improving open sharing mechanism of agricultural scientific data.

3.2 Changing Concepts and Raising Cognition

Researchers always focus on publishing research papers, they believe that scientific and technical literatures are closely correlated to their vital interests, while few of them care about processing and publishing of scientific data. The important thing is that publishing scientific data has nothing to do with the interests of the researchers, which leads to their lack of enthusiasm. In fact, scientific data is not only the basis of scientific research activities, but also the output of scientific research, as scientific literatures do. So by publishing scientific data, academic influence of researchers can also be improved. It is suggested that scientific data can be seen as an indicator, and can be absorbed into personnel performance appraisal and evaluation system of scientific researchers, with the purpose of encouraging agricultural scientific researchers to share data actively and enhancing their sharing awareness. It is essential to connect the scientific data with researcher' academic contribution and to motivate them to pay great importance for processing and publishing of the scientific data, ultimately to change the old conception of advocating papers and ignoring data which has been existing in the mind of researchers[10].

3.3 Strengthening Studying on Data Management and Applied Technology

Sharing of scientific data requires strong technical support, including database management and data mining technology, massive data storage and retrieval, data quality control technology. Processing, storage, management and sharing of agricultural scientific data also depend on the advanced technology. As a resource, agricultural scientific data have the great potential of deep processing and regeneration, it is a necessary way to study on data management and application technologies in order to process and regenerate agricultural scientific data. In the modern information technology, the key technologies to promote scientific data sharing include the interoperability of distributed database, supercomputer environment combined with a database, the combination of mathematical model and the database, standards and specifications of metadata and data documents, data mining , virtual reality and so on. In the field of agriculture, the application of these advanced technologies will be bound to play a crucial role in agricultural scientific data sharing, so study on these technologies must be strengthened.

3.4 Increasing Integration of Data Resource, Improving Service Capabilities of the Data Sharing Platforms

Agricultural Scientific Data Sharing Center as a national platform has integrated some agricultural science data resources at present. Because of the continuity of agricultural research, the continuous generation of new data, meanwhile a large number of agricultural science data resources are scattered in different institutions or departments. These data resources must be integrated timely and efficiently. Therefore it is necessary to strengthen cooperation and communication with other platforms and institutions for implementing data exchange in order to realize adequate integration of the agricultural scientific data resources. In addition, to enhance data sharing service capabilities, the existing platforms will always focus on the needs of users, innovate service means, provide users with various forms, content-rich data-sharing services, including data storage and access to services, data citation services, and data analysis services in a deeper level. Under these service patterns, the platforms will advocate passive service combined with initiative services, paying more attention to increasing user's participation in particular.

3.5 Developing and Training Talent Team of Agricultural Scientific Data Sharing and Management

Scientific data sharing is a complicated systematic project, which must be achieved through collaboration of multi-disciplinary talents. Sharing of scientific data needs a series of steps including data collection, processing, storage, preservation, sharing services and operating maintenance, in addition to the need for talents with expertise. Information management professionals are indispensable for scientific data sharing. Therefore, talent team of agricultural scientific data sharing and management must be constructed in two aspects: On one hand, the information technology professionals and specialists in the field of agriculture and agricultural information management talents are introduced into the data sharing management team; On the other hand, it must speed up to train some compound talents who are proficient in both technologies and management.

4 Conclusions

Over the past decades, sharing and management of scientific data in the field of agriculture have been greatly improved and got some achievements. To some extent, scientific value, economic value and social value of agricultural scientific data have been realized. But there are still a lot of works to be done. The work of data sharing has a long way to go. There are many issues to be further considered and resolved, such as legal boundaries during effective management and sharing of scientific data, and how to use data more availably. Therefore, management and sharing of agricultural scientific data will continue to be a major problem that faced in science communities, which will always run through the entire agricultural science and technology activities. It is becoming an important support to realize agricultural science and technology innovation.

References

[1] SDS/T 1003-2004.Scientific data sharing project technical standard [S]
[2] Hu, H., Liu, S.: Discussion on the Content and Service of"National Agricultural Sci-Tech Data Sharing Platform". Journal of Library and Information Sciences in Agriculture 17(2), 214–217 (2005)
[3] Zhang, L.: Study on the development of Agricultural Scientific Data Sharing in China. Doctoral Dissertation, pp. 19–22. Graduate school of Chinese academy of agricultural sciences, Beijing (2006)
[4] Zhang, X., Li, S.: Study on Agricultural Scientific Data sharing service. Chinese Science and Technology Forum (5), 127–130 (2006)
[5] Wang, Y., Hua, X., Wang, J.: Analysis of the Domestic and Foreign Scientific Data Management and Sharing. Science & Technology Progress and Policy 30(4), 126–129 (2013)
[6] Guo, M.-H., Li, J.-C., Tian, J.-L.: The Management of Scientific Data Sharing in China. Xi'an University of Architecture & Technology (Social Science Edition) 28(4), 84–88 (2009)
[7] Zhao, R.: Prospect and Practice on Agricultural Scientific Data Sharing Center. Agriculture Network Information (6), 4–12 (2009)
[8] Song, L., Meng, X., Zhou, G.: Measures and Suggestions for Information Quality Management in Agricultural Scientific Data Sharing in China. Journal of Agricultural Science and Technology 11(6), 37–42 (2009)
[9] Wang, Q.: Research on the opening Modes, Guarantee Mechanism and Optimization Strategies for Opening and Sharing Scientific Data. Journal of The National Library of China (91), 3–8 (2014)
[10] Liu, C.: Study on Scientific data sharing mechanism in China. Land and Resources Informatization (1), 5–7 (2004)

Research on Building Technology of Aquaculture Water Quality Real-Time Monitoring Software Platform

Yinchi Ma[1,2], Wen Ding[1,2], and Wentong Li[1,2]

[1] Beijing Fisheries Research Institute, Beijing, 100068, China
[2] National Engineering Research Center of Freshwater Fisheries, Beijing, 100068, China

Abstract. At present, the information level of the aquaculture water quality monitoring is relatively backward in China. Building a digital, networked, intelligent real-time dynamic aquaculture water quality monitoring system by the modern electronic information technology, communications technology and wireless sensor network technology will have an important significance for the factory aquaculture technological innovation. The system is not only able to detect the main indicators of the aquaculture water quality (temperature, PH, dissolved oxygen, turbidity, ammonia, etc.) in real-time, but also be able to do the data fusion and data mining through its software platform, and establish the water quality indicators historical database. This software platform can achieve display and analysis of the local or remote monitoring data in real-time and dynamically, which can provide an important technical means and scientific basis to improve the water utilization effectively, and reduce the pollutant emission in aquaculture. This article will focus on the building technology of the system software platform.

Keywords: aquarium, wireless sensor network, software platform, monitoring.

1 Introduction

Wireless sensor network is a wireless network without infrastructure, which combines sensor technology, embedded computing technology, modern network technology, wireless communication technology and distributed intelligent information processing technology. It can work unattended during a long time [1]. In this study, the wireless sensor network technology is used to build an intelligent, network-based aquaculture water quality wireless monitoring system. In addition, we build a water quality monitoring data intelligence analysis software platform used VC++ program language and database technology, which provides an information and visualization interface for the aquaculture technicians. The implementation of this system can fundamentally enhance the aquaculture water quality monitoring level, which is relatively backward [2] ~ [5]. This paper will focus on the design and implementation of data storage module, analysis module and display module within the software platform.

[1] Ma Yinchi (1982 -), Beijing Fisheries Research Institute, Senior Engineer, Master, graduated from Beijing Normal University, State Key Laboratory of Remote Sensing Science, mainly engaged in research of agriculture remote sensing and fisheries information technology.

© IFIP International Federation for Information Processing 2015
D. Li and Y. Chen (Eds.): CCTA 2014, IFIP AICT 452, pp. 87–93, 2015.
DOI: 10.1007/978-3-319-19620-6_11

2 Function and Feature of Software Platform

The System transmits the monitoring data to the monitoring center through the wireless communication technology. It can realize the continuous data recording, analysis, processing and warning in real-time, so that it can remind the aquaculture technician to take the necessary adjustment measures at first time. This Method can effectively guarantee the aquaculture safety under intensive farming condition. The system software platform can realize the following functions:

I. The system can monitor the various water quality parameters automatically during 24-hours in a whole day, and the data is sent to the center server automatically, while the online records and related analysis will be done.

II. Data server will collect each node data, authorized customers can remote login the data server to inquiry the experimental data or have aided analysis at any time through PC installed the client software.

III. The data analysis system can display monitoring data of each sampling node through graphical and tabular form, which provides user with an intuitive water quality trends. At the same time, the collected data will be saved, which will be added to the large-capacity water quality history database. And this history database can provide a scientific basis for production.

The structure design of the system software platform is shown as follow (Fig.1).

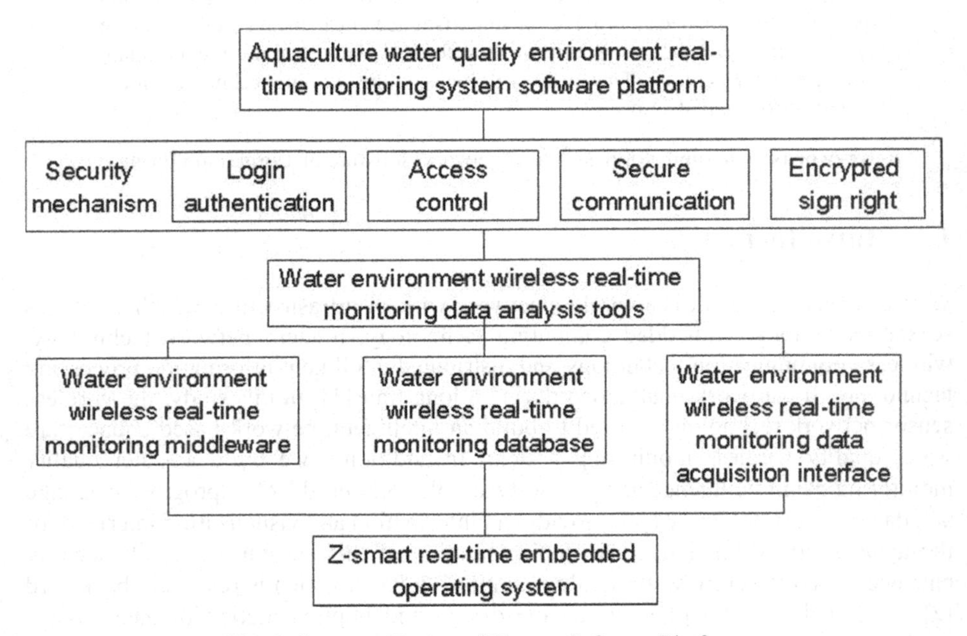

Fig. 1. Structure Design of System Software Platform

The main features of the system software platform include:

I. Remote Real-Time and Intelligent Monitoring

The system combined with advanced embedded computing technology, modern networking and wireless communications technology to automatically collect various sensing parameters to achieve a real-time and intelligent remote monitoring of water quality.

II. Professional and Friendly Interface

The software platform is developed for the application of aquaculture water quality. User can easily view the real-time water quality condition of each aquaculture pond through the software platform interface, as well as the software platform also has a variety of chart and graph display. The Rich display forms of data enhance the usefulness of this system.

III. Powerful Database System

Using the distributed database technology, the platform realizes a multi-functional, highly integrated data service system for the water quality monitoring, which integrated data analysis, data management and data integration display. In addition, the software platform also has a strong storage and query capability for the history data, and it can provide user with comprehensive and reliable data service.

3 Design and Development of Software Platform

The real-time monitoring software platform of aquaculture water quality is self-developed upper application for the water quality monitoring with independent intellectual property rights. The platform is designed for the special needs of the aquaculture industry. User can monitor the aquaculture water environmental parameters such as water temperature, PH, dissolved oxygen, turbidity, ammonia and other data by the on-line monitoring platform, which makes advance warning of the unknown water quality mutation, and effectively curb the risk of aquaculture and significantly improve production and management efficiency. It uses the hierarchical structure technology, which can be adjusted as needed, and dynamically changes the kernel size. In addition, as the software abstract of system hardware, it is primarily responsible for managing and driving the underlying hardware, optimizing network structure, timely detection, controlling the flow of network traffic, ensuring the quality of network application services and extending service hours. In this paper, the network layer protocol stack and middleware technology has been developed based on Z-smart and provided the high-quality application software product for the information processing and service needs of the variety water quality testing applications.

The middleware development is one of focal points of the system. The technology will greatly shorten the development period of the application System.

3.1 Design and Development of Underlying Platform

The real-time embedded operating system called Z-smart by independent research and development is designed for wireless sensor. The middleware is a standard intermediate layer structure with a common property. It is responsible for sensor network system development support for the resource management, dynamic environment analysis and ubiquitous application, such as application change, extension, system upgrades, reused and the other key properties. A key function of the middleware is conversion and agent of heterogeneous networks, such as explaining the data format of sensor network correctly and converting it into TCP / IP data or reverse operation. Therefore, the middleware must have rapid processing capability.

3.2 Design and Development of Application Layer Function Module

For user, the core of wireless sensor network is the perception data instead of the network hardware. User is interested in the data generated by the sensor, instead of the sensor itself. The basic idea of data-centric wireless sensor networks is that the sensor is regarded as the perception of data stream or data source, and the wireless sensor network is regarded as the sensing data space or sensing database. And the data management and processing is regarded as an application target of the network. Based on this idea, the platform application server modules include data loader, data analysis (application analysis) and man-machine interface.

I. Data Loader

The software platform changes data with sensor data collection terminals and data relay node by wireless data interface. In the terminal embedded software development, the system follows the IEEE802.15.4 standard, and software platform communication module is in accordance with the Zigbee Alliance international standards. The complete Zigbee protocol suite is consists of high-level application specification, the application convergence layer, network layer, data link layer and physical layer. The layers apart from the network layer protocol are developed by the Zigbee Alliance, and IEEE Association is responsible for the physical layer and link layer standards.

II. Data Analysis

The system uses data mining technology. User can obtain the needed information from vast amounts of data. In addition, the system also realizes the mining in real-time and mining after the event of the data stream. Data mining refers information processing for decision-making and assessment, in which the large amount of data is automatically analyze and synthesize under certain criteria. In the high-capacity wireless sensor network, a lot of sensing data will be generated in real-time. On the one hand some data is failure data, on the other hand valid data need to be fused together to produce meaningful information. In this case, data mining techniques need to be used. The data is judged and integrated to reduce unnecessary loss of bandwidth and reduce the overall energy consumption so as to provide more meaningful data.

Data mining combines the experience of early information fusion in the commercial applications and military field. According to the center distributed and decentralized distributed judgment ruling theory, tree fusion estimation algorithm is used to realize comprehensive intelligent decision of large amounts of data. The data mining technique of wireless sensor network emphasizes the three core areas of the information integration: First, the information fusion is to complete the process of multi-source information on several levels, where each level representing information abstraction with a different level. Second, information fusion including detection, interconnect, relevant, estimates and information portfolio. Third, the results of the information fusion include the status or identity estimation on the lower level, and the overall decision-making on the higher level.

The system software platform uses multi-sensor data mining technology. And this technology can reduce interference of the invalid data, reduce information redundancy, and improve accuracy. It provides an effective means for the purpose of large amounts of data analysis on the complex system.

III. Man-machine Interface

a. User rights management.

By setting permissions for different users, user can inquiries whether the user is a privilege to observe, manage user permissions, data export privileges, Node Manager or other privileges.

b. Query and graphical display of real-time monitoring point data.

By the list or graphical display of the real-time data, users can view the real-time data of all sensor nodes.

c. Graphic display of node deployment.

Display the location of the node deployment and operation status in monitor area visually.

d. The treelike display of the monitoring points list.

Users can quickly view the real-time data of each node, node status and other information through this feature.

e. Graphic display of wireless sensor monitoring network topology.

Topology displays data transmission path information between the gateway, relay and the sensor node, as shown in Figure 2.

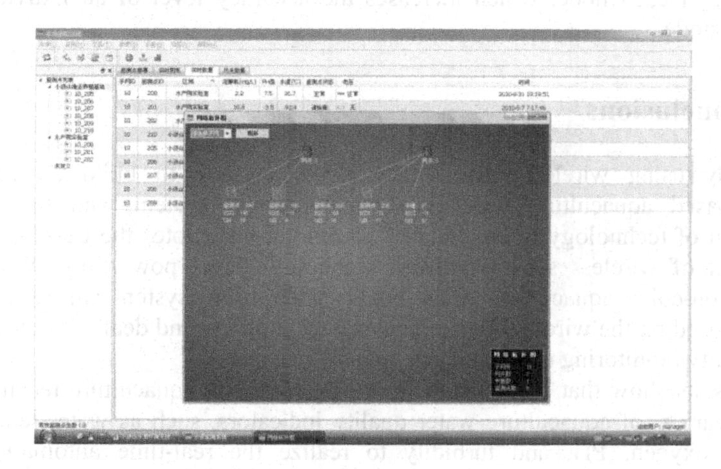

Fig. 2. Display of Topology Diagram

f. Real-time alarm.

Users can set up their own node parameters threshold. When the real-time data of the node exceeds the threshold range, the background color of the node type will be displayed in red, while the node identifier on the node deployment is also displayed in red and the alarm sounds.

g. Query and graphic display of history data.

Historical data can be found all or a specific time period of corresponding node. According to the node number, the sensor type or time period, etc., graphical display can obtain data and showing visual graph, as shown in Figure 3.

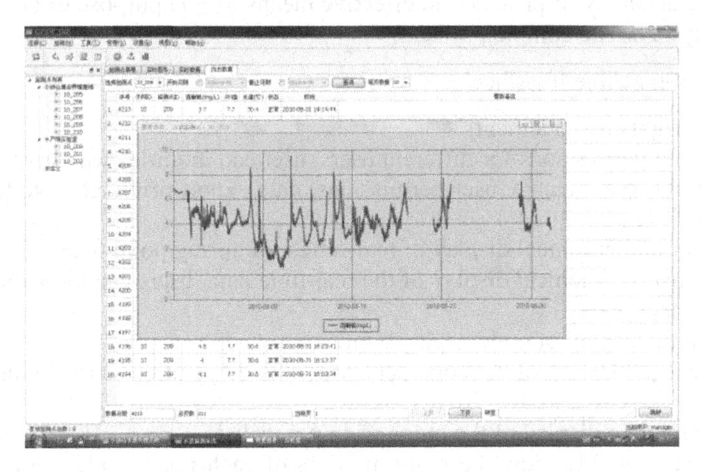

Fig. 3. Graphic Display of History Data

h. Node parameters setting.

Node parameters can be adjusted by users to set send periodic, alias threshold, a coefficient of each node, which increases the accuracy level of an individual data collection node.

4 Conclusions

This study using wireless sensor network technology to build an intelligent, network-based aquaculture water quality monitoring system, and this practical application of technology in the field of aquaculture promotes the development and application of wireless sensor network technology level powerfully. We propose Zigbee protocol a aquaculture water quality monitoring system software platform solution based on the wireless sensor network technology, and deal with the real-time water quality monitoring data management and analysis.

The results show that based on the software platform, aquaculture technician can obtain a variety of aquaculture water quality indicators, such as water temperature, dissolved oxygen, PH, and turbidity to realize the real-time automatic remote monitoring, wireless data transmission, data processing and analysis, intelligent warning, SMS alarm functions. Thereby this technology can maximize the utilization of aquaculture water to protect the suitable environment conditions for aquaculture animal.

References

1. Wu, J., Ma, C., Liu, Y.: Design of Wireless Sensor Network Node for Water Quality Monitoring. Computer Measurement & Control 17(12), 2575–2578 (2009) (In Chinese with English abstract)
2. Du, Z.-G., Xiao, D.-Q., Zhou, Y.-H., Ou, Y.G.-Z.: Design of water quality monitoring wireless sensor network system based on wireless sensor. Computer Engineering and Design 29(17), 4568–4570 (2008) (In Chinese with English abstract)
3. Liu, X., Liu, Z., Wang, P., Miao, L.: Aquaculture security guarantee system based on water quality monitoring and its application. Transactions of the Chinese Society of Agricultural Engineering 25(6), 186–191 (2009) (In Chinese with English abstract)
4. Ma, C.-G., Ni, W.: The design of a factory aquiculture monitor system based on PLC. Industrial Instrumentation & Automation (2), 51–53 (2002) (In Chinese with English abstract)
5. Zhu, W., Ran, G.: Research of Agriculture environment Automatic Monitoring and Control. Freshwater Fisheries 31(1), 60–63 (2001) (Chinese reverse to English)

Mathematical Modeling of Broccoli Cultivation and Growth Period and Yield of Flower Heads

Weiming Liu[1] and Wang En-guo[2]

[1] Taizhou Vocational College of Science & Technology, Taizhou, Zhejiang 318020
[2] Plant Protection Station of Linhai, Linhai 317000, China
15157622288@139.com

Abstract. This study was carried out to evaluate the effects of transplant age, planting density and the level of nitrogen fertilizer on growth period and yield of broccoli. Two mathematical models were constructed for these parameters, they are y_1 =75.1322+1.6361 x_1 -0.0001 x_2 +0.0014 x_3 (R =0.8888**), and y_2 =14999.1852+404.0006 x_1 -0.1986 x_2 +3.6196 x_3 (R =0.6058**), where y_1 and y_2 are growth period and yield, and x_1, x_2, x_3 are transplant age, planting density and dose of nitrogen fertilizer, respectively. These models can be used for the optimization of cultivation techniques to achieve the goal of developing an high yield and high profit competitive broccoli production system.

Keywords: broccoli, transplant age, planting density, nitrogen fertilizer level, growth period, yield, mathematical model.

1 Introduction

Transplant age, planting density and the amount of nitrogen fertilizer are important cultivation factors influencing quality of broccoli and profit of the crop. Previously, these three factors were evaluated for their effects on growth period and yield of broccoli [1-9], but no mathematical model systems were established. This study aimed at developing a simulation modeling system for providing scientific information for decision making in secure high yield and high profit production of broccoli.

2 Material and Methods

2.1 Experimental Design

Broccoli cultivar, Luxiong90, was chosen in this study. There were six treatments of transplant age of 25, 30, 35, 40, 45, 50 days. Planting density had seven treatments of 24000 plants/hm2, 27750 plants/hm2, 30000 plants/hm2, 33000 plants/hm2, 36750 plants/hm2, 42000 plants/hm2 and 48000 plants/hm2. In these treatments, the inter-row and between plant spacing were 0.6×0.7m, 0.6×0.6 m, 0.6×0.55 m, 0.6×0.5 m, 0.6×0.45m, 0.6×0.4m and 0.6×0.35 m. For nitrogen fertilizer treatments, four level treatments using pure N each at a rate of 150, 375, 600 and 825 kg/hm2 were applied

© IFIP International Federation for Information Processing 2015
D. Li and Y. Chen (Eds.): CCTA 2014, IFIP AICT 452, pp. 94–98, 2015.
DOI: 10.1007/978-3-319-19620-6_12

following the schedule in Table 1. In each treatment, three replicates were planted using randomized block design. The plot size was 12m2 for transplant age treatments, 18m2 for planting density treatments, and 15m2 for fertilizer treatments.

Table 1. Experimental design of the nitrogen fertilizer treatment experiment

Total pure N	Schedule of application of nitrogen fertilizer(kg/hm2)				
	First top dressing (Oct.22)	Second top-dressing (Nov. 08)	Flower bud differentiation fertilizer (Nov. 27)	Budding fertilizer (Dec.13).	Heading fertilizer (Dec. 21)
150	15	30	30	45	30
375	37.5	75	75	112.5	75
600	60	120	120	180	120
825	82.5	165	165	247.5	165

2.2 Construction of Mathematical Models and Simulation

Units used in this study were as follows: transplant age (Days), planting density (Plants/hm2), and nitrogen fertilizer doses: kg/hm2; growth period (days); yield (kg/hm2).

Data were subjected to polynomial regression analysis using DPS software [10] to build models for transplant age, planting density, nitrogen fertilizer levels, and their relationship with growth period, and yield. These models were used to simulate the cultivation system for broccoli.

3 Results and Analysis

3.1 The Effects of Transplant Ages on Growth Period and Yield of Broccoli

The growth periods of broccoli varied greatly among those transplant age treatments, it was from 113 to 162 days. In general when using older transplants, the number of days increased during the period from transplanting to floral bud emergence, the duration of vegetative growth post transplanting, floral head growth period, the duration from transplanting to harvest, and total growth period. Statistical analysis identified a significant linear correlation between transplant age (x_1) and growth period (g). The regression model is g =1.8343 x_1 +65.381(r =0.9963**). However, the density of florets on flower heads, the ratio of marketable flower heads and yield of flower head showed a "low-high-low" trend in responses to increasing transplant age. The regression mathematical model for transplant age (x_1) and flower head yield (y) is: y =-40.497 x_1^2 +3089 x_1 -31257(r = 0.9721**).

3.2 Effects of Planting Density on Growth Period and Yield of Broccoli

Results indicate that increasing planting density led to decreases in plants height, canopy width, and the number of leaves per plant. Floral buds emerged prematurely, which resulted in shortened growth period. Statistical analysis indicated that planting density (x_2) had a significant linear correlation with the length of period (days) from transplanting to bud emergence (f), and total growth period (g). The regression mathematical models are: f = 73.867-0.0005 x_2 (r =-0.9662**) and g = 153.71-0.0006 x_2 (r =-0.9663**). Among all the planting density treatments, the highest yield was produced in plots of 36750 plants /hm2, it was 32622 kg/hm2. The yield was 32127.0, 30088.5, 29232.0, 28153.5, 28101.0, and 25357.5 kg/hm2 from treatments of 33000, 42000, 30000, 48000, 27750, and 24000 plants/hm2, respectively. Statistical analysis indicated that planting density (x_2) and flower head yield (y) had an extremely significant binominal correlation. The regression model is: y =-0.00004 x_2^2 +2.8151 x_2 -21740, (r =0.9571**).

3.3 Effects of Nitrogen Fertilizer Level on Growth Period and Yield of Broccoli

Results in Table 2 show that when nitrogen fertilizer was provided at 600 kg/hm2, plants produced the flower heads of the biggest diameter (13.3cm), fresh weight per head (476.85g) and total yield (1440.7 kg/667m2). However, these values started to decline at increasing nitrogen level. Florets became thinner at higher nitrogen level. The width of florets in the high nitrogen level treatments was significantly lower than those supplied with lower doses of nitrogen fertilizer. The highest ratio of marketable flower heads was found in the 375 kg/hm2 of nitrogen fertilizer treatment, it was 89.6%. Nitrogen fertilizer applied at below of above this dose both resulted in reduced ratio of marketable flower heads. The lowest ratio of 76.2% was found in the treatment of 150kg/hm2. Statistical analysis revealed a binomial curve changes in flower head yield (y) with nitrogen fertilizer level (x_3). The regression model is: y =-0.03 x_3^3 +32.986 x_3 +12999(r =0.9870**).

Table 2. Effects of nitrogen fertilizer levels on marketable quality of flower heads of broccoli

Nitrogen fertilizer rate (kg/hm^2)	Diameter of the flower head(cm)	Weight per head(g)	Size of floral buds (cm^2)	Ratio of marketable flower heads(%)	Yield (kg/hm^2)
150	12.5	421.35	76	76.2	17142
375	13.2	450.6	87	89.6	21545
600	13.3	476.85	102	83.8	21611
825	12.8	469.95	116	79.4	19944

3.4 Construction and the Use of Polynomial Decision-Making Mathematical Models in Broccoli Production

3.4.1 Construction of Mathematical Models

Data of growth periods and yield of broccoli from treatments of transplant age, planting density, nitrogen fertilizer doses were analyzed using stepwise regression analysis. The polynomial regression models for total growth period (y_1), transplant age (x_1), planting density (x_2), nitrogen fertilizer doses (x_3) are as follows: y_1 =75.1322+1.6361 x_1 -0.0001 x_2 +0.0014 x_3 (R =0.8888**); and the model for yield (y_2) is: y_2 =14999.1852+404.0006 x_1 -0.1986 x_2 +3.6196 x_3 , (R =0.6058**).

3.4.2 The Application of the Polynomial Decision Making Mathematical Model

Mathematical models can be used to simulate and select the optimum combinations of parameters according to market demand and goals of production. For example, in case of a planting scheme comprised of 130-140 day growth season with targeted yield of 24000-26000 kg/hm^2, the planting density, seedling age and nitrogen fertilizer (pure N) combination should be 35-40 day transplants, at a density of 33000-36750 plants / hm^2, and with the use of 375-600 kg/hm^2 of nitrogen fertilizer. The priority goal of growing broccoli is profit. Because profit is affected by market price, there are differences between yield and profit during some seasons. These modules can be used to arrange production schedule by taking into considerations of predicted market demands and unit price of broccoli.

4 Conclusions and Discussions

4.1 Planting Older Transplants Lengthens Growth Duration, and Planting at Higher Density Shortens the Period of Growth and Development of Broccoli Plants

This study indicates that the growth period of broccoli plants varied significantly under different transplant ages and planting densities. The growth period changed from 113 to 161 days. In general, when older transplants were used, it lengthened the duration from transplanting to floret emergence, the vegetative growth period in the field, flower head development stage, days from transplanting to harvest, and the duration of growth period. There was an extremely significant correlation between treatments of transplant age and the duration of those growth period.

The growth period was 126-142 days in the planting density treatments. Under high planting density, florets emerged soon after transplanting, and the whole growth period became short. These information are very important when making recommendation for a secure production strategy of broccoli.

4.2 The Yield of Flower Heads Has a Binominal Curve Relationship with Transplant Age, Planting Density and Nitrogen Fertilizer Level

Results indicate that the yield of flower heads followed a non-linear "low-high-low" trend in response to increasing levels of transplant age, planting density and nitrogen fertilizer level treatments. Therefore, in cultivation of broccoli crop, it is very important to pay attention to appropriate combinations of transplant age, planting density and the amount of nitrogen fertilizer.

4.3 The Polynomial Mathematical Models Can Provide Information to Decision Making in Designing Schemes for the Production of Broccoli

The polynomial regression models constructed in this study can simulate the effects of transplant age, planting density, nitrogen fertilizer doses on growth period and yield. These models are very useful in designing schemes in response to market demands with specific goals, and guide production management of the broccoli crop.

References

1. Effects of Different Treatment levels of nitrogen Fertilizer on Yield and Quality of Broccoli. Northern Horticulture (1), 6–7 (2006)
2. Effects of application of nitrogen fertilizer at different growth stage on yield and quality of broccoli. Acta Agriculture Shanghai 24(2), 78–80 (2008)
3. Effects of nitrogen and Ca nutrients on quality of broccoli. Journal of Zhejiang Agricultural Sciences (5), 346–348 (2005)
4. A study of sowing date and density on fall broccoli. China Vegetables (3), 35 (1998)
5. A study of planting density and fertilizer on broccoli. Inner Mongolia Agricultural Science and Technology (6), 41–42, 53 (2005)
6. Effect of Main Cultural Techniques on Production and Quality of Broccoli. Chinese Agricultural Science Bulletin 23(6), 545–551 (2011)
7. Effect of cultivar, density and nitrogen fertilizer on yield of broccoli. Vegetables (9), 27–28 (2001)
8. Effect of Main Cultural Techniques on Production and Quality of Broccoli. Chinese Agricultural Science Bulletin 26(15), 274–280 (2010)
9. Studies on the Techniques of the Plug Seedlings in Broccoli. Chinese Agricultural Science Bulletin 26(2), 171–175 (2010)
10. Applied Statistics and SPS Data Processing System. Science Press, Beijing (2002)

Research on the Allocation in the Complex Adaptive System of Agricultural Land and Water Resources of the Sanjiang Plain

Qiang Fu[1,2], Tienan Li[1,2], and Tianxiao Li[1,2]

[1] School of Water Conservancy & Civil Engineering, Northeast Agricultural University, Harbin 150030, China
[2] Institute of Food Science and Engineering, Wuhan Polytechnic University, Harbin 150030, China
fuqiang0629@126.com, 719287576@qq.com, litianxiao.888@163.com

Abstract. As for the improper water and land use structure and difficulty in bringing the system into maximum performance, this research takes the Sanjiang Branch of Heilongjiang Land Reclamation Bureau as an example and uses the concept of complex adaptive system to build an allocation model of the complex adaptive system of agricultural land and water resources of the Sanjiang Plain and to make evolution analysis of optimized allocation of agricultural land and water resources under different groundwater exploitation plans. This research shows that: with the increased exploitation of groundwater, the agricultural production value and the grain yield have increased; the ecological area has decreased within a certain range; with the increased restriction in groundwater exploitation year by year, the efficiently-used land area will decrease, resulting in a lot of land not being effectively utilized and seriously hindering the coupling between the regional water and land resources and the local economy development.

Keywords: Sanjiang Plain, agricultural land and water resources, complex adaptive system.

1 Introduction

Land and water resources are the core resources of agricultural production [1]-[2] and basic materials that human survival and development rely on, playing an important role in the sustainable development of the region [3]. China has a serious shortage of land and water resources. Along with the increase in population and rapid economic development, the scope and intensity of the use of land and water resources have been increasing, and the disputes in fighting for water between various sections and industries become increasingly acute. As agriculture is a big consumer of land and water resources, the proper allocation of land and water resources of is not only related to the social and economic development, but also affects the ecological environment restoration and reconstruction.

© IFIP International Federation for Information Processing 2015
D. Li and Y. Chen (Eds.): CCTA 2014, IFIP AICT 452, pp. 99–106, 2015.
DOI: 10.1007/978-3-319-19620-6_13

However, the previous researches on the optimized allocation of regional resources usually separate land and water resources as independent systems [4]-[6], lay more emphasis on the water resource system and lack an effective coupling with the land resource system, hindering the full overall performance of the regional land and water resource system and easily leading to the system overloaded while running. Therefore, it is necessary to treat land and water resources as a system to make overall optimized allocation to improve the regional land and water resource utilization efficiency, maintain the relative balance of the system, obtain the optimal ecological and economic benefits and achieve sustainable use of land and water resources. Therefore, this research takes the Sanjiang Branch of Heilongjiang Land Reclamation Bureau as an example, treats agricultural land and water resources as a complex adaptive system and make its complex adaptive allocation, so as to provide an important reference for improving the regional agricultural land and water resource utilization efficiency, obtaining higher grain production efficiency and achieving the sustainable use of agricultural land and water resources.

2 Establishment of Complex Adaptive Allocation Model

2.1 Government Subject

The regional government is the controller of the entire region and its ultimate optimization objectives have more than one, which means the adaptability of the government subject is shown by the description of a multi-objective problem. The government subject's adaptability in this model is shown by maximum regional comprehensive benefits; the comprehensive benefits of this model is a single objective that is obtained after the processing of agricultural benefit objective and ecological benefit objective by utility function through the multi-objective analysis method, and that can be used to make a comprehensive evaluation of the system state. The agricultural benefit objective results from the equilibrium between the grain yield objective and the agricultural income objective or, specifically a government's comprehensive benefit objective resulting from the non-dimensionalization of two utility functions:

$$ECOS = \frac{ECOA - ECOAMIN}{ECSOMAX - ECOAMIN} \quad Y = \frac{YA - YSMIN}{YSMAX - YSMIN} \quad M = \frac{MA - MAMIN}{MAMAX - MAMIN} \tag{1}$$

$$AWEL = f(YS, MS) \qquad SWEL = f(AWEL, ECOS) \tag{2}$$

ECOA and ECOS respectively mean standard ecological area (hm^2) and non-dimensionalized standard ecological area; Y and YS respectively mean grain yield (t) and non-dimensionalized grain yield; M and MS respectively mean agricultural production value (RMB) and non-dimensionalized agricultural output value; ECOAMAX means the threshold value of Maximum standard ecological area (hm^2); ECOAMIN means the threshold value of minimum standard ecological area (hm^2); YMAX means maximum grain yield (t); YMIN means minimum grain yield (t); MMAX means maximum grain production value (RMB); MMIN means the minimum grain production value (RMB); AWEL means agricultural benefits; SWEL means comprehensive benefits; f means utility function.

2.2 Agricultural Production Subject

As a sector-level subject, its adaptability is shown by maximum comprehensive benefits of the agricultural sector, including pursuit of grain yield and agricultural production value, and is also a multi-objective problem.

Calculation of grain yield:

$$Y = y_j \bullet s_j \tag{3}$$

Where Y means gross grain yield (t), yield per unit area of the j-*th* crop (t/hm^2), j-*th* crop planting area (hm^2) and j = (1 ... 6) six kinds of crops in the agricultural sector.

Calculation of economic benefits:

$$M = \sum_{j=1}^{6} Y_j \bullet P_j \tag{4}$$

Where M means agricultural income (RMB), yield of the j-*th* crop and unit price of the j-*th* crop (RMB/t).

2.3 Ecological Subject

As a sector-level subject, its adaptability is shown by maximum comprehensive benefits of the ecological sector. This research uses standard ecological area to measure the benefits of the ecological sector. The ecological footprint method [7] is used to cover different ecological types into standard ecological areas through "equivalency factors". This research set the yield factor to 1 and equilibrium factors of pasture, woodland and waters to 0.5, 1.1 and 0.2 respectively which are respectively multiplied by the actual areas to be standard ecological areas.

Standard ecological area:

$$S_{ECO} = \sum_{i=1}^{3} \text{因子} \bullet a_i \cdot \tag{5}$$

Where means standard ecological area, the i-th ecological area and i=(1…3) three ecological types.

3 Model Solving and Analyzing

3.1 Establishment of Boundary Conditions

The raw data come from the Statistical Yearbook of Sanjiang Branch and statistics of water conservation construction in Heilongjiang Province. The data of the sector and the ecological sector of the Sanjiang Branch of Heilongjiang Land Reclamation Bureau in 2008 is show in Table 1 and Table 2 respectively, and the model constraints, as shown in Table 3, are calculated from the data of various years by a certain percentage. The model is initialized by the following data with the establishment of relevant boundary conditions and constraints.

Table 1. Relevant Data of Agricultural Sector

Crop Type	Rice	Wheat	Corn	Barley	Bean	Oil Crops
Sowing area (10^4hm^2)	44.91	0.47	3.62	1.80	8.46	0.68
Yield per unit area (t/hm^2)	8.93	5.25	7.71	5.25	2.37	1.05
Production value per unit area (RMB/ hm^2)	7815	3600	5325	4000	4170	5625
Yield (10^4t)	400.93	2.44	27.88	9.47	20.11	0.71
Production value (RMB10^8)	71.29	0.37	2.90	1.53	6.66	0.56
Irrigation quota (m^3/hm^2)	5000	1280	1360	1230	1330	1370

Table 2. Relevant Data of Ecological Sector

Ecological Type	Forest	Pasture	Waters
Area (10^4hm^2)	16.61	2.35	4.75
Irrigation quota (m^3/hm^2)	150	3000	0
Equilibrium factor	1.1	0.5	0.2

Table 3. Boundary Conditions and Constraints

Constrain name	Threshold value of crop sowing area	Threshold value of ecological area	Groundwater exploitation (108m^3)	Reclaimable wasteland (104hm^2)	Annual precipitation (mm)
Value	50% to 150% of actual area	50% to 150% of actual area	23.93	7.45	6342

3.2 Result Analysis

The Sanjiang Branch has 800000 hm^2 of utilizable agricultural and ecological land. As the region mainly uses groundwater to irrigate, the evolution analysis of optimized allocation of agricultural land and water resources in the region under different groundwater exploitation programs is made as follows. See Table 4 for program numbers and Table 5 and Table 6 for specific water quantity allocation programs and planting structure programs.

Table 4. Groundwater Exploitation of Different Programs

Program number	1	2	3	4	5	6	7	8	9	10
Groundwater exploitation (10^8m^3)	23.80	23.50	23.00	22.47	21.64	21.19	20.94	20.36	19.90	19.47

Table 5. Water Resource Allocation Program

Program	1	2	3	4	5	6	7	8	9	10
Agriculture (10^8m^3)	23.05	22.77	22.19	21.72	20.94	20.46	20.13	19.65	19.16	18.84
Ecology (10^8m^3)	0.74	0.73	0.79	0.75	0.70	0.74	0.81	0.71	0.74	0.63

As can be seen from Table 6, with the reduction in groundwater exploitation, the planting area of rice which consumes more water decreases gradually, the planting area of oil crops which consumes more water but have better economic benefits has no great fluctuation, and the planting areas of other crops which consume less water change within a certain range with no great reduction in the overall sowing areas.

With the actual water consumption in 2008 as the benchmark and other constraints unchanged, simulation calculation is made for each program to obtain the respective relationships between groundwater exploitation and agricultural production value, grain yield, standard ecological area and efficiently-utilized land area, as shown in Figures 3 to 6.

Table 6. Planting Structure Adjustment Program

Program	Rice	Wheat	Corn	Barley	Bean	Oil Crop	Forest	Pasture	Waters
1	44.27	0.06	2.69	0.42	3.15	0.57	22.32	1.34	2.55
2	43.78	0.06	2.80	0.37	2.85	0.47	21.86	1.34	2.46
3	42.46	0.05	3.12	0.47	2.94	0.61	20.71	1.60	2.57
4	41.69	0.06	2.56	0.36	3.11	0.48	19.15	1.54	2.44
5	40.00	0.06	3.03	0.34	3.01	0.55	19.85	1.34	2.39
6	39.05	0.06	2.94	0.35	3.06	0.54	19.86	1.47	2.53
7	38.63	0.06	2.24	0.37	2.89	0.50	19.63	1.71	2.48
8	37.44	0.06	2.75	0.46	2.94	0.70	19.26	1.42	2.43
9	36.61	0.05	2.50	0.43	2.89	0.53	17.27	1.61	2.44
10	35.89	0.06	2.67	0.44	2.85	0.64	16.96	1.25	2.44

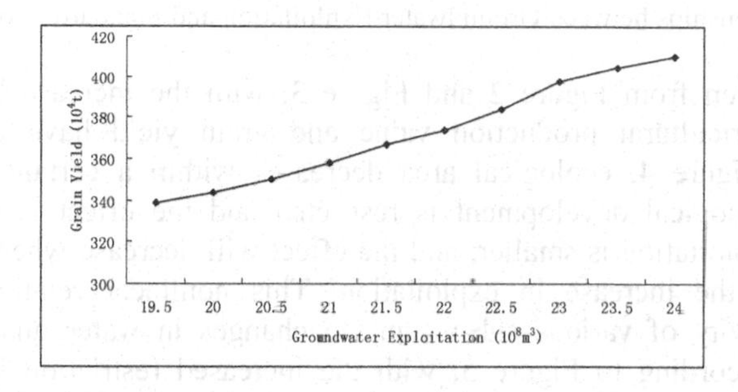

Fig. 1. Relationships between Groundwater Exploitation and Grain Yield

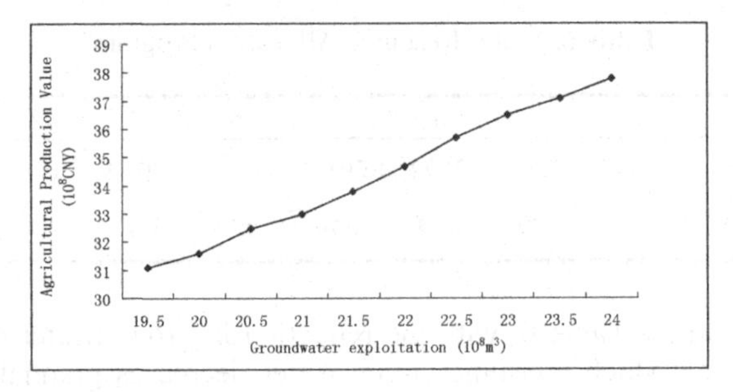

Fig. 2. Relationships between Groundwater Exploitation Agricultural Production Value

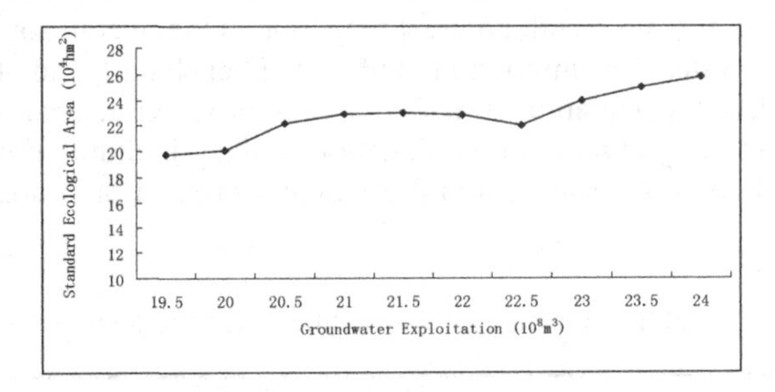

Fig. 3. Relationships between Groundwater Exploitation and Standard Ecological Area

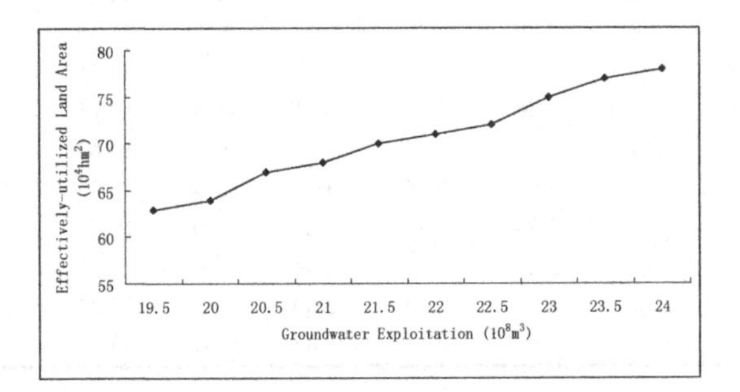

Fig. 4. Relationships between Groundwater Exploitation and Standard Ecological Area

As can be seen from Figure 2 and Figure 3, with the increase in groundwater exploitation, agricultural production value and grain yield have also increased. According to Figure 4, ecological area decreases within a certain range, mainly because the ecological development is restricted and the effect is high when the groundwater exploitation is smaller, and the effect will decrease when the restriction decreases with the increase in exploitation. This nonlinear relationship and the different sensitivity of various subsystems to changes in water quantity show the adaptability. According to Figure 5, with the increased restriction in groundwater exploitation year by year, the efficiently-used land area will decrease, resulting in a lot of land not being effectively utilized and seriously hindering the coupling between the regional water and land resources and the local economy development.

4 Conclusions

According to the characteristics of the agricultural land and water resources of the Sanjiang Plain, this research treat agricultural land and water resources as a complex adaptive system and use the theory of complex adaptive system to put forward an allocation model of the complex adaptive system of agricultural land and water resources of the Sanjiang Plain and to make evolution analysis of optimized allocation of agricultural land and water resources under different groundwater exploitation plans. The evolution analysis has showed that water resources have become a bottleneck restricting the Sanjiang Plain's coordinated economic, social and ecological development. The development and utilization of surface water, groundwater protection and construction of water conservancy projects become effective means to solve the aforementioned problem. This research has important theoretical and practical significance for the development and utilization of agricultural land and water resources of the Sanjiang Plain and the coordinated economic, social, and ecological development.

Acknowledgment. The author would like to thank his collaborators from and support of Natural Science Foundation of China (NO. 51179032、NO. 51279031、NO. 51109036), Ministry of water resources public welfare industry special funds for scientific research project (NO. 201301096), The ministry of education in the new century excellent talents to support plan, Province Natural Science Foundation of Heilongjiang (No: E201241), New Century Talent Supporting Project by Education Ministry, The Yangtze River Scholars Support Program of Colleges and Universities in Heilongjiang Province, Heilongjiang Province Water Conservancy Science and Technology project (No: 201318), Prominent young person of Heilongjiang Province(JC201402).

References

1. Liu, C., Sun, R.: Ecological Aspects of Water Cycle: Progress in the Research on the Water and Energy Balance of the Soil-Vegetation-Atmosphere System. Advances in Water Science 10(3), 251–259 (1999) (in Chinese)
2. Jiang, Q., Fu, Q., Wang, Z., et al.: Space Matching Pattern of the Sanjiang Plain's Land and Water Resources. Journal of Natural Resources 26(2), 270–277 (2011) (in Chinese)
3. Li, T., Fu, Q., Peng, S.: DPSIR-based Evaluation on the Carrying Capacity of Land and Water Resources. Journal of Northeast Agricultural University 43(8), 128–134 (2012) (in Chinese)
4. Sand, J., Liu, B., Xie, X., et al.: Research on the Optimized Allocation of Water Resources Based on Particle Swarm Optimization. Hydroelectric Energy Science 30(9), 33–35, 69 (2012) (in Chinese)

5. Chen, X., Duan, C., Qiu, L., et al.: Application of Particle Swarm-based Large System Optimization Model in the Optimized Allocation of Water Resources in Irrigation Areas. Transactions of the CSAE 24(3), 103–106 (2008) (in Chinese)
6. Dong, P., Lai, H.: Optimized Allocation of Spatial Structure of Land Use Based on Multi-objective Genetic Algorithm. Geography and Geo-Information Science 19(6), 52–55 (2003) (in Chinese)
7. Zhang, J., Zhang, R.Z., Zhou, D.: Evaluation of the Water Resources Carrying Capacity of Shule River Basin Based on Ecological Footprint. Pratacultural Science 21(4), 267–274 (2012) (in Chinese)
8. Fu, Q., Wang, K., Ren, S.: Real-coded and Multi-objective Nested Accelerating Genetic Algorithm Based on Complex System Evolution Optimization. System Engineering Theory and Practice 32(12), 2718–2723 (2012) (in Chinese)

The Effect of Precision Nitrogen Topdressing Decision on Winter Wheat

Guo Jianhua[1,2,*], Meng Zhijun[1,2], Chen Liping[1,2], Ma Wei[1,2], An Xiaofei[1,2], and Yao Hong[1,2]

[1] Beijing Research Center for Intelligent Agricultural Equipment, Beijing 100097, China
[2] National Research Center of Intelligent Equipment for Agriculture, Beijing 100097, China
{guojh,mengzj,chenglp,maw,anxf,yaoh}@nercita.org.cn

Abstract. Topdressing in the growth period is the key to high yield of wheat. So how to use the rapid detection technique to realize scientific and real-time topdressing for lower fertilizer application and higher use efficiency of fertilizer is critical to agricultural development. The aim of this study is to investigate the effect of fertilization as affected by different topdressing recommendations. In this study, field experiments are conducted at Xiaotangshan Precision Agriculture Demonstration Base in Beijing, using commonly-used rapid detection equipment. The experimental results are summarized as follows: Variable rate fertilization based on nitrate nitrogen level in soil and grower's uniform fertilization are basically same in fertilizing amount but the former produces higher yield of 320 kg/ha; high rate application based on low measured NDVI value in the reviving stage leads to highest fertilizer productivity at 19.3 kg/ kg, with an increase of 10.4 kg/ kg compared with grower's uniform fertilization; in terms of profitability, high rate application based on low measured NDVI value in the reviving stage produces highest profit of 12,649 yuan/ha, while high rate application based on high measured SPAD value and high measured NDVI value in the reviving stage produce lowest profits of 9,358 yuan/ha and 9,775 yuan/ha respectively.

Keywords: Topdressing wheat, Precision decision-making, Topdressing recommendation.

1 Introduction

Fertilizer is regarded as material guarantee of sustainable agricultural development and also the key to higher grain yield. It is proven by agricultural development practices that fertilization is the most rapid, efficient and important means of increasing grain yield. China has now become the largest producer and consumer of fertilize (1). The amount of fertilizer applied in China takes up about 1/3 of total amount of global fertilization. Growers tend to spread fertilizer out evenly on the soil surface, which means the rate of fertilizer application is same in one application area or with one grower regardless of varied soil nutrients. As a result, too many fertilizers are applied on high-nutrient soils, leading to low use efficiency of fertilizer.

* Corresponding author.

© IFIP International Federation for Information Processing 2015
D. Li and Y. Chen (Eds.): CCTA 2014, IFIP AICT 452, pp. 107–116, 2015.
DOI: 10.1007/978-3-319-19620-6_14

The average application rate of fertilizer was developed two times the upper limit of the safe use of chemical fertilizers and the average utilization rate of is about 40%. China's arable land accounted for 7% of the world's arable land, but the use of 35% of the fertilizer over the world.(2). Considerable loss of fertilizer has triggered various environmental problems. For example, in some areas characterized by highly intensive agriculture in Beijing, the improper application of fertilizer N has caused excessive nitrate contained in underground water, the content of which even reaches 100% in some areas.

With the development of soil test-based fertilizer recommendation and balanced fertilization technologies, fertilizer's rate of contribution to soil increases gradually and fertilizers applied elevate the grain yield. The traditional testing technique used to be widely applied because of its precision in the test procedures and results. However, due to high workload, time-consuming feature and complicated operation, coupled by limitation in environment, equipment and technology, such testing technique is unlikely to meet the requirements of agricultural modernization. As modern science and technology continue to develop, various sensor-based non-destructive rapid testing equipment and technologies are developed, such as the SPAD Chlorophyll Meter made in Japan. Chlorophyll meter is a tool used to rapidly measure the level of chlorophyll contained in crop leaves, which is now gradually recognized and popularized by experts. At present, it has been applied in N nutrition determination for rice, corn, wheat and other crops [3, 4, 5].

In this study, effective nitrogen level in soils (0-20cm & 20-40cm) is measured along with SPAD value, NDVI value and nitrate content in leaves in the reviving stage for purpose of precision fertilization. Then the decision-making on precision topdressing and the effect of precision fertilization are examined and evaluated with the intent to improve the use efficiency of fertilizer N, maximize ecological economic benefit, and provide guidance for agricultural production.2 The structure of soil parameter distribution map system

2 Experimental Material and Method

2.1 Experiment Design

The experiments were conducted during Mar. 2013 and Jun. 2013 on the fields in Xiaotangshan Precision Agriculture Demonstration Base in Changping District, Beijing. The tested wheat is Jingdong 12, which was sown mechanically on Oct. 3, 2012 at a rate of 13 kg/mu by applying base fertilizer DAP at rate of 15 kg/mu. Sprinkling irrigation is applied.

After entering the reviving stage, the trial zone is delineated through DGPS locating and boundary defining. Then samples are extracted in zigzagging form from the soils at depths of 0-20 cm and 20-40 cm in subzones of the trail zone. Following that, the contents of nitrate nitrogen and water in the soil samples are measured, then SPAD value, NDVI value and nitrate content in leaves. Finally, the amount of fertilizer N to be top dressed is figured out. Meanwhile, field trials of different rates of fertilizer N application are carried out.

Table 1. Topdressing Recommendations

Test Item	Index	Recommended Amount
Nitrate in Leaf	>1400	0
	1000-1400	8
	800-1000	10
	600-800	13
	<600	15
SPAD	<47	15
	47-50	10
	>50	8
GREENSEEK	<0.45	18
	0.45-0.47	15
	0.47-0.5	12
	>0.5	10

2.2 Measurement of NDVI Value

NDVI data is collected using hand-held GreenSeeker NDVI meter produced by U.S. Ntech. The NDVI meter works based on an active remote sensing system, of which the red band length is 671±6 nm, NIR band length 780±6 nm, and spectrum width 0.6 m. The test is conducted in a sunny forenoon in the growth period of wheat. During the test, the top end of remote sensor must be adjusted to be level with the ground and in parallel to wheat canopy at the same height, with a vertical height of 80-120 cm. The typical NDVI value is obtained by averaging the NDVI values of all subzones.

2.3 Measurement of SPAD Value

Leaf SPAD value is measured by SPAD-502 Chlorophyll Meter. The test is carried out on 20 uniformly sized wheat plants randomly selected from the subzones in a sunny forenoon on 9:00 a.m.-12:00 a.m. The SPAD value is obtained by averaging the measured values of mid-part leaf newly generated on the stem.

2.4 Measurement of Leaf Nitrate Content

The instrument used for the test is a German-made nitrate reflectometer. And the test procedures are shown as follows: take a basal part of the stem, and cut into small pieces; use juice squeezer to get the juice out; dilute the juice by a factor of about 10-20; inset the test strip to the liquid to be analyzed; take out the test strip and insert into the reflectometer for matching; and read out the nitrate content (mg/L) directly.

3 Results and Analysis

After the experiments, the growth dynamics and yield of wheat in critical growth stages are analyzed.

3.1 Dynamic Changes of NDVI Value in Different Growth Stages

The experiments show that wheat's N nutrition can be expressed by NDVI value. In the erecting stage, leaves are small, so NDVI value is low; then NDVI value increases at a high rate as wheat grows at a high rate as wheat grows gradually, peaking after entering the booting stage and finally remaining stable thereafter. During the entire experimental period, different fertilizer treatments are consistent with growth changes. NDVI is a vegetation index that can reflect growth trend of crops.

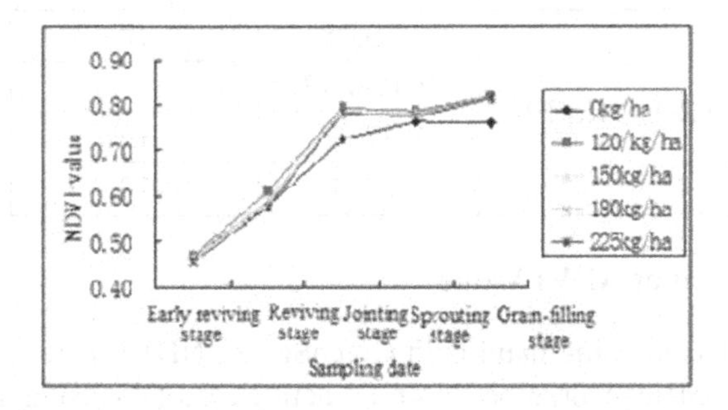

Fig. 1. NDVI Value Changes Affected by Different Fertilizing Amounts

3.2 Correlation of NDVI Value with Different N Application Levels

The correlation of NDVI values obtained by GreenSeeker NDVI meter with different N application levels are analyzed, the result of which suggests that NDVI value is an implication of wheat's N level. It is a useful tool for N nutrition diagnosis. Therefore, by measuring NDVI value, growers can figure out the growth dynamics of wheat. Then, growers can determine the rate of topdressing application based on the growth dynamics and target yield.

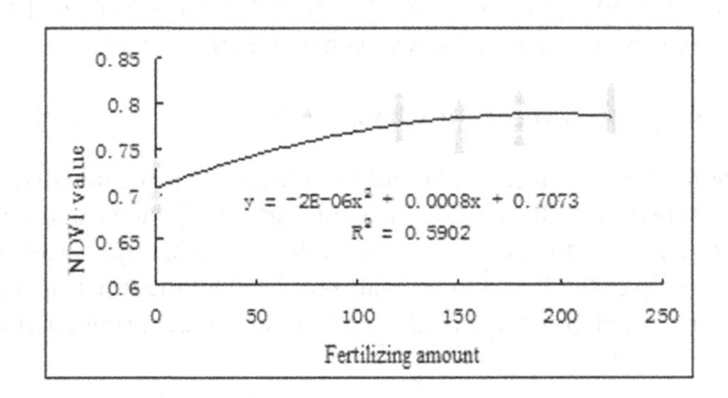

Fig. 2. NDVI Value Changes Affected by Different Fertilizer Treatments

In consideration of differences in wheat species, it is necessary to build a separate control field for determining an appropriate rate of topdressing application in practical production.

3.3 Dynamic Changes of SPAD Value in Different Growth Stages

As can be seen from Fig. 3, during the entire experimental period, SPAD value varies with growth stages, bottoming out in the reviving stage while peaking in the heading stage. Then SPAD value declines gradually and tends to stabilize in the grain-filling stage. Such a change trend is consistent with the change of N nutrition [8]. Leaf N is mostly stored in chloroplast, so N level in leaf is positively correlated with chloroplast content. Chloroplast content and N level increase as a result of fertilization, so is SPAD value. After flowering, the plant starts to age, resulting in a decline in chloroplast and N levels and also SPAD value.

During the sprouting stage, SPAD value is correlated with fertilizing amount, which is especially remarkable in the grain-filling stage.

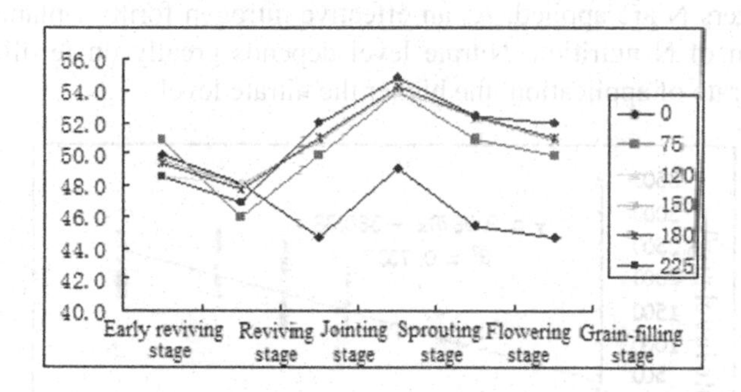

Fig. 3. Measured SPAD Value in Different Growth Stages

3.4 Correlation of SPAD Value with Different N Application Levels

SPAD value is highly correlated with N application level. The correlation coefficients detected on Apr. 22 and May 22 are R2=0.6262** and R2=0.6774** respectively. Leaf N level is positively correlated with chlorophyll level while crop yield is positively correlated with leaf N and chlorophyll levels in specific growth stages.

After topdressing, it takes some time for wheat to uptake the fertilizer N. So wheat is more responsive to nitrogen and SPAD value is more correlated with the rate of topdressing application in the flowing stage.

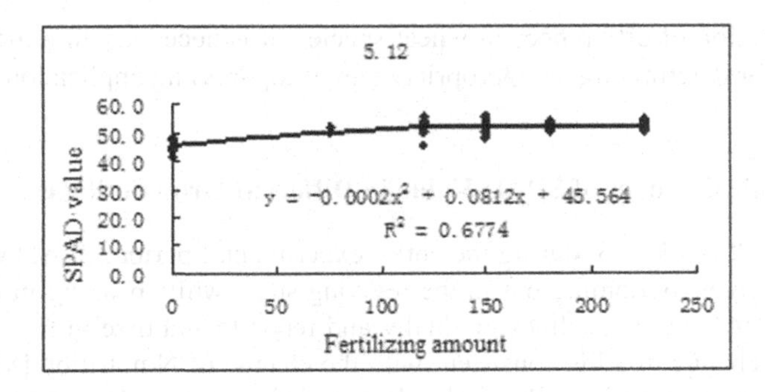

Fig. 4. Correlation of SPAD Values with Different Fertilizing Amounts in Flowering Stage

3.5 Correlation of Nitrate Level with N Application Level

Wheat nitrate level is correlated with fertilizing amount, of which the correlation coefficient is R2=0.7333. Nitrate level varies with N application level and increases as more fertilizers N are applied. As an effective nitrogen form of plants, nitrate is a direct reflection of N nutrition. Nitrate level depends greatly on fertilizing amount. The higher the rate of application, the higher the nitrate level.

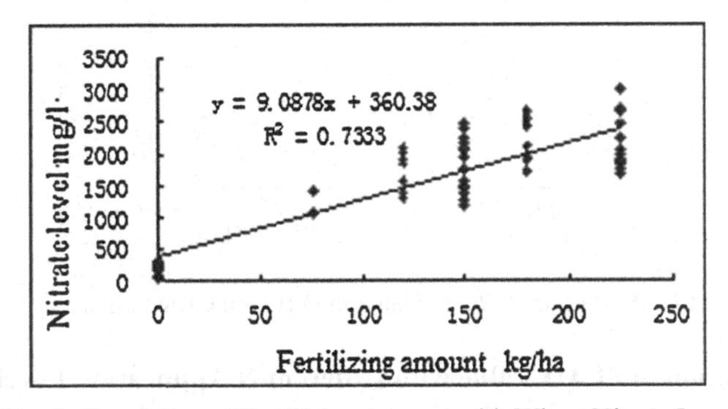

Fig. 5. Correlation of Fertilizing Amount with Wheat Nitrate Level

3.6 Correlation of Chlorophyll Level with Fertilizing Amount

Chlorophyll is vital for photosynthesis, so the amount of photosynthate depends on chlorophyll level. To the extent that is appropriate, higher fertilizing amount produces higher chlorophyll level and more efficient photosynthesis.

During the jointing stage, nitrate level is correlated with yield. The higher the nitrate nitrogen level, the higher the yield.

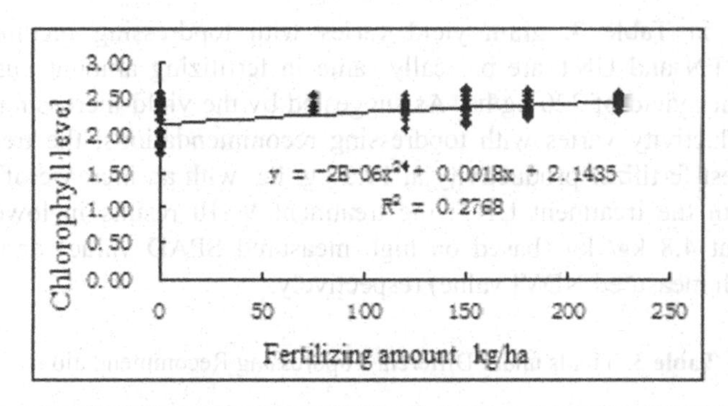

Fig. 6. Chlorophyll Levels Affected by Fertilizing Amount

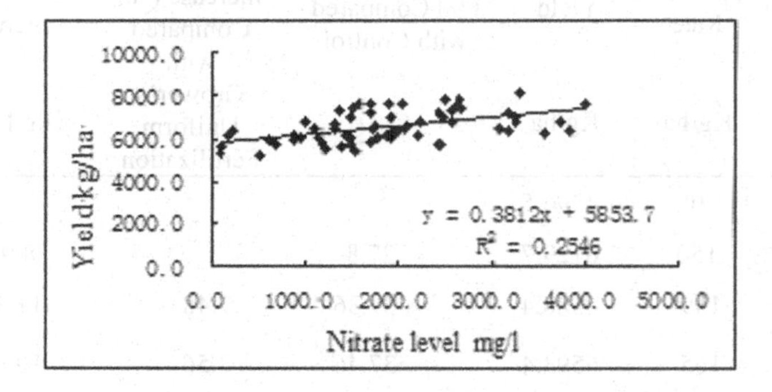

Fig. 7. Correlation of Nitrate Level with Yield in Jointing Stage

3.7 Effects of Different Topdressing Recommendations

In Different topdressing recommendations derived during the reviving stage are put into practice in the field. The trial zone is divided into several 25m2 subzones, in which fertilizers are applied artificially immediately followed by watering.

Table 2. Topdressing Treatment Summary

Treatment	Code	Subzone Qty.
ck	Control	9
Unit	Local grower's uniform fertilization	9
VTN	Variable rate fertilization based on nitrate nitrogen level in soil	18
VS10	Low rate application based on high measured SPAD value in the reviving stage	18
V10	Low rate application based on high measured NDVI value in the reviving stage	18
VPN	Fertilization based on nitrate nitrogen level in wheat	18

As shown in Table 3, grain yield varies with topdressing recommendations. Treatments VPN and UNT are basically same in fertilizing amount but the former produces higher yield of 320 kg/ha. As suggested by the yield-increasing effects, the fertilizer productivity varies with topdressing recommendations; the treatment V10 leads to highest fertilizer productivity at 19.3 kg/ kg, with an increase of 10.4 kg/ kg compared with the treatment UNT; the treatment VS10 results in lowest fertilizer productivity at 4.8 kg/ kg (based on high measured SPAD value) and 5.9 kg/ kg (based on high measured NDVI value) respectively.

Table 3. Yields under Different Topdressing Recommendations

Code	Application Rate Kg/ha	Yield Kg/ha	Yield Increase (%) Compared with Control	Yield Increase (%) Compared with Grower's Uniform Fertilization	Yield Increase kg/kg
CK	0	4796.5			
UNI	150	6129.7	27.8		8.9
VT	159	6588.4	37.36	7.48	11.3
VS10	165	6593.4	37.46	7.56	10.9
V10	132	7346	53.15	19.84	19.3
VPN	153	6449.7	34.47	5.22	10.8

As indicated in Table 4, the treatment V10 produces a profit of 12,649 yuan/ha, with an increase of 2,268 yuan/ha compared with the control (10,381 yuan/ha); the treatments VTN and VS10 both produce lower profits compared with the treatment UNIT.

The treatment VS10 produces lowest profit of 9,358 yuan/ha (based on high measured SPAD value) and 9,775 yuan/ha (based on high measured NDVI value) respectively.

Table 4. Analysis on Profitability of Different Topdressing Recommendations

Code	Application Rate	Average Yield Per Unit	Fertilizer Input	Grain Output Value	Profit
	kg/ha	kg/ha	yuan/ha	yuan/ha	yuan/ha
CK	0	4796.5		8634	
UNI	150	6129.7	652	11033	10381
VT	159	6588.4	691	11859	11168
VS10	165	6593.4	717	11868	11151
V10	132	7346	574	13223	12649

As can be seen from Table 5, the variable coefficient of yield decreases after the treatment VTN. The variable coefficient of the control is highest (20.75%), followed by the Treatment UNIT (18.59%); the variable coefficient of the treatment VTN is only 4.47-14.17%.

Table 5. Yield Variance under Different Topdressing Recommendations

Code	Average Yield (kg/ha)	Standard Deviation	Yield Variance (%) under Different Topdressing Recommendations
CK	4796.5	995.41	20.75
UNI	6129.7	1139.86	18.59
VT	6588.4	395.2	5.96
VS10	6593.4	825.19	12.5
V10	7346	328.38	4.47
VPN	6449.7	917.1	14.17

4 Result and Conclusion

Topdressing is one of critical farming practices to realize high yield of wheat. Scientific topdressing can not only elevate crop yield but also cut down fertilizer input. Grain yield depends partly on topdressing recommendations. Variable rate fertilization based on nitrate nitrogen level in soil and grower's uniform fertilization are basically same in fertilizing amount but the former produces higher yield of 320 kg/ha. As suggested by the yield-increasing effects, the fertilizer productivity varies with topdressing recommendations; high rate application based on low measured NDVI value in the reviving stage leads to highest fertilizer productivity at 19.3 kg/ kg, with an increase of 10.4 kg/ kg compared with grower's uniform fertilization; high rate application based on high measured SPAD value and high measured NDVI value in the reviving stage result in lowest fertilizer productivities at 4.8 kg/ kg and 5.9 kg/ kg respectively. In terms of profitability, high rate application based on low measured NDVI value in the reviving stage produces a profit of 12,649 yuan/ha, and high rate application based on high measured SPAD value and high measured NDVI value in the reviving stage produce lowest profits of 9,358 yuan/ha and 9,775 yuan/ha respectively.

Acknowledgment. This study was supported by Chinese National High Science and Technology Research Project "863" No 2012AA101901 and Ministry of Agriculture nonprofit sector special Project No 201303103.

References

1. Jin, J.Y., Li, J.K., Li, S.T.: Fertilizer and food safety. J. of Plant Nutrition and Fertilizer 12(5), 601–609 (2006)
2. Yan, X., Jin, J.Y., He, P., et al.: Recent advances in technology of increasing fertilizer use efficiency. Scientia Agricultura Sinica 41(2), 450–459 (2008)
3. Chapman, S.C., Barreto, H.J.: Using a chlorophyl meter to estimate specific leaf nitrogen of tropical maize during vegetative growth. Agron. J. 89I, 557–562 (1997)
4. Wu, L.H., Tao, Q.N.: Nitrogen fertilizer application based on the diagnosis of N nutrition of rice plants using chlorophyll meter. J. of Zhejiang Agric. Univ. 25(2), 135–138 (1999)
5. Tang, Y.L., Wang, R.C., Zhang, J.H., et al.: Study on determining nitrogenous levels of barley by hyperspectral and chlorophyll Meter. J. of Triticeae Crops 23(1), 63–66 (2003)

Design and Implementation of WeChat Public Service Platform for the China Research Center for Agricultural Mechanization Development, CAU

Qing Dong and Min-li Yang*

College of Engineering, China Agricultural University, Beijing 100083, China
dqjane@126.com, qyang@cau.edu.cn

Abstract. In order to speed up the effective integration between the agricultural mechanization and information technology, an information service platform based on WeChat (a mobile social network application in China) is designed for scientific research in this paper. By introducing the characteristics of WeChat public service platform and comparing multiple methods of information diffusion, the 4I model of wireless marketing can solve the conflicting information spreading problem with effectiveness, accurateness and personalization. For illustration, a wireless service platform of the China Research Center for Agricultural Mechanization Development, CAU, based on WeChat is proposed designing and establishing. It includes different function modules, such as client information management, intelligent response service and electronic commerce. Some databases have been designed in the platform, such as agricultural mechanization laws and regulations, statistics databases, research institutions and enterprise information, main crops production modes, agricultural machinery products and subsidies, expert system etc. Practical results show that the wireless network application will be used as the final decision in the information diffusion. It is expected to transfer information with lower costs, improve marketing competitiveness and increase the revenue for the enterprises.

Keywords: WeChat public service platform, 4I model of wireless marketing, information diffusion, smart devices, agricultural mechanization.

1 Introduction

The Eighteenth National Congress of the Communist Party of China has put forward to speed up the development of agricultural modernization and effective integration with information technology. Asymmetric information is inevitable in management for modern agriculture industry and is one of the reason block the development successfully. In practice, it is difficult to spread the accurate information of the new equipments and technology and meet the urgent needs of peasants [1, 2]. However, the rapid progress in information technology makes it possible to spread the information

---

* Corresponding author.

© IFIP International Federation for Information Processing 2015
D. Li and Y. Chen (Eds.): CCTA 2014, IFIP AICT 452, pp. 117–129, 2015.
DOI: 10.1007/978-3-319-19620-6_15

via the widespread smart phones for agricultural mechanization technology. According to International Telecommunication Union (ITU) statistics, by the end of 2014, there are 7 billion mobile-cellular users over the world, with 51.4% of them in Asia-Pacific region [3]. Fig.1 shows a strong mobile-cellular user growth since 2005 compared to developed countries and Fig.2 shows a significant increase on share of mobile devices subscriptions by 2000, 2005 and 2014 respectively in developing countries.

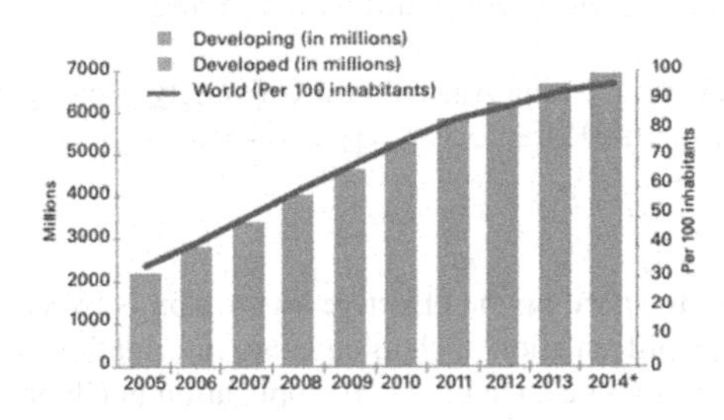

Fig. 1. Mobile-cellular subscriptions, per 100 inhabitants, 2005-2014 (Source: ITU World Telecommunication / ICT Indicators database)

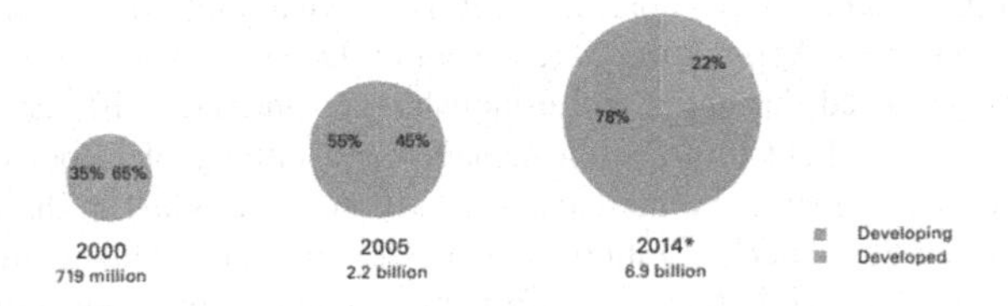

Fig. 2. Share of Mobile-cellular subscriptions, by level of development in 2000, 2005, and 2014 (Source: ITU World Telecommunication / ICT Indicators database)

According to China Internet Network Information Center (CNNIC) statistics, by the end of 2013, there are 0.5 billion mobile user with an annual growth rate of 19.1% in China [4]. Therefore, a smart phone service platform could be established to spread precise information of agricultural mechanization. One of important of McLuhan's media theories is that the really meaningful and valuable information in media is not the content, but themselves, which represents the nature of communication tools and social transformation in this era [5].

Information diffusion consists of three primary parts: sender, channel and receiver. Actually, the channel has various forms, from traditional oral communication, magazines, newspaper, radio and television [6], to newer and innovated technology computer and wireless network. In recent years, mobile media devices, mobile phones and PDA etc. showed close relationship with agricultural machinery. Particularly, mobile phones play an outstanding performance during Trans-regional operations of farm machines (TROFM) in China. Since 2006, the Ministry of Agriculture (MOA) of the People's Republic of China has signed annual mobile phone information service agreement with Foton Lovol for delivering free weather, traffic, agricultural machinery, even oil supply and operation messages. It has sent message of around 133 million pieces via cell phone to people, who participated in the TROFM and enjoyed an experience of convenient and efficiency from wireless information services [7]. However, this kind of information diffusion model is still a traditional 'one-way communication' [8], which is linear, limited and untimely due to a straight line from sender to receiver for informing, persuading or commanding [9]. Even though the users could have dialed to call center for more enquiries, the information is often untimely or inaccurate due to busy line or limited scope of knowledge of operators. This communication model is believed to lack participation, efficient feedback and enough communication in Fig. 3.

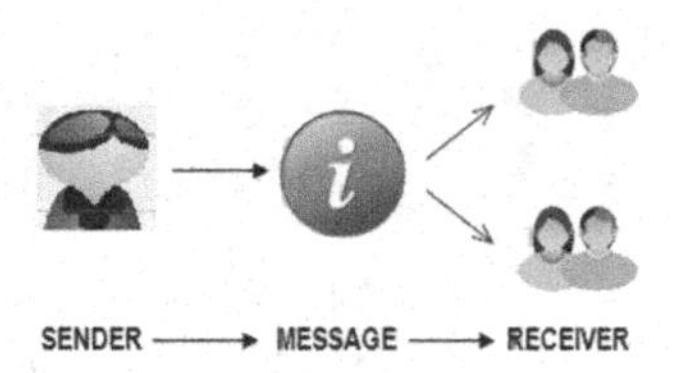

Fig. 3. One-way communication model

Foulger proposed a new ecological model of communication process between information creators, consumers, messages, language and media [10]. The model is demonstrated in Fig. 4, which elaborates Lasswell's five classic communication questions: Who? Say what? In which channel? To whom? What effect? [11]. The answers are: creators for 'Who?' Message for 'says what' , languages and media for 'in which channel' that means information transmit with different media in different languages, consumers for 'to whom' receive the messages, and perspectives, attributions, interpretations and others for 'with what effect'.

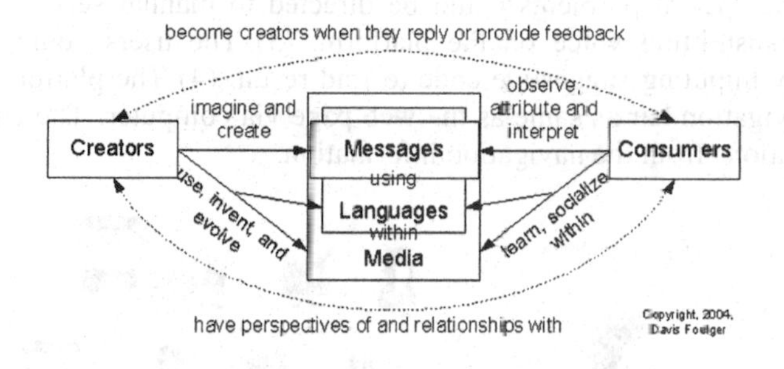

Fig. 4. An ecological model of the communication process

In this paper, although it does not analyze the theory of the five primitives' relationship, which including creators, messages, languages, media and consumers, it proposes a mobile platform based on WeChat application for agricultural mechanization scientific research according to Floulger's theory. Comparing with one-way communication model, it directly forms an interactive network ecological model as shown in Fig. 5.

WeChat, also called 'Weixin' (in Chinese), was created by a Chinese Internet communication conglomerate, named Tencent Company in 2011. It is a free popular mobile instant messaging application, which combines and upgrades the basic functions of SMS (Short Messaging Service) and MMS (Multimedia Messaging Service) in mobile phone [12]. WeChat Public Platform is a social marketing platform that allows government, companies, organizations and others, even sole person to share and market products, ideas, views to concerned groups rather than to individual users. The platform has three particular features: 'semacode subscription', 'message pushing' and 'brand communication' [13].

Fig. 5. Interactive network ecological model

In this paper, we propose a public service platform for agricultural mechanization scientific research based on WeChat public platform that consist of different function modules, such as client information management, intelligent response service and electronic commerce. The platform creates client information files through their registration, LBS (Location Based Service) information or when they click on the 'Follow' button. How the intelligent response service works is shown in Fig. 6. It has four solution channels upon any enquiries from users on this platform: (1) Frequently asked questions can be replied by intelligent response on the displays of mobile phones (2) The special problems could be directed to manual service by inputting message or transmitting voice on the platform. (3) The users could search more information by imputing suggestive code to find result. (4) The platform also has an alternative navigation bar as same as the web page via computer. The end user could find more solutions from the navigation information.

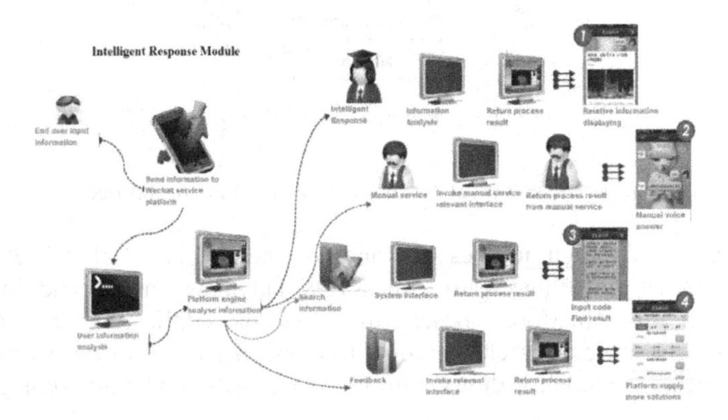

Fig. 6. Intelligent response service

2 Design and Implementation of WeChat Public Service Platform for the China Research Center for Agricultural Mechanization Development, CAU

2.1 The Start-Up Process of WeChat Public Platform

Hardware, smart devices and WeChat application are required to start the public platform. The platform is equipped with a process that is capable of registering a WeChat official account, which may be either a subscription account or a service account via WeChat service website of an account admin platform. Then the system sends information to smart devices, which have installed in WeChat application. To avoid the spam messages, a subscription account only can send one piece of message per day or four pieces of messages of a service account for a month. The process of start-up WeChat public service platform is presented in Fig. 7.

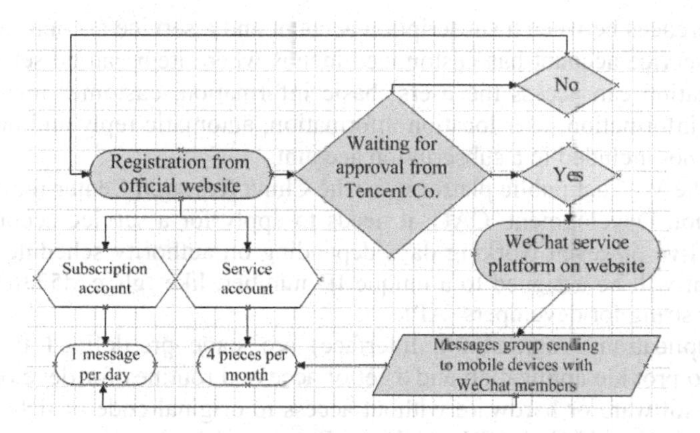

Fig. 7. The process of start-up WeChat public platform

To register on WeChat official website one needs to prepare all relative documents about organizations, then fill in all the blanks or upload the documents after scanning. Options are provided for a service account or a subscription account. The Fig. 8 shows the needed information for registration, including organization type, name, address, contact, business license, business scope, capital, organization code, operator information, certificate of authorization and other supporting documents.

After submitting, the third party of Tencent Company for certification authority would review the documents. If the documents comply with the regulations, the applicant will be able to send grouping messages such as words, voices, pictures, videos and graphics in Chinese or English from the WeChat public service platform to target users, Otherwise, the registration and reviewing have to be performed again.

When searching ‘agricultural mechanization’ from WeChat public service account, there are 34 items with 34 accounts. By contrast, to search ‘Agricultural mechanization’ as keywords, it has 200 results with 200 accounts. However, 99% of them applied subscription account without submitting certification materials, while only few of them owned a service account.

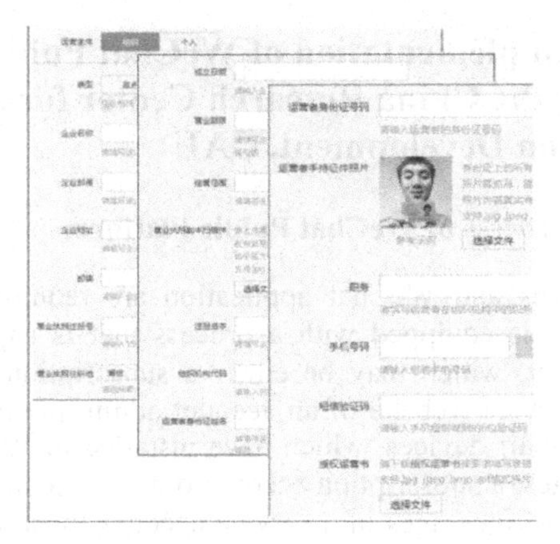

Fig. 8. Registration information

The differences between a subscription account and a service account lie in that an authorized service account has customized menus where item can be set as required, and organization can access the users' basic information, customer interface, voice recognition information, user location information, automatic reply and others. These features are not included in a subscription account.

To start the WeChat public platform for the China Research Center for Agricultural Mechanization Development, CAU, it needs to apply for a service account as well, which take five or seven working days depending on authority schedule. Moreover, each account will be assigned to a unique ID number, like 'gh_e7d52360139a' kind of character string for developers' API.

API (Application Programming Interface) are some pre-defined functions, the purpose is to provide application and a set of accessed routines for developers, which is based on software or hardware without access to original code or detailed internal working mechanism [14]. A URL (Uniform Resource Locator) begins with 'http' and token are needed for the interface in order to represent the right performance of operations.

2.2 The Design and Implementation of the China Research Center for Agricultural Mechanization Development, CAU Based on WeChat Public Service Platform

The China Research Center for Agricultural Mechanization Development, CAU is the first agricultural mechanization research organization in China established in Department of Engineering, China Agricultural University in Jan. 2011 [15]. The main functions of this center are to realize the combinations of soft science and hard technology, research and solve the major development direction, policy issues of agricultural mechanization and agricultural equipment industry. The goals are to improve theory system of agricultural mechanization with China characteristics, to become a technology communication ligament between agricultural mechanization and agricultural equipment industry by carrying the forum and network, and to provide advisory services for the national decision department [16].

The aims of the design of the WeChat platform for upper research center are to deepen the connection with consumers within agricultural mechanization field, build up database of consumers, scientific industry research, and expert information etc. Therefore, a service account is necessary to provide higher quality services rather than a subscription account. However, the approved service account by registration only is a basic WeChat service platform and more advanced functions will be enabled by connecting with third-party developers' API, such as providing customer service, obtaining users' information that includes open ID, nickname, sex, and location etc. generating a semacode with parameters for further promotion, grouping users and managing multimedia file to upload or download.

The customer service interface is used for handling artificial message by posting a JSON (JavaScript Object Notation) data packet that allows developers to provide more quality services to users. For instance, a kind of JSON data packets for sending two graphics with text messages: 'The Press Conference for 18th World Congress of CIGR was hold' and 'Session VII: Information Systems' are represented as follow.


```
{
    "touser":"OPENID",
    "msgtype":"news",
    "news":{
        "articles": [
        {
            "title":" The Press Conference for 18th World Congress of CIGR was hold ",
            "description":"International Commission of Agricultural and Biosystems Engineering",
        "url":" http://www.cigr2014.org/news_and_notices/news/2014/05/37065.shtml ",
        "picurl":"
http://www.cigr2014.org/images/news_and_notices/news/2014/05/22/5C68B1913E5110B8833013461FC4
4C84.jpg "
        },
        {
            "title":"Session VII: Information Systems",
            "description":"discuss and interact among academics, researchers and professions in all
areas of agriculture",
        "url":"http://www.cigr2014.org/about_the_congress/technical_sessions/2014/04/36113.shtml",
        "picurl":"http://www.cigr2014.org/images/bigbanner1.jpg"
        }
        ]
    }
}
```


The design sketch of first graphics with textures messages: 'The Press Conference for 18th World Congress of CIGR was hold', for example, is shown in smart phone as Fig. 9 below.

Fig. 9. The graphics with textures messages: 'The Press Conference for 18th World Congress of CIGR was hold' on smart phone via WeChat

All followed users can read the latest news from pushing message is the same as surfing in website. On the other hand, after users followed the account, information senders could build up a customer database from website background management system in this public service platform.

The fundamental part in this platform is the custom menu which is designed to include some databases, such as agricultural mechanization laws and regulations, statistics databases, research institutions and enterprise information, main crops production modes, agricultural machinery products and subsidies, expert system etc.

According to the WeChat regulation of custom menu, it includes three primary navigations with 8 characters (4 Chinese words) mostly and each of navigation contains at most five secondary menus with 14 characters (7 Chinese words) [17]. More characters will be replaced by '…'. Fig. 10 shows the layout of the design structure and Fig. 11 represents the final sketch draft in smart phone.

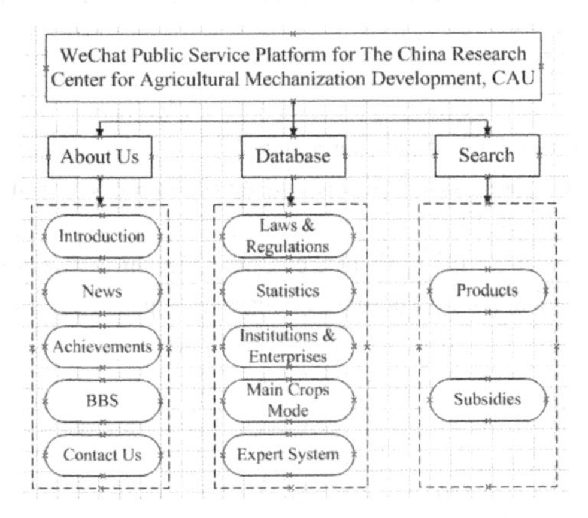

Fig. 10. The structure of WeChat public service platform for the China Research Center for Agricultural Mechanization Development, CAU

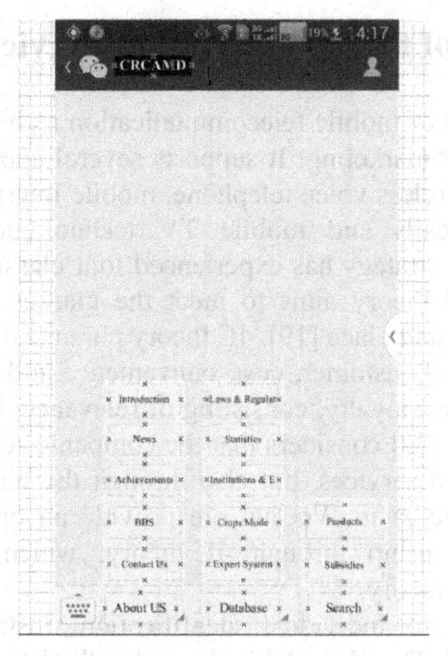

Fig. 11. The final sketch draft of the service platform in smart phone

In Fig. 11, the service account name 'CRCAMD' is the provisional abbreviation name in WeChat public service platform for the China Research Center for Agricultural Mechanization Development, CAU which shows on the top draft. Three primary navigation bars are demonstrated on the bottom line. Moreover, it displays all secondary menu when click each primary navigation bar. After clicking each secondary menu, it will pop up a new html5 page for displaying the final information. If the users need to have a direct conversation with customer service, they should go to the dialog page by clicking the bottom-left keyboard icon. The dialog page shows in Fig. 12, users can send voice, word or picture messages for direct communication.

Fig. 12. The dialog page of the service platform in smart phone

3 The 4I Model of the WeChat Public Service Platform

The third generation (3G) of mobile telecommunication technology brought unlimited business opportunities for marketing. It supports several applications available to 3G devices users, such as: wireless voice telephone, mobile Internet access, fixed wireless Internet access, video calls and mobile TV technologies [18]. Based on 3G characteristics, marketing strategy has experienced four classical theories from 4P, 4C and 4R to 4I theory. 4P theory aims to meet the market demands and stands for product, price, promotion and place [19]. 4C theory pursuits the customer's satisfaction as the goal that is included customer, cost, convenience and communication [20]. 4R tries to establish customers' loyalty, consisting of relevance, reaction, relationship and reward [21]. Peter (2003) [22] considers that the companies compete in the market not only through products and services, but also through the business model. Therefore, 3G and mobile phone media in WeChat era provide an opportunity to discuss the refined marketing relationship through 4I theory, which is gradually replacing traditional 4P, 4C and 4R theory.

The 4I theory refers to individual identification, instant message, interactive communication and I [23]. The 4I model is showed in the Fig. 13.

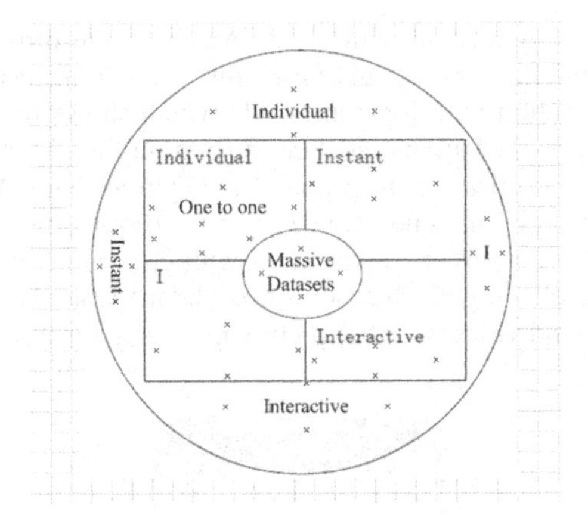

Fig. 13. The 4I model in wireless marketing

From the 4I model, it shows that the 'interactive' is the core of mobile marketing, 'one to one' is intrinsic character, 'I' and 'instant message' belongs to external performance.

(1) Individual Identification. It subdivides the market according to different individual demands and target groups. The WeChat public service platform can subdivides different groups to do business and create brands in accordance with industry, interests, objective, events etc. In order to achieve these, WeChat has dozens of modular units available for doing various tasks, such as micro official website, micro research, micro mart, micro payment, micro membership, and micro community and so on.

(2) Instant Message. It is the distinctive function in WeChat. Each sender can respond in time to meet customers' demands and desires via the instant customer

service. Furthermore, it creates possibilities for deep refine market by dynamic feedback and interactive tracking actions of wireless mobile phone.

(3) Interactive Communication. It emphasizes the customer participation in the marketing. This WeChat public service platform can greatly enhance customer loyalty and satisfaction when they have good communication and timely response experience. For instance, the micro membership improves the user usage and increases the product viscosity.

(4) I. It expresses the personality characteristic and refers to individual needs, exchanges and other personality communications [24].

(5) In this matrix column, it presents a three-dimensional circulation of propagation. Enterprises attract and impact potential customers through WeChat service account, then use html5 to load the information flow. Html5 has more characteristics than html, such as geographical location and local storage. After transforming information flows to download application (App), enterprises manage information and user database through the platform. Among seamless communication of WeChat service account with html5, App and WeChat public service platform, it forms upper referred matrix and own brand media matrix in order to promote the whole media operation.

4 Conclusions

This paper explores the present developing level of information diffusion service in China, particularly the application of social networking service WeChat public service platform which created by Chinese Internet communication company "Tencent". Globally, WeChat public service platform with dozens of advanced modular units and special characteristics is gaining popularity. Governments, organizations, firms even persons can employ this platform to market themselves. Compared with the traditional one-way media communication, the platform provides more convenient, participative and interactive dialogue functions for the general public. Similar to agricultural mechanization companies and organizations, they have their own customers and approaches to promote themselves. As a result of the increasing utilization of WeChat application within smart phone, the public service platform is a new promotional instrument for information diffusion. It contributes to the publicity of product brands, because of high forwarding and transmission rates.

This paper also illustrates the establishment and implement of the WeChat public service platform of the China Research Center for Agricultural Mechanization Development, CAU. It elaborates the detailed processes including service account application, connecting with third-party developers' API for advanced promotion and providing higher quality services in order to enhance the satisfaction and identification of customers.

At the end of this paper, it cites a three-dimensional circulation of propagation matrix from 4I model for analyzing how the WeChat public service platform can satisfy the refined marketing, diversified requirements and personalized service.

Practical results show that the wireless network application will be used as the final decision in information diffusion. It is expected to transfer information in lower costs, improve marketing competitiveness and increase the revenue for the enterprises.

Acknowledgment. A fund for this research was provided by the Beijing Lianhui Digital Technology Limited Company. Meanwhile, the authors thank all senior experts, professors and staffs, who are working for the China Research Center for Agricultural Mechanization Development, CAU for their valuable and constructive comments.

References

1. Zhang, L.: Analysis on the existing problems in agricultural mechanization extension and improvement strategy. Business (4), 253 (2013)
2. Wang, M.: The ideas and countermeasures of agricultural mechanization extension under the new situation. Modern Agricultural Equipments (7), 36–38 (2012)
3. ITU., The world in 2014: facts and figures [EB/OL], http://www.itu.int/en/ITU-D/Statistics/Documents/facts/ICTFactsFigures2014-e.pdf (retrived on April 23, 2014)
4. CNNIC., Statistical Report on Internet Development in China (2014)
5. McLuhan, M.: Marshall McLuhan's Theory of Communication. Global Media Journal-Canadian Edition (2008)
6. Levinson, P.: New New Media. Fudan University Press, Shanghai (2011)
7. Bai, L.: The Information Service Continue to Upgrade in this Summer, China Agricultural Mechanization Herald [EB/OL], http://www.amic.agri.gov.cn/nxtwebfreamwork/detail.jsp?articleId=ff808081377ca6b801377cd96b8402b9 (retrived on May 21, 2014)
8. Scott, A., Le Gall, F., Alexander, R., et al.: The One-Way Communication Complexity of Subgroup Membership. Chicago Journal of Theoretical Computer Science (6), 1–16 (2011)
9. Shannon, C.E., Weaver, W.: The mathematical theory of communication. University of Illinois Press, Illinois (1949)
10. Foulger, D.: An Ecological Model of the Communication Process [EB/OL], http://davis.foulger.info/papers/ecologicalModelOfCommunication.htm (retrived on May 03, 2014)
11. Lasswell, H.: The structure and function of communication in society. Institute for Religious and Social Studies, New York (1948)
12. Sunny, Y.: Weixin–Tencent's Bringing the Mobile IM Revolution to the Mainstream [EB/OL], http://techrice.com/2011/09/21/weixin-tencents-bringing-the-mobile-im-revolution-to-the-mainstream/ (retrived on May 11, 2014)
13. China TMT Daily, Weixin Public Platform. Deutsche Bank Markets Research (2012)
14. Wikipedia, Application programming interface [EB/OL], http://en.wikipedia.org/wiki/Api (retrived on April 25, 2014)
15. The Formal Establishment of The Development and Research Center of China Agricultural Mechanization [EB/OL], http://www.camn.agri.gov.cn/Html/2011_01_17/2_1887_2011_01_17_16052.html (retrived on March 02, 2014)
16. Baidu Encyclopedia, The Development and Research Center of China Agricultural Mechanization [EB/OL], http://baike.baidu.com/link?url=EnIcfpG78MACg4oK_NeVMZtVFWamMk7XPn6XPzdiVwIvPakv8hSHTwLwIBH4VIozjj1elk0QlPddKlTH05t39q (retrived on March 02, 2014)

17. WeChat, Creator's File [EB/OL], http://mp.weixin.qq.com/wiki/index.php?title=19S2qNLlssulpbS0vai907/a (retrived on May 12, 2014)
18. Wikipedia, 3G [EB/OL], http://en.wikipedia.org/wiki/3g (retrived on June 20, 2014)
19. William, D., Perreault, J., Joseph, P.C., et al.: Basic Marketing (18th Revised edn.). McGraw Hill Higher Education (2010)
20. Lauterborn, R.: New Marketing Litany: Four Ps Passé: C-Words Take Over. Advertising Age 41(61), 26 (1990)
21. Schultz, D.E., Tannenbaum, S.I., Lauterborn, R.F.: Integrated Marketing Communications. McGraw-Hill (1993)
22. Drucker, P.F.: Peter Drucker on the Profession of Management - A Harvard Business Review Book. Harvard Business Review Press (2003)
23. Zhu, H.: Wireless Marketing: The Adaptability of the Fifth Media Interaction, Guangdong Economic (2006)
24. Zhu, H.: The essencial marketing of Web2.0: 4I model [EB/OL], http://column.iresearch.cn/u/zhuhaisong/3887.shtml (retrived on May 22, 2014)

Research on Data Sharing Model Based on Cluster

Xiaobin Qiu, Hongqian Chen, and Nan Zhou

Network Center, China Agriculture University,
Beijing 100083, P.R. China
Qxb@cau.edu.cn

Abstract. The data center is the core of digital campus, and the quality of data center depends on the accessing efficiency of data sharing. In this paper we study the advantage and disadvantage of the three current popular sharing model, and propose a sharing model based on cluster according to the reality of the digital campus. By the contrast and analysis of performance, the result show that the model can meet the requirements of the data sharing well.

Keywords: Digital Campus, P2P, Data Warehouse, Middleware, HBase.

1 Introduction

With the rapid development of information technology, the departments of many universities have established their own professional management system, such as management systems for Office Automation(OA), financial information, educational information, scientific research, books and so on. These systems run independently and bring great convenience to people. However, there is insufficient information island, data storage and other problems. This page will discuss the data sharing model that is competent to be applied to the digital campus.

2 Data Share Model

At present, there are three main ways to realize the data sharing, i.e. data warehouse model, sharing middleware model and P2P data sharing model.

2.1 Data Warehouse Model

The data warehouse gets data through dada extraction, cleaning, conversion and loading. And it describes the data of heterogeneous database system through the global schema. The data that shares in different systems stores in the data warehouse. This provides unified data interface, service of data access and data analysis. The model of data integration and sharing based on data warehouse copy the heterogeneous data source to the designated data warehouse[1]. The system only access the copy of data of other system but not the original data. Sharing model as shown in Fig.1:

© IFIP International Federation for Information Processing 2015
D. Li and Y. Chen (Eds.): CCTA 2014, IFIP AICT 452, pp. 130–136, 2015.
DOI: 10.1007/978-3-319-19620-6_16

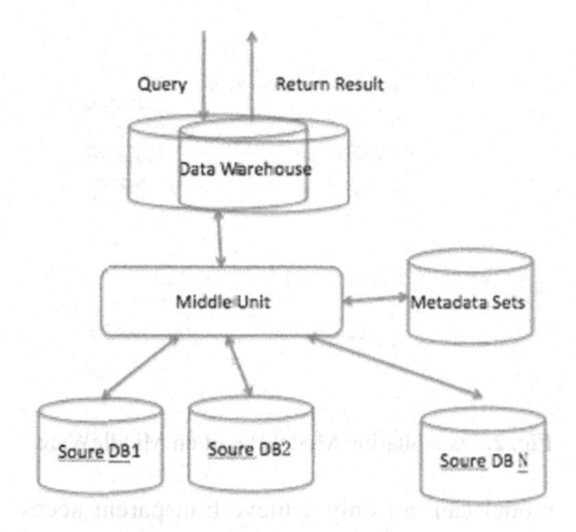

Fig. 1. Data Sharing Model Based on Data Warehouse

The model connects the different systems to realize data sharing by data warehouse. The problems are:

(1) Quality. Data from other database by extraction, conversion, cleaning and other processing from the data source. Once the source database updated, the same semantic extraction data will be inconsistent and lead to data distortion.

(2) Reliability. Once there is the problem in the central database of centralized storage, the sharing data between different application systems will not work, and it will affect the whole integration and sharing system[2][3].

(3) Performance and Real-time. The data joins the data warehouse is by extraction, transformation, cleaning, loading operation and so on, it will take time too much and will affect efficiency of the system of high real-time requirements.

2.2 Middleware Sharing Model

The Middleware sharing mode is based on the global view mode. It realizes the data sharing among systems. The systems access each other by the middleware. There is a unified data logic view in the middle layer that manages the data sharing of source data. It can hide the detail of heterogeneous data to make the distributed source data into a logic whole. The middleware down coordinates the member databases and maintains the view mapping between the member database and middleware. And it up defines and specifies the transmission protocol and interface parameters. The model is focused to establish a global view to realize view mapping between the source data and the middleware that provide the perfect functions and reliable data service. The model shown in Fig.2:

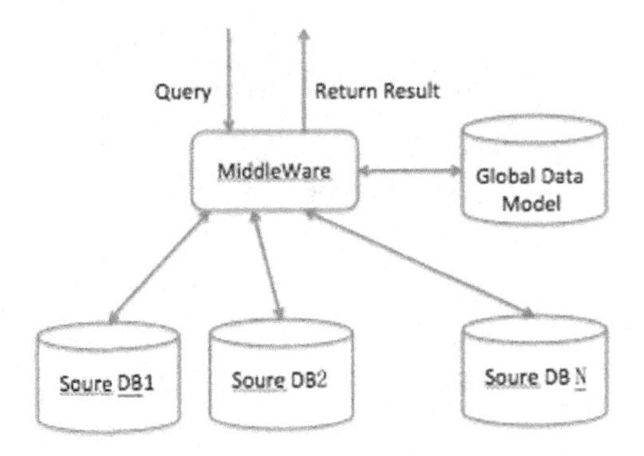

Fig. 2. Data Sharing Model Based on MiddleWare

The middleware model can not only achieve transparent access to heterogeneous data source, but also the management is very simple and the security is good. The main problem is:

(1) The system failure rate increased. When system accesses the data of other system, as the middle layer, the number of links to access increases and the failure probability doubles that leads to the system failure rate increased.

(2)Efficiency. All sharing data transmits by middleware, and the system accesses frequently the middleware or transmits data by middleware that leads to high load. And the bottleneck is the middleware.

2.3 P2P Data Sharing Model

P2P(Peer to Peer Net) data sharing method is developed from the computing technology of P2P. The ideas are derived from the P2P network architecture. It is a distributed loosely coupled data sharing model that is developed from the traditional centralized data management. It accesses directly the database. This is equivalent to between two equal client or server data exchange. Each shares its data equally. The model shown in Fig.3:

P2P (peer-to-peer) data sharing has advantages of source data equivalence, flexible topology and distributed management, the main problems of the:

(1) Bottleneck of System. When the data in one node is shared by multiple system, and the number of accessed increases rapidly, the stress tolerance will be a great challenge, and the note will become the bottleneck.

(2) Node Failure. With the topology expanding and the number of nodes increasing, the complexity of system increases and the possibility of failure of a single node becomes larger. The failure of a core node will affect the operation of the whole system.

(3) Security Flaw. Without QoS quality assurance, anonymous communication and data exchange is unsafe. Without Qos Quality assurance, delay, packet loss, availability and reliability will influence the quality of service.

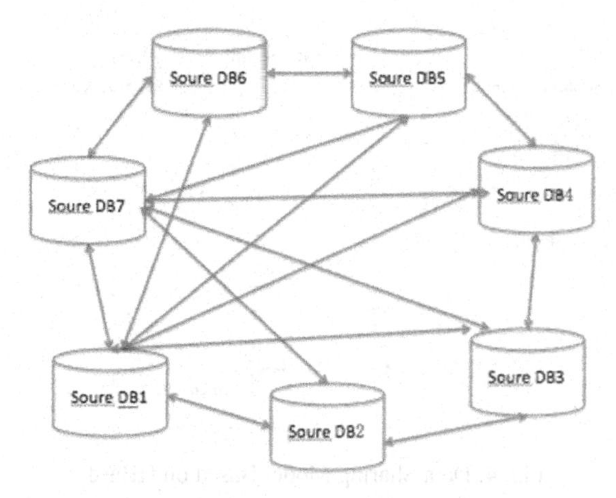

Fig. 3. Data Sharing Model Based on P2P

According to the comparison of above, P2P and middleware tends to involve business system less or high real-time, and data warehouse model is more suitable for complex business. In this paper, according to the actual situation of the digital campus, and combining the advantages and disadvantages of the three kinds of model above, we proposed a data sharing model based on cluster[2][3].

3 Data Sharing Model Based on Cluster

The purpose of digital campus platform is to provide the service of data sharing to the department, and the more important purpose is to build a data warehouse facing theme, and then build a large data to statistic analysis, and provide the data basis for the leading. The connectivity operation and transaction characteristics of relation database, such as Atomicity, Consistency, Isolation, Durability, will make it poor efficiency in scalability and analysis capability. According to the principle of intensive construction of application system integration, the NoSQL database and cloud computing have advantages in data sharing in large-scale. It is high ability to handle large amounts of data and sharing, and it reference to the data warehouse storage sharing idea and concept of cloud storage for all kinds of data types such as structured, semi-structured and unstructured. The relation database is the most widely used before the appearance and development of NoSQL database in recent years. With the rapid processing requirements on large data, NoSQL database gradually shows its advantages. The NoSQL database is the distributed storage system of column oriented that can achieve high performance of concurrent read and write operations, such as Cassandra, Hbase, BigTable.

In this paper, we establish a data sharing model based on the cluster of public databases in a Hadoop HBase database as an example. In the model, it realizes the data sharing and the processing and analysis of data, but it can keep the original private business system. The model shown in Fig.4:

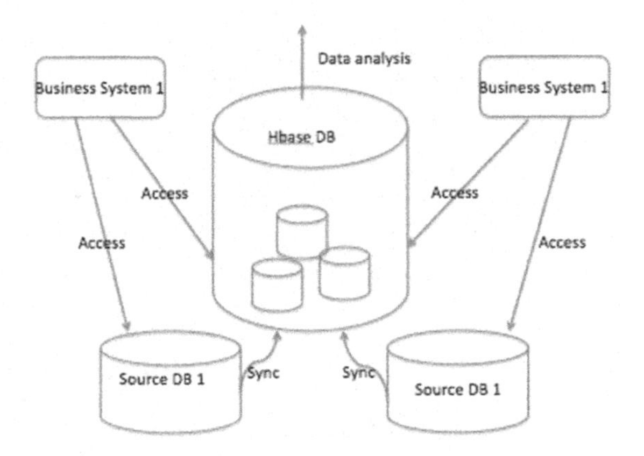

Fig. 4. Data Sharing Model Based on HBase

In the model, the HBase database is the central data warehouse, and synchronizes periodically with the source database. The HBase stores multiple copies for public data to realize sharing efficiently. One system accesses the data of other systems by the mode of accessing public database directly. But the access mode should make corresponding adjustment because HBase is non relational database, and should access database by non relational database mode. This has advantage of the distributed parallel computing(Map/Reduce) and the distributed storage. The different businesses access the different sharing data through the access control[4][5].

4 Performance Analysis and Contrast

The data operation of sharing model can be divided into storage, transmission, query and so no. And the process includes extraction, cleaning, conversion, transmission, processing (add/delete/update), and the extraction, cleaning is a special steps of data synchronization. Now the data query as an example, the time required for each mode as:

Set:

Tc: Time of data converting

Tt: Time of data transmission

Tq: Time of data querying

The time of data warehouse model:

$$T = Tt + Tq \tag{1}$$

The time of middleware model:

$$T = Tc + Tt + Tq \tag{2}$$

Tc is the time of converting of middleware.

The time of P2P model:

$$T = Tt + Tq \tag{3}$$

The time of data sharing mode based on cluster:

$$T = Tt + Tq \tag{4}$$

The operation time is same but mode(2) that have Tc. The Tt is same in the same network, so the key is Tq. mode (1)(2)(3) use relation database and mode(4) uses non relation database.

The following is a group of test data. The key value is used as the query condition in order to ensure the consistency of data because of only key value query in HBase. In 5000000 of data, we query 1 to 100000. The test results in table 1:

Table 1. Query Speed Comparison of HBase and MySql

Number	HBase1	HBase2	HBase3	mysql1	mysql2	mysql3
1	619	602	638	2295	2190	2302
10	738	723	789	2909	2887	2931
100	1102	1124	1210	5232	5219	5214
1000	2522	2612	2460	10213	10207	10211
10000	5532	5510	5628	32012	31018	33101
100000	12092	12030	13021	52351	53121	53462

According to the test data, draw the diagram as Fig.5

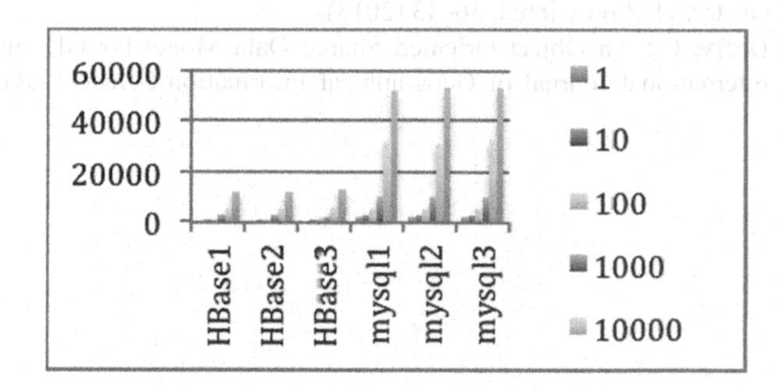

Fig. 5. Query speed comparison chart of HBase and Mysql

Our performance test results show that , the time of HBase and MySql does not increase obviously and, the time range is limited as the quantity of querying data increases,. However, when the volume of data reaches a certain amount, the performance advantage of HBase is obvious.

When the data quantity is growing, HBase can use the cheap PC server to realize the expansion. On the other hand, MySql may also be expanded, for example, multiple copies can be stored in different database servers, and each server has a copy of all the data, but each server can not load balancing automatic. because they serve individually for a part of the user.

5 Conclusion

This paper presents the advantage and disadvantage of the current data sharing model, then proposes a sharing model based on cluster. The performance test confirms that this model has certain advantages. At present, the distributed data is used by many large company such as Google, Baidu, Alibaba and so on. With the development of the large data concept, the data sharing mechanism will be much further improved and widely used in the campus digital construction.

References

1. Lassnig, M., Garonne, V., Dimitrov, G., Canali, L.: ATLAS Data management accounting with hadoop pig and Hbase. In: International Conference on Computing in High Energy and Nuclear Physics, pp. 1–7 (2012)
2. Dykstra, D.: Comparison of the frontier distributed database caching system to NoSQL databases. In: International Conference on Computing in High Energy and Nuclear Physics, pp. 1–5 (2012)
3. Franke, C., Morin, S., Chebotko, A., Abraham, J., Brazier, P.: Efficient processing of semantic web queries in HBase and MySQL cluster. In: The IEEE Computer Society, pp. 36–41 (2013)
4. Franke, C., Morin, S., Chebotko, A.: Efficient processing of semantic web queries in HBase and MySQL cluster. IT Professional, 36–43 (2013)
5. Kumar, M., Duffy, C.: An Object Oriented Shared Data Model for GIS and Distributed Hydrologic. International Journal of Geographical Information Science, IJGIS-2008-0131 (2008)

Agriculture Big Data: Research Status, Challenges and Countermeasures

Haoran Zhang, Xuyang Wei, Tengfei Zou, Zhongliang Li, and Guocai Yang*

School of Computer and Information Science, Southwest University, Chongqing 400715, China
{haoranzhang0715,weixuyang321}@163.com, 670543900@qq.com,
paul.g.yang@gmail.com

Abstract. Agriculture data type and amount in human society is growing in an amazing speed which is caused by the emerging of agricultural Internet of things. The growth of agricultural data volume brought many difficulties in storage and analysis. The realization of the big data and cloud computing technologies provides the solution of these problems. The cloud computing and big data technologies can be applied in agriculture to solve the problems in storage and analysis. This paper elaborates the related concept of agricultural big data briefly, and emphasizes some current researches, the challenges that agricultural big data should face in the future and its countermeasures.

Keywords: agricultural big data, cloud computing, challenge, countermeasure.

1 Introduction

Until now, the agricultural big data has not been clearly defined. We can reach the definition with two parts — big data and agricultural informationization.

The accurate definition of big data also has some disputes. The well-known consulting company McKinsey argues that big data cannot be obtained, stored and managed in a certain time using a traditional database software tools[1]. The definition given by wikipedia shows big data is a blanket term for any collection of data sets which are so large and complex and it becomes difficult to process using on-hand database management tools or traditional data processing applications[2]. there is another view that big data should meet three terms which are Variety, Velocity and Volume[3]. Some people has a different view. They think that three terms can't describe the characteristics of big data, so they come up with four terms. But there is no unified statement on the fouth characteristic. The International Data Corporation IDC believes big data should also has a term of Value[4]. The IBM corporation argues that big data should be considered with Veracity[5]. I think that the term of value makes the definition of big data more accurate. Although the value of big data is sparse, there is great potential inherent value.

Agricultural informationization is a dynamic concept. It improves the comprehensive productivity of agriculture and the overall productivity of agriculture using modern

* Corresponding author.

© IFIP International Federation for Information Processing 2015
D. Li and Y. Chen (Eds.): CCTA 2014, IFIP AICT 452, pp. 137–143, 2015.
DOI: 10.1007/978-3-319-19620-6_17

information technology and information systems[6]. In brief, agricultural informationization makes full use of information technology methods, means and process to achieve the goals.

Agricultural big data uses concepts, techniques and methods of big data and cloud computing to handle a mount of agricultural data. We can obtain the information which is useful to guide agriculture. Wen Fujiang shows that agricultural big data is related to agricultural production in all aspects. It is a multi-professional data mining process[8]. The implementation of agricultural big data technology is a very important component of agricultural informationization. And it also provides new methods and ideas for agricultural research and agricultural business development. Professor Wang refers that we have many datasets of agriculture. As a result, there is a good deal of agricultural data. But the data standards are not uniform and standardized. How to give a reliable and professional decision is an urgent task[9]. In order to solve the problem of the development of agricultural informationization, we should establish a national big data center and make efforts to develop cloud computing and to search agricultural big data mining techniques.

2 Developing Status of Agricultural Big Data

With the rapid rise of the Internet of things and social networks, the amount of data is growing at an unprecedented rate. According to Winter Corporation's survey, the current amount of data is growing at three-time increases every two years[10]. Its growth rate exceeds the growth rate of Moore's Law far. The era of big data is coming. "NATURE" magazine published a special issue named "Big Data: Science in the Petabyte Era"[11]. This paper described the challenges of big data in many aspects,such as Internet technologies, network economics and so on. "SCIENCE" magazine also launched a special issue named "Dealing with Data"[12] in 2011. It illustrated the important scientific value of big data. European Information Society Science and Mathematics Research journal ERCIM News discussed the management and innovative technology issues in their special issue named "Big Data"[13] in 2012. Gartner Company showed that the development of big data has entered the peak period in 2013 Hype Cycle for Emerging Technologies. We showed it in Fig.2-1. Many transnational corporations, such as IBM,ORACLE and FACEBOOK, are making their efforts to develop big data technologies[14]. Academician Professor Li Guojie noted that we should take the redundant data out of raw data in the data explosion era[15]. In December 2013, China Big Data Technology Conference was held and it mainly discussed the technical means and the commercial value of big data applications. Big data has been a intersection of information science, social science, network science, and many other emerging interdisciplinary field. It has became a hot research point.

Recently, big data has been applied to different points of healthcare, manufacturing, transportation, and financial sector in China. With the use of agricultural informationization and Internet of things in agriculture, agricultural big data will become another focus of big data research. Agricultural big data is a application of big data in agriculture. Agricultural big data which is related to all aspects of agricultural production

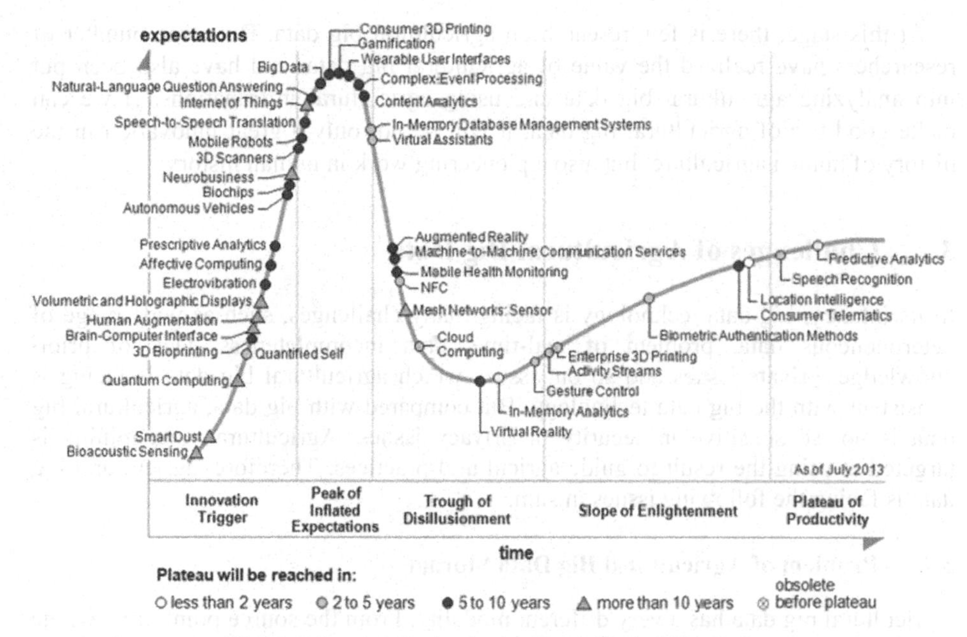

Fig. 1. Emerging Technologies Hype Circle,2013 [16]

is a issue of multi-disciplinary, cross-sectoral analysis and data mining. The first domestic agricultural big data industry technology innovation strategic alliance was established in Shandong Agricultural University in June 2013. Shandong Agricultural University President Wen Fujiang pointed out that in China, the current big data research is just beginning, but agricultural big data research is leading research[8]. Many companies are targeting agricultural big data opportunities. The New York Times reported that in 2012 the Solum company received 17 million dollars fund form Andreessen Horowitz and other companies. Solum company commits to use data analysis techniques to determine the amount of fertilizer in cultivation. It helps farmers improve productivity and reduce costs through the analysis of agricultural big data. Multinational agricultural biotechnology company Monsanto spent 930 million to takeover the insurance accident weather company Climate Corporation [18]. Climate Corporation predicts the weather which may damage agricultural production through analyzing their massive weather data. According to the forecast, farmers can select appropriate agricultural insurance, to reduce the impact of bad weather in agricultural process. The analysis of agricultural big data plays an indispensable role in agriculture.

While agricultural big data is still a fresh vocabularies, but the number of data which is produced in the process of agricultural production is far more than small data. Agricultural data generation method has been changing dramatically. With the wisdom agriculture proposed, widely used sensors promote the development of agricultural Internet of things greatly. Agricultural Internet of things generates a lot of agricultural data. We should use big data technologies to analysis the data. The emergence of cloud computing provide a reliable way to deal with these massive data.

At this stage, there is few research on agricultural big data. But large number of researchers have realized the value of agricultural big data, and have also been put into analyzing agricultural big data and using agricultural informations. If we can make good use of agricultural big data, it will be not only a great innovation in the history of human agriculture, but also a pioneering work in human history.

3 Challenges of Agricultural Big Data

In its infancy, big data technology is facing many challenges, such as wide range of heterogeneous data, problem of real-time, data incompleteness, lack of priori knowledge, private issues and so on. Issues which agricultural big data is facing is consistent with the big data technology. But compared with big data, agricultural big data is not so sensitive in security or privacy issues. Agricultural data mining is targeted at using the result to guide agricultural practices. Therefore, agricultural big data is facing the following issues in sum.

3.1 Problem of Agricultural Big Data Storage

Agricultural big data has a very different modality. From the source point of view, the data comes from Internet of things of the radio equipment, the agricultural information websites, and a variety of advanced mobile terminals. From the content point of view, it includes not only statistical data, but also basic information on agriculture-related economic entities, investment information, import and export information and GIS coordinate information. Data types also include Structured data, semi-structured data and unstructured data. It will be a problem worth studying to store heterogeneous data and the ability to read and write because of the different treatment of different storage hardware devices.

The heterogeneous hardware is also a problem of storing agricultural big data. There will be a very significant performance differences between different machines in the data center. Different hardware devices have different literacy and processing capabilities. Handing equipment will waste a lot of time waitting the slower storage devices. The linear growth of storage devices and servers will not necessarily bring the linear growth of computing power in this case. The "Barrel Effect" restricts the performance of the entire cluster.

3.2 Problem of Agricultural Big Data Analysis

Data analysis is the core of the whole process of agricultural big data, because the value of agricultural big data is produced in the big data analysis process. Currently, there are many problems in agriculture such as food security, soil management, pest forecasting and prevention, soil management and agricultural consumption. We can use the analysis of agricultural big data to solve these problems. Raw data comes from the extraction and integration of agricultural information. We can select all or part of these data to do the research. Conventional analytical techniques are not applicable on processing agricultural big data, such as data mining, machine learning, statistical analysis and other techniques. It is mainly discussed in the following parts.

(1)Traditional data mining algorithms, such as machine learning and other areas, are no longer suitable for agricultural big data. On one hand, algorithms which mine a small amount of data can not be directly applied to big data. On the other hand, agricultural big data has its particularity. The accuracy of the algorithm is no longer the main standard. Algorithm needs to strike a balance between timeliness and accuracy of processing in many cases.

(2)The metrics of the quality of the analysis result is also a major challenge. Big data has complex types. It leads to many problems in designing the indicators to algorithm.

3.3 Problem of Agricultural Big Data Timeliness

As the time elapsing, the inherent value of data keeps attenuating. Therefore, timeliness must be considered during the analysis of agricultural big data. Untimely data analysis may result in the production of agricultural disasters, Especially in the weather, environmental conditions associated with data analysis. For example, the occurrence of "Low grain price hurts farmers" event is the result of managing the cost of production and other information not in time. Therefore, the characteristic of timeliness is particularly important in agricultural big data.

4 Countermeasures of Agricultural Big Data

To solve the problem of agricultural big data, we will give the corresponding countermeasures which will give some guidances for the future work.

4.1 Storage of Agricultural Big Data

Agricultural big data storage affects not only the efficiency of data analysis, but also the cost of data storage. Therefore, we need to come up with high-efficiency and low-cost data storage. One solution is using distributed storage mode. Distributed file system is the physical storage resources in the file system and is not necessarily directly connected to the local node. But the node is connected via a computer network. Many companies which have to deal with huge amount of data have their own distributed file system, such as Google's GFS(Google File System)[19-20], Taobao File System (TFS)[21]. There are some open source distributed file system,such as HDFS[22], NFS[23-24]. There is another solution. The research can be concerned in searching the technologies about the acquisition and integration of multi-source ans multi-modal data. In addition, high-speed storage and creating index are also important aspects.

The general solution to solve the problem of heterogeneous hardware is using different storage devices in different aspects in heterogeneous hardware environments. The problem will become very complicated when the scale of heterogeneous environment extends to thousands of clusters.

4.2 Analysis of Agricultural Big Data

After years of researches and developments, data mining, machine learning, statistical analysis and other information analysis have been proved to have significant effects for small data. These algorithms can be adjusted to accommodate cloud computing system.

But it must be noted that we must consider the characteristics of agricultural big data in the adjustment process of these algorithms. Real-time and predictable characteristics must be considered. It will be a hot spot in the coming period of scientific research.

It is important and difficult to evaluate the results of agricultural big data algorithms. According to the characteristics of agricultural big data, we can use timeliness as a measure standard. Agricultural big data is massive, so we can use the prior knowledge to test the algorithms. It can measure the quality of the algorithms to a certain extent. It can also measure the reliability of the data results.

4.3 Timeliness of Agricultural Big Data

Timeliness is a core demand of agricultural big data analysis. A lot of research are also expanded around this demand. There are three methods to ensure timeliness.

The first method is using stream processing mode. Although streaming mode is suitable to real-time system, its application field is relatively limited. Streaming application model focuses on real-time statistical system, online monitoring. The second method is batch mode. In recent years, the development of batch model real-time system has became a hot topic and has achieved a lot of achievements. MapReduce programming model which Google company made in 2004 is the most representative batch mode. The third method is using a combination of stream processing and batch mode. The main idea is to use the MapReduce programming model to achieve stream processing.

Acknowledgment. This work was supported by the Key Technologies R & D Program(2012BAD35B08), China.

References

1. James, M., Michael, C., Brad, B.: Big data: The next frontier for innovation, competition, and productivity. The McKinsey Global Institute (2011)
2. Big data [EB/OL]. http://en.wikipedia.org/wiki/Big_data/
3. Grobelnik, M.: Big Data Tutorial [EB/OL]. http://videolectures.net/eswc2012_grobelnik_big_data/
4. Barwick, H.: The "four Vs" of Big Data. Implementing Information Infrastructure Symposium [EB/OL]. http://www.Computerworld.com.au/article/396198/iiis_four_vs_big_data/
5. IBM. What is big data? [EB/OL]. http://www901.ibm.com/software/data/bigdata/
6. Zeng, X.J., Ding, C.Y., Wen, H., et al.: Effective Methods for Agricultural Informationization. Agricultural Library and Information Sciences (2), 34–37 (2004)
7. Qian, X.J.: Research of Agricultural Informationization in the Process of China's Agricultural Modernization. China Agricultural University, Beijing (2005)
8. Wen, F.J.: Strategic significance of data and collaboration mechanisms agricultural research. Higher Agricultural Education 11, 002 (2013)
9. Wang, R.J.: Bottleneck of Agricultural Informatization Development in China and the Thinking of Coping Strategies. China Academic Journal Electronic 28(003), 337–343 (2013)

10. WinterCorp. The Large Scale Data Management Experts [EB/OL]. http://www.wintercorp.com/
11. Nature. Big Data [EB/OL]. http://www.nature.com/news/specials/bigdata/index.html/
12. Science. Special Online Collection; Dealing with Data [EB/OL]. http://www.sciencemag.org/site/special/data/
13. ERCIM News. Big Data [EB/OL]. http://ercim-news.ercim.eu/en89/
14. Li, G.J., Cheng, X.Q.: Research status and scientific thinking of big data. Bulletin of Chinese Academy of Sciences 27(6), 647–657 (2012)
15. Li, G.J.: The New Focus of Information Technology Big Data (April 12, 2013). http://www.cas.cn/xw/zjsd/201206/t20120627_3605350.Shtml
16. Fenn, J., LeHong, H.: Emerging Technologies Hype Cycle for 2013: Redefining the Relationship [EB/OL]. http://my.gartner.com/portal/server.pt?open=512&objID=202&mode=2&PageID=5553&showOriginalFeature=y&resId=2546719&fnl=search&srcId=1-3478922244
17. Big Data Across the Federal Government [EB/OL]. http://www.whitehouse.gov/sites/default/files/microsites/ostp/big_data_fact_sheetfinall.pdf
18. Monsanto Acquires The Climate Corporation [EB/OL]. http://www.monsanto.com/features/pages/monsanto-acquires-the-climate-corporation.aspx
19. Ghemawat, S., Gobioff, H., Leung, S.: The google file system. In: The ACM Symposium on Operating Systems Principles, pp. 29–43 (2003)
20. McKusick, M.K., Quinlan, S.: GFS: Evolution on Fast-forward. ACM Queue 7(7), 10–20 (2009)
21. Chucai. TFS Introduction [EB/OL]. http://rdc.taobao.com/blog/cs/
22. Konstantin, S., Hairong, K., Sanjay, R., et al.: The hadoop distributed file system. In: The 2010 IEEE 26th Symposium on Mass Storage Systems and Technologies (MSST), pp. 1–10 (2010)
23. Osadzinski, A.: Network File System(NFS). Computer Standards and Interfaces 8(1), 45–48 (1988)
24. Anderson, T.E., Dahlin, M.D., Neefe, J.M., et al.: Serverless Network File Systems. ACM Transaction on Computer Systems 14(1), 41–79 (1996)

Improved Method for Modeling in Capacitive Grain Moisture Sensor

Yang Liu, Cheng Xinrong*, Mu Haomiao, and Song Yuyao

China Agricultural University, Beijing 100083, China

Abstract. Food is the material basis of human survival, food quality directly related to people's survival and life. Grains as the main food supply, its quality is also essential. Grain moisture of grain stored is an extremely important influence on the main character. However, the ambient temperature and the degree of compaction of grains have a strong nonlinear relationship between the analytic expression and difficult. Therefore the improved BP neural network was used to solve this problem. With improved orthogonal-optimizing method, the study showed that while the RBF nerve network's weighing factors were obtained, the numbers of hidden units could be acquired. This method could avoid too few nerve elements that would result in low accuracy, or too many nerve elements that will result in "over learnt". This method had been approved for its advantages over the ordinary methods with laboratory tests on grains of wheat, rice, corn, etc. Experimental results showed that application of the improved BP model, the measurement accuracy of wheat increased 62.0%, corn increased measurement accuracy of 66.2%, rice increased 66.7% accuracy

Keywords: Grain, Moisture, Improved BP Neural Network, Orthogonal-optimizing Method.

Introduction

Food is the material basis of human existence, food quality is directly related to the survival and life of the people. Our country is a large agricultural country, China's grain output this year to nearly 500 million tons [1-3], in order to ensure the safe storage of grain , grain moisture should be slowed down to safe storage standards by timely grain drying after it is harvested.

There are factors affecting the use of capacitive moisture sensors measure the grain moisture, which is why the causes of error are more complicated. Major source of error for the sample moisture content is uneven, the error ambient temperature, humidity, dust, electromagnetic interference, and the introduction of mechanical vibration generated when the error detection data for the operation. The traditional model of polynomial curve fitting is established through extensive testing and observation, and the need to adopt three more times to fit the experiment, and the adaptability of the model is not strong, fitting accuracy is not high [4-7].

* Corresponding author.

© IFIP International Federation for Information Processing 2015
D. Li and Y. Chen (Eds.): CCTA 2014, IFIP AICT 452, pp. 144–150, 2015.
DOI: 10.1007/978-3-319-19620-6_18

Artificial neural network is a kind of imitation of biological information processing system of brain structure and function, compared with polynomial fitting, and its advantage is that it does not have to learn to be established, including the elimination of non-target parameters including analytic function [9-10]. Neural network learning samples are usually provided by the multi-dimensional calibration experimental data. In recent years, BP artificial neural network algorithms have been used to implement the non-linear fitting of measurement data and temperature compensation.

1 Probe Structure

This paper focuses on the neural network in moisture data acquisition system calibration advantages, you first need to set up the hardware for measurement. The capacitive sensor construction is shown in Fig. 1. In the sensing electrode and the drive electrode of the back surface of the substrate protective electrode is provided, the potential of the guard electrode and the sensing electrode, so as to reduce interference and parasitic capacitance.

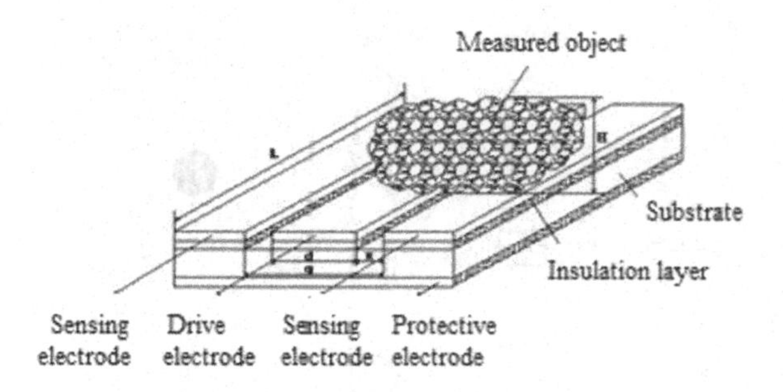

Fig. 1. Flat plate capacitive sensor

Where d is the width of the driving electrodes, referred to as a very wide; q is the distance between the two sensing electrodes, referred to as the pole pitch; drive plate and the distance between the sensor plate, L is the length of driving electrode and the sensing electrode, H cereal measured depth. Sensing electrodes during installation, and grain drying machinery combined into one, thus, it can be considered the width of the sensing electrode is infinite.

Because of the protection of the electrodes, the driving and sensing electrodes between the power line affected by the shape, assuming a power line between the two electrodes of an arc or elliptical arc curve to calculate the curve of the fringe field capacitance of the capacitor, there is clearly a large error. Therefore, for the capacitor shown in Fig. 1, directly calculate the capacitance is very difficult to resolve.

In the experiment, the sensor plate is placed in a grounded container, using the measurement method given volume.

2 Neural Network Design

2.1 BP Neural Network

BP (Back Propagation) network was proposed in 1986 by a team of scientists led by Rumelhart and McCelland. It is a training multilayer feedforward network that was widely used according to a former error back propagation algorithm for. BP neural network topology includes the input layer (input), hidden layer (hide layer) and output layer (output layer).

BP algorithm learning is composed by the forward spread of propagation and reverse spread of propagation. Learning process first began in forward propagation part in the information input by the input layer to the output layer hidden layer processing, this communication process with no influence each layer neurons, each neuron state on its next level only neurons affected state. If the result is not up to the expected output layer value, the learning process into reverse propagation part, even if the output error is returned to the original connection path to modify the weights of each layer neurons, the error signal continues to spread after the minimum [11-14].

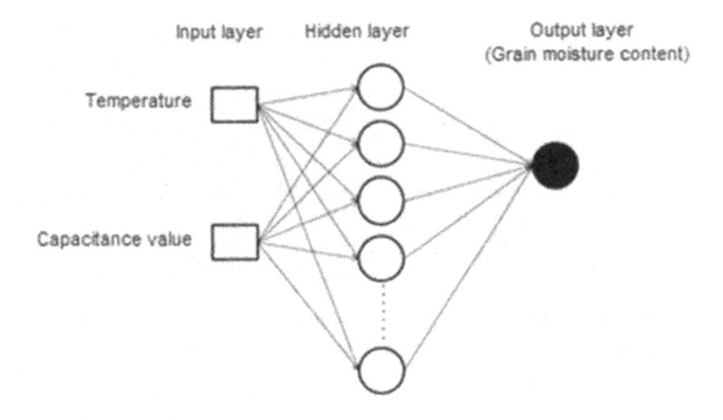

Fig. 2. BP neural network model structure

2.2 BP Neural Network Model

The advantages of the network is that its generalization capabilities BP on nonlinear mapping ability and it has been proven the typical three-layer BP network model can be any rational function approximation theory. So the paper has designed a 3-layer BP network modeling functions to achieve the experimental data approximate relationship that mathematical modeling through BP network, handling the relationship between capacitance values, the standard temperature and moisture values. Model structure is shown in Fig. 2.

This experiment has two inputs (temperature, capacitance value) an output (grain moisture content). So the input layer design two neurons, respectively temperature and capacitance values. The output layer is a neuron for moisture content. For the hidden layer structure, after repeated experiments, the choice of the two hidden layers is better. The transfer function between the layers by Elman network design, the transfer function is Sigmod functions.

The neural network is being learned after the establishment of BP neural network in Matlab. The convergence is achieved after 89 iterations [15-17]. It was shown in Fig. 3.

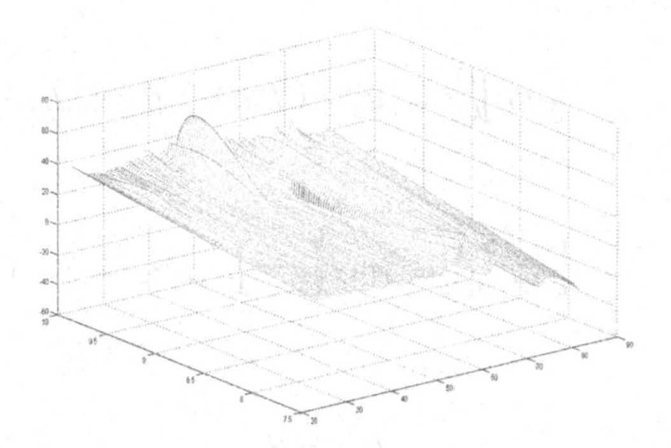

Fig. 3. Neural network fitting surface

At the end of the training, the right network matrix:

$$w\{1,1\} = \begin{bmatrix} -0.2811 & -6.8660 \\ 0.1665 & -0.2048 \end{bmatrix} \tag{1}$$

$$w\{2,1\} = \begin{bmatrix} -4.8079 & -4.8944 \end{bmatrix} \tag{2}$$

Network threshold value:

$$b1 = \begin{bmatrix} 8.1423 & 0.7191 \end{bmatrix} \tag{3}$$

$$b2 = \begin{bmatrix} 7.2273 \end{bmatrix} \tag{4}$$

2.3 OLS Algorithm

Structure optimization of neural networks (ie, the number of hidden nodes selection) is always a difficulty. Hidden nodes numbers in the network structure are commonly large and easily make the network produce over study. OLS method in each orthogonal should use "a contribution to the new interest" criteria Orthogonal, when the OLS meet certain precision, the algorithm terminates, the number of orthogonal vector is then hidden layer nodes. OLS orthogonal method is generally used for the traditional Gram-Schmidt due to the shortcomings of traditional Gram-Schmidt rounding errors large [18-21]. So Gram-Schmidt method to further optimize the modified hidden layer network structure.

Modified Gram-Schmidt method makes the triangular matrix elements are not in columns but by line basis, so rounding error will be smaller.

The calculation steps are:

(1) So that the number of samples equal to the number of hidden nodes N, when t = 1, calculate the

$$[err]_k^{(t)} = g_k^2 / (d^T d) \quad (k=1,2,L,N) \tag{5}$$

(2) Find

$$[err]_{k_1}^{(t)} = \max\left\{[err]_k^{(t)}, 1 \leq k \leq N\right\}$$

to select

$$q_1 = q_k,$$

repeat the cycle when t = h when looking for:

$$[err]_{k_h}^{(t)} = \max\left\{[err]_k^{(t)}, 1 \leq k \leq N, k \neq k_1, \ldots, k \neq k_{h-1}\right\} \tag{6}$$

(3) Select

$$q_h = q_{k_h},$$

when $t=P_s$, if

$$1 - \sum_{i=1}^{P_s} [err]_i < \rho,$$

then ends.

For the hidden layer optimized points. Where in the selected tolerance, it is a parameter related to the degree of accuracy and the final balance of network complexity.

3 Confirmatory Analysis

To verify the data fusion technology and improved BP algorithm proposed to improve the detection accuracy of grain moisture effects on wheat, corn, rice cereal made three common detection comparative experiments.

First make learning samples produced three samples, then using conventional BP neural network, improved BP neural network method for data fusion calculations. The error range is shown in Table 1.

Table 1. The error range of measuring wheat, corn, rice in three methods

Method	Method error range (%)		
	Wheat	Corn	Rice
Data fusion algorithm is not used	-4.9-5	-6-6.8	-4.6-4.8
Regular BP algorithm	-4-3.8	-2.8-5.3	-2.4-3.9
Improved BP algorithm	-1.6-1.9	-2.3-1.3	-1.3-1.6

In rice, for example, the neural network after the training is completed, the drawing of the surface shown in Fig. 3, x-axis represents the temperature, y-axis represents the capacitance value, z-axis represents the water content.

We randomly selected verification test data set 283, and a capacitance value corresponding to the temperature obtained by the sensor device using the standard method from the standard moisture content of grain, and the temperature and the capacitance values into the computer, has been used by BP neural network output, the results shown in Fig. 4.

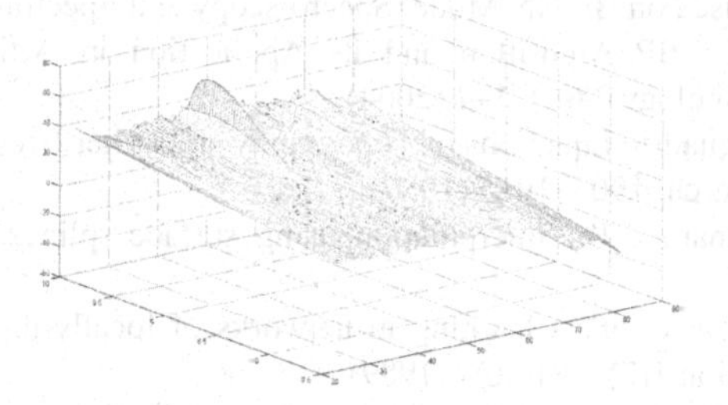

Fig. 4. Neural networks and test point fitting surface

Can be seen from Fig. 4, the dot is measured by standard methods until a point substantially falls on the neural network generated by the fitting surfaces, after the conclusion of the data processing by the neural network after fitting of the aqueous and the moisture content ratio obtained by the standard method with a significant decrease in the average error, improve the measurement accuracy of 62.0% wheat, maize improved accuracy of 66.2%, to improve the accuracy of 66.7% rice, control accuracy can be achieved drying tower requirements.

4 Conclusion

The data confusion method of Radial Basis Function (RBF) nerve network is adopted in this study. With improved orthogonal-optimizing method, the study shows, while the RBF nerve network's weighing factors are obtained, the numbers of hidden units can be acquired. This method can avoid too few nerve elements that will result in low accuracy, or too many nerve elements that will result in "over learnt". This method has been approved for its advantages over the ordinary methods with laboratory tests on grains of wheat, rice, corn, etc.

References

[1] Jianjun, Z., Yuchun, Z.: The Study on the Classification of Maize Drying Characteristics and Moisture Content Intelligent Forecast. Mechanical and Electronic Engineering of Northeastern University (2006)

[2] He, J.: Grain Dry Machinery of Present Situation Development and Tendency in China. Chinese Society of Agricultural Engineering, 252—259 (1993)

[3] Qi, Z., Li, Q.: The Factors Influencing Safety of Grain Storage. Grain Storage 35(1), 3–6 (2006)

[4] Zhang, S.: Measurement Method of Moisture Content in Food Product. Machinery for Cereals Oil and Food Processing (2), 66–69 (2005)

[5] Zhai, B., Bai, Y.: Design of Capacitor Moisture Detector for Cereal. Journal of Liao Ning Institute of Technology 23(1), 34–39 (2003)

[6] Yang, L., Yang, M.: Research on Online Grain Moisture Monitoring System. College of Information and Electrical Engineering, China Agricultural University (2009)

[7] Zhang, B., Lei, Q., Li, G., et al.: Noninvasive Measurement of the Human RBC Concentration Based on BP NN Mode. Spectroscopy and Spectral Analysis 9 (2012)

[8] Hu, J., Tang, X.: BP Algorithm and its Application in Artificial Neural Network. Information Technology 28(4), 1–4 (2004)

[9] Hardy, R.: Multiquadric Equations of Topography and Other Irregualr Sufaces. Journal of Geophysics Research, 1905–1915 (1987)

[10] Harder, R., Desmarais, R.: Interpolation using surface splines. J. Aircraft 9, 189–191 (1972)

[11] Moody, J., Darken, C.: Fast learning in networks of locally-turned processing units. J. Neural Computation 1(2), 281–294 (1989)

[12] Yan, P., Zhang, C.: Artificial Neural Networks and Evolutionary Computing. Tsinghua University Press (2002)

[13] Billings, S., Zheng, G.L.: Radial basis function network configuration using genetic algorithms. Neural Networks 8(6), 877–890 (1995)

[14] Zhi, H., Niu, K., Tian, L., et al.: A Comparative Study on BP Network and RBF Network in Function Approximation. Bulletin of Science and Technology 21(2), 193–197 (2005)

[15] Dong, C.: MATLAB Neural Network and Application. National Defence Industry Press (2005)

[16] Du, X., Li, X., Cui, J.: Application of Subtractive Clustering Method in Medium and Long-Term Eletric Load Forecasting. Journal of Taiyuan University of Technology 36, 165–167 (2008)

[17] Gao, J.: Artificial Neural Network Theory and Simulation Examples, pp. 55-58. China Machine Press (2007)

[18] Zhu, D., Shi, H.: The Principle and Application of Artificial Neural Network, p.129. Science Press (2006)

[19] Xu, L.: Neural Network Control. Publishing House of Electronics Industry (2003)

[20] Huang, H.: Application of RBF Neural Network to Gas Load Forecasting. GuangDong Building Materials 5, 182–184 (2007)

[21] Liu, J.: Study on RBF Neural Network Improvement and its Applications. LanZhou University (2008)

Effects of Reclaimed Water and C and N on Breakthrough Curves in Sandy Soil and Loam

Fangze Shang[1], Shumei Ren[1], Lei Yan[2], Chong Zhang[1], Ganlin Wu[1], Guoya Wang[1], and Chunhuan Zhu[1]

[1] College of Water Resources and Civil Engineering, China Agricultural University, Beijing 100083, China
[2] State Key Laboratory of Water Resources and Hydropower Engineering Science, Wuhan University, Wuhan 430072, China
{shangfangze,renshumei,wuganlin530}@126.com, yanl@whu.edu.cn, h54zhangchong@sohu.com, 18810524293@163.com, shuilizch@cau.edu.cn

Abstract. Long-term irrigation with reclaimed water may change soil physical properties and solute transport rate due to C and N in reclaimed water and the particularity of reclaimed water. Ordinary water, reclaimed water and mixed water which added C and N into reclaimed water were used as background water, then potassium bromide was added to background water and mixed them into three kinds of solutions whose bromide concentrations were all 0.5 mol/L, then soil column breakthrough experiments were conducted. The results showed that bacteria quantity increased both in sandy soil and loam after soil column experiments, and bacteria quantity in sandy soil and loam were all in the following descending order: breakthrough solution using mixed water as background water, breakthrough solution using reclaimed water as background water, and breakthrough solution using ordinary water as background water. However, fungi quantity had no significant difference. Cumulative infiltration in sandy soil and loam can be properly described by power function and logarithm function, respectively. The amount of cumulative infiltration in sandy soil and loam in the same infiltration time were all showed a descending order as: breakthrough solution using ordinary water as background water, breakthrough solution using reclaimed water as background water, and breakthrough solution using mixed water as background water. Breakthrough curves can be well described by CXTFIT 2.1 code, it can be seen from the values of *V* and *D* that reclaimed water and the addition of C and N made solute transport more difficult in soils and increased diffusion coefficient, and these impacts were greater on loam than sandy soil. Reclaimed water and the added C and N increased soil bacteria, complicated soil pore system, and decreased soil hydraulic conductivity.

Keywords: reclaimed water, C and N, cumulative infiltration, bacteria, breakthrough curves, CXFIT 2.1.

1 Introduction

Reclaimed water irrigation is an effective way to resolve the shortage of conventional water resources in agricultural. In China, agricultural water consumption accounts for

© IFIP International Federation for Information Processing 2015
D. Li and Y. Chen (Eds.): CCTA 2014, IFIP AICT 452, pp. 151–159, 2015.
DOI: 10.1007/978-3-319-19620-6_19

more than 60% of the total water consumption [1], and with the development of industrialization and the growth of urban population, conventional water will be more and more used as industrial and domestic water, thus agricultural water will be strictly limited. On the one hand, the quality of secondary reclaimed water is conform to crop irrigation water quality standards [2], on the other hand, Beijing will produce 1.0×10^9 m^3 of reclaimed water per year by 2015, and China will produce more than 680×10^9 m^3 of reclaimed water in 2030[3], thus it is a good choice that irrigation with reclaimed water in large scale in China. However, reclaimed water can be seen as a coexistence system and many researchers studied the specific effects of reclaimed water irrigation from different aspects. C and N in reclaimed water have important effects on soil microorganisms [4], K^+, Ca^{2+}, Na^+ and Mg^{2+} in reclaimed water may cause soil salinity and the change of soil hydraulic conductivity [5-6], suspended solid particles and colloid in reclaimed water may also clog soil pores.

Soil microorganism is one of the most active parts in soils, and it can influence the biochemical reactions of soils, because C and N are the energy sources of microorganisms activities, thus C/N in soil, soil solution and irrigation water have been studied by many researchers [7]. On the one hand, different urban reclaimed water plants produce different quality of secondary reclaimed water, even the same reclaimed water plant produces different quality of secondary reclaimed water in different seasons, leading to the change of C/N in reclaimed water. On the other hand, agricultural measures such as straw returned to field and fertilization can also change C/N value of soils and soil solutions. Breakthrough curve reflects the hybrid displacement characteristics of solute in different mediums, and it also reflects the balance time between solute and mediums, thus the study of breakthrough curve is an effective way to understand solute transport characteristics in soils [8-9]. CXTFIT code which developed by U.S. Salinity Laboratory and based on Levenberg-Marquardt algorithm, has been widely used to calculate solute transport parameters and describe breakthrough curves [10]. Solute breakthrough curves were obtained from soil column experiment, and the parameters were fitted by using CXTFIT code [11-12].

To our knowledge, most soil column experiments seldom consider the actual change of C/N in soils and soil solutions when agricultural measures like crops straw returned to field, fertilization, and reclaimed water irrigation were taken, and more less to study their effects on solute transport, which is important for agricultural management and environmental protection. Thus, our objectives were: (1) to compare the impact of ordinary water, reclaimed water and mixed water on soil microorganisms and cumulative infiltration; (2) to simulate breakthrough curves by using CXFIT 2.1 code; and (3) to explore the effect of C and N on solute transport.

2 Experiments and Methods

2.1 The Properties of Experiment Soils

Soil samples were collected from Beijing Water Science and Technology Institute experiment station in Tongzhou on April 1, 2013. Sandy soil and loam were collected from 120-160 cm and 0-20 cm soil layers, respectively. The percentage of sand, silt, and clay in sandy soil is 88.4%, 10.1% and 1.5%, respectively, and the percentage of sand, silt, and clay in loam is 40.8%, 48.9% and 10.3%, respectively. Physical and chemical properties of soils were shown in Table 1.

Table 1. Physical and chemical properties of experiment soils

Soil texture	pH value	Bulk density ($g \cdot cm^{-3}$)	$EC_{1:5}$ ($\mu s \cdot cm^{-1}$)	TN[a] ($mg \cdot kg^{-1}$)	TP[b] ($mg \cdot kg^{-1}$)	TK[c] ($mg \cdot kg^{-1}$)	Organic C %
Sandy soil	8.9	1.50	120.0	400.5	2.5	15.4	0.17
Loam	8.5	1.45	210.0	1103.4	23.1	98.7	0.45

[a]TN=total nitrogen, [b]TP=total phosphorus, [c]TK=total potassium

2.2 Experimental Design

The reclaimed water was secondary precipitation water taken from Qinghe sewage treatment plant of Beijing, ordinary water was tap water taken from China Agricultural University. Ammonium chloride was added to reclaimed water so that the concentration of total nitrogen up to 80 mg/L, then glucose was added so that the C/N=7 in mixed water. Ordinary water, reclaimed water and mixed water were used as background water, and their chemical characteristics were presented in Table 2. Then potassium bromide was added to background waters, respectively, and finally mixed into three kinds of solutions whose bromide concentration were all 0.5 mol/L. A-1, A-2 and A-3 denoted breakthrough solution using ordinary water, reclaimed water and mixed water as background water to infiltration sandy soil, respectively. B-1, B-2 and B-3 denoted breakthrough solution using ordinary water, reclaimed water and mixed water as background water to infiltration loam, respectively.

Table 2. Important chemical characteristics of the three kinds of background water

Parameters	Background water types		
	Ordinary water	Reclaimed water	Mixed water
pH	7.8	7.6	7.6
EC ($dS \cdot m^{-1}$)	0.72	1.37	1.42
Total nitrogen ($mg \cdot L^{-1}$)	2.0	30.1	90.0
NO_3^--N ($mg \cdot L^{-1}$)	0.1	15.7	15.7
NH_4^+-N ($mg \cdot L^{-1}$)	0.1	0.3	60.3
Organic carbon ($mg \cdot L^{-1}$)	0.5	50.8	630.0
DOM ($mg \cdot L^{-1}$)	3.8	25.0	672.0
TSS[a] ($mg \cdot L^{-1}$)	1.5	6.7	6.7
K^+ ($mg \cdot L^{-1}$)	7.5	8.8	8.8
Ca^{2+} ($mg \cdot L^{-1}$)	48.2	90.3	90.3
Na^+ ($mg \cdot L^{-1}$)	12.3	100.5	100.5
Mg^{2+} ($mg \cdot L^{-1}$)	15.7	38.9	38.9

[a]TSS = total suspended solids

2.3 Soil Column

Soil column and markov bottle were made of organic glass, the inner diameter and height of soil column was 9 cm and 40 cm, respectively, and there was a water inlet and outlet at the top and bottom of soil column, respectively. The inner diameter and height of markov bottle was 9 cm and 50 cm, respectively. Quartz sands whose particle size between 0.2 and 0.4 cm and between 0.8 and 1.5 cm were used as filter layer, and the height was 3 cm, and a piece of filter paper was put on the filter layer, then sandy soil column was packed according to bulk density of sandy soil (1.5 g/cm^3), and the total thickness was 30 cm, loam soil column was packed according to the bulk density of loam (1.45 g/cm^3), and the total thickness was 30 cm, then a little of quartz sand whose particle size between 0.2 and 0.4 cm was added. Infiltration water head was 5 cm for all soil columns (two replications).

2.4 Samples Testing and Data Analysis Methods

Experiments were started on May 1, 2013, continued sampling with 100 ml plastic bottles, and recorded the sampling time of each sample, and the water samples were filtered with 0.45 μm membrane filtration, then stored in 4 ^{0}C fridge. The PIC-10 ion chromatograph was used to measure the concentration of bromide. Fluorescence quantitative polymerase chain reaction (PCR) method was used to measure soil bacteria and fungi quantity. Dates were analyzed using SPSS 17.0 software, and significance level was selected at 0.05.

The convective-dispersive equation can be expressed to [13]:

$$R\frac{\partial C}{\partial t} = D\frac{\partial^2 C}{\partial X^2} - V\frac{\partial C}{\partial X} \tag{1}$$

Where X was distance, cm; t was time, h; R was retardation factor, bromide was a non reactive factor, thus R=1; C was the concentration of bromide, mol/L; D was dispersion coefficient, cm^2/d; V was the average pore water flow velocity, cm/d.

3 Results and Discussion

3.1 Effects of Different Breakthrough Solutions on Soil Microorganisms Quantity

Soil microorganisms including bacteria, fungi, actinomycetes and algae, and the bacteria and fungi account for the largest proportion among them, and C and N are the energy sources for the growth and activity of soil microorganisms. Table 3 showed average soil bacteria and fungi quantity in soil layers after breakthrough experiments. Soil bacteria quantity in sandy soil and loam were all in the following descending order: breakthrough solution using mixed water as background water, breakthrough solution using reclaimed water as background water, and breakthrough solution using ordinary water as background water. However, soil fungi quantity both in sandy soil and loam had no significant difference in different treatments. It is worth mentioning that bacteria and fungi quantity in loam was about 5 and 80 times that of sandy soil, respectively, and bacteria quantity in sandy soil and loam was about 230 and 13 times that of fungi, respectively. Some researchers found that soil microorganisms quantity increased after added fructose and amino acids, and soil structure could also affect soil microorganisms [14-15]. Reclaimed water had higher concentrations of total nitrogen and organic carbon than ordinary water, thus it provided more energy for the growth of microorganisms, and the addition of C and N sources also promoted the activity of soil microorganisms, especially for bacteria. Because solution infiltration in sandy soil was faster than in loam, microorganisms in sandy soil couldn't use C and N effectively, thus microorganisms quantity in sandy soil were less than in loam.

Table 3. Soil bacteria and fungi quantity in sandy soil and loam

Soil microorganisms	In sandy soil			In loam		
types	A-1	A-2	A-3	B-1	B-2	B-3
Bacteria/(10^6 /g soil)	0.04^a	0.13^b	0.75^c	0.98^a	1.32^b	1.74^c
Fungi/(10^5/g soil)	0.01^a	0.02^a	0.01^a	1.04^a	1.11^a	0.99^a

3.2 Effects of Different Breakthrough Solutions on Cumulative Infiltration

The relationship between cumulative infiltration and infiltration time was shown in Figure 1. Three cumulative infiltration curves had similar shape in sandy soil, and the amount of cumulative infiltration in the same infiltration time showed a descending order

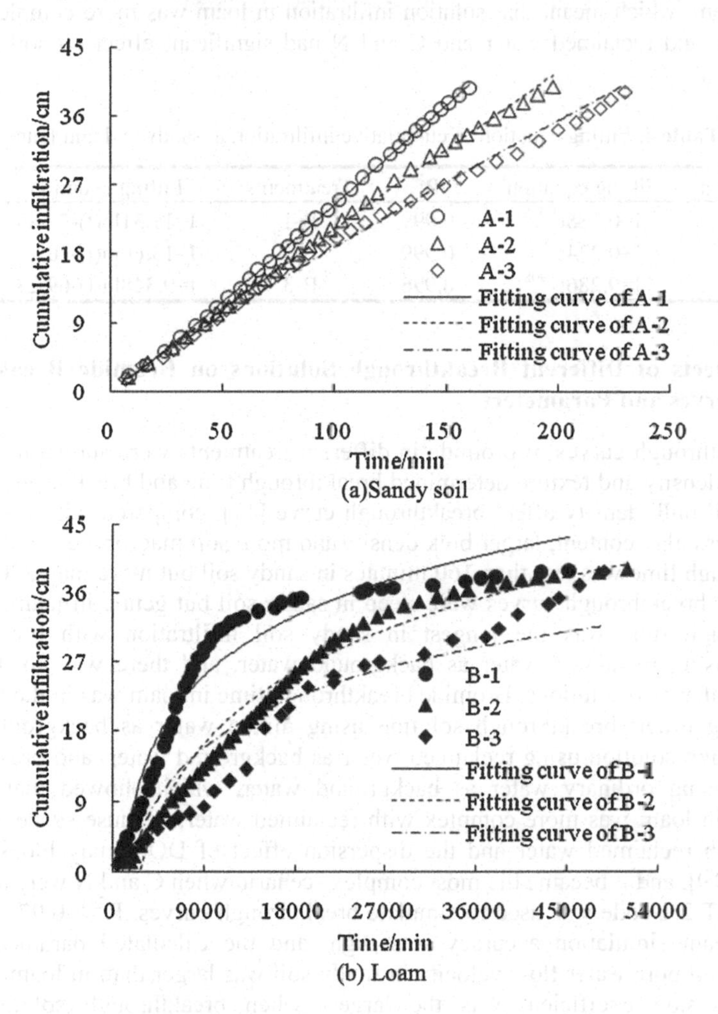

Fig. 1. Soil cumulative infiltration in sandy soil and loam

as A-1, A-2 and A-3. Three cumulative infiltration curves also had similar shape in loam, and the amount of cumulative infiltration in the same infiltration time decreased in the order of B-1, B-2 and B-3. Compared with breakthrough solution using ordinary water as background water, infiltration rate was reduced after infiltration with breakthrough solution using reclaimed water as background water, and infiltration rate was the lowest after infiltration with breakthrough solution using mixed water as background water. Infiltration rate in sandy soil was quicker than in loam when infiltration with the same solution, also the infiltration rate was stable in sandy soil, while the infiltration rate gradually reduced with the increase of infiltration time in loam. Kostiakov put forward that cumulative infiltration could be expressed in power function [16], while in our study, cumulative infiltration could be expressed with power function and logarithm function in sandy soil and loam, respectively (Table 4). The fitting accuracy was higher in sandy soil than in loam, which meant that solution infiltration in loam was more complex than in sandy soil, and reclaimed water and C and N had significant effect on soil hydraulic conductivity.

Table 4. Fitting equation of cumulative infiltration in sandy soil and loam

Treatments	Fitting equation	R^2	Treatments	Fitting equation	R^2
A-1	$I=0.188t^{1.055}$	0.999	B-1	$I=11.31\ln(t)-77.85$	0.939
A-2	$I=0.234t^{0.977}$	0.999	B-2	$I=13.60\ln(t)-107.6$	0.963
A-3	$I=0.286t^{0.908}$	0.996	B-3	$I=9.348\ln(t)-69.24$	0.857

3.3 Effects of Different Breakthrough Solutions on Bromide Breakthrough Curves and Parameters

The breakthrough curves of bromide in different treatments were shown in Figure 2. Soil bulk density and texture determined breakthrough time and breakthrough curve's shape. Soil bulk density affect breakthrough curve [17], compared with loam, sandy soil had less clay content, larger bulk density and more soil macropores, thus bromide breakthrough time was less than 160 minutes in sandy soil but more than 220 hours in loam, also breakthrough curves were steep in sandy soil but gentle in loam. Bromide breakthrough time was the longest in sandy soil infiltration with breakthrough solution using reclaimed water as background water, and there was no difference between other two solutions. Bromide breakthrough time in loam was in the following descending order: breakthrough solution using mixed water as background water, breakthrough solution using reclaimed water as background water, and breakthrough solution using ordinary water as background water, which showed that bromide transport in loam was more complex with reclaimed water, because suspended solid particles in reclaimed water and the dispersion effect of DOM may block the soil pores[18-19], and it became the most complex scenario when C and N were added.

CXTFIT 2.1 code was used to simulate breakthrough curves, $R^2 > 0.97$ (Table 5), which meant simulation accuracy was high, and the calculated parameters were reliable. Soil pore water flow velocity in sandy soil was larger than in loam. In sandy soil, dispersion coefficient was the largest when breakthrough solution using reclaimed water as background water, while in loam, dispersion coefficient was the largest when breakthrough solution using mixed water as background water, which showed that reclaimed water and C and N made the soil pores more complex and increased dispersion coefficient.

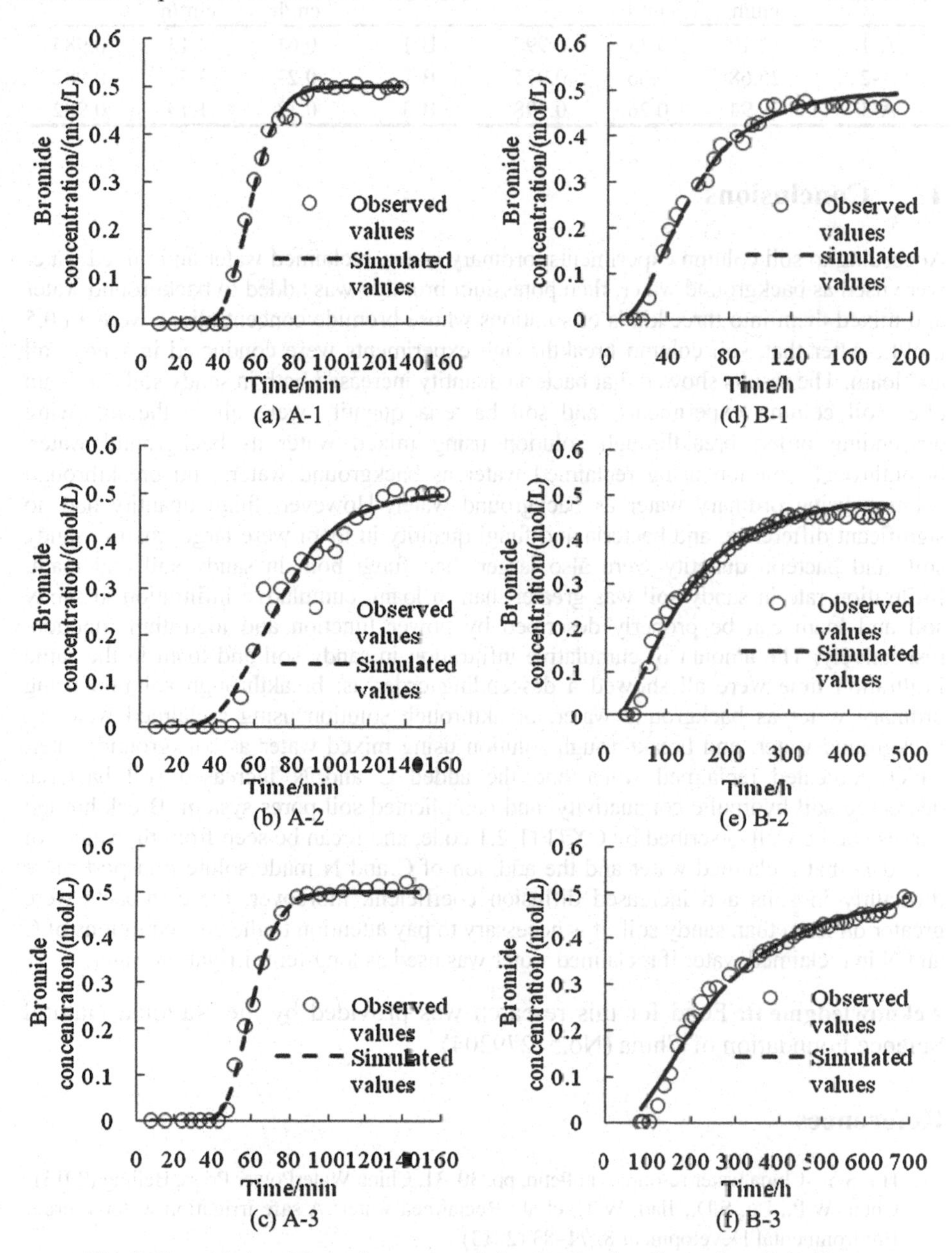

Fig. 2. Bromide breakthrough curves and simulated values in sandy soil and loam

Table 5. Simulation parameters of bromide transport in sandy soil and loam

Treatments	*V*	*D*	R^2	Treatments	*V*	*D*	R^2
	cm/h	cm^2/h			cm/h	cm^2/h	
A-1	32.10	0.33	0.997	B-1	0.61	4.13	0.985
A-2	25.68	0.86	0.975	B-2	0.27	3.20	0.982
A-3	29.94	0.26	0.998	B-3	0.13	1.19	0.982

4 Conclusions

According to soil column experiments, ordinary water, reclaimed water and mixed water were used as background water, then potassium bromide was added to background water and mixed them into three kinds of solutions whose bromide concentrations were all 0.5 mol/L. After that, soil column breakthrough experiments were conducted in sandy soil and loam. The results showed that bacteria quantity increased both in sandy soil and loam after soil column experiments, and soil bacteria quantity were all in the following descending order: breakthrough solution using mixed water as background water, breakthrough solution using reclaimed water as background water, and breakthrough solution using ordinary water as background water. However, fungi quantity had no significant difference, and bacteria and fungi quantity in loam were larger than in sandy soil, and bacteria quantity were also larger than fungi both in sandy soil and loam. Infiltration rate in sandy soil was greater than in loam, cumulative infiltration in sandy soil and loam can be properly described by power function and logarithm function, respectively. The amount of cumulative infiltration in sandy soil and loam in the same infiltration time were all showed a descending order as: breakthrough solution using ordinary water as background water, breakthrough solution using reclaimed water as background water, and breakthrough solution using mixed water as background water, which indicated reclaimed water and the added C and N increased soil bacteria, decreased soil hydraulic conductivity, and complicated soil pores system. Breakthrough curves can be well described by CXTFIT 2.1 code, and it can be seen from the values of *V* and *D* that reclaimed water and the addition of C and N made solute transport more difficultly in soils and increased diffusion coefficient, moreover, these impacts were greater on loam than sandy soil. It is necessary to pay attention to the concentrations of C and N in reclaimed water if reclaimed water was used as long-term irrigation water.

Acknowledgment. Fund for this research was provided by the National Natural Science Foundation of China (No.51279204).

References

1. Hu, S.Y.: China water resources bulletin, pp. 30–31. China WaterPower Press, Beijing (2013)
2. Chen, W.P., Lu, S.D., Jiao, W.T., et al.: Reclaimed water: A safe irrigation water source? Environmental Development 8, 74–83 (2013)
3. Xue, Y.D., Yang, P.L., Luo, Y.P., et al.: Characteristics and driven factors of nitrous oxide and carbon dioxide emissions in soil irrigated with treated wastewater. Journal of Integrative Agriculture 11(8), 1354–1364 (2012)

4. Acea, M.J., Prieto-Fernańdez, A., Diz-Cid, N.: Cyanobacterial inoculation of heated soils: effect on microorganisms of C and N cycles and on chemical composition in soil surface. Soil Biology & Biochemistry 35(4), 513–524 (2003)
5. Muyen, Z., Moore, G.A., Wrigley, R.J.: Soil salinity and sodicity effects of wastewater irrigation in South East Australia. Agricultural Water Management 99(1), 33–41 (2011)
6. Shang, F., Ren, S., Zou, T., Yang, P., Sun, N.: Impact of simulated irrigation with treated wastewater and saline-sodic solutions on soil hydraulic conductivity, pores distribution and fractal dimension. In: Li, D., Chen, Y. (eds.) CCTA 2013, Part I. IFIP AICT, vol. 419, pp. 502–516. Springer, Heidelberg (2014)
7. Wichern, F., Mayer, J., Joergensen, R.G., et al.: Release of C and N from roots of peas and oats and their availability to soil microorganisms. Soil Biology & Biochemistry 39(11), 2829–2839 (2007)
8. Lee, M.J., Hwang, S.I., Ro, H.M.: Interpreting the effect of soil texture on transport and removal of nitrate-N in saline coastal tidal flats under steady-state flow condition. Continental Shelf Research 84, 35–42 (2014)
9. Marín-Benito, J.M., Brown, C.D., Herrero-Hernández, E., et al.: Use of raw or incubated organic wastes as amendments in reducing pesticide leaching through soil columns. Science of the Total Environment 463-464, 589–599 (2013)
10. Toride, N., Leij, F.J., van Genuchten, M.: Th. The CXTFIT code for estimating transport parameters from laboratory or field tracer experiments. Version 2.0, Research Report No. 137. U. S. Salinity Laboratory, Riverside. California (1995)
11. Martínez-Lladó, X., Valderrama, C., Rovira, M., et al.: Sorption and mobility of Sb (V) in calcareous soils of Catalonia (NE Spain): Batch and column experiments. Geoderma 160(3/4), 468–476 (2011)
12. Roger, B., Herbert, J.: Implications of non-equilibrium transport in heterogeneous reactive barrier systems: Evidence from laboratory denitrification experiments. Journal of Contaminant Hydrology 123(1/2), 30–39 (2011)
13. van Genuchten, M.T.: A closed-form equation for predicting the hydraulic conductivity of unsaturated soils. Soil Sciece Society of America Journal 44, 892–898 (1980)
14. Wild, B., Schnecker, J., Eloy Alves, R.J., et al.: Input of easily available organic C and N stimulates microbial decomposition of soil organic matter in arctic permafrost soil. Soil Biology & Biochemistry 75, 143–151 (2014)
15. Jin, V.L., Haney, R.L., Fay, P.A., et al.: Soil type and moisture regime control microbial C and N mineralization in grassland soils more than atmospheric CO_2-induced changes in litter quality. Soil Biology & Biochemistry 58, 172–180 (2013)
16. Kostyakov, A.N.: On the dynamics of the coefficient of water-percolation in soils and on the necessity for studying it from a dynamic point of view for purpose of amelioration (en ruso): Gramingen, Holanda.Transactions, 17–21 (1932)
17. Safadoust, A., Mahboubi, A.A., Mosaddeghi, M.R., et al.: Effect of regenerated soil structure on unsaturated transport of Escherichia coli and bromide. Journal of Hydrology 430-431, 80–90
18. Shang, F.Z., Ren, S.M., Zou, T., et al.: Effects of infiltration and evaporation with treated wastewater and salt solutions on soil moisture and salinize-alkalization. Transactions of the CSAE 29(14), 120–129 (2013)
19. Lado, M., Ben-Hur, M.: Effects of irrigation with different effluents on saturated hydraulic conductivity of arid and semiarid soils. Soil Science Society of America Journal 74(1), 23–32 (2010)

The Design and Implementation of Email Archiving System Based on J2EE

Hu Hongwei, Yu Ping, and Zhou Nan

Net Center, China Agricultural University, Beijing 100083, China
huhongwei@cau.edu.cn

Abstract. With the increasingly widespread use of email and the increasing importance of email data, email archiving has become necessary for mail management. In view of this, we propose a solution of the design and implementation of email archiving system based on J2EE. This solution researches and analysis some key technologies of mail archiving such as mail backup, full-text indexing, email recovery an so on first, and then designs a mail archiving system based on J2EE, resolves the mail real-time backup, full-text search and others. Finally, this system has been applied in actual mail archiving work. Experimental results show that the mail archiving system based on J2EE can achieve real-time email backup and rapid email recovery, this solution can effectively help enterprise users to resolve the problem of email archiving.

Keywords: email archiving, J2EE, Journaling, full text search, email recovery.

1 Introduction

With the rapid development of information technology and the increasingly widespread use of internet, email has become one of the most widely used communication tools in the world because of it's fast, efficient and low cost.

The increasingly widespread use of e-mail makes the email data more important. In modern societ, email data has become critical information assets for government, enterprises and individuals, and plays an increasingly important role in finanical report, commercial negotiations, statistical analysis, decision support and so on.

However, there are some information security risk in using email because of the failures of the email system and the user's faulty operation. In recent years, many problems frequent frequent occurrence while in using email, such as data corruption, data loss and so on, and it Seriously affecting the use of secure e-mail.

The increasing importance of e-mail data and email security incidents occur frequently has raised new demands for the security and integrity of the mail system.

Therefore, building a mail archiving system which can achieve real-time email backup and rapid email recovery has become an important part of modern email management work.

Email archiving system is a solution which can achieve mail data migration, protection and management target, it contains a plurality of functional modules, such as mail backup, full-text indexing, email recovery, Statistical Analysis and so on.

© IFIP International Federation for Information Processing 2015
D. Li and Y. Chen (Eds.): CCTA 2014, IFIP AICT 452, pp. 160–166, 2015.
DOI: 10.1007/978-3-319-19620-6_20

In order to achieve the target of long-term store users' email data, the email backup module backup mail data to mail archiving system from mail server using the mail backup technoloy.

In the full-text indexing module, mail content and attachments can be indexed and added to the index database, which can improve the retrieval efficiency of e-mail data.

The mail recovery module achieves the restore function, which can restroe the data message archiving server to the user's mailbox, or export email data file to the users.

The information of mail server and email archiving server can be analyzed by the statistical analysis module, This information can be displayed in various forms, such as charts, graphics, tables and so on.

Application and research shows that the mail archiving system can resolve some problems, such as offsite backup, long-term storage, category management, real-time retrieval, Rapid Restore and so on. This makes it possible to long-term store and effective use of email data, it also can help manager to analyze email data and make effective decisions.

Generally speaking, email archiving system currently used broadly divided into three classes: email archiving system based on backup technology; email archiving system based on email gateway system and pure software email archiving system. In this paper, we first study and analyze email archiving technology and J2EE paltform, then we design and implement an email archiving system based on J2EE, than can backup, index, restore the email data, and last this system has been tested, compared and analyzed. Application and research show this system can effectively solve backup and restore problem of the mass mail, and it provides security guarantee for e-mail application.

2 Analyze of Email Archiving System

2.1 System Structure Analyze

From a structural point of view, the email archiving system are divided into two main parts, data bankup part and bankuped data manage part. The main function of backup part is to apply the bacukup mechanism to backup the email data, which are received by email server to mail archiving server, and deletes expired backup mail data regularly; the bankuped data manage part achieves the function of bankuped data online query, statistics and restore and so on. The overall structure of the email archiving system is shown in Figure 1.

2.2 System Modules Analyze

From the perspective of functional modules, mail archiving system consists of several major subsystems as follows: backup subsystem, index and backed email query subsystem, restore subsystem, statistical analysis subsystem, user management subsystem and configuration subsystem, as shown in Figure 2. In this paper, we mainly discuss four system: email data backup subsystem, mail index and backed email query system, email restore subsystem, statistical analysis subsystem and user management subsystem.

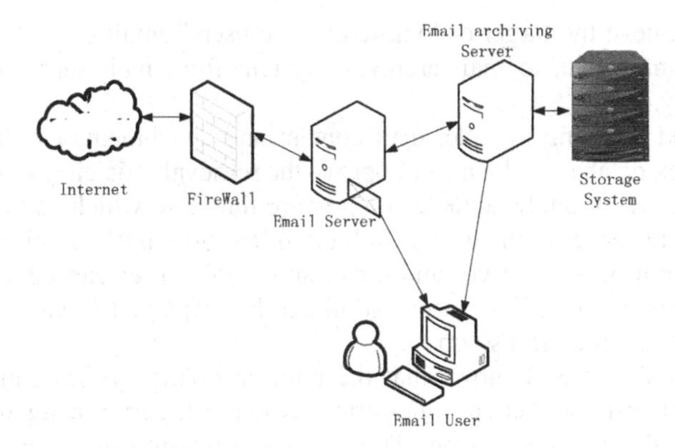

Fig. 1. The overall structure of the email archiving system

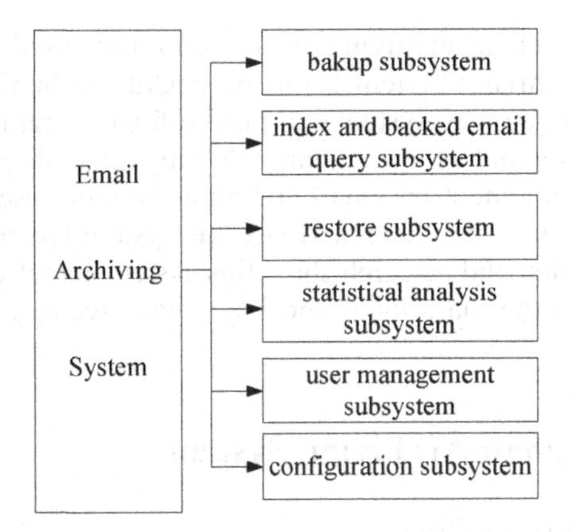

Fig. 2. Email archiving system functions Framework

3 Implementation of Email Archiving System

3.1 Email Backup Subsystem

The mail backup subsystem mainly achieve the function of email backup, and it was dived into three modules, email data acquisition, data store and backed data management.

In this paper, we apply the journaling method to acquire email data from email server. We first change some email manage system parameter to make the mails which are send and received by mail server are all copied to the journaling mailbox. Then the mails which were copied to the journal mailbox will be dumped into the mail archiving server. And last the email which was backuped will be deleted from the journaling mailbox to reduce the pressure of the mail server.

In the mail archiving system, archived email data was stored in a mixed- methods of database and file system. The meta data of the mail was stored in the database, the content and the attachment of the mail was stored in the data file. This method has two advantages. on the one hand, it can improve stored efficiency of archived email and meet the demand of managing massive data, on the other hand, it can achieve fast retrieval of archived messages and impore retrieval efficiency of archived email data.

Taking into account the demand of massive mail data long-term backup and the capacity limitations of the storage system, we conduct a mail backup duplicate checking processing. The email send to multiple users was retain only a single copy. In order to further save storage space, this email archiving system also performed compression process the mail contents and attachments.

In order to achive email backup function, we develop a backup management software which was called "emailbackup", this software apply multiple threads method to read and store email data, and Significantly improves the backup efficiency archived mail.

After setting the parameter of interval time, threads number and backup time of the "mailbackup" software, it will regularly read mail from the journal mailbox and stored data to the email archiving server, and also delete expired backups mail in order to save storage space.

In this paper, we design two methods to set system parameters: the firt method is manual operation, administator can manul set the interval time, threads number and backup time of the "mailbackup" software based on frequencies of email transmission and performance of email archiving server.

3.2 Index and Query Subsystem

Research shows that a prerequisite for the effective management and use of archiving email is to achieve fast retrieval of archived email. However, because the enterprise-class email archiving system stores tens of millions or even billions of email, so it will reduce the efficiency of retrieval archived email if we adopt a common approach to retrieve archived da, and it will not meet the need fast retrieval of archived email.

In this paper, we design a archiving email full-text indexing system based on Luncene in order to achieve fast retrieval of archived email.

Lucene is an open-source full-text search engine toolkit[1], and has been widely used in many files. After analyzing the performance requirements of emal archving system, we design a full-text retrieval subsystem of archived mail basen on Lucene.

This subsustem consists of two parts. The first part is the module of creating and maintaining archived email index; the seconde part is the module of index database retrieving. The structure of the archived mail full-text retrieval subsystem is shown in figure 3.

The establishment of archived email index is procude of adding a index record to the index database.

In this paper, we achieve the function of archived email index combined the process of email backup. The email information will be supplied to the email index system when the email was backed up, and then the email index system will achieve full-text and different attributes index by establish index with email header, sender, recipient, delivery time, email content and attachments.

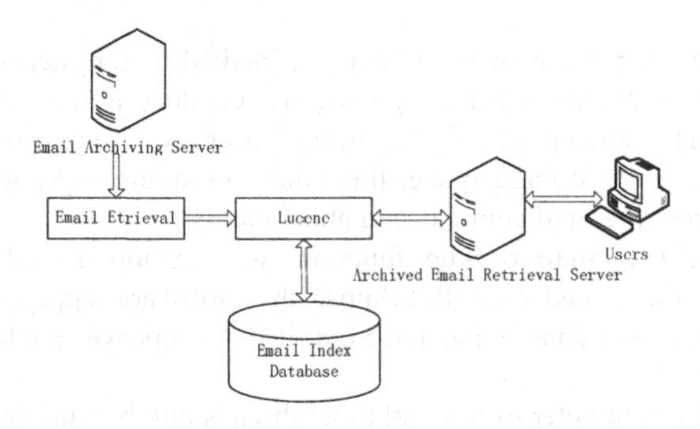

Fig. 3. The structure of the full-text retrieval subsystem

The creation and maintenance of mail index is achieved by three classes of Lucene: indexWriter, document and filed[2].

Index retrieving is the process to retrieve email information from index database based on the web interface supplied by the retrieving system. The retrieving interface of Luene consists of QueryParser, IndexSearcher and Hists[3].

In the full-text retrieval subsystem, we can get the email serial number through email header, sender, recipient, delivery time and other information, than we will get email content and attachment stored in email archiving server.

3.3 Archiving Email Restore and Exprot

Because of the failures of the email system and the user's faulty operation, the user's email often lose, accidentally deleted or damaged issues. Therefore, the restore of the archived email to the email user has to be a major function of email archchiving system.

Tmail archchiving system can restore users' email by sending archived email to the users or export email file to users.

In the process of email restore, we can achieve the email restore by JavaMail interface where retrieving the email informtion.

JavaMail is a java interface usede to send, receive, reade mail which was supplied by Sun Company[4], and it has been widely used in email system[5]. In this paper, we desgin a email restore subsystem used JavaMail, which was called "MailSender". This subsystem achieve the funtion of restoring email to email management from the email archiving system.

The structure of the email restore sybsystem is shown in figure 3.

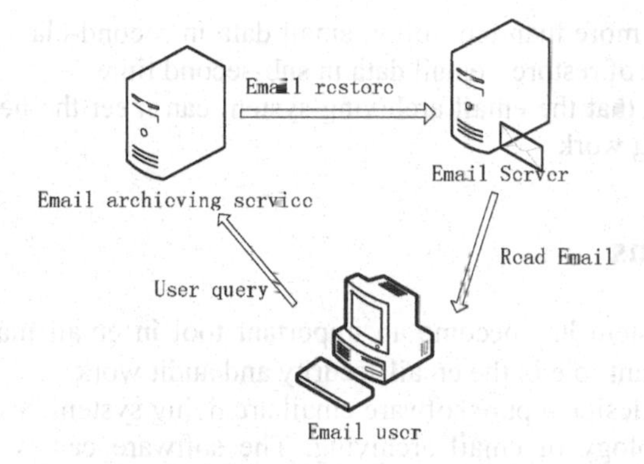

Fig. 4. The structure of the email restore sybsystem

3.4 User Management and Limitation

In the email archiving system, the safe of archieved email is very important. The safe levels of the email archiving system consists of serveral levels, sucn as software level, operation system level, hardware level and network safe level.

In the process of software design, properly setting user limitations is a important work for software level safe.

We dived the users of email archiving system into two categories, comman user and administrator user.

The comm user is matched to the email user, he can query and restore the archived email.

The administrator user is the administrator of the email archiving system, administrator is dived into three categories: the system operator, email auditor and the super administrator.

The responsibilities of system operator is to manage common users and query email basic information, he can manage uses that are from one or some email groups. The system operator can only read email's basic information, such as header, sender, recipient, delivery time and so on, but he can't read email's content.

The email auditor can read email content and audit email informaiton within the scope of relevant laws and regulations, the audit users was assigned the super administrator.

The responsibilities of super administrator is to manage the system operators and email auditors, and he wil set proper parameter information of the email archiving system.

4 Test and Analyze

In order to test performance of the email archiving system, we carryid out test on a university anda company. The parameters of email archiving system was set below, the threads number of backup system is three and the interval time is thirty seconds.

The thirty day's test shows that the email archiving system can achieve the demand of email real-time bacup, and the full-text index subsystem can accurately retrieve mail from more than ten milion email data in second-class time, and it also achieve the function of restore email data in sub-second time.

Analysis showed that the email archiving system can meet the need of enterprise-class email archiving work

5 Conclusions

Email archiving system has become an important tool in email management work, and plays an important role in the email security and audit work.

In the paper, we design a pure sofware email archiving system based on J2EE after studying the technology of email archiving. The software can work on windows, linux, unix and other operating systems, and can applied to a variety of databases, such as Mysql, Oracle, Sql Server and so on.

The email archiving system designde by this paper can meet the demand of email archiving in the the premise of reducing the dependence on hardware and software environment, it also improves the applicability of email archiving products and takes a new way for the development and promotion of email archiving system.

References

[1] Li, X.-L., Du, Z.-L.: Investigation on Personalized Search Engine Based on Lucence. Computer Engineering 16(19), 258–260 (2010)

[2] Guan, J.-H., Gan, J.-F.: Design and implementation of web search engine based on Lucene. Computer Engineering and Design 28(2), 489–491 (2007)

[3] Guohua, B.K.G.: Study And Application Of Vertical Search Engine Based On Lucene And Heritrix. Computer Applications and Software 26(1), 258–260 (2009)

[4] Zhu, A.-Q., Wang, H.-J., Gao, H.-F.: Design and implementation of message system inMIS based on J2EE. Computer Engineering and Design 26(10), 2775–2777 (2005)

[5] Zhong, L., Liu, L., Xia, H.-X.: Development Research of Web Mail System Based on JavaMail AP. Journal of Wuhan University of Technology 28(6), 84–86 (2006)

Stimulating Effect of Low-Temperature Plasma on Seed Germination Characteristics of *Trifolium repens*

Nandintsetseg Munkhuu[1], Changyong Shao[1], Decheng Wang[1], Liangdong Liu[1], Imtiaz Muhammad[3], Changbin He[1], Shanzhu Qian[1,2], Ru Jia[1], and Jinkui Feng[1]

[1] College of Engingeering, China Agricultural University, Beijing 100083, China
[2] College of Mechanical and Electrical Engineering, Inner Mongolia Agricultural University, Hohhot 010018,China
[3] College of Agronomy and Biotechnology, China Agricultural University, Beijing 100093, China

{Nandia_0011,mtz_yousafzai}@yahoo.com,
{shaochangyong68,fengjinkui09}@163.com, wdc@cau.edu.cn,
{lld62336512,changbin_he}@126.com,
{719725846,455084047}@qq.com

Abstract. The low temperature plasma (LTP) technology a new technology, uses for pre-sowing treatments. In this research work we used the new LTP technology equipment "low-temperature plasma modified instrument" and applied different doses in LTP technology to explore its effect on seed germination and other related characteristics on Trifolium repens crop seeds. The seed germination, germination vigor, plant fresh weight, plant dry weight, plant height and root length were significantly affected by LTP doses. The maximum seed germination (73.75%) recorded at LTP 120W treatment followed by 72% at 160W. In addition early germination was observed on same treatment, while minimum seed germination was recorded at 140 Watts LTP treatment. Maximum plant height (8.16mm) at 160W and maximum root length (74.14mm) was recorded at 180W followed by root length (73.89mm) at 160W. From our present work results, it is concluded that low temperature plasma dose 160W is an optimum dose to get high germination rate and healthy seedlings of Trifolium repens crop seeds.

Keywords: Low-temperature plasma, trifolium repens, germination force, germination rate.

1 Introduction

To secure uniform and early germination of seeds and to break seed dormancy, different methods of pre-sowing treatments have been developed to enhance germination rate of different plant species, mainly include, coat breaking, medicine, chemical mixing pile, medicament seed treatments and so on, used to enhance seed germination and seeds sterilization [1, 2]. They played an important role but still have many limitations, such as chemical processing method can't really improve seed itself to adapt to the environmental conditions, the overall disease resistance, cold resistance, early maturity, seed

© IFIP International Federation for Information Processing 2015
D. Li and Y. Chen (Eds.): CCTA 2014, IFIP AICT 452, pp. 167–174, 2015.
DOI: 10.1007/978-3-319-19620-6_21

performance, causing the destruction of the soil environment. Along with the development of modern science and technology, represented by mechanical, electrical and thermal physics technology to deal with the method of seed research and application, starting to get use of crop seed with mechanical, electrical and thermal physical properties, by low temperature plasma treatment technology, magnetic treatment technology, biological spectrum technology, modern physics engineering technology such as solar irradiation technology processing crop seeds, the seeds can activate the endogenous substances, rejuvenate seeds, and stimulate and improve the seed itself ability to adapt to external environment, improve crops resistance, thus improve crop yield. At the same time reduce the risk of damage to the environment and pollution [3-7]. The low temperature plasma (LTP) technology is a new technology, which uses for pre-sowing treatments.

This research article is about the application of low temperature plasma technology to study, the effect of low temperature plasma treatments on seed germination and seedlings characteristics of Trifolium repens, its mechanism is by gas discharge, resulting from the gas ionization discharge, gas molecules and atoms are broken down and ionized. The negatively charged electrons and positively charged ions, uses for formation of plasma. When seeds by plasma glow are passed through discharge zone by light, the light interacts with seed surface layer, the light absorption and scattering occurs. Absorption of energy in form of light forced vibration and converts into heat energy. When material molecule absorbed photons, the electron will transitioned from a low energy state (ground) to a high energy (excited) state, molecular absorption of energy caused by the energy level transition, namely from the ground state transition to the excited level and cause the energy state transition of material, so that the seeds, enhance the vitality of life. A large number of experimental results show that low temperature plasma seed treatment technology has been in crop seed treatment before planting on significant results have been achieved. Through special equipment for processing, through germination experiment, pot experiment and field plot test, showed that the plasma activation seeds to produce a variety of biological effects, improved the germination potential, germination rate, root development, drought resistance and disease resistance [8]. The seedlings grow strong, improve quality, promote early maturity and also increase yield.

Low-temperature plasma seed treatment is a modern eco-agricultural technology that stimulates plant growth. It is based on non-ionizing low-level radiation, which can activate the vitality of seeds but without causing gene mutations, and is quite different from space breeding or mutation breeding by particle beam [8]. Zivkovic *et al.* reported *that* cold air plasma pretreatment significantly improved the germination of Paulownia tomentosa[9] and Jiang JF also discovered that Low-temperature plasma seed treatment has been reported to improve the growth and yield of wheat[10].

In the present work to test of data acquisition and intelligent monitoring system can realize the test conditions, test process and test results of key parameters, such as monitoring, control and acquisition[11]. For the development of digital processor intelligent seeds and development conditions, the results will guide the physical engineering seed processor output dose of key process parameters, such as maximizing the growth of the seed germination, improve the rejuvenation ability of seeds, at the same time for development, can be used for large-scale production of seed processor to provide technical support and application prospect, so the research is of great importance [12].

2 Materials and Methods

2.1 Experimental Set-Up

Low temperature plasma seed treatment mechanism is low temperature plasma in the vacuum ultraviolet energy crop seeds of biological macromolecules of transition, the transition from the ground state to the excited state, which makes the seed produce positive biological effects. Non-ionizing radiation effect on organisms living tissue has been studied in recent years. The different non-ionizing radiation can produce different biological effects.

As the mechanism of low-temperature plasma effect on seeds is not very clear, and the LTP processing device is quite simple now (Figure 1), it becomes impossible to use this technology and devices in a large scale. However, this question also make low-temperature plasma treatment before seed sowing become into a very attractive project to explore, a large number of lab tests have been done in lab and small open field by the authors, all results prove that plants show better characteristics in terms of germination, root system, resistance to drought and diseases, early maturity and yield after their seeds well treated by a suitable dose of low-temperature plasma (LTP).

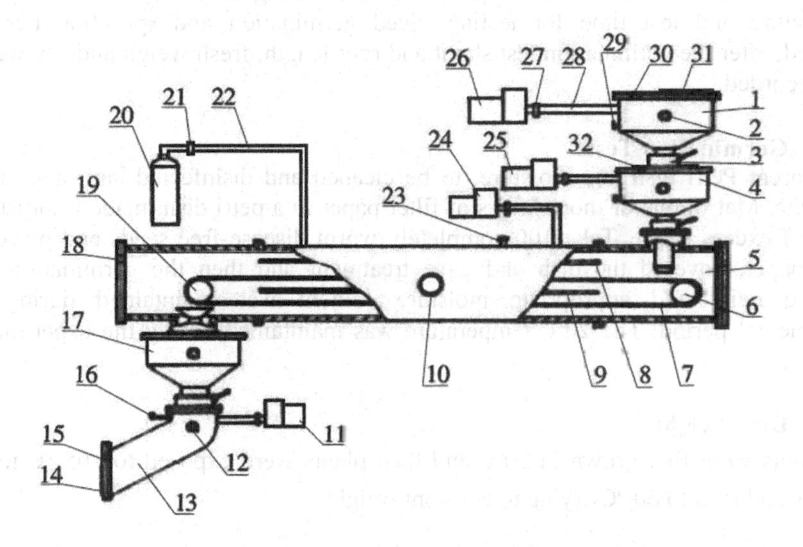

Fig. 1. Diagram of low-temperature plasma test stand device

1, Feeder compartment I; 2, Detector tube I; 3, Butterfly gate; 4, Feeder compartment II; 5, Feeder compartment III; 6, Driven wheel; 7, Transmission belt; 8, Top crown; 9, Bottom crown; 10, Detector tube II; 11, Vacuum air pump I; 12, Detector tube III; 13, Feeder compartment IV; 14, Discharge port; 15, Thermocouple gauge tube I; 16, Deflating valve I; 17, Feeder compartment V; 18, Thermocouple gauge tube II; 19, Driver wheel; 20, Gas tank; 21, Gate; 22, Inlet pipe; 23, Deflating valve II; 24, Vacuum air pump II; 25, Vacuum air pump III; 26, Vacuum air pump IV; 27, Electromagnetic relief valve; 28, Vacuum tube; 29, Deflating valve III; 30, Thermocouple gauge tube III; 31, Feed cap; 32, Thermocouple gauge tube IV.

2.2 Sample Seeds

Germination tests were carried out under laboratory conditions with Trifolium repens seeds purchased from the Royal Barenbrug Group.

2.3 Experimental Design

2.3.1 LTP Treatment

Using this technology the seeds of Trifolium repens were exposed to different treatments such as 0W (1), 20W (2), 40W (3), 60W (4), 80W (5), 100W (6), 120W (7), 140W (8), 160W (9), 180W (10), 200W (11), 220W (12), 240W (13), 260W (14) and 280W (15).

The experimental work was conducted in control conditions, 100 seeds were used for each treatment and the experiment was replicated four times. Data were recorded for the inspection of Trifolium repens seeds, referring to the corresponding test temperature and test time for testing. Seed germination and sprouting per day recorded, after the germination test shoot and root length, fresh weigh and dry weight were recorded.

2.3.2 Germination Tests

Transparent Petri dish and moisture, to be cleaned and disinfected and make them non-toxic. Mat in one or more layers of filter paper in a petri dish, made it moist and drain off excess water. Take 100 completely worm disease-free seeds and placed in moist paper, covered the dish and gave treatment and then the germination was recorded daily and appropriate moisture content was maintained during the experimental period. The 20°C temperature was maintained during the experimental work.

2.3.3 Dry Weight

The plants were first grown in 20°C and then plants were exposed to 105 °C for 15 minutes and then to 60 °C drying to constant weight.

2.4 Statistical Analysis

The data was analyzed by Microsoft office excel 2007 and statistix-8.1. As the experimental work conducted in controlled environmental conditions, so the statistical design CRD (completely randomized design) was used in the analysis. The LSD (least significant different) test was used for means comparison at 5% probability level.

3 Results and Discussion

3.1 Germination

The seeds of Trifolium repens were subjected to different low temperature plasma treatments to explore the effect of different low temperature plasma doses control, 20, 40, 60, 80, 100, 120, 140, 160, 180, 200, 220, 240, 260 and 280 W on germination rate and force of Trifolium repens seeds.

Table 1. Germination rate of Trifolium repens seeds at different LTP treatments

Treatment	CK	20 W	40 W	60 W	80 W	100 W	120 W	140 W	160 W	180 W	200 W	220 W	240 W	260 W	280 W
%Germination	66.5	67.75	68.75	69	66	69	73.75[a]	64.5	72	66.75	70.25	69.25	64.75	71	72.25
%Increase	0	1.25	2.25	2.5	-0.5	2.5	7.25	-2	5.5	0.25	3.75	2.75	-1.75	4.5	5.75
Relative to CK (%)	0	1.88	3.38	3.76	-0.8	3.76	10.90	-3.01	8.27	0.38	5.64	4.14	-2.63	6.77	8.65

Note: [a]Optimum treatment

Table 2. Germination force of Trifolium repens seeds at different LTP treatments

Treatment	CK	20 W	40 W	60 W	80 W	100 W	120 W	140 W	160 W	180 W	200 W	220 W	240 W	260 W	280 W
%Vigor	64	65	67.75	66.25	62.5	68	71.75[a]	63	70.25	64	67.5	66.25	64	69	66.5
%Increase	0	1	3.75	2.25	-1.5	4	7.75	-1	6.25	0	3.5	2.25	0	5	2.5
Relative to CK (%)	0	1.6	5.9	3.5	-2.3	6.3	12.1	-1.6	9.8	0.0	5.5	3.5	0.0	7.8	3.9

Note: [a]Optimum treatment

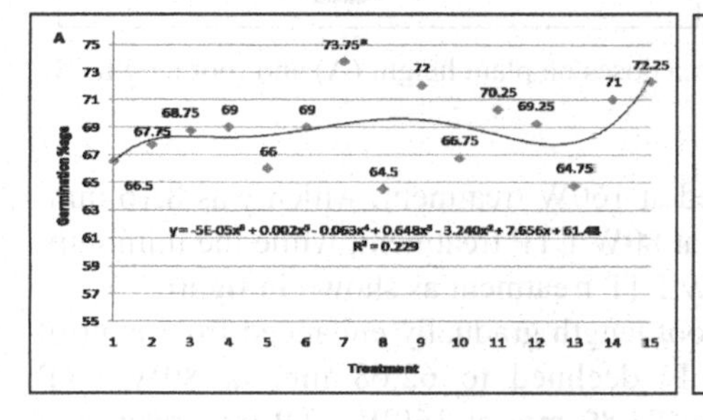

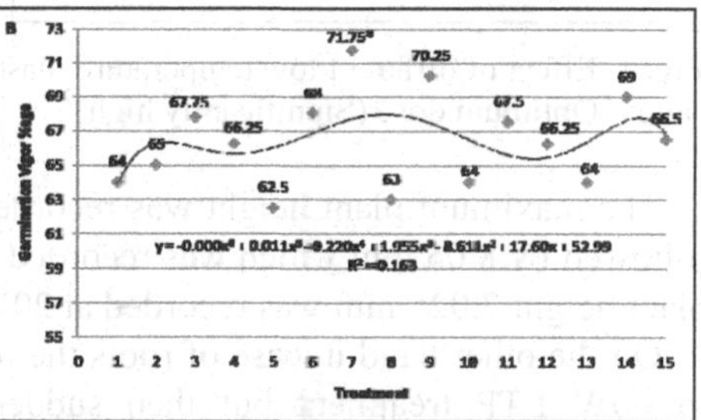

Fig. 2. Effect of different low temperature plasma doses on germination rate (A) and germination force (B)

Note: *Optimum dose (Significantly high)

The germination rate 66.5% was recorded on control. With 20W, 40W and 60W doses the germination rate non-significantly increased to 67.75, 68.75 and 69% respectively, then declined to 66% on 80W and then increased to 69% with 100W low temperature plasma treatment, then with further the germination rate significantly increased to 73.75% recorded at 120W, and then with further increase to 140W the germination rate significantly declined to 64.5% and then again significantly increased to 72% with 160W dose. Then 66.75, 70.25, 69.25, 64.75, 71 and 72.25% germination rate was recorded for 180, 200, 220, 240, 260 and 280W respectively (Table 1, Figure 2A).

The germination force was also significantly affected by LTP doses and almost similar trend was recorded as observed for the germination rate. The maximum germination force was recorded at 120W which was 71.75% followed by 70.25% recorded at 160W LTP treatment, while the lowest germination force was found at 80W LTP treatment (Table 2, Figure 2B).

3.2 Plant Height and Root Length

The data was recorded for Plant height and root in response to different doses control, 20, 40, 60, 80, 100, 120, 140, 160, 180, 200, 220, 240, 260 and 280 W of low temperature plasma. The LTP doses significantly affected the plant height as well as plant root length (Fig.3).

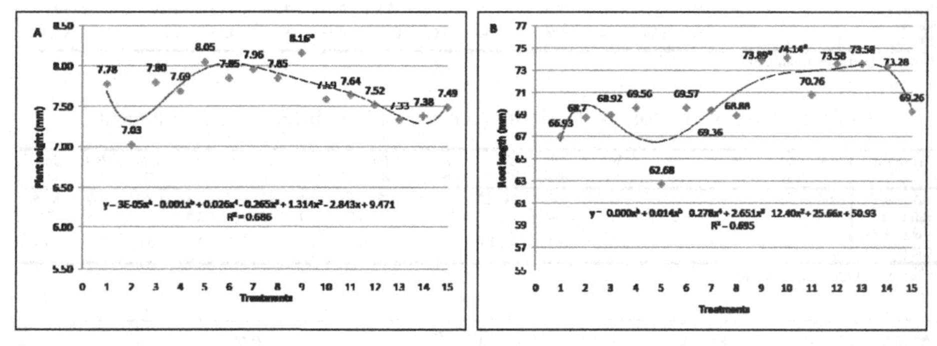

Fig. 3. Effect of different low temperature plasma doses on plant height (A) and root length (B)
Note: *Optimum dose (Significantly high)

The maximum plant height was recorded at 160W treatment, which was 8.16 mm, followed by 8.05 mm which was recorded at 80W LTP treatment, while the minimum plant height 7.025 mm was recorded at 20W LTP treatment as shown in figure 3A.

On the other hand in case of roots the root length gradually enhanced from control to 60W LTP treatment but then suddenly declined to 62.68 mm at 80W LTP treatment. And then gradually increased to 73.89 mm at 160W LTP treatment, and further improved to 74.14 mm at 180W LTP treatment. While further increase to 200W LTP treatment the root length declined to 70.76 and then increased to 73.58 mm at both220 and 240 W LTP treatments, then declined (Figure 3B).

3.3 Plant Fresh and Dry Weights

The plants fresh weight as well as plants dry weight was recorded for the treated plants. The plant fresh weight enhanced with low temperature plasma treatments as given in Fig. 4.

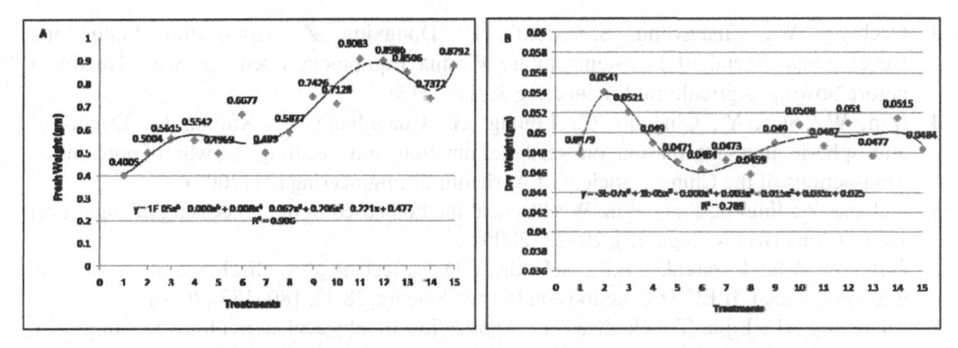

Fig. 4. Effect of different low temperature plasma doses on plant fresh weight (A) and dry weight (B)

The maximum plant fresh weight 0.9083 g was recorded for the 200W LTP dose, followed by 0.8986 g and 0.8792 g at LTP doses 220W and 280W respectively, while minimum fresh weight 0.4005 g was recorded for the control plants, and 0.4969 g for the LTP dose 80W (Figure 4A).

On the other hand similarly minimum dry weight was recorded for the 140W LTP dose, which was 0.0459 g, while maximum plant dry weight was recorded at 20W, which was 0.0541g, followed by 0.0521 g and 0.0515 g at LTP doses 40W and 260W respectively as shown in Figure 4B.

4 Conclusions

In the present work we observed that LTP doses significantly affected Trifolium repens seed germination, germination vigor, plant height and root length. The maximum Trifolium repens seed germination percentage and germination vigor percentage were recorded at 120W followed by 160 W LTP treatment, while maximum plant height and root length were recorded at 160W LTP dose.

Acknowledgment. This study is funded by China Agriculture Research System (CARS-35).

The study was supported by The ministry of Education basic business expenses special fund for scientific research projects, low temperature plasma treatment on the influence of seed germination seedling characteristics research 2013QJ019.

References

1. Azad, S.M., Manik, M.R., Hassan, M.S., Matin, M.A.: Effect of different pre-sowing treatments on seed germination percentage and growth performance of Acacia auriculiformis. Journal of Forestry Research 22(2), 183–188 (2011)
2. Dhoran, V.S., Gudadhe, S.P.: Effect of Plant Growth Regulators on Seed Germination and Seedling Vigour in Asparagus sprengeri Regelin. International Research Journal of Biological Sciences 1(7), 6–10 (2012)

3. Decheng, W., Changyong, S., Xianfa, F., Dongxing, Z.: Application Status and Development Trend of Low-temperature Plasma Equipment Used for Seed Treatment before Sowing. Agricultural Engineering 3(z1) (2013)
4. Min, W., Size, Y., Qingyun, C., Lihong, G., Guangliang, C., Xiujun, L.: Effects of atmospheric pressure plasma on seed germination and seedling growth of cucumber. Transactions of the Chinese Society of Agricultural Engineering (2) (2007)
5. Yuhang, Z., Jinglou, Z., Qingfa, W.: Physical methods used in the processing of sugar beet seeds. China Beet & Sugar (2), 20–22 (2005)
6. Dubinov, A.E., Lazarenko, E.R., Selemir, V.D.: Effect of glow discharge air plasma on grain crops seed. IEEE Transactions on Plasma Science 28(1), 180–183 (2000)
7. Lianglong, H., Lijia, T., Zhichao, H.: Application of physical agriculture techniques in cleaned seeds treatments. Journal of Anhui Agricultural Science 35(13), 3778–3779 (2007)
8. Jiang, J., Lu, Y., Li, J., Li, L., He, X., Shao, H., Dong, Y.: Effect of Seed Treatment by Cold Plasma on the Resistance of Tomato to Ralstonia solanacearum (Bacterial Wilt). Plos One (May 19, 2014). doi: 10.1371/journal.pone.0097753
9. Zivkovic, S., Puac, N., Giba, Z., Grubisic, D., Petrovic, Z.L.: The stimulatory effect of non-equilibrium (low temperature) air plasma pretreatment on light-induced germination of Paulownia tomentosa seeds. Seed. Sci. Tecnol. 32, 693–701 (2004)
10. Jiang, J.F., He, X., Li, L., Li, J.G., Shao, H.L., et al.: Effect of cold plasma treatment on seed germination and growth of wheat. Plasma Sci. Tech. 16, 54–58 (2014). doi:10.1088/1009-0630
11. Changyong, S., Decheng, W., Xin, T., Lijing, Z., Yan, L.: Stimulating effects of magnetized arc plasma of different intensities on the germination of old spinach seeds. Mathematical and Computer Modeling 58, 814–818 (2013)
12. Changyong, S., Xianfa, F., Xin, T., Lijing, Z., Lili, Z., Decheng, W.: Effects of Low-temperature Plasma on Seed Germination Characteristics of Green Onion. Transactions of the Chinese Society for Agricultural Machinery 6(44), 201–205 (2013)

Research and Implementation of Modeling Grid DEM Based on Discrete Data

Jing Wang, Kai Xia, and Xuegong Chen

Software School, Central South University, Changsha 430070, China
weyure@126.com, xiakai05@163.com, cxgcsu@csu.edu.cn

Abstract. Interpolation method is used to construct regular grid DEM (Digital Elevation Model) based on discrete data. The key to DEM interpolation algorithm lies in the efficiency of data search around the points to be operated and the precision of DEM interpolation by selecting an appropriate interpolation function model. At present, the point by point interpolation is very common. Based on the comparison and analysis of characteristics and application scope of various DEM interpolation algorithms, this paper proposed a new interpolation model which combines distance-weighted average method and least squares fitting method. Results show that proposed method can improve search speed and interpolation accuracy remarkably and has good application value for large area.

Keywords: DEM interpolation, digital elevation model, distance-weighted average method, least squares fitting method.

1 Introduction

With the rapid development of GIS (Geographic Information System), DEM (Digital Elevation Model), as a digital description and simulation of the surface topography, has become an important component of spatial data infrastructure and "Digital Earth". The accuracy of DEM has a significant impact on terrain visualization, GIS analysis and credibility of the decision-making [1, 2]. It is necessary to improve the accuracy of modeling DEM. Most of original data are discrete data. The key issue of DEM is how to construct regular grid DEM by interpolating with discrete data. Therefore, research on improving the DEM accuracy is of great significance.

To construct Grid DEM, first we divide the research region into two-dimensional grid to form spatial structure of grid that covers the entire area. Then discrete points surrounding the interpolated grid points are used to calculate its elevation values, we can obtain the grid data of this region with a certain format. Interpolation is the core issue of modeling DEM, which involved in DEM in all aspects of production, including quality control, accuracy assessment and analysis applications. According to the distributed range of interpolation points, DEM data interpolation algorithm generally can be divided into overall interpolation, block interpolation and point by point interpolation. The point by point interpolation is widely used because its calculation method has many advantages, such as simple, flexible block, high

© IFIP International Federation for Information Processing 2015
D. Li and Y. Chen (Eds.): CCTA 2014, IFIP AICT 452, pp. 175–184, 2015.
DOI: 10.1007/978-3-319-19620-6_22

precision and so on. The point by point interpolation can be divided into the distance-weighted average method, moving surface fitting method, least squares fitting method, etc. Every interpolation method has some drawbacks. For example, the distance-weighted average method is less precise and poor in smoothness, and only suitable for points which are well-distributed and sampled from even terrain; Moving surface fitting method is of high precision and flexible in calculation, but imposes high requirements on the sampling point; Least squares fitting method is suitable for most of terrain but time consuming [3-8]. Different interpolation algorithms have their own advantages, disadvantages and application scope, so we should select the appropriate interpolation algorithm according to a certain problem.

Simply using one method of interpolation to construct DEM in entire region have limitations, such as high requirements on the sampling point, time consuming and so on, which will affect the applicability and accuracy of model. People put forward some combination method, for example, integration of moving surface fitting method with distance-weighted average method, distance-weighted average method with quadratic surface method. Both of these methods only suitable for points which are well- distributed and sampled from even terrain. The method that combines moving surface fitting method with least squares fitting method does not consider moving surface method has a high requirement on accuracy of sampling data [9-13].

In response to above problems, taking all conditions into account and after analysis of the application of various interpolation methods, this paper presents a method that integrates the distance-weighted average method with the least- squares fitting to solve problems arising in DEM modeling and improve accuracy and efficiency of the interpolation method.

The following of the paper is organized as follows. Section 2 presents the proposed approach. Section 3 describes an experiment and shows the analysis. Section 4 concludes the paper.

2 Data and Methods

2.1 Data

These elevation data used in this paper are sampled from three areas, such as alpine areas, glacier areas and even areas. Each area contains more than 50000 sampling points. Using the distance-weighted average method, least squares method, moving surface fitting method, moving surface fitting method and the proposed combination method to construct DEM based on discrete data, we analyze and evaluate the accuracy of the model.

Set aside 100 sampling points as the checkpoints that do not participate in DEM modeling. Using different methods to model DEM, comparing the checkpoints original elevation values and calculated elevation values through four interpolation algorithm to calculate max error and RMSE (root mean square error), then we analyze and evaluate the results.

2.2 Distance- Weighted Average Method

To use the distance-weighted average method, we need to select these discrete points nearest to the interpolation point. The easiest way is to directly calculate the distance between interpolation points and the discrete points, then choose discrete points according to distance and interpolate, but this interpolation method need large time and space to compute, which has a bad impact on interpolation speed [10]. So we choose certain search areas according to requirement of mathematical model for the interpolation number and adjust the search scope.

The distance-weighted average method is as follows:

$$h = \frac{\sum_{k=1}^{m} w_k h_k}{\sum_{k=1}^{m} w_k} \tag{1}$$

h_k is the point to be interpolated, w_k is the weight of discrete point k, and h_k is elevation values of sampling point k.

Currently weight calculated in the following ways:

$$P_i = d_{ki}^{-u} \tag{2}$$

$$P_i = \left(\frac{R - d_{ki}}{d_{ki}} \right)^2 \tag{3}$$

$$P_i = e^{\left(\frac{d_{ki}}{m} \right)^2} \tag{4}$$

R is a search radius; m is a constant, P_i is the weight of discrete points. In this paper, we choose the first method, and set u = 2.

A problem arises when there are few reference points. When the reference points are very far from each other or distributed unevenly, if we still use these reference points that has little effect on the interpolated points to interpolate, it will cause disturbance of interpolation points instead of improve the accuracy of the interpolation points, which make the interpolation results not ideal.

To improve the efficiency of distance-weighted average algorithm, this paper adopts the idea of gradually expanding the search area and restricted orientation search. The idea is as follows:

First, classify grid in accordance with the actual needs.

Second, determine the number of discrete points needed to be calculated.

Third, search discrete points around the interpolated grid points with the restricted orientation. If there are enough discrete points in each quadrant, use the distance-weighted average method for interpolation. If not, expand search area gradually. When search area reaches threshold, then stop.

2.3 Least Squares Fitting Method

When the reference points are very far from the interpolation points or distributed unevenly, distance-weighted average method has great error. Least squares fitting method, which combines the advantage of distance-weighted average method and trend surface method, is suitable for most of terrain, and can effectively eliminate noise data [12].

Weighted least squares fitting method assigns all discrete points that have influence on interpolation points a certain influence coefficients, and assumes the coefficients is consistent with a quadratic polynomial as follows:

$$\mathrm{p(x, y)} = \mathrm{c}_1 + \mathrm{c}_2 x + c_3 y + c_4 xy + c_5 x^2 + c_6 y^2 \tag{5}$$

$$\mathrm{Q} = \sum_{\mathrm{i}=2}^{\mathrm{n}} [\mathrm{p}(\mathrm{x}_i, y_i) - z_i] w\left[(x_i - a)^2 + (y_i - b)^2\right] \tag{6}$$

When we use least squares fitting method for interpolation, these points that are nearer to the interpolation point are allocated a higher weight than these points which are far away. As the number of discrete points the least squares fitting method requires is more than 6, we need to find coefficients ci (i= 1, 2, … , 6) to make (6) minimum. $w\left[(x_i - a)^2 + (y_i - b)^2\right]$ is weight and is a function of the distance. Take $w\left[(x_i - a)^2 + (y_i - b)^2\right] = \frac{1}{\left[(x_i - y_i)^2 + (y_i - b)^2\right]^2}$, when (a, b) close to (x_i, y_i), it has a large value and a small value while (a, b) is far from (x_i, y_i). To get minimal Q, according to the principle of least squares $\frac{\partial \mathrm{Q}}{\partial \mathrm{c}_\mathrm{i}} = 0$, so we can get the following equation:

$$\frac{\partial \mathrm{Q}}{\partial \mathrm{c}_1} = \sum_{\mathrm{i}=1}^{N} 2(P(x_i, y_i) - z_i)w = 0 \tag{7}$$

$$\frac{\partial \mathrm{Q}}{\partial \mathrm{c}_2} = \sum_{\mathrm{i}=1}^{N} 2(P(x_i, y_i) - z_i)wx_i = 0 \tag{8}$$

$$\frac{\partial Q}{\partial c_3} = \sum_{i=1}^{N} 2(P(x_i, y_i) - z_i) w y_i = 0 \tag{9}$$

$$\frac{\partial Q}{\partial c_4} = \sum_{i=1}^{N} 2(P(x_i, y_i) - z_i) w x_i y_i = 0 \tag{10}$$

$$\frac{\partial Q}{\partial c_5} = \sum_{i=1}^{N} 2(P(x_i, y_i) - z_i) w x_i^2 = 0 \tag{11}$$

$$\frac{\partial Q}{\partial c_6} = \sum_{i=1}^{N} 2(P(x_i, y_i) - z_i) w y_i^2 = 0 \tag{12}$$

The equations from (7) to (12) can be solved to obtain coefficients of p (x_i, y_i). Use these coefficients in equation (5); we can obtain value of grid points. Solving these equations needs large time. When there are large discrete points, there is a sharp decline in speed. When the reference points are very far from the interpolation points or distributed unevenly, least squares fitting method has better accuracy than distance-weighted average method. So we combine least squares fitting method and distance-weighted average method to improve the accuracy and efficiency of interpolation.

2.4 Proposed Method

When the reference points are very far from each other or distributed unevenly, the result of using weighted average algorithm would not be ideal. This paper brings up the idea of gradually expanding the search area and restricted orientation search. As the number of discrete data points the distance-weighted average method requires is from 4 to 8, we adopt four quadrants search method, so there should be at least one point in each quadrant. The least squares fitting method has no special requirements on the distribution of discrete points, but it needs at least six discrete points. When the number of discrete points is over six and the distribution does not satisfy the requirements of distance-weighted average method, we adopt least squares fitting method.

In addition, in order to accelerate the computing speed and improve the efficiency of modeling, we judge distance between point to be interpolated and discrete point firstly. For example, if the nearest distance between point to be interpolated and discrete point is less than 0.5 meter, then we use value of the elevation of discrete point directly and don't do next step, if not, use the proposed algorithm for interpolation. Algorithm is as follows:

First, for point P to be interpolated, search discrete points in grid with number 1 around the point to be operated with the restricted orientation searching. If each of four quadrants has more than one discrete point, use distance-weighted average method for interpolation;

Second, if not, expand search area with number 2. Check if there is more than one discrete point in each quadrant, if yes; use distance-weighted average method for interpolation;

Third, if discrete points in one or more quadrants are less than one, but a total of discrete points are more than six, use least squares fitting method to interpolate.

Fourth, if discrete points are less than 6, expand the search area with number 3 and use least squares fitting method to interpolate if the requirement is satisfied ;

Sixth, if not, don't do interpolation and process the next grid point.

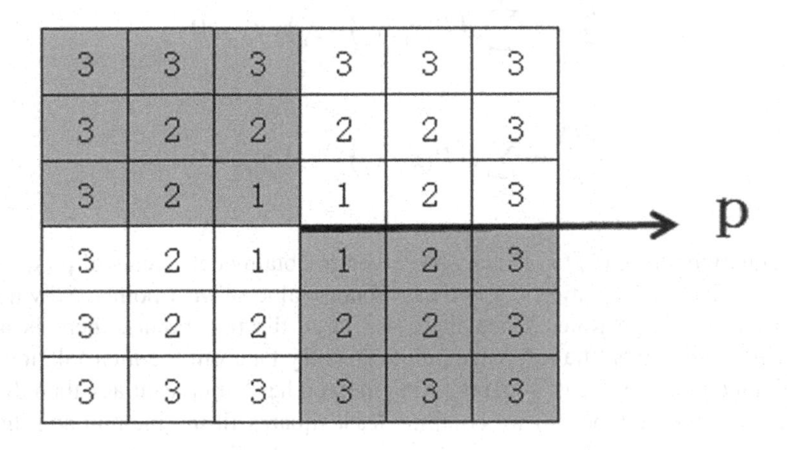

Fig. 1. Grid model searching

3 Experiments and Discussion

3.1 Experiments

Using distance-weighted average method, least squares fitting method, moving surface fitting method and proposed method to model DEM. Set aside 100 sampling points as the checkpoints that do not participate in DEM modeling, if the check point distribute unevenly, we may get an inaccurate evaluation of interpolations, in this paper we make the check point distribute evenly in order to obtain an accurate evaluation. We compare the original elevation values and calculated elevation values through four interpolation algorithms to calculate RMSE and max error. In order to directly display the results, we randomly selected 10 sampling points from 100 sampling points to compare. Fig. 2, Fig. 3 and Fig. 4 show absolute value of error in alpine areas, glacier areas and even areas respectively. Table 1, table 2 and table 3 show comparison of various interpolation precision in alpine areas glacier areas and even areas respectively. The results are as following:

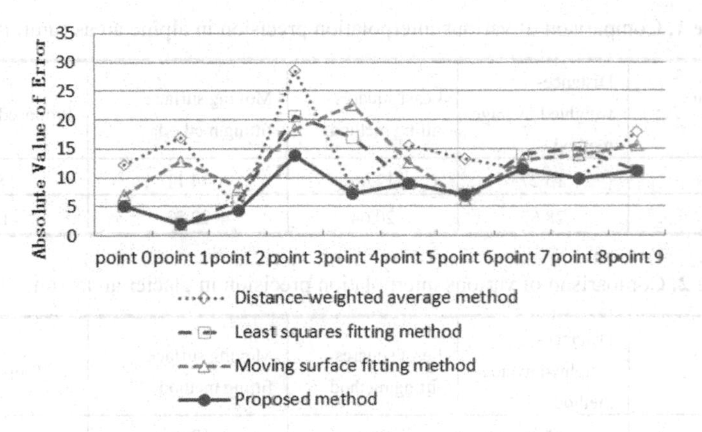

Fig. 2. Absolute value of error in alpine areas

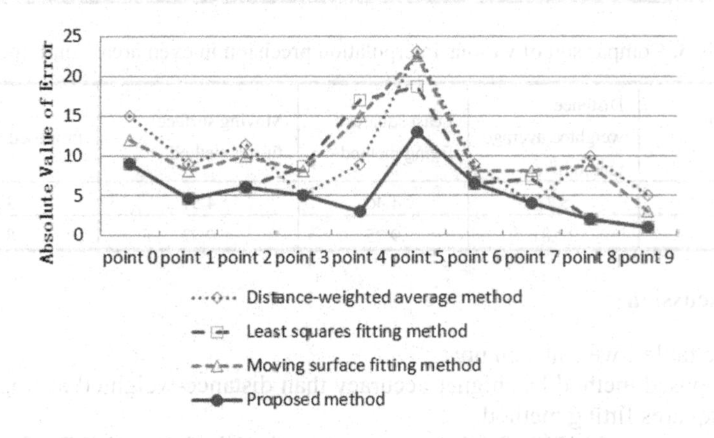

Fig. 3. Absolute value of error in glacier areas

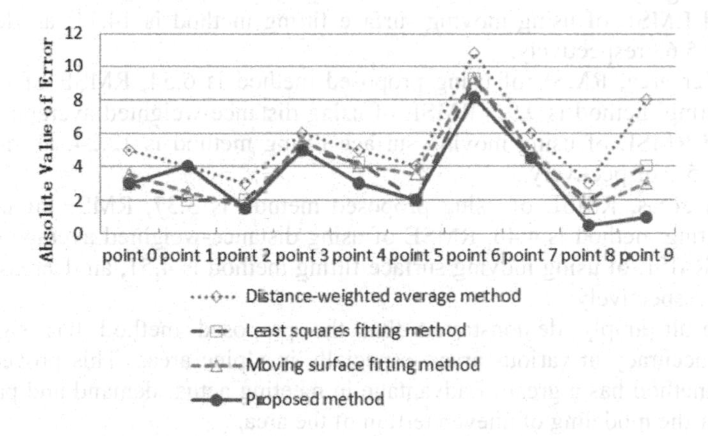

Fig. 4. Absolute value of error in even areas

Table 1. Comparison of various interpolation precision in alpine areas (unit: meter)

Interpolation method	Distance-weighted average method	Least squares fitting method	Moving surface fitting method	Proposed method
RMSE	15.27	12.06	14.11	8.48
Max error	28.67	20.64	22.53	13.77

Table 2. Comparison of various interpolation precision in glacier areas (unit: meter)

Interpolation method	Distance-weighted average method	Least squares fitting method	Moving surface fitting method	Proposed method
RMSE	11.36	9.91	12.24	6.54
Max error	23.13	18.71	27.15	12.92

Table 3. Comparison of various interpolation precision in even areas (unit: meter)

Interpolation method	Distance-weighted average method	Least squares fitting method	Moving surface fitting method	Proposed method
RMSE	5.78	4.46	4.51	3.37
Max error	10.81	9.25	9.33	8.16

3.2 Discussion

From these tables, we can sum up:

The proposed method has higher accuracy than distance-weighted average method and least squares fitting method:

In alpine areas, RMSE of using proposed method is 8.48, RMSE of using least squares fitting method is 12.06, RMSE of using distance-weighted average method is 15.27, and RMSE of using moving surface fitting method is 14.11, an decrease of 3.58, 6.79, 5.63 respectively.

In glacier area, RMSE of using proposed method is 6.54, RMSE of using least squares fitting method is 9.91, RMSE of using distance-weighted average method is 11.36, and RMSE of using moving surface fitting method is 12.24, an decrease of 3.37, 4.82, 5.7 respectively.

In even areas, RMSE of using proposed method is 3.37, RMSE of using least squares fitting method is 4.46, RMSE of using distance-weighted average method is 5.78, and RMSE of using moving surface fitting method is 4.51, an decrease of 1.09, 2.41, 1.14 respectively.

This result amply demonstrated that the proposed method has significantly improved accuracy in various areas, especially in alpine areas. This proved that the proposed method has a great of advantage in meeting actual demand and particularly suitable for the modeling of uneven terrain of the area.

The distance-weighted average method and moving surface fitting method has weak accuracy in alpine areas and glacier areas, better accuracy in even areas. The least squares fitting method has higher accuracy than distance-weighted average method in each kind of areas, especially in undulating areas. But in even area, accuracy is only a few higher. Taking the computing time into account, the least squares method is suitable for terrain that changes significantly.

4 Conclusions

Using distance-weighted average method, least squares fitting method, moving surface fitting method and proposed method to model DEM, according to the experimental results, after analysis, we can get the following conclusions.

The proposed method which combines distance-weighted average method and least squares fitting method has adopt both advantages and improved accuracy in each kind of terrain; it is suitable for the modeling of large areas. The experiment has demonstrated that the proposed method has improved efficiency of interpolation and prevented generating unwanted interpolation points.

The thinking of distance-weighted average method takes large ground gradient as a continuous process. But when the terrain changes significantly, such as ridges or rifts, the accuracy is not particularly ideal. So this method is suitable for even areas.

The least squares fitting method has high accuracy and time consuming, so it is suitable for terrain changes significantly.

Moving surface fitting method is suitable for points which are well- distributed and sampled from even terrain. When the reference points are very far from each other or distributed unevenly, this method would not be ideal, because it has a high requirement on accuracy of sampling data.

It is difficult to find a model that fully fit the complex topography of the surface. Therefore, in practical applications, how to make better use of space to fit the data and actual ground is an important direction for future research.

References

1. Li, Z., Zhu, Q.: Digital Elevation Model, 2nd edn., pp. 6–7. Wuhan University Press, Wuhan (2003)
2. Twito, R.H., Mifflin, R.W., McGaughey, R.J.: MAP Program: Building the Digital Terrain Model. Pacific Northwest Research Station, US (2000)
3. Tang, G., Li, F., Liu, X.: Digital Elevation Model, 2nd edn., pp. 76–85. Science Press, Wuhan (2005)
4. Charleux Demargne, J., Puech, C.: Quality assessment for drainage networks and watershed boundaries extraction from a digital elevation model (DEM). In: Proceedings of the 8th ACM International Symposium on Advances in Geographic Information Systems, vol. 5, pp. 89–94 (2011)
5. Aguilar, F.J., Agüera, F., Aguilar, M.A., et al.: Effects of terrain morphology, sampling density, and interpolation methods on grid DEM accuracy. Photogrammetric Engineering & Remote Sensing 7, 805–816 (2005)
6. Heritage, G.L., Milan, D.J., Large, A.R.G., Fuller, I.C.: Influence of survey strategy and interpolation model on DEM quality. Geomorphology 112(3), 333–344 (2009)

7. Lazzaro, D., Montefusco, L.B.: Radial basis functions for the multivariate interpolation of large scattered data sets. Journal of Computational and Applied Mathematics 140(1–2), 521–536 (2002)
8. Li, J., Chen, C.S.: A simple efficient algorithm for interpolation between different grids in both 2D and 3D. Mathematics and Computers in Simulation 58(2), 125–132 (2002)
9. Pan, P., Wang, G., Zhang, H.: Study and practice of DEM generation. Journal of Zhengzhou Institute of Surveying and Mapping 24(1), 57–60 (2007)
10. Li, Y., Yang, W., Yang, R., et al.: Improvement of DEM interpolation algorithms based on moving surface fitting and distance-weighted. Map 33(4), 168–171 (2010)
11. Wang, J., Yang, X., Yao, J.: Integration of the two methods to interpolate grid DEM. Journal of Taiyuan University of Technology 5(39), 63–69 (2008)
12. Jie, H.U.: The study of DEM modeling method based on regular grid. Geomatics and Spatial Information Technology 35(11), 138–140 (2012)
13. Zhang, J., Guo, L., Zhang, X.: Effects of interpolation parameters in inverse distance weighted method on DEM accuracy. Journal of Geomatics Science and Technology 1, 51–56 (2012)

Effect of Different Nitrogen Fertilizers with Reclaimed Water Irrigation on Soil Greenhouse Gas Emissions

Ning Ma, Shumei Ren, Peiling Yang, Yanbing Chi, and Dawei Gao

College of Water Resources and Civil Engineering,
China Agricultural University, Beijing 100083
{ningma90,chiice,18810395806}@163.com,
{renshumei,yangpeiling}@126.com

Abstract. In order to investigate the effect of using different nitrogen fertilizer with reclaimed water irrigation on the emissions of soil greenhouse gases (CO_2 and N_2O), plot experiments were conducted using clean water and reclaimed water combined with different nitrogen fertilizer (urea, ammonium sulfate, slow release fertilizer) for irrigation. No significant differences in CO_2 and N_2O emission flux was observed between the treatments irrigated with clean water and ones with reclaimed water. The soil CO_2 emission flux had no significant relationship with the application of nitrogen fertilizer, whereas the soil emission flux increased significantly as applied with nitrogen fertilizer. The N_2O emission flux reached its maximum in 2-5 days after irrigation.

Keywords: Reclaimed water, Nitrogen fertilizer, CO_2, N_2O, Emission flux.

1 Introduction

Global warming is a critical issue challenging the security of the whole human society today. The main trigger of global warming is the excessive emission of greenhouse gas (i.e. CO_2, CH_4 and N_2O) and the contribution of these three to the greenhouse effect rate are up to 80% [1]. CO_2 makes the greatest contribution to adding greenhouse effect, accounting for about 60%, which is the most important greenhouse gas [2]. N_2O is long-lasting greenhouse gas and not only it has a warming effect, but also can damage the ozone layer. The N_2O warming effect is 296~310 times of CO_2 [3]. In the process of agricultural production, agricultural greenhouse gas emissions is one of the important causes for global warming, about 5%-20% CO_2, 15%-30% CH_4 and 80%-90% N_2O derives from soil every year, so agricultural emission reductions allows of no delay[4].

China is one of the 13 countries under acute water shortage and per capita water resource is only about 1/3 of the world average level. In China, especially in the northern water shortage area, the use of reclaimed water has become an inevitable trend. The reclaimed water refers to the industrial wastewater or domestic sewage treated to non-potable water, which achieves a certain quality standard and can be repeatedly used in a certain extent. It is not affected by climate, convenient to use, stable, reliable and high guaranteed. It can not only reduce the discharge of sewage, but also reduce demand on freshwater resources. The reclaimed water has become an important supplemental source

© IFIP International Federation for Information Processing 2015
D. Li and Y. Chen (Eds.): CCTA 2014, IFIP AICT 452, pp. 185–192, 2015.
DOI: 10.1007/978-3-319-19620-6_23

of water for agricultural irrigation production [5]. Reclaimed water as a kind of irrigation water can increase soil fertility in some degree, and the application of reclaimed water irrigation has a certain degree of security in heavy metal [6]. Water shortage will be expected to reach 1.3 × 1010 m^3 and the available quantity of reclaimed water will reach $7.67 \times 10^{10} m^3$ by the year 2030[7]. However, because the reclaimed water contains high concentration of nutrient, salinity, bacteria, organic matter and suspended solid particles, using reclaimed water will change soil micro environment and it may cause environmental problems including soil salinization, decreasing soil fertility and increasing agricultural greenhouse gas emission if used inappropriately [8]. In this case, the objective of this study is to investigate the effect of different nitrogen fertilizer with reclaimed water irrigation on soil greenhouse gas emissions, which may be used as a guideline for greenhouse gas emissions control in long-term sustainable use of reclaimed water.

2 Materials and Methods

2.1 Experimental Material

The experimental soil was vadose zone soil collected from the southeast suburb of Beijing irrigation area where the soil was loam at the upper 80 cm depth and sandy at the lower 40 cm. Basic physical characteristics of soil are shown in Table 1.

Table 1. The basic physical parameters of experimental soils

Soil depth	Particle size composition（%）			Soil texture	Bulk density	Field capacity	Initial water content
	Clay	Silty sand	Sand				
cm	<0.002mm	0.02~0.002mm	2~0.02mm		$g{\cdot}cm^{-3}$	$cm^3{\cdot}cm^{-3}$	%
0~80	9.11	37.85	54.04	Loam	1.40	0.36	16.1
80~120	3.39	8.13	88.49	Sand	1.40	0.33	11.2

The reclaimed water was secondary stage sedimentation effluent from the Qinghe sewage treatment plant. The basic parameters of irrigation water are shown in table 2.

Table 2. Irrigation water quality parameters

Water quality		Clean water	Reclaimed water
pH		7.56	7.97
EC	$\mu S \cdot cm^{-1}$	494	1010
BOD_5	$mg \cdot L^{-1}$	<0.5	3.8
COD_{Cr}	$mg \cdot L^{-1}$	<5	16.5
Organic nitrogen	$mg \cdot L^{-1}$	0.710	1.02
Total organic carbon	$mg \cdot L^{-1}$	1.4	5.6
Total nitrogen	$mg \cdot L^{-1}$	1.00	5.18

2.2 Experimental Design

The experiment started in the water conservancy test hall in College of Water Resources and Civil Engineering, China Agricultural University from July 29, 2013. The soil tank area was 1.2m × 1.2m, 1.5m high. The irrigation method was surface flood irrigation. There were six treatments which were: CK, CK-APSU, RW-APSU, RW-U, RW-APS, RW-SPC (Table 3).

According to efficient water-saving irrigation system of Beijing Plain area for summer maize flat water years, the irrigation quota was 600m^3/ hm^2, simulating summer maize irrigation. Irrigation was conducted three times in the experiment. The first irrigation amount was 40L, the second and the third irrigation amount was 20L for each treatment. Nitrogen fertilizer (300 kg/hm^2) was applied with water at the first irrigation. Test used urea with 46% nitrogen content, ammonium sulfate nitrogen content of 21% and slow-release fertilizer nitrogen content of 46%.

Table 3. Fertilization treatments

Treatments	Urea (g/m^2)	Ammonium sulfate (g/m^2)	Slow-release fertilizer (g/m^2)
Clean water (CK)			
Clean water with ammonium sulfate and urea (CK-APSU)	32.61	71.43	
Reclaimed water with ammonium sulfate and urea (RW-APSU)	32.61	71.43	
Reclaimed water with urea (RW-U)	65.22		
Reclaimed water with ammonium sulfate (RW-APS)		142.86	
Reclaimed water with slow-release fertilizer (RW-SPC)			65.22

2.3 Experimental Methods

The basic principle of sampling static chamber [9] is to use the sealed bottomless boxes (made from a chemically stable material) to cover up the surface to be measured for a certain period of time and extract the gas inside. Determine trace gas concentration using chromatograph machine and then calculate the gas exchange rate of surface quilt-trace gases between ground and atmosphere according to the changing time rate of gas concentration.

Temperature and soil moisture are all important factors affecting trace gases. Therefore, in the determination of fluxes of greenhouse gas emissions, it is also necessary to measure the temperature and soil moisture in observation point. This experiment used TRIME tube which had been laid in the soil already, measuring water content in different soil depths (0-10cm, 10-20cm, 20-40cm, 60-80cm, 80-100cm, 100-120cm) for each treatment. Domestic JM624 portable digital thermometer was used for measuring air temperature and soil temperature in observation points.

Gas sampling is generally collected at 9:00-11:00 am, for soil temperature during this period is closest to average daily temperature. Before covering the box, the tank was filled with 1/2 water. After covering the boxes, we extracted gas in 0,10,20,30 min with 40mL polypropylene medical syringes which have three-way valve. Sample collection was tested by Chinese Academy of Agricultural Sciences Institute of Agricultural Environment and Sustainable Development Analysis Test Center and the analysis is completed within 24h. After the initial time and the last sample extraction were completed, the temperature value would record. After the gas samples was collected, using TRIME tube to measure water content in different soil depths (0-10, 10-20, 20-40, 60-80, 80-100, 100-120 cm) of different treatments. Samples were collected before irrigation and at 1 d, 2 d, 5 d, 10 d, 18 d after irrigation.

2.4 The Data Processing Principle

Flux refers to the amount of substance of unit area per unit time [10]. According to the calculation formula of sampling static chamber:

$$F = \frac{M}{V_0}\frac{P}{P_0}\frac{T_0}{T}H\frac{dc}{dt} \tag{1}$$

For the specific target compounds, the type M is the molar mass, P_0 and T_0 are ideal gas pressure and temperature (1013.25 hpa and 273.15 K) under standard conditions, V_0 is the target compound molar volume in the standard state, i.e. 22.4L·mol-1, H is the sampling box chamber top space height, P and T is the actual sampling gas pressure and temperature, dc/dt is regression curve slope of the time changing target gas concentration.

The parameters required in calculation are the actual height of sampling box H, the sampling pressure P, sampling temperature T in the box, and regression curve slope of the time changing target gas concentration dc/dt.

2.4.1 The Determination of Actual Height of Sampling Box (H)

The vertical distance from the bottom depth of the soil surface is defined from the water at the bottom of the base when measuring, then

$$\text{The actual height} = \text{sampling box height} - \text{bottom depth} \tag{2}$$

In this experiment, the sampling box height is 50cm, the water depth is -3cm, the actual height of sample box is 53cm, namely H=0.53m.

2.4.2 The Determination of Pressure (P)

Pressure of ideal gas under standard state is 101.3 kPa. The testing area pressure all year round maintains a stable value with small changes. Take a stable pressure value 100.5 kPa as the pressure sampling value P_0.

2.4.3 The Determination of Temperature (T)

The temperature T_0 is 273.15K under standard conditions. The temperature value is the average temperature of the sampling process. Then use the formula T=t+273.15 to convert Celsius temperature into thermodynamic temperature.

2.4.4 The Determination of Regression Curve Slope of the Time Changing Target Gas Concentration dc/dt

Detect the collected target gas concentrations, combine with the time recorded in the sampling process, and use excel to analyze regression curve slope of the time changing target gas concentration dc/dt. The dc/dt of CO_2 and N_2O can be calculated. Data were analyzed using SPSS 17.0 software.

3 Results and Discussion

3.1 Effects of Different Nitrogen Fertilizers with Reclaimed Water Irrigation on Soil CO_2 Emissions

Figure 1 showed CO_2 emission flux over time under the conditions of clean water irrigation, reclaimed water irrigation with urea, reclaimed water irrigation with ammonium sulfate and reclaimed water irrigation with slow release fertilizer. Irrigation and drainage was applied at August 7th, September 8th, September 27th. As shown in Figure 1, treatments of CK, RW-U, RW-APS had consistent gas emissions trend. Obtained from the test results, CO_2 emissions increased gradually to peak at 2-5 days after each irrigation, but treatment of slow-release fertilizer peaked emission flux in the first tenth day after irrigation, later than the other three treatments. Thus, compared to CK water treatment, reclaimed water with urea, ammonium sulfate treatment and slow-release fertilizer had relatively higher total CO_2 fluxes. However, CO_2 flux emissions had a certain relationship with the soil moisture. Figure 2 showed CO_2 emission flux changes with time after the clean water and reclaimed water irrigation with both ammonium sulfate and urea. In the two treatment conditions, the law of CO_2 emissions was close, but RW-APSU CO_2 emission flux was slightly higher than that of CK-APSU. Emission flux peaked on the second day after the first irrigation. CO_2 emissions and soil moisture are clearly related. Emission peak occurred after irrigation and the first emission peak was higher than the second and third irrigation.

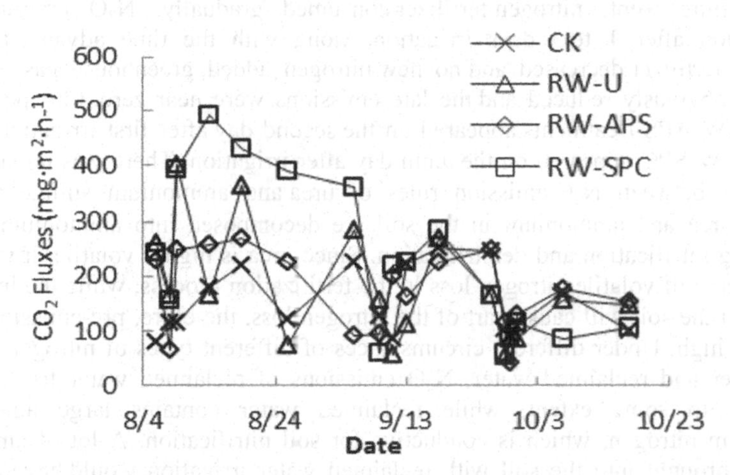

Fig. 1. Greenhouse gas emission fluxes of CO_2 with the date change curve of CK, RW-U, RW-APS, RW-SPC

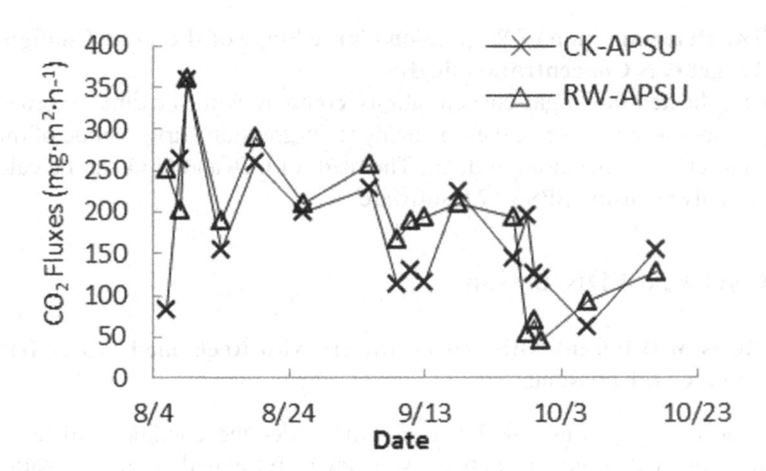

Fig. 2. Greenhouse gas emission fluxes of CO_2 with the date change curve of CK-APSU, RW-APSU

3.2 Effects of Different Nitrogen Fertilizers with Reclaimed Water Irrigation on Soil N_2O Emissions

Figure 3 was N_2O emission flux over time under the conditions of clean water irrigation, reclaimed water irrigation with urea, reclaimed water irrigation with ammonium sulfate and reclaimed water irrigation with slow release fertilizer. The overall curve trend of four treatments in N_2O emissions flux variation with time was consistent. August 7th, September 8th, September 27th was the three irrigating dates. As shown in fig.3, N_2O emission increased gradually after irrigation. At the beginning of the experiment, the N_2O had a high emission flux mainly because the basic fertilizer was brought into a lot of nitrogen, promoting the production of N_2O. As the time went, nitrogen fertilizer consumed gradually, N_2O emissions had a promotion after 1 to 3 days irrigation. Along with the time advanced, because the basic fertilizer decreased and no new nitrogen added, greenhouse gas emissions flux was obviously reduced and the late emissions were near zero. Flux peak in the RW-U, RW-APS treatments appeared on the second day after first irrigation, and the peak of RW-SPC appeared on the tenth day after irrigation. There was no significant difference between N_2O emission rules of urea and ammonium sulfate treatment, because urea and ammonium in the soil are decomposed into ammonium ions by promoting nitrification and denitrification. Since urea is highly volatile, it will cause some degree of volatile nitrogen loss in the fertilization process, while the hydrolysis process in the soil will cause part of the nitrogen loss, therefore, pre-emission flux of urea was high. Under different circumstances of different types of nitrogen fertilizer with water and reclaimed water, N_2O emissions of reclaimed water treatment had increased to some extent, while reclaimed water contains large amounts of ammonium nitrogen, which is conductive for soil nitrification. A lot of ammonium nitrogen brought into the soil with reclaimed water irrigation would have a certain role in promoting soil N_2O emissions.

Figure 4 showed N_2O flux variations with time of the clean water and recycled water irrigation with ammonium sulfate and urea. Under both treatment conditions, the law of N_2O emissions was nearly the same, fluxes of the reclaimed water was slightly higher than clean water, emission flux peak appeared on the second day after first irrigation. The figure showed that N_2O emissions had significant relationships with nitrogen fertilizer applied. And reclaimed water had complex ingredient, it interacted with nitrogen leading to a slight increase in N_2O Daily emission flux.

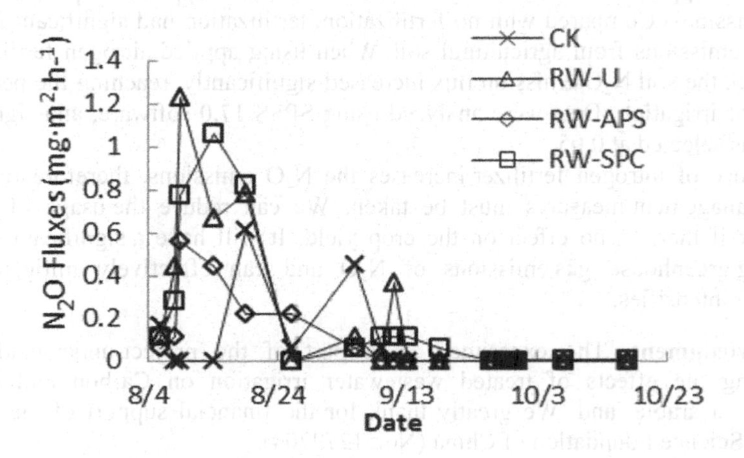

Fig. 3. Greenhouse gas emission fluxes of N_2O with the date change curve of CK, RW-U, RW-APS, RW-SPC

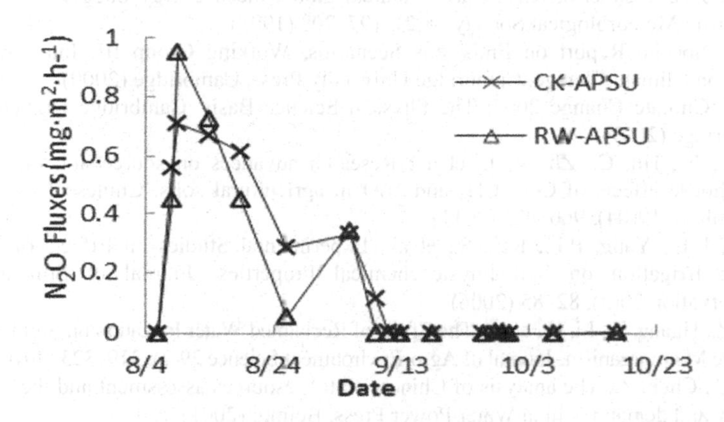

Fig. 4. Greenhouse gas emission fluxes of N_2O with the date change curve of CK-APSU, RW-APSU

4 Conclusions

Soil greenhouse gas emissions caused by reclaimed water had a small increase compared with clean water, but the increase was not significant. Therefore the reclaimed water irrigation had no significant effect on greenhouse gas emissions.

Different nitrogen fertilizer had no significant effect on soil CO_2 emissions but there was some relation with soil water content.

Soil greenhouse gases fluxes of N_2O has a significant correlation with the nitrogen fertilizer applied or not, the nitrogen fertilizer application promoted soil N_2O emissions. Compared with no fertilization, fertilization had significant influence on N_2O emissions from agricultural soil. When using applied nitrogen fertilizer after irrigation, the soil N_2O emission flux increased significantly, reaching the peak at 2-5 days after irrigation. Data were analyzed using SPSS 17.0 software, and significance level was selected at 0.05.

The use of nitrogen fertilizer increases the N_2O emissions, therefore reasonable field management measures must be taken. We can reduce the usage of nitrogen fertilizer if there is no effect on the crop yield. It will have a significant effect on reducing greenhouse gas emissions of N_2O and can effectively mitigate global warming intensifies.

Acknowledgment. The experiment is a part of the project-understanding and modelling the effects of treated wastewater irrigation on Carbon and Nitrogen Circling in arable land. We greatly thank for the financial support of the National Natural Science Foundation of China (No.51279204).

References

1. Kiehl, J.T., Trenberth, K.E.: Earth's annual global mean energy budget. Bulletin of the American Meteorological Society 78(2), 197–208 (1997)
2. IPCC. Special Report on Emissions Scenarios, Working Group III, Intergovernmental Panel on Climate Change. Cambridge University Press, Cambridge (2000)
3. IPCC. Climate Change 2007: The Physical Science Basis. Cambridge University Press, Cambridge (2007)
4. Zhang, Y., Hu, C., Zhang, J., et al.: Research advances on source/sink intensities and greenhouse effects of CO_2, CH_4 and N_2O in agricultural soils. Chinese Journal of Eco-Agriculture 19(04), 966–975 (2011)
5. Yang, L.L., Yang, P.L., Ren, S., et al.: Experimental Studies on Effects of Reclaimed Water Irrigation on Soil Physicochemical Properties. Journal of Soil and Water Conservation 20(2), 82–85 (2006)
6. Jiao, Z., Huang, Z., Li, Y., et al.: The Effect of Reclaimed Water Irrigation on Soil Performance and the Microorganism. Journal of Agro-Environment Science 29(2), 319–323 (2010)
7. Liu, C., Chen, Z.: The analysis of Chinese water resources assessment and the tendency of supply and demand. China Water Power Press, Beijing (2001)
8. Toze, S.: Reuse of effluent water - benefits and risks. Agricultural Water Management 80(1-3), 147–159 (2006)
9. Yao, H.: Gas exchange of carbon and nitrogen – fom the experiment to the model, pp. 77–82. China Meteorological Press, Beijing (2003)
10. Wang, Y., Wang, Y., et al.: Carbon exchange of observations between Chinese terrestrial and freshwater lakes and atmosphere. Science Press, Beijing (2008)

Research on Spatial Variability Characteristics of Black Soil Unfrozen Water in Songnen Plain during Freezing-Thawing Period

Zilong Wang[1,2,3,4], Qiang Fu[1,2,3], Jun Meng[4], Qiuxiang Jiang[1,2,3,4], and Xianghao Wang[1]

[1] College of Water Conservancy and Architecture, Northeast Agricultural University, Harbin 150030

[2] Collaborative Innovation Center of Grain Production Capacity Improvement in Heilongjiang Province, Harbin 150030

[3] Key Laboratory of Water-saving Agriculture, College of Heilongjiang Province, Harbin 150030

[4] Postdoctoral Mobile Research Station of Agricultural and Forestry Economy Management, Northeast Agricultural University, Harbin 150030

wzl1216@163.com.cn

Abstract. To address the difficulties of the complexity and the quantitative description of spatial non-homogeneity in the process of freezing and thawing of soil moisture migration, taking the unfrozen water as the research object, and using geostatistics to study the spatial variability characteristics of unfrozen soil water in different freezing and thawing stages. The results show that different freezing and thawing stages of unfrozen soil water has good spatial structure and strong spatial correlation, and it has a strong spatial redistribution effect on soil moisture. The application of geostatistics provides a new train of thought for in-depth study on freezing-thawing process of soil moisture transport mechanism.

Keywords: Songnen Plain, black soil, soil unfrozen water, spatial variability.

1 Introduction

Freezing and thawing of soil water movement, as an important part of the water cycle in nature, agriculture, water resources, environmental engineering, which occupies an extremely important position. To carry on an in-depth investigation into the variations of the freezing and thawing period soil moisture, master the laws of its motion, will not only help to promote the development of the theory of soil water dynamics, but also provide the theoretical basis for water resources accurately evaluation, efficient utilization of soil water, and a reasonable determination of farmland irrigation technical parameters, etc [1].

Study on the theory of soil water dynamics have made considerable progress, along with the changes of soil moisture from the morphology point of view to the energy

© IFIP International Federation for Information Processing 2015
D. Li and Y. Chen (Eds.): CCTA 2014, IFIP AICT 452, pp. 193–199, 2015.
DOI: 10.1007/978-3-319-19620-6_24

state theory [2],the research methods also vary from the qualitative description [3-4] based on soil physics test to the quantitative research [5-6] based on the numerical simulation. However, dramatic phase transitions and spatial variability of soil moisture in the freezing thawing process, that lead to the existing methods in the analysis of soil moisture variation during freezing and thawing period insufficient. Geostatistics is a theory, which study on the spatial variability of the properties of a system. The theory has been applied in the study on spatial variability of in unfreezing soil water [7].But it has fewer applications on the freezing and thawing soil moisture variation.

Therefore, in this paper, the methods of geostatistics were used to analysis the spatial variability of black unfrozen soil water in Songnen Plain in the freezing-thawing period, expecting to provide scientific basis for reasonable regulation and utilization of black soil water in Songnen Plain.

2 Materials and Methods

2.1 Study Area and Test Scheme

Field experiments were measured in a place of dry land in the city of Harbin, soil type is black soil. Harbin city is located in the southeast of Songnen Plain, has a temperate continental monsoon climate with its severe cold winter, dry and windy spring, has no perennial frozen soil layer, the annual average temperature is 3.6°C, the average January temperature is -20.3°C, the average frost free period is 141 days, the average annual precipitation is 523.3mm, the average annual snowfall is 63.1mm, the annual average frost depth is about 1.75m.

Field experiment treats the soil unfrozen water in freezing-thawing stages as the observation object, and uses TRIME-T3 time domain reflectometry to monitor soil unfrozen water during the soil thawing period of 2008 to 2009. In the same vertical profile every 2m buried 1 TDR probe, altogether 10 probes was buried, each TDR probe sets 8 measuring points, respectively 20cm, 40cm, 60cm, 80cm, 100cm, 140cm, 180cm, 220cm and 260cm.

2.2 Research Methods

Geostatistics is the most important method to study the spatial variability characteristics of soil. Since 1963, the famous French statistician Matheron G has put forward the concept of geostatistics [8], through decades of development, geostatistics has been widely applied in soil science, ecology, geology, hydrology, meteorology, environment and many other fields.

Since geostatistics methods and theory can not only keep the autocorrelation of space variables, but also can make the simulation value of measured point be equal to the measured value, it can use a few sampling point to simulate many unknown information and spatial distribution, which greatly reduced the field sampling and observed workload, enhanced visibility [9] of data analysis.

3 Results and Analysis

3.1 Semivariance Analysis

Based on the soil unfrozen water content observational data in different freezing-thawing stages, using the software of GS+ 7 for the semi variance analysis, as shown in Table 1.

Table 1. Semivariance analysis of soil unfrozen water content in different freezing-thawing stages

Date	Fitting model	Nugget	Sill	Nugget/Sill(%)	Range(m)	R^2
08-12-19	Sphere	0.228	1.323	17.2	14.41	0.827
09-01-10	Sphere	0.001	1.443	0.1	14.8	0.899
09-03-16	Sphere	0.001	1.41	0.1	14.2	0.876
09-03-27	Sphere	0.001	1.392	0.1	13.91	0.873
09-04-23	Sphere	0.216	1.081	20.0	9.82	0.737
09-05-19	Sphere	0.238	1.078	22.1	6.47	0.912

Results of the Semi variance analysis of soil unfrozen water content in different freezing-thawing stages in the experimental areas show that: soil unfrozen water content in different freezing-thawing stages has good spatial structure, their theoretical semivariogram models were all spherical model, as the same as the theoretical semivariogram models of unfrozen soil total water content in different freezing-thawing stages in previous research [10].The spatial structure of the soil moisture test area should adopt the spherical model to describe. Model fitting precision is lower than the total soil moisture content's, but still can meet the precision requirement. As is shown from the unfrozen soil water content Nugget/Sill in different freezing-thawing stages, the spatial correlation of most unfrozen soil water content is stronger than the corresponding position of the unfrozen soil water content [11].

According to the horizontal time span comparison, the spatial correlation of the initial freezing period (December 19, 2008) and the melt season (April 23 and May 19, 2009) is relatively weak. Soil has strong spatial correlation in stable freezing period. Effect of freezing process on soil unfrozen water content spatial correlation was in contrast to the situation in soil water content.

Investigation of the range of soil unfrozen water content: range of soil unfrozen water content in every freezing stages of the freezing period (before April 23, 2009) was about 14m,the range of melting period decreased by 50%.It shows that the freezing-thawing processes also have impact on the spatial correlation distance of unfrozen soil water content, but relatively small.

3.2 Analysis of Spatial Variation Characteristics

Using the Geostatistical Analyst module in ArcGIS 9 Software to carry on the spatial local estimation of the unfrozen soil water content in different freezing-thawing stages and to map the spatial distribution, as shown in Fig.1.The left side data in the spatial distribution of the unfrozen soil water content in different stages in Fig.1 represents the depth of the measuring points, the unit is cm; the upper marks represents each TDR probe.

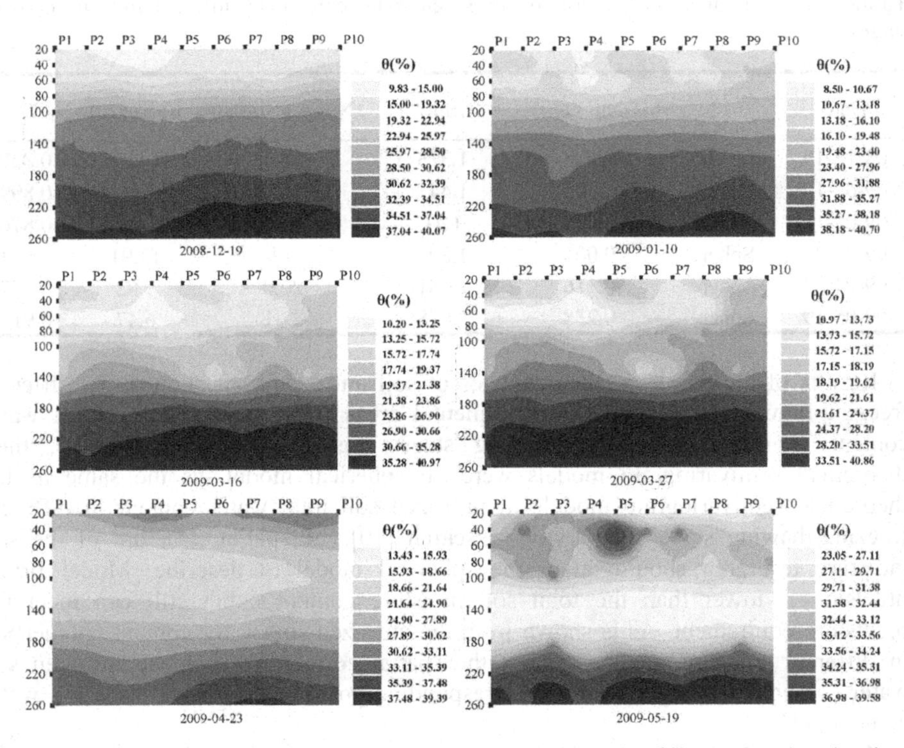

Fig. 1. Spatial distribution maps of soil unfrozen water content in different freezing-thawing stages

From Fig.1 we can see, in the process of freezing-thawing stages, the two dimensional spatial distribution of the unfrozen soil water content in the profile has certain regularity. The unfrozen soil water content and soil the negative temperature has an extremely close relationship [12-13]. In December 19, 2008, the soil belongs to the initial freezing period, freezing depth is 65cm, the equivalent zone of soil unfrozen water content present banded rules, it shows that the phase change process of water plays an important role in the redistribution of soil moisture. For the part above 60cm measuring point in the freezing layer and the part below 100cm measuring point in the unfrozen layer, the banded equivalent zone of soil unfrozen water content is wide, that is to say soil unfrozen water in these region changes little in the vertical direction. However, the equivalent zone of soil unfrozen water content at the measuring points between 60cm and 100cm is narrow, it changes heavily in the vertical direction. Because of the existence of soil freezing front, the unfrozen soil water below the location of the freezing front migrate to the frozen zone, this phenomenon dovetails the fact of the freezing depth of 65cm.

Based on the same principle, the freezing depth is about 100cm in January 10, 2009, 160cm in March 16, 2009, and in March 27, 2009 the freezing depth reached the maximum 162cm, in the adjacent soil layer below measuring points for 100cm and the soil layer between 140cm to 180cm between the unfrozen water content dramatic changes in the vertical direction, which were associated with the development of freezing depth or the depth of the freezing front, that's to say that the drastic change in the vertical direction of soil unfrozen water content in each frozen stages was caused by soil freezing. Therefore, the location of the freezing soil can be inferred according to the spatial distribution chart of the soil unfrozen water content in the freezing period. The unfrozen soil water content in freezing period showed a trend of increasing gradually from top to bottom. At the same time, notably in the freezing period of soil, for the frozen layer, the spatial distribution of the unfrozen water content presents regularly banded structure as the same as in the initial freezing period. It shows that when the soil is frozen, the unfrozen water content decreases rapidly, and continues to decline with the temperature's further reducing, but the freezing effect on the movement of the unfrozen water in frozen layer is small, the possible reasons to this question mainly are the atmospheric environmental conditions, the boundary of micro topography and the spatial variability of the soil water conductivity characteristics.

For the soil thawing period, the soil began to melt from two directions [14-15], while the soil unfrozen water content also increased from two directions. The spatial distribution situation of the soil unfrozen water content in the whole soil profile in April 23 and May 19, 2009 were shown in Fig.1. Frozen soil region is the low value area of unfrozen water content in soil, such as the measuring point between 40cm to 140cm in April 23, 2009 and the measuring point between 100cm to 140cm in May 19, 2009. At the same time, according to the spatial distribution of soil unfrozen water content in soil thawing period melting position between the upper and lower soil layer could be roughly inferred. For the soil surface in melting period, the initial melting period was the main driving force for migration of the unfrozen soil water content. Thus, the distribution of the soil surface unfrozen water content was regularly banded, and so was the situation in initial soil melting period. The soil unfrozen water content showed strong spatial variability after the surface soil began to thaw for a period of time. This is because in this period soil unfrozen water content is mainly affected by the different position of soil evaporation and the spatial variability of the water conductivity characteristics [16]. The lower soil also melted, but was less affected by external factors, so the unfrozen water content showed no spatial variability characteristics.

In order to analyze the severity that the different soil unfrozen water content vary with time in soil freezing-thawing period, spatial distribution maps of soil unfrozen water content coefficient of variation were drawn, as shown in Fig.2.

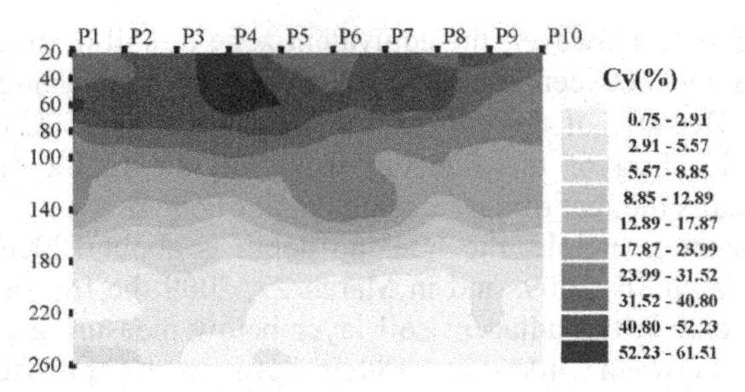

Fig. 2. Spatial distribution map of coefficients of variation of soil unfrozen water content in freezing-thawing period

We can see from Fig.2, during the whole freezing-thawing period, coefficient of variation of the upper layer of the soil unfrozen water content is large, the lower layer of the soil unfrozen water content on the contrary. This is due to the upper layer of the soil unfrozen water content is not only the influence of the freezing-thawing effect, but also restricted by other factors. The coefficient of variation of bottom soil unfrozen water content is very small, below 5.57%. The phase change of the soil moisture in freezing period did not occur in this part of soil layer, so there were many other factors that bring about the slight variation of its soil unfrozen water content with time changes.

4 Conclusions

Different stages of freezing and thawing soil not frozen water content has good spatial structure and strong spatial correlation, but has relatively small influence on the spatial correlation distance of the soil unfrozen water content. In the whole process of freezing-thawing, the severe variation of soil water content in vertical direction happened in the position of soil freezing front. The spatial distribution of soil unfrozen water content was mainly influenced by the effect of freezing-thawing in the initial freezing-thawing period, and was influenced by the environment and the variability of the soil itself in the late period of freezing-thawing period.

Using geostatistical theory and method is a good way to analyze different stages of freezing-thawing soil spatial variation features of unfrozen water content, also can visually depict the spatial distribution state's variation of soil unfrozen water content in different freezing-thawing period, at the same time, it can infer the freezing-thawing position and the variability of the whole soil profile according to the spatial distribution of the unfrozen soil water content in the freezing-thawing period, so as to provides a new thought and method for further research on soil moisture migration mechanism of Songnen plain's black soil region at the freezing-thawing period.

Acknowledgment. The study was supported by the National Natural Science Foundation of China (51209039, 51279031), Specialized Research Fund for the Doctoral Program of Higher Education (20122325120016), China Postdoctoral Science Foundation (2012T50321, 2011M500632), Program for New Century Excellent Talents in University, Science Foundation for Distinguished Young Scholars of Heilongjiang

Province (JC201402), Natural Science Foundation of Heilongjiang Province (E201241), Foundation of Heilongjiang Educational Committee (12521016), Heilongjiang Postdoctoral Fund (LBH-Z11231) and Doctoral Foundation of Northeast Agricultural University (2012RCB13).

References

1. Cheng, G.: Recent Development of Geocryological Study in China. Acta Geographica Sinica 45(2), 220–223 (1990)
2. Lei, Z., Yang, S., Xie, S.: Soil water dynamics. Tsinghua University Press, Beijing (1988)
3. Bloomsberg, G.L., Wang, S.J.: Effect of moisture content on permeability of frozen soils. In: Proc. American Geophysical Union Pacific Northwest Regional Meeting, Portland, OR, October 16-17, 1969
4. Qiu, G., Wang, Y.: Freezing Point of Loess in Lanzhou, China. Journal of Glaciology and Geocryology 12(2), 105–115 (1990)
5. Shoop, S.A., Bigl, S.R.: Moisture Migration during Freeze and Thaw of Unsaturated Soils: Modeling and Large Scale Experiments. Cold Regions Science and Technology 25(1), 33–45 (1997)
6. Zheng, X., Fan, G.: Numerical Emulation on Simultaneous Soil Moisture and Heat Transfer under Freezing and Thawing Conditions. Journal of System Simulation 13(3), 308–311 (2001)
7. Jiang, Q., Fu, Q., Wang, Z.: Research on Spatial Variability of Soil Water Characteristics In Western Semi-arid Area of Heilongjiang Province. Journal of Soil and Water Conservation 21(5), 118–122 (2007)
8. Matheron, G.: Principles of Geostatistics. Economic Geology 58, 1246–1266 (1963)
9. Chen, Y., Shi, H., Wei, Z.: Forecast Theory and Conditional Simulation of Soil Water and Salt Information's Spatial Variability. Science Press, Beijing (2005)
10. Wang, Z., Fu, Q., Jiang, Q., Li, T., Wang, X.: Spatial Variability of Soil Moisture Profile in Seasonal Frozen Soil Region in Different Stages. Scientia Geographica Sinica 30(5), 772–776 (2010)
11. Cambardella, C.A., Moorman, T.B., Novak, J.M., et al.: Field-scale Variability of Soil Properties in Central Iowa Soils. Soil Sci. Soc. Am. J. 58, 1501–1511 (1994)
12. Xu, X., Deng, Y.: Experimental Study on Water Migration in Frozen Soil. Science Press, Beijing (1991)
13. Zhou, Y., Guo, D., Qiu, G.: Chinese Frozen Soil. Science Press, Beijing (2000)
14. Guo, Z., Jin, E.: Analysis on the characteristics of soil moisture transfer during freezing and thawing period. Advances in Water Science 13(3), 298–302 (2002)
15. Siberia Branch of the Soviet Academy of Sciences Institute of frozen soil. General Geocryology. Science Press, Beijing (1988)
16. Shi, C., Li, S.: Agricultural Hydrology. Agriculture Press, Beijing (1984)

Comprehensive Evaluation of Land Resources Carrying Capacity under Different Scales Based on RAGA-PPC

Qiuxiang Jiang[1,2,3,4], Qiang Fu[1,2,3], Jun Meng[4], Zilong Wang[1,2,3,4], and Ke Zhao[1]

[1] College of Water Conservancy and Architecture, Northeast Agricultural University, Harbin 150030
[2] Collaborative Innovation Center of Grain Production Capacity Improvement in Heilongjiang Province, Harbin 150030
[3] Key Laboratory of Water-saving Agriculture, College of Heilongjiang Province, Harbin 150030
[4] Postdoctoral Mobile Research Station of Agricultural and Forestry Economy Management, Northeast Agricultural University, Harbin 150030
jiangqiuxiang914@163.com

Abstract. Land carrying capacity is an important indicator of the regional population, resources and environment evaluation of sustainable development. In this study, based on fully considering the influencing factors of land resources carrying capacity, land resources carrying capacity evaluation indicator system is constructed, composed of four sub target, i.e. the utilization of land resources, social development, economic and technology and the ecological environment. Real-coded accelerating genetic algorithm is used to optimize projection pursuit model to carry on comprehensive evaluation of land resources carrying capacity under different scales. Results show that national resources carrying capacity level of large-scale land falls under grade III under national scale; Heilongjiang Province and Sanjiang Plain belong to grade II under medium scale; The city of Jixi, Hegang, Shuangyashan and Jiamusi fall on grade I and the city of Qitai River, Muling and Yilan belong to grade II under small scale. The measures of effective utilization of land resources can be further proposed combined with the regional land resource evaluation results.

Keywords: land resources, carrying capacity evaluation, projection pursuit evaluation, Sanjiang Plain.

1 Introduction

Land is the material basis for human survival and development, human basic necessities of life all cannot do without the land. As land resources are limited, and land resources are non-renewable, but with the rapid increase in population, socio-economic growth, the destruction of land resources, the conflict between human demand and the limited land resources escalates and the irrational land use has threatened the sustainable development of human society and economy. Therefore, scientific evaluation of the land resources carrying capacity and rational development, utilization and protection of land, integrative distribution of land resources, make sustainable development of land resources possible.

© IFIP International Federation for Information Processing 2015
D. Li and Y. Chen (Eds.): CCTA 2014, IFIP AICT 452, pp. 200–209, 2015.
DOI: 10.1007/978-3-319-19620-6_25

At present, most of scholars at home and abroad concentrate on the exploitation of land resources [1], sustainable utilization [2] optimal allocation [3], and the ecological environment evaluation [4] and so on, the research on comprehensive evaluation of land resources carrying capacity is very limited. However, comprehensive evaluation is the basis of water-related research aforementioned, Sustainable utilization of land resources can only be achieved by management within its carrying capacity. Land Resources Carrying Capacity (LRCC) is an important indicator for evaluating the regional population, resources and environment and sustainable development [5], the research has important significance on the development strategy of regional economic development, population growth, land use planning and layout of production. In view of this, land resources not only contains the natural attributes, but also contains the complex social, economic and ecological attributes. Real-coded accelerating genetic algorithm- projection pursuit model was applied to carry on comprehensive evaluation of land resources carrying capacity of Sanjiang Plain, and the comparative study was done under both national and regional scale to form scientific understanding and judgment on land carrying capacity levels. The research results can provide reference for the sustainable use of land resources and the development of society.

2 Study Area

Sanjiang Plain is part of the plains of the Northeast, is China's largest freshwater marsh distribution area, including Ken River, Muling River plains, Xingkai lake plains and the triangular confluence region of Heilongjiang, Songhua and Ussuri. The Sanjiang plain is an important commodity grain production base, and also the key to the realization of Heilongjiang Province billion kilogram of grain production engineering and food safety in China. Over the years, excessive pursuit of economic benefits results in the increasing of plant area in Sanjiang plain [6], single plant structure. Coupled with the absence of appropriate government, Soil quality degraded severely, resulting in the increase of the proportion of low yielding fields. The utilization of land resources in Sanjiang Plain is regarded to be unreasonable, a series of problems has appeared, such as the low utilization efficiency and the decreasing of the carrying capacity, which has imposed a serious threat to the sustainable use of resources in Sanjiang Plain and sustainable socio-economic development. Therefore, in the premise of ensuring national food security and regional ecological security, scientific evaluation of land resources carrying capacity, the rational allocation of land resources, improving the land resource sustainable support ability in industrial and agricultural production and the development of social economy, seems to be basis of the realization of Sanjiang plain soil resources sustainable utilization.

3 Research Methods

This study will combine real coding based Accelerating Genetic Algorithm (RAGA) and Projection Pursuit Classification Model (PPC) organically. RAGA is used to optimize the parameters of projection direction in PPC, conversing the high-dimensional data to a low dimensional space, that is to say, multiple evaluation indicators are synthesized into an integrated one, then sort by projection value and recognition [7].

3.1 PPC Model

PPC model modeling steps are as follows [8-10]:

Step 1: Indicators normalized. The evaluation sample set is $\{x^*(i,j) \mid i=1,2,\ldots,n, j=1,2,\ldots,p\}$, where $x^*(i,j)$ is the jth indicator value of ith sample, n is the number of the samples、p is the number of the indicators. In order to eliminate the variation range of each indicator value dimension and unified the indicator value, the type of normalization, using the following formula for normalized.

The bigger the better indicators (forward): $x(i,j) = \dfrac{x^*(i,j)}{x_{\max}(j)}$ (1)

The smaller the better indicators (reverse): $x(i,j) = \dfrac{x^*(i,j)}{x_{\max}(j)}$ (2)

Where $x_{\max}(j)$、$x_{\min}(j)$ are the biggest and smallest value of the jth indicator; And $x(i,j)$ is the normalized sequence.

Step 2: Constructing projection indicator function $Q(a)$. Projection pursuit method is process of projecting the P dimension data $\{x(i,j) \mid j=1,2,\ldots,p\}$ into one dimension projection value $z(i)$ in the direction of $a = \{a(1), a(2), a(3), \ldots, a(p)\}$

$$z(i) = \sum_{j=1}^{p} a(j)x(i,j) \quad (i=1,2,\ldots,n) \tag{3}$$

Where a is the unit vector. Then do the classification according to the one dimensional scatter diagram of $\{z(i) \mid i=1,2,\ldots,n\}$. When projecting the indicators, the scattered feature of indicator $z(i)$ should be as densely as possible for region projection point while as dispersed as possible for the whole projection point. Therefore, the projection indicator function can be expressed as:

$$Q(a) = S_z D_z \tag{4}$$

Where S_z is the standard deviation of indicator $z(i)$, D_z is the regional density of $z(i)$, that is

$$S_z = \sqrt{\frac{\sum_{i=1}^{n}(z(i)-E(z))^2}{n-1}} \tag{5}$$

$$D_z = \sum_{i=1}^{n}\sum_{j=1}^{n}(R - r(i,j)) \cdot u(R - r(i,j)) \tag{6}$$

Where $E(z)$ is the average value of $\{z(i) \mid i = 1,2,...,n\}$, R is the window radius of the regional density, the usual value is $0.1S_z$; $r(i,j)$ is the distance between samples, $r(i,j) = |z(i) - z(j)|$; $u(t)$ is the unit step function, when $t \geq 0$, it is 1, when $t < 0$, it is 0.

Step 3: Optimizing the projection indicator function. When the sample set of indicators is given, the projection indicator function only varies with projection direction a. The direction of projection reflects the different characteristics of different data structure, the best projection direction projection direction can maximize the feature structure of high dimensional data, so we can estimate the best projection direction through solving the maximum of the projection indicator function, that is:

$$\text{Objective function:} \quad \max : Q(a) = S_z \cdot D_z \tag{7}$$

$$\text{Constraints:} \quad s.t.: \sum_{j=1}^{p} a^2(j) = 1 \tag{8}$$

This is a complex nonlinear optimization problem, in which $\{a(j) \mid j = 1,2,...,p\}$ is the decision variables. Traditional optimization method is difficult to handle this problem, so in this paper, accelerated real-coded genetic algorithm is used to address the high dimensional global optimization problems.

Step 4: Comprehensive Evaluation. The best projection direction a^* obtained by 3 was put into formula (3) to acquire projection value $z^*(i)$ of each sample, i.e. the comprehensive evaluation value of land resources carrying capacity.

3.2 Accelerated Real-Coded Genetic Algorithm

The genetic algorithm, put forward by Professor Holland of the University of Michigan's America, is a kind of adaptive global optimization probability search algorithm simulating heredity and evolution of biology in the natural environment. It mainly includes selection, crossover and mutation operations. The specific steps are as follows [7, 11]:

Step 1: randomly generate N groups' random variables.
Step 2: calculate the objective function value, sorted by size, from largest to smallest.
Step 3: calculate the evaluation function based on ranking.
Step 4: selection operation, generating new population.
Step 5: crossover the new population generated in step 4.
Step 6: mutate the new population generated in step 5.
Step 7: evolutionary iteration.
Step 8: take the variable interval of outstanding individual generated in step 1 and 2 as the variable initial change interval, then run step 1 step 7, bringing about accelerated process, until the function value of best individual less than the default value or run number reaches the predetermined value. At this time, the best individual in the current population is treated as RAGA results.

4 Comprehensive Evaluation of Land Resource Carrying Capacity

4.1 Construction of Evaluation Indicator System

Land resources carrying capacity evaluation indicator system (Fig.1) was constructed according to the principle of constructing evaluation indicator system, comprehensive analysis on the regional land resources and influence factors, the experience of other scholars' research achievements [12-14] and the status of the ecological environment. Evaluation system is divided into the total objective layer, sub objective layer and indicator layer, where the indicator layer contains 29 indicators.

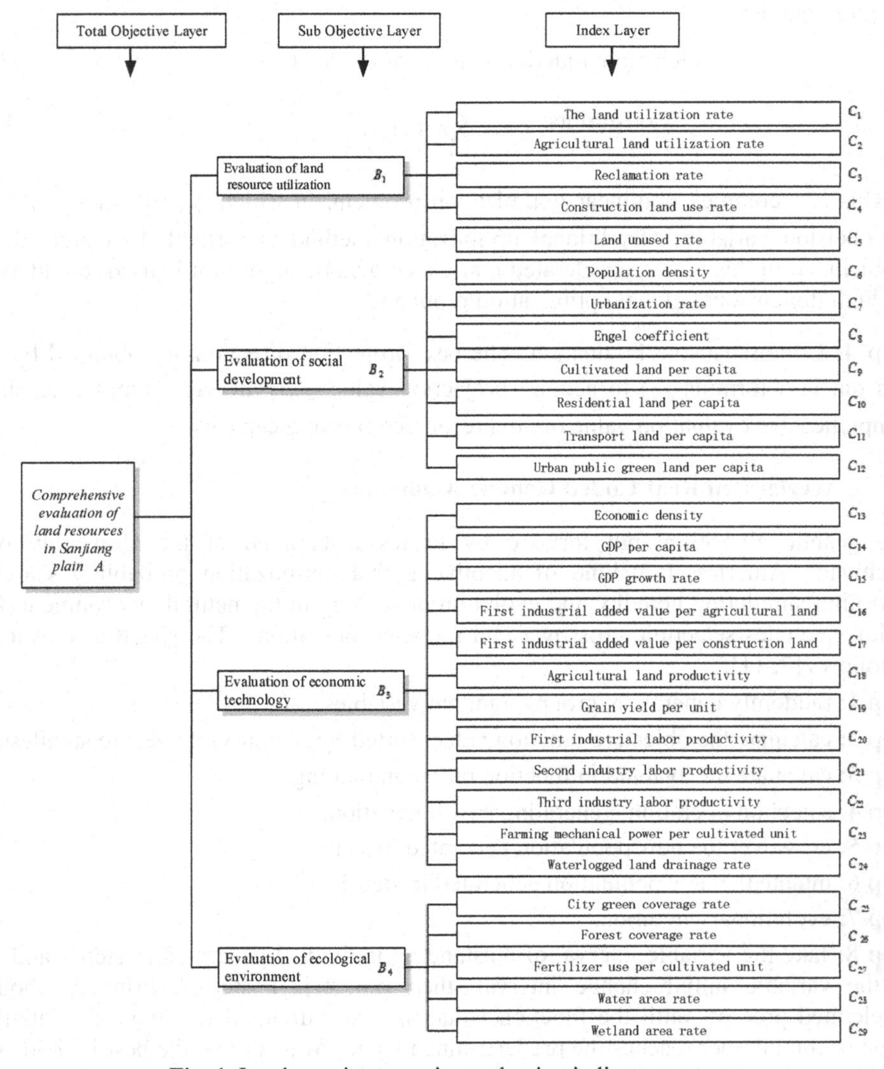

Fig. 1. Land carrying capacity evaluation indicator system

Table 1. The comprehensive evaluation indicator value of land resources carrying different scale in 2012

Indicator	Unit	Type	Small scale							Medium scale		Large scale
			Jixi	Hegang	Shuang-yashan	Jiamusi	Qitahe	Muling	Yilan	Sanjiang Plain	Heilongjiang province	China
C_1	%	F	84.28	90.78	92.86	87.22	96.97	98.44	95.41	89.81	87.1	72.21
C_2	%	F	79.8	86.52	88.59	82.78	91.41	90.06	90.53	85.11	83.8	68.43
C_3	%	F	43.74	37.61	48.05	62.91	45.26	23.82	52.89	48.74	26.16	12.68
C_4	%	F	4.36	4.09	4.16	4.29	5.53	2.01	4.65	4.2	3.3	3.44
C_5	%	R	0.13	0.33	0.06	0.1	0.09	0.07	0.14	0.13	0.09	0.24
C_6	people/km^2	R	83	74	68	74	149	44	88	78	85	141
C_7	%	F	63.91	80.74	64.3	50.08	58.87	49.48	42.26	60.12	56.9	52.57
C_8	%	R	33.82	34	29.8	35.1	38.21	39.61	35.2	33.35	36.9	36.2
C_9	hm^2/ capita	F	0.53	0.51	0.7	0.85	0.3	0.55	0.6	0.63	0.31	0.09
C_{10}	m^2/ caplta	F	303	294	311	287	222	270	286	288	255	156
C_{11}	m^2/ capita	F	173	206	216	287	110	168	197	196	195	118
C_{12}	m^2/ capita	F	15.4	26	15.3	16.2	26.1	8.2	28.2	18.5	19.3	12.3
C_{13}	Yuan/km^2	F	259	244	256	206	483	221	260	251	303	541
C_{14}	10^4 Yuan/ capita	F	3.1	3.3	3.7	2.8	3.2	5.1	3	3.5	3.6	3.8
C_{15}	%	F	13.6	13.5	13.5	13.8	8.3	17	12.3	13.2	10.1	9.7
C_{16}	10^4 Yuan /km^2	F	92.03	83.15	94.65	75.04	54.08	43.43	84.95	80.68	55.74	79.73
C_{17}	10^4 Yuan /km^2	F	4249	4217	4148	3348	7840	9040	3937	4344	7760	14108
C_{18}	10^4 Yuan /hm^2	F	1.11	0.59	0.81	1.14	0.84	2.12	1.8	1.06	1.96	3.86
C_{19}	kg/hm^2	F	3305	2061	2648	4123	3131	2679	7445	3522	4867	4843
C_{20}	Yuan/ capita	F	5.88	10.42	8.56	3.58	2.77	4.53	5.68	5.39	2.65	2.03
C_{21}	Yuan/ capita	F	12.12	11.28	18.39	11.55	11.65	26.86	8.24	13.16	16.13	10.12
C_{22}	Yuan/ capita	F	5.88	4.58	5.18	6.68	5.02	4.29	5.47	5.63	7.3	8.36
C_{23}	kw/hm^2	F	1.93	1.49	1.47	1.65	2	1.48	2.53	1.7	3.85	8.43
C_{24}	%	F	56.9	82.2	73.9	50.3	61.8	60.4	80.2	65.8	75.3	51
C_{25}	%	F	12.77	19.26	10.44	11.95	39.01	3.57	11.69	13.81	16.31	24.67
C_{26}	%	F	32.6	45.9	37.7	16.6	43.8	65.7	35.1	33.4	42.6	20.4
C_{27}	kg/hm^2	R	46.6	76.1	58.7	99.2	48.8	58.3	90.8	74.8	203	479.7
C_{28}	%	F	11.69	6.8	5.97	10.02	2.55	0.88	4.41	7.89	1.91	1.72
C_{29}	%	F	15.59	8.89	7.07	12.69	2.94	1.48	4.45	10.05	9.53	4.01

Note: 1, the data on cities (counties) in Sanjiang plain and Heilongjiang province evaluation indicators are in reference [15]; 2, national evaluation indicator data are in reference [16]; 3, F stands for forward, R stands for reverse.

4.2 Comprehensive Evaluation of Land Resources Carrying Capacity

In order to be more scientific and reasonable on the evaluation of the land resources carrying capacity in Sanjiang plain, this study selected 10 land samples for carrying capacity evaluation, including the national large-scale, the Heilongjiang Province and Sanjiang plain under medium scale and seven cities in Sanjiang plain under small scale. Value table of land resources in different scale statistics 2012 carrying capacity evaluation indicator. The comprehensive evaluation indicator value of land resources carrying different scale in 2012 is shown in Table 1.

Matlab is used to program RAGA-PPC model to evaluate the total target and four sub goals of land resource carrying capacity under different scale. The total goal, the best projection value (i.e. Z*) of sub goal and its ranking are shown in Table 2.

On the basis of accumulation and dispersion characteristics of land resource carrying capacity value, coupled with the comprehensive evaluation of the land resources carrying capacity in Heilongjiang Province, land resources carrying capacity evaluation level can be divided into three grade: grade I $Z_A^* \geq 3.30$, land resources carrying capacity is strong, at stress-free state; grade II $2.6 < Z_A^* < 3.30$, land resources carrying capacity belongs to the medium level, at stable state; grade III $Z_A^* \leq 2.6$, land resources carrying capacity is weak, in a state of stress. Comprehensive evaluation result of land resource carrying capacity under different scale in 2012 is shown in Table 2.

Table 2. Comprehensive evaluation result of land resource carrying capacity under different scale in 2012

Evaluation Objection		Small scale							Medium scale		Large scale
		Jixi	Hegang	Shuangyashan	Jiamusi	Qitahe	Muling	Yilan	Sanjiang Plain	Heilongjiang province	China
B_1	Z_1^*	1.32	1.12	1.66	1.63	1.56	1.18	1.48	1.40	1.16	0.74
	Ranking	6	9	1	2	3	7	4	5	8	10
B_2	Z_2^*	1.43	1.50	1.71	1.87	0.95	1.54	1.49	1.55	1.22	0.70
	Ranking	7	5	2	1	9	4	6	3	8	10
B_3	Z_3^*	1.18	1.02	1.11	1.12	1.30	1.36	1.44	1.16	1.61	2.51
	Ranking	6	10	9	8	5	4	3	7	2	1
B_4	Z_4^*	1.66	0.98	0.90	1.28	0.60	0.37	0.62	1.09	0.59	0.32
	Ranking	1	4	5	2	7	9	6	3	8	10
A	Z_A^*	3.47	3.34	3.49	3.35	2.76	2.77	3.11	3.23	2.67	2.12
	Ranking	2	4	1	3	8	7	6	5	9	10
Evaluation grade		I	I	I	I	II	II	II	II	II	III

4.3 Land Resources Carrying Capacity Evaluation and Regional Differences

Land resources carrying capacity evaluation on national large-scale shows that in 2012 the national land resource carrying capacity is much lower than that in Heilongjiang province and the Sanjiang plain, the land resources utilization degree is low; Because of population pressure, land resources support capacity for social development and ecological environment protection is low, in the lower level of land resources carrying capacity (grade III), at a high stress level. But the national economic and technological level is high (economic and technical evaluation ranks 1st), hence, we can control the population while increasing funding for land resources for land resources and technology development and management improvement, hence enhance the overall land resources carrying capacity, reducing the social development pressure on land resources.

Land resources carrying capacity evaluation on medium scale shows that because of the rich land resources in Heilongjiang province and Sanjiang plain and the low population density, the evaluation of land resource utilization, the supporting capacity of land resources for social development and ecological environment protection, the comprehensive land resources carrying capacity are higher than the national level, the land resources carrying capacity belongs to middle grade (III) and the use of land resources and social development are at the equilibrium state. Because of the characteristics of geographic location and natural resources, the production activity of Heilongjiang province and Sanjiang plain dominated by agricultural production, industrial structure is irrational, backward in economy. Therefore, in the economic and technical evaluation, the economic and technical level of Heilongjiang province and Sanjiang Plain are lower than the national average level, restricting the advantage role of land resources in this area.

Especially as the commodity grain production base, the Sanjiang plain should strengthen transformation of low yielding fields, put the cultivation techniques of high yield and high yield varieties into practice to comprehensively promote the agricultural production level, improve agricultural land output efficiency, grain production capacity, at the same time, efforts should be made to expand the area of basic farmland protection, improve the role and status of agricultural product base, the formation of high yielding agriculture industry clusters and the industrial chain as soon as possible, hence ensure the use and coordination of economic and technical development of land resources.

Because of the different distribution of natural resource, land resources carrying capacity in Sanjiang plain under small scale varies significantly. The land resources carrying capacity of Jixi City, Hegang City, Shuangyashan city and Jiamusi city are at a high level (grade I), in a stress-free state. The area is rich in land resources, the available land resources quantity is enough to ensure the social and economic development. Future utilization of land resources should be focused on the depth development resources, the rational allocation of land resources, at the same time it should combine with the characteristics of regional resources to adjust the industrial structure, improve the efficiency of resources, to ensure the benign circulation of land resources.

The land resources carrying capacity of Qitaihe City, Muling City and Yilan County in Sanjiang plain belongs to the medium level (grade II), the use of the land resource and social economic development scale is in equilibrium. The area is the area of Sanjiang plain in the highest population density in the area, with the expansion of social development scale, shortage of land resources will become a limiting factor to the social and economic development. So in the future, population density should be controlled while promoting the rational distribution of industrial land, strengthening mining land reclamation, increasing the utilization of barren hills and wasteland, expanding the available land area to reduce the pressure on land resources. The efforts should also be put on the environment protection, the rational development of forest and wetlands, and all the exploitation should be on the premise of ecological environment.

5 Conclusions

(1) In the full analysis of land resources carrying capacity and regional characteristics, established the regional land resource carrying capacity evaluation indicator system, we set up the regional land resource carrying capacity evaluation indicator system. This system contains one total goal, four sub goals and 29 indicators, evaluating the regional land resource carrying capacity from four perspectives (i.e. land resources utilization, social development, and economy, technological and ecological environment).

(2) The paper constructed comprehensive evaluation model of land resources carrying capacity and acquired the comprehensive evaluation value of land resources carrying capacity under large scale (nation), medium scale (Heilongjiang province, Sanjiang plain), small scale (6 cites and 1 county in Sanjiang plain) in 2012.

(3) After full consideration of the comprehensive evaluation value of land resources carrying capacity, we put land resources carrying capacity into three levels. Among them, the national land resources carrying capacity falls on grade III; Heilongjiang province and Sanjiang plain under medium scale belongs to grade II; Jixi city, Hegang city, Shuangyashan city and Jiamusi city under small scale belong to grade I, while Qitahe city, Muling city and Yilan city belongs to fall on grade II. The result concluded in the paper, coupled with consideration of the characteristics of the regional social and economic development could support the policy making.

Acknowledgment. The study was supported by the National Natural Science Foundation of China (51179032, 51279031, 51209038), Cultivation Plan of Excellent Talents in New Century of Ministry of Education of Heilongjiang (1155-NCET-004), Ministry of Water Resources' Special Funds for Scientific Research on Public Causes (201301096-0204), Scientific Research Fund of Heilongjiang Provincial Education Department (12531009), Doctor Research Fund of Northeast Agricultural University (2012RCB58) and Heilongjiang Postdoctoral Fund (LBH-Z13049).

References

1. Liu, Y.: Optimal allocation of regional land use. Academic Press, Beijing (1999)
2. Hu, Y., Yang, Z.: Method for ecological benefit assessment of rural land consolidation. The Chinese Society of Agricultural Engineering 20(5), 275–280 (2004)
3. Wang, X., Bao, Y.: Study on the methods of land use dynamic change research. Progress in Geography 18(1), 81–87 (1999)
4. Food and Agriculture Organization of the United Nations, FESLM: An International Framework for Evaluating Sustainable Land Management. World Soil Resources Report 73 (1993)
5. Slesser. M., Hounam, I.: Carrying Capacity Assessment. Report to UNESCO and FAO, Rome (1984)
6. Jiang, Q., Fu, Q., Wang, Z., Jiang, N.: Spatial matching patterns of land and water resources in Sanjiang plain. Journal of Natural Resources 26(2), 270–277 (2011)
7. Fu, Q., Zhao, X.: Principle and application of projection pursuit model. Science Press, Beijing (2006)
8. Fu, Q., Wang, Z., Lian, C.: Applying PPC model based on RAGA in evaluating soil quantity variation. Journal of Soil and Water Conservation 16(5), 109–110 (2002)
9. Xiang, J., Shi, J.: The statistical method of data processing in nonlinear system. Science Press, Beijing (2000)
10. Zhang, X.: Projection pursuit and its application in hydrology and water resources. Sichuan University, Chengdu (2000)
11. Zhou, M., Sun, S.: The principle of genetic algorithm and its application. National Defense Industry Press, Beijing (2000)
12. Wang, S., Mao, H.: Design and evaluation on the indicator system of land comprehensive carrying capacity. Journal of Natural Resources 16(3), 248–254 (2001)
13. Zhu, Y.: Study on the integrated assessment system of land resources carrying capacity for Chongming Island. East China Normal University (2007)
14. Ying, J., Li, J.: Evaluation of land resources carrying capacity in the process of urbanization. The Construction of Small Towns (5), 31–37 (2010)
15. Heilongjiang Provincial Bureau of Statistics. Statistical yearbook of Heilongjiang Province. Chinese Statistics Press, Beijing (2013)
16. The people's Republic of China National Bureau of Statistics. China statistical yearbook. China Statistics Press, Beijing (2013)

Software Design of Distribution Map Generation for Soil Parameters Based on VC++

Xiaofei An[1,2], Zhijun Meng[1,2], Guangwei Wu[1,2], and Jianhua Guo[1,2]

[1] Beijing Research Center for Intelligent Agricultural Equipment, Beijing 100097, China
[2] National Research Center of Intelligent Equipment for Agriculture, Beijing 100097, China
{anxf,mengzj,wugw,guojh}@nercita.org.cn

Abstract. Software system on distribution map generation of soil parameters was developed based on VC++. With the help of MapX control, soil spatial data could be edited, analyzed, generated distribution maps and so on. Soil parameter scatter graph was displayed and classified by equal count method, equal range method, natural break method and standard deviation method. In the software system soil parameter statistics data was also calculated and a new perimeter and area algorithm was developed. After comparison of different interpolation methods, Inverse Distance Weighted (IDW) method was selected to generate the soil parameter contour map and soil parameter iso-surface. At the same time the system gave the legend color and drawing method. These maps could also be saved as vector graphics. Finally a distribution map generation of soil parameters for soil moisture and total nitrogen was conducted by the soil samples from Xiangtang apple orchard in Beijing suburb, including scatter graph, contour map and iso-surface. The errors of perimeter and area were less than 3.72% and 3.01%, respectively. The result shows that the system has perfect function and stable performance.

Keywords: Soil parameters, Distribution map, MapX control, Precision agriculture.

1 Introduction

Soil is the basis of agricultural production, human has continued to research and analysis soil in the recent years. According to the theory of soil science, soil has two main functions: the first function is used to store water and supply crops; the second is to maintain the healthy growth of crop root. In order to describe the two functions, human use several parameters such as soil moisture, soil total nitrogen, soil fertility and many other parameters [1-4]. At the same time, domestic and foreign scholars have obtained a lot of results on the rapid detection of soil parameters [5-10].

As a kind of new management ideas and modern agriculture, precision agriculture requires precise scientific fertilization and nutrient management [11-12]. In order to achieve the crop production in the process of scientific regulation measures of investment and management decisions, it must implement the precise fertilization decision-making. As a result it is particularly important to generating high precised soil parameter distribution map.

© IFIP International Federation for Information Processing 2015
D. Li and Y. Chen (Eds.): CCTA 2014, IFIP AICT 452, pp. 210–217, 2015.
DOI: 10.1007/978-3-319-19620-6_26

Currently, combinations of GIS tool software and visualization technology integrated have become the mainstream of the GIS application [13, 14]. This article discussed the development of soil parameter distribution map system based on VC ++. With the help of MapX control [14-16], it could display soil data, scatter plot and generate soil parameter distribution map.

2 The Structure of Soil Parameter Distribution Map System

2.1 Experimental Samples

According to the system requirement, soil parameter distribution generation system mainly included the following modules: basic operation module, data import module, data statistics module and parameter map generation module. Figure 1 is the overall hierarchy diagram of soil parameter distribution generation system. In VC + + 6.0 environment, the system has realized the import of soil parameter information, statistics and mapping, and other functions with MapX control.

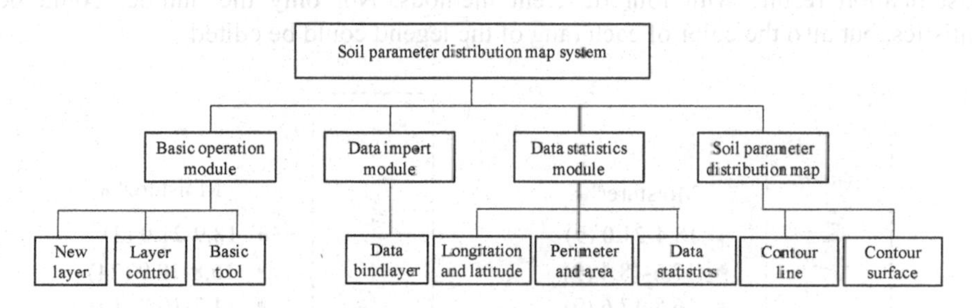

Fig. 1. Overall structure of system

3 Development of Soil Parameter Distribution Map System

3.1 Data Binding

Soil parameter information included soil moisture value, soil total nitrogen and soil GPS coordinates. Data binding is the process of soil parameter data from the data source into MapX [15]. In the MapX control, every map was corresponded to the multiple layers (the layers), and each layer (layer) have datasets, which contains a dataset object. Datasets with some properties and methods were mainly used to add and remove the dataset object in the collection.

Soil parameter data mainly included GPS location information of soil and soil parameters attribute values. Then the binding layer type was set to miBindLayerTypeXY, which could realize to WGS54 coordinates X and Y coordinates of points. Figure 2 shows the imported the scatter plot after binding the soil moisture data.

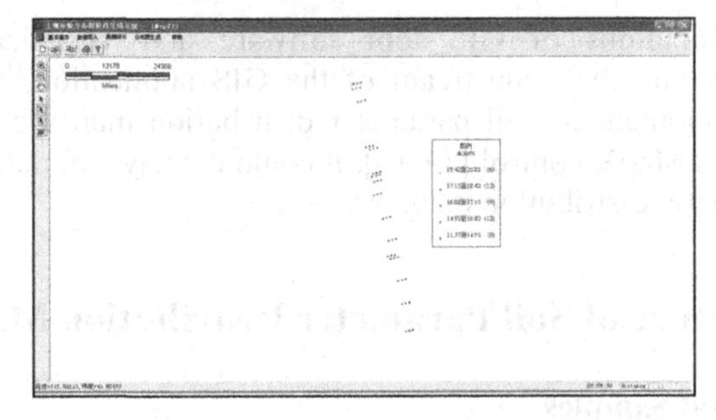

Fig. 2. Data binding result

3.2 Data Classification and Statistics

After the soil data was binded to the soil layer, it could be classified by four methods, such as the counting method, the same range method, the nature devision method and the standard devision method. Figure 3 (a), (b), (c) and (d) shows the data classification results with four different methods. Not only the number could be statistics, but also the color of each rang of the legend could be edited.

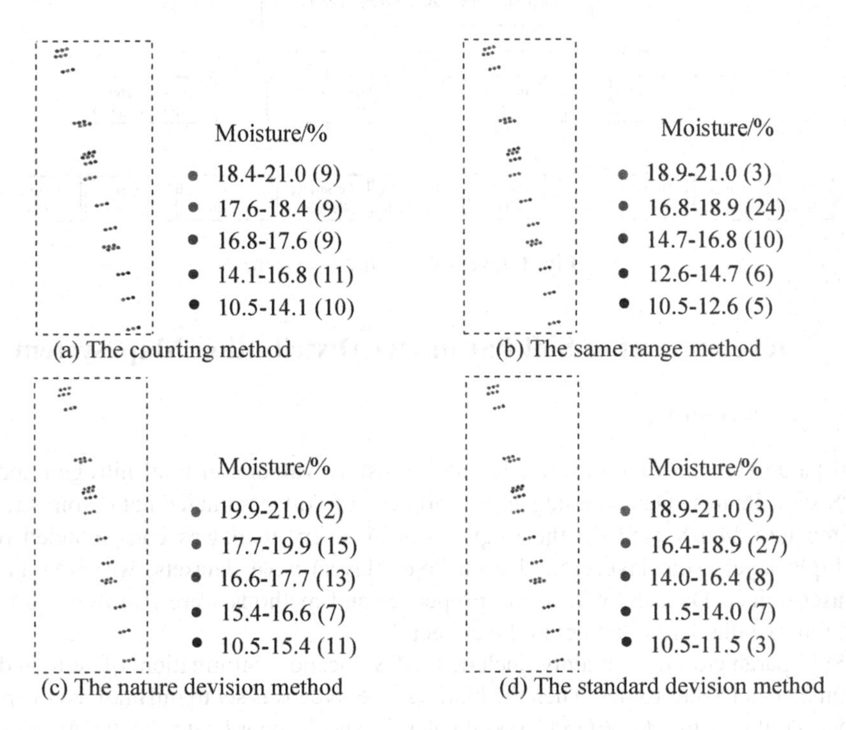

Fig. 3. Data classification method

It was said that soil and crop information always has a certain statistical regularity, although sometimes it does not get the distribution function of random variables. In this system the soil parameters could be calculate by maximum, minimum, the statistics of the mean and standard deviation. The mean value and mean square error of the calculation formula were shown in (1) and (2).

$$\bar{v} = \frac{1}{n} \times \sum_{k=0}^{n-1} v_k \tag{1}$$

$$\sigma = \sqrt{\frac{1}{n} \times \sum_{k=0}^{n-1} (v_k - \bar{v})^2} \tag{2}$$

Where: σ *was* the data standard deviation value, $\bar{v}$ was the average of the data, n was the number of data points, k was the data point number, v_k was the first k data of soil parameters.

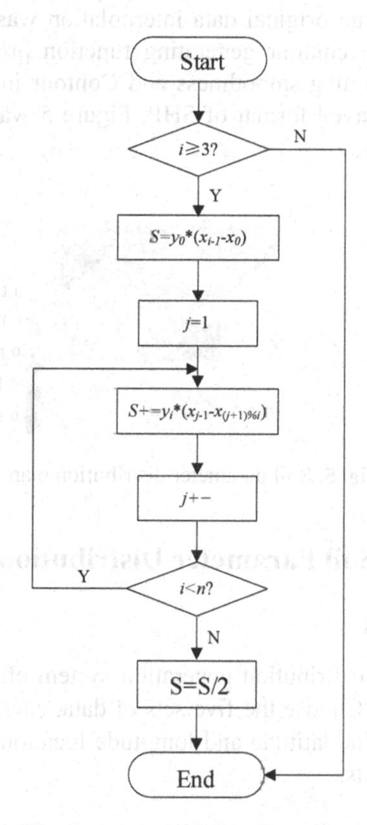

Fig. 4. Area calculation flow chart

3.3 Perimeter and Area Calculation

This system designed an algorithm to calculate perimeter and area with the direct use of GPS handheld device data. And projection method was directly to gaussian zoning projection. The center was east longitude 116 °, projection of 40°.

MapX control provided map.distance method. This method could be used to calculate the linear distance between two points. Along the land border manually choose some key points, and then map.distance method was used repeatedly in sequence. Finally the perimeter was obtained accumulation. The system adopts the offset compensation method to calculate polygon area. Figure 4 was area calculation flow chart.

3.4 Soil Parameter Distribution Map Generation

With the steps of data binding, scatter plot, interpolation, contour map, soil parameter distribution map could be generated. By comparing the different types of spatial data interpolation methods, inverse distance weighting spatial data interpolation algorithm was finally chosen. After the original data interpolation was completed, the contour map was generated by the contour generating function provided by Contour.OCX control. Both the corresponding smoothness and Contour interval could be set. And finally the map could be saved format of SHP. Figure 5 was the distribution of soil parameters generated dialog.

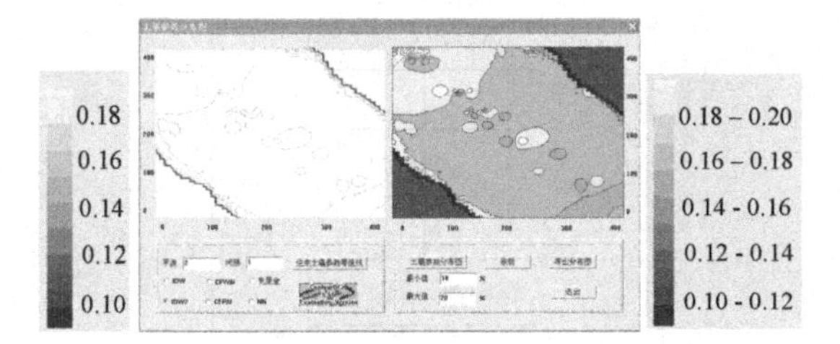

Fig. 5. Soil parameter distribution map

4 Application of Soil Parameter Distribution Map System

4.1 Efficiency Analysis

To test the soil parameter distribution generation system efficiency, different sets of import test data was used. Choose the five sets of data, each group data included the number, soil parameter value latitude and longitude location information. Figure 6 is data import speed test results.

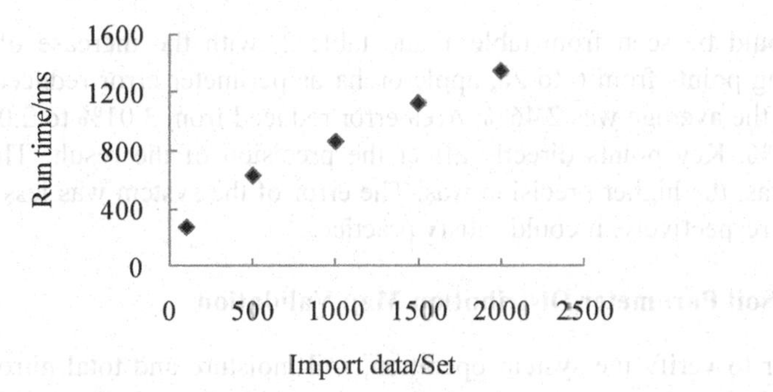

Fig. 6. Data import speed result

It could be seen from the figure 6, when import external data, this system increased with the amount of data. The consumption of time required to import the data were linearly increasing trend. When the data points was 2000 sets (the amount of data with 8000), the system took 1344 ms. From the perspective of the function of angle and speed of system, functions could be fully realized, moderate speed in data quantity could meet the needs of practical application.

4.2 Perimeter and Area Validation

In order to test this system the perimeter area of the accuracy of the algorithm, experiments were carried out in Beijing changping apple orchards with GPS handheld instrument G738CM and the system. After sample the apple orchard with the device, the perimeter and area were calculated by the system. Table 1 and table 2 were the perimeter and area calculation error, respectively.

Table 1. Perimeter calculation error

Number	Key point	System perimeter/m	System perimeter with GPS device/m	error/%
1	6	247.00	238.14	3.72
2	8	246.81	238.00	3.70
3	12	242.90	239.47	1.43
4	20	245.93	241.06	2.02
5	28	245.58	242.14	1.42

Table 2. Area calculation error

Number	Key point	System area/m^2	System area with GPS device /m^2	error/%
1	6	2160.62	2097.40	3.01
2	8	2079.50	2035.34	2.17
3	12	2110.55	2087.28	1.11
4	20	2231.84	2180.40	2.36
5	28	2248.96	2203.98	2.04

It could be seen from table 1 and table 2, with the increase of the number of sampling points from 6 to 28, apple orchards perimeter error reduced from 3.72% to 1.42%, the average was 2.46%; Area error reduced from 3.01% to 2.04% , an average of 2.14%. Key points directly affect the precision of the result. The more the key point was, the higher precision was. The error of the system was less than 3.72% and 3.01%, respectively. It could satisfy practice.

4.3 Soil Parameter Distribution Map Validation

In order to verify the system operation, soil moisture and total nitrogen distribution maps of changing apple orchard were generated respectively, including the scatter plot, contour map and level surface. Inverse distance weighting spatial data interpolation method was used on the original data interpolation. The results showed that the system had stable performance. Figure 7 (a), (b), (c) and figure 8 (a), (b), (c) were the distributions of soil moisture and soil total nitrogen.

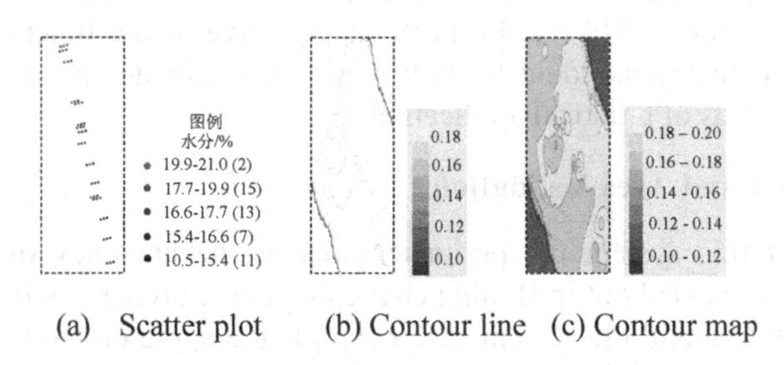

(a) Scatter plot (b) Contour line (c) Contour map

Fig. 7. Soil moisture distribution map

5 Conclusions

Soil parameter distribution map system was developed based on VC++. The system could realized the scatter plot, contour map, contour surface generation and vector graph save function. Through the analysis of the system efficiency and perimeter area computation verification test it showed that when the data points was 2000, the time-consuming needs 1344 ms, calculation error within 3.72% and 3.01% respectively; The results showed that the system was stable and reliable enough. It provided a new means to guide the variable rate fertilization.

Acknowledgment. This paper was supported by the 863 programmar (2012AA101901) and 948 programmar(2014-G32).

References

1. Li, M.: Spectral analysis and application. Science Press, Beijing (2006)
2. Zheng, L.: Real-time Sensing of Soil Parameters Based on Spectroscopy. China Agricultural University, Beijing (2007)
3. Guo, J., Zhao, C., Wang, X., et al.: Research advancement and status on crop nitrogen nutrition diagnosis. Soil and Fertilizer Sciences in China (4), 10–14 (2008)

4. Zhang, W., Li, X., Huang, W., et al.: Comprehensive assessment methodology of soil quality under different land use conditions. Transactions of the CSAE 26(12), 311–318 (2010)
5. Luo, X., Zang, Y., Zhou, Z.: Research progress in farming information acquisition technique for precision agr iculture. Transactions o f the CSAE 22(1), 167–173 (2006)
6. He, D., Yang, C., Yang, Q., et al.: Research progress of real-time measurement of soil attributes for precision agriculture. Transactions of the CSAE 28(7), 78–85 (2012)
7. Mouazen, A.M., Kuang, B., De Baerdemaeker, J., et al.: On-line measurement of some selected soil properties using a VIS-NIR sensor. Soil & Tillage Research 93(1), 13–27 (2007)
8. Christy, C.D.: Real-time measurement of soil attributes using on-the-go near infrared reflectance spectroscopy. Computers and Electronics in Agriculture 61, 10–19 (2008)
9. Adamchuk, V.I., Hummel, J.W., Morgan, M.T., et al.: On-the-go soil sensor for precision agriculture. Computers and Electronics in Agriculture (44), 71–91 (2004)
10. Li, M., Pan, L., Zheng, L., et al.: Development of a portable SOM detector based on NIR diffuse reflection. Spectroscopy and Spectral Analysis 30(4), 1146–1150 (2010)
11. Wang, M.: Precision agriculture with the practice of agricultural science and technology innovation. China Soft Science (4), 21–25 (1999)
12. Wu, C.: Research on the variable rate fertilization and economic analysis. Jilin University (2003)
13. Jin, H., Liu, H., Miao, B.: Secondary development of GIS application software functions based on MapX controls. Computer & Digital Engineering 39(1), 61–63 (2011)
14. He, Y., Fang, H., Feng, L.: Information processing system for precision agriculture based on GPS and GIS 18(1), 145–149 (2002)
15. Yin, X., Zhang, W.: MapX technology based on viscal C++. Metallurgical Industry Press, Beijing (2009)
16. Yan, H., Wu, X., Zhu, G.: The development of GIS Application software on the basis of MapX control in VC++. Journal of Kunming University of Science and Technology 26(6), 12–17 (2001)

RBF Neural Network Based on K-means Algorithm with Density Parameter and Its Application to the Rainfall Forecasting

Zhenxiang Xing[1,2,3], Hao Guo[1], Shuhua Dong[4], Qiang Fu[1,2,3], and Jing Li[1]

[1] College of Water Conservancy & Civil Engineering, Northeast Agricultural University, Harbin 150030, China
[2] Collaborative Innovation Center of Grain Production Capacity Improvement in Heilongjiang Province, Harbin 150030, China
[3] The Key Lab of Agricultural Water Resources Higher-Efficient Utilization of Ministry of Agriculture of PRC, Harbin 150030, China
[4] Heilongjiang Province Hydrology Bureau, Harbin, 150001
zxxneau@hotmail.com,
{ghneau,shdshuiwen,fuqiangneau,lijingneau}@sina.com

Abstract. The Radial Basis Function (RBF) neural network is a feed-forward artificial neural network with strong approximation capability. A K-means algorithm based on density parameter was introduced to determine clustering center aimed to improve the training rate of the RBF. It could reduce sensitivity of traditional K-means algorithm for initial clustering centers. A rainfall forecasting model of RBF based on K-means algorithm was built, which was applied to forecast monthly rainfall in Shuangyashan City during the flood season, aiming to test the effectiveness of this model. The case study showed that the mean relative error of rainfall forecasting in flood season (from June to September) of the year 2006, 2007 and 2008 was 10.81%, and the deterministic coefficient was 0.95. It demonstrated a higher forecasting accuracy comparing to a RBF model based on a standard K-means algorithm and BP (Back Propagation) model, and the rainfall forecasting results satisfied the requirements of hydrologic prediction.

Keywords: Rainfall forecasting, Radial Basis Function Neural Network, Density parameter, K-means.

1 Introduction

The rainfall is an important process with higher uncertainty in natural water cycle. A rainfall forecasting method with high-accuracy could predict the change amount of precipitation, which could provide important significance to decision-making of flood control and disaster reduction. There are a lot of methods for rainfall forecasting, such as the regression analysis [1], the grey prediction [2], the fuzzy prediction [3], and artificial neural network [4.5] and so on.

© IFIP International Federation for Information Processing 2015
D. Li and Y. Chen (Eds.): CCTA 2014, IFIP AICT 452, pp. 218–225, 2015.
DOI: 10.1007/978-3-319-19620-6_27

The artificial neural network has many advantages of rainfall forecasting, which has strong ability to deal with nonlinear problem and high generalization ability. So the BP neural network and the RBF neural network is widely used network model. Compared with the BP network, the number of hidden layer of RBF network can be adaptive adjusted in training phase. In addition, input layer and hidden layer of RBF use linear connection instead of weights. It seems that the RBF has advantages compared with BP, which can greatly improve the convergence speed of network. RBF also has better approximation capability of nonlinear function. In this paper, a RBF model was trying to be used to forecast the precipitation in Naolihe catchment in Sanjiang plain.

2 Radial Basis Function Neural Network

The Radial Basis Function Neural Network (RBF) is a three-layer feed forward network with single hidden layer [6], such as input layer, hidden layer, output layer. The network also can approximate any continuous function with arbitrary precision theoretically.

In RBF network is, the hidden layer space is constructed by the RBF, so an input vector can be directly (do not need the weights) mapped into hidden space. The mapping relationships between hidden layers to output layers were described as a linear function and the outputs of network are linear weighted sum of hidden unit output [7].

Gaussian function is commonly used as the radial basis function in RBF network. The expression of activation function is:

$$\phi(x_p - c_i) = \exp\left[-\frac{1}{2\sigma^2} \| x_p - c_i \|^2\right] \tag{1}$$

where φ is activation function; $\| x_p - c_i \|$ is the European norm; $x_p = (x_1^p, x_2^p, \ldots, x_m^p)^T$ is the pth input samples; $p=1,2,\ldots,P$ (P is the total number of samples); c_i are a center of Gaussian function; σ are a variance of Gaussian function.

The outputs of network are obtained through the RBF structure

$$y_j = \sum_{i=1}^{h} w_{ij} \exp\left[-\frac{1}{2\sigma^2} \| x_p - c_i \|^2\right] \tag{2}$$

where w_{ij} denote weights between the hidden layer to output layer; $i=1,2,\ldots,h$ (is number of nodes in a hidden layer); y_j denote actual outputs from the jth of output corresponding with input sample; and other symbols have the same meaning as above.

3 K-means Algorithm Based on Density Parameter

In an RBF neural network, three parameters are needed to be solved, which are the centers of basis function in hidden layer, the variance and the weights of the hidden layer to output layer. The key to build a good network is to select an adaptive basis function center. There are many methods, such as the randomly selected algorithm, the self-organization selected algorithm, the clustering analysis algorithm and the orthogonal least squares algorithm to do this job. Among the methods above, the K-means clustering algorithm is one of the fairly effective learning algorithms.

The K-means algorithm is a clustering algorithm based on distance, i.e., a distance is the assessment criteria of the comparability. The traditional K-means algorithm is easy to understand, which is easier for programming. However, it is obvious that K-means algorithm is sensitive to the initial clustering center and clustering results fluctuate when given different initial input. It will infect the final characteristics of sample groups. The K-means algorithm based on density parameters can reduce the influence caused by initial clustering centers to clustering results comparing to a traditional algorithm. Therefore, the K-means algorithm based on density parameter of RBF is applied to forecast the rainfall in this paper.

3.1 The Concept of Density Parameter

The aggregation of sample data: S= $\{x_1, x_2, \ldots, x_n\}$, the initial clustering centers: $z_1, z_2, \ldots, z_k$.

Define 1 an Euclidean distance between each two samples

$$d(x_i, x_j) = (|x_{i1} - x_{j1}|^2 + |x_{i2} - x_{j2}|^2 + \ldots + |x_{ip} - x_{jp}|^2)^{1/2} \tag{3}$$

where $x_i=\{x_{i1}, x_{i2}, \ldots, x_{ip}\}$ and $x_j=\{x_{j1}, x_{j2}, \ldots, x_{jp}\}$ are samples with p-dimension.

Define 2 an average distance between samples

$$MeanDist = \frac{1}{C_n^2} \times \sum d(x_i, x_j) \tag{4}$$

where n is the total number of samples.

Define 3 the density parameter [8].

In the density space, neighborhood of any point is defined as the region which point p is the center and *MeanDist* is the radius. The number of points in region is known as the density parameter based on *MeanDist*, called *density* (p, *MeanDist*).

3.2 K-means Algorithm Based on Density Parameter

The Euclidean distance is used as similarity measure of the K-means algorithm. The mutual farthest k data objects are more representative than random k data. It is considered that noisy data are often mixed in practical data. In order to avoid selecting the noisy points, we can get k points in the high density area as the initial clustering center by following steps [8]:

(1) Calculate the distance between any two data according to formula (3), called as $d(x_i, x_j)$;

(2) Calculate the average distance of total data according to formula (4), called as *MeanDist*;

(3) Calculate the density parameters of all data, named as *density* (*p*, *MeanDist*), and composing a data set named *D*.

(4) Find the maximum value from D, $z_k = \max\{density(p_i, MeanDist) \mid i \in (1,2\ldots,n) density\ (p_i, MeanDist) \in D\}$ .If $d\ (p_i, z_k) < MeanDist$, the *density* (*p*, *MeanDist*) is deleted from D; z_k is the kth of cluster center.

(5) Repeat the step (3) and step (4) until finding cluster centers, and the number of cluster centers is k.

Therefore, cluster centers can be obtained in accordance with the method, that is, the final center of the basis function of RBF network.

4 The Rainfall Forecasting Model Based on RBF

4.1 The Structure of RBF Network

4.1.1 The Number of Input Layer Neurons

It has much significance to choice the number of input neurons for RBF network. The unreasonable choice will affect training ability of the network, and even lead to model crash. Therefore, the node number of input neurons of RBF network is determined by autocorrelation analysis technology in this article.

4.1.2 The Number of Hidden Layer Neurons

The hidden layer neurons are used to store connection weights and thresholds between input layer and hidden layer. It is characterized in terms of reflecting inherent law between training sample and expected output. There are no accurate and scientific methods to determine the hidden layer neurons at present, and as well as more determined by empirical formula and numerical test. The traditional method is also used in this paper, and the empirical formulas are [9]:

$$n_2 = \sqrt{n_1 + m + a} \tag{6}$$

$$n_2 = \log 2^{n_1} \tag{7}$$

where n_2 is the number of hidden layer neurons, n_1 is the number of output layer neurons, m is the number of input layer neurons, a is a constant between [0, 1].

4.2 The Rainfall Forecasting Model

The network model is built by RBF tool box of MATLAB, the program language as follows:

```
net =newrb(p,t,goal,spread,mn,df)
```

where *p* and *t* are the input vector and the output vector of sample respectively; *goal* is the mean square error for the purpose of preventing excessive fitting. In this paper, *goal* was set as 0.001; *mn* is the maximum number of neurons; *df* is the number of neurons to add between displays; *spread* is the width of the radial basis function, and in this paper it is determined by the distance between each center of hidden layer neurons, and the formula is

$$spread = b \times d_i \tag{8}$$

where b is the overlap coefficient, and usually this value are set as an integer greater than 1[10]; d_i is the minimum distance of basic function centers of the hidden layer, the basis function center is obtained by K-means algorithm based on density parameters.

5 Case Study

5.1 The Experimental Data

The rainfall forecasting model of RBF based on K-means algorithm was built, which was taken precipitation in Shuangyashan City of Sanjiang plain in each flood season from 1955 to 2005 (June to September) as an example. The precipitation of the year from 1955 to 2003 was used for network training, and that from 2004 to 2005 was used for network testing, forecasting the rainfall of the year 2006, 2007 and 2008.

5.2 Data Preprocessing

Input and output data of RBF network should be within [0, 1], so the precipitation need to be normalized before training model according to following formula,

$$y_i = \frac{x_i - x_{\min}}{x_{\max} - x_{\min}} \tag{9}$$

where x_i and y_i is the rainfall before and after normalizing respectively; x_{min} and x_{max} are the minimum and maximum values of all rainfall.

The processed data will be used as the samples of the RBF neural network model.

5.3 Identifying and Training RBF Network

The key point to build RBF neural network is to determine number of nodes in input layer and hidden layer, which is based on the characteristics of rainfall. The correlation analysis of rainfall series in flood season in Shuangyashan was built, and we can estimate that the rainfall series is suitable for the AR (p) model. The order of Markov process should be four by analyzing the AIC criterion. That is to say, the input layer neurons of RBF network are four. According to the empirical formulas and trial calculation, number of nodes in hidden layer is eight. Therefore, the structure of network in this paper is 4:8:1.

To achieve converging, the rainfall of the year from 1955 to 2003 is trained 172 times by using neural network which has been constructed above, and the fitting error of network is 0.001. The rainfall of the year 2004 and 2005 is used to test and verify the generalization ability of the RBF network model. The mean prediction relative error is 10.9736% and the deterministic coefficient is 0.95. In conclusion, the constructed RBF network could be applied to forecast rainfall in Naolihe catchment.

5.4 Results and Analysis

The calculation in Table 1 show that the mean relative error of rainfall forecasting in 2006, 2007 and 2008 is 10.81% by using established RBF network model, the deterministic coefficient is 0.95; the relative error using K-means algorithm is 14.30%, and the deterministic coefficient is 0.86. The mean relative error of BP network model is 14.38%, the deterministic coefficient is 0.87.Meanwhile, fitting time and convergence times of the established RBF network model are reduced, which are 80s and 172 times respectively; and the standard K-means algorithm's fitting time and convergence times are 97s and 235 times, while the BP network model's are 125s and 302 times. Therefore, compared with RBF model of standard K-means and BP model, the computing speed of the RBF network that has been constructed in this paper are increased by 18% and 36% respectively, and the mean relative error are reduced by 24% and 25%.The fitting figure between observed rainfall and forecasted rainfall in training period, testing period and forecasting period was shown in Fig 1. (The first 182 series is training period, series of183th to 190th is testing period, and other series is forecasting period).

Table 1. The results of rainfall forecasting compared impoved RBF network with standard K-means RBF network and BP network

Time	Observed rainfall /mm	a		b		c	
		Forecas-ted rainfall /mm	relative error /%	Forecas-ted rainfall /mm	relative error /%	Forecas-ted rainfall /mm	relative error /%
0606	127.9	105.76	-17.31	106.68	-16.59	99.86	21.92
0607	148.7	130.88	-11.98	125.03	-15.92	120.86	18.72
0608	64.9	70.00	7.86	72.05	11.01	70.14	8.08
0609	17.2	20.07	16.70	18.80	9.30	20.00	16.29
0706	65.9	70.22	6.56	75.01	13.82	73.03	10.83
0707	35.8	41.04	14.64	40.37	12.77	42.08	17.53
0708	156.1	142.37	-8.79	132.50	-15.12	140.66	9.89
0709	39.6	35.40	-10.60	45.09	13.87	43.53	9.92
0806	109.9	100.56	-8.50	95.65	-12.97	104.40	5.00
0807	48.4	52.37	8.21	42.38	-12.44	40.07	17.21
0808	101.0	110.20	9.11	118.00	16.84	117.84	16.68
0809	29.3	32.05	9.39	35.46	21.03	35.30	20.49
MRE	—	—	10.81	—	14.30	—	14.38
DC	—	0.95	—	0.86	—	0.87	—

a. The improved RBF network that has been built in this paper;
b.RBF network based on standard K-means; c. BP network;
MRE is the mean relative error; DC is the deterministic coefficient

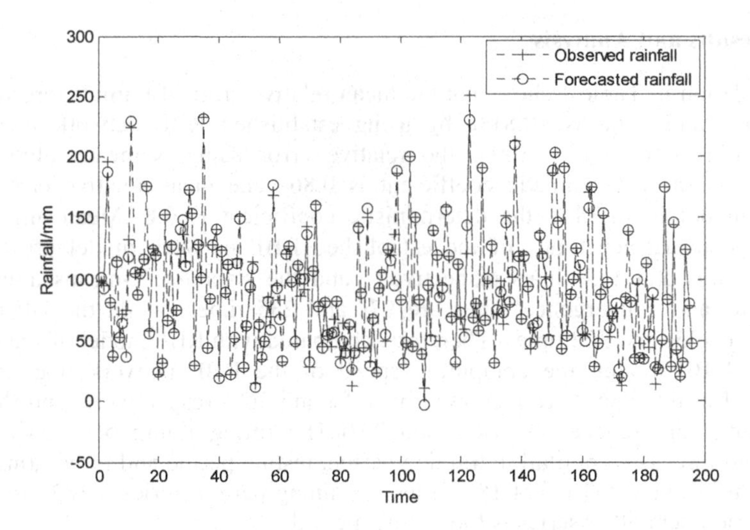

Fig. 1. The fitting figure between observed rainfall and forecasted rainfall

6 Conclusions

The K-means algorithm is commonly used as learning algorithm in RBF neural network. The initial clustering centers are selected randomly based on the traditional K-means algorithm, which result in different convergence abilities. The method of density parameters was trying to be used to reduce the sensitivity of selecting randomly for clustering centers in this paper, and to calculate the width of basis function of the RBF network. And then a rainfall forecasting model of RBF based on K-means algorithm was built, which was applied to forecast monthly rainfall in Shuangyashan City in Heilongjiang Province. The case study results and conclusions were shown as follows:

1) Comparing with the traditional K-means algorithm, the RBF network model based on the density parameter had its advantages for determining the number of clustering results and the centers of basis function. And the model had higher convergence rate and forecast accuracy.

2) The accuracy of RBF network model on account of rainfall forecasting during flood season in Shuangyashan City had been improved greatly compared with BP network model. The mean relative error of monthly rainfall in 2006, 2007 and 2008 were reduced by 24% and the deterministic coefficient was increased by 10%. So the model built in this paper could be used as an effective method for rainfall forecasting or other time series forecasting with strong uncertainties.

3) In this paper, number of nodes in the network was determined by empirical formula and numerical test. In order to further increase the calculation precision of model forecasting, some intelligent optimization algorithms such as genetic algorithm could be adopted to optimize number of nodes in hidden layers in the future research.

Acknowledgment. Funds for this research was provided by the national natural science foundation (51109036; 51179032), the specialized research fund for the doctoral program of higher university of the Ministry of Education (20112325120009) the special scientific research funds for Ministry of water resources public welfare industry (201301096), The leader talent echelon back-up headman funding projects of Heilongjiang province (500001), the studying funds of postdoctoral in Heilongjiang province (LBH-Q12147), the science and technology project of Heilongjiang Provincial Education Department(11541022).

References

1. Chang, Q., Zhang, X., Cai, H., et al.: A Precipitation Forecast Model Based on Regression Analysis and Time Series Analysis. Journal of Soil and Water Conservation Bulletin 29(1), 88–91 (2009)
2. Li, C., Gu, Y.: Grey Prediction Model in the Application of the Upper Yangtze River Basin Surface Rainfall Forecast. Journal of Climate Science 31(4), 223–225 (2003)
3. Sun, C., Lin, X.: Research on fuzzy Markov Chain Model with Weights and its Application in Predicting the Precipitation State. Journal of Systems Engineering 17(4), 294–299 (2003)
4. Feng, Y., Wu, B., Cui, L.: Study on Forecasting Typhoon Rainfall Based on BP Neural Network. Research of Soil and Water Conservation 12(3), 289–293 (2012)
5. Li, X., Cheng, C., Lin, J.: Bayesian Probabilistic Forecasting Model Based on BP ANN. Journal of Hydraulic Engineering 37(03), 354–359 (2006)
6. Yi, Y., Lu, W., Zhang, Y., et al.: Study on the Surrogate Model of Groundwater Numerical Simulation Model Based on Radial Basis Function Neural Network. Research of Soil and Water Conservation 19(4), 265–269 (2012)
7. Zhang, D.: Neural Network Application Design of Matlab. China Machine Press, Beijing (2009)
8. Zhang, J.: Research and Application of K-means Clustering algorithm. Wuhan University of Technology (2007)
9. Zhu, M.: Data Mining. Hefei: Press of University of Science and Technology of China (2002)
10. Su, M.: Research on Radial Basis Function Neural Network Learning Algorithms. Soochow University, Suzhou (2008)

Design of Transplanting Mechanism with B-Spline Curve Gear for Rice Pot Seedling Based on UG

Yanjun Zuo[1], Huixuan Zhu[1], and Peng Cao[2]

[1] College of Engineering, Northeast Agricultural University, Harbin 150030, China
[2] The 49th Institute of China Electronics Technology Group Corporation, Harbin 150001, China
{42648135,108091835,36125922}@qq.com

Abstract. In allusion to the existing rice pot seedling transplanter exists the defects of complex structure, high cost and low efficiency, the transplanting mechanism with b-spline curve gear for rice pot seedling was designed based on parametric modeling software UG NX 8.0. Firstly, the 3-D models of part were built, and then assembled. Secondly, the interference detection of assembly model was done. Finally, relevant 2-D engineering drawings were generated for manufacture. It is provides the reference for the design of rice pot seedling transplanter, and preparation for further simulation and analysis with softwares of Adams and Ansys.

Keywords: Rice pot seedling transplanting, B-Spline, UG, Virtual design.

1 Introduction

Rice is one of the main food crops. and perennial planting area is about 30 million hm2 in China. The area account for national grain and world's rice planting areas are about 30% and 20%, it only less than India and is the second largest in the world[1-2]. Therefore, our rice production has a pivotal position to food security not only for China but also for world. There are two modes of rice planting (direct seeding and transplanting) at present[3], but the Asia is given priority to rice seedling transplanting, the mechanization levels of Japan and Korea are highest and more than 98% during Asia. The cultivation mode of rice transplanting is also divided seedling transplanting and pot seedling transplanting, comparing with seedling transplanting, the pot seedling transplanting has the advantages of seedling no hurting and no need recovering, production increasing and so on[4]. Since Japan invents the first transplanter for rice pot seedling in 1975, each large agricultural machinery company has producted their own models in succession, such as Yanmar, Iseki and so on. The material structure and processing technology have improved through the forty years development, but the its basic structure and working principle are still the same[5-10].

* The paper is supported by the National Natural Science Fund Projects (Project number is 51175073), the Special Fund for Agro-scientific Research in the Public Interest (Project number is 201203059-01), the Northeast Agricultural University Doctor Startup Fund (Project number is 2012CRB56).

© IFIP International Federation for Information Processing 2015
D. Li and Y. Chen (Eds.): CCTA 2014, IFIP AICT 452, pp. 226–232, 2015.
DOI: 10.1007/978-3-319-19620-6_28

The transplanting machines of rice pot seedling use 3 devices to achieve the 3 actions of taking, transportation and planting separately, which gives rise to complex structure, high cost and low efficiency, so the promotion area is limited[7,11-12]. The rice transplanter is main except rice pot seedling transplanting machine with double cranks, which is invented by Zhao Yun et al. The double cranks transplanting machine can use a mechanism to achieve 3 actions orderly for rice pot seedling, has relatively simple mechanism, but the disadvantages of large vibration and low efficiency are existent because of the characteristics of bar mechanism[13]. Therefore, the study on transplanting mechanism for rice pot seedling to efficient, light simplified and full-automatic, is very necessary.

In the study, UG NX 8.0 was applied for the design of transplanting mechanism with b-spline curve gear for rice pot seedling. The 3-D models of each part can be built and assembled easily, and interference detection of assembly models was done through the software. Therefore, virtual design can promoted standardization, normalization and serialization to physical prototype based on virtual manufacturing, assembly and test.

2 Characteristics of UG NX 8.0

Software of UG NX 8.0 is developed by SIEMENS, it is a flagship digital software for solution of NX. The software is comprehensive 3-D modeling software which gathers CAD, CAE and CAM, it sets up a new standard for machinery industry and is favoured by insider. In order to meet the requirements of users which can develop new product quickly, UG has more extensive product design module, such as modeling, simulation, assembly, bending and so on, and also has high performance of charting capability, so it is used widely in machinery, automobile and other manufacturing industries. In addition, UG can realize seamless joint with other virtual simulated softwares, such as Adams and View 2010, the model that is built in UG, can be operated directly after import[14].

3 Results and Discussion

3.1 Working Principle

The sketch of transplanting mechanism for rice pot seedlings is shown in figure 1, it is composed by transmission component with non-circular gear and transplanting component. The sun gear, middle gear and planet gear are non-circular gear. Planet carrier and sun shaft are solid joint, planet gear and transplanting arm(including cam, shifting fork, spring, spring place, seedling-pushing rod and seedling clip) are solid joint. Power income from sun shaft, planet carrier rotates uniformly, transplanting arm rotates reversely through middle gear, shifting fork is swaied by combined effect of cam and spring, the movement of seedling-pushing rod is formed by sway of shifting fork, the open and close of seedling clip are realized by movement of seedling-pushing rod. The whole transplanting process is achieved for rice pot seedling by the above actions.

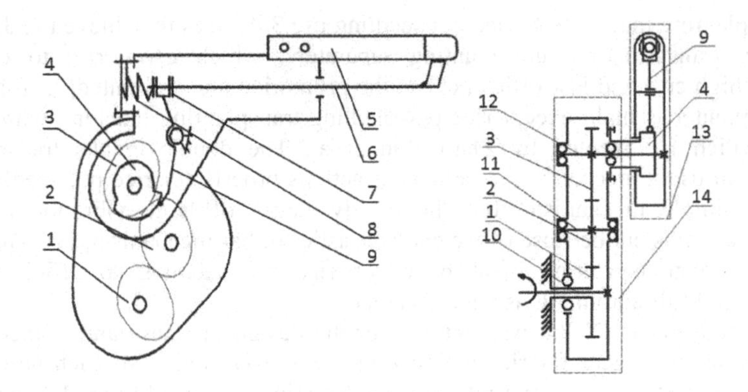

Fig. 1. The sketch of transplanting mechanism for rice pot seedlings 1. Sun gear 2. Middle gear 3. Planet gear 4. Cam 5. Seedling clip 6. Seedling-pushing rod 7. Spring place 8. Spring 9. Shifting fork 10. Sun shaft 11. Middle shaft 12. Planet shaft 13. Transplanting arm 14. Planet carrier

3.2 Part Design

The characteristics of criterion, voxel and basic shape design that were needed, could be built by 3-D capabilities in UG, such as base level, base coordinate, sphere, cylinder, square body, cone, stretching, rotating, draft, and slightly, convex platform, slotting and so on. Because the pitch curve of non-circular gear was irregular, the pitch curve was imported using interface of UG, and then stretched. The 3-D model of non-circular gear was obtained as shown in figure 2a, the other five more important or distinctive parts were all simple, the 3-D modeling process was not need to special instruct, the models were shown in figure 2.

3.3 Assembly Design

Based on section of 3.2, using the assembly module which is the built-in modules of UG, the assembly was done through adding the relevant constraints. There are tow types of assembly in the whole machine assembly, they are step by step assembly and whole assembly. The whole assembly can be used under the condition of less mechanism and simple, all the parts are imported into the assembly drawing and imposed constraints in turn. The characteristics of transplanting mechanism are that components and parts are more, the structure and positional relationship are all relatively complicated, so the type of step by step assembly would be used in this paper. The transmission component with non-circular gear and transplanting component were assembled into sub assembly by the top- down order, and shown in figure 3. The whole machine assembly of transplanting mechanism has been assembled by adding the relevant constraints continuely after the tow components completing, the 3-D virtual prototype model was shown in figure 4.

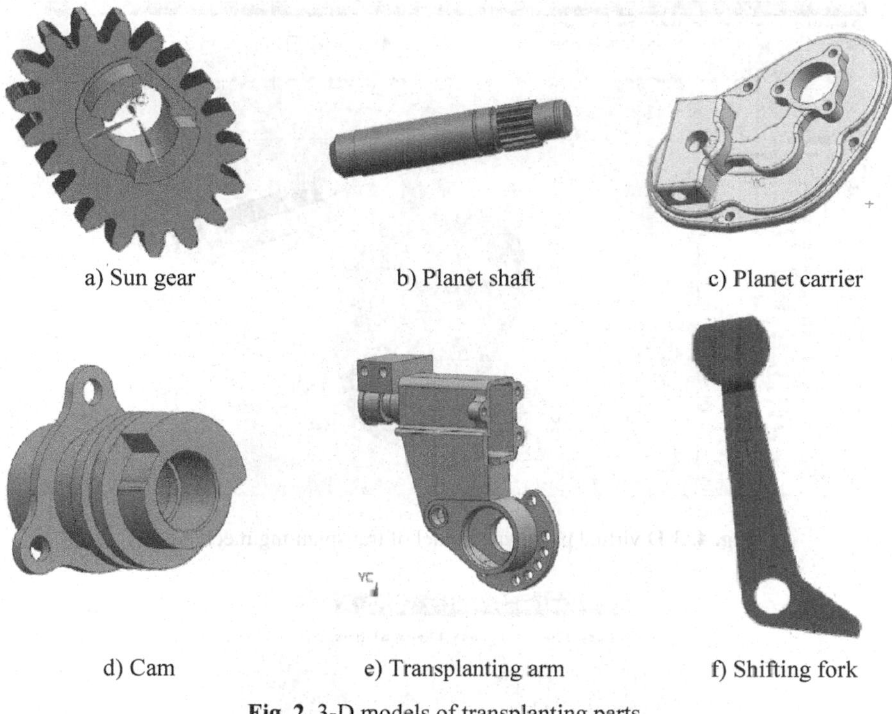

a) Sun gear b) Planet shaft c) Planet carrier

d) Cam e) Transplanting arm f) Shifting fork

Fig. 2. 3-D models of transplanting parts

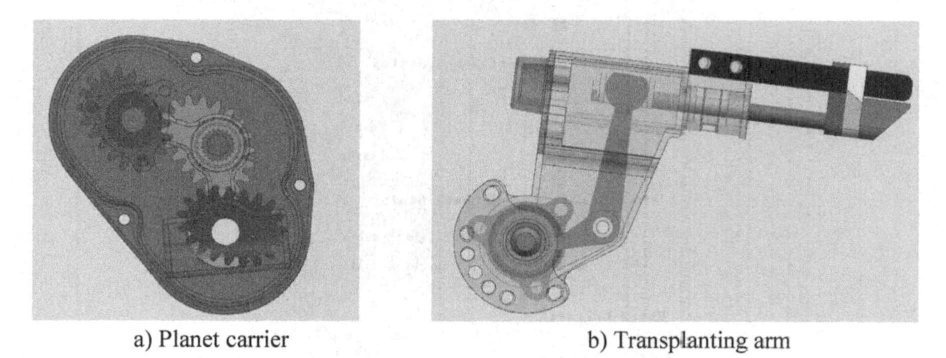

a) Planet carrier b) Transplanting arm

Fig. 3. Sub assembly of transplanting mechanism

3.4 Interference Detection

Based on section of 3.3, each parts and components were related by adding the relevant constraints, but its relationship of mutual cooperation must be checked. In UG the specific methods of interference detection was that, Menu → Analysis → Examine geometry, check geometrical characteristics of each part in the pop-up menu, and then the system would analyze automatically for transplanting mechanism assembly, the interference detection results were obtained and shown in figure 5. It is know that the whole mechanism without interference from the figure 5.

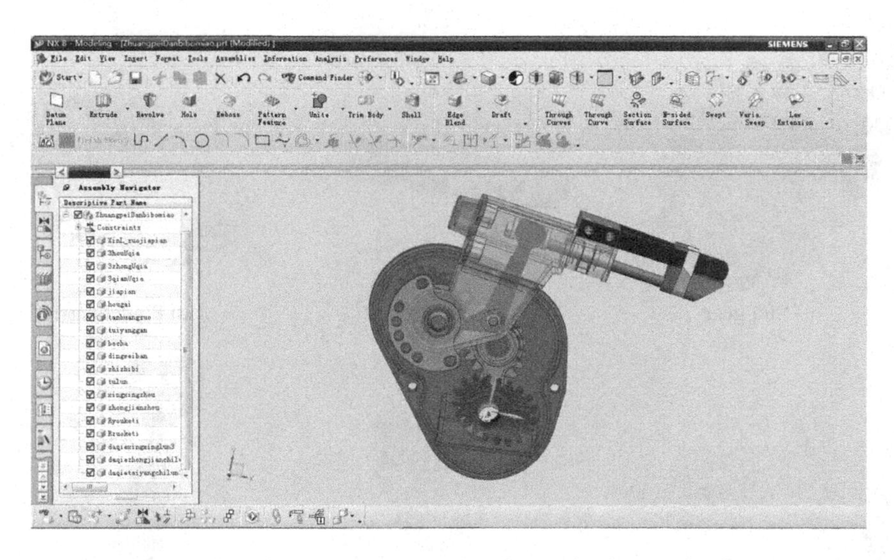

Fig. 4. 3-D virtual prototype model of transplanting mechanism

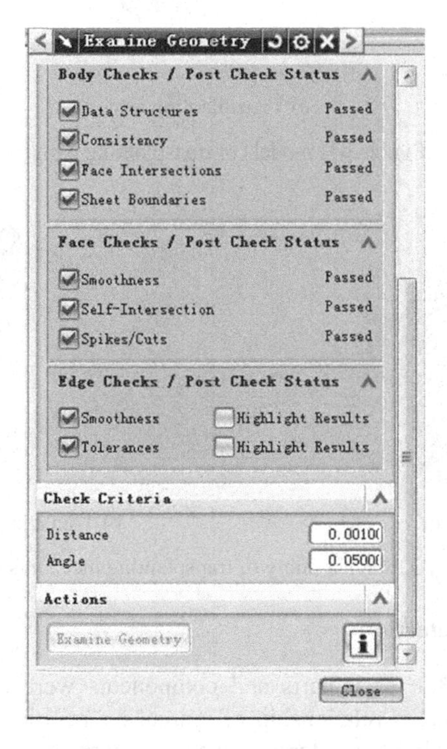

Fig. 5. The interference detection results of virtual prototype

3.5 Generation of 2-D Engineering Drawings

Based on above sections, 2-D engineering drawings were generated from corresponding parts and assemblies in the drawing module, and automatic dimensioning was done in each drawing. 3-D models and 2-D engineering drawings were related with each other, namely any modification of dimensions made in 3-D part and assembly module would be reflected in drawing module and vice versa, which improving the efficiency. The different angle of projection views, local amplification figure, sectional view and so on, would be produced according to needs in the process of generating 2-D engineering drawings. Some necessary annotations were inserted into drawings as these were required for manufacture, such as weld symbol, geometric tolerance, surface finish symbol, BOM (Bill of Material) and so on. The 2-D assembly drawing was obtained after further processing to the figure 4, and shown in figure 6, the 300mm is the rice row spacing.

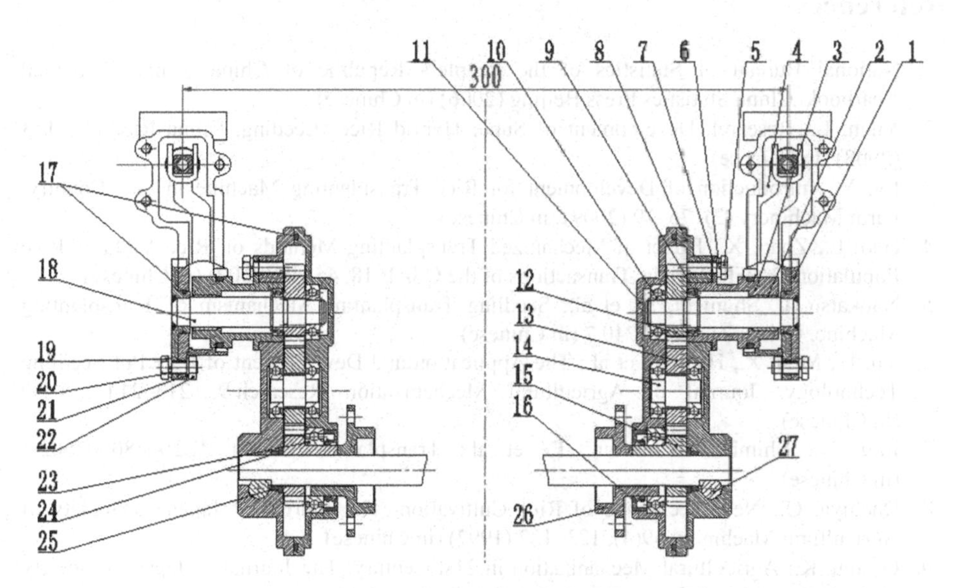

Fig. 6. 2-D assembly drawing of transplanting mechanism 1. Positioning board 2. Cam bushing 3. Cam 4. Transplanting arm 5. Screw 6. Spring washer 7. Packing paper of cam 8. Right shell of planet carrier 9. Circlip 10. Deep groove ball bearing 11. Left shell of planet carrier 12. Planet gear 13. Middle gear 14. Middle shaft 15. Sun gear 16. Skeleton seal ring 17. O seal ring 18. Planet shaft 19. Bolt 20. Elastic washer 21. Flat washer 22. Hex nut 23. Deep groove ball bearing 24. Flat washer 25. O-rubber seal ring 26. Flange 27. Locking pin

4 Conclusions

(1)In allusion to the existing rice pot seedling transplanter exists the defects of complex structure, high cost and low efficiency, the necessity of efficient, light simplified and full-automatic was put forward to study on transplanting mechanism for rice pot seedling.

(2)UG NX 8.0 was applied to accomplish parts design, assemblies design, interference detection and generation of 2-D engineering drawings. Results showed that the design was reasonable and feasible.

(3)The created parts and assemblies will be preparation for further simulation and analysis with softwares of Adams and Ansys, it is provides the reference for the design of rice pot seedling transplanter.

Acknowledgment. The paper is supported by the National Natural Science Fund Projects (Project number is 51175073), the Special Fund for Agro-scientific Research in the Public Interest (Project number is 201203059-01), the Northeast Agricultural University Doctor Startup Fund (Project number is 2012CRB56).

References

1. National Bureau of Statistics of the People's Republic of China. China Statistical Yearbook. China Statistics Press Beijing (2006) (in Chinese)
2. Yuan, L.: Research Development of Super Hybrid Rice Breeding. China Rice (1), 1–3 (2008) (in Chinese)
3. Lv, Y.: Introduction of Development for Rice Transplanting Machine in Our Country. Farm Machinery (2), 76–79 (2009) (in Chinese)
4. Gao, L., Zhao, X.: Effect of Mechanized Transplanting Methods on Rice Yield & Rice Population Growth Trends. Transactions of the CSAE 18, 45–48 (2002) (in Chinese)
5. Naokatsu, I., Shunichi, S., et al.: Seedling Transplanting Mechanism of Transplanting Machine, China: ZL 99118740.7 (in Chinese)
6. Yu, L., Mou, X., Han, X., et al.: The Application and Development of Rice Pot Seedling Technology. Journal of Agricultural Mechanization Research 9, 213–214 (2011) (in Chinese)
7. Itou, N., Shimizu, S., Wata, T., et al.: Transplanter. China: ZL200480007602.4 (in Chinese)
8. Dachiya, O.: New Technique of Rice Cultivation. The Journal of Japanese Soeiety of Agrieultural Maehinery 59(4), 123–127 (1997) (in Chinese)
9. Osamu, K.: Agricultural Mechanization in 21st Century. The Journal of Japanese Soeiety of Agrieultural Maehinery 59(6), 4–9 (1997) (in Chinese)
10. Hu J.: The Research of Auto-transplanter's Picking Seedling Mechanism and Seeding Mechanism[EB/OL]. `http://www.docin.com/p-280100256.html#documentinfo.2011-10-31` (in Chinese)
11. Bing, Y., Chen, Z.: Development and Application of Mechanization Technology for Rice Pot Seedling Transplanting. Agriculture Machinery Technology Extension (4), 52 (2011) (in Chinese)
12. Qin, Y.: Rice Practical Cultivation. Chinese Agricultural Press, Beijing (1984) (in Chinese)
13. Zhao, Y., Xin, L., Feng, J., et al.: Transplanting Mechanism of Pot Seedling Transplanting Machine with Double Cranks. China: ZL 201220206603.3 (in Chinese)
14. Ma, Q.: The Article of UG-CAE. China Machine Press, Beijing (2002) (in Chinese)

The Catchment Water-Based System Health Evaluation Based on the TOPSIS Model

Zhenxiang Xing[1,2,3], Jing Li[1], Meixin Liu[1], Qiang Fu[1,2,3], and Hao Guo[1]

[1] College of Water Conservancy & Civil Engineering,
Northeast Agricultural University, Harbin 150030, China
[2] Collaborative Innovation Center of Grain Production Capacity Improvement
in Heilongjiang Province, Harbin 150030, China
[3] The Key Lab of Agricultural Water Resources Higher-Efficient Utilization
of Ministry of Agriculture of PRC, Harbin 150030, China
zxxneau@hotmail.com,
{lijingneau,liumeixinneau,fuqiangneau,ghneau}@sina.cn

Abstract. The health status of Naolihe Basin has already varied significantly affected by the climate change and human activity. Therefore, it is reasonably much important to analyze and evaluate the health status of river, which could provide basis for the comprehensive development and protection in the Naolihe Basin. So, this paper takes Baoqing country as the study region, uses the stable state, the harmonious degree and evolution rate to describe the health status, and establishes the health evaluation index system of water-based system in Baoqing country based on the concept of water-based system. The entropy method is used to the weight vector of steady state index and harmonious degree index, the TOPSIS was applied to analysis the health status of water-based system of Baoqing country during 2005 to 2009. The comprehensive evaluation result shows that the water-based system of Baoqing country is in the state of comparative healthy, the main reason is that water resources of the study area are relatively insufficient, and drinking water security and water consumption cannot obtain the complete guarantee, which leads to lower steady state.

Keywords: Naolihe Basin, water-based system, health assessment, entropy weight method, TOPSIS.

1 Introduction

Water-based system refers to habitats carrying system that is made up of water and related wading medium and engineering in a certain hydrological scale and spatial dimensions. Space for different hydrological scale, water-based system has different characteristics. The system covers the water, air, soil, biological, social economy and many other aspects, which is a comprehensive on river health, the development and utilization of drainage basin water resources and regional water resources carrying capacity, etc. It is a huge system on real significance [1-2].

At present, the study of water-based system is still in its primary stage, with few related research and application. Zhou Huicheng etc. [3-4] put forward the concept of city water system, which established water system health index of Dalian city. Xiang Biwei etc. [5] established the water-based system of Dongjiang River

© IFIP International Federation for Information Processing 2015
D. Li and Y. Chen (Eds.): CCTA 2014, IFIP AICT 452, pp. 233–239, 2015.
DOI: 10.1007/978-3-319-19620-6_29

Basin and constructed the index system of ecosystem health assessment of the Dongjiang River Basin Water System based on the water-based system. The rough set theory and set pair analysis were applied to the water-based system in Dongjiang River Health Assessment by Huang Guoru etc [6].

BaoQing is located in the northeast of Heilongjiang province, belongs to Shangyashan city. BaoQing's economy keeps developing fast and rich in resources [7]. It is an important commodity grain base. However, economic development have done much harm to the resource environment and threatened the regional Health.

2 Regional Health Assessment Index Based on Water-Based Systems

To consider the influence of the consistency of the system variables, contradictory, and time variability on the system, we need to describe the characteristics of the system in water base system. When we evaluated the health status of water-based systems, we used the three characteristic properties of water-based systems to characterize the state of water-based systems, namely stability, harmony and evolutionary. Meanwhile we used the stable state, the harmonious degree and evolution rate to describe. Because the economy of studied areas was dominated by agriculture, we put forward table 1 and 2 through the study of theories related and the combination of natural and social characteristics of the study area. Table 1 represents the index and standard value of steady state and 2 represents the index and standard value of harmonious degree. The evolution rate represents the degree of change of water-based systems on the time scale to track and characterize the stability and harmony of water-based system. So evolution rate can be divided into steady state evolution rate and the harmonious degree evolution rate. By using the calendar feature values of steady-state and harmony to draw the break point diagrams of steady-state and harmony, we select the slope of the process line as a characteristic value of evolution rate. Index and standard value of evaluation rate are shown in table 3.

Table 1. Index and standard value of steady state

rank	Un	Pd/per˙ $(km^2)^{-1}$	Pswla/ (%)	Pcse/ (%)	Cf kg˙$(km^2)^{-1}$	Cp /t˙$(km^2)^{-1}$	Ga /m^{2}˙ per^{-1}
I	<40	<30	<5	>90	<100	<0.5	>20
II	40~50	30~60	5~10	80~90	100~150	0.5~2	11~20
III	50~60	60~70	10~15	70~80	150~225	2~3	6~11
IV	>60	>70	>15	<70	>225	>3	<6

Note:a, I ~ IV level respectively: very stable, stable, basically stable and unstable;
Un: urbanization progress; Pd: population density (per˙ $(km^2)^{-1}$);
Pswla: proportion of soil and water loss area (%);
Pcse: proportion of controlling soil erosion (%);
Cf: consumption of fertilizer; Cp: consumption of pesticides; Ga: green area.

Table 2. Index and standard value of harmonious degree

rank	Lwcpc /L·(per·d)$^{-1}$	Rufiw /%	Iwd /%	Hwccpp /%	Pwa /%	Prdws /%	Pw /%	Peia /%
I	>300	>60	>90	<30	>10	>90	>50	>90
II	200~300	50~60	80~90	30~40	5~10	80~90	50~30	80~90
III	150~200	40~50	70~80	40~50	3~5	70~80	20~30	60~80
IV	<150	<40	<70	>50	<3	<70	<20	<60

Note: I ~ IV level respectively: very harmonious, harmonious, basic harmonious and disharmony; Lwcpc: life water consumption per capita; Rufiw: repeating utilization factor of industrial water; Hwccpp: high water consumption crop planting proportion; Iwd: industrial water disposal; Pwa: proportion of water area; Prdws: proportion of rural drinking water standards; Pw: proportion of woodland; Peia: proportion of effective irrigation area

Table 3. Index and standard value of evaluation rate

rank	steady state evolution rate	harmonious degree evolution rate
I	<-0.02	<-0.02
II	-0.02~0	-0.02~0
III	0~0.2	0~0.2
IV	>0.2	>0.2

3 Comprehensive Evaluation of the Health of Water-Based Systems

TOPSIS model is a method for multi-objective-decision-making, also known as approaching ideal solution [8~11]. According to nature of the index itself and index data, two contrast regimens are calculated, namely virtual positive ideal solution and virtual negative ideal solution. The formula as follows:

$$Z^*=\left\{(\max Z_{ij} / j \in J), (\min Z_{ij} / i \in J') \quad i=1,2,...,n\right\}=\left\{Z_1^*, Z_2^*, ..., Z_n^*\right\} \tag{1}$$

$$Z^-=\left\{(\max Z_{ij} / j \in J), (\min Z_{ij} / i \in J') \quad i=1,2,...,n\right\}=\left\{Z_1^-, Z_2^-, ..., Z_n^-\right\} \tag{2}$$

where: Z_{ij} is the element of the standardized weighted matrix, J、J' stand for index of the bigger the optimal type and index of the smaller the optimal type.

Calculate positive ideal reference solution S_i^+ and negative ideal reference solution S_i^- of each monitoring site by using the Euclidean distance coefficient.

$$S_i^+=\sqrt{\sum_{i=1}^{n}(Z_{ij}-Z_j^*)^2}\,(i=1,2,...,n) \tag{3}$$

$$S_i^-=\sqrt{\sum_{i=1}^{n}(Z_{ij}-Z_j^-)^2}\,(i=1,2,...,n) \tag{4}$$

Once the distance from each monitoring site was determined, the relative closeness coefficient C_i^* of each monitoring site can be computed by equation (5)

$$C_i^* = S_i^- / (S_i^* + S_i^-), 0 \le C_i^* \le 1 \quad (i = 1,2,...,n) \tag{5}$$

If an index value of the monitoring site is same with the positive ideal reference solution, then $C_i^*=1$; If an index value of the monitoring sites is same with the negative ideal reference solution, then $C_i^*=0$.

TOPSIS method was used for water-based system health evaluation. First of all, we confirm the characteristic values of stable state and the harmonious degree of water-based systems. Then we obtain the evolution rate by the characteristic values. In the end, the stable state, the harmonious degree and evolution rate are assessed comprehensively.

3.1 Determine the Index Weight

There are currently a lot of methods to confirm index weight in the multi-index comprehensive evaluation, such as the Delphi method and Analytic Hierarchy Process (AHP), Mean square deviation method, Membership frequency method and Principal Component Analysis (PCA) [12]. Entropy method can obtain the weight by mining data own information and eliminate the influence of subjective judgment of the index weight [13-14], and make the index weight results more objective. The index of steady state and index of harmonious degree are calculated by the Entropy method.

$$\alpha_{稳} = [0.1266, 0.1171, 0.1059, 0.1449, 0.1076, 0.1303, 0.1300, 0.1376]$$

$$\beta_{和} = [0.1218, 0.1219, 0.0916, 0.1218, 0.1225, 0.1025, 0.1025, 0.0927, 0.1226]$$

3.2 Steady State and Harmonious Degree Evaluation

3.2.1 Determine the Characteristic Value of Steady State Rank and Harmony Degree Rank

We select the standard values to establish matrix in table 1 and 2. Though the normalized processing and weight vector, weighted normalized matrix is obtained. Finally we calculated the Euclid approach degree of each level, namely, the standard value of steady state rank and harmony degree rank, shown in table 4.

3.2.2 Calculate the Characteristic Value of Steady State and Harmony Degree

Characteristic matrix is composed of the index value and standard values of steady state and harmony degree of study area. By taking the characteristic matrix to the TOPSIS model, we calculate the approach degree to obtain the characteristic value of steady state and harmony degree in Baoqing during 2005-2009. Then by comparing the characteristic value of steady state rank and harmony degree rank, we obtain steady state rank and harmony degree rank, shown in table 5.

Table 4. Grade standards of steady state and harmonious degree

steady state rank	standard deviation	harmony degree rank	standard deviation
Ⅰ	>0.66	Ⅰ	>0.63
Ⅱ	0.48~0.66	Ⅱ	0.36~0.63
Ⅲ	0.24~0.48	Ⅲ	0.18~0.36
Ⅳ	<0.24	Ⅳ	<0.18

Table 5. Grades of steady state and harmonious degree

years	characteristic value	steady state rank	years	characteristic value	harmony degree rank
2005	0.48	Ⅲ	2005	0.55	Ⅱ
2006	0.45	Ⅲ	2006	0.50	Ⅱ
2007	0.49	Ⅲ	2007	0.42	Ⅲ
2008	0.49	Ⅲ	2008	0.48	Ⅲ
2009	0.41	Ⅲ	2009	0.52	Ⅱ

3.3 Assessment of Evolution Rate

According to the characteristic value of stable state and the harmony degree in table 5, we calculate the evolution rate of steady state and harmony degree. Combining the table 3, we obtain the evolution rate of water-based systems during 2006-2009, shown in table 6.

3.4 Comprehensive Assessment of the Health of Water-Based Systems

Based on the analyzing the stable state, the harmonious degree and evolution rate respectively, we can a comprehensive evaluation for the health of water-based systems. According to the actual situation of the study area, we assigned the weight of stable state, evolution rate of steady state, harmony degree and evolution rate of harmony degree. Finally, the weight was 0.4, 0.4, 0.1, 0.1. In addition, due to the smaller numerical of the evolution rate, we cannot directly combine the evolution rate with stable state and harmony degree. Therefore, we need to normalized calculation for evolution rate, and then make a comprehensive evaluation. If the value ≥ 0.62, system is in healthy state; If the value0.42~0.62, system is in healthier state. The evaluation results listed in table 7.

Table 6. Grades of evaluation rate

years	2006	2007	2008	2009
evolution rate of steady state	-0.01	0.03	-0.003	-0.08
rank	Ⅳ	Ⅲ	Ⅳ	Ⅳ
evolution rate of harmony degree	-0.05	0.08	0.05	0.04
rank	Ⅳ	Ⅲ	Ⅲ	Ⅲ

Table 7. Assessment result of Baoqing water-based system

years	2005	2006	2007	2008	2009
characteristic value	0.514	0.472	0.466	0.558	0.462
rank	healthier	healthier	healthier	healthier	healthier

3.5 The Evaluation Results Analysis

We can see from Table 5, the steady state of Baoqing was in grade Ⅲ. The main reason lies in the study area less water, leading to safe drinking water and water consumption cannot get effective guarantee. Therefore, when efforts to increase water resources development, it is easy to cause instability of water-based system. The level of harmony degree is superior to the steady state level, in the majority with Ⅱ level. This is mainly because the study area is mainly dominated by agriculture, while the study about renovation and destruction of water-based systems is relatively less. Comprehensive health state of water-based systems was in a relatively healthy state and has also been picking up in 2007.

4 Conclusions

In this paper, a water-based system evaluation model is introduced to catchment health assessment, and the health evaluation index system of water-based system was established according to the actual situation of Baoqing country of Naoli river basin, and then the TOPSIS model was used to evaluate catchment health status. Through the study, several conclusions were drawn as follow:

(1)The health status of water-based system of Baoqing country was relatively lower, the main reason was that water resources of the study area are relatively insufficient, and drinking water security and water consumption cannot be guaranteed completely, which lead to the lower steady state.

(2)In order to improve the health status of water-based system of Baoqing country, relies on improving the rate of water resources should be improved, increasing the irrigation area should be developed, and the awareness of savings water should be raised, and so on.

(3)The evaluation index system for Baoqing was not mature completely due to the limitation of data and research methodology, and certain limit of determination of evaluation indexes cause agricultural characteristics of Baoqing. A further research should be considered with more data of industrial and environmental index.

Acknowledgment. Funds for this research was provided by the national natural science foundation (51109036; 51179032), the specialized research fund for the doctoral program of higher university of the Ministry of Education (20112325120009) the special scientific research funds for Ministry of water resources public welfare industry (201301096), The leader talent echelon back-up headman funding projects of Heilongjiang province (500001), the studying funds of postdoctoral in Heilongjiang province (LBH-Q12147), the science and technology project of Heilongjiang Provincial Education Department(11541022).

References

1. Liu, N.: Study on the concept, connotation and evolvement of the base-system of water. Advances in Water Science 16(4), 475–481 (2005) (in Chinese)
2. Liu, N., Du, G.: Integrating the hydrological technology for analyzing the base-system of water. Advances in Water Science 16(5), 696–699 (2005) (in Chinese)
3. Zhou, H., Cong, F.: Multi-objective fuzzy assessment of health of Dalian water-based system. Journal of Dalian University of Technology 48(5), 720–725 (2008) (in Chinese)
4. Zhou, H., Cong, F.: Evaluation of Dalian water-based system based on index weight reliability analysis. Journal of Liaoning Technical University (Natural Science) 27(5), 770–773 (2008) (in Chinese)
5. Xiang, B., Huang, G., Feng, J.: Fuzzy Comprehensive Assessment of Water-based System Health for Dongjiang Basin Based on AHO Method. Water Resources and Power 29(10), 1–4 (2011) (in Chinese)
6. Huang, G., Wu, C., Xiang, B.: Health state assessment of water-based system for Dongjiang basin based on rough set theory and set pair analysis. Systems Engineering-Theory & Practice 33(1), 1–7 (2013) (in Chinese)
7. Kuang, W., Zhang, S., Hou, W., et al.: Baoqing County of Sanjiang Plain Land Use Change TuPu Analysis. Journal of the Graduate School of the Chinese Academy of Sciences 23(2), 242–250 (2006) (in Chinese)
8. Pan, N., Zhou, S.: Improved Entropy Weight-based TOPSIS Model and Its Application. Yunnan Water Power 23(5), 8–12 (2007) (in Chinese)
9. Hu, Y.: The Improved Method for TOPSIS in Comprehensive Evaluation. Mathematics In Practice and Theory 32(4), 572–575 (2002) (in Chinese)
10. Li, X., Liu, Z., Peng, Q.: Improved Algorithm of TOPSIS Model and Its Application in River Health Assessment. Journal of Sichuan University (Engineering Science Edition) 43(2), 14–20 (2011) (in Chinese)
11. Wang, T.C., Chang, T.H.: Application of TOPSIS in evaluating initial training aircraft under a fuzzy environment. Expert Systems with Applications 33, 870–880 (2007)
12. Ma, Y.: The fuzzy synthetic evaluation of water resource sustainable development and utilization in Xi'an city based on AHP. Chang' an University, Xi'an (2008) (in Chinese)
13. Hasan, M.M., Dunn, P.K.: Entropy, consistency in rainfall distribution and potential water resource availability in Australia. Hydrological Processes 25, 2613–2622 (2011)
14. Wang, Y.: Matter-element Model Based on Coefficients of Entropy for Comprehensive Evaluation of Irrigation Water Quality. Journal of Irrigation and Drainage 28(3), 73–76 (2009) (in Chinese)

Efficiency Evaluation of Agricultural Informatization Based on CCR and Super-Efficiency DEA Model

Xu Han, Li Wang, Hui Wang, and Shuqin Wang

School of Information Engineering,
Minzu University of China, Beijing 100081, China
hanxu@muc.edu.cn, wlry8812@sohu.com,
{wanghui_610919,cd-wx}@163.com

Abstract. In this research, we want to evaluate the efficiency of input / output in agricultural informatization (AI) and the redundancy of AI by using DEA method. An index system evaluating the input and output in agricultural informatization was built with the support of CCR Model and Super-efficiency DEA Model, which contains 9 indices. We processed the data of agricultural informatization in Huaihua and Xiangxi, two areas in Hunan province, with DEMP and EMS software and analyzed the efficiency of agricultural informatization in different 5 years and figure out the tendency of developing status in AI from 2009 to 2013. The results show that the CRSTE, VRSTE, SCALE efficiency of AI in two areas is efficient and the inputs and outputs about AI in 2009 and 2011 have spaces of improving. In general, the developing tendency of AI efficiency in two areas is stable in these years even though they are not developed areas in agricultural informatization. The index system of AI evaluation in this research could be better and reliable in future if we get more data and add more indices in it, but it is hard for us now to get more data because of the limitation of incomplete statistical data released by local governments.

Keywords: DEA, CCR, efficiency evaluation, informatization.

1 Introduction

With the development of information communication technologies (ICTs), the trend of informatization is spreading fast on the world. Agricultural informatization is an important component in our information society, and it is the inevitable outcome that modern information technology combines with agricultural interior demand.

Everett Rogers (2000) defines informatization as the process through which new communication technologies are used as means for furthering development as a nation becomes more and more an information society [1]. As a description of the development, informatization refers to the extent of a geographical area, economy or society continues to develop based information and information communication technologies, in other words, it refers to the degree of workforce size enhanced based on the information and information communication technologies [2]. Agricultural Informtization refers to the information communication technologies (ICTs) are comprehensively applied in the fields of agriculture, and penetrating into agricultural production, marketing, consumption, and rural social, economic, technological, and other specific aspects of the whole process [3].

© IFIP International Federation for Information Processing 2015
D. Li and Y. Chen (Eds.): CCTA 2014, IFIP AICT 452, pp. 240–246, 2015.
DOI: 10.1007/978-3-319-19620-6_30

The efficiency evaluation of AI in this research refers to the production inputs benefit which is achieved by calculating indices about the ICTs supporting agriculture and the output in agriculture. The indices needed in evaluation are different from other researches because of the different research aims.

Methods in informatizaiton measurement and evaluation abroad [4] usually are Machlup method, Porat method, Informatization Index method. The methods above aims to mainly figure out the informatization level of certain society or economy unit, however they do not focus on the efficiency and can't tell whether it is reasonable in input of the indices related with the AI.

Methods in informatizaiton measurement and evaluation in China are Delphi method, Analytic Hierarchy Process (AHP) method, The method for the measurement of informatization level used in China, raised by National Information Center, combines Machlup method and Porat method, and includes 6 factors and 21 indices. [5] These methods above applied in domestic rural or agricultural informatization research mostly play extra effort on measurement of AI and service ability.

Comparing the researches on evaluating the level of AI, development status of AI and service ability of AI, the research on the efficiency of AI is rare.

Data Envelopment Analysis (DEA) method is seldom used in researches on evaluation of AI, but Machlup Method, Porat Method, AHP method, and Factor Analysis method are common methods in such researches. DEA method needs lesser indices than other methods discussed above to evaluate the efficiency of AI and overcome the shortage of data collecting difficulty.

2 Methodology

2.1 CCR Model

Data envelopment analysis (DEA) is a nonparametric method in operations research and economics for the estimation of production frontiers [8]. It is used to empirically measure productive efficiency of decision making units (or DMUs). Non-parametric approaches have the benefit of not assuming a particular functional form/shape for the frontier, however they do not provide a general relationship (equation) relating output and input [9].

CCR Model is one basic DEA model. CCR Model is proposed by Charnes, Cooper and Rhodes in 1978.

$$
(D1)\begin{cases}
min\theta = V_{D_1} \\
s.t. \\
\sum_{k=1}^{n} X_k\lambda_k + s^- = \theta X_t \\
\sum_{k=1}^{n} X_k\lambda_k - s^+ = Y_t \\
\sum_{k=1}^{n} X_k\lambda_k - s^+ = Y_t \\
s^+ \geq 0, s^- \geq 0
\end{cases}
$$

In D1, set $X_l = (x_{1l}, x_{2l}, \cdots, x_{ml}), Y_l = (y_{1l}, y_{2l}, \cdots, y_{sl})$.
Description of CCR Efficiency Model:

1) Suppose there are *n DMUs: DMU₁, DMU₂,..., and DMUₙ*.
2) Suppose m input items and s output items are selected with the properties.
3) Let the input and output data for DMU_j be $\{x_{1j}, x_{2j}, \cdots, x_{mj})$and $(y_{1j}, y_{2j}, \cdots, y_{sj})$
4) The calculation of total efficiency of DMU_l can be changed to linear programming problem in next slide.
5) Let $Xl=(x_{1l}, x_{2l}, \cdots, x_{ml})$, $Yl= (y_{1l}, y_{2l}, \cdots, y_{sl})$
6) We assume the return scale in CCR Model is not changed.
7) θ is the efficiency rate ,θ is at most 1($0 \leq \theta \leq 1$)

Economy Meaning of θ: when the output Y is alternative by k DMUs output with a linear combinations, θ is the compressibility of its input X, θ is also known as the efficiency rate value.

1) *If* θ=1, the DMU being examined is the point of being efficient frontier surface, and it is efficient status.
2) *If* θ<1, the DMU being examined is invalid status, 1-θ is the more input of the DMU being examined than efficient status.

DEA efficient economic meaning: output can't be any increased unless one or more inputs to increase or reduce other types of output. Under the same condition, input can't be any decreased.

2.2 Super-efficiency DEA Model

When there are many DMUs which are efficient and their efficiency of evaluation is 1(θ=1), CCR model can't tell which is better among these DMUs. Anersen (1993) proposed the super-efficiency DEA model to sort efficient DMUs and give them ranking.

$$(D_2)\begin{cases} min\theta = V_{D_2} \\ s.t. \\ \sum_{\substack{k=1 \\ k\neq 1}}^{n} X_k\lambda_k + s^- = \theta X_t \\ \sum_{\substack{k=1 \\ k\neq 1}}^{n} X_k\lambda_k - s^+ = Y_t \\ \lambda_k \geq 0, k = 1,2,\cdots,n \\ s^+ \geq 0, s^- \geq 0 \end{cases}$$

2.3 Enovation in Super-efficiency DEA Model Compare with CCR Model

Super-efficiency DEA Model: in the process of evaluating DMU_l efficiency, the inputs and outputs of DMU_l are alternated by the linear combination of inputs and outputs of the other DMUs excluding themselves.

CCR model: in the process of evaluating DMU_l efficiency, all inputs and outputs, including the inputs and outputs of themselves as part of the calculation.

An efficient decision-making unit can make it into a pro-rata increase, while the efficiency of the value does not change, the proportion of its investment increased is the super-efficiency evaluation value.

3 Data

Using the DEA and Super-efficient DEA model to evaluate the inputs and outputs of agricultural informatization (AI) in Huaihua and Xiangxi based on the data from 2009 to 2013 collected from statistical bulletin of national economic and social development in Hunan Province, We can analysis the data of inputs and outputs of agricultural informatization in two areas, and find out the aspects which can be improved in AI investment, also we can figure out the tendency of AI in two areas.

We built an index system containing 9 indices in concern of the feasibility and accuracy, and the data is collected from statistical bulletin of national economic and social development published by Statistical Bureau of Hunan Province.

In the index system, Y1 means "Added Value of Agriculture(Billion Yuan)", Y2 means "Farmers' Per Capita Income(Yuan)", X1 means "Main Business Volume of Post(Billion Yuan)", X2 means "Electricity for Rural Use(Million kWh)", X3 means "Rural Telephone Subscribers(Million)", X4 means "Cell phone Subscribers(Million)", X5 means "Rural TV Coverage(%)", X6 means "Rural Radio Coverage (%)", X7 means "Internet Accounts(Million)".

Data about invest and output of AI in Huaihua and Xiangxi is listed in table 1:

Table 1. Data about invest and output of Agricultural Informatization in Huaihua and Xiangxi

City	Year	Y1	Y2	X1	X2	X3	X4	X5	X6	X7
Huaihua	2009	4.602	2905	4.246	0.184	0.808	1.6246	97.1%	91.5%	0.145
	2010	5.291	3520	5.833	0.211	0.7217	2.048	97.5%	91.5%	0.2144
	2011	5.52	3573	2.511	0.228	0.602	2.335	98.0%	92.8%	0.242
	2012	8.16	4701	2.743	0.3127	0.575	2.6395	98.5%	93.1%	0.3215
	2013	8.3	5849	3.001	0.3441	0.4834	3.0664	98.5%	93.3%	0.3604
Xiangxi	2009	6.23	2858	2.797	0.2492	0.3178	0.9856	92.0%	89.0%	0.0993
	2010	6.51	3173	3.421	0.2604	0.2882	1.1153	93.8%	72.9%	0.138
	2011	6.934	3674	1.382	0.27736	0.272	1.0114	94.5%	73.5%	0.1702
	2012	7.22	4229	1.525	0.2888	0.2537	1.229	95.5%	87.0%	0.1996
	2013	7.45	5260	1.64	0.298	0.24	1.293	95.7%	97.0%	0.231

4 Empirical Results and Discussion

The data was processed by DEAP (Data Envelopment Analysis Program, Version 2.1) using CCR model, the results are listed in table2. And we also used EMS software (Efficiency Measurement System, Version 1.3) using to process the data with Super-efficiency CCR model, the results are listed in table3.

Table 2. DEA value of AI in Huaihua and Xiangxi (2009-2013)

Year	CRSTE		VRSTE		SCALE		SCALE MERIT	
	Huaihua	Xiangxi	Huaihua	Xiangxi	Huaihua	Xiangxi	Huaihua	Xiangxi
2009	0.999	1.000	1.000	1.000	0.999	1.000	increase	-
2010	1.000	1.000	1.000	1.000	1.000	1.000	-	-
2011	0.968	1.000	1.000	1.000	0.968	1.000	increase	-
2012	1.000	1.000	1.000	1.000	1.000	1.000	-	-
2013	1.000	1.000	1.000	1.000	1.000	1.000	-	-
Average	0.993	1.000	1.000	1.000	0.993	1.000		

In table 2, CRSTE means “technical efficiency from CRS DEA”, VRSTE means “technical efficiency from VRS DEA”, SCALE means “scale efficiency = CRSTE / VRSTE”.

Table 3. Super-efficiency DEA value of AI in Huaihua and Xiangxi (2009-2013)

Year	Huaihua	Rank	Xiangxi	Rank
2009	128.58%	2	133.00%	1
2010	101.97%	4	104.12%	5
2011	99.76%	5	116.01%	3
2012	114.02%	3	107.65%	4
2013	203.68%	1	131.48%	2

Based on the data in table 3, we can figure out the tendency of agricultural informatization in Huaihua and Xiangxi areas.

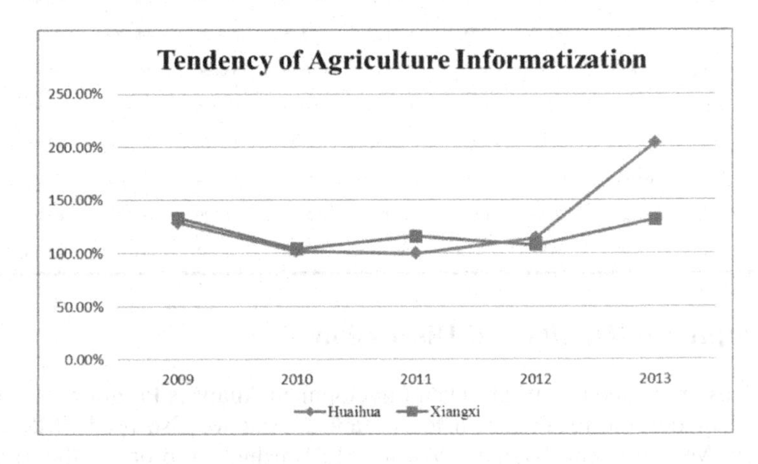

Fig. 1. Tendency of AI in Huaihua and Xiangxi

4.1 Results Analysis in Table 2

1) The results of CRSTE, VRSTE and SCALE are 1 in 2010, 2012 and 2013, which means the inputs and outputs about AI in these years are reasonable, and they are in efficient status.
2) The CRTES values of Huaihua in 2009 and 2011 are below 1, but greater than 0.9, which means the inputs and outputs about AI in these years have spaces of improving.
3) The scale merit of Huaihua increased in 2009 and 2011, which means the technical efficiency from CRS DEA is increased by expand the scale of AI investment.

4.2 Results Analysis in Table 3 and Figure 1

1) In 2011, it is inefficient considering the inputs and outputs in AI in Huaihua.
2) From 2010 to 2013, the efficiency of AI in Huaihua is gradually increased, but the efficiency in Xiangxi is relatively gentle and does not changed so much.
3) The efficiency from 2009 to 2010 in both areas is decreased, but the reasons are different. The increase of electricity for rural use leads to the decrease of efficiency in Huaihua, however, the decrease of rural radio coverage leads to the decrease of efficiency in Xiangxi.

4.3 Discussion

The efficient status (θ=1) does not means developed status or developing rapidly of AI, but only tell us that considering the output in those years, the inputs is not redundant.

Even the efficiency status of AI in Xiangxi is efficient, the developed degree in this area is lower than Huaihua, which can be seen from the data in table 1.

If we want to describe the development status of AI in undeveloped areas, more indices should been take into account in statistical bulletin.

5 Conclusions

The CCR model and Super-efficiency CCR Model are suitable to describe whether there any redundancy comparing inputs with outputs in AI, but can't figure out the development level of AI.

The index system can be modified better if more indices data is collected, and describe the efficiency of inputs and outputs comprehensively.

Even the efficiency status of AI in the two areas is efficient, the developing level of AI is still low, especially in the popularity rate of Internet.

Acknowledgment. Fund for this paper was provided by MOE (Ministry of Education in China) Project of Humanities and Social Sciences (Project No. 12YJA630123). Research in this paper is also supported by School of Information Engineering, Minzu University of China, Beijing.

References

1. Rogers, E.M.: Informatization, globalization, and privatization in the new millennium. The Asian Journal of Communication 10(2), 71–92 (2000)
2. http://wiki.mbalib.com/wiki (browsed in May10, 2014)
3. Zhang, L., et al.: Evaluation of the rural informatization level in four Chinese regions- A methodology based on catastrophe theory. Mathematical and Computer Modeling 58, 868–876 (2013)
4. Liu, Y., Ge, J.: Informatization Measurement Theories Comparative Study in Domestic and Foreign. Information Studies: Theory & Application 2, 144–147 (2004) (in Chinese)
5. Wei, Z., Yang, Z.: Study on Rural Information Undertaking Public Satisfaction Index Model. Information Science 2, 278–283 (2008) (in Chinese)
6. Chen, L., Ye, J.: Analysis on the Agricultural Expenditure of County Finance in the Transformation of Agricultural Modernization. Journal of Jilin Agricultural University (2), 230–236 (2012)
7. Liang, C.: The construction of Agricultural Information Service Performance Evaluation System. Library Theory and Practice 9, 31–35 (2012) (in Chinese)
8. Charnes, A., Cooper, W., Rhodes, E.: Measuring the Efficiency of Decision-making Units. European Journal of Operational Research 2, 429–444 (1978)
9. Aristovnik, A., et al.: Relative efficiency of police directorates in Slovenia: A non-parametric analysis. Expert Systems with Applications: An International Journal 40(2), 820–827 (2013)

A Portable Impedance Detector of Interdigitated Array Microelectrode for Rapid Detection of Avian Influenza Virus

Xiaohong Wang[1], Zhuo Zhao[2], Yuhe Wang[2], and Jianhan Lin[1,*]

[1] MOA Key Laboratory of Agricultural information acquisition technology (Beijing), China Agricultural University, Beijing 100083, China

[2] Modern Precision Agriculture System Integration Research Key Laboratory of Ministry of Education, China Agricultural University, Beijing 100083, China

wxh16@163.com.cn, {1992zhaozhuo,cauyuhe}@gmail.com, jianhan@cau.edu.cn

Abstract. Impedance biosensors are featured with fast detection, easy operation and low-cost, and have been reported in the detection of avian influenza viruses, foodborne pathogens and pesticide residues. Based on our previous research on impedance biosensors for rapid detection of avian influenza virus, a portable impedance detector was redesigned using an S3C2440AL ARM9 microprocessor and an improved AD5933 impedance converter to meet the higher requirements on frequency and magnitude in impedance measurement from the new interdigitated array microelectrodes, which were redesigned for reducing the cost and improving the produce quality. The impedance measurement range is 50 Ω-1MΩ and the frequency response range is 100 Hz-100 kHz. Compared to commercial E4980A precision LCR meter on the measurements of solid-state resistor, solid-state equivalent circuit of the electrode and avian influenza virus test, this impedance detector showed a relative error of less than 5% at the characteristic frequency of 100 Hz and a standard deviation of less than 5% in parallel tests. Besides, a linear relationship between impedance change ΔZ and the concentration of avian influenza virus ranging from 2^{-1} HAU/ 50 μl to 2^{4} HAU/ 50 μl was found.

Keywords: impedance analysis, biosensor, interdigitated array microelectrode, avian influenza detection.

1 Introduction

Avian influenza (AI) is an infectious disease of birds caused by influenza A virus. According to WHO statistics, animals in 62 countries and people in 15 countries have been infected by AI H5N1 virus since 2003, and the mortality rate of human is nearly 60% [1]. Highly pathogenic AI H5N1 has caused huge economic losses and has become a serious threat to public health and safety.

* Corresponding author.

© IFIP International Federation for Information Processing 2015
D. Li and Y. Chen (Eds.): CCTA 2014, IFIP AICT 452, pp. 247–256, 2015.
DOI: 10.1007/978-3-319-19620-6_31

Rapid screening of suspected cases is quite critical to preventing and controlling the spread of AI H5N1 virus. Conventional methods for detection of H5N1 virus mainly include viral isolation culture and RT-PCR. However, these methods are not suitable for in-field screening of AI H5N1 due to their time-consuming procedures or complex sample pretreatment and requirement on specialized facilities. As an alternative, impedance biosensors have shown their potentials to offer a rapid, simple and sensitive detection [2] and have been reported in literatures for detection of biological and chemical targets, including foodborne pathogens [3-6], pesticide residues [7-8] and AI virus [9-10]. Interdigitated array (IDA) microelectrodes are often used in the fields of electrochemical analysis [11-13] with advantages of high sensitivity, low detection limit and high signal to noise ratio [14-17] and has been reported in the studies on detection of AI virus and foodborne pathogens [18-23].

In our previous studies, an impedance biosensor was developed with advantages over conventional methods in detection of avian influenza, such as high specificity, high sensitivity and fast detection [9-10]. This impedance biosensor used the antibodies against AI virus immobilized onto the IDA microelectrode to capture the viruses and this resulted in an impedance increase of the electrode because the viruses blocked the electron transfer between the electrode and the electroactive probes. The impedance change was measured using a commercial impedance analyzer and analyzed to determine the concentration or presence of AI viruses. Meanwhile, an impedance detector based on an 89C51 single-chip microprocessor was developed for in-field measurement of impedance change of the IDA microelectrode with the measuring frequency of 10 kHz and the impedance measurement resolution of 1kΩ [24]. However, the IDA electrode was recently redesigned for the purpose of reducing the cost and improving the quality. The characteristic response frequency was found to shift to 50-100 Hz from original 1-10 kHz and the impedance values of the electrode were measured with smaller lower-limit-amplitude of ~100 Ω from original ~1 kΩ. Both the characteristic response frequency and the lower-limit-amplitude were beyond the range of the previously-developed impedance detector. Therefore, this study intended to redesign and develop a portable impedance detector to meet the higher requirements on frequency and amplitude in impedance measurement from the newly-redesigned IDA microelectrode and evaluated using solid-state circuits and AI virus.

2 Development of Impedance Detector

2.1 Hardware Redesign

Since ARM microprocessors generally have obvious advantages over 51 single-chip microprocessors, such as faster processing speed and more peripheral resources, the impedance detector was redesigned using an S3C2440AL ARM9 microprocessor instead of original 89C51 single-chip microcontroller and an improved AD5933 impedance converter with an external clock and a signal conditioning circuit. As shown in Fig.1, the hardware includes: (1) an impedance measurement module to measure the impedance amplitude and phase angle of the electrode, (2) a power-supply module using an AMS1117-3.3 low dropout linear regulator to provide with accurate 3.3 voltage from 5 VDC power adaptor, (3) a data storage module using a K9F2G08UXA flash memory to save the testing data, (4) a serial communication module using an SP3232EEN chip for program debugging, and (5) a USB communication module to send the testing data to computer, and (6) a WXCAT35-TG3 TFT-LCD touch screen to display the data and operate the detector.

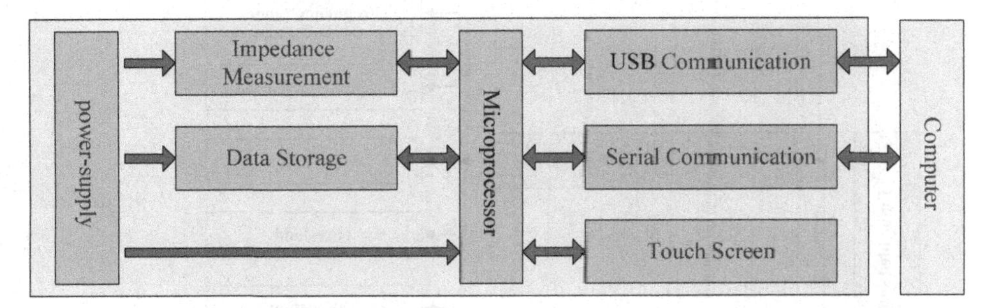

Fig. 1. Schematic diagram of the hardware structure of the impedance detector

Impedance measurement is vital for the development of the impedance detector. The impedance measurement module was redesigned based on our previous prototype with an AD5933 impedance converter, a high-pass filter, a voltage follower, a current-voltage converter and an impedance measuring range adjustor.

The AD5933 impedance converter obtained from Analog Device used Discrete Fourier Transform (DFT) to convert the data collected in the time domain into the frequency domain and obtained the real part (*R*) and the imaginary part (*I*), which could be accessed through the I^2C interface. However, its frequency response range was 1-100 kHz, which didn't cover the characteristic frequency of the newly redesigned electrode, and impedance measuring range was 1 kΩ-1 MΩ, which didn't cover the impedance amplitude of the electrode for detection of AI virus. Thus, an external clock with a frequency of 125 kHz generated by an internal timer of the microprocessor was employed and provided to the AD5933 converter.

Since the excitation signal with a peak-to-peak voltage of 0.198V from AD5933 applied on the electrode had a DC bias of 0.173 V, which didn't match that of the signal receiver in AD5933 with a DC bias of 1.650 V, the electrode might be polarized and the impedance measuring accuracy was greatly affected due to the saturation of analog-digital converter in AD5933. Thus, the excitation signal was sequentially processed by a first order high-pass filter with a cut-off frequency of 1 Hz, a voltage follower using 1/2 AD8606 to make the DC bias of the excitation signal 1.650 V, and a current-voltage converter using 1/2 AD8606 to convert the current passing the electrode into voltage. Besides, an impedance measuring range adjustor using ADG811 was used to automatically adjust the feedback resistor (four optional: 100 Ω, 1 kΩ, 10 kΩ and 100 kΩ) to make the ratio of the feedback resistor to the electrode between 0.1 and 1.0, indicating that the analog-digital converter in AD5933 was working in linear range.

2.2 Embedded Software Development

The embedded software of the impedance detector was developed using Linux C for driver programming and QT4 for application programming under Linux environment. As shown in Fig.2, the embedded software mainly included three dialog boxes for system setting, impedance measurement and analysis, and data display and storage.

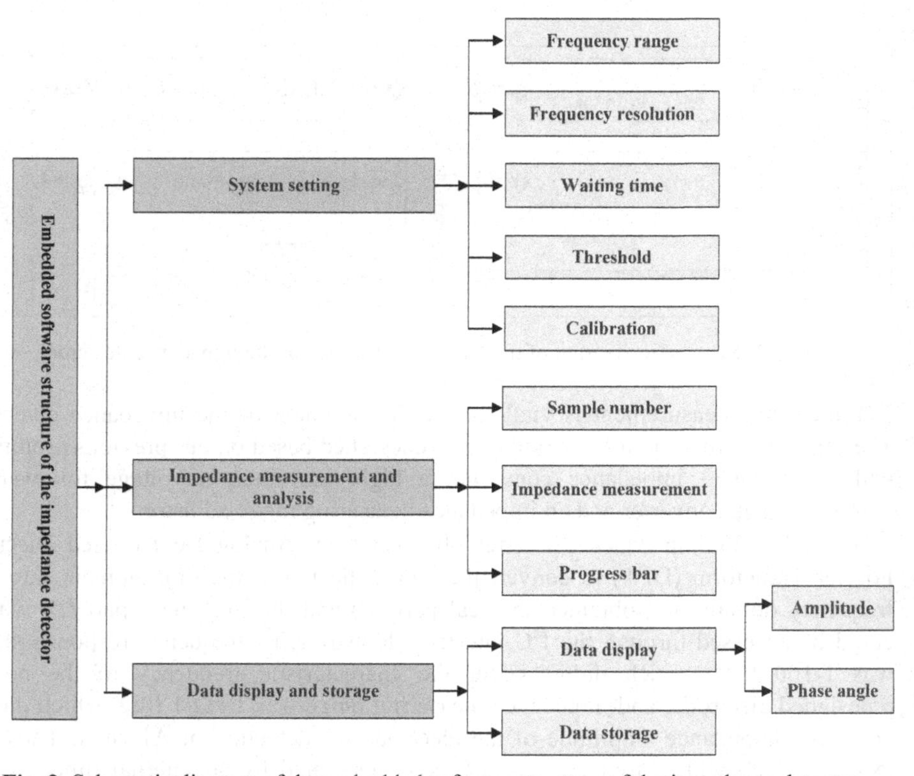

Fig. 2. Schematic diagram of the embedded software structure of the impedance detector

In the dialog box for system setting, system parameters, such as the scanning frequency range and resolution for impedance measurement, the threshold for determining the presence of AI virus, and the waiting time for AI virus incubation, could be set using QcomboBox and QspinBox classes. Besides, system calibration was achieved with the use of standard resistors.

In the dialog box for impedance measurement and analysis, the number of testing samples was automatically set and increased by one after the sample was done. Also, a progress bar using QprogressBar was used to display the progress of the measurement. As shown in Fig. 3, the impedance data were obtained by the following procedures: (1) initialize AD5933, (2) set the parameters in the dialog box for system setting, (3) collect impedance data including real value (*R*) and imaginary value (*I*) and calculate impedance amplitude (*M*) and phase angle (*A*), (4) judge data validity,

(5) save valid data and repeat seven measurements, (6) bubble sort the seven valid impedance data, (7) average the three middle impedance data as the result (M_A) for impedance measurement.

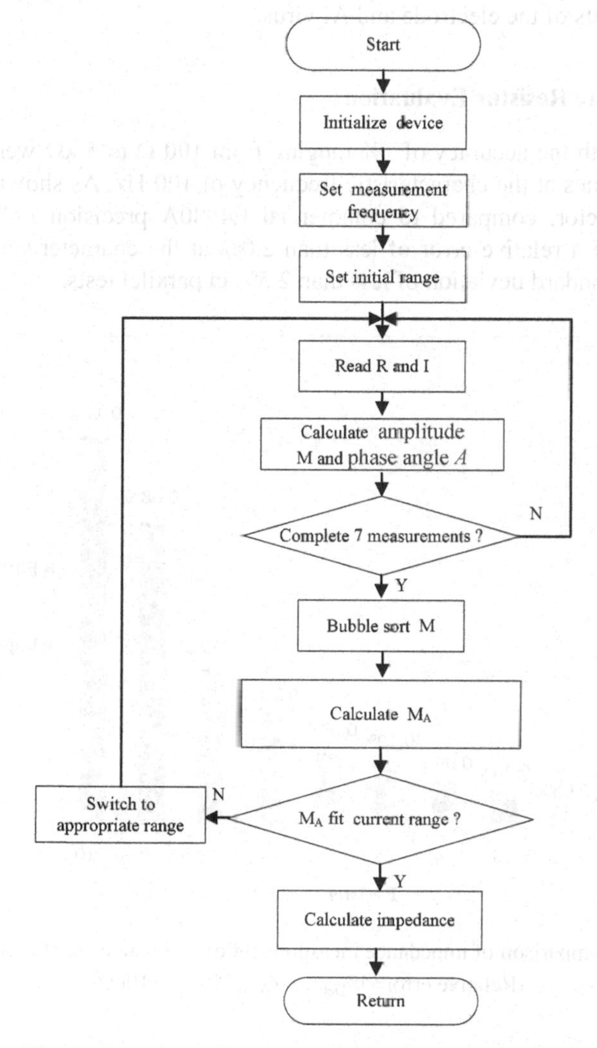

Fig. 3. Flow chart of impedance measurement subprogram

In the dialog box for data display and storage, the impedance amplitudes and phase angles measured before and after AI viruses were incubated with the anti-H5 antibodies immobilized on the surface of the electrode, the changes of impedance amplitudes and phase angles, and the testing results of positive or negative, were displayed and could be stored into a flash disk via USB port.

3 Evaluation of Impedance Detector

The impedance detector was evaluated using solid-state resistors, solid-state equivalent circuits of the electrode and AI virus.

3.1 Solid-state Resistor Evaluation

Ten resistors with the accuracy of 1% ranging from 100 Ω to 5 kΩ were respectively measured six times at the characteristic frequency of 100 Hz. As shown in Fig.4, this impedance detector, compared to commercial E4980A precision LCR meter from Agilent, showed a relative error of less than 2.0% at the characteristic frequency of 100 Hz and a standard deviation of less than 2.5% in parallel tests.

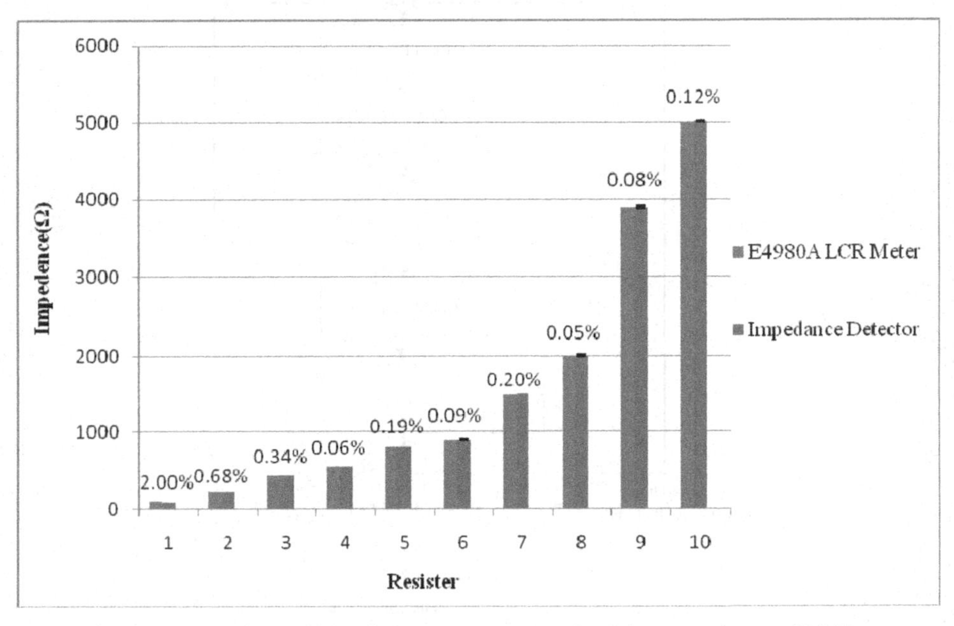

Fig. 4. Comparison of impedance measurement of solid-state resistor at 100 Hz (Relative error= $|Z_{Detector}-Z_{Meter}|/Z_{Meter}$ *100%)

3.2 Equivalent Circuit Evaluation

The impedance biosensor developed in our previous study could be simulated using Randle model shown in Fig. 5. The equivalent circuit included the electrolyte resistance (R_S) representing the resistance of the bulk electrolyte solution, the electric double layer capacitor (C_{dl}) representing the effect of ions near the surface of the electrode, and the electron transfer resistance (R_{et}) representing the resistance of the electron transfer of the redox probes .

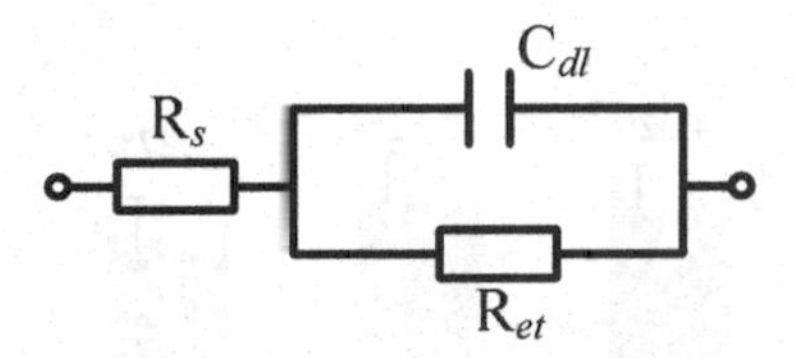

Fig. 5. Equivalent circuit of the impedance biosensor

Four equivalent circuits consisted of solid-state resistors and capacitors with impedance amplitude ranging from 100 Ω to 2 kΩ were respectively measured six times at the characteristic frequency of 100 Hz. As shown in Fig.6, this impedance detector, compared to E4980A meter, showed a relative error of less than 1.0% at the characteristic frequency of 100 Hz and a standard deviation of less than 2.0% in parallel tests.

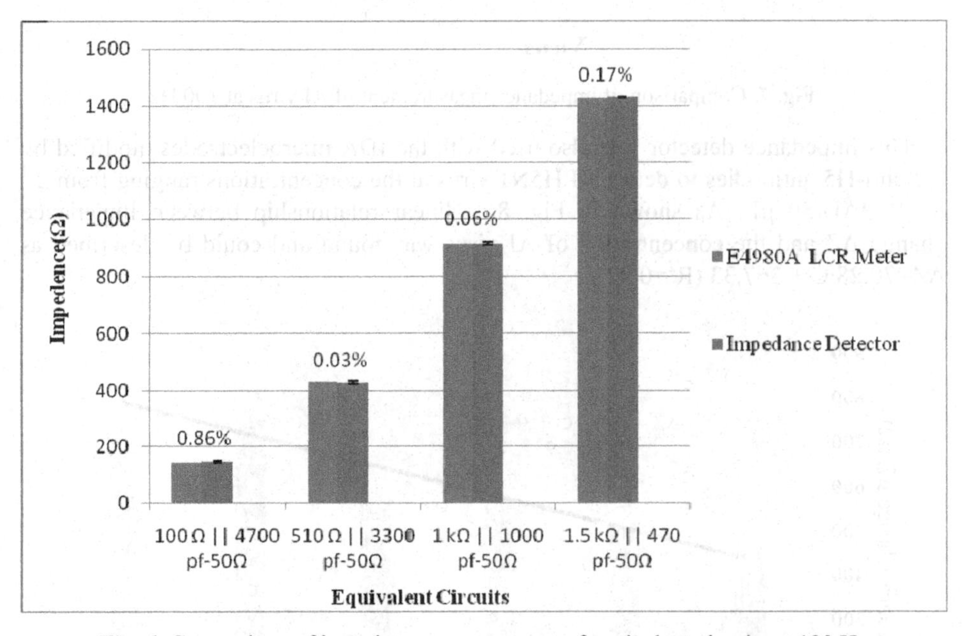

Fig. 6. Comparison of impedance measurement of equivalent circuits at 100 Hz

3.3 AI Virus Evaluation

The evaluation by AI virus was conducted using IDA microelectrodes fabricated by wet-etching process in the Institute of Semiconductors, Chinese Academy of Sciences. The anti-H5 antibodies were immobilized onto the surface of the electrode by Protein A method [9-10] and impedance measurements of AI H5N1 virus with a concentration of 2^4 HAU/50 μL at 100 Hz were performed at the presence of 10 mM $[Fe(CN)_6]^{3-/4-}$ using E4980A meter and this developed impedance detector in parallel.

As shown in Fig.7, this impedance detector showed a relative error of less than 4.1% at the characteristic frequency of 100 Hz and a standard deviation of less than 4.5% in parallel tests.

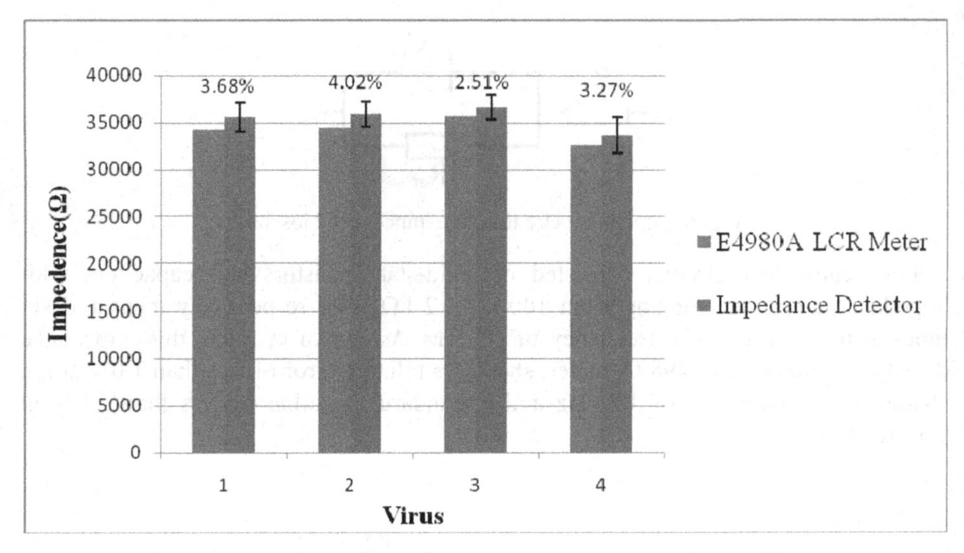

Fig. 7. Comparison of impedance measurement of AI virus at 100 Hz

This impedance detector was also used with the IDA microelectrodes modified by the anti-H5 antibodies to detect AI H5N1 virus at the concentrations ranging from 2^{-1} to 2^4 HAU/50 μL. As shown in Fig. 8, a linear relationship between impedance change ΔZ and the concentration of AI virus was found and could be described as $\Delta Z = 70.28 \cdot C + 367.33$ ($R^2 = 0.92$).

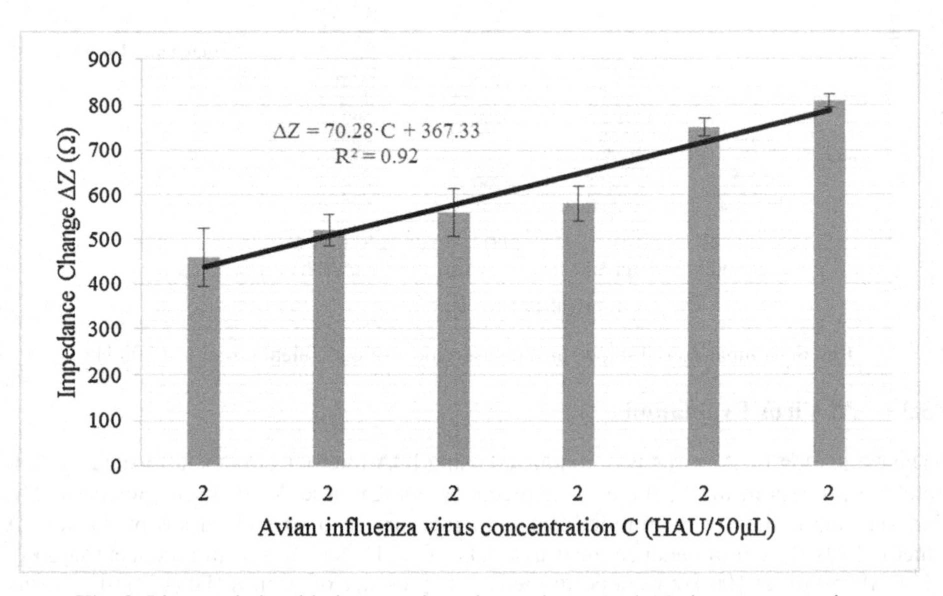

Fig. 8. Linear relationship between impedance change and AI virus concentration

4 Conclusions

A portable impedance detector was redesigned and developed using an S3C2440AL ARM9 microprocessor and an improved AD5933 impedance converter with a comparable accuracy and stability in impedance measurements of both solid-state circuits and AI virus with commercial E4980A precision LCR meter, which could meet the requirements of the newly-redesigned IDA microelectrode for rapid detection of AI virus. This improved impedance detector is an important promotion of developing a simple, rapid, robust, cost-effective, and reliable detection method for in-field screening of avian influenza.

Acknowledgment. This study was supported by Special Fund for Agro-scientific Research in the Public Interest (No. 201303045), Chinese Universities Scientific Fund (No. 2014RC013) and Program of the Co-Construction with Beijing.

References

1. WHO, Human infection with avian influenza A(H5N1) virus – update (2014), `http://www.who.int/csr/don/2014_01_09_h5n1/`
2. Miao, L., Liu, Z., Zhang, S.: Research Evolvement of Electrochemical Immunosensor. Chinese Journal of Medical Physics 23(2), 132–134 (2006)
3. Xiao, F., Zhang, N., Gu, H., et al.: A monoclonal antibody-based immunosensor for detection of Sudan I using electrochemical impedance spectroscopy. Talanta 84(1), 204–211 (2011)
4. Li, Z., Wang, Z., Sun, X., et al.: A sensitive and highly stable electrochemical impedance immunosensor based on the formation of silica gel–ionic liquid biocompatible film on the glassy carbon electrode for the determination of aflatoxin B_1 in bee pollen. Talanta 80(5), 1632–1637 (2010)
5. Geng, P., Zhang, X., Meng, W., et al.: Self-assembled monolayers-based immunosensor for detection of *Escherichia coli* using electrochemical impedance spectroscopy. Electrochimica Acta 53(14), 4663–4668 (2008)
6. Li, D., Wang, J., Gai, L., Ying, Y., et al.: Rapid detection of *Escherichia coli* O157:H7 using electrochemical impedance immunosensor. Chinese Journal of Sensors and Actuators 21(5), 709–714 (2008)
7. Shi, H., Sun, X., Wang, H., et al.: Development of an Electrochemical Impedimetric Immunosensor for Determination of Microcystin-LR in Water. Journal of Instrumental Analysis 28(6), 633–637 (2009)
8. Sun, X., Li, Z., Cai, Y., et al.: Electrochemical impedance spectroscopy for analytical determination of paraquat in meconium samples using an immunosensor modified with fullerene, ferrocene and ionic liquid. Electrochimica Acta 56(3), 1117–1122 (2011)
9. Yan, X., Wang, R., et al.: Impedance Immunosensor Based on Interdigitated Array Microelectrodes for Rapid Detection of Avian Influenza Virus Subtype H5. Sensor Letters 11(6-7), 1256–1260 (2013)
10. Yan, X., Wang, M., Wen, X., et al.: Rapid detection of avian influenza virus using immunomagnetic separation and impedance measurement. Applied Mechanics and Materials 239, 367–371 (2013)

11. Ciszkowska, M., Stojek, Z.: Graphite multi-micro-disc based mercury film electrode: Comparison of experimental and theoretical anodic stripping results. Journal of Electroanalytical Chemistry and Interfacial Electrochemistry 191(1), 101–110 (1985)
12. Cai, H., Lee, T.M.-H., Hsing, I.-M.: Label-free protein recognition using an aptamer-based impedance measurement assay. Sensors and Actuators B: Chemical 114(1), 433–437 (2006)
13. Kim, Y.S., Niazi, J.H., Gu, M.B.: Specific detection of oxytetracycline using DNA aptamer-immobilized interdigitated array electrode chip. Analytica Chimica Acta 634(2), 250–254 (2009)
14. Min, J., Baeumner, A.J.: Characterization and optimization of interdigitated ultramicroelectrode arrays as electrochemical biosensor transducers. Electroanalysis 16(9), 724–729 (2004)
15. Ehret, R., Baumann, W., Brischwein, M., et al.: Monitoring of cellular behaviour by impedance measurements on interdigitated electrode structures. Biosensors and Bioelectronics 12(1), 29–41 (1997)
16. Chang, B.W., Chen, C.H., Ding, S.J., et al.: Impedimetric monitoring of cell attachment on interdigitated microelectrodes. Sensors and Actuators B: Chemical 105(2), 159–163 (2005)
17. Berggren, C., Bjarnason, B., Johansson, G.: Capacitive biosensors. Electroanalysis 13(3), 173–180 (2001)
18. Yang, L., Li, Y., Erf, G.F.: Interdigitated Array Microelectrode-Based Electrochemical Impedance Immunosensor for Detection of *Escherichia coli* O157: H7. Analytical Chemistry 76(4), 1107–1113 (2004)
19. Yang, L.: Electrical impedance spectroscopy for detection of bacterial cells in suspensions using interdigitated microelectrodes. Talanta 74(5), 1621–1629 (2008)
20. Radke, S.M., Alocilja, E.C.: A high density microelectrode array biosensor for detection of *E. coli* O157: H7. Biosensors and Bioelectronics 20(8), 1662–1667 (2005)
21. Valera, E., Ramón-Azcón, J., Rodríguez, Á., et al.: Impedimetric immunosensor for atrazine detection using interdigitated μ-electrodes (IDμE's). Sensors and Actuators B: Chemical 125(2), 526–537 (2007)
22. Wang, R., Wang, Y., Lassiter, K., et al.: Interdigitated array microelectrode based impedance immunosensor for detection of avian influenza virus H5N1. Talanta 79(2), 159–164 (2009)
23. Wang, R., Li, Y., Mao, X., et al.: Magnetic bio-nanobeads and nanoelectrode based impedance biosensor for detection of avian influenza virus. In: Proceedings of IEEE Nano/Molecular Medicine and Engineering (NANOMED), pp. 214–217 (2010)
24. Lin, J., Lum, J., Wang, R., et al.: A portable impedance biosensor instrument for rapid detection of avian influenza virus. In: Proceedings of IEEE SENSORS 2010, pp. 1558–1563 (2010)

Application of Information Technology on Traceability System for Agro-Food Quality and Safety

Xue Xia, Yun Qiu, Lin Hu, and Guomin Zhou

Agricultural Information Institute (AII) of the Chinese Academy of Agricultural Sciences (CAAS), Beijing 100081, China

Abstract. In the near past, the increasing demand of produce supply and the frequent foodborne disease have turned the issue of agro-food quality and safety into the key point of concerned all over the world. The approach of agro-food traceability is being under the exploration in lots of communities in the world for controlling the occurrence of quality and safety accidents, restoring consumer confidence. In this paper, we first introduce the implication of the traceability system for agro-food quality and safety, and the development situation of the system in China and overseas have been described. Then, the application of information techniques for agro-food traceability has been elaborated base on the summary of key techniques in recent produce traceability. At last, the existence problem in the construction of traceability system for agro-food quality and safety is pointed out so as to provide a reference to optimize and improve the traceability system of produce quality and safety.

Keywords: Traceability System, Information Technology, Agro-Food, Quality and Safety.

1 Introduction

Since the inception of the new century, quality and safety incidents of agro-food were frequently exposed both in China and abroad. Many foodborne diseases caused by agricultural products are emerging in an endless (as shown in Table 1 and Table 2).

These agro-food safety incidents not only cause a great loss to the customer and the country, but led to national consumption panic. There is a certain worrying from the consumers for agro-food safety. Meanwhile, constant exposure of agro-food safety issues attracts extensive attention socially as well.

Achieving traceability for the Agro-food is an irresistible trend of developing agriculture in the world. The establishment of the traceability system has become an important direction for the world's agricultural development. This article mainly introduces the development of the traceability system for the agro-food safety, and focused on the research and application of the information technology in the traceability system for agro-food safety.

© IFIP International Federation for Information Processing 2015
D. Li and Y. Chen (Eds.): CCTA 2014, IFIP AICT 452, pp. 257–269, 2015.
DOI: 10.1007/978-3-319-19620-6_32

Table 1. The quality and safety incidents of agro-food in abroad over the years

Year	Country	Incidents
2006	United Kingdom	Cadburys' contaminated chocolate events led to 42 people poisoning
2008	Canada	The listeria infection events cause 23 people death
2009	United States	The tainted peanut butter events sickens 636 people
2011	Germany	The poison cucumbers events give rise to at least 16 people death
2012	Japan	The raw meats poisoning events cause at lowest 4 people death

Table 2. The quality and safety incidents of agro-food in China over the years

Year	Incidents
2003	The ham dichlorvos events brought negative influences on society
2006	The clenobuterol hydrochloride poison events sickened 336 people in Shanghai
2008	The melamine milk events sickened hundreds of thousands of children
2011	The Shanghai "lean" events undermine the whole meat industry in China
2013	The dead pig floating events in Shanghai influences domestic pork consumption

2 Implication of the Traceability System for Agro-Food

So far, there is no uniform recognition on the concept of food traceability in the world. The definition of food traceability given by International Standard Organization (ISO) is the capacity that can trace the produce in terms of the history, the usage and the location by using the recorded markers. The Codex Alimentarius Commission (CAC) defines the traceability as the ability of food tracing at every designated section in the producing, processing and circulating. According to the EU General Food Law (Regulation 178/2002), the definition of traceability is the ability to trace and follow a food, feed, food-producing animal or substance intended to be, or expected to be incorporated into a food or feed, through all stages of production, processing and distribution[1].

The traceability system for agro-food safety can be defined as a continuous guarantee system for information stream in production, processing and circulation of agricultural products. In other words, it is a quality safeguard system to record and save related information of agricultural products in their entire supply chain [2]. This traceability system aimed at strengthening information transmission of the produce safety, preventing foodborne hazards and protecting consumers' interests.

3 Present Development and Conditions of the Agro-Food Traceability System

3.1 Present Development and Conditions in Overseas Agro-Food Traceability System

European Union (EU) is the first place to research and apply the agro-food traceability. A series of agro-food safety accidents in Denmark and Scotland make EU pay more attention to the safety of food and produce. In January 2000, the EU Council released “White Paper on food Safety” in which emphasized the responsibilities of food enterprises for food quality and safety. Moreover, The Hazard Analysis and Critical Control Point (HACCP) has introduced as a safe manufacture practice. In January 2001, EU officially enacted the law no.178/2002, which stipulate the produce companies and food companies must provide relevant measures and data so that the food safety and traceability can be ensured in the whole supply chain [3-5]. In 2006, EU implemented new legislation- EU legislation for food and feed hygiene. It involves the entire food supply chain, and implemented the whole seamless connection from the production of raw material, the launch of ready-for-sale to the customer service of the produce quality and safety [6]. The Seventh Framework Programme (FP7) has been launched by EU in 2007, which requires establishing the Good Traceability Practice, designing the extensible markup language for produce and food traceability-TraceCoreXML, and the specification of GS1, EPC, RFID, E-barcode label [7].

In the United States, the agro-food traceability system can be divided into three parts: the traceability system for agricultural production; the traceability system for packaging and processing; the traceability system for circulation and sales process [8]. The tracing objects are mainly livestock products. The Public Health Security and Bioterrorism Preparedness Response Act (PHSBPRA) had issued by the US government in 2002. It clearly specifies that the government intend to exercise mandatory management for the food safety, and request food firms to build up the relevant records of the whole distribution process [9]. In 2009, the US government passed the Food Safety Enhancement Act in which the U.S. Department of Agriculture was required to establish the food traceability system and develop new traceability technologies and techniques. The integrated family tree of food traceability could be established by means of the linked traceability record from production to distribution. In January 2011, the US government released the Food Safety Modernization Act (FSMA) which specified the food traceability. The act required the high risk food enterprises retaining the record in terms of production, storage and transportation in order to trace the destination and prevent the food security crisis caused by various issues [10].

In Japan, the establishment of the agro-food traceability system synchronizes with the improvement of the produce-related laws, and the established institution of food traceability was early begun to explore. In 2002, seven major institutions, backed by the Ministry of Agriculture, Forestry and Fisheries of Japan, set out to develop the food traceability system. The system record and pass relevant food information in production, processing and circulation with the Internet technology, the barcode technology and the sensor technology. In April 2003, the Japanese government released the Guide for Food Traceability System on the basis of continuous improvement of the laws and regulation relation to food safety. After that, it became a mainstay for enterprises to establish food traceability system. In May 2003, the Food Safety Basic Law was implemented in Japan. The law presented that the research should be launched gradually in both technical and economic perspectives, and comprehensively advances the whole process traceability system from the food production to the market circulation. In 2009, the Japanese government released the Rice Traceability Law, which requested domestic rice production companies, processing companies and distributors have to keep the rice trading records while providing the origin information of rice and batcher [11].

3.2 Present Development and Conditions of Agro-Food Traceability System in China

The Chinese government payed great attention to the research and application of the traceability technology for agro-food, and a series of laws and regulations had been enacted to manage and regulate the produce market. In 2002, The Ministry of Agriculture released the Animal immunity marking rules, which stipulated that the animal of compulsory immunization such as pigs, cattle and sheep should equipped immune ear tags and set up the immune-managed file. In December 2007, No. 11 Decree of the Ministry of Agriculture published the Measures for the Management of geographical symbol of agricultural products. Its implementation means geographical origin traceability of produce based on location was possible. At the same time, Beijing launched the capital food safety traceability system of the Olympic Games in order to ensure the food safety. In 2009, the Ministry of Agriculture published the General Principles of Operation Rules for Produce Quality and Safety, which formulated seven operating standards for food safety tracing covered fruits, grains, tea, meats, vegetable and flours. There were fully demonstrated the request in this standards that production recordable, logging query able and quality traceable [12]. In October 2011, the Ministry of Commerce issued the Construction Guidance of the Traceability System for Facilitating the Circulation of Meets and Vegetable during the 12th Five-Year-Plan Period. The guidance claimed that to accelerate the construction of the impeccable traceability system for meets and vegetable and explored the thorough management model and implements the function for fast tracking so as to boost the construction of the city traceability system [13].

4 Key Technologies in Tracing the Quality and Safety of Agro-Food

4.1 Marking and Identification Technology

At present, the common data carrier technologies in marking system are Barcode technology and Radio Frequency Identification (RFID) technology. The Barcode can be divided into one-dimension barcode and quick response (QR) barcode. The one-dimension barcode coding produce feature information as a code by using a series of letters and numbers, and obtain the produce information from the Internet through read those codes [14]. The quick response barcode can save information by use of the two-dimensional space with better abilities of disturbance and correction and stronger information capacity and density, which can be serving for the purpose of marking and coding for a large number of products [15].

The RFID technology is an automatic identifying technology with contactless mode. The basic RFID system is composed of the electronic tags, the readers and the antennas. The RFID system transport information by using the space coupling of radio wave and the entity objects attributes can be identified with that information [16]. The RFID technology has a series of advantages which are handy information reading and writing, longer reading distance, higher precision, and less susceptible to the effects of harsh conditions.

4.2 Environmental Factors Surveillance Technology

The well growing of crops relies on an appropriate growth environment. Environmental factors, such as air temperature and humidity, soil temperature and moisture, rainfall, wind speed, soil total nitrogen and PH value, play an important role in crop growth. Appropriate transportation and storage environment determine the quality of produce. Through informationize the environmental factors collected by sensors, the crops' environmental conditions can be perceived directly so as to accumulate the essential environment data for the traceability system for agro-food quality. With the introduction and continuous improvement of the Wireless Sensor Network (WSN) technology, an Ad Hoc Network System is formed in wireless mode. All sensors are worked with each other, and perception, acquisition and processing, which make the monitoring of environmental factors for crops and produce more efficient, energy saving, safety and eco-friendly[17].

4.3 Data Transmission Technology

The data transmission technology is the bridge to link all of produce traceability information from scattered individual famers, collective farms, processing firms, transport firms and dealerships. The data transmission technology contains short-distance transmission technology and long-distance transmission technology. Short-distance transmission technologies, which include Bluetooth, ZigBee and Infrared, have the features of handy transport, lower cost and higher safety despite its inconspicuous transfer rate. The long-distance transmission technology contains GSM technology, GPRS technology and 3G technology that features longer transmission range and faster transmission rate. On the other hand, they have relatively high cost. Therefore, the long-distance transmission technology is fit for data collection and sharing with long distance places. All of these transmission technologies providing a better basis of data exchange for the agro-food traceability system.

4.4 Geoscience Information Technology

At present, the technologies of Geographic Information System (GIS) and the General Purpose Radar (GPS) have been widely used in the fields of industry, agriculture, transportation and architecture. GIS refers to the computer system for collection, storage, query, analysis and representation of geography spatial data. It has a top ability in synthetically analysis to administrate the spatial data. GPS can be used as accurately positioning for field information and operation equipment. Moreover, combining with the different distribution of the elements in soil, GPS can assist with irrigation, fertilization, spraying, weeding in agricultural filed operation [18].

5 Application of Information Technology on the Traceability System of Agro-Food Safety and Quality

To ensure the customers' interests, the agro-food traceability system is designed to effectively pinpoint the source of the produce safety accidents with the help of the original records of the faulty products and deal with those faulty products in time in the case of a problem on the agro-food quality and safety. In recent years, the related scholars and research institutes take some particular produces as the major study objects to explore and research the traceability system of agro-food quality and safety with advanced information technology.

5.1 Application of Information Technology on Tracking of Livestock Safety and Quality

In abroad, Fröschle et al. [19] had study on the barcode of readability and optimal print position. In the study, some barcodes were printed on the livestock's mouths and legs with inkjet, then scanning the barcodes to get information, and finally obtained the optimal printing and reading location through statistical analysis. Tomeš et al. [20] proposed a RFID system combined with biological traceability databases-"RFID -Biotrack database system" which kept a lot of livestock's' ID information in each RFID label. Through this ID information, the original of the animal products could be recognized accurately. Farag Sallabi et al. [21] designed and constructed a digital document system for livestock production process based on mobile electronics. The system was divided into four sections: front-end process subsystem; data transmission subsystem; background process subsystem and application administration subsystem. By using the handheld devices like cell phones and PDAs, the production process of the livestock could be logged concisely and precisely. The collected data would be transmitted in real time to the database system with wireless network in order to track down the specific source rapidly by means of the E-documents of production history during the avian disease outbreak.

In China, synthesizing animal identification technology, GPRS technology and 3G technology, Xiong et al. [22-23] developed pork traceability system with PDA that served as the reader device. The system established the electronic files in pig breeding process, and the detailed information of pock data could be queried. At present the system had been applied widely in Tianjin city. Based on the analysis of the key information in breeding beef cattle, Kanget al. [24] developed the PDA-based traceability system for beef cattle quality and safety in feeding process. The data source of the system consists of both the fattening processing information collected by PDA and the feeding information collected by PC. In view of existence question in meats sales-term, Renet al. [25] constructed the meat-productions traceability system by using of RFID technology, EPC technology, GIS technology and IOT technology. The application of the system facilitates the quality and safety tracing in meat-products sales. Panget al. [26] united with RFID technology and WSN technology designing a scheme of acquisition and transmission for dairy traceability information. Furthermore, the reader module had been developed to achieve information convergence for the data fusion problem of the RFID device and WSN device.

5.2 Application of Information Technology on Tracking of Fisheries Safety and Quality

In abroad, K Seineet al. [27] came up with a construction method for the traceability system of marine products by use of QR code and Internet. In this system, the marine products would be coded with a unique ID number and saved the ID number into database, and then the paper or plastic label that printed with QR code would be covered directly on the marine products. The producers, distributors and retailers could record the fishery operation activities respectively via the Internet, and the customers would acquire the historical information of the relevant marine products by scanning the QR code of the products with their cell phones. Thompsonet al. [28] proposed a traceability solution for marine products based on ERP model. In this case, using EDI and ODBC as the basis of data interchange, and portable data collector, label reading device, RFID device, electronic scale, fish classifier, timer and thermometer as tools, the traceability system of seafood had been discussed. Thakuret al. [29] conducted a method study for building the traceability system model of the frozen mackerel which took the framework agreement of EPCIS as the basis and UML state diagram as a tool.

In China, based on TTT (Temperature, Time, Tolerance) theoretical model, Zhang et al. [30] unite with RFID technology, GPRS technology and mobile technology, and developed the traceability system of frozen tilapia with thermal management method. The refrigerated storage temperature of the frozen tilapia in the cold chain could be real-time monitoring. And the information technology, such as RF wireless communication, GPRS, database and multimedia, could be used to provide the remote support for the monitoring management of tilapia cold-chain. Through query the traceability system, the customers could easily get the temperature information in the tilapia cold-chain. Ren et al. [31] unite with the key point of the ACCP system and the practical experience, analyzed the tilapia breeding process and the essential information of the process, and the web-based traceability system had been structured with .NET platform for the tilapia breeding quality and safety. The system could effectively monitor the key link that impacts the tilapia quality, and provided an appropriate operating platform to control the quality and safety in tilapia breeding management. Qi et al. [32] developed a traceability system for recirculation aquaculture by means of wireless sensor network. In the system layer of remote monitoring, the environmental and operation data collected by WSN and RFID, and entered those data with WLAN into the database in the data service layer. By using the visualized desktop and GUI, the users of the client application layer could easily get the information about aquaculture field in order to exchange information among fisheries managers, breeders and consumers. Yang et al.[33], using USB Key as the basis and QR code as the traceability information carrier of the aquatic product, constructed an aquatic product supervisory system with embedded technology and .NET technology targeted at the enterprise certificate authority. The system implemented a new approach of the effective administration between fishery firms and government agencies.

5.3 Application of Information Technology on Tracking of Fresh Products Safety and Quality

In abroad, Arima et al. [34] took strawberry as the study object, and designed acquisition system with robotics and traceability technology for tracing the information in the process of crop production. The system through automatic logging the strawberry's production information with versatile robots in the links of spraying, harvest and classification, and took this production information as the source data of traceability information. Hertog et al. [35] took the Belgium tomato as the study object, and gathered the tomato's temperature information from the planting process to the sales process by using the RFID label which integrated temperature sensors. A quality changed model of the tomato would be made based on the temperature information in order that the users could identify the origin of tomatoes with RFID label and the quality change model. Porto et al. [36] proposed a design method of the computer-based information systems for certified plant traceability by analyzing the production process and specifications of the citrus nursery. This system was considered to be the ideal tool to curb the diffusion of the plant diseases in citrus nursery supply chain.

In China, Yang et al. [37-38], united with the Extensive Markup language and the traceability data model of the vegetables from the aspect of information technology, constructed the Vegetables Traceability Information Markup Language (VTML) and designed the Schema model of the VTML, which enabled seamless traceability information exchange. Moreover, a safe production and quality traceability system for vegetable had been developed with database technology, network technology, product coding technology and early-warning technology. Li et al. [39] took fresh cucumbers as the study object, designed the record-keeping and decision-support system for traceability in cucumber production by utilized PDA device decision support system and geographic information system. The agricultural activities would be better guided by use of the fertilizing reference model and pest warning system. By analyzed the agricultural operations in mango production and mobile capture technology, Liu et al. [40] united with the PDA capture technique and QR code label technique to design a mobile collection system for traceability information, and applied this system to the export system of the Taiwan mango. Based on the original geography codes as index and the production information, Deng et al. [41] constructed a system for safe producing management and product tracking of fresh vegetables with Flex technique, Web GIS technique and Web Services technique, and realized the tracking from the production base to the consumers.

5.4 Application of Information Technology on tracking of Other Produces Safety and Quality

In abroad, Pérez-Aloe et al. [42] discussed the performances of various electronic labels in different environment in cheese traceability. Papetti et al. [43] took cheese as the study objects as well, proposed an electronic traceability system for nondestructive quality analysis. Serrano et al. [44] discussed how to use the data of the GIS as references to enhance the efficiency of traceability control for high quality honey, and a web geographical information system had been developed apply to script multiple features. Thakur et al. [45] proposed a relational database model that was used for internal traceability for grain management. The grain relational database model had been designed, which includes necessary basis information such as product information and quality characteristics, circulation information, gathering and distribution information and destroy information. The customers would trace back to details in grain circulation by using the system.

In China, Li et al. [46] united with network technology, GIS technology, GPS technology, universal coding technology to constructed a management information system for bee products traceability by analyzing the factors that influenced the quality and safety of domestic bee products, and realized the forward tracking and reverse tracking for bee products' quality and safety. Zheng et al. [47] took grain and oil products as the object, and studied a hardware terminal for multi-platform traceability. By using of UCC/EAN-128 coding techniques, a traceability platform for the quality and safety of grain and oil products had been constructed with multi-level and multi-role. Zhang et al. [48] gather the information in exploration origin, environment detection and planting process into the data source of traceability in the light of the cultivation standards of pollution-free potato, and a platform system for potato management and traceability had been developed in .NET platform with QR code technology and IOT technology, and realized the traceability barcode remote identifying and querying traceability information.

6 Problems in Construction of the Traceability System for Agro-Food Safety and Quality with Its Techniques

6.1 Lack a Uniform Traceability Information Platform

Currently, there was lacking a nationwide traceability information platform. The traceability system presents a fragmented state, and the efficient information sharing among systems was not achieved, the diverse traceability systems could track and manage produce just in a given area. In addition, the incompatibility issues among traceability system also had more serious, which could not only cause inconvenience for consumers to query the traceability information of agro-food, but affected the government's demands to supervise the quality and safety of the produce by agro-food traceability system.

6.2 Rapid Quality-Checking System Needs to be Urgently Developed

The quality of agricultural products is affected by temperature, humidity, atmospheric environment and soil constituent. Thus, to protect the quality and safety of produce, the relevant studies in sensing technologies are required to be lunched. Meanwhile, the identification technology of low-cost items is one of basic condition in construction of the agro-food traceability system. For adapts to the development demands of agricultural production with scale, precision and installation, the portable hardware systems with good environment adaptability will be researched and developed to rapid detect the produce's quality.

6.3 Comprehensive Traceability Information is Difficult to Obtain

Due to a late start of research for agro-food traceability in China, the techniques of package, labeling and traceability are incomplete and the quality management systems are inadequate, which give a rise to the information traceability for agro-food quality just addressed part of the produce safety issues in individual regions. Meantime, since agricultural features in China that decentralized production and independent operation, a situation that separated peasant business have been presented in process of purchase and sale, which makes it hard to uniform the produce's quality and difficult to acquire traceability information.

7 Conclusions

It is an effective way for each country in the world to deal with the food safety by means of the quality and safety traceability for agro-food which is an inevitable trend for the world agriculture and food industry with the sound development. In China, the construction of traceability system for agro-food quality and safety is still in the exploratory stage. Compared with western developed countries, in China, the approaches and techniques of traceability are not mature. The management systems in agriculture and food industry need to be perfected. The recognition of the traceability system for agro-food quality and safety is still obviously insufficient for consumer groups. Therefore, as for pushing the traceability system of agro-food quality and safety, the availability of technology, international compatibility, economic affordability and efficiency of reasonable implementation will be taken into account synthetically based on the experience of agro-food traceability in developed countries [49] and a traceability system of quality and safety for agro-food need be studied according with national situations. Meanwhile, the laws and regulations of produce and food must be enacted and improved to further constrains and regulates the market orders of agricultural products.

References

1. Petter, O., Melania, B.: How to define traceability. Trends in Food Science & Technology 29, 142–150 (2013)
2. Zhang, H.L., Sun, X.D., Liu, Y.D., et al.: Research on the Feedback System for Quality and Safety of Agricultural Products. Hubei Agricultural Sciences 49(12), 3220–3223 (2010)
3. Schwagele, F.: Traceability from European perspective. Meat Science 71(1), 164–173 (2005)
4. Mousavi, A., Sarhadi, M., Lenk, A., et al.: Tracing and traceability in the meat processing industry: a solution. British Food Journal 104(1), 7–19 (2002)
5. Bertolini, M., Bevilacqua, M., Massini, R., et al.: FMECA approach to product traceability in the food industry. Food Control 17(2), 137–145 (2006)
6. Wang, J.Z.: Existing Problems and Countermeasures for Emergency Treatment of Agricultural Product Quality Safety Accident at Domestic. Journal of Anhui Agricultural Sciences 41(6), 2680–2682 (2013)
7. Babot, D., et al.: Comparison of visual and electronic identification devices in pigs: On-farm performances. Journal of Animal Science 9(84), 2575–2581 (2006)
8. Xíng, W.Y.: System Construction of the Agricultural Products Traceability in U.S. World Agriculture 4(324), 39–41 (2006)
9. Smith, G.C., Tatum, J.D., Belk, K.E., et al.: Traceability from a US perspective. Meat Science 71(1), 174–193 (2005)
10. Gao, Y.S., Huan, P., Hu, D.G., et al.: Interpret and Appraise of the US FDA Food Safety Modernization Act. Journal of Inspection and Quarantine 21(3), 71–76 (2011)
11. Zhao, R., Chen, S., Qiao, J.: Tracing and Supervision System of Food Quality and Safety of America, EU and Japan and the Relevance to China. World Agriculture 3, 1–4 (2012)
12. Wu, M.Y., Song, Y., Ma, C., et al.: Fruit processing industry development report. China Fruit & Vegetable 2(2013), 3–5 (2012)
13. Fang, R.J.: Study on different stakeholders' transition behaviors and supervisory system in food quality and safety traceable information. Shenyang Agricultural University (2012)
14. Lu, J.H., Guan, J.F., Min, W.J.: A Tracing System Based on One-dimensional Electronic Code for Quality Safety of Green Fruits. Science & Technology Review 21, 59–62 (2010)
15. Fang, Z.K., Sun, M., Zhao, F.: Two-Dimensional Barcode Technology and Recent Applications on Agriculture Production Quality Traceability System. In: Conference of CSAE (2009)
16. Li, J., Ma, M.Y., Qin, X.Y., et al.: Research and advances in quality safety control and traceability technology for animal products. Transactions of the CSAE 24(suppl. 2), 337–342 (2008)
17. Yao, S.F., Feng, C.G., He, Y.Y., et al.: Application of IOT in Agriculture 33(7), 190–193 (2011)
18. Li, D.L.: Introduction to Internet of Things in Agriculture. Science Press, China (2012)
19. Fröschle, H.-K., et al.: Investigation of the potential use of e-tracking and tracing of poultry using linear and 2D barcodes. Computers and Electronics in Agriculture 66(2), 126–132 (2009)
20. Tomeš, J., Lukešová, D., Machá, J.: Meat traceability from farm to slaughter using global standards and RFID. Agricultura Tropica et Subtropica 42(3), 98–100 (2009)

21. Sallabi, F., et al.: Design and implementation of an electronic mobile poultry production documentation system. Computers and Electronics in Agriculture 76(1), 28–37 (2011)
22. Xiong, B.H.: A solution on pork quality traceability from farm to dinner table in Tianjin city, China. Agricultural Sciences in China 9(1), 147–156 (2010)
23. Xiong, B.H., Luo, Q.Y., Yang, L., et al.: Development on mobile traceability system of feeding process of pigs and quality safety of its meat products based on 3G technology. Transactions of the Chinese Society of Agricultural Engineering 28(15), 228–233 (2012)
24. Kang, R.J., Fu, Z.T., Tian, D., et al.: Design and implementation of beef cattle breeding traceability system based on PDA. Microcomputer Information 5, 50–52 (2010)
25. Ren, S.G., Xu, H.L., Li, A., et al.: Meat-productions tracking and traceability system based on internet of things with RFID and GIS. Transactions of the CSAE 26(10), 229–235 (2010)
26. Pang, C., He, D.J., Li, C.Y., et al.: Method of traceability information acquisition and transmission for dairy cattle based on integrating of RFID and WSN. Transactions of the CSAE 27(9), 147–152 (2011)
27. Seine, K., et al.: Development of the traceability system which secures the safety of fishery products using the QR code and a digital signature. In: Proc. IEEE TECHNO-OCEAN 2004, Kobe, Japan, November 9-12 (2004)
28. Thompson, M., Sylvia, G., Morrissey, M.T.: Seafood traceability in the United States: Current trends, system design, and potential applications. Comprehensive Reviews in Food Science and Food Safety 4(1), 1–7 (2005)
29. Thakur, M., et al.: Managing food traceability information using EPCIS framework. Journal of Food Engineering 103(4), 417–433 (2011)
30. Zhang, J., et al.: Development of temperature-managed traceability system for frozen and chilled food during storage and transportation. Journal of Food, Agriculture & Environment 7(3-4), 132–135 (2009)
31. Ren, X., Zhang, X.S., Mu, W.S., et al.: Design and implementation of tilapia breeding quality safety traceability system based on web. Computer Engineering and Design 30(16), 3883–3890 (2009)
32. Qi, L., et al.: Developing WSN-based traceability system for recirculation aquaculture. Mathematical and Computer Modeling 53(11), 2162–2172 (2011)
33. Yang, X.T., Wu, T., Sun, C.H., et al.: Design and Application of Aquatic Enterprise Governance Traceability System Based on USB Key 43(8), 128–133 (2012)
34. Arima, S., et al.: Traceability based on multi-operation robot; information from spraying, harvesting and grading operation robot. In: Proceedings of the 2003 IEEE/ASME International Conference on Advanced Intelligent Mechatronics, AIM 2003, vol. 2. IEEE (2003)
35. Hertog, M.L.A.T.M., et al.: Smart traceability systems to satisfy consumer expectations. Acta Horticulturae 768, 407–415 (2008)
36. Porto, S.M.C., Arcidiacono, C., Cascone, G.: Developing integrated computer-based information systems for certified plant traceability: Case study of Italian citrus-plant nursery chain. Biosystems Engineering 109(2), 120–129 (2011)
37. Yang, X.T., Qian, J.P., Zhao, C.J., et al.: Construct ion of information description language for vegetable traceability based on XML and its application to data exchange. Transactions of the CSAE 23(11), 201–205 (2007)
38. Yang, X.T., Qian, J.P., Sun, C.H., et al.: Design and application of safe production and quality traceability system for vegetable. Transactions of the CSAE 24(3), 162–166 (2008)
39. Li, M., et al.: A PDA-based record-keeping and decision-support system for traceability in cucumber production. Computers and Electronics in Agriculture 70(1), 69–77 (2010)

40. Liu, Y.C., Hong, M., Gao, X.L., et al.: Development and Application of Mobile Traceability Data Construction for Agriculture. In: Proceedings of the 8th Asian Conference for Information Technology in Agriculture and World Conference on Computer in Agriculture Tianmu Convention Center Taipei City, Taiwan, September 3-6 (2012)
41. Deng, X.F., Huang, X.H., Ren, Z.Q., et al.: Geocoding- based technology of safety and traceability for fresh vegetables. Acta Agriculturae Zhejiangensis 1, 120–124 (2012)
42. Perez-Aloe: Application of RFID tags for the overall traceability of products in cheese industries. In: 2007 1st Annual RFID Eurasia. IEEE (2007)
43. Papetti, P., et al.: A RFID web-based infotracing system for the artisanal Italian cheese quality traceability. Food Control 27(1), 234–241 (2012)
44. Serrano, S.: GIS design application for "Sierra Morena Honey" designation of origin. Computers and Electronics in Agriculture 64(2), 307–317 (2008)
45. Thakur, M., Bobby, J.M., Charles, R.H.: Data modeling to facilitate internal traceability at a grain elevator. Computers and Electronics in Agriculture 75(2), 327–336 (2011)
46. Li, S.J., Zhu, Y.P., E, Y., et al.: Status quo of quality safety of bee products and construction of whole-recess traceability system. Transactions of the CSAE 24(2), 293–297 (2008)
47. Zheng, H.G., Liu, S.H., Meng, H., et al.: Construction of Traceability System for Quality Safety of Cereal and Oil Products. Scientia Agricultura Sinica 42(9), 3243–3249 (2009)
48. Zhang, J.T., Liu, X., Shi, Z.: Design and Application of Potatoes Safety Management and Traceability Information System. Agriculture Network Information 12, 46–48 (2011)
49. Wang, L.F., Lu, C.H., Xie, J.F., et al.: Review of traceability system for domestic animals and livestock products. Transactions of the CSAE 21(7), 168–174 (2005)

Computer Computing and Simulation—In View of the Leaves' Categories, Shapes and Mass

Jiahong Li, Heng Li, and Qiang Fu

School of Hydraulic and Construction Engineering,
Northeast Agricultural University Heilongjiang Harbin, 150030, China
liquanhui1992@163.com, mr.liheng@gmail.com, fuqiang@neau.edu.cn

Abstract. Leaf is the important part of a tree. This paper mainly studied its categories, shapes and mass, then established four mathematic models to describe and analyze them. Firstly, this paper analyzed the reasons why leaf shapes are different by the theory of mechanics. The mechanics model explained the structure and forming principles of a leaf. Secondly, in order to classify the leaves accurately, this paper selected digital image processing and built two classified criterion: ratio of leaf area to its perimeter and ratio of maximal length to maximal width. After edge extraction and Fourier fitting. Thirdly, this paper utilized the connection between illumination intensity and leaf area to explain how the distribution of leaves and branches affect leaf area. According to the stem height at each branch tip, this paper got the relation between the branches, then established a model of leaf area to proof that the leaf shape has a relation with the tree profile and the branches structure. Finally, this paper set up leaf mass model to compute the total leaf mass, based on the leaf area index. It was calculated by choosing digital image processing technique. Using the leaves mass, this paper analyzed the relevancy of the mass and the size characteristics of the tree.

Keywords: computer, simulation, leaves, mathematic model.

1 Introduction

Leaves are the important parts of a tree. The classification, shapes and mass of leaves have big relations with a tree. The leaves are the places where plants can absorb the carbon and keep the carbon-water balance; they are also the survival foundation of the plant, so the form and physiology characteristics of the leaves will become important growth indexes of the tree undoubtedly. The leaves have the important denotation to the update of the vegetation, the community and the ecological system [1]. The types, shapes and mass of the leaves contact to the climate and its growth closely. Through the leaves, we can also learn about the relationship between plants and the global carbon cycle, so its significance is far-reaching. Nowadays, someone researched tree structure, obtaining the tree model based on L system and real-time rendering technology conclusion; someone imitated the dynamic model of veins, describing the importance

© IFIP International Federation for Information Processing 2015
D. Li and Y. Chen (Eds.): CCTA 2014, IFIP AICT 452, pp. 270–284, 2015.
DOI: 10.1007/978-3-319-19620-6_33

of veins to leaf characteristics; someone applied physical methods to mimic the change process of the leaf shapes. But few has a systematic, integrated and comprehensive research about shape and mass of leaves. In this paper, problems which we need to solve is the diversity of the shape, the effect of the distribution of leaves and branches to the leaf shape, the relation between the leaf shape and the outline of the tree and the structure of the branches and how to calculate the weight of the leaves, we should establish the corresponding models to describe leaf shape and the mass of the tree.

There are the main steps in this paper,

(1) This paper use some mechanics knowledge which related to the growth of the leaf to describe the leaf shape, then analyze the reasons of shape diversity about the leaves;
(2) This paper simulation the leaf shape according to the Fourier series, then classified the kinds of the leaves according some index in the final figures;
(3) This paper utilized the expression about the light interception ratio (LIR) and the leaf area index (LAI), explaining the blade shape is effected by the distribution of the leaves and tree; By considering the sun light (photosynthesis) affects the leaf shape, we established the model to show the shape has some relations with the outline of tree and the structure of the branches;
(4) To measuring relative parameters easily, this paper selected digital image processing to calculate the leaf area index. So that we can weigh the mass of leaves by leaf area index and projected area. Confirming relations between the leaf mass and the mass, height, volume of the tree, we took correlation analysis and qualitative analysis to accomplish them.

To achieve the goal of simulating the relation between parameters, we did simulating experiment several times.

2 Analyses and Modeling

2.1 The Leaves Categories

2.1.1 The Leaf Variety

There are no two leaves exactly the same. The same tree's growths of the leaves are not completely the same. This paper mainly studied the leaf shape's diversity of a tree. Through understanding the structures of the leaves and the influence of environment to the leaves, we use the opinion of mechanics [2] that the force can be divided into the internal force and the external force. It is similar to the growth of the leaves which is effected by the external environment and the inherent factors itself. So the leaves stand its weight and external force produced by the environment. Then the petiole will bend and twist, so it can be regarded as a cantilever beam mechanical operation approximately. So we establish the model:

Nomenclature

$F = mg$ The gravity expression of a leaf

$f(x)$ It is leaf quality

We can Deduced from the Following Expressions:

$M = \int_0^m \frac{m}{x} gxd_x = mgl$ The moment of force which is received by the leaves

$\sigma_{11} = \frac{F_1 l^2}{I_Z}$ The bending stress which is received by the leaf roots (expressed by force)

$\sigma_{21} = \frac{Ml}{I_Z}$ The bending stress which is received by the leaf roots (Expressed by moment of force)

$\sigma_{31} = \frac{F_2}{S}$ It stands for pressure stress

$I_Z = \frac{\pi d^4}{64}$ Moment of inertia

$\tau_{\max} = \frac{4FS}{3A}$ Maximum shear stress

$\sigma_{\max} = \frac{\sigma_x + \sigma_y}{2} + \sqrt{\frac{(\sigma_x - \sigma_y)^2}{2} + \tau_x^{\ 2}}$ The maximum of normal stress which is received by body of the cell

$\sigma_{\min} = \frac{\sigma_x + \sigma_y}{2} - \sqrt{\frac{(\sigma_x - \sigma_y)^2}{2} + \tau_x^{\ 2}}$ The minimum of normal stress which is received by body of the cell

g Gravity acceleration

l The vertical distance of point to the main branch

x Integral variable

F_1 F_2 Blade stress

S Stress area

d Related inertia length of the Leaves

A Cross-sectional area

σ_x σ_y Stress component

τ_x Shear stress

The expressions above can be shown in the figures:

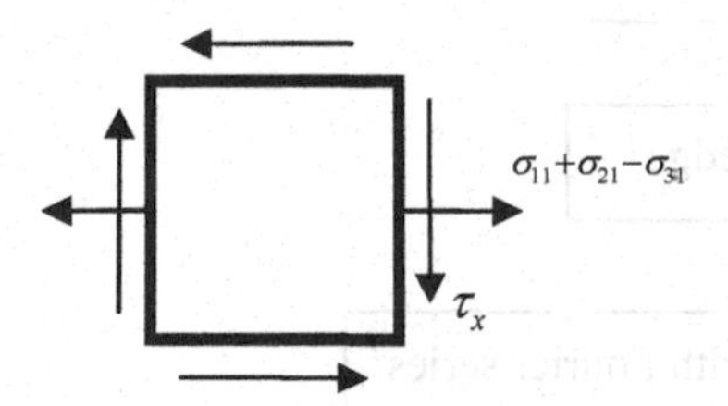

Fig. 1. The cell of the leaf body

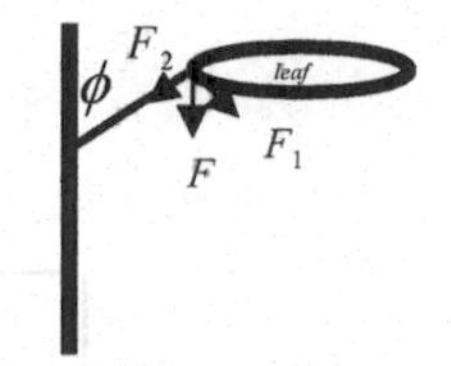

Fig. 2. Simulation tree

$\sigma_x = \sigma_{11} + \sigma_{21} - \sigma_{31} \qquad \sigma_y = 0 \qquad \tau_x = \tau_{max}$

then,

$$\sigma_{max} = \frac{\sigma_x}{2} + \sqrt{\left(\frac{\sigma_x}{2}\right)^2 + \tau_x} \quad \sigma_{min} = \frac{\sigma_x}{2} - \sqrt{\left(\frac{\sigma_x}{2}\right)^2 + \tau_x}$$, thus

$\sigma_1 = \sigma_{max}, \sigma_2 = 0, \sigma_3 = \sigma_{min}$

According to the third strength theory $\sigma = \sigma_1 - \sigma_3 < [\sigma]$, in the cell of the body, when the total maximum stress minus minimum stress, the result is less than the allowable stress. Considering the effect of the load from the outside, we should change the third strength theory into the following expression:

$$\sigma = k(\sigma_1 - \sigma_3) < [\sigma] \tag{1}$$

Due to the leaf must meets the strength theory to keep standing on the branch, we can get:

$$f(x) = F(\phi, L, k, [\sigma], S) \tag{2}$$

So, the mechanics of the relevant knowledge, analyzes the specific reason for the diversity of leaf shapes. The differences of blade shapes are effected by $\varphi, L, k, [\sigma], S$ and other factors.

2.1.2 Image Processing

Boundary Extraction

There is a very practical operation called boundary in the image processing. After extracting the boundary of image, we can do the further operation, such as image segmentation, extraction of the location and the skeleton extraction and so on. In this paper, we extracted an image of the leaf, then extract the edge. The process can be described as follow:

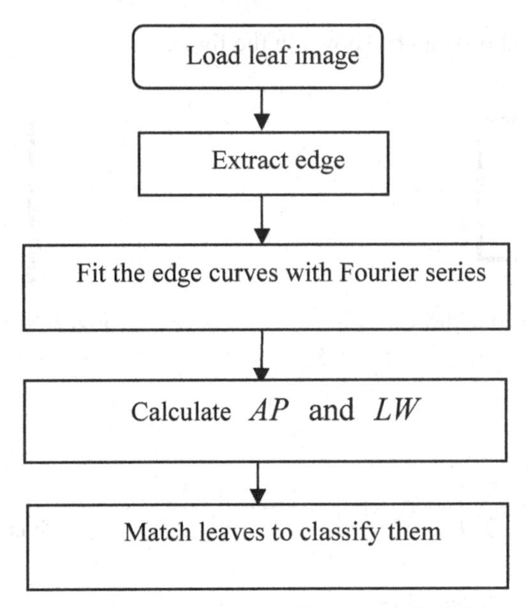

Fig. 3. The main steps of image processing

Therein, in the image after introduction, we extracted the edge of the leaf image to meet the requirements of leaf image. Then the graphics morphology processing is based on Mathematical Morphology (Mathematical Morphology) set theory method. IN this model, we applied to "corrosive" and "inflation" which are the morphological processing method [3]. Image processing the rapid expansion of the corrosion, morphology method is characteristic: it USES the rectangular element structure on the additional two sliding window that is vertical and horizontal two directions of the sliding window, with rectangular structural elements through traverse the whole image move corrosion, inflation in the process of operation, through the level and the vertical sliding window a record has been compared with local information, so that in the future can be directly used in calculating the extraction, in order to achieve reduce and eliminate the comparison of the calculation of the repeat redundant, save the computational time, and presents the existing technology results which can compared with experimental. Thus, we will get the leaves of the main context extracted.

2.1.3 Fourier Series Fitting

After the boundary extraction, we use the Fourier series [4] image analysis fitting out the boundary of the blade, the coefficients of the normalization of the leaves to final simulation blade shape relevant model. Among them, Fourier series fitting has successfully used by many scholars to achieve closed boundary characteristics, elliptic Fourier descriptor is using the stack to approximate the object boundary curves.

Because the blade of the image is a continuous border and have closed cycle, so Fourier series can be used to approximate the boundary to a closed boundary, the border in the x y direction of Fourier series could start for as follows:

$$x(t) = A_0 + \sum_{n=1}^{\infty}\left(a_n \cdot \cos\frac{2n\pi t}{T} + b_n \cdot \sin\frac{2n\pi t}{T}\right) \tag{3}$$

$$y(t) = C_0 + \sum_{n=1}^{\infty}\left(c_n \cdot \cos\frac{2n\pi t}{T} + d_n \cdot \sin\frac{2n\pi t}{T}\right) \tag{4}$$

In the mathematical expressions

$$a_n = \frac{T}{2n^2\pi^2}\sum_{p=1}^{k}\frac{\Delta x_p}{\Delta t_p}\left(\cos\frac{2n\pi t_p}{T} - \cos\frac{2n\pi t_{p-1}}{T}\right) \tag{5}$$

$$b_n = \frac{T}{2n^2\pi^2}\sum_{p=1}^{k}\frac{\Delta x_p}{\Delta t_p}\left(\sin\frac{2n\pi t_p}{T} - \sin\frac{2n\pi t_{p-1}}{T}\right) \tag{6}$$

$$c_n = \frac{T}{2n^2\pi^2}\sum_{p=1}^{k}\frac{\Delta y_p}{\Delta t_p}\left(\cos\frac{2n\pi t_p}{T} - \cos\frac{2n\pi t_{p-1}}{T}\right) \tag{7}$$

$$d_n = \frac{T}{2n^2\pi^2}\sum_{p=1}^{k}\frac{\Delta y_p}{\Delta t_p}\left(\sin\frac{2n\pi t_p}{T} - \sin\frac{2n\pi t_{p-1}}{T}\right) \tag{8}$$

$$A_0 = \frac{1}{T}\sum_{p=1}^{k}\left[\frac{\Delta x_p}{2\Delta t_p}\left(t_p^{\,2} - t_{p-1}^{\,2}\right) + \xi_p\left(t_p - t_{p-1}\right)\right] \tag{9}$$

$$C_0 = \frac{1}{T}\sum_{p=1}^{k}\left[\frac{\Delta y_p}{2\Delta t_p}\left(t_p^{\,2} - t_{p-1}^{\,2}\right) + \delta_p\left(t_p - t_{p-1}\right)\right] \tag{10}$$

$$\Delta t_p = \left(\Delta x_p^{\,2} + \Delta y_p^{\,2}\right)^{\frac{1}{2}} \tag{11}$$

$$t_p = \sum_{i=1}^{p}\Delta t_i \quad \xi_p = \sum_{j=1}^{p-1}\Delta x_j - \frac{\Delta x_p}{\Delta t_p}\sum_{j=1}^{p-1}\Delta t_j \quad \delta_p = \sum_{j=1}^{p-1}\Delta y_j - \frac{\Delta x_p}{\Delta t_p}\sum_{j=1}^{p-1}\Delta t_j \tag{12}$$

$$\xi_1 = \delta_1 = 0 \tag{13}$$

In the expressions

n——Elliptic represents the order number, $n > 0$

k——the number of points on the boundary

T——period

P——the boundary point serial number

a_n、b_n、c_n、d_n——Elliptic Fourier coefficient

Δt_p、Δt_j——Two boundary point between the lines

Δx_p、Δx_j——In the direction of x incremental

Δy_p、Δy_j——In the direction of y incremental

A_0 And C_0 the harmonic dc component, from the border, it represents the center point of the border, at the same time it is a harmonic elliptical center.

Every n form of 4 coefficients a_n、b_n、c_n、d_n represents a elliptic, n order elliptic also is n time harmonic Use type above, After the treatment of boundary to get data Fourier transformation calculation, get the elliptic boundary the Fourier descriptor as shown in figure 4:

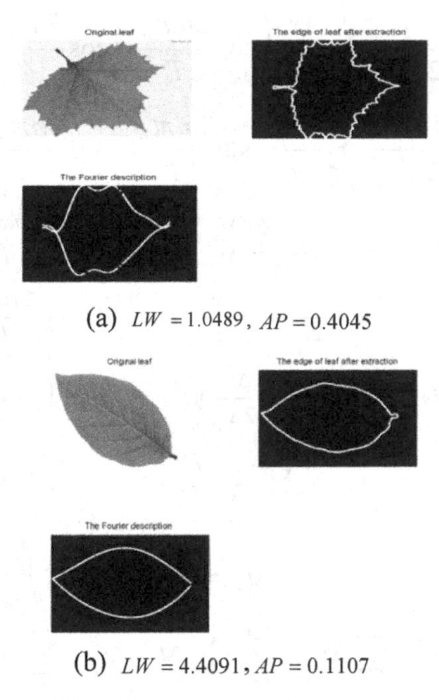

(a) $LW = 1.0489$, $AP = 0.4045$

(b) $LW = 4.4091$, $AP = 0.1107$

Fig. 4. Leaves simulation process

2.1.4 Analysis of the Model

For every descriptor, hope it has scale transform rotation transformation and the starting point of the transformation invariance, So gets the elliptic Fourier descriptor for starting point and rotation of the size of the normalized the arbitrary starting point get elliptical Fourier descriptor for the record a_n、b_n、c_n、d_n. When starting point along the boundary clockwise movement λ units, and when the original x、y coordinate counter-clockwise ψ angle to coordinate u、v, get new elliptic coefficient a_n^{**}、b_n^{**}、c_n^{**}、d_n^{**}, there is

$$\begin{bmatrix} a_n^{**} & b_n^{**} \\ c_n^{**} & d_n^{**} \end{bmatrix} = \begin{bmatrix} \cos\psi_1 & \sin\psi_1 \\ -\sin\psi_1 & \cos\psi_1 \end{bmatrix} \bullet \begin{bmatrix} a_n & b_n \\ c_n & d_n \end{bmatrix} \begin{bmatrix} \cos(n\theta_1) & -\sin(n\theta_1) \\ \sin(n\theta_1) & \cos(n\theta_1) \end{bmatrix} \tag{14}$$

Where $\theta_1 = \frac{1}{2}\arctan\left(\frac{y_1^*(0)}{x_1^*(0)}\right) = \arctan\frac{c_1^*}{a_1^*}$ (15)

$$(0 \le \psi_1 < 2\pi)$$

$$\begin{bmatrix} a_1^* & b_1^* \\ c_1^* & d_1^* \end{bmatrix} = \begin{bmatrix} \cos\theta_1 & \sin\theta_1 \\ -\sin\theta_1 & \cos\theta_1 \end{bmatrix} \begin{bmatrix} a_1 & b_1 \\ c_1 & d_1 \end{bmatrix} \tag{16}$$

At the same time, half the size of the long axis for

$$E^*(0) = \left(a_1^{*2} + c_1^{*2}\right)^{\frac{1}{2}} \tag{17}$$

Through the type Calculated about starting point and the rotation Angle of the normalized coefficient; Then to scale to normalization, to get the coefficient of half the size of the divided by long axis can get the result; On to the translation of the normalization, just ignore dc component A_0 and C_0. The connection between errors of Fourier fitting and series is shown in figure 5:

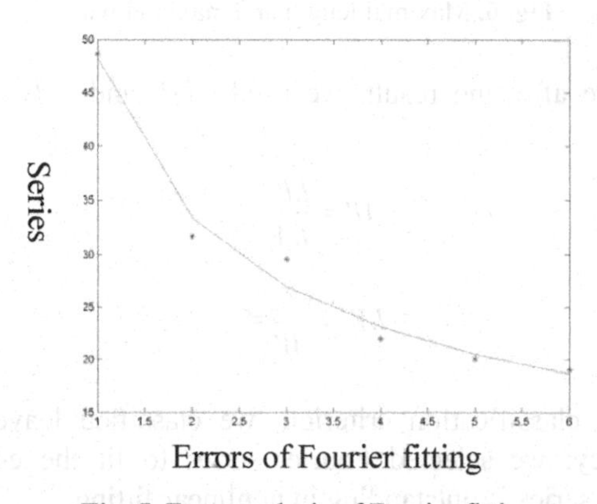

Fig. 5. Error analysis of Fourier fitting

notes:

(1) With elliptic Fourier description method for describing the shape of need only a descriptor can complete complex shape description
(2) When used for reconstruction of the harmonic times are over 10 times, errors can be ignored, and can be accurately rebuild original shape [5]

2.1.5 Classification Criterion

Many criterions for classifying leaves used digital morphological features, and the main criterions is ratio of leaf area to its perimeter AP and ratio of maximal length to maximal wide LW [6]. Leaf area, its perimeter, maximal length and maximal wide both can't measure leaves shape alone. Since they are affected by leaf size [7]. But their ratio can do. With them, classification leaves became quickly and accurate.

Leaf area (LA): the value of leaf area is easy to calculate, if only counting the mount of pixels of binary value 1 on smoothed leaf image.

Leaf perimeter (LP): leaf perimeter is calculated by counting the number of pixels consisting leaf edge.

Maximal length ($L_{\max}$): maximal length is the maximal distance between arbitrary two point on leaf edge.

Maximal wide ($W_{\max}$): it is the longest line which is perpendicular to the main vein. Maximal length and maximal wide is shown in figure 6.

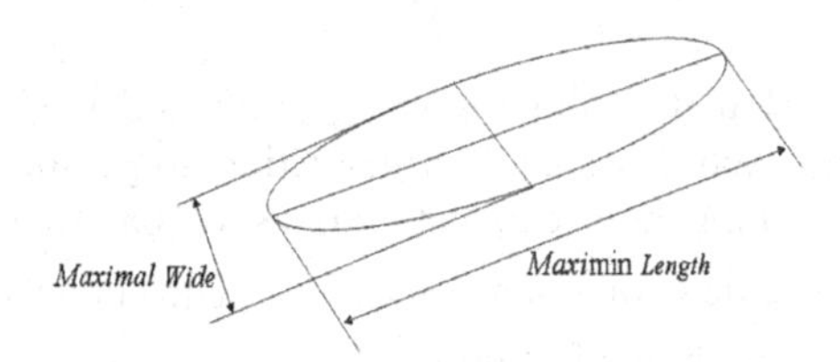

Fig. 6. Maximal length and maximal wide

To avoid leaf size affecting result, we used AP and LW as criterion. They expressed as follows:

$$AP = \frac{LP}{LA} \tag{18}$$

$$LW = \frac{L_{\max}}{W_{\max}} \tag{19}$$

According to the classification criterion, we classified leaves by program. To improve the accuracy, we selected Fourier series to fit the edge of leaves. The advantage of Fourier series is outstanding in nonlinear fitting.

2.2 The Leaves Shapes

2.2.1 Minimum Overlapping

From the angle of the light energy, we considered the situation that the leaves have the overlapping, then we introduce the light interception ratio (LIR) [8] which is the proportion of incoming of irradiation intercepted by the canopy, is generally computed by using the turbid medium analogy which supposes a uniform distribution of leaf area in the canopy. This analogy has been proved to be very robust since the Beer-Lambert equation is fairly insensitive to violations of the uniform assumption. See for example [9] the light interception ratio is thus classical given by equation

$$LIR = 1 - \exp(-kLAI) \tag{20}$$

Where,

LAI The leaf area index whose original definition is the total one-sided area of photosynthetic tissue per unit ground surface area.

k Is a fixed numerical value, standing for the extinction coefficient for the Beer Law, related to the leaves' orientation.

Where,

PA Stands for projection area of leaves and it equals to the leaf area (la) multiplied by leaf sum (ls). So

$$PA = leaf\ \ area * leaf\ \ \ \ sum = la * ls \tag{21}$$

FS Stands for floor space

AOE Stands for absorbed optical energy

TOE Stands for total optical energy

Then,

$$LAL = \frac{\mathrm{PA}}{\mathrm{FS}}\ \ LIR = \frac{\mathrm{AOE}}{\mathrm{TOE}} \tag{22}$$

we can obtain

$$\frac{\mathrm{AOE}}{\mathrm{TOE}} = 1 - \exp(-k\frac{\mathrm{PA}}{\mathrm{FS}}) = 1 - \exp(-k\frac{la * ls}{FS}) \tag{23}$$

In the expression, TOE, *k*, *ls* and FS are permanent only AOE and la are variation. So there are some relations between them. When the *la* is bigger, the shape turns to be large and the AOE also becomes plentiful. So the sentence that "shapes "minimize" overlapping individual shadows that are cast, so as to maximize exposure" was right.

2.2.2 Distribution of the Leaves and the Branches

The leaves will get more energy through the exposure in the sun if necessary. The distribution of the leaves and the branches will decide whether the leaves need energy. When the distribution is sparse, the leaves will expose mostly, then the leaves needn't change their shapes to receive the sunshine; when the distribution is dense, the leaves need to change too small to receive the sunlight. So when the leaves change to receive the sunlight, the shapes of the leaves will change as follow. Through the opinion, we can answer to the question 2 that the distribution of leaves within the "volume" of the tree and its branches effect the shape.

2.2.3 Outline of the Tree and the Structure of the Branches Leaf Areas

The leaves area distribution for the absorption of light by function have very important influence, so the outline and the branches of the tree structure will affect the growth of the leaves, from cross-sectional area of tree branches and leaves area and branch prediction model is out of the length:

$$h_x = h_0 + \cos\theta \times l_{br} \tag{24}$$

Where

h_x The stem height at each branch tip

h_0 Height of branch emergence

θ Branch angle from the vertical

l_{br} Branch length

Crown length of each tree was divided into 0.1-m sections and branch leaf area within each section expressed as a proportion of total leaf area. These data were fitted to a two-parameter cumulative Weibull function using the NLIN procedure of the SAS software package (SAS Institute Inc., Cary, NC)[10]. The cumulative foliage distribution function took the form:

Based on (5) the cumulative proportion of leaf area model can be reached

$$l_{ac} = 1 - \exp\left[-\left(\frac{rh}{\beta}\right)^{\alpha}\right] \div \left[1 - \exp\left[\left(\frac{1}{\beta}\right)^{\alpha}\right]\right] \tag{25}$$

Where

l_{ac} Cumulative proportion of leaf area

rh Relative crown height

α β Estimated parameters

For $1 < \alpha < 3.6$, the probability distribution is mound shaped and positively skewed; if $\alpha = 3.6$, the distribution is approximately normal and if $\alpha > 3.6$, the distribution becomes negatively skewed. The β parameter describes the scale of the distribution and has been interpreted as leaf area density or leaf area per unit height (Gillespie et al. 1994). Which reflects the outline of the outline of tree branches and structure and crown have relationship, the shape of the crown has influence for leaves' shape.

The proportion of leaf area at a given crown location was estimated from the fitted Weibull [11] function:

$$la_x = \frac{la\left\{\exp\left[-\frac{x-w}{\beta}\right]^{\alpha} - \exp\left[-\frac{x+w}{\beta}\right]^{\alpha}\right\}}{1-\exp\left[-\left(\frac{1}{\beta}\right)^{\alpha}\right]} \tag{26}$$

Where

la_x The predicted leaf area

la Total tree leaf area

Crown volume was estimated by calculating the horizontal projection length of each branch at its maximum stem height, which is related to the branches of the tree. The branches of the different Angle will influence the leaves accept photosynthesis. Photosynthesis decided to the leaf to sunshine absorption and their own manufacturing organic matter, for the leaves the veins which influence the shape of leaves are the influence of photosynthesis determined to the movement of nutrients and carbon dioxide absorption.

So the outline of the tree branches and the distribution of the shape have influence for the leaves.

2.3 Section 3 the Leaves Mass

2.3.1 Calculating the Mass

Improving the production of orchard, protecting trees and the ecological environment require us to monitoring trees growth condition. One of monitoring indexes of trees is the mass of leaves [12]. The mass usually calculates by the total leaves area (PA) and its surface density (SLM), just is

$LM = PA * SLM$

Since SLM relates to the tree species and its environment of its location. So SLM needs to measure when counting LM. Calculating PA needs leaf area index (LAI) and projection area of crown (PA) [13]. Projection zone is usually treated as ellipse. So PA is easy to gain.

2.3.2 Based on the Digital Cover Photography

There are many ways to measure LAI. Among those ways, graphic processing has many advantages, such as no damage to trees and high precision. Using graphic processing to measure LAI, we should define crown porosity η.

$$\eta = 1 - \frac{IM_{PA}}{IM_C} \tag{27}$$

Where,

IM_{PA} Is foliage cover in image. It counts by the fraction of pixels not in any gaps;

IM_C is defined as the ground fractional of the vertical projection of solid crown which include the porosities.

LAI can be expressed as follow:

$$LAI = -IM_C \frac{\ln(\eta)}{k} \tag{28}$$

Where, k leaf inclination angle distribution.

2.3.3 Analysis of the Model

After LM calculated, we confirm the connection between LM and the size characteristics of the tree by Correlation analysis. The result is showed in table1. According it, we can know the mean crown radius correlation obvious. And then we inferred the qualitative relationship by fitting (fig. 7.).

Table 1. The connection between leaf mass and some characteristics of tree

	Tree height	Ground-crown distance	Mean crown radius	Trunk circum. breast	Sapwood rings
LM	0.447	0.0582	0.8394	0.2168	0.5816

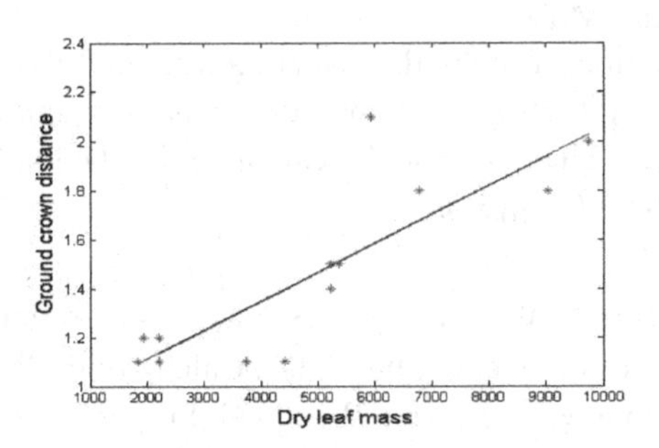

Fig. 7. The relationships between dry leaf mass and ground crown distance

2.3.4 Result of the Model

Through the calculation expression, we can get the mass of the leaves. Next, we based on the Digital Cover Photography and the Analysis of the model, we can draw the leaves quality and the radius of the crown, and the tree height, the volume of the tree have some relationship.

3 Conclusion

(1) There are many reasons that lead to the difference between leaves. Some of them are primary causes, another are not. In order to search for the primary causes, we selected the mechanics structure of leaf. In the mechanical angle, we established the mechanism model based on leaf forming and analyzed reasons which affects leaf shape, then we found out the influence factors. To classify leaves exactly, we defined the classification criterions which are chosen from image processing technology to accomplish the goal.

(2) Leaf incidence and internodes can affect the rate of leaf coverage, enlarging the exposure. At the same time, the rate of leaves coverage should make production of photosynthesis equal or large to the production of transpiration. So leaf area and distribution of leaves and structures are affected each other.

(3) The mass of leaves was calculated. With the mass, we analyzed and confirmed the relativity between mass and some of the size characteristics of the tree.

(4) The cover rate of the leaves will content the photosynthesis and the transpiration for the final purpose. The use of light and carbon for leaves influence the leaf shape. The distribution of leaves and branches influence the light absorbed by the leaves, so the shapes of the leaves are affected by the distribution of leaves and branches.

(5) The mass of the leaves is determined by the canopy height, volume and radius, so they have the certain relations.

Acknowledgment. Funds for this research was provided by the special scientific research funds for Ministry of water resources public welfare industry (201301096), the studying funds of Key laboratory of water saving in Heilongjiang province (2012KF503), the development potential of groundwater resources in Heilongjiang provincial water resources bureau(201318).

References

1. Niklas, K.J.: A mechanical perspective on foliage leaf form and function (1998)
2. Chen, S.Y.Y., Lestrel, P.E., Kerr, W.J.S., McColl, J.H.: Describing shape changes in the human mandible using Elliptic Fourier functions. Eur. J. Orthod. 22, 205–216 (2000)
3. Daliri, M.R., Torre, V.: Robust symbolic representation for shape recognition and retrieval. Pattern Recognition 41(5), 1782–1798 (2008)
4. Zheng, X.: Leaf Vein Extraction Based on Gray-scale Morphology (2010)
5. Royer, D.L., Meyerson, L.A., Robertson, K.M., et al.: Phenotypic plasticity of leaf shape along a temperature gradient in Acer rubrum. PLoS One 4(10), e7653 (2009)
6. Sarkar, D., Srimany, A., Pradeep, T.: Rapid identification of molecular changes in tulsi (Ocimum sanctum Linn) upon ageing using leaf spray ionization mass spectrometry. Analyst 137(19), 4559–4563 (2012)
7. Franz, E., Gebhardt, M.R., Unklesbay, K.B.: Shape description of completely visible and partially occluded leaves for identifying plants in digital images. Trans. ASAE 34(2), 1991 (1991a)

8. Sato, Y., Kumagai, T., Kume, A., et al.: Experimental analysis of moisture dynamics of litter layers—the effects of rainfall conditions and leaf shapes. Hydrological Processes 18(16), 3007–3018 (2004)
9. Li, Y.F., Zhu, Q.S., Cao, Y.K., Wang, C.L.: A Leaf Vein Extraction Method Based On Snakes Technique. In: International Conference on Neural Networks and Brain 2005, October 13-15, vol. 2, pp. 885–888 (2005)
10. Gunawardena, A.H., Greenwood, J.S., Dengler, N.G.: Programmed cell death remodels lace plant leaf shape during development. The Plant Cell Online 16(1), 60–73 (2004)
11. Mori, S., Hagihara, A.: Crown profile of foliage area characterized with the Weibull distribution in a hinoki (Chamaecyparis obtusa) stand. Trees 5, 149–152 (1991)

Numerical Simulation of Regulating Performance of Direct-Operated Pressure Regulator for a Microirrigation Lateral

Chen Zhang and Guangyong Li

College of Water Resources and Civil Engineering,
China Agricultural University, Beijing 100083, China
swzhangchen@163.com, lgyl@cau.edu.cn

Abstract. A lateral inlet direct-operated pressure regulator is a novel device for microirrigation system that ensures the equal operating pressure of the lateral inlet required for high uniformity. This study develops a computational fluid dynamics (CFD) model in combination with inlet pressure and regulation assembly displacement to analyze the outlet pressure of the pressure regulator. The model is validated by a comparison of experimental measurements, and the predicted results show good agreement. The effects of the regulation assembly displacement and geometrical structure (regulation assembly inlet height) on regulating performance are investigated. Results show that the magnitude of the regulation assembly movement affecting by inlet pressure, preset pressure, and flow rate significantly changes in the beginning of the regulation range and then changes slowly. The spring parameters can be designed according to the force–displacement characteristic (the F–L_v curve) of the T-shape regulating plunger. A greater regulation assembly inlet height corresponds to a lower preset pressure and less sensitivity of pressure loss to the movement of the regulation assembly. The pressure distribution through the regulator provides an improved understanding of the pressure difference in the regulating plunger with various displacements. The CFD model can reflect the motion characteristics of the regulation assembly and reveal the key factor of the regulator design. The results form the sound basis for future design and performance optimization of pressure regulator.

Keywords: microirrigation, pressure regulator, pressure regulating performance, CFD analysis.

1 Introduction

A microirrigation system located in a hilly and mountainous area or in a large submain unit will have varying lateral inlet pressure considerably along the submain because of terrain slope and hydraulic friction loss. Pressure regulators perform a critical function in ensuring the lateral inlet pressure required for high uniformity.

Bernuth and Baird [1] tested three brands of agricultural irrigation pressure regulators to characterize their performance. A line segments separate and linear

© IFIP International Federation for Information Processing 2015
D. Li and Y. Chen (Eds.): CCTA 2014, IFIP AICT 452, pp. 285–303, 2015.
DOI: 10.1007/978-3-319-19620-6_34

regression computer program was developed to determine the preset pressure and regulation range. Their work serves as a reference for testing and performance evaluation. According to Tian et al. [2], the main factors affecting pressure regulator performance are spring stiffness coefficient, spring length, upriver and downriver areas of the regulating part, and space between the regulating part and block cap. An orthogonal experiment was conducted to quantificationally analyze the effects of those factors on preset pressure. The results show that a linear relationship exists between the four factors and the preset pressure.

The development of computational fluid dynamics (CFD) has increased the accuracy of predicting the fluid characteristics and flow field distribution inside a fluid machinery. Many researches have been made on the flow prediction of relief valves and safety valves. Mokhtarzadeh-Dehghan et al. [3] conducted a finite element study on the flow of oil through a hydraulic pressure relief valve of the differential angle type. The results show that the total force on the plunger increases approximately linearly with increasing plunger lift. Depster et al. [4] and Depster and Elmayyah [5] reported a 2D representation by using Reynolds-averaged Navier–Stokes equations and standard k–ε turbulence model to investigate the force-lift and flow-lift characteristics of a spring-operated safety relief valve. Han et al. [6] presented the steady and transient characteristics of a Contra-push check valve (CPCV) simulated by CFD codes. Results show that the size of the gap between plug and sleeve is a key factor in CPCV design, another one is annular area of the plug. Chattopadhyay et al. [7] presented the simulating investigations of the flow through a spool type pressure regulating valve at different opening and different pressure drop. Song et al. [8] developed a simplified dynamic model to simulate the dynamic characteristics of a conventional pressure relief valve. The lift force coefficient of the valve at several fixed lifts was calculated using the static CFD analysis and was then imported into the dynamic model as the inherent characteristic of the valve to predict the plug lift during the reclosing process. Shahani et al. [9] performed dynamic simulation models to analyze the regulator performance of a high pressure regulator. The regulator behavior consist of the output pressure change versus time, the displacement of the moving parts versus time, and the regulator mass flow rate versus time were obtained. Beune et al. [10] developed a multi-mesh numerical valve model to analyze the opening characteristic of high-pressure safety valves. Song et al. [11] developed a numerical model including moving mesh techniques to investigate the fluid and dynamic characteristics of a direct-operated safety relieve valve. The effects of design parameters on flow forces acting on the disc and the lift of the valve are analyzed and compared.

For a lateral inlet direct-operated pressure regulator, the regulation assembly moves under different inlet pressure and flow rate conditions. However, the previous experimental studies pay no attention to the motion characteristic of the regulation assembly and its influence to the regulating performance. In this paper, a mathematical model combined with regulation assembly displacement and geometrical structure is developed to simulate the hydraulic performance of pressure regulator by CFD approaches. Basing on steady CFD analysis of the pressure regulator at several fixed displacements, the relationship among regulator inlet pressure, regulation assembly displacement, and the outlet pressure are obtained. The results of this study can be used as reference for the design and the performance optimization of the pressure regulator.

2 Working Principle of a Pressure Regulator

A lateral inlet direct operated pressure regulator mainly consists of eight parts (Fig. 1). The regulation assembly plays a part in regulating the outlet pressure, which consists of the regulation assembly inlet, spring, and T-shape regulating plunger. When water flows across the pressure regulator, head loss occurs in the gap between the regulation assembly inlet and the top surface of the T-shape regulating plunger. Thereafter, the hollow plunger transmits the water to the outlet.

When the inlet pressure is below the preset pressure, the outlet pressure increases with the inlet pressure increasing. The T-shape regulating plunger stay stationary for the pressure difference induced force on the plunger is less than the pre-stressed spring force (Fig. 2a). As the inlet pressure is higher than the preset pressure, the force acting on the plunger that overcomes the preloading of the spring forces the plunger to move against the flow direction (Fig. 2b). Therefore, the gap becomes narrow, and the friction loss increases. Consequently, the outlet pressure decreases and ensures the regulator to maintain its constant outlet pressure. When the inlet pressure decreases from a high value to the preset pressure, the movement process of the regulation assembly becomes the opposite.

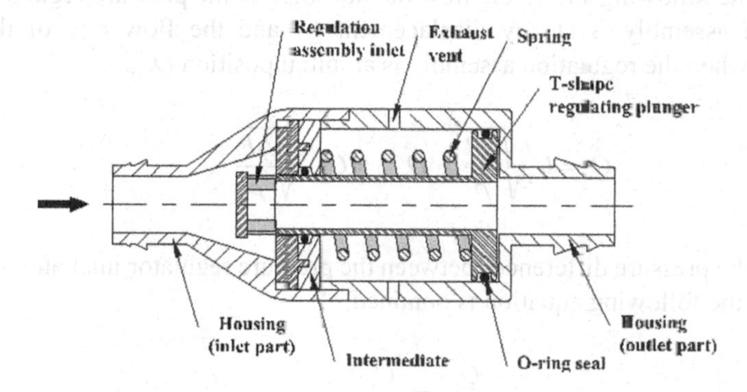

Fig. 1. Schematic cross-section of a pressure regulator

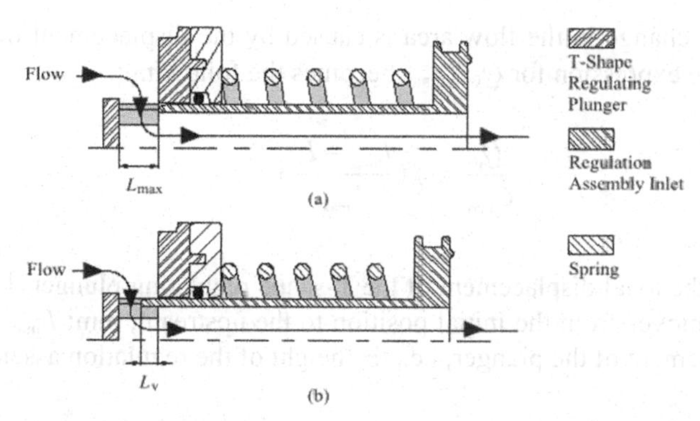

Fig. 2. Regulating process of a pressure regulation assembly (the axisymmetric part) (a) Before regulation range and (b) regulation range

3 Numerical Methods

3.1 Mathematical Models

Outlet Pressure and Displacement Expression

On the basis of the flow rate formula, the flow rate of the pressure regulator is expressed as follows:

$$Q = C\sqrt{\frac{\Delta P}{\rho}}, \tag{1}$$

where Q is the flow rate throughout the pressure regulator, L/h, and is calculated as $Q=vA$, where v is the average velocity in the pipe, m/s and A is the nominal cross-section, mm^2; ΔP is the pressure difference between the pressure regulator inlet and the outlet, Pa; C is the flow coefficient; and ρ is the water density, 1000 kg/m^3.

The flow area of the regulation assembly inlet changes with different pressure drops (Fig. 2).The following shows the flow rate throughout the pressure regulator when the regulation assembly is at any displacement Q_v and the flow rate of the pressure regulator when the regulation assembly is at initial position Q_{max}:

$$Q_v = C_v\sqrt{\frac{\Delta P}{\rho}},\quad Q_{max} = C_{max}\sqrt{\frac{\Delta P}{\rho}}.$$

When the pressure differences between the pressure regulator inlet and the outlet are the same, the following equation is obtained:

$$\frac{Q_v}{Q_{max}} = \frac{C_v}{C_{max}}. \tag{2}$$

When the change in the flow area is caused by the displacement of the regulation assembly, the expression for Q_v/Q_{max} becomes the following:

$$\frac{Q_v}{Q_{max}} = f\left(\frac{L_{max} - L_v}{L_{max}}\right), \tag{3}$$

where L_v is the axial displacement of the T-shape regulating plunger, i.e., the distance the plunger moves from the initial position to the upstream, mm; L_{max} is the maximal axial displacement of the plunger, i.e., the height of the regulation assembly inlet, mm.

From Equation (1), the pressure difference between the pressure regulator inlet and outlet ΔP is expressed as follows:

$$\Delta P = \frac{\rho Q^2}{C^2} \quad (4)$$

By combining Equations (2) to (4), Equation (4) can be written as

$$\Delta P = \frac{\rho Q^2}{f(\frac{L_{max} - L_v}{L_{max}})^2 C_{max}^{\ 2}} \quad (5)$$

The outlet pressure equation of the pressure regulator can be written as

$$P_{out} = P_{in} - \Delta P = P_{in} - \frac{\rho Q^2}{f(1-\frac{L_v}{L_{max}})^2 C_{max}^{\ 2}}, \quad (6)$$

where P_{in}, P_{out} are the inlet and outlet pressures of the pressure regulator, respectively, Pa. The function expressions are provided by the results from the numerical simulation in the following section.

Equations of the Mechanical Models

Fig. 3 shows the force analysis of the T-shape regulating plunger. Friction force is assumed to be negligible. A force balance equation of the plunger is formulated as

$$F_s + P_1 A_1 = P_2 A_2, \quad (7)$$

where the spring force F_s is given by

$$F_s = K_s (L_0 + L_v) \quad (8)$$

The force acting on the T-shape regulating plunger F is induced by the pressure difference between the top and bottom surfaces of the plunger. The expression for force F then becomes

$$F = P_2 A_2 - P_1 A_1, \quad (9)$$

where K_s is the spring stiffness, N/mm; L_0 is the pre-stressed spring length, mm; P_1, P_2 are the pressures on the top and bottom surfaces of the T-shape regulating plunger, respectively, Pa; A_1, A_2 are the areas of the top and bottom surfaces of the plunger, respectively. The values of A_1, A_2 are 12.7and 510mm^2, respectively.

Governing Equations and Turbulence Models

To describe the 3D flow phenomenon in a pressure regulator, uncompressible Navier–Stokes equations are numerically solved in the commercial software package ANSYS FLUENT.

Conservation of mass is expressed as

$$\frac{\partial}{\partial x_i} u_i = 0 \ . \tag{10}$$

Conservation of momentum is expressed as

$$\frac{\partial}{\partial t}(\rho u_i) + \frac{\partial}{\partial x_i}(\rho u_i u_j) = -\frac{\partial p}{\partial x_i} + \frac{\partial \tau_{ij}}{\partial x_j} + \rho g_i + F_i \ , \tag{11}$$

where u_i is the fluid velocity component in the i direction, g_i is the acceleration of the gravity component in the i direction, $\tau_{ij} = \left[\mu_{eff} \left(\frac{\partial u_i}{\partial x_j} + \frac{\partial u_j}{\partial x_i} \right) \right] - \frac{2}{3} \mu_{eff} \frac{\partial u_l}{\partial x_l} \delta_{ij}$, and δ_{ij} is the Kronecker symbol.

The turbulent kinetic energy, k, and its rate of dissipation, ε, are obtained from the following transport equations in the standard k–ε model:

$$\frac{\partial}{\partial t}(\rho k) + \frac{\partial}{\partial x_i}(\rho k u_i) = \frac{\partial}{\partial x_j}\left[\left(\mu + \frac{\mu_t}{\sigma_k} \right) \frac{\partial k}{\partial x_j} \right] + G_K - \rho\varepsilon \ , \tag{12}$$

and

$$\frac{\partial}{\partial t}(\rho\varepsilon) + \frac{\partial}{\partial x_i}(\rho\varepsilon u_i) = \frac{\partial}{\partial x_j}\left[\left(\mu + \frac{\mu_t}{\sigma_\varepsilon} \right) \frac{\partial \varepsilon}{\partial x_j} \right] + C_{1\varepsilon} \frac{\varepsilon}{k} G_k - C_{2\varepsilon} \rho \frac{\varepsilon^2}{k} \ , \tag{13}$$

where μ is the coefficient of dynamic viscosity; G_K is the generation of turbulence kinetic energy due to the mean velocity gradients; and the model constants C_μ, $C_{1\varepsilon}$, $C_{2\varepsilon}$, σ_k, and σ_ε have the following value: C_μ=0.09, $C_{1\varepsilon}$=1.44, $C_{2\varepsilon}$=1.92, σ_k=1.0, and σ_ε=1.22.

Equation (6) shows that the outlet pressure P_{out} is calculated by L_v, and Q_v. Equations (7) – (9) indicate that the spring stiffness K_s and pre-stressed spring length L_0 are determined by L_v, and F. However, the value of L_v in a certain inlet pressure cannot be obtained directly from the experiment. Hence, a series of discrete models at different displacements are established, and then the F–L_v curve is fitted with the CFD results at different positions. The procedure for the F–L_v curve calculation of the pressure regulator is shown in Fig. 4.

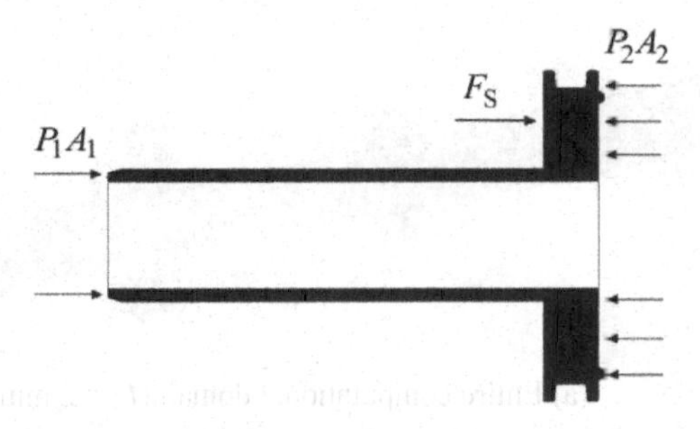

Fig. 3. Force acting on the T-shape regulating plunger

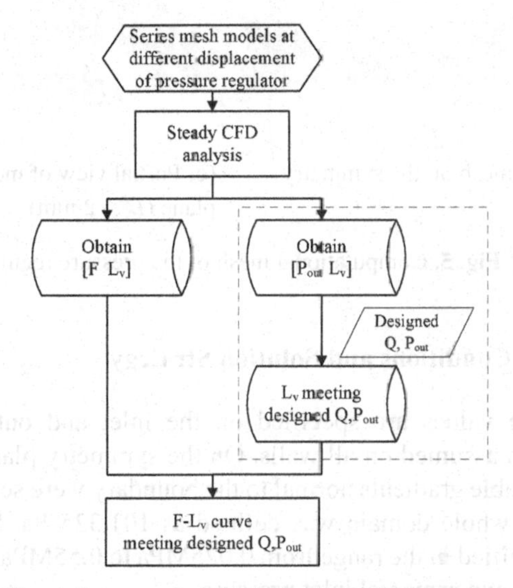

Fig. 4. Strategy for obtaining the F–L_v curve

3.2 Computational Mesh

To reduce computational time, the computational domain is reduced to half of the full 3D pressure regulator model. Considering the fully developed fluid, the pipeline was extended to 5 times the nominal diameter in the inlet and outlet. The full 3D structured hexahedral mesh is generated by using the commercial mesh generator ICEM CFD (Fig. 5). To obtain good quality and to represent the details of the regulation assembly inlet, a domain decomposition method that splits the entire complex pressure regulator interior domain into several subdomains is applied [12]. The interface boundary is set to the corresponding zones of the subdomains. Mesh quality has been rigorously checked for parameters, such as skewers and aspect ratio. The mesh quality metrics in ICEM CFD shows that the mesh quality of the model is greater than 0.65, which is sufficient for this case. Mesh independence for a 3D simulation is undertaken by using 4 different mesh levels composed of approximately 100,000, 200,000, 500,000, and 800,000 nodes, respectively. The mesh with 500,000 nodes is selected for the numerical analysis. The number of mesh models range from 490,000 to 540,000 for the different displacement models.

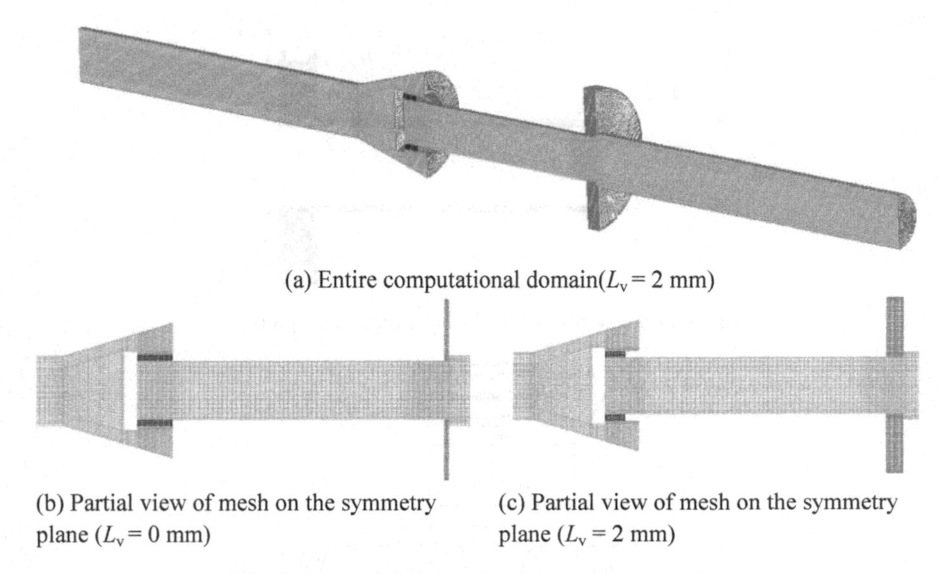

(a) Entire computational domain(L_v = 2 mm)

(b) Partial view of mesh on the symmetry plane (L_v = 0 mm)

(c) Partial view of mesh on the symmetry plane (L_v = 2 mm)

Fig. 5. Computational mesh of the pressure regulator

3.3 Boundary Conditions and Solution Strategy

Different pressure values are specified on the inlet and outlet. No-slip boundary condition has been assumed on all walls. On the symmetry plane, the scalar variables and the scalar variable gradients normal to the boundary were set to zero. The reference pressure over the whole domain was defined as 101,325 Pa. Several values of inlet pressure were specified in the range from 0.075MPa to 0.55MPa, which were contained in the range of the experimental inlet pressure.

The second-order upwind scheme was used for descretization of momentum and the SIMPLE algorithm was applied for coupling of the pressure and velocity, the standard k-ε model for turbulent flow. The continuity equation and momentum were considered converged when the residuals for each of variables reached to the order 10^{-5}.

4 Results and Discussion

4.1 Outlet Pressure and Displacement Expression

Fig. 6 shows the predicted flow rate relative to the maximum flow rate in each fixed displacement for the pressure regulators of different regulation assembly inlet heights. The regression equation is obtained as follows:

$$\frac{Q_v}{Q_{\max}} = a\ln(1-\frac{L_v}{L_{\max}})+b \quad . \tag{14}$$

The values of coefficients for different regulation assembly inlet heights are given in Table 1. The maximum flow coefficient is obtained by Equation (1). By substituting Equations (14) into Equation (6), the following form can be obtained:

$$P_{out} = P_{in} - \frac{\rho Q^2}{[a\ln(1-\frac{L_v}{L_{max}})+b]^2 C_{max}{}^2} \tag{15}$$

Table 1. a, b, and C_{max} values of the pressure regulator for different L_{max}.

L_{max} /mm	4	5	6	7
a	0.3474	0.3442	0.309	0.2939
b	1.0368	1.0567	1.0756	1.0994
C_{max}	4.80E-05	4.93E-05	5.00E-05	5.01E-05

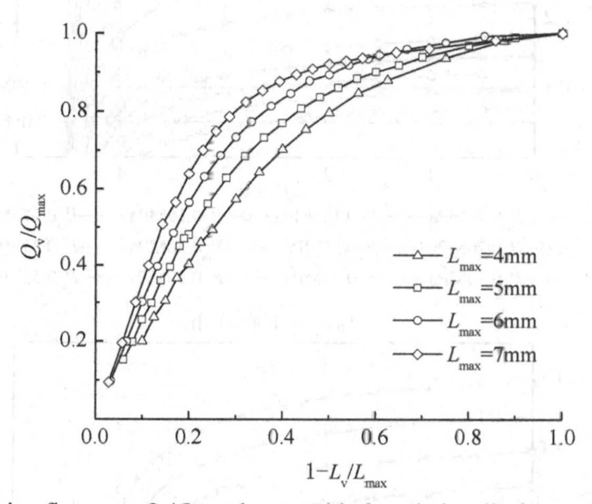

Fig. 6. Relative flow rate Q_v/Q_{max} change with the relative displacement L_v/L_{max}

Fig. 7 shows the outlet pressure change in different displacements under different inlet pressures as calculated by Equation (15). The outlet pressure slowly changes when the displacement is small. When the displacement increases, the outlet pressure rapidly decreases. A higher inlet pressure corresponds to a greater variation of outlet pressure when the displacement is large. The flow rate affects the outlet pressure.

Under the same inlet pressure, the comparison between Figs. 7(a) and (b) shows that a higher the flow rate corresponds to a lower outlet pressure. An ideal regulator can maintain a constant outlet pressure regardless of the inlet pressure or flow rate, provided that the inlet pressure is above the preset pressure. The constant preset pressure 0.05 and 0.1MPa, which represent the two straight lines, are plotted in Fig. 7. The intersection of the straight lines and outlet pressure curves provide the value of displacement that the T-shape regulating plunger will reach under that inlet pressure (Fig. 8).

The data shows the law of movement of the regulation assembly. The displacement of the T-shape regulating plunger increases rapidly as the inlet pressure increases at the beginning of the regulating process. When the inlet pressure is greater than 0.2MPa (Fig.8 a), the distance increases slightly from 4 mm to 4.5 mm. Comparing the two motions with the same flow rate but different preset pressure show that the regulation assembly set low preset pressure moves first. However, the displacements are almost the same when the inlet pressure is high. A decrease in flow rate tends to advance the motion of the regulation assembly and increases displacement (Fig. 8).

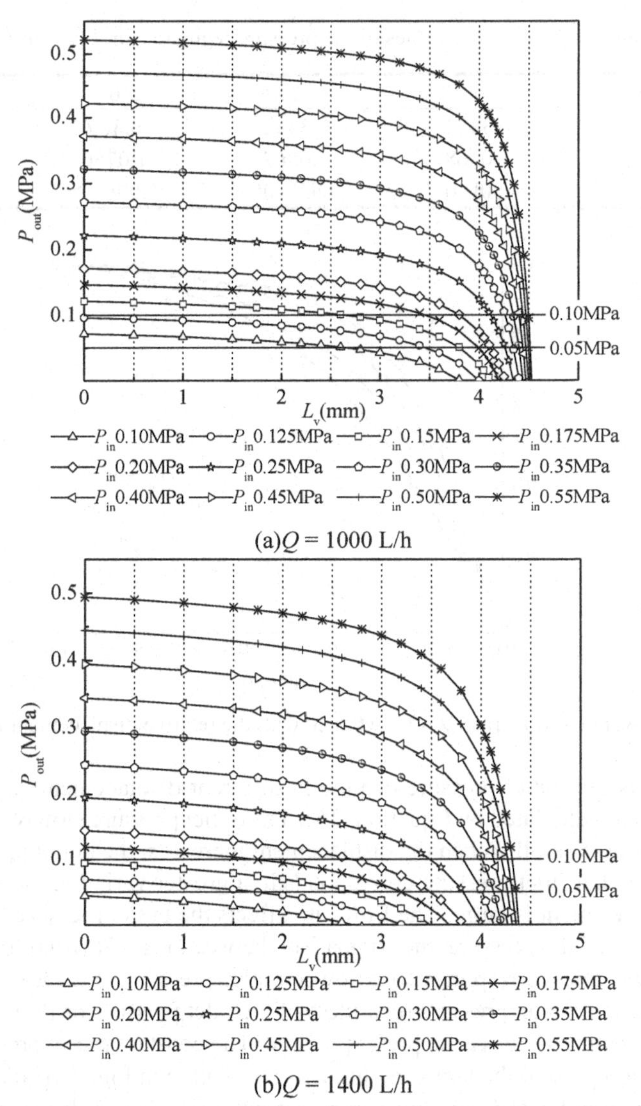

Fig. 7. Outlet pressure change with different displacements at different inlet pressures (L_{max} = 5 mm)

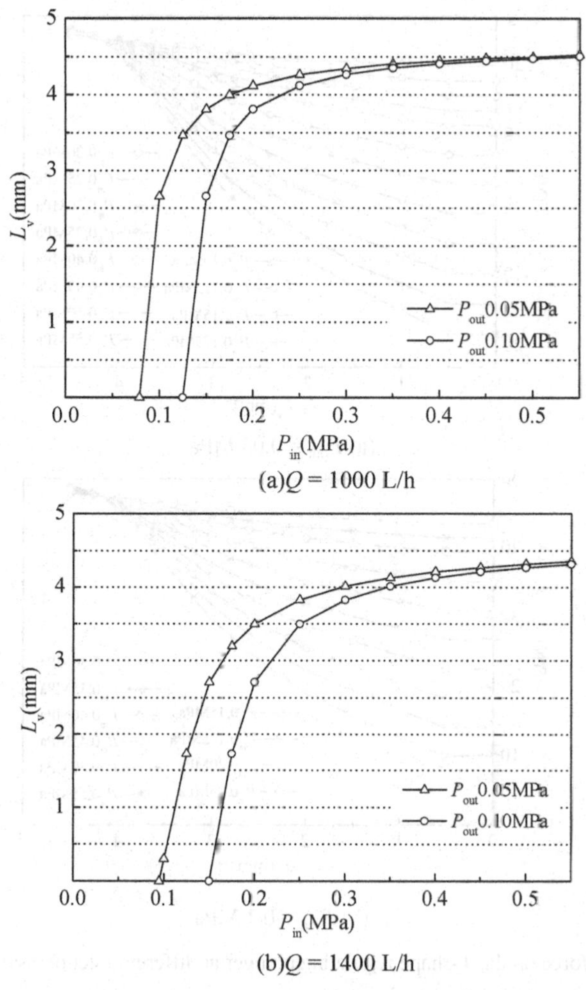

Fig. 8. Displacement of the T-shape regulating plunger at different inlet pressures(L_{max} = 5 mm)

4.2 Displacement-Force Characteristics

Fig. 9 shows the axial force induced by the pressure difference on the T-shape regulating plunger for two preset pressure conditions. When the pressure difference between the inlet and outlet is the same, the force acting on the plunger increases with increasing displacement. A quadratic polynomial functional relationship is analyzed between the displacement and the force. A high preset pressure leads to a large force magnitude. Therefore, a large spring per-stressed force should be exerted for the pressure regulator with a high preset pressure.

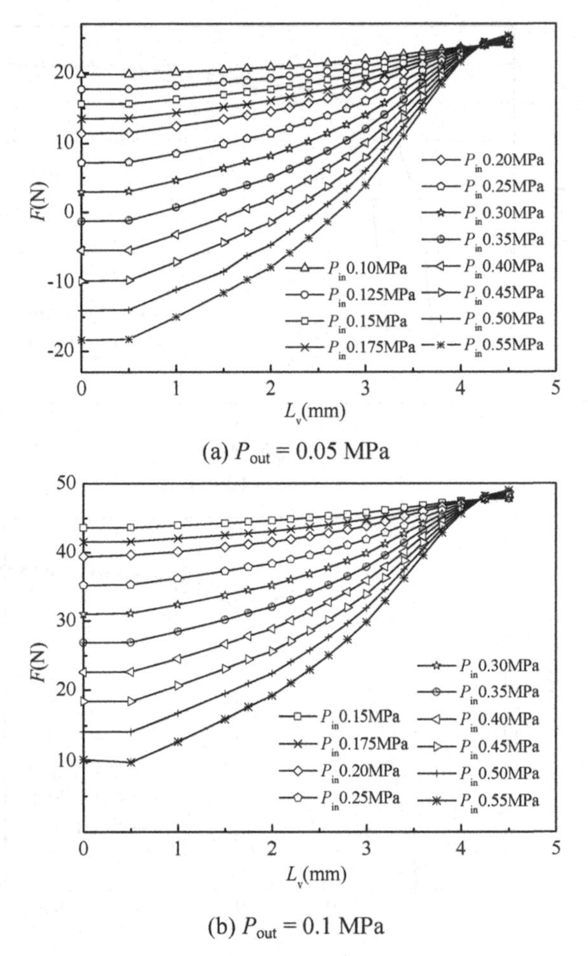

(a) $P_{out} = 0.05$ MPa

(b) $P_{out} = 0.1$ MPa

Fig. 9. Predicted force on the T-shape regulating plunger at different inlet pressures ($L_{max} = 5$ mm)

By mapping Figs. 8 to 9, the relationship between the force acting on the T-shape regulating plunger (F) and the displacement (L_v) meeting the preset pressure and flow rate requirements was obtained Fig. 10 shows the F–L_v curves for the different flow rates with preset pressures of 0.05 and 0.1MPa. The pressure induced force has a nonlinear relationship with the displacement. When the displacement is smaller than 3 mm, the force acting on the plunger slightly changes. The force increases sharply when the plunger increases up to the maximum displacement. Note that, as the spring force satisfies the Hooke's Law, the spring force (the F_s–L_v line) will intersect with the F–L_v curve but cannot coincide with the curve. Therefore, the reasonable spring parameters, i.e., spring stiffness and pre-stressed spring length, can be designed by providing the minimum deviation between the F_s–L_v line and the F–L_v curve.

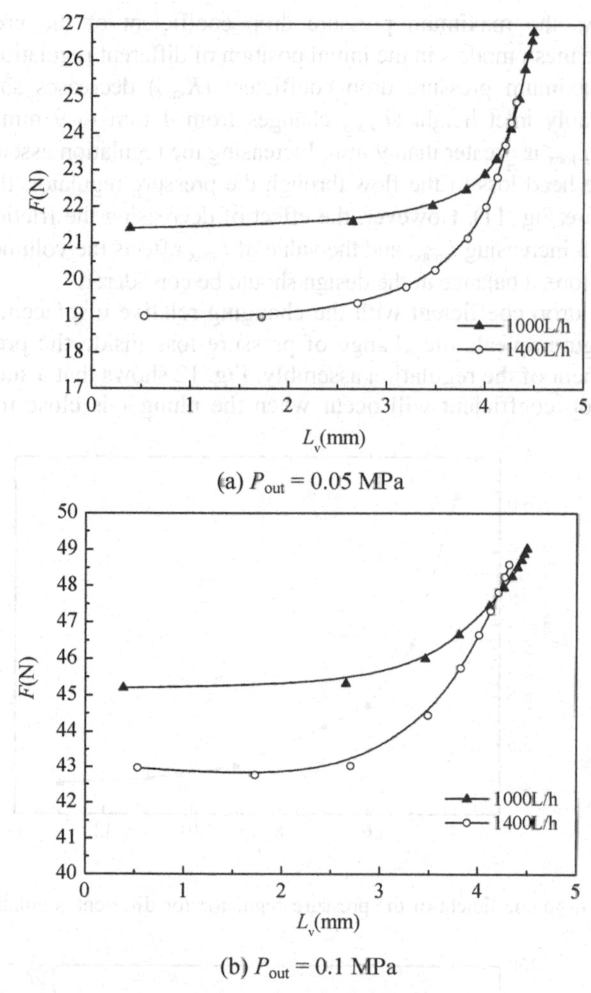

(a) P_{out} = 0.05 MPa

(b) P_{out} = 0.1 MPa

Fig. 10. Fitted F– L_v curves (L_{max} = 5 mm)

The flow rate of the pressure regulator is not a constant value but a range of values, which means that the outlet pressure variation caused by the flow rate variation does not exceed the limit within the flow rate range. In this paper, a certain flow rate that meets the requirement is considered as the upper limit of the discharge.

4.3 Effect of the Regulation Assembly Inlet Height

The hydraulic behavior of the pressure regulator is characterized by the local pressure drop introduced by the regulator itself [13]. This pressure drop is made nondimensional by the mean dynamic pressure in the regulator. The pressure drop coefficient is expressed as follows:

$$K = \frac{2\Delta P}{\rho v^2}. \tag{16}$$

Fig. 11 shows the maximum pressure drop coefficient of the pressure regulator calculated by the mesh models in the initial position of different regulation assembly inlet heights. The maximum pressure drop coefficient (K_{max}) decreases sharply when the regulation assembly inlet height (L_{max}) changes from 4 mm to 9 mm K_{max} decreases gradually when L_{max} is greater than 9 mm. Increasing the regulation assembly inlet height can decrease the head loss of the flow through the pressure regulator, thus reducing the operating pressure(Fig. 11). However, the effect of decreasing the friction loss becomes insignificant with increasing L_{max}, and the value of L_{max} affects the volume of the pressure regulator. Therefore, a balance in the design should be considered.

The pressure drop coefficient with the changing relative displacement is shown in Fig. 12. This figure reveals the change of pressure loss inside the pressure regulator with the movement of the regulation assembly. Fig. 12 shows that a sudden increase in the pressure drop coefficient will occur when the plunger is close to the maximum

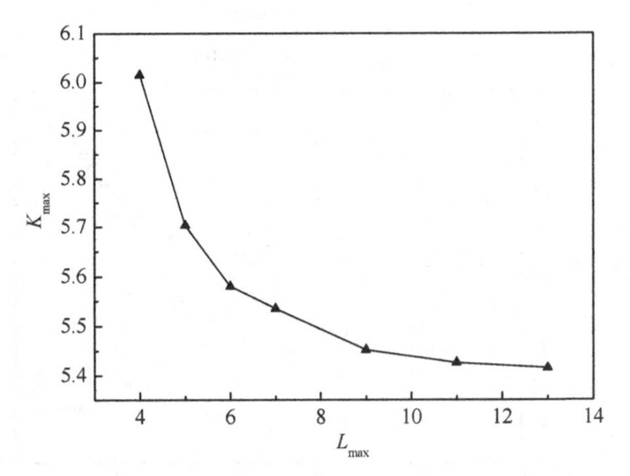

Fig. 11. Pressure drop coefficient of the pressure regulator for different regulation assembly inlet heights

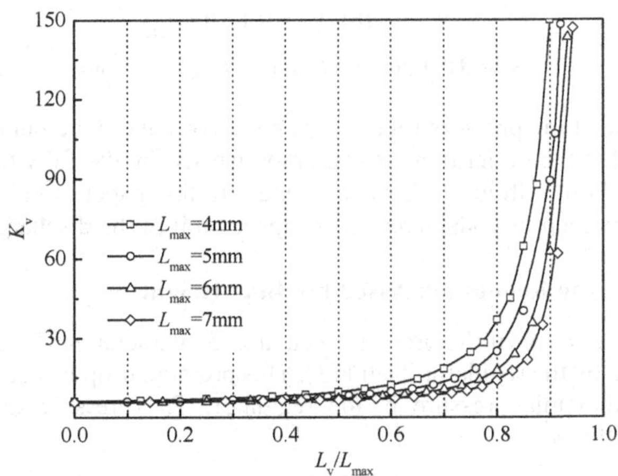

Fig. 12. Pressure drop coefficient changes with the relative displacement for different regulation assembly inlet heights

displacement. Furthermore, curve L_{max}=4 mm shows that when the displacement is over 60% of the regulation assembly inlet height, the pressure drop coefficient will begin to increase rapidly. However, the curve for L_{max}=7 mm shows a sudden increase point of approximately 80% of the regulation assembly inlet height. Therefore, increasing the regulation assembly inlet height can decrease the sensitivity of pressure loss to the movement of the regulation assembly and ensure a stable outlet pressure.

4.4 Pressure Distribution

The pressure distribution in the same inlet and outlet pressure conditions at six different displacements are shown in Fig. 13. The pressure contours are distributed densely in the regulation assembly inlet and upstream of the plunger, thus indicating that the pressure loss mainly occurs at those regions. As the T-shape regulating plunger changing its position towards the regulator inlet, the high pressure stagnates in the outer region of the regulation assembly inlet. The largest pressure drop position becomes centralized to the region between the top surface of the plunger and the regulation assembly inlet. The movement of the regulation assembly results in changes in the pressure distribution of the pressure regulator, thus, the force produced by the pressure difference between the top and bottom surfaces of the plunger is affected significantly.

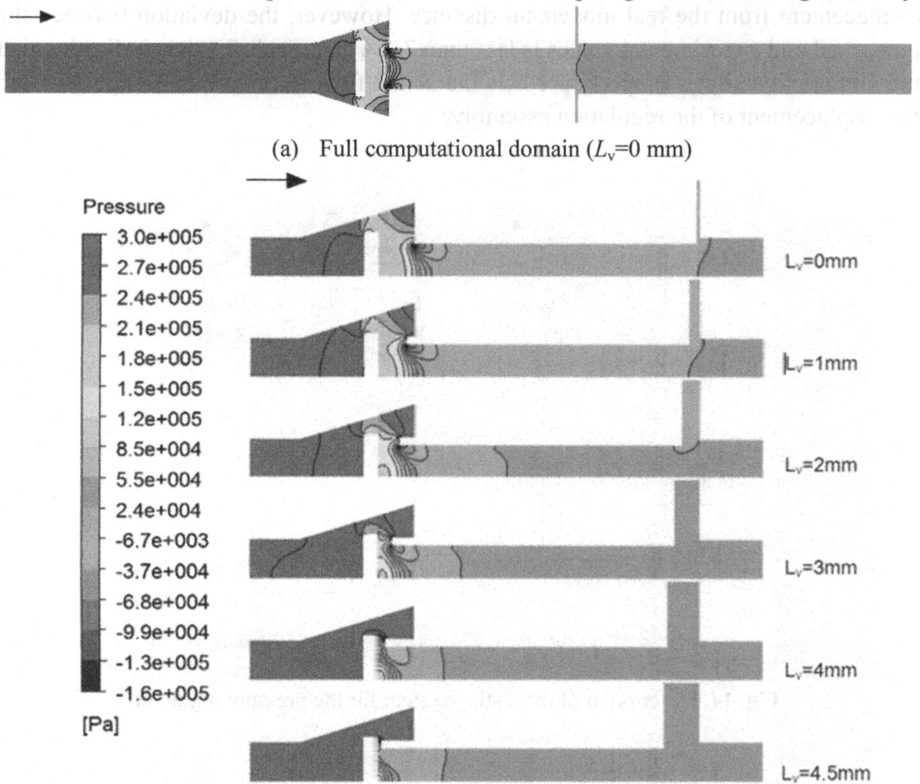

(a) Full computational domain (L_v=0 mm)

(b) Upstream and downstream regions of the T-shaped regulating plunger

Fig. 13. Pressure distribution on the symmetry plane (L_{max} = 5 mm, P_{in} = 0.3 MPa, P_{out} = 0.05 MPa)

4.5 Validation of the Computational Model

To validate the predicted results, several experiments were applied to the pressure regulator (5mm regulation assembly inlet height, 0.05MPa preset pressure, 2.31 N/mm spring stiffness, and 5.5mm pre-stressed spring length). A layout and schematic drawing of the experimental setup is shown in Fig. 14. The setup is created to measure the outlet pressure and the flow rate through the regulator in a series of inlet pressures. By substituting the tested pressure and flow rate data into Equation (15), the displacement of the T-shaped regulating plunger can be obtained (Fig. 15).

The accuracy of the numerical model was verified by modeling the geometry with the calculated displacement and repeating the CFD calculations. Fig. 16 shows a comparison of flow rates between the CFD model and experimental results. Table 2 shows the deviation of the numerical and experimental flow rates. A low flow rate is predicted before the regulation range (inlet pressure of less than 0.05MPa) (Fig. 16). This error can be accounted for by the difference in the geometrical model and testing production. Within the regulation range, the difference between the predicted and experimental flow rate is caused by the additional deviation of the calculated displacement from the real movement distance. However, the deviation between the numerical and experimental values is less than 7.3% (see Table 2), thus indicating that the numerical analysis properly predicts the actual flow in the pressure regulator and the displacement of the regulation assembly.

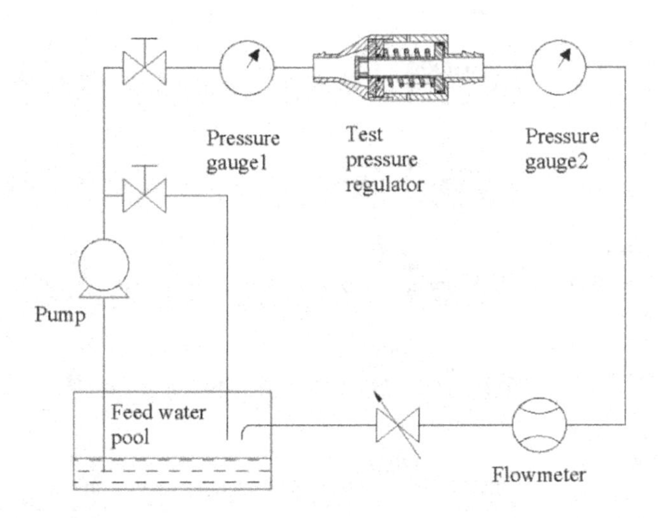

Fig. 14. Description of the testing system for the pressure regulator

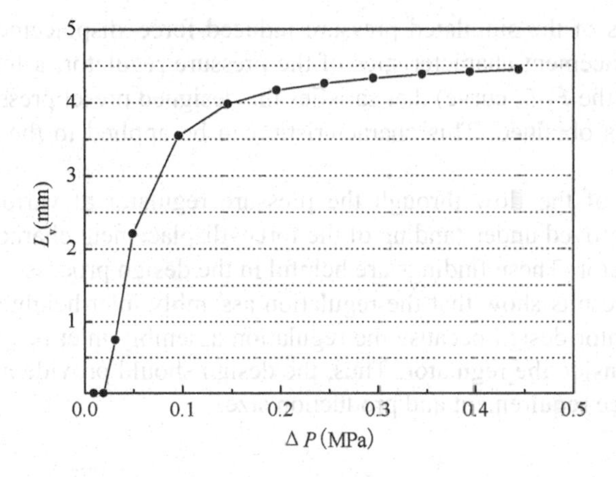

Fig. 15. Calculated displacement of the testing pressure regulator (L_{max} = 5 mm)

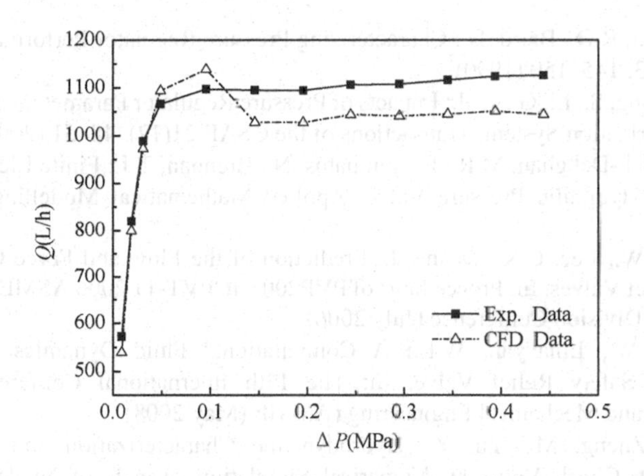

Fig. 16. Comparison of the numerical and experimental flow rates (L_{max} = 5 mm)

Table 2. Deviation of the numerical and experimental flow rates

ΔP (MPa)	0.01	0.02	0.03	0.05	0.1	0.15	0.2	0.25	0.3	0.35	0.39	0.44
Deviation (%)	-5.9	-2.4	-1.5	2.7	3.9	-6.1	-6	-5.6	-6.1	-6.4	-6.4	-7.3

5 Conclusions

The flow process inside a lateral inlet pressure regulator has been investigated by commercial ANSYS FLUENT code. A mathematical model that features the inlet pressure and displacement to the outlet pressure is presented by the CFD approach. This model reveals the movement of the regulation assembly inside the regulator and provides the value of displacement that the experiment cannot directly provide.

On the basis of the simulated pressure induced force–displacement and the outlet pressure–displacement characteristics of the pressure regulator, a force–displacement characteristic (the $F–L_v$ curve) that satisfies the designed preset pressure and flow rate requirements is obtained. This characteristic can be applied to the spring parameter design.

The details of the flow through the pressure regulator at various displacements provide an improved understanding of the force–displacement characteristic inside the pressure regulator. These findings are helpful in the design process.

Numerical results show that the regulation assembly inlet height is a key factor in pressure regulator design because the regulation assembly inlet height determines the pressure loss inside the regulator. Thus, the design should provide a balance between the performance requirement and production size.

References

1. Bernuth von, R.D., Baird, D.: Characterizing Pressure Regulator Performance. Transaction of ASAE 33, 145–150 (1990)
2. Tian, J., Gong, S., Li, G., et al.: Impacts of Pressure Regulator Parameters on Preset Pressure in Micro-irrigation System. Transactions of the CSAE 21(12), 48–51 (2005)
3. Mokhtanadeh-Dehghan, M.R., Ladommatos, N., Brennan, T.J.: Finite Element Analysis of Flow in a Hydraulic Pressure Valve. Applied Mathematical Modelling 21(7), 437–445 (1997)
4. Dempster, W., Lee, C.K., Deans, J.: Prediction of the Flow and Force Characteristics of Safety Relief Valves. In: Proceedings of PVP2006-ICPVT-11 2006 ASME Pressure Vessels and Piping Division Conference (July 2006)
5. Dempster, W., Lmayyah, W.E.: A Computational Fluid Dynamics Evaluation of a Pneumatic Safety Relief Valve. In: The 13th International Conference on Applied Mechanics and Mechanical Engineering (AMME (May 2008)
6. Han, X., Zheng, M., Yu, Y.: Hydrodynamic Characterization and Optimization of Contra-push Check Valve by Numerical Simulation. Annals of Nuclear Energy 38(6), 1427–1437 (2011)
7. Chattopadhyay, H., Kundu, A., Saha Binod, K., et al.: Analysis of Flow Structure Inside a Spool Type Pressure Regulating Valve. Energy Conversion and Management 53(1), 196–204 (2012)
8. Song, X., Park, Y., Park, J.: Blowdown Prediction of a Conventional Pressure Relief Valve with a Simplified Dynamic Model. Mathematical and Computer Modelling 57(1-2), 279–288 (2013)
9. Shahani, A.R., Esmaili, H., Aryaei, A., et al.: Dynamic Simulation of a High Pressure Regulator. Journal of Computational and Applied Research in Mechanical Engineering 1(1), 17–28 (2011)
10. Beune, A., Kuerten, J.G.M., Heumen van, M.P.C.: CFD Analysis with Fluid-Structure Interaction of Opening High-Pressure Safety Valves. Computers & Fluids 64(15), 108–116 (2012)
11. Song, X., Cui, L., Cao, M., et al.: A CFD analysis of the dynamics of a direct-operated safety relief valve mounted on a pressure vessel. Energy Conversion and Management 81, 407–419 (2014)

12. Song, X., Wang, L., Park, Y.: Transient Analysis of a Spring-Loaded Pressure Safety Valve Using Computational Fluid Dynamics (CFD). Transactions of the ASME 132, 054501/1–054501/5 (2010)
13. Jeon, S.Y., Yoon, J.Y., Shin, M.S.: Flow Characteristics and Performance Evaluation of Butterfly Valves Using Numerical Analysis. In: IOP Conference Series Earth and Environmental Science 12(012099), vol. (1), pp. 1–7 (2099)

Analysis and Research of K-means Algorithm in Soil Fertility Based on Hadoop Platform

Guifen Chen[1], Yuqin Yang[1], Hongliang Guo[1], Xionghui Sun[1], Hang Chen[1,2], and Lixia Cai[1]

[1] Jilin Agricultural University, Changchun 130118, China
[2] Jilin provincial science and technology information Research Institute, Changchun130033, China
guifchen@163.com, 1172126066@qq.com

Abstract. In order to study the K - means algorithm for evaluation of soil fertility, solve the large amount of calculation and high time complexity of the algorithm, this paper proposes the K-means algorithm based on Hadoop platform. First, K-means algorithm is used to cluster for Nongan town soil nutrient data for nine consecutive years; clustering results show that: the accuracy rate increased year by year, and consistent with the actual situation. Then for the K-means clustering algorithm in processing large amounts of data has the disadvantages of high time complexity, This paper uses the K-means algorithm Based on Hadoop platform to realize the clustering analysis of soil fertility of large amounts of data; the results show that: compared with the traditional serial K-means algorithms, improves the operation speed. The above analysis shows that, K- means algorithm is an effective soil fertility evaluation method; Based on Hadoop platform of parallel K-means algorithm has great realistic meaning to analysis of large amount of data of soil fertility factors.

Keywords: K-means algorithm, Hadoop platform, MapReduce model, Soil fertility.

1 Introduction

With the wide application of agricultural information technology, 3S technology (GPS, GIS, RS),the Internet of things technology and Expert System (ES) technology are applied extensively in precision agriculture, so that the rapid growth of the agricultural sector data [1]. K - means algorithm based on Hadoop platform, can quickly and accurately to the large amount of data of soil fertility carries on the comprehensive evaluation and correct analysis, is of great significance to guide farmers reasonable fertilization. Xiong Chunhong for Jiang Xi tea area soil fertility problems and the status of the heavy metals of fresh tea leaves, she uses mining techniques were analyzed and predicted, creating favorable conditions for soil fertility analysis and comprehensive evaluation [2]. Li Lianghou proposed application of clustering anslysis in classifying site type and evaluating soil fertility[3].The research on weighted space fuzzy dynamic clustering algorithm by Chen Gui-fen, proved the effectiveness of soil fertility

© IFIP International Federation for Information Processing 2015
D. Li and Y. Chen (Eds.): CCTA 2014, IFIP AICT 452, pp. 304–312, 2015.
DOI: 10.1007/978-3-319-19620-6_35

evaluation[4].The traditional clustering algorithm in processing large-scale data in terms of efficiency of real-time or from the angle of system resources, are not well resolved. In the clustering algorithm, K-means algorithm is the most widely used clustering algorithm based on partition, it has the advantages of rapid and simple; but it has the disadvantage of large amount of calculation and sensitive to the initialization center. In contrast, this paper studies the K-means clustering algorithm based on Hadoop platform of MapReduce parallel programming method, and relevant experiments were carried out for the large-scale soil fertility data .

2 Key Technologies and Algorithms Introduced

2.1 MapReduce Model

MapReduce is a distributed parallel computing model proposed by Google Labs, its basic idea[5,6]:(1)The MapReduce database to user program input data set is divided into several small data set, tthen fork copies user processes to other machines in the cluster;(2)A copy of user program has a called master is responsible for scheduling, allocation of jobs to idle worker(Map worker or Reduce worker); (3)After the worker is assigned a Map job, read data from the input data and extract key value pairs, performs Map computation tasks and generate the intermediate key value pairs, and then cached in memory;(4)The middle of the cache key value to be written to a local disk regularly, and is divided into R districts, each corresponding to a Reduce worker, master is responsible for forwarding the middle of the key positions to Reduce worker;(5)Reduce worker read the intermediate key value pairs and to sort them out, so that the same keys of the key to gather together;(6)After sorting the intermediate key value to be Reduce worker traversal, and then for each unique keys, reduce worker will pass key and values associated with the key to the Reduce function, the output generated by the Reduce function will be added to the output file in the partition;(7)When all the Map and Reduce operations are completed, master wakes genuine user program, MapReduce function call returns user program code.[6] MapReduce operation process shown in Fig. 1[7,8]:

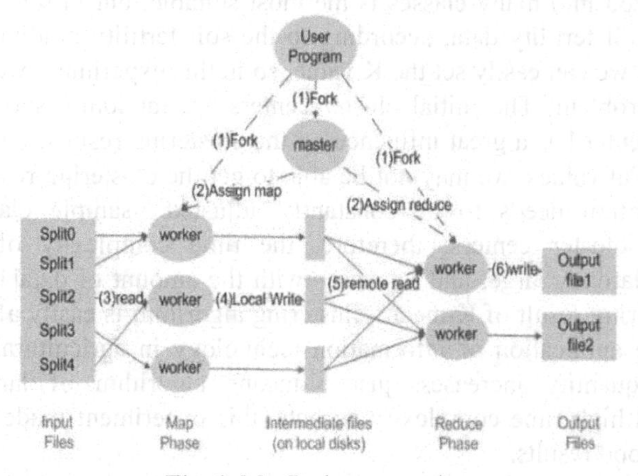

Fig. 1. MapReduce operation

2.2 K-means Algorithm

K-means algorithm is a clustering algorithm based on partition method, it is one of the earliest proposed classic clustering algorithm. The main idea of the K-means algorithm is the n object into the k class (cluster) ($k \leq n$).In this experiment we refer to the "National Survey of cultivated land fertility and quality evaluation of technical regulations" in the grading standard, Nong'an basic farmland is divided into six grades [9]. First of all, in the whole soil fertility data set we randomly selected k objects, each object represents an initial cluster center or the initial average. For each remaining object, according to its distance from the center of each cluster, assign it to the nearest cluster. And then calculate the average of each cluster, each object in the database is compared with the average value of each cluster, the object is assigned to the most similar clusters. This process is repeated until the cluster objects are "similar", while objects in different clusters are "dissimilar", namely the criterion function converges and make the minimum square error function.

K - means algorithm steps:

Algorithm: K - means. Division of the K- means algorithm based on the average of the objects in the cluster.

Input: the number of clusters k and the database contains n objects.

Output: k clusters, make the least square error criterion.

Methods:

(1) Choose k objects as the initial cluster centers;

(2) Repeat;

(3) According to the average of the objects in the cluster,each object (again) is assigned to the most similar cluster;

(4) Update the average of the cluster, namely calculating the average of objects in each cluster;

(5) To calculate the clustering criterion function;

(6) Until criterion function will not change.[10]

The disadvantage of K-means algorithm: When using K-means clustering algorithm, k values are given in advance. Normally, We don't know a given data set should be divided into many classes is the most suitable, but in this experiment, we are aiming at soil fertility data, according to the soil fertility grading standards and past experience we can easily set the K value, so in this experiment we do not need to consider this problem; The initial cluster centers are randomly selected, select the initial cluster center has a great influence on the clustering results, once the selection is not good initial values, we may not be able to get the clustering results effectively; K-means algorithm needs to be constantly adjusted sample classification and calculate new cluster centers; therefore, the time complexity of the K-means algorithm is relatively large, and increases with the amount of data[11];At the same time, the clustering result of K-means clustering algorithm is easily affected by noise data.; With the application of information technology in agricultural field, the soil fertility data quantity increases, just K-means algorithm of large amount of calculation and high time complexity aspects, this experiment made improvements, and achieved good results.

3 K-means Clustering Algorithm to Achieve MapReduce

Datasets of MapReduce processing should have such characteristics: it can be broken down into many small data sets, and each of the small data set can be completely processed in paralle [12,13], soil fertility data is well positioned to meet this demand. The basic idea is that K-means clustering algorithm actualizes MapReduce: each iteration start a MapReduce process, MapReduce to complete data records to the cluster center distance calculation and the new clustering center calculation. Fig. 2 describes the K-means clustering algorithm MapReduce parallel implementation[14]. According to the computing needs of MapReduce, preprocessing data records in the form of rows are stored, so that pre-treatment data can be fragmented by row, and no correlation between the pieces of data, fragmentation process is completed by the MapReduce model, without writing code.

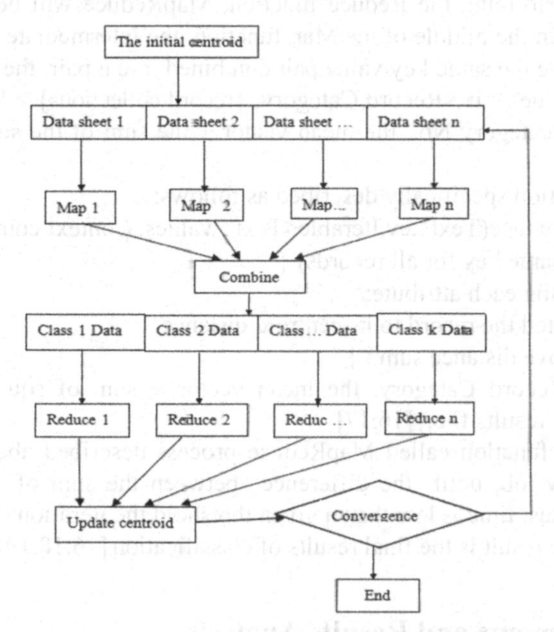

Fig. 2. K - means clustering algorithm MapReduce parallel implementation

3.1 Map Function Design

The task of the Map function is progressive to read data of soil fertility from the Hadoop Distributed File System (HDFS) file,, calculate the distance from the center to each record and re-mark the new category it belongs to the cluster. The input of Map function is the original data set and the last round of iteration (or initial clustering) clustering center , input data records <key,value>, ie <row number of rows>; output of intermediate results <key, value> that <record category, record> [15,16].

Map function specifically described as follows:
Public void map(Object key, Text value ,Context context){
To calculate the distance the record to each center of mass;
Compare the above distance;
The record boils down to the class of the nearest centroid;
To write <Record Category, Record> into the intermediate file ;
}[7,16]

3.2 Reduce Function Design

The task of Reduce function is based on the output of the Map function, update the clustering center, for the next round of the Map function. Meanwhile, calculating the value of standard measurement function, for the main function whether iteration is over. Before performing the Reduce function, MapReduce will be merged with the result of output in the middle of the Map function, the intermediate results in multiple <key,value> have the same key value pair combined into a pair. the Reduce function's output <key,value> is <Record Category, {record collections} > [15,16]. The output <key,value> is<category No., the mean vector + the sum of the squared error of the class>.

Reduce function specifically described as follows:
Public void reduce(Text key,Iterable<Text> values, Context context{
for(The same key for all records) {
Averaging each attribute;
Calculated the record to its centroid distance
The above distance sum ; }
The <Record Category, the mean vector + sum of square error for each class > write the results file;}[16,17]

In the main function called MapReduce process described above, each iteration apply for a new job, until the difference between the sum of the squares of the errors and the last time is less than a given threshold,the iteration ends. Map function last intermediate result is the final results of classification [16,18,19].

4 Experiments and Results Analysis

4.1 Experimental Datasources

In recent years, with the application of information technology in agricultural field, we also get a lot of correlated with soil fertility data. The experimental data is mainly from the national “863” plan- “research and application of corn precise operating system” [20] project demonstration base-NongAn country of JiLin Province for many years conducted aprecision fertilization after soil fertility data applied research. This paper select the representative soil nutrient data to integrated analysis, such as, Alkaline hydrolysis nitrogen, Available potassium and Available phosphorus. From the town of NongAn、WanShun and other towns during 2005 to 2013years.

Table 1. Part of thesampling data

Town name	Alkaline hydrolysis nitrogen (mg/kg)	Available phosphoru s (mg/kg)	Available potassium (mg/kg)	Latitude	Longitude	Elevation
SanGang	139	17	118	44.07632	124.78503	213
SanGang	87	14.5	115	44.07774	124.78433	213
SanGang	80	13.6	126	44.07768	124.78444	217
SanGang	153	13.2	115	44.07841	124.7742	214
SanGang	87	12	120	44.07709	124.77288	220
SanGang	139	13.1	123	44.07123	124.7769	216

4.2 Application of K-means Algorithm in the Analysis of Soil Nutrients

K-means clustering algorithm is one of the classic algorithm. First, K-means algorithm is used to cluster for Nongan town soil nutrient data for nine consecutive years (2005-2013),according to the test zone N, P, K data, hierarchical clustering analysis of soil fertility. The experimental results shown in Fig. 3:

The above clustering results showed that: after continuous precision fertilization, three comprehensive similarity of entire plots of soil nutrient data that are Alkaline hydrolysis nitrogen、Available phosphorus and Available potassium increased year by year, soil fertility tends to be balanced. Experimental results consistent with the actual situation, indicating that K- means algorithm is an effective soil fertility evaluation method. However, with the passage of time, the data of soil nutrients in Nongan county is increasing year by year, environmental factors associated with the soil fertility is considered, the disadvantages of K-means algorithm which are the large amount of calculation and high time complexity becomes more obviously , So we propose the K-means algorithm based on Hadoop platform.

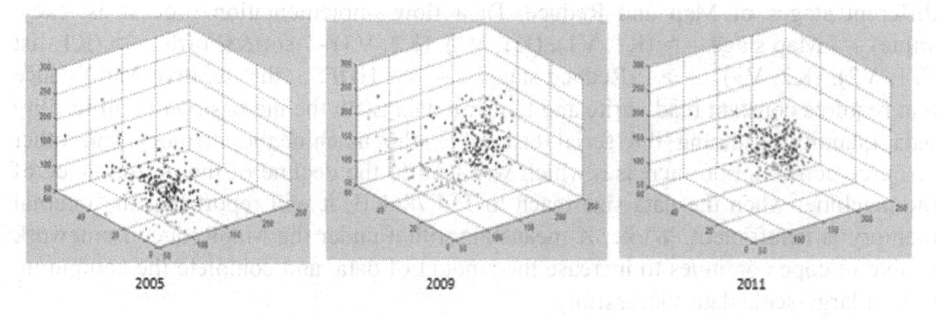

Fig. 3. Clustering results

4.3 K-means Algorithm Based on Hadoop Platform to Achieve

In the Hadoop environment, the data file stored in Hadoop Distributed File System(HDFS) will be automatically divided into a plurality of data block, each size is 64MB. In Map stage , each data block is processed by a Map task, the whole data files can be assigned to different nodes. Assume that each node can perform m tasks, there are n nodes involved in parallel computing, parallel K-means algorithm's time complexity is n*k*t*o/ (m*n), Compared with the time complexity of the serial algorithm n * k * t * o ,parallel K-means algorithm can significantly improve the operation efficiency .

To analyze the advantages of Hadoop platform in large amounts of data processing, in the case of the same hardware configuration environment, the same size of data, We performed serial K-means algorithm and Hadoop platform K-means algorithm computation time comparison. The experimental results are shown in Tab 2, where T1 is the operation time of serial algorithm, T2 is the operation time of the algorithm under the framework of MapReduce.

Table 2. Comparison of experimental results

Serial	File(KB)	Records(105)	t1 (s)	t2(s)
1	1,248	0.32609	4.09	248.81
2	4,991	1.30436	35.13	280.44
3	19,963	5.21744	157.11	461.43
4	39,926	10.43488	246.08	690.97
5	79,851	20.86976	331.45	1217.25
6	109,794	28.69592	979.76	1641.88
7	114,786	30.00028	Out of memory	2507.39

As seen from Tab 2, when the data size is less than 109,794KB, the efficiency of the serial K-means algorithm is higher than K-means algorithm under the framework of single MapReduce, this is because, MapReduce task divides one task into two different stages of Map and Reduce. Data flow implementation process is: (key, value) -- >Map stage -- > (K1, V1), (K1, V2), (K2, V3) - >sort&shuffle -- > (K1, list (V1, V2), (K2, V3) -- > Reduce stage -- > HDFS, the iterative MapReduce that requires constant read, write and transmit data ;with the increasing of soil fertility data quantity, Running the serial task of the machine,the memory and other resource consumption increases,which will lead to the decline of the performance of the machine; when the data size reach to 114,786KB, it will report that the internal memory is insufficient. While, K-means algorithm under the MapReduce framework is able to cope continues to increase the amount of data, and complete the computing task of large-scale data successfully.

5 Results and Discussion

Research on the NongAn town 2005 to 2013 soil nutrient data clustering showed that when dealing with large amounts of data, Based on Hadoop platform of parallel K-means algorithm have good operational results and practical significance .

(1)K-means algorithm is used to cluster for Nongan town soil nutrient data for nine consecutive years(2005-2013).Experiments show that after continuous precise fertilization ,comprehensive similarity of entire plots of soil nutrient data increased every year ,the soil fertility tends to equilibrium .Experimental results consistent with the actual situation, indicating that K- means algorithm is an effective soil fertility evaluation method.

(2)K-means algorithm based on Hadoop platform, When dealing with small amount of data, Serial K-means algorithm efficiency is better than under the MapReduce framework parallel K-means algorithm. Because when a small amount of data , in the MapReduce framework of K-means algorithm for each iteration, restarting a new JobTracker, the startup and interactive process will consume some resources.

(3)K-means algorithm based on Hadoop platform, when dealing with large amount of data , Hadoop platform can be well done the calculate of the large amounts of the soil fertility information data , which fully reflects the capabilities and advantages of the Hadoop's processing big data .Running the serial task of the machine,the memory and other resource consumption increases and causes the machine performance degradation,which lead to report insufficient memory and this can not meet the needs of data growth.

(4)With the increase of the amount of data, the Value that the difference of running time under the MapReduce framework divided by the total time tend smaller ,which reflects the stability and reliability of Hadoop platform.

Acknowledgements. This work was supported by the national "863" project (2006AA10A309), National Spark Plan (2008GA661003) and Shi Hang of Jilin province projects (2011- Z20).

References

1. Turner, B.L., Meyer, W.B.: Land use and land cover in global environmental change: considerations for study. Int. J. Soi. Sci. 130, 669–680 (1991)
2. Xiong, C.: Assesment and Prediction on Heavy Metals Status for Soils and Fresh Tea Leaves in Jiangxi Major Tea Regions Based on GIS Data Mining Technology. Nanchang University, Nanchang (2011)
3. Li, L., Li, J.: Application of Clustering Analysis in Classifying Site Type and Evaluating Soil Fertility. In: 2010 Third International Conference on Education Technology and Training (ETT 2010), pp. 468–471 (2010)
4. Chen, G., Cao, L., Wang, G.: Application of Weighted Spatially Fuzzy Dynamic Clustering Algorithm in Evaluation of Soil Fertility. Scientia Agricultura Sinica 42(10), 3559–3563 (2009)

5. Qian, Y.: Research and implementation of large-scale dataclustering techniques. University of Electronic Science and Technology, ChengDu (2009)
6. Li, L.: Research and implementation of Hadoop +Mahout intelligent terminal based cloud application recommendation engine. University of Electronic Science and Technology, Chengdu (2013)
7. Li, J., Cui, J., Wang, R.: Review of MapReduce parallel programming model. Journal of Electronic 39(11), 2635–2641 (2011)
8. Liu, P.: Open the shortcut leading to the actual Hadoop cloud computing, pp. 60–74. Electronic Industry Press, Beijing (2011)
9. Chen, G.: Research and Application of Spatial Data Mining Technology for Precision Agriculture. Jilin University, ChangChun (2009)
10. Pan, W.: Research and Application of Mining parallel K-means based on meteorological data cloud. Nanjing Information Engineering University, Nanjing (2013)
11. Naldi, M.C., Campello, R.J.G.B.: Evolutionary K-means for distributed data sets. Neurocomputing 127(15), 30–42 (2014)
12. Liu, P.: Open the shortcut leading to the actual Hadoop cloud computing, pp. 60–74. Electronic Industry Press, Beijing (2011)
13. Wang, L., Tao, J., Ranjan, R., Marten, H., Streit, A., Chen, J., Chen, D.: G-Hadoop: MapReduce across distributed data centers for data-intensive computing. Future Generation Computer Systems 29(3), 739–750 (2013)
14. Zhou, T., Zhang, J., Luo, C.: Realization of K-means clustering algorithm based on Hadoop. Computer Technology and Development 23(7), 18–21 (2013)
15. Srirama, S.N., Jakovits, P., Vainikko, E.: Adapting scientific computing problems to clouds using MapReduce. Future Generations Computer Systems 28(1), 184–192 (2012)
16. Xie, G., Luo, S.: Research on Application of Hadoop based on MapReduce model. Software World 29(8), 4–7 (2010)
17. Zhao, W., Ma, H., Fu, Y., Shi, Z.: Parallelk-means clustering algorithm designed Hadoop cloud-base computing platform. Computer Science 38(10), 166–167 (2011)
18. Li, Y., Deng, S., Wen, Y.: PageRank algorithm block matrix under Hadoop-MapReduce. Computer Technology and Development 21(8), 6–9 (2011)
19. Jiang, X., Li, C., Xiang, W., Zhang, X., Yan, H.: K-means clustering algorithm MapReduce parallel realization. Huazhong University of Science and Technology (Natural Science) 39(suppl.), 120–124 (2011)
20. Chen, G., Ma, L., Chen, H.: Research status and development trend of precision fertilization technology. Jilin Agricultural University 35(3), 253–259 (2013)

Application and Prospect of New Media in Forecast of Plant Pests

Zhiwei Zhao, Feng Qin, and Haiguang Wang*

Department of Plant Pathology,
China Agricultural University, Beijing 100193, China
wanghaiguang@cau.edu.cn

Abstract. Forecast of plant pests is a long-term foundation work for plant protection and it is of great significance for sustainable management of plant diseases, insects and other plant pests. Traditional media have played important roles in plant pest forecasting. However, with the rapid development of science and technology, various forms of new media are very suitable for plant pest forecasting. In this study, the limitations of the traditional media were analyzed and the advantages of the new media were revealed. Present situation of the applications of the traditional media in pest forecasting was introduced. And in the forecast of plant pests, the applications of the new media such as mobile phone short message, microblogging and WeChat, were also presented. The potential applications of the new media in plant pest forecasting in the future were prospected and some problems existing in the application of the new media that should be solved were discussed.

Keywords: plant pest, forecast, new media, plant disease epidemiology.

1 Introduction

With the rapid development of computer technology and information technology, more and more people in the world can get access to networks. According to a statistical report on internet development in China from China Internet Network Information Center (CNNIC), up to December, 2013, the number of net citizens in China has reached 618 million and the popularity rate of internet was 45.8%. The number of mobile net citizens in China has reached 500 million. In particular, the number of rural net citizens has reached 177 million and the ratio of mobile net citizens among rural net citizens was 84.6%. The popularization and application of the networks provide the convenient conditions for agricultural informatization and agricultural knowledge dissemination. And this will play a significant role in solving 'the last mile' problem in agricultural informatization.

Various forms of new media are emerging along with the rapid development of digital technology, communication technology and internet technology, and have been becoming an indispensable part of people's daily life. Meanwhile, the new media

* Corresponding author.

© IFIP International Federation for Information Processing 2015
D. Li and Y. Chen (Eds.): CCTA 2014, IFIP AICT 452, pp. 313–323, 2015.
DOI: 10.1007/978-3-319-19620-6_36

influence people's lifestyles, behaviors and thinking modes. In this context, plant protection personnel should think about how to use the new technologies and the new media in plant pest management and plant pest forecasting, and should think about how to avoid the limitations of the traditional media in forecasting information release, forecasting information dissemination and forecasting information feedback. The rapid collection and analysis of the data and information that required for plant pest forecasting and the timely release and dissemination of forecasting results are very important for the control of plant pests. In general, the critical periods of plant pest prevention and control are short. Therefore, the timely dissemination of the information of plant pests is especially important. Compared with the traditional media, the new media have the characteristics, such as incomparable information transfer rate, excellent interaction, openness and sharing, etc. Therefore, the new media provide an excellent solution for plant protection personnel to conduct effective data collection and information release.

2 Traditional Media

Generally, the traditional media include newspapers, periodicals, radio and television. The information released by the traditional media always is collected, screened and then released to the audiences by the professionals from newspaper offices, periodical offices, radio stations, TV stations, etc, and is highly professional and reliable. However, the released information is transmitted in the one-way mode without timely feedback from information audiences. Thus it cannot form an effective information transmission circuit, so the transmission and the use of the information are discounted. Although some traditional media also set feedback part for the audiences, the feedback mode is generally very cumbersome and the feedback cycle is very long. It always cannot meet the timely feedback requirements in our work. So the feedback information cannot play an effective role in improving the level of our work. After interpretation of the received information, individual audiences with different knowledge, life experience, personality, habits, etc, may get different understanding of the information. Sometimes the real information will be misinterpreted. Therefore, timely communication between the audiences and the information sources is very important. The new media with interactive feature can provide a good solution for these problems.

3 New Media

The term 'new media' originated in the United States and was spread to other countries and regions. In fact, the so-called 'new media' is relative, and only in a specific period, some media are called as new media. Wired broadcasting appeared in 1893 and it became the new medium at that time compared with newspaper. Subsequently, the radio technology was developed, and the wireless broadcasting then became the new medium instead of the wired broadcasting. In view of the current 'new media', many definitions were put forward. Basically, the new media at the present stage refer to the media based on the digital technology and internet technology that can realize many-to-many communication. Now the new media include some media based on internet such as blog, podcast, microblogging, WeChat, network chat tools (e.g., Tencent QQ and MSN), search engine, websites, network television (web TV), internet protocol television (IPTV), online journals and the Internet of things, and some media based on mobile phone network such as mobile phone short messages, mobile phone MMS (multimedia messaging service), mobile phone TV, mobile phone newspaper, WeChat and mobile phone chat tools (e.g., Tencent QQ). Compared with the traditional media, the new media have timely, massive and interactive features, and the information can be disseminated in text, image, sound and multimedia. Thus the new media provide a bridge for the audiences to accurately get access to specific information in time and to perform real-time communication with the people who release the information. Through the new media, personalized information service can be provided, and the needs of the people at the present stage for personalized information can be met. The new media have played important roles in many fields, such as in education [1-5]. In particular, the new media have also been applied in agriculture [6], such as the dissemination of agricultural science and technology [7]. The new media can be used for forecasting information release, plant protection knowledge popularization and technology extension, reporting information and information feedback, timely information dissemination and communication, and prompt correction of plant protection information that can cause the social public security problems and crisis. The application of the new media in plant pest forecasting has very important significance.

4 Forecast of Plant Pests

Plant pest forecast is a prerequisite for the implementation of pest management measures and plays an important role in the modern integrated management of plant pests. Plant pest forecasting provides services for control measure making and pest management. According to the accurate forecast of plant pests, good preparations can be made for pest control and the risk of plant pests can be reduced; various control techniques can be applied more reasonable to improve the pest control effect and the pest control benefit; the unnecessary control cost and the environmental pollution caused by pesticide abuse also can be reduced [8]. Information collection and dissemination of forecasting results are the important parts of plant pest forecasting.

At present, plant pest forecasting in China mainly focuses on the major crop pests such as rice blast, wheat stripe rust, potato late blight, rice planthopper, corn borer. Generally, the information and the data used for plant pest forecasting in China are collected via the observation using forecast light, system investigation in pest observation field and field survey by local plant pest forecasting personnel. After the collection of the data and information by the personnel, it usually will take a long time before super administrative department can get the data and information. Then the data and information is analyzed to predict the trend of plant pests in the future, and finally the forecasting information is released through reports on plant pests, newspaper, radio, TV and agriculture websites. In recent years, forecasting methods of plant pests are developing in the visual direction. China Central Television (CCTV) and local television stations release the forecasting information of some major plant pests at irregular intervals. Some television stations establish special TV columns of plant pest forecast. To guide the pest control in agricultural production, they regularly release the forecasting information of plant pests and provide the corresponding prevention and control measures. A large number of agricultural websites can also publish some forecasting information of plant pests. However, the agricultural websites often cover everything and contain a large amount of information, and pest forecasting information cannot be systematically released in time. Special agricultural websites focusing on plant pest forecasting should be established. The problem-solving agricultural science and technology knowledge service system can be developed and provide convenient and efficient one-stop problem solving services [9]. Some web-based plant pest forecasting systems have been established [10-14] and some of them have already played important roles in practice [10-12].

Although the limitations of newspaper, radio, television and other traditional media have been discussed above, only the limitations of agriculture websites were discussed here. The agriculture websites are run on the internet, but the influence of these websites is low. As well as the traditional media, the release mode of plant pest forecasting information is still one-to-many communication in the absence of interaction.

The collected information about plant pests and the released forecasting information of plant pests via the traditional media have played very important roles, but based on the previous discussion, the effects of plant pest forecasting should be influenced by the traditional media's congenital deficiency such as the traditional media's lack of good interactive feature. To make full use of forecasting information and to provide more timely and higher quality of plant pest forecasting information services for farmers, the new media should be used to collect and disseminate information. Thus the work on forecasting plant pests will be greatly improved. The new media have be applied in plant pest forecasting and control and have played important roles in the management of some plant pests [15, 16].

5 New Media and Plant Pest Forecasting

At present, the new media applied in plant pest forecasting mainly include mobile phone SMS, mobile phone newspaper, the Internet of things and Wechat, etc. Wireless Application Protocol (WAP) is a protocol standard for accessing information and advanced value-added services over a mobile wireless network using a mobile terminal. Agricultural information network can be constructed using WAP in combination with internet technology and mobile phone technology, and no matter when and where, users can get access to the network resources using mobile phone terminals. It will be useful for the solution of 'the last mile' problem in the agriculture informatization. Pest information and forecasting information, the knowledge about control technology can be quickly spread to the farmers. Thus the farmers can learn about the incidence of plant pests in time and can effectively master the control measures. Fig. 1 showed the information transfer process in plant pest forecasting using new media. Compared with the traditional media, there are many differences in the information transfer process in plant pest forecasting using new media. As shown in Fig. 2 and Fig. 3, the relationship between the different subjects including the government department of plant protection, forecasting technical personnel, experts and farmers, is more complex than that when the forecasting information is disseminated based on the traditional media. Some management service departments in China have developed agricultural information service using mobile phone text messages for local farmers. Farming issue information is released to guide the agricultural production. Among these applications, mobile phone short message and mobile phone newspaper are relate to plant pest forecasting, but plant pest forecasting is not a major service. Since the coverage of mobile phone in China is increasing, the new media based on mobile phone network will have many new applications in plant pest forecasting. Blog, microblogging, Wechat and the Internet of things will play important roles in plant pest forecasting. In recent years, extensive attention has been paid to the Internet of things. Radio frequency identification (RFID) and sensor network are used more and more widely in agriculture [17].

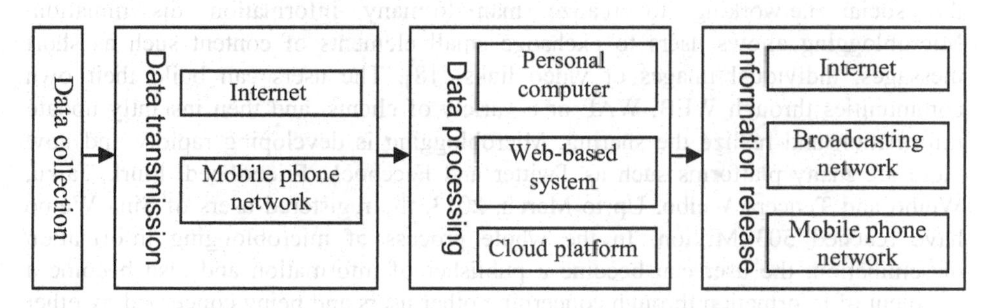

Fig. 1. Information transfer process in plant pest forecasting using new media

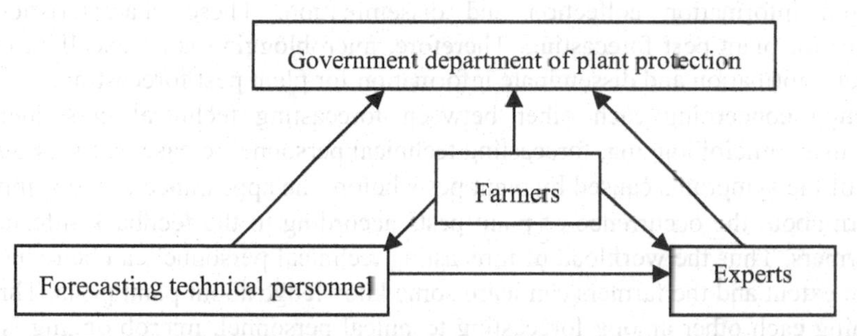

Fig. 2. Information transfer between different subjects in data collection for plant pest forecasting using new media

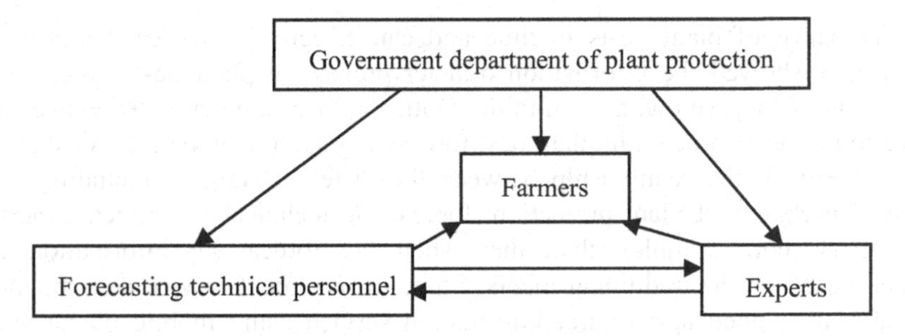

Fig. 3. Information transfer between different subjects in information release for plant pest forecasting using new media

5.1 Application of Microblogging in Plant Pest Forecasting

Microblogging, namely microblog, is a broadcast medium that exists in the form of blogging, but its content is typically smaller than a traditional blog. A microblog contains up to 140 Chinese characters words including punctuations. Microblogging is a kind of new medium after blog and it relies on the relationships similar to that of the social networking to realize many-to-many information dissemination. Microblogging allows users to exchange small elements of content such as short messages, individual images or video links [18]. The users can build their own communities through WEB, WAP or a variety of clients, and then instantly update information and realize the sharing. Microblogging is developing rapidly and now there are many platforms such as Twitter and Facebook, FriendFeed, Plurk, Jaiku, Weibo and Tencent Weibo. Up to March, 2013, the registered users of Sina Weibo have reached 503 Million. In the whole process of microblogging information dissemination, the user can become a publisher of information and also become a recipient of information through concerning other users and being concerned by other users. This is so-called We Media. For microblogging, the threshold to get access is very low. The characteristics of microblogging are wide coverage, strong interaction and rapid information collection and dissemination. These characteristics are necessary for plant pest forecasting. Therefore, microblogging is an excellent means to collect information and disseminate information for plant pest forecasting.

Through concerning each other between forecasting technical personnel and farmers using microblogging, forecasting technical personnel release text messages or images of the symptoms caused by plant pests before the appearance of the symptoms and learn about the occurrence of plant pests according to the feedback information from farmers. Thus the workload of forecasting technical personnel can be reduced to a certain extent and the farmers can learn some knowledge about plant pests. Through concerning each other among forecasting technical personnel, microblogging is very helpful for the exchange of their experiences and the improvement of their forecasting level. In particular, during the critical periods of plant pests, information communication between forecasting technical personnel in adjacent areas is very helpful to improve the accuracy of forecasting. Mutual concerns between farmers also can help to break the geographical limitations and to realize the exchange of the farmers' experiences in agricultural production. Similarly, mutual concerns between forecasting technical personnel and experts, that between farmers and experts and that among experts can play an important role in the collection and dissemination of forecasting information.

Microblogging has the properties of We Media, so the government department of plant protection, experts, forecasting technical personnel and farmers all can issue forecasting information on their microblogs. Concerning the reliability and extensive dissemination of the information, the official microblog opened by the government department of plant protection should be preferred to release forecasting information of plant pests. The plant pest forecasting information can be released by the official microblog opened by the government department of plant protection, and then the forecasting information can be rapidly communicated among experts, forecasting technical personnel and farmers through the network established based on mutual concern. Then experts, forecasting technical personnel and farmers can disseminate the information to others through their own networks. Thus the forecasting information can be disseminated widely. Therefore, if the reliability and accuracy of plant pest forecasting information released by the government departments can be guaranteed, the immediate spread of the information will no longer be a problem. In addition, to realize timely communication among experts, forecasting technical personnel and farmers and to quickly feedback the questions concerned by forecasting technical personnel or farmers, the official microblogging should be monitored in real time by the related professional personnel. Using the official microblog opened by government department of plant protection to release the forecasting information, the spreading width and the extent of application of the information will reach a high level that is hard to reach by other media.

5.2 Application of the New Media Based on Mobile Phone in Plant Pest Forecasting

Now mobile phone short messages have been applied in plant pest forecasting. However, in the application processes, there are still three problems that need to be solved. First, the problem that the farmers need personalized information service is still not completely solved. Second, the information interaction between the information promulgator and the audiences is not as good as expected. Third, the publishing form of the forecasting information is still simple. Therefore, mobile phone short message services in plant pest forecasting in the future should made efforts to solve the above three problems. With large community of users and high rural coverage, the mobile phone short messages can have more extensive application in plant pest forecasting if the above three problems can be solved satisfactorily. In the same way, mobile phone TV and mobile phone newspaper are facing with some problems in the application in plant pest forecasting.

Now there are many kinds of smart mobile phones with video call function. Thus, a bridge has been provided for the remote diagnosis of plant pests and face-to-face communication among farmers, forecasting technical personnel and experts. The absence of technical personnel, their uneven ability to work, heavy workload of the field survey, short critical period for pest control and the lack of knowledge about plant pests by farmers, etc, lead to the collection of first-hand data becoming the first problem met in plant pest forecasting. Through mobile phone video call with related experts, not only the remote diagnosis of plant pests can be realized, but also farmers can make real-time communication with experts and experts can solve the practical problems that forecasting technical personnel or farmers meet in the agricultural production in time. This will provide great supports for plant pest forecasting and plant pest control.

As a mobile text and voice messaging communication service, WeChat developed by Tencent in China was first released in January, 2011. It is an instant messaging client supported on Wi-Fi, 2G, 3G, and 4G data networks. The app is available on many operating systems including Android, iPhone, Windows Phone, etc, and multiple languages are supported including traditional/simplified Chinese, English, Spanish, Japanese, Italian, Thai, Korean, Indonesian, Russian, etc. Voice message, videos, pictures and words can be sent using the app. The app supports group chat and provides the functions of public platforms and friends circles. Now WeChat has won a large number of users in the world. Up to August 15, 2013, the users of WeChat outside China have reached over 100 million. In October, 2013, WeChat total users reached 600 millions worldwide. WeChat can be applied to exchange the information of plant pest information and pest control experiences inside the agricultural cooperation organizations through group chat or friends circles. Some experts, companies and plant protection personnel have used WeChat for the extension of plant protection technologies and the release of plant pest information. WeChat public platforms also can be built by the government department of plant protection. The forecasting information can be released on the public platforms and can be instantly transferred to users. Users also can feedback information whenever and wherever possible. As well as microblogging, the information on the WeChat public platforms should be monitored in real time.

5.3 Application of Other New Media in Plant Pest Forecasting

The forecasting information of plant pests can be released and disseminated via QQ group. In practice, many QQ groups focusing on regular crop advisory, plant protection and plant pest management have been built. Because of the limited capacity of group members, different QQ groups can be built for different individuals. Blogs and podcasts can be used to assist microblogging in plant pest forecasting. Compared with microblogging, blogs and podcasts have larger spaces for data uploading. They can be extended reading materials for users such as experts, forecasting technical personnel and farmers to get a further understanding of how the plant pests occur and develop and how to effectively prevent and control the plant pests. The Internet of things should be made full use to collect data and information about the occurrence of plant pests and then to provide more timely and accurate data supports for plant pest forecasting.

6 Prospect of Application of New Media in Forecast of Plant Pests

Now the application of the new media in plant pest forecasting is very limited because of the low rural net coverage, high internet access fees, farmers' lack of network knowledge, etc. With the development of economic in China and the improvement of rural living conditions, the new media will be increasingly used by farmers. It is very important to overcome the limitations of the traditional media in plant pest forecasting and to widely use the new media to carry out plant pest forecasting. Thus it will greatly promote the development of plant pest forecasting. Among many new media, the Internet of things will play an important role in the information collection of plant pests, and microblogging and WeChat will play important roles in the dissemination of forecasting information. The application of the new media in plant pest forecasting will promote the development of digital forecasting system of plant pests and the informatization of plant protection in China. It is expected that new media will be more widely applied in the plant pest forecasting in the future.

Based on internet, broadcasting network and mobile phone network, the national integrated information platform for plant pest forecasting can be developed using the advanced technologies including digital technology, internet technology, database technology, multimedia technology, communication technology, etc. The basic framework of integrated information platform for plant pest forecasting based on the new media was shown in Fig. 4. Now in the Big Data Era, cloud platform services should be made full use in plant pest forecasting. Then the coverage of the forecasting information can be improved, and the information can be more quickly transfer to users. According to their own needs, users can make information retrieval and screening at any time; they can subscribe or unsubscribe the required information at any time; and they also can perform the information feedback and effective communication at any time. Therefore, sustainable control of plant pests can be achieved.

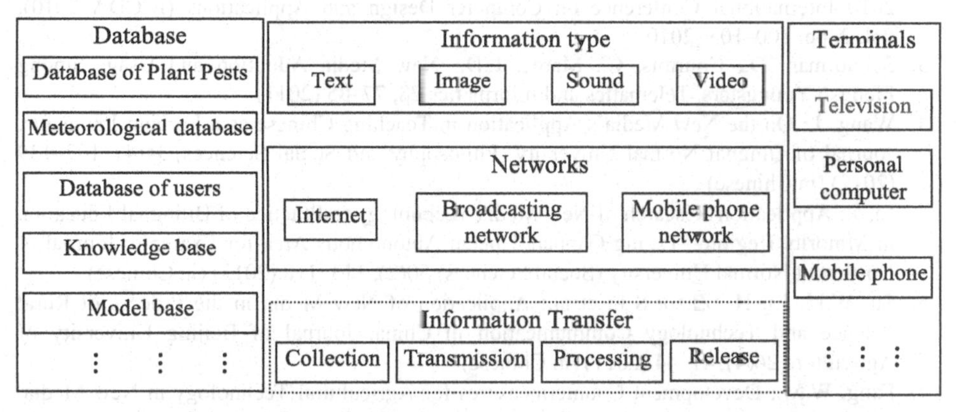

Fig. 4. The basic framework of integrated information platform for plant pest forecasting based on new media

In the information transmission via the traditional media, as passive recipients, the audiences are usually in passive position. In the information transmission via the new media, the initiative of the audiences is enhanced, and they can conduct active searches and receive information of interest to them. They also can independently release information on the internet. To avoid the users' misunderstanding or wrong direction of public opinion caused by released incorrect forecasting information, it is critical to ensure that the released information is true and legal. The information verification is strict required for the release of forecasting information. The government department of plant protection as the 'gatekeeper' should strictly be responsible for information verification. The government supervision and legislation should be conducted to guarantee the correctness and effectiveness of the information.

The ability of audiences or users to use the services via new media should be improved. Farmers should be trained how to use mobile phone or computer to obtain forecasting information and to seek help through uploading information. The rural popularity and the rural coverage of mobile phones, computers and internet should be increased through financial subsidies and other means. Experts and forecasting technical personnel should learn to process the information used for plant pest forecasting and forecasting information using the new media.

Acknowledgment. This work was supported in part by National Key Technologies Research and Development Program (2012BAD19BA04) and the National Natural Science Foundation of China (31101393).

References

1. Zeng, Y.: Discussion on the Teaching of Applying Course of Computer Network in the Circum Stances of New Media. Education and Teaching Research 23(8), 102–103, 106 (2009) (in Chinese)
2. Su, L.H., Jin, B.Y.: About the Application of New Media Technology to Education. In: 2010 International Conference on Computer Design and Applications (ICCDA 2010), vol. 2, pp. 100–103 (2010)
3. Schuurman, D., Courtois, C., Marez, L.D.: New Media Adoption and Usage among Flemish Youngsters. Telematics and Informatics 28, 77–85 (2011)
4. Wang, J.: On the New Media's Application in Teaching Chinese as a Foreign Language. Journal of Qinghai Normal University (Philosophy and Social Sciences) 34(4), 137–139 (2012) (in Chinese)
5. Li, J.: Application Research of New Media Technology to Practice of Bilingual Education in Minority Region—Taking Gannan Tibetan Autonomous Area for Example. Journal of Northwest Normal University (Social Sciences) 50(2), 113–118 (2013) (in Chinese)
6. Su, W.L., Li, H., Zhao, S.J., et al.: Application of New Media in the Process of Rural Science and Technology Communication in China. Journal of Beijing University of Agriculture 26(4), 41–44 (2011) (in Chinese)
7. Fang, W.M.: Development Countermeasures for Agricultural Technology in New Media. Guangdong Agricultural Sciences 41(14), 203–205 (2012) (in Chinese)
8. Xiao, Y.Y., Ji, B.H., Yang, Z.W., et al.: Epidemic and Forecast of Plant Diseases, 2nd edn., pp. 78–84. China Agricultural University Press, Beijing (2005) (in Chinese)

9. Wan, M., Meng, X.X.: Design and Construction of Problem-solving Agricultural Science and Technology Knowledge Service System. Journal of Agricultural Science and Technology 15(2), 33–38 (2013) (in Chinese)
10. Forrer, H.R., Steenblock, T., Fried, P.M.: Monitoring of Potato Late Blight in Switzerland and Development of PhytoPRE+ 2000, an Internet Based Decision Support System. Journal of Agricultural University of Hebei 24, 38–43 (2001)
11. Kelly, N.M., Tuxen, K.: WebGIS for Monitoring "Sudden Oak Death" in Coastal California. Computers, Environment and Urban Systems 27, 527–547 (2003)
12. Hu, T.L., Zhang, Y.X., Wang, S.T., et al.: Construction and Implementation of China-light: a Monitoring and Warning System on Potato Late Blight in China. Plant Protection 36, 106–111 (2010) (in Chinese)
13. Pavan, W., Fraisse, C.W., Peres, N.A.: Development of a Web-based Disease Forecasting System for Strawberries. Computers and Electronics in Agriculture 75, 169–175 (2011)
14. Kuang, W., Liu, W., Ma, Z., Wang, H.: Development of a Web-based Prediction System for Wheat Stripe Rust. In: Li, D., Chen, Y. (eds.) Computer and Computing Technologies in Agriculture VI, Part I. IFIP AICT, vol. 392, pp. 324–335. Springer, Heidelberg (2013)
15. Luvisi, A., Panattoni, A., Triolo, E.: Electronic Identification-Based Web 2.0 Application for Plant Pathology Purposes. Computers and Electronics in Agriculture 84, 7–15 (2012)
16. Pande, A., Jagyasi, B.G., Choudhuri, R.: Late Blight Forecast Using Mobile Phone Based Agro Advisory System. In: Chaudhury, S., Mitra, S., Murthy, C.A., Sastry, P.S., Pal, S.K. (eds.) PReMI 2009. LNCS, vol. 5909, pp. 609–614. Springer, Heidelberg (2009)
17. Yan, M.J., Xia, N., Wan, Z., et al.: The Application of the Internet of Things in Agriculture. Chinese Agricultural Science Bulletin 27(8), 464–467 (2011) (in Chinese)
18. Kaplan, A.M., Haenlein, M.: The Early Bird Catches the News: Nine Things You Should Know About Micro-blogging. Business Horizons 54(2), 105–113 (2011)

Modeling the Drivers of Agricultural Land Conversion Response to China's Rapidly Rural Urbanization: Integrating Remote Sensing with Socio-Economic Data

Yiqiang Guo[1], Jieyong Wang[2,*], and Chunxian Du[3]

[1] Key Laboratory of Land Consolidation and Rehabilitation, Land Consolidation and Rehabilitation Center, Ministry of Land and Resources of China, Beijing 100035, China
wjy@igsnrr.ac.cn

[2] Institute of Geographic Sciences and Natural Resources Research, Chinese Academy of Sciences, Beijing 100101, China
guoyiqiang2002@126.com

[3] National Agro-Technical Extension and Service Center, Ministry of Land and Resources of China, Beijing 100026, China
chunxyan@163.com

Abstract. Agricultural land has changed remarkably response to the rapidly rural urbanization in China since the further market-oriented reform of early 1990s. Using accurate remote sensing images and socioeconomic data, this paper identified the temporal pattern of agricultural land conversion and estimated their driving forces during 1992-2010 in Yueqing City. Agricultural land change can be summarized to three trends: Firstly, the conversion between non-agricultural land and agricultural land was very remarkable, the non-agricultural land increased obviously from 1992 to 2010, in verse farmland decreased continuously; Secondly, the loss agricultural land mostly were the higher quality cropland. The expansion of settlements and industrial park was almost completely at the expense of fertilize farmland in flat area, but hillside and tideland were reclaimed for planting grain to keep the dynamic balance of total agricultural land amount. Thirdly, the changes of agricultural land distributed regularly and according to the distance to the No.104 national highway. In addition, the stepwise regression analysis indicated that both the booming of rural private enterprises and increasing of population was main driving force of agricultural land conversions. According to the results, we discussed the negative impact of fertile farmland shrink and gave some feasible advices in the end.

Keywords: agricultural land conversion; modeling socio-economic driving forces; rural urbanization; Yueqing City.

1 Introduction

Global Land Project (GLP) was initiated by IGBP and IHDP in September 2005. Its central objective is to retain the sustainable development of human and natural system

* Corresponding author.

© IFIP International Federation for Information Processing 2015
D. Li and Y. Chen (Eds.): CCTA 2014, IFIP AICT 452, pp. 324–336, 2015.
DOI: 10.1007/978-3-319-19620-6_37

and mitigate its frangibility in the globalization, by means of measuring, simulating and understanding the socio-environment coupled system, and interpreting both the changes of surface process and its social, economic and political results (GLP, 2005; Turner *et al.*, 2007). China is the most rapidly developing nation of the world and home to 1.3 billion people. Since the early 1980s, the unprecedented combination of economic and population growth has led to a dramatic land transformation across the nation (Chen, 1999; Lin, 2001; Liu *et al.*, 2005; Tian *et al.*, 2003). As a result of rapid urbanization, farmland conversion is dramatically frequent in China. In the eastern coast of China, the development process has been characterized by rural industrialization and urbanization. As a result, many coastal regions of China such as the Yangtze River Delta region, Pearl River Delta region and Beijing metropolitan areas experienced dramatic economic and spatial restructuring, which resulted in tremendous farmland conversion (Lin, 2001; Weng, 2002; Li and Yeh, 2004; Long *et al.*, 2007a). Agricultural land conversions, while restricted by physical conditions, are mainly driven by socio-economic factors, which are tightly inter-related with human production activities.

In the late 1980s, some strict measures were adapted to control the conversion of agricultural land into non-agricultural use (Lin and Ho, 2005; Erik and Ding, 2008). The policies of "Dynamic balance of Total Farmland (*gengdi zongliang dongtai pingheng*)" and "Basic Farmland Protection Regulations" have relieved the rapid loss of farmland to some extent, but they did not have resulted in massive loss of fertilize farmland (Yang and Li, 2000). Actually, the absolute net change of farmland can not reflect the true relationship between farmland conversion and socio-economic development.

In this paper, through a detailed study of Yueqing, Zhejiang province, attempts to provide a better understanding of the agricultural land conversion and its driving forces response to rapid rural industrialization and urbanization. First, land use conversion over a period of rapid rural economic growth was described. Second, the relationships between agricultural land conversion and socio-economic development are discussed. Thirdly, the characteristics of f agricultural land conversion and the mechanism of socio-economic driving forced were concluded. In the end, several key problem for land use management in rapid rural industrialization were discussed.

2 Study Area

Yueqing City is located in the southeast of the Zhejiang province and faced to the East China Sea (Figure 1), with an area of coverage of 1251 km^2 and a population of 1389,300. The territory is composed of two main areas from south-east to north-west. The south-eastern area is occupied by plain and platform where is the bread basket of the city region, and the north-western area is a part of *Yandang Mountains* where suitable for fruit, citrus, tea, and edible fungi production.

Yeqing has been regarded as the origin place of the Wenzhou Model of China's economic development at the beginning of the reform and opening-up policy (Zhang and Li, 1990). The Wenzhou Model is famous for its domestic private investment in traditional manufacturing and bottom-up rural urbanization (Zhang and Li, 1990; Liu, 1992). Before the reform and opening up, Yueqing used to be one of the poorest regions in eastern China, it has a very limited amount of arable land at 0.52 mu (15 mu=1 hectare) per capita, about one-third of the national average. Meanwhile, Yueqing's proximity to Taiwan made it a likely war frontline in the planned economy era from 1949 to the late 1970s. Neither the central or provincial (Zhejiang) governments were inclined to spare their limited capital for Yueqing's infrastructural and industrial development (Tsai, 2002). With limited land resource, poor road access to major cities, and little supporting policy from the central government, it seemed to lack all the conditions necessary for economic growth. However, over the past three decades, Yueqing has been one of the most dynamic economical regions and one of the fastest urbanization cities all over China. To a large extent, it is representative of bottom-up rural urbanization all over the China. Therefore, it might provide some insights into the agricultural land-use change response to China's rapid urbanization through this case study.

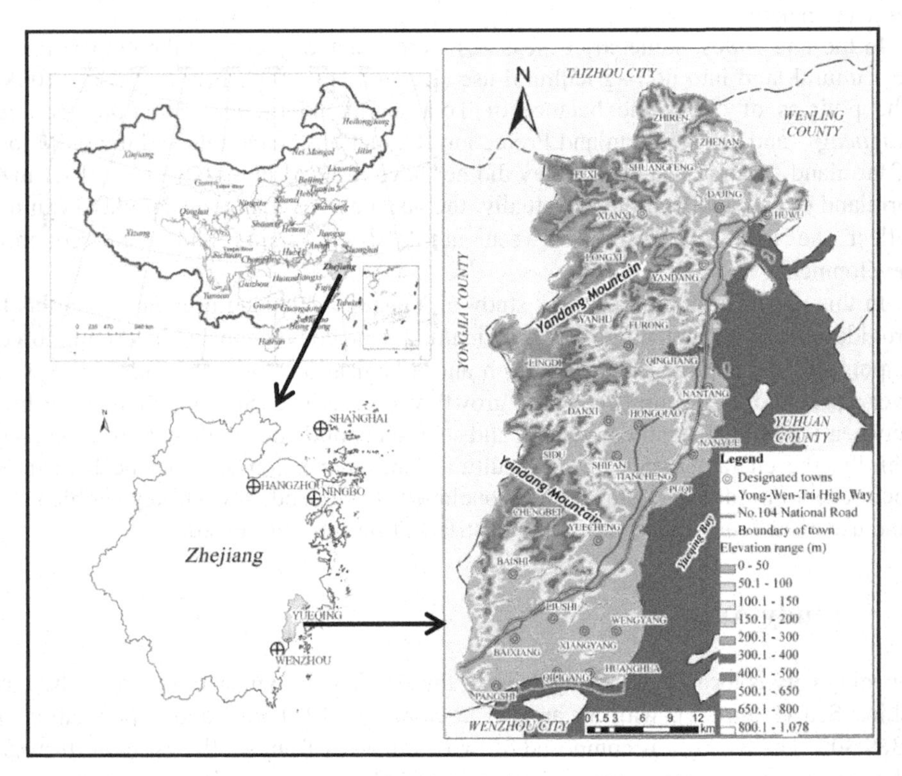

Fig. 1. Study area: Yueqing, Zhejiang province, China

3 Materials and Methods

3.1 Data Source and Pre-processing

Detailed spatial data of agricultural land conversion need to be produced for this study. We selected high-resolution remote sensing imagery as the basic data resource, and land use data, topographic maps, and socio-economic statistical data were also collected for further data processing.

Agricultural Land Use Data

Four historical Landsat satellite images (TM data taken in October 1992, and ETM+ data in October 2010; (Path 118 (orbit), Row 40-41 (scene center)) were downloaded from the International Scientific Data Service Platform, and the Landsat ETM+ SLC-off image was processed with a self-adaptive local regression model for multi-temporal imagery. All of the images were clear and nearly free of clouds (total cloud cover less than 5%). In order to assist image interpretation, two land use maps were collected from Land and Resources Department of Yueqing City. They were mapped by the two National Land Surveys, in 1990 and in 2008, respectively. The slope map was derived from ASTER Global Digital Elevation Model (ASTER GDEM) data which is acquired by a satellite-borne sensor "ASTER" to cover all the land on earth. It was downloaded from its official website. In September 2011, a field survey, including 90 sampling-points, was carried out by using a global positioning system to identify present land-use types and trace land-use histories. These points were evenly distributed in the study area, and at the same time land-use types and transportation accessibility were taken into consideration.

Twenty well distributed ground control points (GCPs) were selected for geometric correction of remote sensing imagery according to the Second Land Survey map. Two periods of images were geo-referenced to the Xi'an 1980 Coordinate System. The root mean square error (RMSE) was less than 1 pixel. A first order polynomial fit was applied and all the data were re-sampled to a spatial resolution of 30 m using nearest neighbor method. In order to improve the visual interpretability of images, a color composite (Landsat TM Bands 4, 5, and 3) was prepared and its contrast was stretched using the Gaussian distribution function. The 3 × 3 high pass filters were applied to the color composite to further enhance visual interpretation of linear features, e.g., rivers and vegetation features.

Socio-Economic Data

In addition, a history time-series socioeconomic data were collected from the local governments and published statistical yearbooks (Yueqing County Statistical Bureau, 1993- 2011). They are used to analyze potential driving forces resulting in agricultural land conversion in the study area. The selection of data items took into consideration of the literature of China scholars (Xie and Costa, 1991; Ma and Xiang, 1998; Kirkby, 2000; Marton, 2000; Lin, 2001) and listed in Table 1. Data were compiled for townships (31 in total). Besides, with the recognition that many research questions cannot be addressed if analysis is based solely on official data, thorough field observations and unstructured interviews were conducted in field studies during Nov. 5-18, 2011. A wide range of people in Yueqing were interviewed, including government officials, enterprise managers, villagers, migrant workers, and town residents.

Table 1. List of the selected and socioeconomic variables

Abbreviation	Description	Unit	label
T-HH	Total households	INH[a]	x_1
T-POP	Total population	IND[b]	x_2
R-POP	Total population registered in rural area	IND	x_3
U-POP	Total population registered in urban area	IND	x_4
E-POP	Exotic Population	IND	x_5
PU	Percentage of population registered in urban area	%	x_6
R-INC	Rural per capita net income	Y[c]	x_7
GIO	Gross industrial output value	MY[d]	x_8
TVA	Total output value of agriculture	MY	x_9

[a]INH: the count of households; [b]IND: the individual count of population; [c]Y: Chinese Yuan; [d]MY: Million Chinese Yuan

3.2 Methods

Land Use Classification

In order to detail the land use/cover changes, it was classified i into 9 types: Farmland (FM), Forestland (FL), Garden plots (GP), Urban Settlements (US), Rural Settlements (RS), Water Body (WB), Industrial land (IL), Transportation land (TL), and unused land (UL). (see Table 2). The training areas were established according to more than 90 field-sampling points and the two National Land Use Surveys maps by the software Erdas 9.0. Then, the two images were interpreted by the method of supervised classification using Maximum Likelihood Classifier (MLC), visual interpretation involved the use of image characteristics such as texture, pattern, and color to translate image into land use. A 3 × 3 majority-neighborhood filter was used to adjust the classified images.

Land Use Change Matrix

A land use/cover change matrix reveals the internal variations of land use changes during the study period, showing the detailed information on cropland conversions. The transition matrix of land use/cover was conducted through the spatial overlay of the two-phase land use maps interpreted from Landsat data.

The two land-use grid maps were mainly used to detect the internal variations of land-use change from 1992 to 2010. For each pair of grid datasets, a change matrix A_{ij} was constructed. Then, B_{ij} and C_{ij} were calculated based on matrix A_{ij} accounting for inner change of all land-use types. Where row *i* is the land-use types of 2006, and column j is the land-use type of 1992. B_{ij} is the percentage of land-use type j that converted to i, C_{ij} is the percentage of land-use type j that came from i.

$$B_{ij} = A_{ij} \times 100 / \sum_{i=1}^{10} A_{ij} \tag{1}$$

$$C_{ij} = A_{ij} \times 100 / \sum_{j=1}^{10} A_{ij} \tag{2}$$

4 Results

4.1 Agricultural Land changes in Yueqing

Land use has changed significantly over the period from 1992 to 2010 in Yueqing City (Table 2, Figure 2, Figure 3). From 1992 to 2010, farmland decreased by 22.2%. In contrast, urban settlements, rural settlements, industrial land and transportation land increased by 503.4%, 23.4%, 51.9%, and 424.7%. To a large extent, land-use change from 1992 to 2010 in Yueqing County was characterized by a replacement of agricultural land with urban settlement, rural settlements, and industrial land.

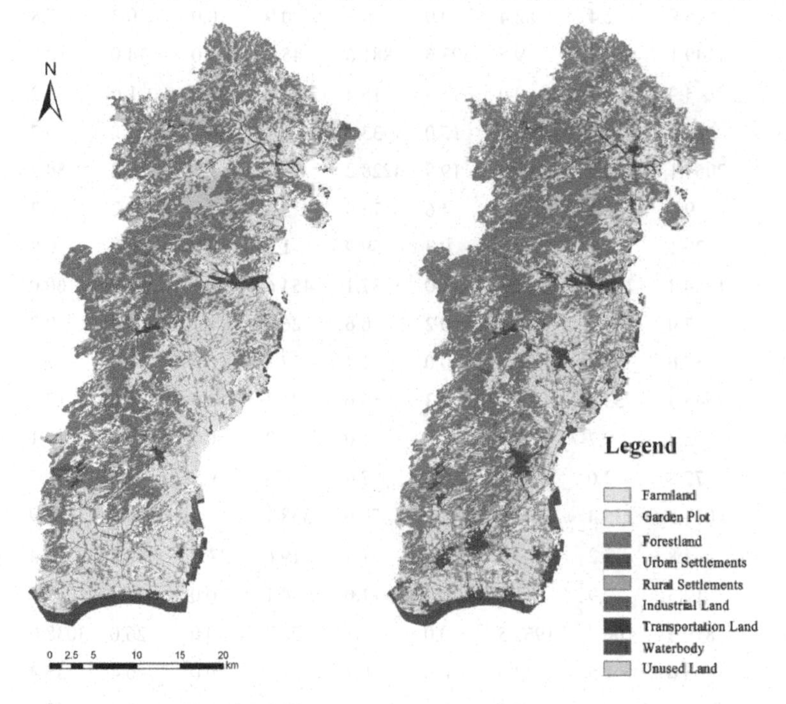

Fig. 2. Land use map of Yueqing in 1992

Fig. 3. Land use map of Yueqing in 2010

Exchanges of land use were analyzed from two aspects during the period of 1992 to 2010. One was how many percentages of the later land-use type came from the former ones, the other was how many percentages of the former land-use type changed into the later ones. The expansions of urban settlements are at the expense of occupying great amounts of farmland and rural settlement, 43.8 % and 33.6% of urban settlements came from farmland and rural settlement respectively. Rural settlements increases are at the expense of occupying lots of farmland, 29.1% of rural settlements came from farmland and 72.3%% of transportation land came from farmland.

Table 2. the internal matrix of Land use changes of Yueqing City during 1992 - 2010 (ha, %)

1992 2010		FM	GP	FL	US	RS	IL	TL	WB	UL	Total
	A	27741.7	1068.8	1399.7	6.4	181.2	136.1	0.4	423.9	306.6	31264.6
FM	*B*	134.0	14.7	2.6	1.5	3.2	7.9	2.6	5.9	3.6	
	C	173.0	8.0	11.8	0.0	1.2	1.0	0.0	2.5	2.4	
	A	1760.7	3819.7	984.4	0.1	74.0	19.4	0.5	69.9	228.3	6957.0
GP	*B*	10.2	52.5	1.8	0.0	1.3	1.1	3.4	1.0	2.7	
	C	25.3	54.9	14.1	0.0	1.1	0.3	0.0	1.0	3.3	
	A	3518.2	1456.2	49073.1	0.6	174.0	540.3	0.2	150.8	4656.0	59569.4
FL	*B*	21.5	20.0	90.8	0.1	3.0	31.3	1.2	2.1	54.3	
	C	5.9	2.4	82.4	0.0	0.3	0.9	0.0	0.3	7.8	
	A	1149.1	49.7	9.5	393.6	881.6	45.9	2.0	74.0	18.7	2624.0
US	*B*	4.9	0.7	0.0	90.5	15.4	2.7	14.8	1.0	0.2	
	C	43.8	1.9	0.4	15.0	33.6	1.7	0.1	2.8	0.7	
	A	2053.1	254.0	191.9	19.9	4228.2	93.3	1.1	156.8	58.2	7056.5
RS	*B*	9.7	3.5	0.4	4.6	73.9	5.4	7.7	2.2	0.7	
	C	29.1	3.6	2.7	0.3	59.9	1.3	0.0	2.2	0.8	
	A	1509.4	238.9	227.5	1.0	37.1	454.6	1.2	91.9	60.0	2621.5
IL	*B*	7.0	3.3	0.4	0.2	0.6	26.3	8.7	1.3	0.7	
	C	57.6	9.1	8.7	0.0	1.4	17.3	0.0	3.5	2.3	
	A	545.3	53.1	25.8	4.3	55.6	21.1	6.9	30.0	12.5	754.7
TL	*B*	2.4	0.7	0.0	1.0	1.0	1.2	50.4	0.4	0.1	
	C	72.3	7.0	3.4	0.6	7.4	2.8	0.9	4.0	1.7	
	A	1011.5	235.3	146.3	8.9	79.9	338.9	1.5	6197.8	202.9	8222.9
WB	*B*	4.8	3.2	0.3	2.0	1.4	19.6	11.2	85.8	2.4	
	C	12.3	2.9	1.8	0.1	1.0	4.1	0.0	75.4	2.5	
	A	879.3	106.8	1963.5	0.0	7.1	76.3	0.0	26.0	3038.0	6097.0
UL	*B*	5.6	1.5	3.6	0.0	0.1	4.4	0.0	0.4	35.4	
	C	14.4	1.8	32.2	0.0	0.1	1.3	0.0	0.4	49.8	
Total		40168.2	7282.5	54021.6	434.9	5718.7	1725.9	13.7	7221.0	8581.1	125167.6

Note: *FM-Farmland; GP-Garden Plots; FL-Fores land; IL-Industrial Land; TL – Transportation Land WB-Waterbody; UL- Unused Land

4.2 Spatial Distribution Characteristics of Land Use Conversions

The road system in rural China is the key factor to promote rural industrialization and agricultural economy (Rozelle and Jiang, 1995). There are few investments to infrastructure by the state and local government in Yueqing until the end of 1980s, because it is close to Taiwan. Since the No. 104 national highway was finished in 1992, it has been the lifeline of rural development in Yueqing County. The road shaped a banded spatial pattern of urban and rural expansion. six buffers were built along the No. 104 national highway at intervals of one kilometer to exanimate the spatial characteristics of rural land use change. The changes of land-use in these buffered belts were reported in Figure 4.

The No. 104 national highway played a primordial role in rural socio-economic development and its power of influence diminished outwards. For example, the increase of urban settlement, rural settlement, construction land and transportation land were highest within the 1 km buffer; respectively, the decrease of paddy field and dry field were highest. Besides, the main types of land use conversion were calculated over 6 buffered belts that include farmland to urban settlement (FMUS), farm land to rural settlement (FMRS), farm land to transportation land (FMCL), and rural settlement to urban settlement (RSUS), the trend of land use conversions were depicted in Figure 4.

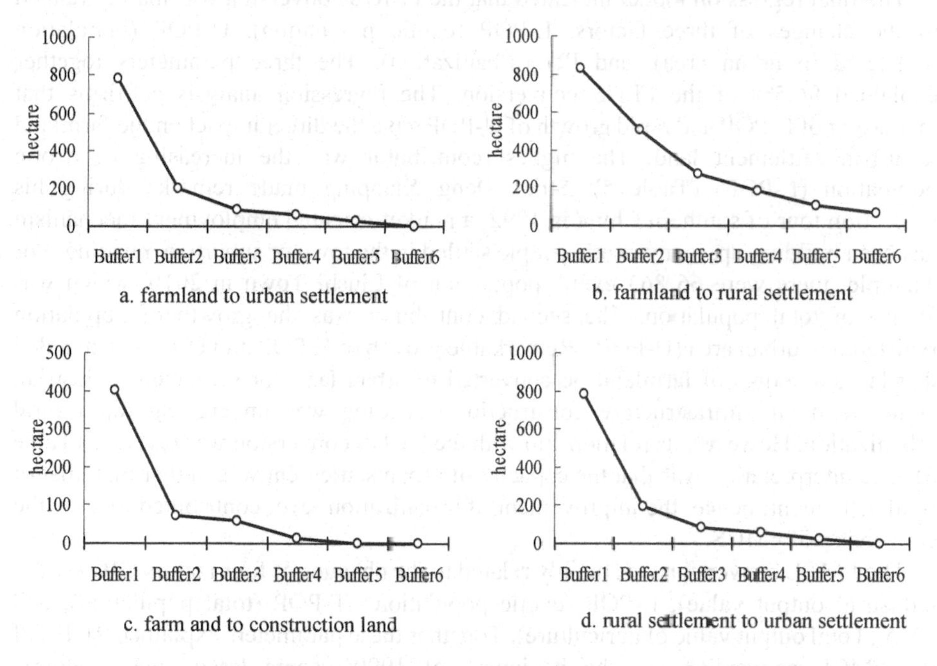

Fig. 4. Main land use conversions along the buffered belts from No. 104 national highway

4.3 Modeling Socio-Economic Driving Forces of Agricultural Land Conversion

Using socioeconomic data of 31 towns from statistical yearbooks and land use conversion data derived from land-use maps, the interactions between land-use conversions and socioeconomic driving forces were examined. The quantitative changes between 1992 and 2010 for the socioeconomic variables were computed, and four main land-use conversion types were counted for all towns that including FMUS (farmland to urban settlement), FMRS (farmland to rural settlement), FMCL (farmland to construction land), FLFM (forest land to farmland) and ULFM (unused land to farmland). The regression analyses were conducted between the farmland conversion and socioeconomic variables (Table 3).

Table 3. Regression final model of dependent variables

Dependent variables	Final regression model	Adjusted R^2
FMUS	$y_1 = 0.305x_5 + 1.752x_6 - 4358.819x_4 + 495.778$	0.965
FMCL	$y_2 = 0.039x_8 + 0.791x_5 + 1.780x_2 + 431.808$	0.911
FMRS	$y_3 = 0.034x_8 + 4.022x_7 + 604.522$	0.869
FLFM	$y_4 = 0.949x_9 + 700.593$	0.371
ULFM	$y_5 = 0.418x_4 + 107.178$	0.425

The final regression model indicated that the FMUS conversion was mainly related to the changes of three factors: E-POP (exotic population), U-POP (population registered in urban area), and PU (urbanization). The three parameters together explained 96.5% of the FLUS conversion. The regression analysis confirms that increasing of E-POP and rapid growth of U-POP gave the direct impact on the farmland to urban settlement land. The highest contributor was the increasing of exotic population (E-POP) (Table 5). Since Deng Xiaoping made remarks during his inspection tour of southern China in 1992, a market-oriented employment mechanism has been building up, many exotic people settled in the town or suburb in Yueqing. For example, there were 66 863 exotic population of Liushi Town in 2010, which was 39.3% of total population. The second contributor was the growth of population registered in urban area (U-POP). Remarkable growth of E-POP and U-POP demanded that large amounts of farmland be converted to urban land for settlement industrial, commercial and infrastructure construction. Yueqing was undergoing rapid rural urbanization. However, its relationship with the FMUS conversion was reversed (Table 6). The interpretation was that the capacity of urban settlement was higher than that of rural settlement; hence, the improvement of urbanization level contributed to slow the conversion of FMUS.

The FMCL conversion was mainly related to the changes of four factors GIO (Gross industrial output value), E-POP (exotic population), T-POP (total population), and TVA (Total output value of agriculture). Together these parameters explained 91.1% of the FMCL conversion. At the beginning of 1990s, some large area economic development zones (jingji kaifa qu) were built, a part of private household enterprises moved out of rural settlements by reason of enlarging production scale, which settled in economic development zones, and lots of exotic population worked in these areas. As it shown in table 6, the largest contributor was the increasing of GIO, rural industrialization caused farmland converting to construction land. On one hand, rural industrialization attracted a lot of exotic population from central and western regions of

China; on the other hand, enlarging production scale contributed to enlarging workshop place through occupying farmland. For instance, the build-up part of Zhejiang Provincial Yueqing Economic Development Zone covered 6.05km^2 with more than 160 enterprises, and two thirds of labors came from outside .

The conversion of FMRS can be explained 87.7% by the increasing of GIO and growth of R-INC. Rural private enterprises were widely regarded as the primary form of rural industrialization for Wenzhou Model. As Lin and Ho (2003) pointed out, the expansion of construction land was largely a result of rural industrialization in China. Numerous factories such as buttons, ball pen and electric accessory can be easily found in the residential backyards of the villagers in rural Yueqing. The increasing of rural income levels spured peasants to extend their dwelling space. Usually, peasants in China want to build another new multifunctional, more comfortable or spacious house if they become affluent, although they did not really need. Peasants preferred to build their houses closing to roads and other available infrastructure. Besides, another universal phenomenon was found during interviews with town officials and village cadres across Yueqing, it was pointed out that many private businessmen managed enterprise all over China, but they built luxury and spacious dwelling in their hometown, for example, there were two brothers in Daixia village Hongqiao town, each of them spent more than 400 000 Yuan (RMB) on building new rural villa, but they only lived few days during the holiday of Chinese New Year Festivities.

FLFM and ULFM contributed to balance the farmland loss by urbanization and industrialization. The conversion of FLFM mostly distributed in the mountain and hill area of western Yueqing, it can be explained 37.1% by the increasing of TVA. There were many under development villages in the western Yueqing, they were far away from urban area, central towns and arterial road. Agricultural production was the principal source of peasants' income in this area; therefore, peasants were willing to cultivate land on slope to plant crops, vegetables and other cash corps. The conversion of ULFM mostly distributed along the coastline in the eastern Yueqing, it can be explained 42.5% by the growth of U-POP. To guarantee the goal of the dynamic balance of the total farmland, the government was planning to reclaim unused land such as tideland, marshlands, and barren land.

5 Discussion

China's rapid industrial growth has received wide attentions in the literature (Allen et al., 2005). As the most dynamic private economics in China, Yueqing provides a good case for us to examine the agricultural land conversion and its driving forces behind China's fast rural industrialization. Then, several lessons can be drawn from the results of our research.

This study examines the temporal and spatial pattern of land use change as a consequence of rural industrialization. Using accurate spatial data from two land survey and history statistic data of 31 towns, the study clearly identifies a temporal pattern of land use change during 1992-2010 based on GIS. Land-use change in Yueqing can be characterized by three major trends: First, the non-agricultural land (urban and rural settlements, construction land, and transportation land) increased substantially from 1992 to 2010, thus causing farmland to decrease continuously, land use conversion between non-agricultural land and farmland were very remarkable. Secondly, the conversion of farmland in Yueqing was a very serious issue. On one hand, expansion of urban and rural settlements, and construction land was almost completely at the expense of farmland, on the other hand, lots of fertilize farmland lost, but hillside and tideland were reclaimed for planting grain. Thirdly, there are clear spatial patterns of land use change geographically, which are crafted by the No.104 national highway built in 1992. The buffer analysis indicated that the distance is the nearer to highway road, the integrated degree is higher.

Our study has also highlighted the driving forces of farmland conversions through method of regression step by step. E-POP (exotic population), U-POP (population registered in urban area), and PU (urbanization) were the main driving forces of farmland to urban settlement; GIO (Gross industrial output value), E-POP (exotic population), T-POP (total population) have driven farmland to construction land dramatically; the conversion of FMRS can be explained 87.7% by the increase of GIO and growth of R-INC; TVA and U-POP can respectively explain forest land and unused land to farmland. In general, both the increase of rural private enterprises including GIO and R-INC growth and population development including population structure change for E-POP and U-POP increase are main driving forces of farmland conversions.

The rapid industrialization and urbanization in Yueqing stimulated the demand of farmland converting to non-agricultural land, which led the expansion of rural settlements and local environmental pollution; however, there were such few areas of farmland that they cannot suppose this unsustainable developing model. Many private household enterprises scattered in rural area, accordingly, which resulted in the low efficiency of land production and the disorder expansion of urban. For an instance, we investigated a private household enterprise in Beibaixiang Town that occupied 0.31 ha land and employed 4 workers; accordingly, the efficiency of land use was much lower than urban areas. In the long run, the government should combine the rural industrialization and current urbanization process, and encourage private household enterprises to move to industrial centralization region. The countermeasures that established a scientific land utilization ethics and economical, intensive use of land, the model of the urban and rural integration reform measures and so on were put forward from the angle of saving, intensive use and sustainable development.

Rural economic development is very inequality in Yueqing. There were 9 less developed towns located in the hill region which the net income per peasant was less than 4 000 Yuan (RMB); however, there were some developed towns which the net income per peasant has exceeded 10 000 Yuan (RMB) such as Liushi, Hongqiao and Beibaixiang. There was 12.7% of total population, but only 1.5% of total gross industrial output value and 7.5% of total output value of agriculture in the less developed towns. Developed towns have occupied lots farmland in the past three decades, and the farmland already reduced to the limit by 0.017ha per capita in Yueqing County. The state had no choice but to adapt the strictest farmland protection system against the increasing threat of food safety. There is few surplus of farmland for the less developed towns to develop industries; moreover, to guarantee the goal of the dynamic balance of the total farmland, some hillside areas were reclaimed for planting grains that was easy to soil erosion. In interviews with town officials and village cadres of Lingdi and Yandang Town, two less developed towns, they pointed out that the shortage of land for industrial and housing development were the biggest obstacle of rural development. Therefore, the government of Yueqing County should work out a feasible development plan for less developed region, which emphasized at least four point as follow: (1) to take advantage of its ecological resources and topographical benefits to develop specialty products and industries; (2) to increase budgetary funds and loans for poverty alleviation, balance the development of different rural region; (3) to provide much more employment opportunities in the developed towns, and give priority to underdeveloped rural labors ; and (4) to implement ecological emigration plan, encourage peasants to emigrate from remote hill villages to developed region.

Acknowledgements. This work was supported by the National Natural Science Foundation of China (41401201).

References

Allen, F., Qian, J., et al.: Law, finance, and economic growth in China. Journal of Financial Economics 77(1), 57–116 (2005)

Chen, B.: The existing state, future change trends in land-use and food production capacities in China. Ambio 28(8), 682–686 (1999)

Ding, C.R.: Land policy reform in China: assessment and prospects. Land Use Policy 20, 109–120 (2003)

Lichtenberg, E., Chengri, D.: Assessing farmland protection policy in China. Land Use Policy 25(1), 59–68 (2008)

Global Land Project (GLP), Science Plan and Implementation Strategy. IGBP Report No. 53/IHDP Report No. 19. IGBP, Stockholm (2005)

Li, P., Li, X., Liu, X.: Macro analysis on the driving forces of the land use change in China. Geographical Research 20(2), 129–138 (2001)

Li, X., Yeh, A.G.O.: Analyzing spatial restructuring of land use patterns in a fast growing region using remote sensing and GIS. Landscape Urban Plann. 69, 335–354 (2004)

Lin, Ho, S.P.S.: The state, land system, and land development processes in contemporary China. Annals of the Association of American Geographers 95, 411–436 (2005)

Lin, G.C.S.: Metropolitan development in a transition socialist economy: spatial restructuring in the Pearl River Delta, China. Urban Studies 38, 383–406 (2001)

Lin, C.: Evolving spatial form of urban-rural interaction in post reform China: a case study of the Pearl River Delta. Professional Geographer 53, 56–70 (2001)

Liu, Y.-L.: Reform from Below: The Private Economy and Local Politics in the Rural Industrialization of Wenzhou. The China Quarterly 130, 293–316 (1992)

Liu, J.Y., Zhan, J.Y., Deng, X.Z.: Spatio-temporal patterns and driving forces of urban land expansion in China during the economic reform era. Ambio 34, 450–455 (2005)

Long, H.L., Liu, Y.S., Wu, X.Q., Dong, G.H.: Spatio-temporal dynamic patterns of farmland and rural settlements in Su–Xi–Chang region: Implications for building a new countryside in coastal China. Land Use Policy 26(2), 322–333 (2008)

Long, H.L., Tang, G.P., Li, X.B., Heilig, G.K.: Socio-economic driving forces of land-use change in Kunshan, the Yangtze River Delta Economic Area of China. J. Environ. Manage. 83(3), 351–364 (2007a)

Ma, L.: Urban transformation in China, 1949–2000:a review and research agenda. Environment and Planning A 34, 1545–1569 (2002)

Rozelle, S., Jiang, L.: Survival strategies and recession in rural Jiangsu. Chinese Environment and Development 6, 43–84 (1995)

Sonobe, T., Hu, D., Otsuka, K.: From inferior to superior products: an inquiry into the Wenzhou model of industrial development in China. Journal of Comparative Economics 32, 542–563 (2004)

Sonobe, T., Hu, D., Otsuka, K.: From inferior to superior products: an inquiry into the Wenzhou model of industrial development in China. Journal of Comparative Economics 32(3), 542–563 (2004)

Tian, H., Melillo, J.M., Kicklighter, D.W., Pan, S., Liu, J., McGuire, A.D., et al.: Regional carbon dynamics in monsoon Asia and its implications for the global carbon cycle. Global and Planetary Change 3, 201–217 (2003)

Tsai, K.S.: Back-alley banking: private entrepreneurs in China. Cornell University Press, Ithaca (2002)

Turner II, B.L.: The emergence of land change science for global environmental change and sustainability. The National Academy of Sciences of the USA 104(52), 20666–20671 (2007)

Xu, W.: The changing dynamics of land-use change in rural China: a case study of Yuhang, Zhejiang Province. Environ. Plann. A 36, 1595–1615 (2004)

Xu, W., Tan, K.C.: Impact of reform and economic restructuring on rural systems in China: a case study of Yuhang. Zhejiang Journal of Rural Studies 18(1), 65–81 (2002)

Weng, Q.H.: Land use change analysis in the Zhujiang Delta of China using satellite remote sensing, GIS and stochastic modelling. J. Environ. Manage. 64, 273–284 (2002)

Shen, X.P., Ma, L.J.C.: Privatization of rural industry and de facto urbanization from below in southern Jiangsu, China. Geoforum 36, 761–777 (2005)

Xie, Y., Costa, F.: The impact of economic reforms on the urban economy of the People's Republic of China. The Professional Geographer 43, 318–335 (1991)

Xie, Y.C., Batty, M., Zhao, K.: Simulating emergent urban form using agent based modeling: desakota in the suzhou–wuxian region in China. Ann. Assoc.Am. Geogr. 97(3), 477–495 (2007a)

Xie, Y.C., Fang, C.L., Lin, G.C.S., Gong, H.M., Qiao, B.: Tempo-spatial patterns of land use changes and urban development in globalizing China: a study of Beijing. Sensors 7, 2881–2906 (2007b)

Xie, Y.C., Yu, M., Tian, G.J., Xing, X.R.: Socio-economic driving forces of arable land conversion: a case study of Wuxian City, China. Global Environ. Change 15, 238–252 (2005)

Yang, H., Li, X.: Cultivated land and food supply in China. Land Use Policy 17(2), 73–88 (2000)

Yueqing County Statistical Bureau, Statistical Yearbook of Yueqing County, 1990. Yueqing, Zhejiang Province, China (1993)

Yueqing County Statistical Bureau, Statistical Yearbook of Yueqing County, 1990. Yueqing, Zhejiang Province, China (2011)

Zhang, R., Li, H.: Wenzhou model research. China Social Sciences Publishing House, Beijing (1990)

Study on the Detection and Warning System of Rice Disease Based on the GIS and IOT in Jilin Province

Guogang Zhao[1,2], Haiye Yu[1,2], Guowei Wang[1,2,3], Yuanyuan Sui[1,2], and Lei Zhang[1,2]

[1] College of Biological and Agricultural Engineering, Jilin University, Changchun 130022, China
[2] Key Laboratory of Bionic Engineering, Ministry of Education, Changchun 130022, China
[3] School of Information Technology, Jilin Agricultural University, Changchun 130118, China
{zhaoguogang2000,41422306}@qq.com, {z_lei,haiye}@jlu.edu.cn, suiyuan0115@126.com

Abstract. Rice disease which has intimate connection with climate has a tremendous impact on the output and quality of rice. The present paper intends to construct a web to monitor the climate environment of rice growing by making use of IOT(Internet of things) and to exactly spot the regions where related statistics are collected and warned by taking advantage of GIS and then by means of C# language and SQL server, to realize the visuality of system and database management. Thereafter, under fuzzy clustering algorithm based on the collected environmental data to judge whether the rice disease will happen and need warning in case that unnecessary lose is caused.The finding reveals that by practicing the present system in Rice Disease Department of Academy of Agricultural Science of Chang Chun, the system functions well in detecting the serious disease.

Keywords: GIS, Internet Of Things, rice disease, monitor, warning.

1 Introduction

Jilin Province is advantageous in grain production and is also one of important commercial commodity grain base. Rice is one of three kinds of corps in jilin province and thus rice disease has influenced the output and quality of rice. Hense, to construct the disease detection system is necessary. Currently, the previous studies on rice disease are as follows: Su Mingyong with his fellows collected the monitoring data and analyzed the data in order to spread the information and put forward the scientific motivation for treatment [1,2]. Yuan Tao with his co-workers realized the prior warning by means of case inference technology based on the symptoms of disease [3]. However, the above studies obviously just focused on the disease rather than preventing before the disease. Xiao Deqin studied wireless water conductor collecting the data to analyze the water in the rice field, which innovate the previous studies lying in the wireless water conductor that is able to realize to conduct water on a wider range and more remote distance. Besides [4], there are more researches which

© IFIP International Federation for Information Processing 2015
D. Li and Y. Chen (Eds.): CCTA 2014, IFIP AICT 452, pp. 337–343, 2015.
DOI: 10.1007/978-3-319-19620-6_38

are related to rice disease based on specialist system and database [5-9]. However, the approaches above are limited on the range of database to infer rather than realize the real-time monitoring based on the latest statistics and then make the policy.

The paper will build a wireless conductor net based on IOT mainly being reflected by water temperature, air moisture and daylight illumination of rice field to collect data on which the calculating ways are ensured and then days among monitoring which are mot the same as the set temperature humidity level and illumination are recorded to forecast whether there will be rice disease in the chosen rice field. At this time, GIS will work to present the current number for the above items and their corresponding location. If there needs warning, then the mark will be made in the map. Meanwhile, the message will be sent to the farmer by mobile phone in order to make exact measurement to save rather than lose. So the significance of the present study lies in the above innovation.

2 System Design

The c/s construction model is adopted to realize the serving way of multiply customer enjoying the data. The framework of the system is as follows: (see fig. 1)

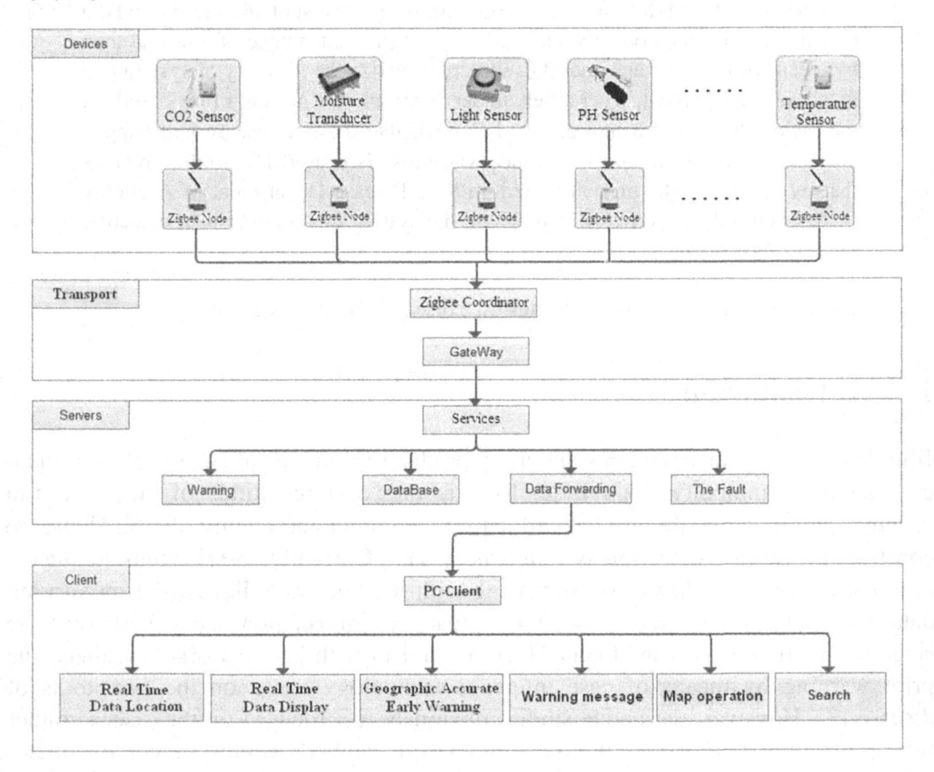

Fig. 1. The whole framework of system

2.1 Data Communication Agreement

Data communication agreement is a series of agreement which aims to guarantee the communicators to effectively and reliably communicate within the data communication web. In the IOT, communication between server and hardware, between sever and various clients needs to be realized. Therefore, the present communication agreement is the key to ensure the function of system, which adopts statistics agreement as the follows:"data-node number-electricity-air-temperature—air humidity-illumination-water temperature."When the system receives the message from the monitoring equipment, the clipping is made according to the communication format to store in the right database.

2.2 Data Collection

Data is collected by Socket programmer within C# language. The working procedure is as fig.2.

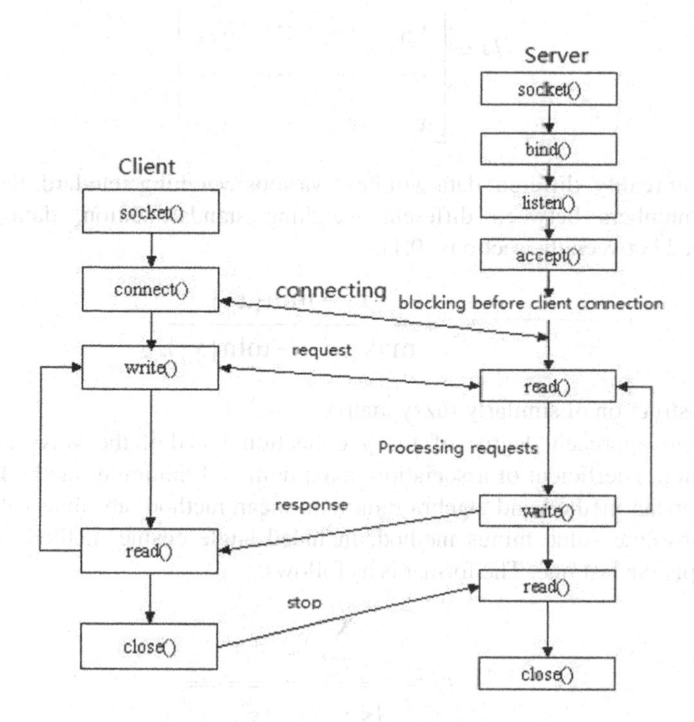

Fig. 2. Working procedure of Socket

2.3 Function of GIS

GIS not only presents where the data is from but also exactly tells the disease regions when the warning message is got, which facilitates the rice planter to prevent the disease beforehand. The GIS in present system takes advantage of MapObjects composition of ESRI Company in America. It consists of 46 parts, which constitutes 5 categories, individually being items of data operation (10), targets of map emergence(17), objects of projection(8), objects of graphs(7) and items of address match (4). Among this kinds of C# system, when any item is selected, the corresponding function can be made.

2.4 Precautionary Model

The model is realized by fuzzy clustering algorithm based on the data digging. The working procedure is as follows:

① The standardization of data weighing

To construct a primitive database D owing to the issues , among which x_{ij} means the attributive value of j in the i sequence.

$$D = \begin{bmatrix} x_{11} & x_{12} & \cdots & x_{1m} \\ x_{21} & x_{22} & \cdots & x_{2m} \\ \cdots & \cdots & \cdots & \cdots \\ x_{n1} & x_{n2} & \cdots & x_{nm} \end{bmatrix}$$

However, in reality, different data will have various weighing standard. So in order to compare numbers between different weighing standardization, data should be standardized between the sections [0,1].

$$x'_{ij} = \frac{x_{ij} - \min\{x_{ij}\}}{\max\{xi_j\} - \min\{x_{ij}\}} \tag{1}$$

② Construction of similarly fuzzy matrix

To calculate approach degree of fuzzy collection i and j, the ways are amounts accumulation, coefficient of association, maximum and minimum method, arithmetic minimum mean method and algebra minimum mean method, absolute value pointing method, absolute value minus method, included angle cosine method. The present paper adopts the last one. The format is as follows:

$$r_{ij} = \frac{\sum_{k=1}^{m} x_{ik} \cdot x_{jk}}{\sqrt{\sum_{k=1}^{m} x_{ik}^2} \cdot \sqrt{\sum_{k=1}^{m} x_{jk}^2}} \tag{2}$$

Based on the formula in the above, the fuzzy matrix $R = (r_{ij})_{n \times n}$ is got.

③ Flowing cluster procedure

Fuzzy cluster analysis needs a equal fuzzy matrix which should be self-contradictory, symmetry and transitivity. Common fuzzy matrix has got the first two qualities.

Thus we can make use of self-square blending to revise matrix R as matrix $t(R)$, that is,

$$R \to R^2 \to R^4 \to \cdots \to R^{2(k+1)} = R^{2k} \tag{3}$$

$$R^2 = R \circ R = (r'_{ij})_{m \times m},$$

$$r'_{ij} = \max_{k=1}^{m}(\min(r_{ik}, r_{jk})) \tag{4}$$

$$t(R) = R^{2k} = (t_{ij})_{m \times m} \tag{5}$$

After the t(R) matrix has been got, the different number elements will be listed from the maximum to the minimum. λ chooses the data according to the sequence to form the mobile cluster map which can be used to construct various models when different warning information is spotted.

3 System Realization

3.1 Data Collection in Real Environment

GIS and Socket functional programming will be made by use of C# language and according to the definition of communication protocols, data capture will be received and obtain data collection sites and corresponding environmental parameter information will be obtained. The collected information is displayed in the system with the result in the fig.3:

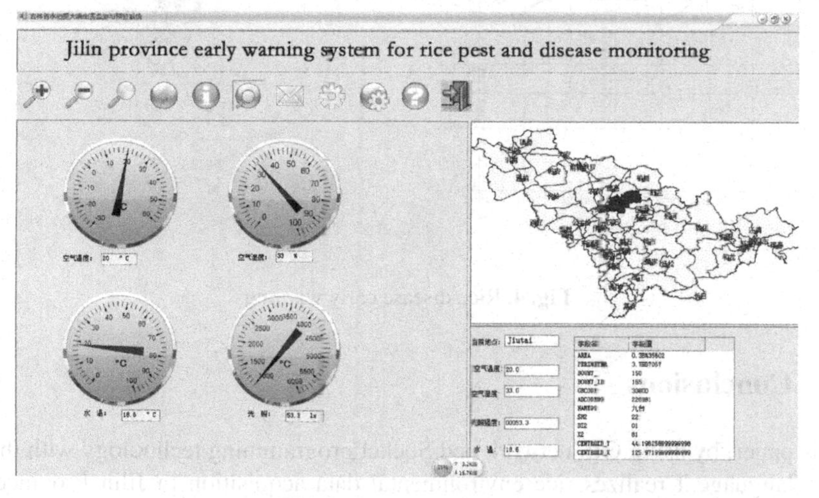

Fig. 3. The data collection

3.2 Early Warning of Rice Disease

Through nearly twenty years of history data of Jilin Province are analyzed by fuzzy cluster analysis, according to the number of high temperature and high humidity weather, there are 4 categories: the first category includes no disease with high temperature and high humidity weather number less than 1 days; the second category involves 65% of rice having lighter diseases and 35% medium disease condition; the third kind in high temperature and high humidity weather lasting 4~6 days, 45% rice has mild disease, 36% medium condition and 19% heavy; fourth kinds of high temperature and high humidity weather more than 6 days, the rice disease occurred with 50% lighter, 23% moderate and 27% heavy.

When there is data acquisition, the system will judge according to the early warning model. If there is indeed warning, it will be marked on the map with red, at the same time, the early warning information in the form of text messages will be sent to mobile phone users; if there is no warning, no treatments. Fig. 4 shows the warning at the interface:

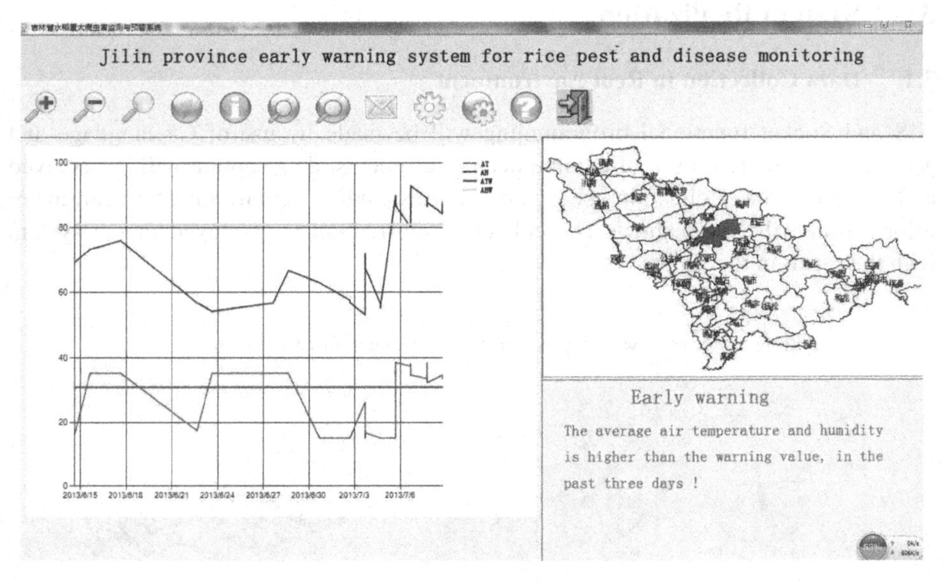

Fig. 4. Rice disease early warning

4 Conclusions

In this paper, by using GIS network and Socket programming technology with the use of C# language,it realizes rice environmental data acquisition in Jilin Province, the display of geographic coordinates, the rice disease early warning and the geographical position of early warning and short message sending functions. Meanwhile, it presents a classification of occurrence of rice diseases so that the system in a timely manner to make judgments, suggesting that early warning information.

The system was conducted with a year of operation in Changchun City Rice Research Institute, effectively to give advanced warning against rice blast effect and improve the quality and yield of rice, making project objectives.

Acknowledgment. Funds for this research was provided by National 863 subjects (2012AA10A506-4, 2013AA103005-04), Jilin province science and technology development projects(20110217), China Postdoctoral Science Foundation the 54th surface funded(2013M541308), Jilin University Young Teachers Innovation Project (450060491471).

References

1. Su, M.Y., Liu, J.: Construction of rice diseases and insect pests prevention and control system. Rural Economy and Science-Technology (9), 162–163 (2013)
2. Feng, Y.: Policy research of the rice pest control system in guangdong province. South China Rural Area 28(7), 4–17 (2012)
3. Yuan, T., Chen, X., Ma, C., et al.: Research on the Comprehensive Prevention and Control System of Rice Diseases and Insect Pests in Yangtze River Farm. Chinese Agricultural Science Bulletin 29(27), 182–186 (2013)
4. Xiao, D., Gu, Z., Feng, J., et al.: Design and experiment of wireless sensor networks for paddyfield moisture monitoring. Transactions of the CSAE 27(2), 174–179 (2011) (in Chinese with English abstract)
5. Chen, X., Wang, X., Yuan, T., et al.: Construction of information service platform of integrated pest control of rice based on service-oriented architecture (SOA). Acta Agriculturae Shanghai 28(4), 127–131 (2012)
6. Sun, M., Luo, W., Feng, W., et al.: A Web-based expert system for diagnosis and control management of diseases in vegetable crops cultivated under protected conditions. Journal of Nanjing Agricultural University 37(2), 7–14 (2014)
7. Lai, J.C., Ming, B., Li, S.K., et al.: An Image-Based diagnostic expert system for corn diseases. Agricultural Sciences in China 9(8), 1221–1229 (2010)
8. Gonzalez-Andujar, J.L.: Expert system for pests, diseases and weeds identification in olive crops. Expert Systems with Applications 36(2), 3278–3283 (2009)
9. Abu-Naser, S.S., Kashkash, K.A., Fayyad, M.: Developing an expert system for plant diseases diagnosis. Journal of Artificial Intelligence 1(2), 78–85 (2008)

Study of Plant Animation Synthesis by Unity3D

Yanna Jiang[2], Boxiang Xiao[1,*], Baozhu Yang[1], and Xinyu Guo[1]

[1] Animation Department, Beijing Research Center for Information Technology in Agriculture, Beijing 100097, China
{xiaobx,yangbz,guoxy}@nercita.org.cn
[2] College of Information Engineering, Capital Normal University, Beijing 100048, China
jiangyannacs@126.com

Abstract. Virtual plant is a novel research issue in computer applications, and realistic plant animation is widely used in many fields. The research of precise, efficient and photorealistic plant animation synthesis is one of the important issues that the current virtual plant facing with. This paper using maize achieves real time and realistic plant animation, based on the platform of Unity3D. The virtual maize model is generated by use of interactive parameterized modeling method. By setting a good interactive interface, users can dynamically adjust the parameters to realize different animation effects. Finally, concluding and analyzing this method. In addition future research and application direction is discussed in order to achieve better plant animation.

Keywords: Unity3D, Plant animation, Animation synthesis.

1 Introduction

1.1 Plant Modeling and Animation

In recent years, plant modeling and animation synthesis become a novel research, achieving wide attention of domestic and foreign researchers ,generating a number of representative studies [1-6]. As for the issue of plant platform simulation, Zhao Xing et al. [1] from the aspect of botany, established a two-scaled automaton model of virtual plant growth based on the growth mechanism of plants. By restoring camera movement and tree's three-dimensional point clouds, Tan et al. [2] constructed realistic 3D models of natural trees based on real plant image. On this basis, for plant dynamic virtual simulation and animation synthesis problem, Feng Jinhui et al. [3] using nonlinear mechanics and wave theory to solve the motion of different branches, simulating tree swaying in the wind. Habel et al. [4] using geometry and dynamic model which is physically based, through a two-step nonlinear deformation method to simulate the interaction of branches and the external force acting on them, realizing the simulation of branches bending and swaying in the wind with a high sense of reality. Diener [5]

* Corresponding author.

© IFIP International Federation for Information Processing 2015
D. Li and Y. Chen (Eds.): CCTA 2014, IFIP AICT 452, pp. 344–350, 2015.
DOI: 10.1007/978-3-319-19620-6_39

using the coupling of video analysis and hierarchical clustering offers as an alternative to physical simulation. Feature tracking is applied to the video footage, allowing the 2D position and velocity of automatically identified features to be clustered. Extracting the motion of real plant in the wind from video, mapping it to 3D plant models, forming the plant animation. Xiao Boxiang et al. [6] presented a motion capture data-driven animation approach ,reconstructed plant motion by real plant scanned data and motion capture data. Based on the virtual maize model, implemented data-driven plant animation. Existing research methods provide a technique reference to achieve high precise and high realistic animation synthesis. At the same time, the development tool oriented game production also provide a good platform for realistic simulation of the plant.

1.2 Unity3D Simulation Platform

As a fully integrated, cross-platform professional game engine, Unity3D [7] is a hierarchical comprehensive development tool, can be used for interactive contents ,such as real-time three-dimensional animation,3D video games, architectural visualization and so on. It provides a visual authoring environment, uses C# and JavaScript as the scripting language, has a detailed property editor and a dynamic game preview. Unity3D [8] has a highly optimized graphics rendering pipeline for DirectX and OpenGL. With excellent advanced rendering effects and user customization support, Unity3D can sufficiently handle large volumes of three-dimensional model in real time. It has a very powerful physics engine, can realize realistic interactive feel , create high-quality 3D simulation system [9] and real visual effects to simulate physical phenomena of the real world. Unity3D provides a highly perfect light and shadow rendering system with the function of soft shadows and baking, can release the game to Windows, Mac, Android or iPhone platform. Using the Unity web player plugin, which is only 500K in size, can release web games, supporting Mac and Windows'web browser. A complete Unity3D program is a combination of several scenes. Each scene including many models, through the script can dynamically control their behavior. With the Unity as virtual reality development platform, the efficiency of development is high, the effect is realistic, the interactive capabilities is powerful, the amount of data is small.

2 Key Technique Research and System Development

This paper uses Unity3D as development platform. First, three-dimensional model of typical plant is generated by use of interactive parameterized modeling method. Then, getting and setting the attribute value of WindZone using C# scripting language. Interactive interface is made by use of NGUI, users can adjust the parameters to dynamically modify the value of wind main, wind turbulence, wind pulse magnitude, and wind pulse frequency. The overall architecture of the system is shown in Figure 1:

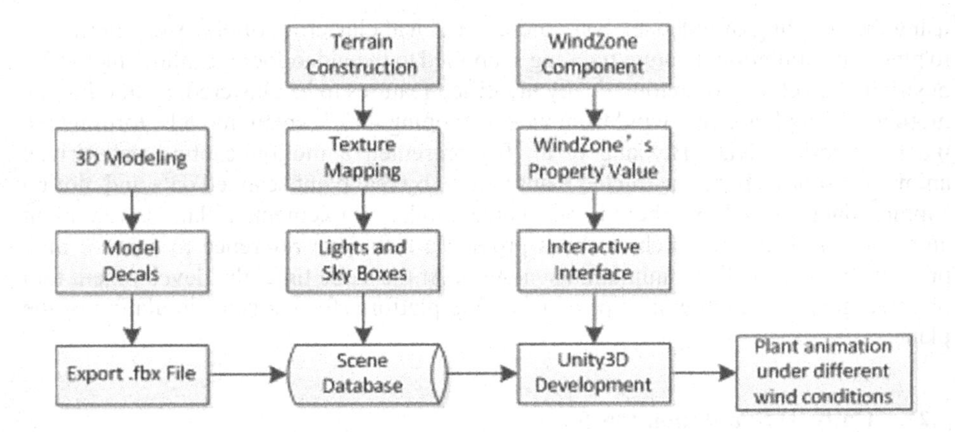

Fig. 1. The framework of plant animation system

2.1 Interactive Plant Parameterized Three-Dimensional Modeling Method

Plant three-dimensional modeling method is varied. For different modeling purposes and application requirements, different modeling methods can obtain different modeling results. Parameterized modeling method is the method which starts earlier, is more influential and widely used. This kind of method uses a set of parameters to represent the main organs and three-dimensional morphological structure of the plant. Usually in an interactive way to realize parameters' dynamic adjustment and editing, in order to achieve the purpose of modifying the three-dimensional shape of models [10]. This paper adopts an interactive parametrized modeling approach. First collecting three-dimensional morphology structure data of maize in field with the three-dimensional instrument Immersion G2LX. Then constructing the three-dimensional model of maize, by defining the veins curve and leaf blade margin curve to achieve the virtual modeling of leaf characteristics. Finally generating maize plant and population model in accordance with the plant type structure of typical maize.

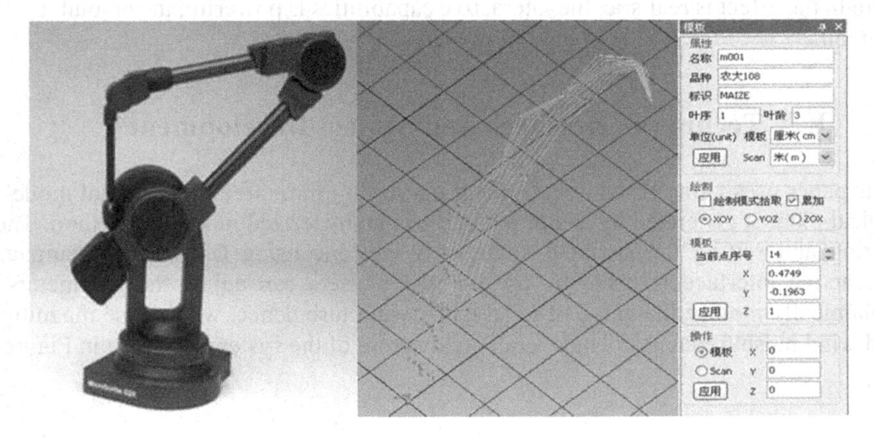

Fig. 2. Immersion G2LX Digitizer, Leaf Template and Modeling Parameters

2.2 Key Technique Research

2.2.1 Model Import

There are two methods for constructing three-dimensional model in the scene. One is the use of built-in functions of Unity3D. Another way is to import the 3D model as the same way of importing resources, where the models are constructed in the modeling software. This paper uses 3DMax as the modeling software to construct the three-dimensional plant model. Converting the format of three-dimensional plant to .fbx, which is the compatible format in Unity3D. The imported model information in Unity3D including spatial location, shader name, etc. Then, in the inspector panel of Unity3D to set the textures. Though directional light is added in the scene, there still exists the phenomenon that the imported model appears partial dark. This problem is solved by setting and adjusting point light.

2.2.2 Interactive Technique

A good user interface is produced by use of NGUI plugin. Through clicking the button, users can switch the browse modes. By dynamically input parameters to achieve different levels of animation effects. There are three cameras in the scene. The NGUI camera is to display the constructed two-dimensional user interactive interface. By defining two camera objects in the scene, dynamically controlling the state of the scene camera and first-person camera, browse mode is switched. By importing the first-person controller in the scene and setting its position and viewing angle, users through mouse, keyboard roaming in the scene, and then from different aspects to observe animation effects.

2.2.3 Plant Animation

This paper realizing the plant animation by adding the component of WindZone, as shown in figure 3.And then getting and setting the component's each property value in scripts, to achieve the plant's swaying effects under different wind conditions. Unity3D comes with the tree component is SpeedTree, thus the tree supporting swing, WindZone just provides wind condition data to swing. For external imported plant model, there is still no wind effect although the WindZone component is added. Initially using brush to draw the plant model onto the terrain, when choosing the model needs to set the value of Bend Factor greater than 0, as shown in Figure 4. Bend Factor is used for setting the camber of plant, thus plant model can have oscillating effects in the scene. As plant model is drawn directly onto the terrain, the attribute value of WindZone can not be dynamically obtained and modified by scripts. Thus importing Unity tree package and creating tree model. Then in the inspector panel uses the imported model and texture to replace that tree's mesh and texture. Adding WindZone for plant model in the way of adding component, binding the written C# script on to the plant model. Finally solving the problem of external plant model swaying and wind conditions'dynamic change.

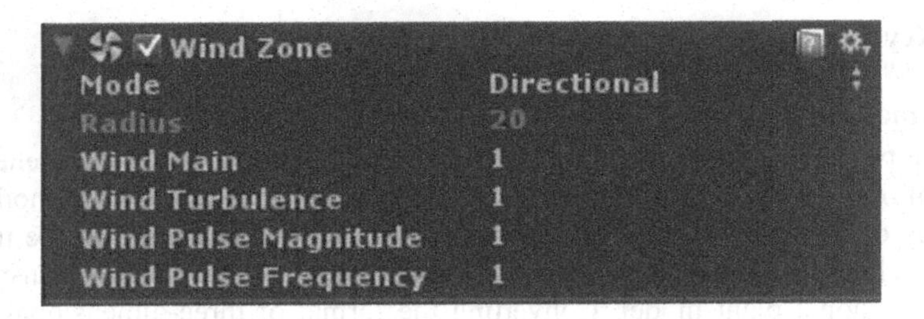

Fig. 3. Properties of WindZone

Fig. 4. Selected plant model and camber setting

Fig. 5. (a) Maize canopy at static status; (b) Motion of canopy by influence of wind

2.3 System Development

Based on Unity4.2 using maize as the study object implemented plant population animation. This paper is mainly divided into the following sections. First is the scene integration [11], including creating terrain, setting textures and adding light and sky boxes in the scene. Second, importing the plant model and trees package .Creating a tree model and setting the model's specific location through translation, rotation and scaling. Letting the imported external model and textures to replace tree's mesh and textures. Adding WindZone component for the plant model, binding a C# script which can access and change the wind conditions. Interactive interface using NGUI plugin to implement, by use of input box through event-trigger to get the parameters dynamically setted by users. Changing the property values of WindZone to achieve the plant animation under different wind conditions. The implemented animation effects is shown in Figure 5.

3 Conclusions

By study of the above key techniques, a simple and low-cost plant animation system is built. This system achieves a certain degree of realism and interactivity. Through the interactive interface, users can switch browse mode and dynamically change the wind conditions. In the mode of first-person, users can roam in the scene by use of mouse and keyboard, and can from different angles to observe the plant's motion effects caused by different wind conditions. In the follow-up work, we intended to implement a plant animation system with a more perfect function, more friendly man-machine interactive interface. To further enhance the realism and interactivity of user experience, data glove can be also used, thus allowing users to interact with the scene more efficiently.

Acknowledgement. This work is supported by National Natural Science Foundation of China (Grant No.61300079);by Beijing Municipal Natural Science Foundation Project “Study of plant dynamic modeling and simulation based on kinematic analysis”(Grant No. 4132028); by Special Fund for S&T Innovation of Beijing Academy of Agriculture and Forestry Sciences (Grant No. KJCX201204007).The authors thank the reviewers for their valuable comments and suggestions.

References

1. Zhao, X., de Reffye, P., Xiong, F., et al.: Dual-scale automaton model for virtual plant development. Chinese Journal of Computers 24(6), 608–615 (2001)
2. Tan, P., Zeng, G., Wang, J., et al.: Image-based tree modeling. ACM Transactions on Graphics (TOG) 26(3), 87 (2007)
3. Feng, J., Chen, Y., Yan, T., Wu, E.: Going with wind-physically based animation of trees. Chinese Journal of Computers 21(9), 769–773 (1998)
4. Habel, R., Kusternig, A., Wimmer, M.: Physically guided animation of trees. Computer Graphics Forum 28(2), 523–532 (2009)

5. Diener, J., Reveret, L., Fiume, E.: Hierarchical retargetting of 2D motion fields to the animation of 3D plant models. In: Proceedings of the 2006 ACM SIGGRAPH/Eurographics Symposium on Computer Animation, pp. 187–195. Eurographics Association (2006)
6. Xiao, B., Guo, X., Zhao, C.: An Approach of Mocap Data-driven Animation for Virtual Plant. IETE Journal of Research 59(3), 258–263 (2013)
7. Zhu, Z.: Design and application research of virtual experimental system based on Unity3D. Master's thesis (May 2012)
8. Chen, H., Ma, Q., Zhu, D.: Research of interactive virtual agriculture simulation platform based on Unity3D (3), 184–186 (2012)
9. Xuan, Y.: Unity3D game development. The People's Posts and Telecommunications Press (2012)
10. Xiao, B., Guo, X., et al.: An interactive digital design system for corn modeling. Mathematical and Computer Modelling 51(11-12), 1383–1389 (2010)
11. Finney, K.C.: 3D Game Programming All in One. Tsinghua University Press, Beijing (2005) Qi, B., Xiao, Y. (transl.)

A New Algorithm of Bayesian Model Averaging Based on SCE - UA Collection Averaging

Liu Jun-Hua[1,2], Zhang Hong-Qin[1], Zhang Cheng-Ming[1,*], Zhao Tianyu[3], and Ma Jing[1]

[1] Shandong Agricultural University, College of Information Science and Engineering, Taian, Shandong

[2] Institute of Agricultural standards and Detecting Technology SAAS, Jinan, China

[3]. Shandong South-North Water Diversion Corporation Limited, Jinan, China

Abstract. Bayesian Model Averaging (BMA) is a statistical method used for multi-model ensemble forecast system. Firstly, the likelihood function of BMA is improved by eliminating the explicit constraint, that the sum of weights is 1, and use SCE-UA for the minimization of its, which presents a new method for solving the Bayesian model averaging, that is the BMA-SCE-UA method. With three land surface models of soil moisture simulation test of multiple numerical model. By comparing the common Expectation Maximization (EM) method with the SCE-UA method, the results show that: SCE-UA method can improve the simulation performance of soil moisture in a large extent, and the soil moisture obtained by the BMA collection simulation and observation matches well, no matter from the amplitude variation and seasonal variability, which makes it possible that generating high accuracy data set of soil moisture with the method of BMA-SCE-UA and using multiple land surface models.

Keywords: Bayesian Model Averaging, SCE-UA, soil moisture, optimization algorithm.

1 Introduction

The model prediction is an import method that has been used in many areas. It contains the physical mechanism between the variables that can be more deeply reveal the changes and regulations of variables. But the prediction performance is not satisfied because there are many uncertainties in the models. Bayesian Model Averaging, a statistical probability forecast post-processing method that put forward by Rafery et al. This method obtains more reliable prediction results by giving the model forecast probability distribution some weights which is after the deviation correction .The weights are posterior probability, which represent the contribution of each model for

---

* Corresponding author.

© IFIP International Federation for Information Processing 2015
D. Li and Y. Chen (Eds.): CCTA 2014, IFIP AICT 452, pp. 351–358, 2015.
DOI: 10.1007/978-3-319-19620-6_40

the ensemble forecast. There are many research results have been already existed by using the BMA method show that the prediction performances are accurate, reliable and with incomparable advantages over other approaches [1-3].

The key of BMA is accurately to estimate the weights of each model. Now the Expectation Maximization (EM) method is widely used. EM algorithm is easy, low calculation cost and requires that the weights are nonnegative and the sum of weights is 1. EM algorithm is used effectively and can generate relatively stable weights and variances [1,4]. Unfortunately, this method cannot guarantee to get the global optimal solution, especially in high-dimensional case. In addition, EM algorithm assumes the variables are normal distribution. However, it is difficult to satisfy the assumption in many cases.

SCE-UA algorithm is not only an efficient optimization method to solve the nonlinear constraint problems but also can obtain the global optimal solution. Calculating the weights and variances are aimed at finding a set of optimal values to meet the constraints of the objective function. At this time, the weights can be as parameters, and rating by SCE-UA method. Structuring the objective function used for estimating weights is the key of BMA method, therefore proposes the new method for solving the Bayesian model averaging (BMA-SCE-UA) [2,5,6]. The BMA-SCE-UA method is different from the EM algorithm without the assumption that forecast variables are normal distribution. This article selects three models (CLM, VIC and Noah) as the prediction models of soil moisture to test the performance. Comparing the results obtained by the EM, BMA-SCE-UA, the observation and the prediction of three models shows that in the calculation precision aspect, the BMA-SCE-UA method is better than EM algorithm.

2 Experiments and Methods

SCE-UA method is an efficient optimization method to solve the nonlinear minimization constraint problems, which is combining the deterministic complex search technology and the principle of biological evolution in nature. It combines with deterministic methods, stochastic methods, competition evolution methods and complex mixture methods to ensure the flexibility, global, consistency, and validity of the SCE-UA algorithm. The core of SCE-UA algorithm is the Competition Complex Evolutionary algorithm (CCE) [7-9].In the CCE method, each vertex is the potential father and is likely to produce the next generation.

There are multiple parameters in SCE-UA algorithm, mainly including: the vertices of each complex polygon m ($m \geq n+1$), the vertices of each child complex number q($2 \leq q \leq m$), the complex number of participating in the evolution p ($p \geq 1$), the sample size s ($s = p*m$), consecutive reproduce complex y ($y = 1$) of each child, each child complex evolution steps z ($z = 2n + 1$), the number of the waiting for optimization parameters n .The m value should not be too large, otherwise the calculation time will be too long and efficiency also will not be high yet [10-12].

Bayesian Model Averaging needs to estimate the weights and variances of each member of the mode of conditional probability function. Firstly, this paper improves the likelihood function of BMA by eliminating the explicit constraint, that the sum of weights is 1, and minimizes the likelihood function by the SCE-UA method. To solve the Bayesian model averaging, we put forward a new method that is the BMA-SCE-UA method.

Actually, estimating the weights and variances is to maximum the objective function as follows:

$$\max\left(\sum_{s=1}^{S}\sum_{t=1}^{T}\log\left(\sum_{k=1}^{K}\omega_k g_k\right)\right) \tag{1}$$

Here $0 \le \omega_k \le 1$, and $\sum_{k=1}^{K}\omega_k = 1$. In order to get rid of the restrain $\sum_{k=1}^{K}\omega_k = 1$, this paper introduces an intermediate variable and normalizes. Make $\omega_k = \frac{\alpha_k}{\sum_{k=1}^{K}\alpha_k}$ to eliminate the constraint $\sum_{k=1}^{K}\omega_k = 1$, and at the same time, the above maximum problem (1) is converted into minimum problem by inversing, namely:

$$\min\left(\frac{1}{\left(\sum_{s=1}^{S}\sum_{t=1}^{T}\log\left(\sum_{k=1}^{K}\frac{\alpha_k}{\sum_{k=1}^{K}\alpha_k}g_k\right)\right)}\right) \tag{2}$$

$$(0 \le \alpha_k \le 1)\ (k = 1, \cdots, k) \tag{3}$$

The flowchart of solving the minimization problem by using SCE-UA algorithm is as following figure 1:

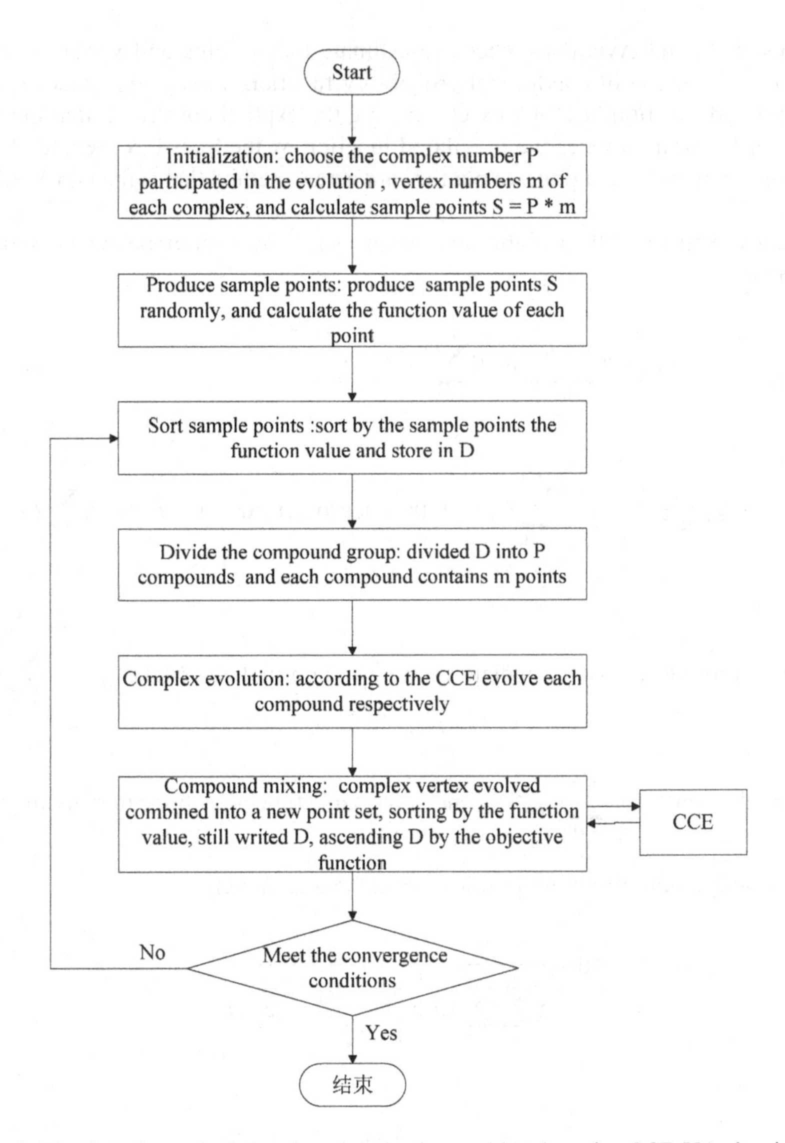

Fig. 1. The flowchart of solving the minimization problem by using SCE-UA algorithm

3 Experiment Data

This paper uses three land surface models (CLM3.5, Clm-Vic, Noah) with the atmospheric forcing data drive models to generate the soil moisture data to test the effectiveness of BMA-SCE-UA method. We choose the north region of China (110 ° ~ 120 ° E, 40 ° to 45 ° N) containing the observation of 14 sites from 1993 to 2006 to train the BMA. Each site is observed at 8, 18, 28 monthly. Observation level is divided into three layers: 0~10 cm, 40~50 cm and 90~100 cm. Due to observation at 40~50 cm and

90~100 cm is relatively scarce, the experiment is considered only 0~10 cm level. We can analysis the calculation accuracy by comparing the EM with SCE-UA method,.

4 Experiment Process and Analysis

Figure 2 shows the soil moisture data results of three models prediction (CLM3.5, CLM - VIC, Noah), average prediction and observation from 1993 to 2006. As can be seen from figure 2, the soil moisture is changing with season regulation and the trend of change each year. At the same time, three model forecast precisions still need to be improved.

Figure 2 shows the forecast results of three models ensemble mean have no better than a single mode. On one hand, each model prediction results can have equal coefficient by using simple arithmetic average, and without considering the performance of each model. On the other hand, it could be that the number of used models (only three) is less, which makes the simple average simulation results to be uncertainty.

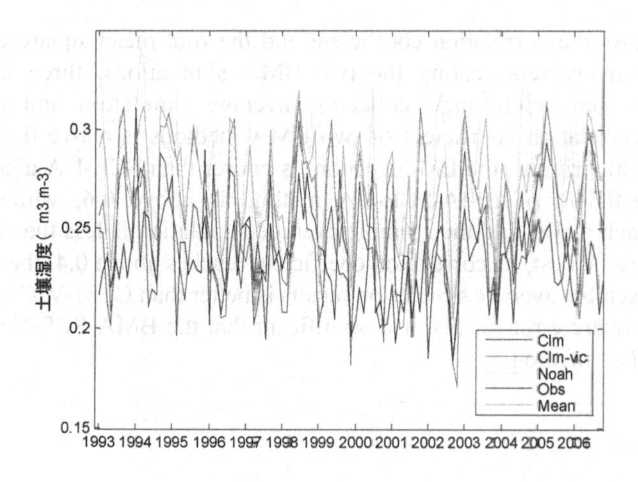

Fig. 2. Soil moisture time series

Figure 3 depicts the soil moisture time series results simulated by two BMA methods and observations. As can be seen from figure 2, the simulation results of two BMA methods have obvious improvement in comparison with the results of each model and model averaging. This is mainly because the BMA method can give full consideration to the prediction ability of each model and endow the different coefficient according to their respective contributions. Their weight coefficients are the posteriori probability representing the prediction techniques in training model, which will improve the simulation precision in a large extent. Firstly a simple linear regression is used for the model forecast results in BMA method, which is actually error correction and bound to improve the simulation precision in certain extent.

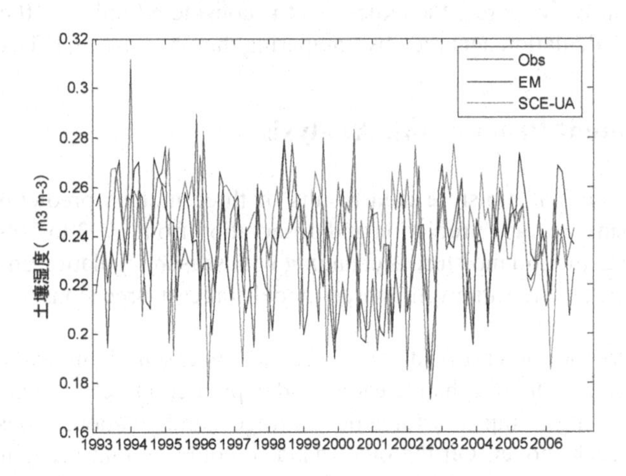

Fig. 3. Simulated time series of two methods of BMA

Figure 4 shows the correlation coefficient and the root mean square error of soil moisture respectively representing the two BMA simulations, three land surface models directly simulation, their collection average simulation and observation. Generally the correlation coefficient of two BMA methods is above 0.5. Compared with two BMA algorithms, the EM algorithm is poorer than SCE-UA algorithm. The correlation coefficient of BMA-SCE-UA method is above 0.6, while the direct simulation of each model and their simple average simulation is less than 0.5, and the effect of Noah is the best, its correlation coefficient reaches above 0.4, the other is less than 0.4, the ensemble average simulation results is better than CLM-VIC model. From the root mean square error results, it is significant that the BMA-SCE-UA method is superior to the EM method.

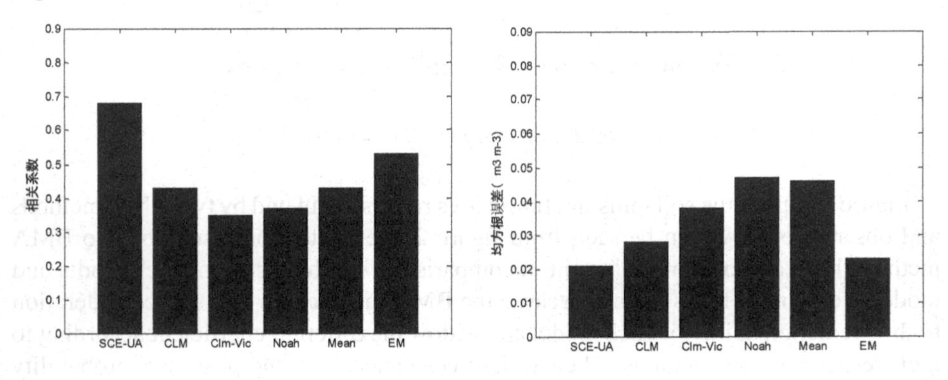

Fig. 4. The correlation coefficient and the root mean square error of soil moisture time series of different methods of simulation results and the observation

5 Discussions

Firstly, the BMA likelihood function is improved in this paper, getting rid of the restraint that sum of weights is 1 and transforming into minimization problem, then solving by the SCE-UA. In this thing called BMA-SCE-UA method. The soil moisture multi-model numerical simulation test using three land surface models shows that the BMA-SCE-UA method is better than the EM algorithm in calculation precision. In addition, the BMA-SCE-UA method is rid of the assumption that forecast variables are normal distribution, and broadening the application. Numerical experiments show that the BMA-SCE-UA method can largely improve the simulation effect, the BMA collection simulation of soil moisture and the observation of soil moisture match well, no matter from the amplitude variation and seasonal variability, which makes it possible that makes use of multiple land surface process models combining the BMA-SCE-UA method to generate higher accuracy of soil moisture data integration. It should be pointed out that, due to the nature of the BMA multi-model prediction method is a model training method, so we must use observed data in the collection each forecast model for training and testing to determine the coefficient of the BMA. But as for some regions with no observation data, this method is invalid.

Acknowledgments. National High Technology Research and Development Program of China (863 Program) (2013AA122003), Foundation for Outstanding Young Scientist in Shandong Province (BS2011DX031), and Science and Technology Develop Project in Shandong Province (2012GSF11713), Funded by Key Laboratory of Geo-informatics of State Bureau of Surveying and Mapping (201319).

References

1. Tian, X.J., Xie, Z.H., Wang, A.H., et al.: A new approach for Bayesian model averaging. China Earth Sci. (2011), doi:10.10007/s11430-011-4307-x
2. Ciccone, A., Jarociński, M.: Determinants of Economic Growth: Will Data Tell? American Economic Journal: Macroeconomics (2010); Zeugner, S.: Bayesian Model Averaging with BMS (2011) (forthcoming)
3. Vrugt, J.A., ter Braak, C.J.F., Diks, C.G.H., Robinson, B.A., Hyman, J.M., Higdon, D.: Accelerating Markov chain Monte Carlo simulation byself-adaptive differential evolution with randomized subspace sampling. Int. J. Nonlinear Sci. Numer. Simul. (2008a) (in press)
4. Vrugt, J.A., ter Braak, C.J.F., Clark, M.P., Hyman, J.M., Robinson, B.A.: Treatment of input uncertainty in hydrologic modeling: doing hydrology backwards with Markov Chain Monte Carlo simulation. Water Resour. Res. (2008b), doi:10.1029/2007WR006720
5. Wöhling, T., Vrugt, J.A.: Combining multi-objective optimization and Bayesian model averaging to calibrate forecast ensembles of soil hydraulic models. Water Resour. Res. (2008), doi:10.1029/2008WR007154
6. Ajami, N.K., Duan, Q., Sorooshian, S.: An integrated hydrologic Bayesian multimodel Combination framework: Confronting input, parameter, and model structural uncertainty in hydrologic prediction. Water Resour. Res. 43, W01403 (2007), doi: 10.1029/2005 WR004745

7. Raftery, A.E., Gneiting, T., Balabdaoui, F., et al.: Using Bayesian model averaging to calibrate forecast ensembles. Mon. Weather Rev. 133, 1155–1174 (2005)
8. Ajami, N.K., Duan, Q., Gao, X., Sorooshian, S.: Multi-model combination techniques for hydrological forecasting: application to distributed model intercomparison project results. J. Hydrometeorol. 8, 755–768 (2006)
9. Rajagopal, R., del Castillo, E.: Model-Robust Process Optimization Using Bayesian Model Averaging. Technometrics 47(2), 152–163 (2005)
10. Xie, H., Eheart, J.W., Chen, Y., et al.: An approach for improving the sampling efficiency in the Bayesian calibration of computationally expensive simulation models. Water Resour. Res. 45, W06419 (2009), doi:10.1029/2007WR006773
11. Tian, X.J., Xie, Z.H., Dai, A.G., et al.: A dual-pass variational data assimilation framework for estimating soil moisture profiles from AMSR-E microwave brightness temperature. J. Geophys. Res. 114, D16102 (2009), doi:10.1029/2008JD011600.
12. Tian, X.J., Xie, Z.H., Dai, A.G., et al.: A microwave land data assimilation system: Scheme and preliminary evaluation over China. J. Geophys. Res. 115, D21113 (2010), doi:10.1029/2010JD014370.

Research on Social Risk Evolution and Control of the Large Hydraulic Project Construction Based on Society Burning Theory*

Bo Wang[1,2], Dechun Huang[1,2], Haiyan Li[3], Chang Zheng Zhang[1,2], and Zheng Qi He[1,2]

[1] Business School of Hohai University, Nanjing 211100, China
[2] Industrial Economics Institute of HHU, Nanjing 211100, China
[3] College of Mathematics and Information Science, Beifang University of Nationalities, Yinchuan 750021, China

{Sensitivebaby,hhu2007}@126.com,
{huangdechun66,lihaiyanmath}@163.com,
751952329@qq.com

Abstract. The construction of large hydraulic project improves the efficiency of water using and it will bring the social risk which has the dynamic evolution characteristic. In order to research the evolution process of social risk of large hydraulic project construction, a social risk evolution model is established based on society burning theory to depict the risk dynamic evolutionary process quantitatively. At last, an early warning control model is constructed. At the same time, the rational comprehensive mechanism of interests, the informational processing communication mechanism and the comprehensive emergency control system are put forward to reduce the risk level from the point of view of guarding and controlling the social risk.

Keywords: large hydraulic project, society burning theory, social risk evolution model, risk control.

1 Introduction

In order to solve the urgent needs for social and economic development, China constructs a number of large hydraulic projects, such as the Three Gorges Project, the Gezhou Dam and South-North Water Transfer Project which are under construction. The successful implementation of major projects could bring profound influence in politics and economics [1].However, there are some complex issues need to solve in the process of construction and operation; even improper handling way may lead to social conflicts. At present, China under the rapid social transformation, all kinds of social contradictions and problems occur frequently because of rapid social changes. Social risks associated with the social structure changes bring a major challenge to maintain social stability.

* Fund supported: Special Program For Key Program for International S&T Cooperation Projects (2012DFA60830).

© IFIP International Federation for Information Processing 2015
D. Li and Y. Chen (Eds.): CCTA 2014, IFIP AICT 452, pp. 359–370, 2015.
DOI: 10.1007/978-3-319-19620-6_41

At present, academic literatures in this field focus, mainly, on the social impact assessment, social risk and stability of hydraulic project. D.M Rosenberg [2-3] considers the problems of economic and environmental benefits of large hydraulic project construction respectively. Fu Peng gave a quantitative research through establishing an index system on hydraulic conservancy project [4]. The researches about social risk of water conservancy project construction mainly focus on immigrant. Li Hua and Jiang Hualin[5] pointed out the social integration of immigrants is at the heart content of Three Gorges Project on Migration and social stability. Chen Yan analyzed the social risk from immigrants' resettlement and made qualitative analysis according to the risk Analysis Questionnaire [6].Yang Fan and Yu Jianxing analyzed the risk factors for resettlement project and applied the fuzzy mathematics method to make a risk analysis and the quantitative analysis [7]. In addition, Zhang Jie and Wang Huimin[8] analyzed the social risks brought about by South-North Water Transfer Project from the point of view of risk society theory. Yu Wenxue discussed social problem and countermeasure for construction of water conservancy projects from the perspective of the history and development [9].Judging from combing the literature, we find out the current studies of the risks and social influences, triggered by water conservancy project construction, mainly focus on static assessment and lack of dynamic and systematic research. However, construction of water conservancy is system engineering. The risk produced from that is a dynamic evolutionary process which gradually accumulated until the outbreak. How to research the social risk evolution process of large hydraulic project from the point views of the risk evolution, we should use the society burning theory as an important tool and method. Fan Zemeng and Niu Wenyuan developed a novel mechanism model for controlling the stability of the social system based on society burning theory. They saw the stability of the social system as the balance of four forces from the systematic perspective [10].

In this paper, we introduce the force balance principle and concept of energy and seen the social risk evolution of large hydraulic project as the acceleration proceeds of object movement. Then, a social risk evolution model, based on the framework of society burning theory, was developed. According to this model, we give a quantitative description with respect to the social risk evolutionary progress and construct the risk control model which helps us to study how to guard and control the social risk. The workings in this paper maybe offer a reference for quantitatively analysis and control the social risk of large hydraulic project.

2 Definition of Social Risk and Theoretical Foundation

2.1 Definition of Social Risk about Large Hydraulic Project

Risk is loss of uncertainty which throughout the development of the society, it always could not be eliminated [11]. In the progress of large hydraulic project construction, there are many stakeholders involved, moreover, all the factors of large hydraulic project construction influence with each other. All of these result in particularity social risks, so we should make a definition for social risk. Feng Biyang deemed that the social risk is social losses uncertainty from abroad view of definition; at the same time, social risk refers to a risk like political risk, economic risk, cultural risk, financial risk and decision-making risk [12]. For the risk of large hydraulic project, the definition of that means the uncertainty of loss about politics, engineering, environment and economy. Associated with that, we called the narrow definition of social risk of large hydraulic project is a possibility which could result in the unrest and instability of the local community. The main reasons give rise to social risk is the contradictions produced by compensation and allocation of large hydraulic project under different time and space between people and nature. In this paper, our research focuses on the narrow definition of social risk.

2.2 Theoretical Basis for Social Risk Evolution Analysis of Large Hydraulic Project

The society burning theory likens the social disorder, instability and turmoil to burning material. This theory shows disharmony between people, people and nature will bring the negative influences (burning material), when these influences accumulated to a certain degree, it will form a certain population density and geographical spatial scale under wrong public opinion (Oxidizer); at the same time, it will make the society out-of-balance, disorder so far as to breakdown(Ignition temperature)[13]. On the other hand, construction of large hydraulic project involves so many people and interest group, uneven interest compensation and redistribution can lead to social conflicts and disharmony easily. It can produce underlying social risk because of "burning material". So, on the one hand, we can regard all the social conflicts and unstable factors. The spread of rumors between people and people can be seen as the "Oxidizer" which accelerate the social contradictions and make the risk of instability gathered. When the contradictions become the common interest expression, it will form a certain population density and geographical spatial scale, the social risk level will close in on the critical range of social unrest. Under the incitation of the emergent mass incident, it will emerge the "Ignition temperature" of social unrest which may result in the "society burning" if the influences exceed the maximum safe bearing capacity. So, we draw a conclusion that the society burning theory gives a reasonable tool for analysis the risk evolution of large hydraulic project.

3 Model of Society Risk Evolution of Large Hydraulic Project Based on Society Burning Theory

3.1 Idea for Model Building

Fan Zemeng and Niu Wenyuan [14] regard emergencies as the process of the material movement from starting to stopping. We consider this idea accords with research on the social risk evolution. So, social conflicts, risk material, the volume of social contradictions and risk material determine the dynamic evolution of social risk. Social risk of large hydraulic project is a process of accumulation, acceleration and breaking out. Therefore, we introduce the object quality, acceleration and kinetic energy with analog modeling method for social risk evolution.

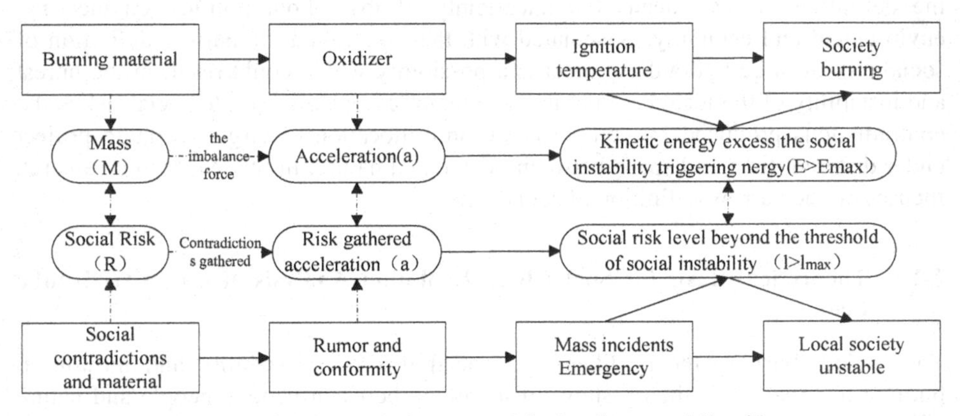

Fig. 1. Social risk evolutionary model of analogy modeling ideas

3.2 Model Assumption

We assume that the model satisfies the following assumptions:

(H1): The stability of the local community of large hydraulic project construction depends on the balance between $\vec{F_N}$ and $\vec{F_P}$. Where, $\vec{F_P}$ denotes the synergy forces which power depends on the ability of assure large hydraulic project local social stability. $\vec{F_N}$ denotes the impact force which could lead to social instability which power depends on the damage by large hydraulic project construction.

(H2): The amount of society contradictions and social risk materials $M_B(t)$ is a dynamic variable over time. The direction of movement depends on the direction and power of the resultant force between $\vec{F_N}$ and $\vec{F_P}$ .The power of the resultant force can be measured in a period.

(H3): In the process of large hydraulic project construction, social risk level can be reflected by the amount of society contradictions and social risk materials accurately.

(H4): When $\vec{F_N} > \vec{F_P}$, social risk will close in on critical range of social unrest, at this time, we need social control force $\vec{F_C}$ to maintain social stability, which power depends on emergency mechanism and safeguard stability in large hydraulic project construction.

3.3 Representation Model

Newton's laws of classical kinematics believe the state of motion of the object is determined by its balance of the force. When the object showed accelerated motion, its direction of motion is determined by the direction of the resultant force. Social risk has dynamic evolutionary feature. We denote $M_B(t)$ is social contradictions and risk material and $\vec{a_t}$ is "oxidizer" at time t, so we get the acceleration model of social risk evolution.

$$\vec{a_t} = (\vec{F_N} - \vec{F_P}) / M_B(t) = \frac{(\vec{f}(y_1, y_2, \cdots, y_i, t) - \vec{h}(z_1, z_2, \cdots z_j, t))}{m(x_1, x_2, \cdots x_k, t)} \tag{1}$$

Where $\vec{F_N} - \vec{F_P}$ denotes the resultant force, $M_B(t) = m(x_1, x_2, \cdots x_k, t)$ is social contradictions and risk material at time t. Here $(x_1, x_2, \cdots x_k, t)$ is a vector of social contradictions and risk material. We denote the impact force is $\vec{F_N} = f(y_1, y_2, \cdots, y_i, t)$, here $(y_1, y_2, \cdots, y_i, t)$ is a negative factors vector of large hydraulic project construction; $\vec{F_P} = h(z_1, z_2, \cdots z_j, t)$ represents synergy forces, here $(z_1, z_2, \cdots z_j, t)$ is a vector of large hydraulic project construction for compensation.

We denote the initial velocity of society contradictions and social risk materials is $\vec{V_0}$, the model of the kinetic energy generates at time t by $M_B(t)$ as follow:

$$\vec{V_t} = \vec{V_0} + \int_{t_0}^{t} \vec{a_t} dt \tag{2}$$

$$E(t) = M_B(t) \cdot \vec{V_t}^2 / 2 = M_B(t) \cdot (\vec{V_0} + \int_{t_0}^{t} \vec{a_t} dt)^2 / 2 \tag{3}$$

Here, we define the maximum bearing capacity of local society as the triggering energy of social instability from stable to unstable.

$$E_{max} = S(w_1, w_2, \cdots w_s) \tag{4}$$

Here, $(w_1, w_2, \cdots w_s)$ represent a vector of society development which level depends on the system of laws, military and police, the wealth of society, emergency security system etc.

When the social risk level will close in on the critical range of social unrest, social control force contributes to lower levels of social risk. So, we have the relation among "fire extinguishing agent" $\vec{a_C}$, fire extinguishing speed $\vec{V_{C,t}}$ and the social control energy $E_C(t)$.

$$\vec{a_C} = \vec{F_C} / M_B(t) = \frac{\vec{g}(c_1, c_2, \cdots, c_l, t)}{m(x_1, x_2, \cdots x_k, t)} \tag{5}$$

$$\vec{V}_{C,t} = \vec{V}_{C,0} + \int_t^{t_4} \vec{a}_c dt \tag{6}$$

$$E_C(t) = M_B(t) * \vec{V}_{C,t}^{\,2} / 2 = M_B(t) * (\vec{V}_{C,0} + \int_t^{t_4} \vec{a}_c dt)^2 / 2 \tag{7}$$

Where $\vec{V}_{C,0}$ is the initial speed of social control force, opposite direction to $\vec{V}_t$, and the direction and magnitude of "fire extinguishing agent" $\vec{a}_C$ is decided by the social control force $\vec{F}_C$.

4 Social Risk Evolution of Large Hydraulic Project

The society development direction of large hydraulic project depends on the balance effect between impact force $\vec{F}_N$ and synergy force $\vec{F}_P$ .Synergistic effect will promote the society development and vulnerable effect will result in the possibility of social unrest. For convenience, we denote magnitude of impact the force and synergy force of large hydraulic project like $|\vec{F}_N|$ and $|\vec{F}_P|$ respectively.

4.1 Social Risk Evolution Dominated by Synergy Effects

(1) when $|\vec{F}_N| \cong |\vec{F}_P|$,the two forces into balance, the $|a_t| \cong 0$ and $|V_t| \cong 0$ .The negative effects from large hydraulic project construction into balance with positive effects from compensation and investment of project construction. So the amount of society contradictions and social risk materials $M_B(t)$ and social risk $l(t)$ at lower levels. Kinetic energy $E(t) = M_B \vec{V}_0^{\,2} / 2$ produced by $M_B(t)$ is lower than triggering energy; the stable state will sustain stability and balance, see fig. 2.

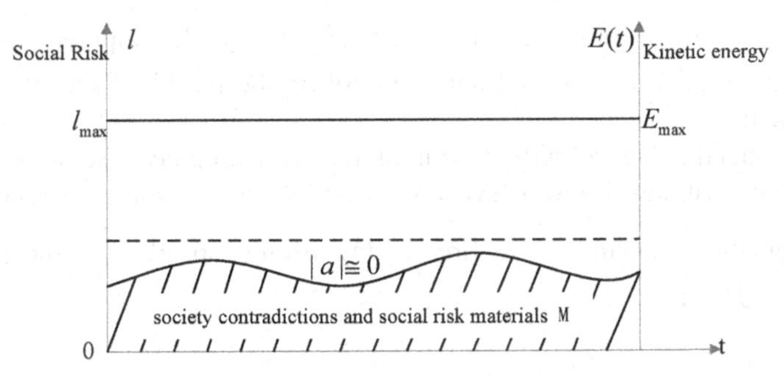

Fig. 2. Social risk evolutionary trend when $|F_N| \cong |F_P|$

(2) When $|\vec{F}_N| < |\vec{F}_P|$, interests compensation makes the positive effects greater than negative effect in large hydraulic project construction process. The acceleration of social contradictions and risk material $|a_t| < 0$ has the same direction with $|\vec{F}_P|$. This means that the amount of society contradictions and social risk materials $M_B(t)$ decreases gradually and social risk $l(t)$ has a lower level. The velocity of social contradictions and risk material $\vec{V}_t$ is less than initial velocity $\vec{V}_0$, $\vec{V}_t$ has the opposite direction with $\vec{V}_0$. At the same time, the kinetic energy $E(t)$ reduces gradually, see fig. 3.

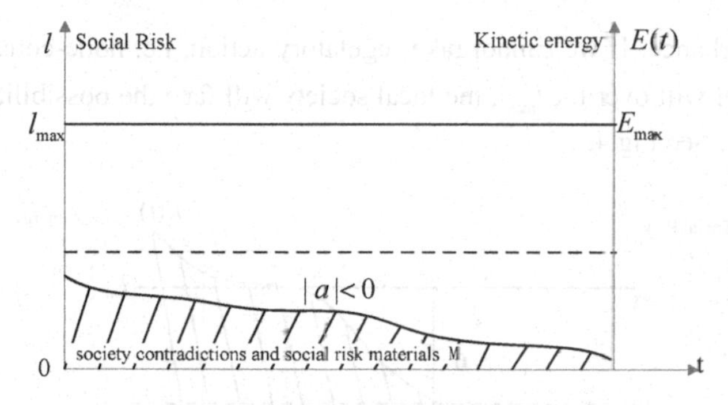

Fig. 3. Social risk evolutionary trend when $|F_N|<|F_P|$

4.2 Social Risk Evolution Dominated by Vulnerability Effects

Social risk evolution is a process which includes risk production, development and diffusion until calm. Fu Yun(2008) give a model which describes the group unexpected incidents as four stages :latent, development, climax, attenuation. With the accumulation of social contradictions and risk material, the potential social risks will show the different levels in different stages. Therefore, we can divide social risk evolution into four stages, like the initial stage, acceleration stage, explosion stage and Calm stage.

4.2.1 Social Risk Evolution with None-control Force

At the initial stage of large hydraulic project construction process, $t \in (t_0, t_1)$. There is a balance between $\vec{F}_N$ and $\vec{F}_P$ and the amount of society contradictions and social risk materials $M_B(t)$ is small and social risk has a lower level. This mainly thanks to the society contradictions and risks coordinate with reasonable investor compensation scheme, such as environmental protection, immigration resettlement, fund compensation and job placement etc., so the social risk lie at a lower level.

At the acceleration stage, $t \in (t_1, t_2)$, the policy, fund and other compensation measure will show up same shortages, such as the investigation discovered that some immigrants recognize their life quality lower and the final-period support for the migrants failed to reach its budget targets[15]. There is going to emerge some phenomenon like "industry hollowness", "return of immigrant", "psychosocial environment fragility" and conformity behavior increasing etc [16].That will make a scale effect of risks communication under inaccurate reports and spreading of rumors. At this stage, the acceleration of society contradictions and social risk materials $|a_t| > 0$, which has the same direction with $\vec{F_N}$. $M_B(t)$ move with $\vec{V_t}$ and its kinetic energy will close in on maximum bearing capacity. At the same time, the social risk level $l(t)$ will accelerate to l_{max}. Social risks break out intensively ignited by mass disturbance. If we cannot take regulatory action, i.e. none-control force $\vec{F_C}$, the social level will over the l_{max}, the local society will face the possibility of disorder and instability. See Fig.4.

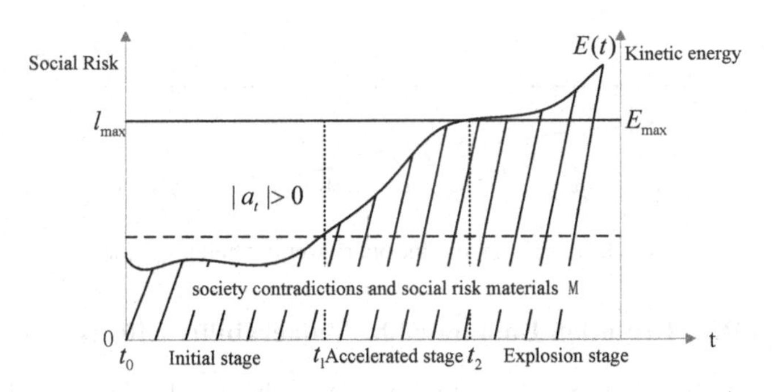

Fig. 4. Social risk evolutionary trend with none-control force $\vec{F_C}$

4.2.2 Social Risk Evolution with Control Force

In the process of large hydraulic project construction, social risk at a lower level under social development dominated by synergy effects and the possibility of disorder is small. When social risk level is close to the l_{max}, the local society maybe comes up instability. Quantities of social conflicts, disorder action and public order cases go up. In this situation, disclose the information, emergency measures were carried out until society restore stability and normal order as soon as possible.

Seen as fig. 5, when $t \in (t_2, t^*)$, the mount of society contradictions and social risk materials gathers with acceleration $|a_t| > 0$ and social force $\vec{F_C}$ will reduce the $M_B(t)$ with acceleration $\vec{a_c}$. Social control energy $E_C(t)$ is increasing and social risk level will lower. There is a peak value in this stage about $l(t)$ and $E(t)$ at time

t^*.In $t \in (t^*, t_3)$, the social risk level will reduce under $\vec{F_C}$, the amount of $M_B(t)$ saw the area of the dotted line form in fig. 5 and when $t \in (t_3, t_4)$,the social risk level will lower at the reasonable range and society order can be controlled.

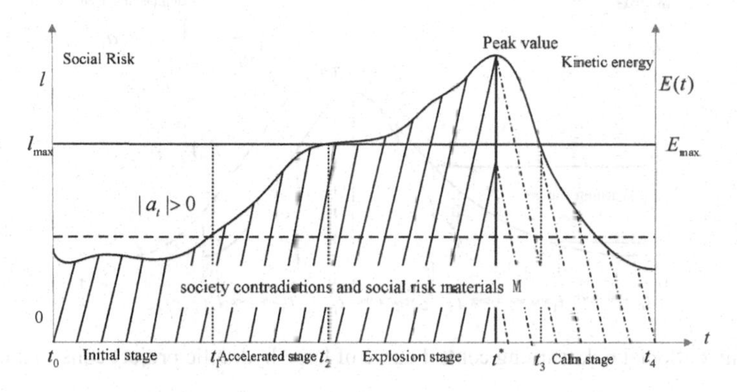

Fig. 5. Social risk evolutionary trend with control force $\vec{F_C}$

5 Social Risk Control of Large Hydraulic Project Construction

5.1 Social Risk Warning Mode of Large Hydraulic Project

Social risk warning is an important link for cut off the risk spreading. Social risk will often breakout suddenly, but the emergence and development of risk is a process. So only build the early warning mechanism, can we prevent loss from social risk [17].When social risk level approach and exceeds the risk threshold, the local society will be in a critical state of disorder and $E(t)$ approach E_{max} too. At the stage (t_1, t_2) ,the comprehensive risk warning system takes an important role in identifying risks and lowering the possibility of society instability and disorder. At the time t , under the effect of resultant force between $(\vec{F_N} - \vec{F_p})$ and $\vec{F_C}$ .Correspondingly, $|\vec{a_t}| < |\vec{a_{C,t}}|$ and $|\vec{V_t}| < |\vec{V_{C,t}}|$ .These changes are shown in Fig. 6.

In figure 6, the full line shows the velocity of social risk and dotted line shows the velocity of social control force. Here t_0^* is the initial time of social control force involving, (t_0^*, t_1^*) and (t_1^*, t_2^*) are accelerated stage and reducing stage of "fire extinguishing" velocity $\vec{V_{C,t}}$. $\vec{V_{C,t}}$ will reach the peak value at t_1^*.The area R that $\vec{V_t}$ curve connects with the ordinate of a_t named warning range. At initial stage, here $|\vec{a_t}| > |\vec{a_{C,t}}|$, social risk is in the spreading state. In (t_0^*, t_1^*) stage, the power of social control force increase gradually, accordingly $|\vec{a_t}| < |\vec{a_{C,t}}|$ and $|\vec{V_t}| < |\vec{V_{C,t}}|$. Social risk level and velocity will lower with time.

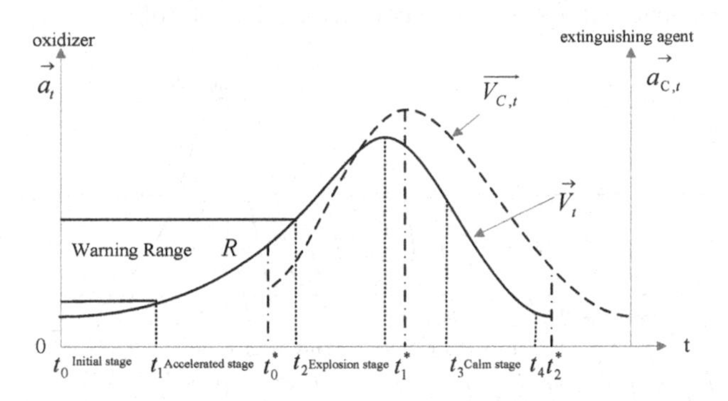

Fig. 6. Social risk warning control curve of large hydraulic project construction

5.2 Measures of Social Risk Prevention and Control of Large Hydraulic Project Construction

(1) Establishing Reasonable Benefits Compensation Mechanism

The reasonable benefits compensation mechanism and measures can reduce the society contradictions and social risk materials. Large hydraulic project construction involves the extensive interest groups and has the widespread influence. Project construction leads to social interest recombined and distribution pattern changed. But beyond that, definition of interest groups and the strength of compensation also affect the "synergy effects". For a long time, definition of an interest group and compensation is planned by the government, so that also lack of the actual investigation. But policy carry out maybe neglect the interested parties' real demand. So we must consider the different group and overall planning by different level and region. Taking measures according to local conditions and adjusting policy is the necessary choice. At the same time, we should build the benefits and ecology compensation mechanism by steps and stages.

(2) Establishing Information Processing and Communication Mechanism

The one way to control the spread of social risk is controlling the acceleration of society contradictions and reducing social risk materials. The "oxidizer" in society burning theory can appear public opinions and make the add fuel to the flame function [18].The construction of large hydraulic project construction often makes the folks feel anxious, group psychology. These factors will make the social psychological environment fragile and fake information spreading. So we need to establish the information processing mechanism. On the one hand, we should build the public opinion monitor mechanism for identifying information and publish to public, which can hold back the folk information from source. On the other hand, when the social psychology no way to appealed, feelings of anxiety and inadequacy will spread in the chain of acquaintances and form the scale effect in the short period.

So, we should build the multiple communication mechanism between government and the grassroots and build mutual trust.

(3) Building Perfect Social Emergency Control System

As the main measures for social risk control, the social emergency control system provides an important guarantee for risk control. The social control force decides the efficiency of risk control. For the past few years, social emergency event, such as SARS and Wenchuan earthquake, reflects the shortage of the social emergency control system. So, first of all, we should strengthen the construction of social emergency ability and change the oneness and limitations of previous emergency measures. At the same time, we should build the multi-departments interaction mechanism and information sharing mechanism to guarantee the timeliness of emergency control measures. Secondly, mass disturbances are the triggered factor has the sudden and dynamic feature so that we should take the dynamic action for it [19]. At last, we need to guide and encourage non-governmental organization and the public welfare social group to take part in building the social emergency control system. Meanwhile, we should play the functions of civil organizations to guide public opinion reasonably thereby reducing the social risk level.

6 Conclusions

Large hydraulic project construction brings the economic effectiveness, but it results in the social risk which has the evolutionary character. In this paper, we build the social risk evolution model based on society burning theory and analyze the evolutionary process from the balance between synergy effects and vulnerability effects. By quantitative description, we draw conclusions to control the social risk and avoid risk distorting and diffusion. First of all, establishing the perfectible benefits compensation mechanism to reduce the amount of society contradictions and social risk materials; secondly, establishing a good information processing and communication mechanism; at last, building the comprehensive social emergency control system to improve level of social risk emergency security .

In this paper, we describe the social risk evolution process and propose control measures according to social risk evolution mode. But how to measure and calculate the variables of the model and make the quantitative calculation is another work in future research.

Acknowledgment. Funds for this research was provided by Special Program For Key Program for International S&T Cooperation Projects (2012DFA60830), the National Social Sciences Fund(11BGL088) and the graduate student research innovation project for Jiangsu province university, 2011(CXLX12_0260).

References

1. Gao, L., Liu, J.: National Key Projects and National Innovation Ability. China Soft Science (4), 17–22 (2005)
2. Rosenberg, D.M., Bodaly, R.A., Ushe, P.J.: Environmental and social impacts of large scale hydro-electric development: who is listening? Global Environmental hange 5(2), 127–148 (1995)
3. Rosenberg, D.M., Berkes, F., Bodaly, R.A., Hecky, R.E., Kelly, C.A., Rudd, J.W.M.: Large-scale impacts of hydroelectric development. Environmental Review (5), 27–54 (1997)
4. Fu, P., Chen, K., Xie, Y., Zhang, D.: Method for environment impact assessment of hydropower development considering social factors. Journal of Hydraulic Engineering 40(8), 1012–1018 (2009)
5. Li, H., Jiang, H.L.: On Social Amalgamation and Social Stabilization of Three Gorges Project Resettlement. Journal of Chongqing University (Social Sciences Edition) 9(2), 37–40 (2002)
6. Chen, Y., Chen, S., Wang, S.: Social risk evaluation of reservoir immigrants. Journal of Economics of Water Resources 23(2), 62–64 (2005)
7. Yang, F., Yu, J.: The risk analysis and evaluation of Water conservancy project resettlement work. Journal of Hydraulic Engineering 36(10), 1258–1262 (2005)
8. Zhang, J., Wang, H.: Society risk analysis of South-to-North Water Society. Yangtze River 4(2), 18–19 (2006)
9. Yu, W.: Social issues caused by water resources project and Countermeasures. Journal of Hohai University 7(4), 54–56 (2005)
10. Liu, Z., Niu, W.: Social stability mechanism model. System Engineering Theory and Practice (7), 69–76 (2007)
11. Hu, A., Wang, L.: A Study of the Measurement Methods and Experiences Concerning the Risks from Social Transition (1993~2004). Management World (6), 46–54 (2006)
12. Feng, B.: Social risk: perspective, contents and factor. Tianjin Social Sciences (2), 72–77 (2004)
13. Niu, W.: The Social Physics and the Warning System of China Social Stability. Bulletin of the Chinese Academy of Sciences (1), 16–20 (2001)
14. Liu, Z., Niu, W., Gu, J.: Social emergency response model and control pattern analysis. China Soft Science (8), 85–92 (2007)
15. Huang, D.: Can it take root After landing?–investigation and thought about rural migrants placed from the Three Gorges. World of Survey (5), 28–30 (2009)
16. Sun, Y.: The Three Gorges Reservoir Area “after the migration period” major social problems analysis of regional social problems reasons and Countermeasures. China Soft Science (6), 24–33 (2011)
17. Song, L.: Social risk warning system design and operation. Journal of Southeast University 1(1), 70–76 (1999)
18. Shan, Y., Gao, J.: Social Physical Interpretation of Reasons for Mass Disturbances: The Introduction of Social Combustion Theory. Journal of Shanghai University of Finance and Economics 6(12), 26–33 (2010)
19. Tong, X., Zhang, H.: Group events and its governance–further consideration under the integrated framework social risk and public crisis. Academic Circles (2), 35–45 (2008)

Exploring the Influential Factors of e-Banking Satisfaction in Rural Areas in China

Mengyu Ren, Yan Li*, Yu Wang, and Zihao Zhao

International College, China Agricultural University, Beijing 100083, China
icbliyan@cau.edu.cn, {76914325,371489197,2485847508}@qq.com

Abstract. Online banking has become a new and popular business in recent years. What factors influence the success of online banking and how can online banking get a larger development have drawn much research interest of people. This paper mainly focuses on the research of online banking standing at the users' view. It will explore the factors that influence the satisfaction of customer. Through this paper, we would not only get the new awareness about the factors affecting the users' satisfaction about online banking, but also gain the profound reflection of the future way regarding the online banking.

Keywords: online banking, satisfaction, rural areas.

1 Introduction

With rapid growth and development of information and network technology, banks have been deeply affected and they are going through unprecedented changes. Those traditional banking businesses are now far behind and left by those online banking businesses since online banking has been bringing people much more convenience [1]. More precisely, with the help of online banking, people can access to banking transactions instead of going to banking institutions [2].

In recent years, along with the economic growth and the popular usage of computer and cell phone in rural areas, the Internet has created tremendous need of using e-banking services in suburban in China [3]. However, restricted by the traditional thought of have the financial activities in the banks and the limited layout of bank site, the development of online banking service is still facing with difficulties. For example, there are only 150,000 people among 505,000 have their online banking accounts in Yutian county, Xinjiang province. The results of former research on rural people showed that 30% of the rural people opened their online banking [4]. The reason of the acceptance of online banking services is that they were recommended, not for their personal needs. Besides, half of them choose to use online banking because there are less commission charges. According to the previous study, the main difficulties that online banking service has faced with are, for example, some people are not capable with using the computer and the Internet; rural population are not committed to financial products; online banking products and services lack innovation.

---

* Corresponding author.

© IFIP International Federation for Information Processing 2015
D. Li and Y. Chen (Eds.): CCTA 2014, IFIP AICT 452, pp. 371–378, 2015.
DOI: 10.1007/978-3-319-19620-6_42

Therefore, determining the main factors influencing customers' satisfaction can be valuable for service providers to make better improvements. In other words, to increase the popularity of using online banking among rural population, banks have to put more emphasis on understanding the needs of rural customers. In this study, we develop a research model basing on reviewing previous literature and test out the relative importance between customer satisfaction level and the main five influential factors, which are system quality, information quality, service quality, trust and switching cost.

2 Theoretical Background

2.1 System Quality

System quality is the measure of information processing system itself [5]. System quality can be measured using ease of learn, ease of use, efficiency, stability and security. Ease of learn denotes a feeling of customer that the system is easy to access. Ease of use can provide convenience to customer when using the system and let people prefer to use the system as well. Efficiency often relates to response time, the less the response time the more customer can be satisfied. Stability means the quality or state of being steady, which creates a sense of reliable from consumers. Security is another important factor that affects consumers' satisfaction. Sathye found that 70% of customers expressed their concerns on security [6]. Hence, it is assumed that system quality positively affects the consumer's satisfaction.

H1: There is positive relationship between system quality and satisfaction.

2.2 Trust

"Willingness to rely on an exchange partner in whom one has confidence" is what trust means [7]. Trust in online banking is a new and emerging area of interest in research. Researchers found that trust will affect satisfaction in the long term [8]. A customer's satisfaction will be enhanced over time, if his or her feeling of faith in the provider is satisfied [9]. Hence, it is assumed that trust effects positively on consumers' satisfaction of online banking system.

H2. There is a positive relationship between trust and satisfaction.

2.3 Service Quality

Service quality refers to the overall support delivered by the service providers. Dabholkar, Shepherd, and Thorpe stated that since service quality had sub-dimensions of reliability and responsiveness, it would lead to customer satisfaction [10]. Researchers maintain that perceived service quality is cognitive and thus followed by satisfaction [11]. With the development of online banking, increasing managers realize that service play an extreme role in the banking. Several empirical studies confirmed that a higher level of service quality was related to a higher level of customer satisfaction. SERVQUAL theory is an important theory to give the way of measuring the service quality [12]. The theory elaborated that quality included five dimensions: reliability, tangibles, responsiveness, assurance, and empathy. And we will design our questions of service quality based on these metrics.

H3. There is a positive relationship between service quality and satisfaction.

2.4 Information Quality

Information quality is a measure of information system output. Information quality refers relevance, usefulness, informativeness, timeliness, completeness and clarity to the users' satisfaction level [12]. Previous studies have demonstrated the significant effects of relevance, which includes relevant depth and scope that can affect bank expertise. Usefulness denotes a feeling of customer that the information is helpful and useful to learn. Informativeness guarantees the users to gain as much information as they should know. Timeliness refers to whether the information is the latest or not, the more often banks update their information, the more their customers can be satisfied. Completeness and clarity are other two key factors. The degree of completeness and clarity is positively related to users' satisfaction level.

H4. There is a positive relationship between information quality and satisfaction.

2.5 Switching Cost

Switching cost is the costs that the consumer incurs by changing one service provider to another [13]. It can be a barrier to change service providers so that the online banking providers are typical beneficiaries. Thus it is a way for improving customer loyalty as well as the customer satisfaction [14]. As a direct effect on customer satisfaction, switching cost offers many advantages for service providers. For instance, it weakens customers' sensitivity to price and satisfaction of the product brand [15]. Specifically, if the customers know the risks of switching their current online transaction way to other ways, such as the trouble in building a new contact relationship, the difficulty in using an alternative service, a long and complex proceed to learn, as well as the time waste of using other transaction method, it will increase the probability that they keep the relationship with the current online banking transaction method. We assume that there is a positive and strong relationship between switching cost and satisfaction. In this paper, we will test their concrete relationship regarding online banking.

H5. There is a positive relationship between switching cost and satisfaction.

3 Research Model

Based on the theoretical background discussed above, this study establishes a research model which suggests five links involved in customer satisfaction regarding the online banking. As showed in Fig. 1, consumers' satisfaction is positively influenced by system quality, trust, service quality, information quality and switching cost. Table 1. shows the main 24 attributes of measuring the five factors.

Table 1. Measurement of five factors

Factors	Items of measurement
System quality	Online banking is easy to learn. Online banking is easy to use. Online banking system can quickly deal with my business. Online banking will not reveal my information and privacy without allowance. Online banking system is stable, no error.
Information quality	Online banking is able to give users financial news. Online banking is able to provide me the various information. Online banking is able to provide me with information that I need. Online banking is able to provide me information of the latest financial products. Online banking are able to provide me with complete information. Online bank's website is designed clear and easy to read.
Service quality	Some related department of bank will give me help when I meet with difficulties while using online banking. The service representative of online banking is polite and worth trust. Online banking provides personal service according to users' needs.
Trust	I think banks that provide online banking service are with high reputation. I think banks give much care to its users. I think banks that provide online banking service is reliable.
Switching cost	Going to the counter gives me much more trouble rather than using online banking. Going to the counter is more expensive compared with using online banking. Going to the counter requests me to learn how to manipulate. Going to the counter wastes my time compared to using online banking.
Customer satisfaction	I will keep using online banking. I will recommend others to use online banking. In general I am satisfied with using online banking.

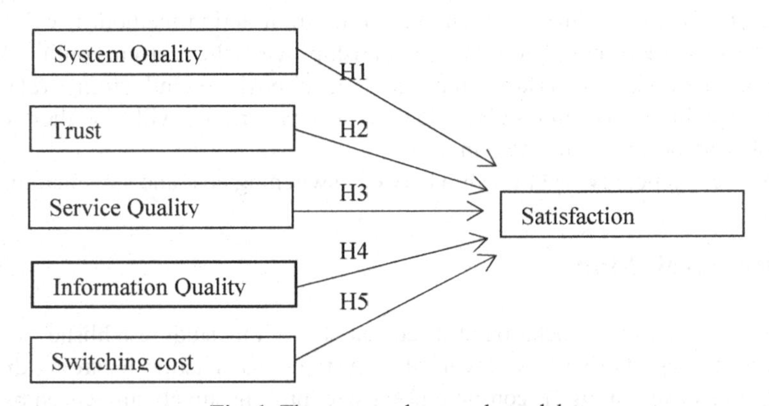

Fig. 1. The proposed research model

4 Research Methods

In June 2013, after pretested by some online banking users and reviewed by three professional experts in related area, our questionnaire was revised based on their feedback and suggestions. 200 questionnaires were collected in Chang Ping County in Beijing and hand out. After eliminating insincere and incomplete responses through data filtering, we got a total number of 150 usable responses. All data will be analyzed by using SPSS and LISREAL software.

The descriptive statistics of the sample are listed in Table 2. Of the 150 participants, 53% are males, 47% are females, and more than half of respondents are below 35 years old. Most of them are young people. The enterprise staff and farmers are the main research respondents. As for disposable income, the majority of respondents have disposable income between 1,000 and 3,000RMB. As for family structure, typical "three in one family" is still the most common structure while there are many big families with more than 3 family members. For the network-using frequency, with the popularity of computer, most people can use Internet every day. It is also an important factor contributing to the popularity of online banking.

Table 2. The profile of sample demographics

Variable		Count	%
Gender	Male	79	53%
	Female	71	47%
Age	<20	36	24%
	21-35	54	36%
	36-50	50	33%
	>50	10	7%
Occupation	Government staff	24	16%
	Enterprise staff	36	24%
	Small business owners	27	18%
	Teachers	5	3%
	Farmers	36	24%
	Students	22	15%
Disposable income per month	<1000	12	8%
	1000-1500	58	39%
	1500-3000	40	28%
	3000-4500	20	13%
	4500-6000	11	7%
	>6000	9	5%
Family members	1	10	7%
	2	20	13%
	3	70	47%
	>3	50	33%
Network-using frequency	More than 6 hours everyday	24	16%
	Less than 6 hours everyday	68	45%
	Several times per week	35	23%
	Several times per month	18	12%
	Several times per year	5	34%

5 Results and Discussion

As shown in Fig. 2, Hypothesis H1 that system quality positively influences users' satisfaction (β=0.511) was supported. System quality was partially the determinant of customer satisfaction. Almost 90% of respondents believe that online banking is easy to use and is easy to learn and 31.3% of them strongly agree with this. 82.6% of respondents agree that online banking can easily deal with business. This result demonstrates the importance of system quality in building the customers' satisfaction in online banking system. Similar to Sathye's finding, almost 70% of customers expressed their concerns on satisfaction [6]. Especially in suburban, the education level is not high, easy to learn and easy to use the system is extremely important to make people be satisfied, which is consistent with the result.

Hypothesis H2 that trust positively influences users' satisfaction (β=0.526) was supported. 74.6% of respondents agree that the bank which provides online banking service has high reputation. 60% of respondents think that banks which provide online banking service is reliable. That implies a large number of online banking users trust that banks are able to provide high quality service in online banking. This result is consistent with earlier finding conducted by Kim, Ferrin and Rao (2009) that trust will affect satisfaction in the long term [9]. The data collected indicates that reliability of online banking service providers is a key to enhance satisfaction. We thought there is no difference between suburban area and urban area on the term that trust affects satisfaction of customers on online banking.

Hypothesis H3 that service quality positively influences users' satisfaction (β=0.247) was not strongly supported. A result can be explained by the characteristic of the system. The conception of "e-service" emerged on the growth of the Internet. Initially, firms developed an online presence due to the cost reductions that could be gained from automation [10]. The online banking system emphasizes on self-using which means people can easily finish the transaction or other financial activities themselves by using the Internet instead of going to the bank personally or communicate with the bank representatives. Consequently, the service provided by departments and representatives of bank has a weak correlation with users' satisfaction. In the meanwhile, the ease of use of the online banking system is still a large barrier of the online banking usages. The bank should consider improving the service quality through the services provided by both the bank staff and the online systems.

Hypothesis H4 that information quality positively influences users' satisfaction (β=0.288) was not strongly supported. It explains the purpose of users to use online banking system. According to the survey, most people use the online banking system in order to finish transfer transaction and currency exchange conveniently. They are not focusing on the financial information the online banking system provided because merely a small portion of people use online banking system to buy financial product.

Hypothesis H5 that switching cost positively influences users' satisfaction (β=0.680) was strongly supported. This result tells us that customers are willing to use the system if it can save time and money. Especially in suburban, the salary level is not very high. The lower the cost, the more customers will be satisfied.

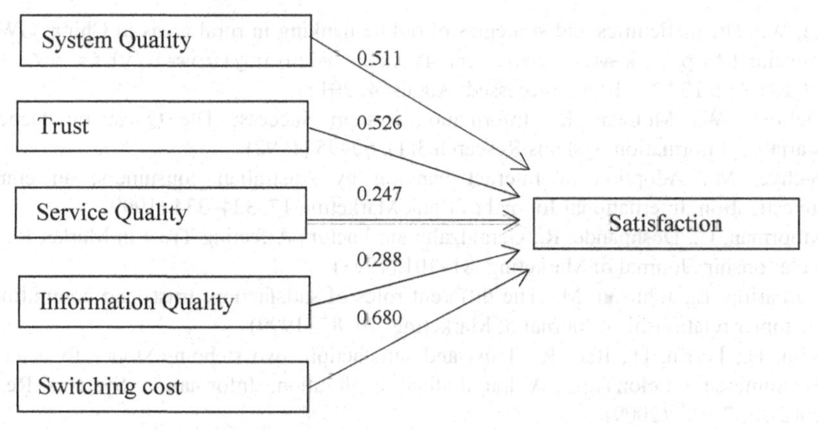

Fig. 2. Results of structural modeling analysis

6 Conclusion

E-banking is fast becoming norm in China, and e-banking system has been adopted by more and more users in rural areas in China. It forces most financial institutions to re-examine their systems and practices and to look for new ways to deliver their services over the Web. E-banking system evaluation does not have committed guidelines and models. Although the past researches evaluate the system effectiveness from different aspects, it was considered more suitable to evaluate the system in terms of users' satisfaction. A research model was developed basing on reviewing previous literature and test out the relative importance. As a result, this research model considers more aspects and help system designers and managers find out the system problems and factors needing improvement. Among all the influential factors, switching cost, trust and system quality are the dominant effects on the satisfaction. As the e-banking system is in the introduction period in rural market, users are more sensitive to the cost, the ease of use of the system and the trust to the bank, rather than the quality of the information and service that the system can provide to them. Therefore, the banks may develop the appropriate strategies copying with the users' current demand and changes in the future. .The main limitations should be noted are the single site of sampling and the criteria of the assessment.

References

1. Yuan, X., Lee, H.S., Kim, S.Y.: Present and Future of Internet Banking in China. Journal of Internet Banking and Commerce 15, 3–9 (2010)
2. Kurnia, S., Peng, F., Liu, Y.R.: Understanding the Adoption of Electronic Banking in China. The University Of Melbourne Press (2009)
3. Chen, X.: Farmers become netizens after the Internet access in rural areas in China. Chinese Financial News (June 20, 2013), `http://www.financialnews.com.cn/dfjr/jyjl/201306/t20130620_35152.html` (accessed: August 4, 2013)

4. Li, W.: The difficulties and strategies of online banking in rural areas in China. GWYOO website, `http://www.gwyoo.com/lunwen/yinhanglunwen/yhfzlw/201209/540137.html` (accessed: August 4, 2013)
5. DeLone, W., McLean, E.: Information System Success: The Quest for Dependent Variable. Information Systems Research 3(1), 60–95 (1992)
6. Sathye, M.: Adoption of Internet banking by Australian consumers: an empirical investigation. International Journal of Bank Marketing 17, 324–334 (1999)
7. Moorman, C., Deshpande, R., Geraldzaltman: Factors Affecting Trust in Market Research Relationship. Journal of Marketing, 81–101 (1993)
8. Garbarino, E., Johnson, M.: The different roles of satisfaction, trust, and commitment in customer relationships. Journal of Marketing, 70–87 (1999)
9. Kim, D., Ferrin, D., Rao, R.: Trust and satisfaction, two stepping stones for successful e-commerce relationships: A longitudinal exploration. Information Systems Research 20(2), 237–257 (2009)
10. Loonam, M., Loughlin, D.: An observation analysis of e-service quality in online banking. Journal of Financial Services Marketing 13(2), 164–178 (2008)
11. Dabholkar, P., Shepherd, D., Thorpe, D.: A comprehensive framework for service quality: An investigation of critical conceptual and measurement issues through a longitudinal study. Journal of Retailing 76(2), 139–173 (2000)
12. Oliver, R.: Whence consumer loyalty? Journal of Marketing 63(4), 33–44 (1999)
13. Tsung, F.C., Li, T.H.: A Study of Assessment of Library E-resource Metasearch System. Journal of Library and Information Studies, 111–142 (2008)
14. Lee, J., Feick, L.: The impact of switching costs on the customer satisfaction-loyalty link: Mobile phone service in France. Journal of Service Marketing 15(1), 35–48 (2001)
15. Dick, A., Basu, K.: Customer loyalty: Toward an integrated conceptual framework. Journal of the Academy of Marketing Science 22, 99–113 (1994)

Study on Key Technology for the Discrimination of Xihu Longjing Tea Grade by Electronic Tongue

Bolin Shi, Houyin Wang, Lei Zhao, Ruicong Zhi, Zhi Li,
Lulu Zhang, and Nan Xie

Food and Agriculture Standardization Institute, China National Institute of Standardization,
Beijing 100191, P.R. China

Abstract. Electronic tongue has the characteristics of sensitivity and instability. However the technical specification for its research has still not been formed. In this paper, it was introduced the key technology of electronic tongue qualitative discriminant analysis to micro-difference samples. The research objects were four different grades of Xihu Longjing Tea with less difference within small producing area in Hangzhou of Zhejiang Province in China. According to the research, the stability of equipment had not been shown until the fifth repetition for the same sample by the electronic tongue. Finally, the signal from seventh repetitive test was selected to represent the intelligent taste fingerprint for this sample. The fingerprints of electronic tongue collected at different days showed linearity drifting for the same sample. The special tea samples were set into each series of experiments to be considered as the reference sample. All samples' signals were calibrated with the difference value between the reference sample fingerprint of corresponding test to the designated reference sample. Principal component analysis (PCA) results showed that the same grade samples were clustered, while the different grades samples were more dispersion and non-overlapping. Through Mahalanobis Distance and Residual Method, the four abnormal tea samples were rejected from highest and 1st grade respectively. The electronic tongue's Longjing Tea Grade models were built by soft independent modeling of class analogy (SIMCA). The discrimination accuracy for tea sample grade were both 100% for correction set and prediction set. Through this study, it was established the technical specification and flow for the quick detection of tea by electronic tongue, which including the determination on intelligent taste spectrum repeatability performance, system error calibration for spectrum drifting, rejection for abnormal tea sample based on taste spectrum and establishment for the quality judgment model . The technical specification provides the theoretical base for reasonable use of electronic tongue.

Keywords: Electronic Tongue, Longjing Tea, Grade, Shifting Calibration, Repeatability, Abnormal Sample.

1 Introduction

The 30~48% contents of tea are water-soluble substances, which represents the taste of tea soup and directly reflect the tea quality and grade (Rahim etc., 2014). Currently, the commonly used analysis technologies for flavoring matter of tea are

© IFIP International Federation for Information Processing 2015
D. Li and Y. Chen (Eds.): CCTA 2014, IFIP AICT 452, pp. 379–392, 2015.
DOI: 10.1007/978-3-319-19620-6_43

liquid chromatography, spectroscopy, mass spectra and nuclear magnetic resonance method as well as the combination and cooperation between them. However, the flavoring ingredients are extremely complex. The whole quality of tea is not composed by certain or some kinds of taste matters but the comprehensive performance by dozens, even by over hundreds kinds of flavoring matters. The researchers were difficult to detect these ingredients one by one, and they can only figure out some ingredients with main flavor, which makes it difficult to test the taste of tea comprehensively (Shi Bolin etc., 2009).

The electronic tongue is a kind of electronic intelligent identification system developed by simulating human body's taste mechanism and is a kind of novel food analysis, identification and testing technology developed within recent years. Being different from the common analyzer, the result obtained by the electronic tongue is not the qualitative or quantitative for the certain kind or some kinds of ingredients in the tested sample but is the overall taste information of sample which is also called as "Taste Fingerprint" data (Jiang Sha etc., 2009). IVARSSON etc. utilized the electronic tongue to appraise nine different kinds of tea and obtained the satisfied results through combining the pattern recognition method with multivariate analysis and principal component analysis. Lvova etc.(2003) used the electronic tongue microsystem with full solid status to quantitative determine the multiple ingredients of green tea from Korea. Chen Quansheng etc. (2008) utilized the electronic tongue technology to perform the classification and identification study for pan-fired green tea with different grades through combining the identification method for K nearest neighbor domain and neutral network mode. He Wei etc. (2009) applied the electronic tongue technology into the grading and classification study for Pu'er tea to study the correlation with sensory evaluation result.

In the former tea quality study, the electronic tongue was employed to test the difference among different types of tea (such as, red tea, black tea and green tea) (Wu Jian etc., 2006; Liu Shuang etc., 2014) or the difference among the same kind of tea with various geographic origins (such as, green tea classification in Zhejiang, Fujian, Anhui, Jiangxi Province) (Wu Xinyu etc., 2007; Runu Banerjee etc., 2012). Since the quality difference was greater among samples in the these studies, it was easy to obtain the satisfied prediction result. Although these studies were specially taken into account the excavation of electronic tongue signal and the application of electronic tongue, the corresponding technical specification and systematic key technology solution on electronic tongue application were not formed in accordance with the electronic tongue technology advantages and the existing technology level. In order to conduct the rapid detection research of electronic tongue for different grades of Xihu Longjing Tea, three key elements should be taken into account. They were repeatability performance analysis of intelligent taste spectrum for electronic tongue, error calibration for spectrum drifting system and rejection for outlier tea samples based on taste spectrum. It was helpful to explore the solution on repeatability and reproducibility of electronic tongue instrument as well as the importance of selecting representative samples in classified model. This paper revealed the key technology and technical specification in the rapid detection of electronic tongue and reflected the value of electronic tongue test for micro-difference sample (such as Xihu Longjing Tea with different grades) in the practical application.

2 Experiments and Methods

2.1 Sample Preparation

Xihu Longjing Tea Sample from 2013 year with four grades (highest grade, special grade, 1st Grade and 2nd Grade) were collected by Zhejiang Hangzhou Standardization Research Institute from Xihu Longjing Tea production area. The tea samples were sub-packaged in 3 g/ bag individually by the aluminum foil bag with good sealing performance (Beijing Huadun Plastics Co., Ltd., 10cm*10cm, food grade, avirulent and insipidity), and were stored in the refrigeration house below −4℃ (Xu Yanjun etc., 2004). According to the experimental dosage, several small bags were conveniently taken each time.

The preparation for tea soup used for electronic tongue detection was as follows. 1.00 g tea sample was brewed into 150 mL boiling ultrapure water. Covered the watch glass, the tea sample was put into 100°C boiling water batch for continuous digestion (DK-98-11A water bath kettle was made by Tianjin Taisite Instruments Co., Ltd.), and was stirred every 10 min at a time. After 45 min, the sample was suction filtration. And the sample was tested by electronic tongue two hours later (Xue Dan etc., 2010).

2.2 Intelligent Taste Acquisition Method

ASTREE electronic tongue made by Alpha MOS in France was adopted, mainly including automatic liquid sampler LS16, electrochemical transducer array (including 7 chemical sensors and 1 reference electrode) and data acquisition unit. Prior to tea sample testing, the electronic tongue was subjected to such procedures as self-inspection, activation, training and calibration so as to ensure the equipment reliability and stability.

Every fingerprint of sample was collected according to the flow of “Tea Soup Sample (120s) Cleaning Fluid No. 1 (10s) Cleaning Fluid No. 2 (10s)”. Each sample was repeatedly tested for seven times and provided with two same cleaning fluids (Figure 1). After all samples were detected, the probe of electronic tongue sensor was stopped in the ultrapure water. Finally, the sensors were completely cleaned based on the setting cleaning procedure. The detection parameters of electronic tongue were presented in Table 1.

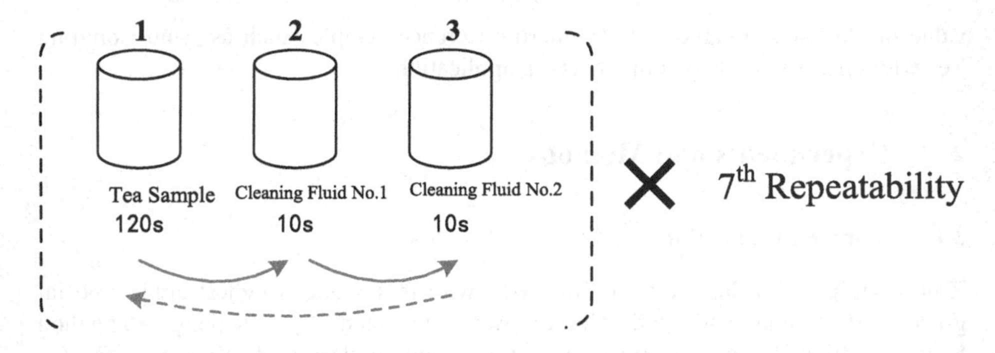

Fig. 1. The steps of sample test

Table 1. The detection parameters of electronic tongue

Parameter Items	Parameter Value
Activation, Training and Calibration Solution	0.0lmol/L HCl
Cleaning Solution	Ultrapure Water
Cleaning Time	10 s
Stirring Speed	1 r/s
Volume of Sample	25 mL
Volume of Sample Cup	40 mL
Sampling Time	120 s
Repeated times of Determination	7 times
Sampling Temperature	25°C for Room Temperature

2.3 Multivariate Statistic Method

Mahalanobis was widely applied in the outlier judgment, because the method was used to measure whether the sample affects the whole samples set (Shi Bolin etc., 2010). If the Mahalanobis value of certain sample was too high, the regression model would have greater dependency on this sample, which would affect the stability of model. In other words, this sample was an outlier sample.

Principal component analysis（PCA）was used for data dimension reduction (Shi Bolin etc., 2012). PCA is a statistical technique that is used to analyze the interrelationships among a large number of variables and to explain these variables in terms of a smaller number of variables, called principal components, with a minimum loss of information. The principal component scoring not only reflected the similarity and peculiarity among tea samples, but also revealed the internal characteristics and clustering information of samples. It represented that whether the sample had greater difference in various categories of sample set, and whether the automatic clustering phenomenon was formed in accordance with the quality characteristics among samples.

Soft independent model classification analysis (SIMCA) was adopted to establish the classification identification model with different grades for Xihu Longjing Tea (Shi Bolin etc., 2011). The foundation of SIMCA was principal component analysis (PCA). Each class of samples had an independent model which was trained by PCA. Meanwhile the residuals of each class were generated. The unknown sample was brought into every class to compare the residuals in different models.

All algorithms were programmed by Matlab 7.0.

3 Results and Discussion

3.1 Selection of Stable Tea Spectrum for Electronic Tongue

The response value that the electronic tongue obtained was the difference value between electric potential of tested sample and electric potential of reference electrode (R-R0, where R0 was electric potential value for reference electrode, and R was electric potential value of sensor of tested sample). Figure 2 was the response curve of certain sample on seven sensors within 120s, from intense response originally to stability finally. The response on the final stage included the whole information for tea sample. Moreover, the steady value at 120 s reflected a minimum relative standard deviation (RSD) for the same sample, and a maximum distinction for the different samples. Therefore, the steady value at 120 s was the characteristic response signal for subsequent study.

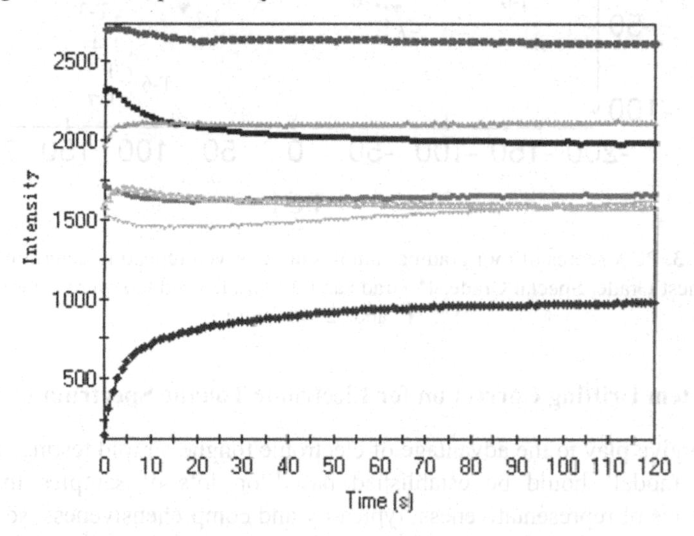

Fig. 2. The response curve for electronic tongue of a certain Sample

For electronic tongue equipment, it was required to guarantee the repeatability of detection signal, otherwise, the reliability of testing data may be queried. The animal's gustation can perfectly catch the taste characteristics of sample after taste training continuously. Similarly, the taste fingerprint from electronic tongue would be more reliable after repeated testing. In the previous researches, the fingerprint information of electronic tongue was used by only one time test result or the mean response signal after three-time repetition. In this study, 7-time tests were repeated simultaneously for each sample. Figure 3 was the PCA scoring for four different grades samples after 7-time repeated tests. The figure showed that the signal repeatability of electronic tongue was poor in the former three-time testing. However, with the increase of repetition time, the testing results were tended to be stable continuously. Especially, the testing results for final three times showed good repeatability to the same sample and great difference among the various grades samples. Therefore, the electric potential value at 120s acquired during 7th repetition was selected to be the original information of intelligent taste for Xihu Longjing Tea. In the repeatability test of electronic tongue, the signal choosing was solved.

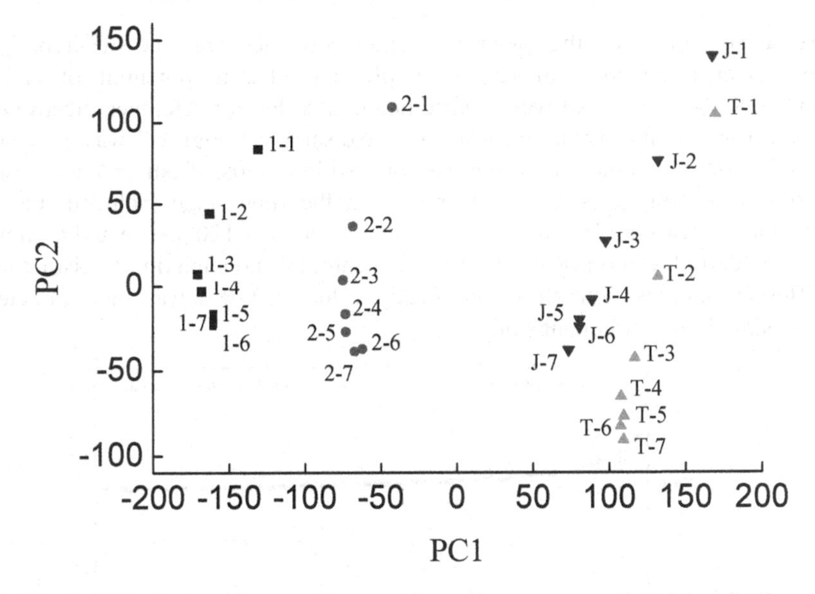

Fig. 3. PCA scores of four grading samples under seven repeating determination (The Highest Grade, Special Grade, 1st Grade and 2nd Grade Products were expressed as "J", "T", "1" and "2" in figure.)

3.2 System Drifting Correction for Electronic Tongue Spectrum of Tea

In order to give play to the advantage of electronic tongue's rapid testing, a long-term prediction model should be established based on lots of samples that had the characteristics of representativeness, typicality and comprehensiveness, so as to make sure of the model robustness and stability. These tea samples were usually detected for several days by electronic tongue. In the same day, the 7th repeated result was used as the sample fingerprint to solve the repeatability. Due to influence by various complicated factors such as environment, the spectrums of the same sample were shifted seriously in different days test, leading to bad reproduction. Therefore, the results would be meaningless if the original data was used directly and the accurate conclusion would be difficult to be made in this case.

The same experimental operation was adopted on each tea soup sample during the whole test period, lasting for a month and acquiring one time every week. The principle analysis was applied to deal with the original data of 4 grades obtained in 4 different days. Figure 4 showed that the samples belonging to the same grade did not gather together properly and but had great dispersion; while for the samples of different grades, there was obvious trend of intersection and serious overlapping, which would not be helpful to the discrimination. Besides, contribution ratios of first principle component and second principle component were only 50% and 25%, showing that the cohesiveness of original data was worse and not suitable for the tea grade discrimination.

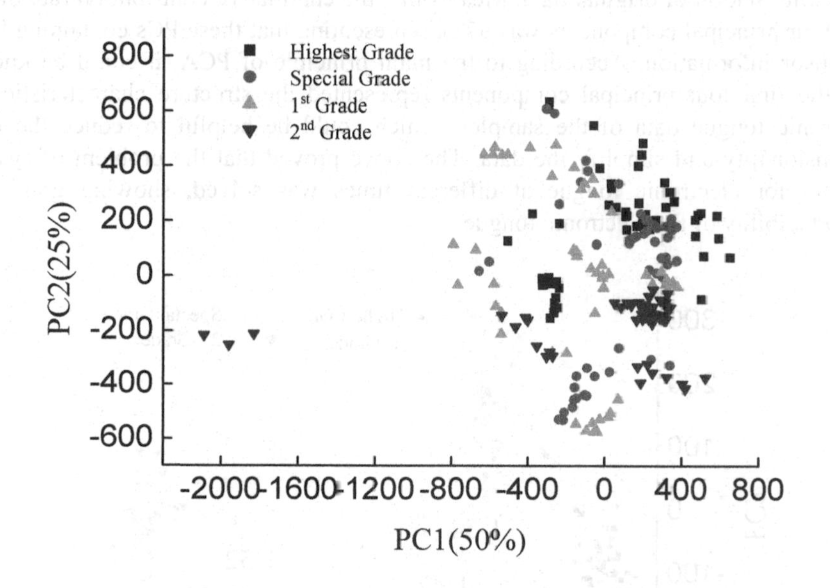

Fig. 4. PCA Scores of Original Date from Four Grading

In order to solve the problem of error at different times, the reference sample was introduced during the data collecting. 1st Grade tea soup was chosen as the reference sample whose preparation method and acquisition method kept the same as other experimental samples. The reference sample was arranged in each sequence experiment, which means for each experiments, there was always a corresponding reference existing. It was necessary to conduct the signal drifting treatment to remove the drifting by utilize the value of reference sample gained in each measurement before modeling. The method was shown below: firstly, fixed the response value of reference sample on certain sequence as the reference value(REF), then figured out different value (Δi=refi-REF) of response value of other sequence reference samples with REF, and then subtracted the value of the corresponding reference samples from the original data of the samples. In this way, the drift could be minimized as largely as possible and data by corrected in this way could be used to build the reliable prediction mode.

Figure 5 was the PCA scoring figure of the samples calibrated by the correction. When compared with the results before the correction, data after correction showed that the cohesiveness of the same grade of tea samples was enhanced and the discreteness between different grade samples was improved, which brought more significant difference between different grads of samples. By analyzing the value range of the horizontal and vertical coordinates of the figure before and after calibration, we could find that after the correction, the values of PC 1 and PC 2 were reduced. But the contribution rates of the first principal component and the second principal component were improved largely, whose contribution rates were 78% and 12%respectively, and the cumulative contribution rate reached 90%, representing the main information of original data. Meanwhile, the cumulative contribution rate of the first four principal components was 97%, representing that these PCs containing 97% of sensor information. According to the main principle of PCA, it could be known that the first four principal components represented the structure characteristics of electronic tongue data of the samples, which could be helpful to reduce the data dimensionality and simplify the data. The above proved that the problem of system drifting for electronic tongue at different times was solved, showing good test reproducibility of the electronic tongue.

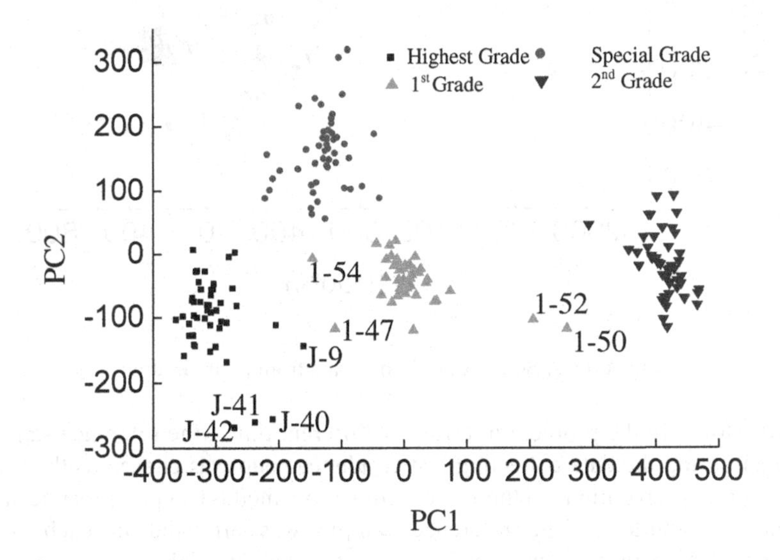

Fig. 5. PCA Scores Under Sensors Modification

3.3 Rejection for Abnormal Sample Point of Tea

The sample set used for pattern recognition was required to be provided with representativeness and correctness of original data, and without anomalism. The existence of the outlier would affect, even change the distribution trend of original data leading to the effect of the prediction model accuracy. From figure 5 it could be found there outliers in highest grade and 1st grade marked by the separation from the cluster. By analyzing (Figure 6) the samples of four grades with Mahalanobis distance in combination with residual errors, No. 54 sample was doubted as the outlier considering its largest Mahalanobis value, meaning that it was apart from the set. Meanwhile, the residual value of No. 50 was largest, which meant the prediction for this sample was inaccurate. In other words, the model cannot be able to explain the sample properly, so it should be treated as abnormal sample, too. However, there was not similar phenomenon for 2nd Grade and Special Grade samples.

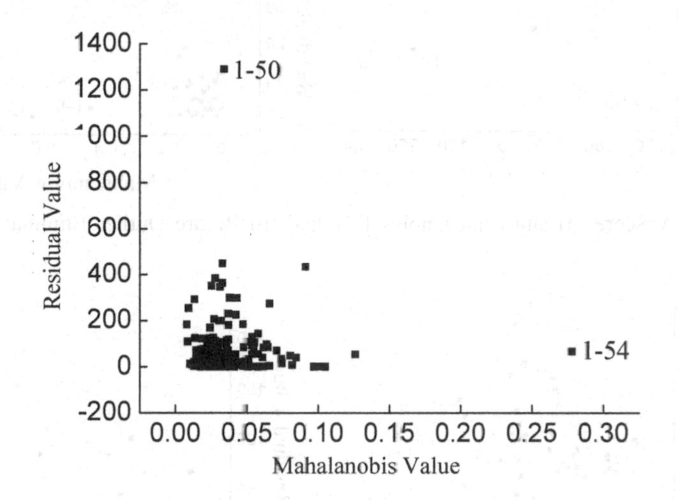

Fig. 6. PCA Scores (a) and Mahalanobis Residual (b) before Outlier Elimination

In order to comprehensively analyze the abnormal sample point, the principal component analysis and calculation of Mahalanobis distance value for the 1st Grade samples and highest Grade samples were conducted separately. From Figure 7, it could be found that the samples of No.47, No.50 and No. 54 were the three maximum Mahalanobis distance samples in 1st Grade, being considered to be the abnormal sample points. However, No. 52 sample not only detached from the samples set in the score figure but also had larger Mahalanobis distance value and residual value, also being of abnormal sample point. From Figure 8, it could be found that the Mahalanobis distance of No.9 sample in the competitive product sample set was maximum, being of abnormal sample point, but the samples of No.40, No.41 and No.42 were far from the samples set in the scoring chart, also being judged to be of abnormal sample points. So the quantity of abnormal sample points to be removed was 8 in total, including four primary samples respectively: 1-47, 1-50, 1-52 and 1-54; and four competitive product samples respectively: J-9, J-40, J-41 and J-42.

Table 2 was the number of the samples before and after outlier elimination for each grade, and the final division of the sample set. The quantity of sample used for discriminating the grade of tea were 209 in total, among which, two-thirds were randomly chosen to be served as calibration set samples to establish the qualitative classification model so as to enable the modified samples to not only have good representativeness but also broaden the prediction range of the model to strengthen the adaptive capacity of the model. The rest one-thirds were used for prediction set samples to inspect the veracity and reliability of the established model.

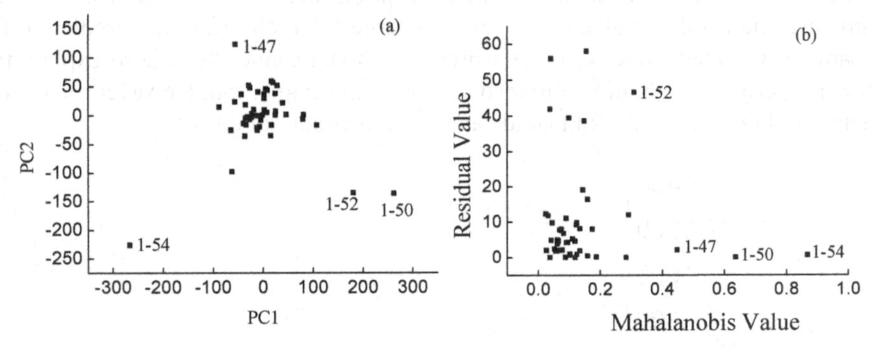

Fig. 7. PCA Score (a) and Mahalanobis Residual (b) Before Outlier Elimination of Grade-I Sample

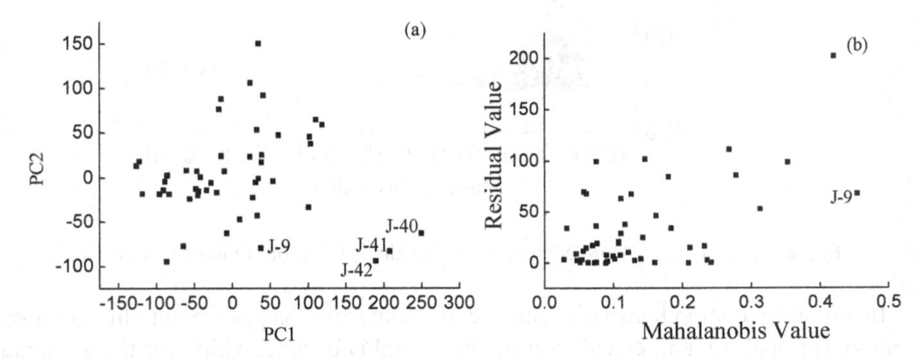

Fig. 8. PCA Score (a) and Mahalanobis Residual (b) before Outlier Elimination of Competitive Product Grade Samples

Table 2. Quantity for Different Sample Sets of Graded Model

Name of specimen clustering	After Rejection			Before Rejection
	Sample quantity of training set	Sample quantity of prediction set	Total sample amount	
Highest Grade	33	17	50	54
Special Grade	36	18	54	54
1st Grade	34	16	50	54
2nd Grade	36	19	55	55
Quantity of all graded sample	139	70	209	217

3.4 Grading Model Establishment of Xihu Longjing Tea Based on Electronic Tongue

Although the individual sensor was provided with high sensitivity and selectivity for identifying the different grades of tea samples, but the component of tea soup was complex, whose information were inevitably mixed together. Through choosing repetitive signal of electronic tongue, after calibrating signal drift at different time and eliminating outlier tea samples, the performance of PCA for extracting the independent information from large numbers of data was improved effectively. According to Figure 9, the score chart of PC 1 and PC 2 could make the distinction between different grade of samples more obvious (the cumulative contribution rate of the first two principal components had been up to 90%), among which the performance of PC 1 was the greatest, being able to distinguish different grade samples obviously. With the decrease of grade level, the score value in PC 1 increased gradually; while the variance of score value between principal component 3 and principal component 4 was very small (the cumulative contribution rate of principal component 3 and principal component 4 was 6%).

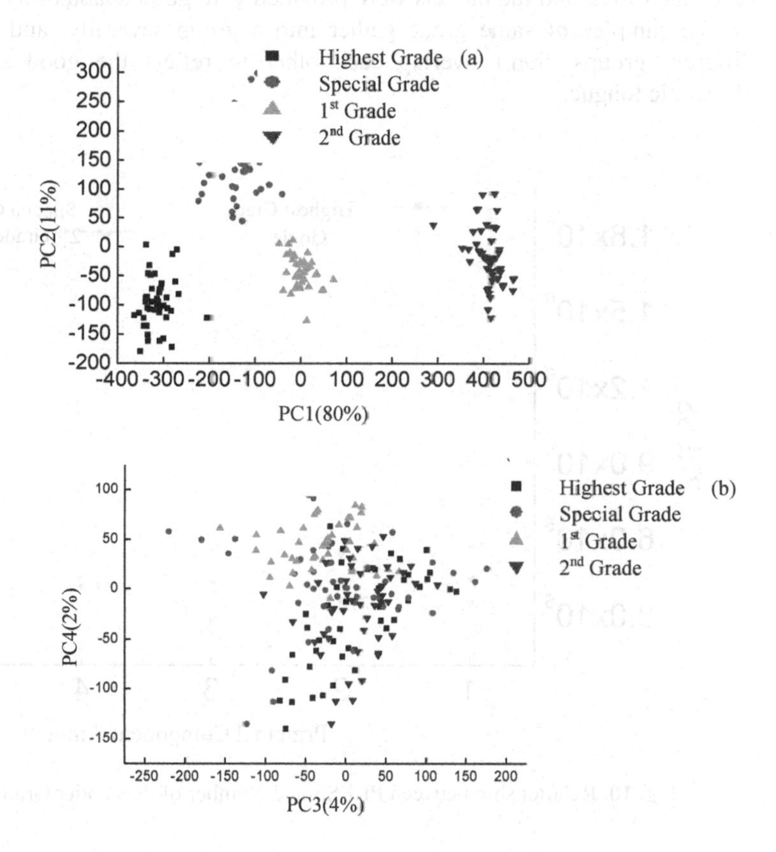

Fig. 9. PCA Scores between Different PCs after Outlier Elimination: PC1-PC2, (b) PC3 -PC4

SIMCA discrimination model was built based on the basis of PCA analysis, where the choice of principal component numbers was very important to SIMCA modeling. The best principal component number of models for the different grades was determined by cross-validation. Figure 10 was the relationship between the sum of square of prediction residual of different grade samples model and the number of principal components. In the case of a little change of SIMCA value, selecting the relatively less principal components, and finally the number of principal components of SIMCA modeling for Highest Grade, Special Grade, 1st Grade and 2nd Grade were selected as 2, 1, 2, 3, respectively. The SIMCA tea grade discrimination model was established according to the selected number of principal components of each grade of samples. Table 3 was the final model result. As what was found in the table, the electronic tongue sensor could distinguish the four grades of Xihu Longjing Tea effectively, not only for cross-validation for 139 modeling samples(33 Pcs. of Highest Grade, 36 Pcs. of Special Grade, 34 Pcs. of 1st Grade, 36 Pcs. of 2nd Grade) reaching 100%, but also for 70 prediction samples with the unknown grade level (17 Pcs. of Highest Grade, 18 Pcs. of Special Grade, 16 Pcs. of 1st Grade, 19 Pcs. of 2nd Grad) reaching 100%, and the models were provided with good adaptability and robustness. So the samples of same grade gather into a group severally, and the samples of different groups didn't overlap each other to reflect the good susceptibility of electronic tongue.

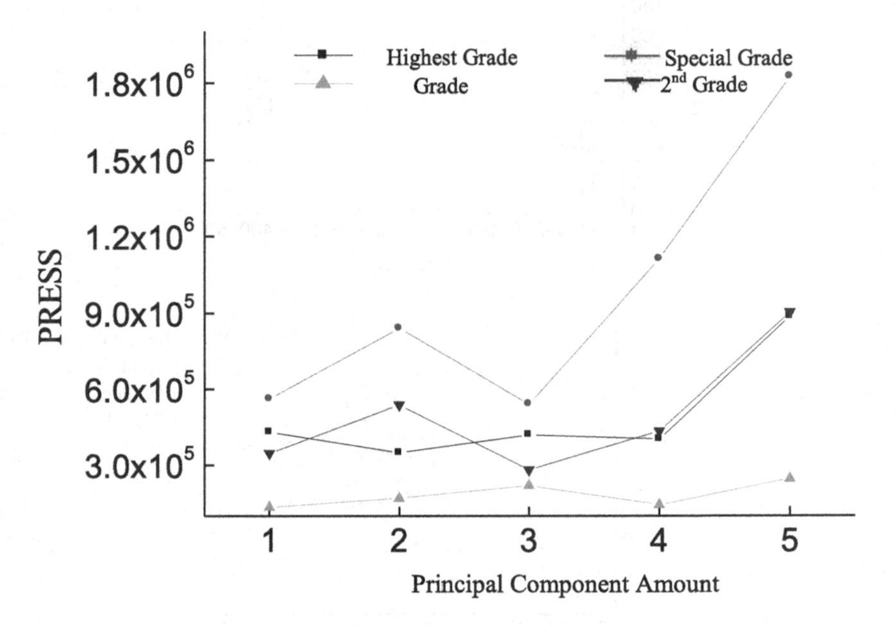

Fig. 10. Relationship between PRESS and Number of PCs under Grading Model

Table 3. Results of SIMCA Modeling for Four Grades of Samples

Model Type	Number of PC	Integral Identification Ratio(%) Correction Set	Prediction Set
Highest Grade, Special Grade, 1st Grade, 2nd Grade	2,1, 1, 3	100	100

4 Conclusions

The study chose the graph of samples in the seventh repetition to stand for the intelligent taste fingerprint of the sample since in this case the stability and authenticity of the tea soup could be guaranteed effectively. The corresponding reference on each sequence was adopted to be served as the basis of calibration to solve the problem of system error generated by linear drifting which guaranteed the stability of the graph signal, making Mahalanobis distance and residual method could be run quickly, factually and accurately, eliminating the abnormal samples in tea sample set. Finally, we used SIMCA modeling with of the selected number of principal components, and the accuracy of discrimination for Xihu Longjing Tea reached 100%. In conclusion, the study focusing on three key points of electronic tongue: repeatability, reproducibility and representative. By handling these three points we could obtain signal graph with high signal to noise ratio, and the authenticity of the characteristics of the samples. The study developed the process of confirming the repeatability, correcting graph drift system error, eliminating the outlier samples and building the discrimination model with technical specifications and procedure. All these will provide the theory basis for application of electronic tongue in other fields.

Acknowledgment. This study had been sponsored by Beijing-sponsored programs for excellent talents training (No. 2012D009999000001), National High Technology Research and Development Plan (863 Plan) for the Project of No. 2011AA1008047.

References

1. Rahim, A.A., Nofrizal, S., Saad, B.: Rapid tea catechins and caffeine determination by HPLC using microwave-assisted extraction and silica monolithic column. Food Chemistry 147, 262–268 (2014)
2. Shi, B., Zhao, L., Wang, H., Yin, J.: Application of intelligent sensory technique in evaluation of tea quality. Food Science 30(19), 351–355 (2009)
3. Jiang, S., Chen, Q., Hu, X., Yang, Y., Ni, Y.: Application of electronic tongue on black tea beverage discrimination. Transactions of the CSAE 25(11), 345–348 (2009)
4. Patrik, I., Susanne, H., Nils-Eriker, H., et al.: Discrimination of tea by means of a voltammetric electronic tongue and different applied waveforms. Sensors and Actuators B 76, 449–454 (2000)

5. Lvova, L., Legin, A., Vlasov, Y., et al.: Multicomponent analysis of Korean green tea by means of disposable all-solid-state potentiometric electronic tongue microsystem. Sensors and Actuators B 95, 391–399 (2003)
6. Chen, Q., Jiang, S., Wang, X.: Discrimination of tea's quality level based on electronic tongue and pattern recognition. Food and Machinery 24(1), 124–126 (2008)
7. He, W., Hu, X., Zhao, L., Liao, X., Zhang, Y., Wu, J.: Application of electronic tongue in the Pu-er tea quality grade analysis. Science and Technology of Food Industry 30(11), 125–127 (2009)
8. Wu, J., Liu, J., Fu, M., Li, G.: Classification of Chinese green tea by a voltammetric electronic tongue. Chinese Journal of Sensors and Actuators 19(4), 963–965 (2006)
9. Liu, S., Tan, J., Lin, Z., Lu, C., Wang, X.: Application of electronic tongue in green tea sensory estimation and grade evaluation. China Tea (5), 19–20 (2014)
10. Wang, X., Chen, Q.: Discrimination on grades of roasted green tea using electronic tongue. Journal of Anhui Agricultural Science 35(28), 8872–8873 (2007)
11. Roy, R.B., Tudu, B., Shaw, L., Jana, A., Bhattacharyya, N., Bandyopadhyay, R.: Instrumental testing of tea by combining the responses of electronic nose and tongue. Journal of Food Engineering (110), 356–363 (2012)
12. Xu, Y., Chen, W., Yin, B., Huang, L.: Green tea store and preservation technology. Agriculture Machinery Technology Extension 9, 29 (2004)
13. Xue, D., Shi, B., Zhao, L., Yin, J.: Classification of tea's grade level based on electronic tongue. Food Science and Technology 35(12), 278–281 (2010)
14. Bolin, S., Lei, Z., Wen, L., Houyin, W., Dazhou, Z., Ingyuan, Y.J.: Outlier sample analysis on near infrared spectroscopy determination for apple interior quality. Transactions of the CSAM 41(2), 132–137 (2010)
15. Bolin, S., Lei, Z., Ruicong, Z., Xingjun, X., Dazhou, Z.: Quality recognition of Xihu-Longjing tea based on intelligent olfactory. Transactions of CSAM 43(12), 130–135 (2012)
16. Shi, B., Zhao, L., Zhi, R., Xi, X., Zhu, D.: Aroma quality discrimination of Xihu-Longjing tea by electronic nose. Transactions of the CSAE 27(Supp. 2), 302–306 (2011)

Research on Pattern Recognition Method for Honey Nectar Detection by Electronic Nose

Ningjing Liu[1], Bolin Shi[2,*], Lei Zhao[2], Zhaoshen Qing[1], Baopin Ji[1], and Feng Zhou[1]

[1] College of Food Science and Nutritional Engineering, China Agricultural University, Beijing 100083, P.R. China
[2] Food and Agriculture Standardization Institute, China National Institute of Standardization, Beijing 100191, P.R. China

Abstract. Electronic nose (e-nose) utilizes the gas sensors array to absorb the volatile organic compounds (VOCs) of samples to classify them into different clusters, and it is noted by the sensitive of the sensors. However, limited by the methodologies of the pattern recognition, this kind of advantage had not been exploited fully. The research studied on different types of pattern recognition method, and selected the optimum method for the detection of samples with little nuance, exemplified by honey nectar detection for including rape honey, linden honey and acacia honey. It was found that support vector machine (SVM, non-linear prediction model) showed better performance than linear discriminate analysis (LDA, linear prediction model) for the classification of tiny different samples, especially when multi-group detection was involved. After the optimized method was selected, the key points of the SVM model were analyzed, and two key parameters were displayed, which were kernel parameter and penalty parameter. Three algorithms, including grid searching (GS), particle swarm optimization (PSO) and genetic algorithm (GA) were applied to find the appropriate parameter values. The results showed parameters optimized by genetic algorithm (kernel parameter and penalty parameter is 0.11 and 14.38 respectively) led to the optimal model, whose training accuracy was 98.78% and prediction accuracy was 97.5%. The results suggested that in the method of SVM with parameters selected by GA, e-nose could handle well the discrimination of similar samples like honey nectar detection.

Keywords: electronic nose, nectar detection, pattern recognition, support vector machine.

1 Introduction

Electronic nose is a kind of a newly developing technology detection system. It utilizes gas sensors array, signal processing and pattern recognition to imitate the olfactory system of human beings. Gas sensors array, usually constituted by metal-oxide semiconductor sensors, absorbs the volatile organic compounds (VOCs) of samples, and the absorption reaction of the VOCs and sensors leads to the change of

* Corresponding author.

© IFIP International Federation for Information Processing 2015
D. Li and Y. Chen (Eds.): CCTA 2014, IFIP AICT 452, pp. 393–403, 2015.
DOI: 10.1007/978-3-319-19620-6_44

surface electron intensity [1]. This kind of current change will be transformed by signal processing to digital signals. After the digital signals obtained, they are analyzed by the pattern recognition system, and make the final discriminant results.

Unlike other detection methods, such as gas chromatography and high efficiency liquid chromatography, e-nose detection is based on the overall responses from all sensors, combining with the characterization from various angle, not only analyzing by one or two feature ingredients, nor one or two characteristic signals. In this case, focusing on one or two sensors results tends to be meaningless. So pattern recognition is brought into the system to deal with the overall detection [2]. Pattern recognition run by using training samples to cultivate the system and foster its ability of discriminate the special items like a person [3]. It has been applied into lots of research areas, like military, finance, medicine, industry and agriculture. Combining with different chemo metrics algorithms, it has shown the great potential in formulation optimization, production control and discriminate detection.

Pattern recognition plays an important role in the e-nose system, and affects the detection result directly, which means the performance of e-nose system largely depends on the method chosen for the pattern recognition. For the e-nose detection, the advantage of sensitive is presented when similarities are analyzed, which brings an advanced requirement to the pattern recognition [4]. However, some pattern recognition methods which were being used were not accurate enough for this level of sensitiveness. Linear discriminate models, like linear discriminate analysis (LDA), discriminate partial least-square (DPLS) etc., had been widely used in e-nose analysis, and turned to be more and more mature. But this kind of models might show weakness in nuance since in this case, it was not guaranteed that all the samples could be divided accurate by the linear classifier. Focusing on this, the research study the appliance of non-linear discriminate model, support vector machine (SVM). By mapping the signal data into higher dimension space, data points in the space tended to be more disperse, which made it easier for the model to classify the samples into different clusters. Besides, SVM is based on the principle of structural risk minimization, which will enhance the robust of model significantly [5]. However the complex of SVM not only brought the steady and robust of the model, but also the difficulty of parameter selecting. Under the different value of parameters, the model showed various degree of performance [6]. The study chose different chemo metrics algorithms to optimize the penalty parameter and kernel parameter of SVM so that the best classifier could be obtained.

2 Experiments and Methods

2.1 Experimental Samples

Three different kinds of botanical origin honey were chosen, including rape honey (76 units), linden honey (55 units) and acacia honey (113 units). To ensure the authenticity of the samples, all the honey were collected from beekeepers directly by the members of the group. Considering the florescence of different nectar honey was not the same, samples collected in different time were stored in refrigerator in -18°C until the collection had been completed.

2.2 Instrument

The e-nose system was FOX 4000, made by Alpha MOS in France. It was constituted by 18 metal-oxide semiconductor gas sensors, distributing in 3 chambers. The e-nose was installed with HS100 head space auto-sampler, which contained 2 pallets with 64 head space bottles (10ml).

2.3 Methodology

To avoid the disturb of crystals which was formed during storage, water bath was employed to heat the honey before the test under the temperature of 40°C for 15 minutes [7]. The parameters of the detection were shown in Table 1, which had been optimized by the orthogonal test [8].

Table 1. Parameters of e-nose detection

Parameters	Value
Head Space Time(S)	300
Head Space temperature (℃)	40
Inject Volume (ul)	500
Inject Speed(μL/s)	500
Inject Temperature(℃)	45
Acquisition Time(S)	240
Delay Time(S)	1080

Each sample obtained a 18*120 matrix, 18 lines for 18 sensors, and 120 columns for 120s detection time. Independent Component Analysis (ICA) was used to extract the feature information in the data matrix, which generated a new matrix of 8*120 for each sample. Afterwards, genetic algorithm was utilized to select 20 characteristic points for each unit, which represented the characterization of samples. According to the ratio of 2:1, all the samples were divided into 2 parts, training data (164 samples) and prediction data (86 samples), to build and validate the model.

2.4 Chemometrics Algorithms

2.4.1 Prediction Model

Two different kinds of methods were selected to make a comparison, non-linear model and linear model. Non-linear model exemplified by support vector machine (SVM). SVM was after the principle of structural risk minimization, which was great helpful for the model to avoid over training of the model and especially suitable for the small margin data detection [9]. It utilized the non-linear kernel function to map the data into a higher dimension space, and built a classified plane to distinguish the data sets into different clusters. To guarantee the minimum structural risk, the plane should be kept as far as possible from the data of both clusters. The linear model was built by the linear discriminate analysis (LDA), which made the classification through mapping the data to a specific direction where all the data could be separated from each other as disperse as possible [10]. Fisher function was normally applied to find this specific direction.

2.4.2 Parameter Optimization

The approaches for parameter optimization included grid searching, genetic algorithm (GA) and particle swarm optimization (PSO).

Grid searching was a kind of exhaustive searching method. By presetting the searching range and searching step length (usually the logarithm of 2), the algorithm could find the optimum solution as long as the range was suitable and step length was precise [11]. In this study, the search range was from 2-4 to 210, and the length of step was log2.

GA, simulating the process of natural evolutionary, could search multi-direction with retaining a population of candidate solution. The algorithm coded the candidates by binary system and run the iteration including choosing, crossing and variation until the optimal item obtained [12]. In this study, the accuracy of the model was set as the fitness function, and the number of population was 20 with the max iteration being 100.

PSO was similar to the GA, but it replaced the crossing and variation by calculating the difference between the fitness value of items and the fitness value of optimum in the population. Compared with GA, PSO was only affected by the optimal item, not the whole population, which would bring a faster convergence rate [13]. Same with GA, the number of population was 20 with the max iteration being 100.

3 Results and Discussion

3.1 Comparing Different Kinds Model

The research had comparing the performance of two different kinds of discriminate model, LDA and SVM, standing for linear model and non-linear model respectively. The accuracy of the models was shown in table 2.

Table 2. Accuracy of different discriminate models

Model Methods	Training Set				Prediction Set			
	Accuracy %	Clusters			Accuracy %	Clusters		
		Rape	Linden	Acacia		Rape	Linden	Acacia
LDA	89.6	45/51	27/38	75/75	83.8	17/25	12/17	38/38
SVM	95.7	50/51	35/38	72/75	92.5	24/25	14/17	36/38

From Table 2, it could be found that linear discriminate classifier showed excellent ability in acacia honey detection, the accuracy of training set and prediction set could reach 100% at the same time. However, this kind of performance was absent upon the detection of other two kinds of honey, especially for linden honey. This phenomenon illustrated that it was difficult for linear model to find out an appropriate classifier to discriminate the rape honey and linden honey, and these two kinds of honey confused badly in linear space. However, this matter did not occur in SVM model. Compared with LDA, SVM could distinguish three kinds honey well. The main reason of this was that SVM utilized the kernel function to map the data into higher dimension space, where the data sets got more disperse and it was great helpful for the classifier to find the difference among three clusters. Simultaneously, SVM was after the principle of structural risk minimization, which demanded the separating hyperplane to keep as far as possible from the data sets, and this ensured great probability of generalization, reflected in the higher value prediction accuracy.

For the further analysis, three bipartition model were built by these two methods, referring to Table 3 to Table 5.

Table 3. Discriminate model of Rape honey and Linden honey

Methods	Training Set			Prediction Set		
	Accuracy%	Clusters		Accuracy%	Clusters	
		Rape	Linden		Rape	Linden
LDA	97.6	51/51	36/38	78.6	17/25	16/17
SVM	92.1	50/51	32/38	86.0	21/25	15/17

Table 4. Discriminate model of Rape honey and Acacia honey

Methods	Training Set			Prediction Set		
	Accuracy%	Clusters		Accuracy%	Clusters	
		Rape	Acacia		Rape	Acacia
LDA	100.0	51/51	75/75	100.0	25/25	38/38
SVM	100.0	51/51	75/75	100.0	25/25	38/38

Table 5. Discriminate model of Linden honey and Acacia honey

Methods	Training Set			Prediction Set		
	Accuracy%	Clusters		Accuracy%	Clusters	
		Linden	Acacia		Linden	Acacia
LDA	98.2	38/38	73/75	98.2	17/17	37/38
SVM	100.0	38/38	725/75	100.0	17/17	38/38

The above three tables illustrated that compared with tripartition model, LDA tended to adapt to bipartition classification models, especially for the rape-acacia discrimination and linden-acacia discrimination. In these two models, there was no capability performance difference between LDA method and SVM method. Furthermore, in the case of rape-linden models, training accuracy of LDA was higher than that of SVM, despite of poor performance in prediction accuracy of LDA. However, neither LDA nor SVM could make efficient division of rape and linden honey, which meant these two kinds of honey had similar properties, but the prediction model of SVM still kept in higher level than LDA. It was because that when dealing with the similar samples, SVM model did not pursuit of training accuracy blindly, but minimized the risk of mistaken as lower as possible to maintain the generalization ability of the model [14].

Meanwhile, by comparing the triparition model and bipartition models, it could be proved that SVM had great advantages on multi-classification. It was mainly because that SVM utilized the kernel function to map the data into higher dimension where the data got more separate than in the lower dimension, which was great helpful for generating the classifier. This kind of improvement showed significantly in multi-classification. However, as to the linear model, especially when the micro-difference samples were detected, it was hard for the model to search a well-fit classify plane, which led to the phenomenon that when this case occured, neither training accuracy nor the prediction accuracy tended to be acceptable. The principle of SVM, structural risk minimization, allowed SVM model obtain the least diversity between the accuracy of training and the accuracy of prediction. However, LDA model, in some cases, demanded the high value of training accuracy too excessively to ignoring the generalized ability of the model, leading to the over training of the model.

The results above showed that when dealing with the similar samples of multi-clusters, SVM model showed the better performance, while as to the samples with large difference, and for the bipartition classifier, LDA model was also a good choice, considering its less complex model structure.

3.2 Effects of SVM Parameters

Although SVM model received better discriminate results, it was limited by the choice of different parameter values. Among all the parameters, the most three important factors were kernel function, kernel parameter value and penalty value. Since the task of kernel function was mapping the data into higher dimension space, as long as the function met the basic requirements of the kernel function, it did not make any significant difference [15]. While the other two factors, values of kernel parameter and penalty parameter were the key points of an excellent discriminate model. Figure 1 showed the accuracy of training sets under different parameter values by grid searching. It could be found that different value groups led to the different performance. Specifically, the figure did not illustrate an upward trend for the accuracy with the increase or decrease of the value, but tended to be a wave curve, which could be illuminated clearly by analysis the risk of SVM principle. Models with the kernel parameter (r) of 2^0 to 2^{-5} and penalty parameter (c) of 2^3 to 2^6 showed higher accuracy, and when r=0.11 and c=16, it reached the peak.

Where R was real risk, R_{emp} was empirical risk, n was the number of the samples, h was VC dimension, η was confidence level, and Φ was confidence interval.

In SVM, the higher dimension was the key point of detection, which was decided by the kernel parameter. In non-linear mapping, the kernel function was just like a kind of tool which could influence how the data mapping into the space, while the kernel parameter could confirm which space the data mapping into. In the high dimension space, the dimension determined the Vapnik-Chervonenkis (VC) Dimension of the space, namely the capability of the plane classification [16]. The relationship between the real risk and VC dimension was shown in formulation 1 (h stands for the VC dimension).

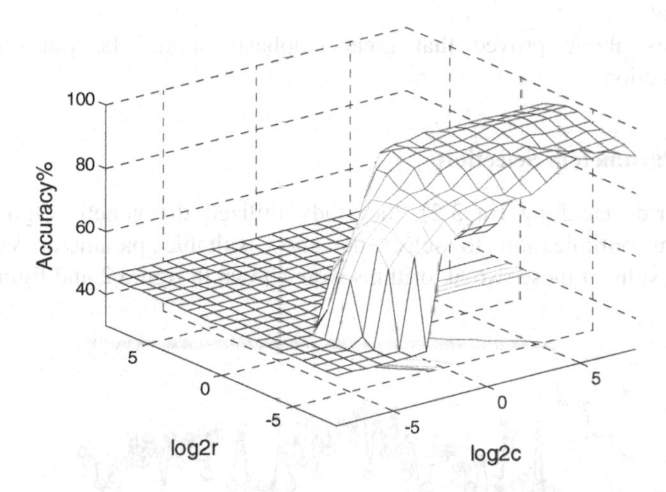

Fig. 1. Grid search graph for the optimization for the parameters of SVM

$$R(W) < R_{emp}(W) + \sqrt{\left|\frac{h\left[\ln\left(\frac{2n}{h}\right)+1\right]-\ln\left(\frac{\eta}{4}\right)}{n}\right|} \tag{1}$$

$$R(W) < R_{emp}(W) + \Phi \tag{2}$$

When the dimension got higher, the structure of the space tended to be complex, and the data sets became more diffuse, leading to the increase of VC dimension, enhancing the ability of the classifier, meaning the lower empirical risk. Along with that, confidence interval got wider, which brought larger disparity between the real risk and empirical risk. It could be shown by the phenomenon of higher training accuracy and lower prediction accuracy. In contrast, when the dimension got lower, the sets tended to be more concentrate. Although the classification would be complicate, the gap between the effectively risk and empirical risk would not be two wider, resulting in the little difference between training accuracy and prediction accuracy. For the detection requirements, the empirical risk and confidence interval needed to be minimized at the same time if it was possible, so that the final performance of the model could be acceptable.

Apart from the kernel parameter, penalty parameter also played a great importance role in model classification. Limited by the characteristics of sets, even in higher space, there still be the possibility of indivisibility. When this occurred, the overall classification should be considered in order the entire results obstructed. This could be obtained by setting the penalty parameters to ignore the individual samples to maintain the total classification. By changing the weight of penalty, the demanding proportion of real risk and empirical risk could be confirmed. When the penalty got larger, the model pursued lower empirical risk, but brought wider confidence interval, while the penalty got smaller, it turned reverse. Hence the penalty parameter was also an important role.

The analysis above proved that great emphasis should be put upon these parameters selection.

3.3 SVM Parameters Selection

Besides the grid searching (in 3.2), the study utilized the genetic algorithm and particle swarm optimization to select the most suitable parameter value. The optimization results of these two algorithms were shown in figure 2 and figure 3.

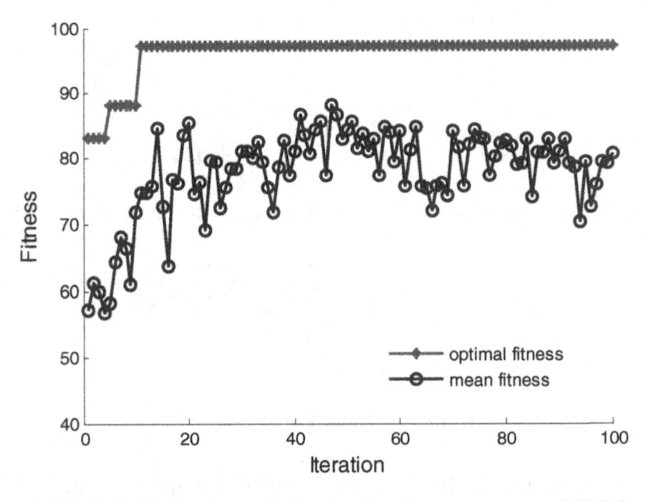

Fig. 2. PSO for the optimization for the parameters of SVM

Figure 2 illustrated the process of PSO. From the figure, it could be found that PSO got faster convergence rate. Fitness function became plane and stable in 18th generation. As to the mean fitness, the value increased wavy and in th 40th it reaches at approximate 80%. However, when it generated to 65th iteration, the mean value dropped a little, but it regained into 80% soon. It was because some bad items were generated but these items were weed out quickly to keep the main trend in normal direction. It proved that PSO had great ability of resistance for the accidental fault to avoid influence upon the final results. Finally, the optimization results was r=20.02, c=0.09.

Compared with PSO, the GA had a slower convergence rate. From the figure 3, it could be found that GA got stable in 44th generation. The difference in convergence rate was mainly led by the difference of algorithm theories. PSO put more energy on excellent individual, while GA needed an overall evaluation of all the items in one generation. Finally, when c=14.38, r=0.11the model received the best results.

After that, parameters optimization results by three methods are shown in table 6.

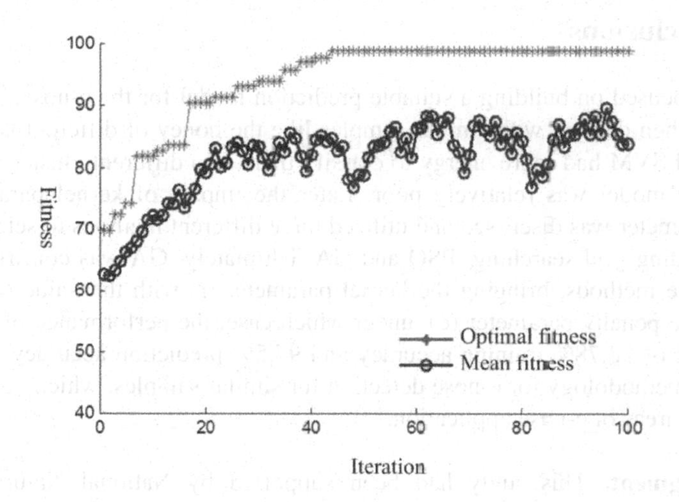

Fig. 3. GA for the optimization for the parameters of SVM

Table 6. Predict accuracy of SVMs optimized by different methods

Optimization Methods	c	R	Training Set Accuracy %	Training Set Clusters Rape	Linden	Acacia	Prediction Set Accuracy %	Prediction Set Clusters Rape	Linden	Acacia
Grid Searching	16.00	0.11	97.56	50/51	37/38	73/75	95%	24/25	16/17	36/38
PSO	20.02	0.09	96.95	50/51	36/38	73/75	93.75%	24/25	15/17	36/38
GA	14.38	0.11	98.78	51/51	37/38	74/75	97.5%	25/25	16/17	37/38

The results referring to Table 6 showed that there was little different between the kernel parameter selected by different methods, while the value of penalty parameter seemed more significant. It was mainly because the model largely relied on the structure of higher space, which was determined by the kernel parameter, so the optimization results of different methods got close to each other; while penalty only presented the tolerance of the error, not as critical as the kernel parameter, which led to fluctuate in different methods.

As to the capability of the model built by different parameter values, GA got the highest scores. Grid searching limited by the step length which was setting artificially, may not correspond with best searching length, being easily to jump the optimum if the length was too long; while for the PSO, considering the fast convergence rate and little improvement in the later stage, it could be doubted that the PSO had been fallen into local minimum, leading to the ignorance of globally optimal solution. If the algorithm fell into this trap, the process would be limited by searching the solution around local minimum repeatedly, and could not jump out of this [17]. In the study, the fast convergence rate and the poor optimization of PSO was identical to it. Ultimately, GA was applied to select the optimized values, and in the case of r=0.11 and c=14.38, the SVM got the best performance with the training accuracy of 98.78% and prediction accuracy of 97.5%.

4 Conclusions

The study focused on building a suitable prediction model for the e-nose. The results show that when dealing with similar samples like the honey of different nectar, non-linear model SVM had more energy to classify them into different cluster accurately, while linear model was relatively poor. Later the impact of kernel parameter and penalty parameter was discussed and utilized three different methods to select the best value, including grid searching, PSO and GA. Ultimately, GA was confirmed as the most suitable methods, bringing the kernel parameter (r) with the value of 0.11 and 14.38 for the penalty parameter (c), under which case, the performance of the model got the peak of 98.78% training accuracy and 97.5% prediction accuracy. The study provided a methodology for e-nose detection for similar samples, which could widely broaden the areas of e-nose application.

Acknowledgment. This study had been supported by National Natural Science Foundation of China (31101292).

References

1. Mirasoli, M., Gotti, R., Di Fusco, M., Leoni, A., Colliva, C., Roda, A.: Electronic nose and chiral-capillary electrophoresis in evaluation of the quality changes in commercial green tea leaves during a long-term storage. Talanta 129(1), 32–38 (2014)
2. Singh, H., Bhasker Raj, V., Kumar, J., Mittal, U., Mishra, M., Nimal, A.T., Sharma, M.U., Gupta, V.: Metal oxide SAW E-nose employing PCA and ANN for the identification of binary mixture of DMMP and methanol. Sensors and Actuators B: Chemical 200, 147–156 (2014)
3. Meyer-Baese, A., Schmid, V.: Statistical and Syntactic Pattern Recognition, pp. 151–196. Academic Press, Oxford (2014)
4. Rodriguez-Mendez, M.L., Apetrei, C., Gay, M., Medina-Plaza, C., de Saja, J.A., Vidal, S., Aagaard, O., Ugliano, M., Wirth, J., Cheynier, V.: Evaluation of oxygen exposure levels and polyphenolic content of red wines using an electronic panel formed by an electronic nose and an electronic tongue. Food Chemistry 155(15), 91–97 (2014)
5. Karaçalı, B., Ramanath, R., Snyder, W.E.: A comparative analysis of structural risk minimization by support vector machines and nearest neighbor rule. Pattern Recognition Letters 25(1), 63–71 (2004)
6. Lin, S.-W., Lee, Z.-J., Chen, S.-C., Tseng, T.-Y.: Parameter determination of support vector machine and feature selection using simulated annealing approach. Applied Soft Computing 8(4), 1505–1512 (2008)
7. Ajlouni, S., Sujirapinyokul, P.: Hydroxymethylfurfuraldehyde and amylase contents in Australian honey. Food Chemistry 119(3), 1000–1005 (2010)
8. Liu, N., Shi, B., Zhao, L., Qing, Z., Baopin, J., Zhou, F., Tian, W.: Optimizing an Electronic Nose for Analysis Honey from Different Nectar Sources. Sensor Letters 11, 1145–1148 (2013)
9. Ahmad, A.S., Hassan, M.Y., Abdullah, M.P., Rahman, H.A., Hussin, F., Abdullah, H., Saidur, R.: A review on applications of ANN and SVM for building electrical energy consumption forecasting. Renewable and Sustainable Energy Reviews 33, 102–109 (2014)

10. Sharma, A., Paliwal, K.K., Onwubolu, G.C.: Class-dependent PCA, MDC and LDA: A combined classifier for pattern classification. Pattern Recognition 39(7), 1215–1229 (2006)
11. Liu, R., Liu, E., Yang, J., Li, M., Wang, F.: Optimizing the Hyper-parameters for SVM by Combining Evolution Strategies with a Grid Search. Ruiming 344, 712–721 (2006)
12. Tam, S.M., Cheung, K.C.: Genetic algorithm based defect identification system. Expert Systems with Applications 18(1), 17–25 (2000)
13. Subasi, A.: Classification of EMG signals using PSO optimized SVM for diagnosis of neuromuscular disorders. Computers in Biology and Medicine 43(5), 576–586 (2013)
14. Corani, G., Gatto, M.: VC-dimension and structural risk minimization for the analysis of nonlinear ecological models. Applied Mathematics and Computation 176(1), 166–176 (2006)
15. Vapnik, V.: The Nature of Statistical Learning Theory. Data Mining and Knowledge Discovery 6, 1–47 (1999)
16. Koiran, P., Sontag, E.D.: Vapnik-Chervonenkis dimension of recurrent neural networks. Discrete Applied Mathematics 86(1), 63–79 (1998)
17. Nitta, T.: Local minima in hierarchical structures of complex-valued neural networks. Neural Networks 43, 1–7 (2013)

Study on Cloud Service Mode of Digital Libraries Based on Sharing Alliance

Xiaorong Yang[1,2], Dan Wang[1,2], Lihua Jiang[1,2], Jian Ma[1,2], and Hui Xie[1,2]

[1] Agriculture Information Institute, Chinese Academy of Agriculture sciences, Beijing, P.R. China
[2] Key Laboratory of Agricultural Information Service Technology (2006-2010), Ministry of Agriculture, The People's Republic of China

Abstract. As a new service model, cloud services have become the growth point of the service innovation of digital libraries. By analyzing and summarizing the cloud service mode of the digital libraries at home and abroad, this paper presents three-layer cloud service model of digital libraries which includes data layer, automated management layer and service layer from bottom to top. Based on this model, this paper puts forward the union mechanism established by some libraries which joint provide intelligence research service, knowledge push service, reference service and so on.

Keywords: Cloud services, Digital library, Service mode, Sharing alliance.

1 Introduction

With the development of cloud computing technology, the cloud service application research becomes a hot issue. The operating mode and technical architecture of cloud service are closely related to Google and Amazon. Google put forward a complete set of distributed parallel cluster infrastructure according to the characteristics of the large scale network data. And Google provides a series of SaaS for individual users and enterprises including the Google search engine, Google maps, photos and videos sharing, social networks, Gmail, Google calendar, Google Apps Market place and so on. In order to make full use of idle IT infrastructure, Amazon provided storage servers, bandwidth and CPU resource to the third party by Amazon Web Services in 2002. In 2006, the network service platform based on cloud computing were provided to enterprises so that they do not need to undertake the underlying work such as hardware maintenance and can focus on their own business.

Information industry is always an important field of information technology application. To apply cloud computing technology, a library can make full use of its resources and technical strength. The service efficiency can be enhanced obviously by innovating service mode.

© IFIP International Federation for Information Processing 2015
D. Li and Y. Chen (Eds.): CCTA 2014, IFIP AICT 452, pp. 404–410, 2015.
DOI: 10.1007/978-3-319-19620-6_45

2 Cloud Service Status of Domestic and Abroad Libraries

2.1 Cloud Service Application Research Status of Abroad Libraries

Online Computer Library Center (OCLC) is a profitable and the largest library cooperation organization all over the world. In 2009, OCLC launched the WorldCat Local based on cloud computing technology which provides a full set of cloud computing information management service for libraries such as WorldCat.org, WorldCatLocal, Questionpoint, CONTENTdm and so on. As a library OCLC firstly applied the cloud computing technology to provide services for users.

National Digital Information Infrastructure and Preservation Program (NDIIPP) is a joint action of digital resources preservation initiated by the American Congress. In 2009, NDIIPP and Duraspaces company started the experimental project-DuraCloud together. DuraCloud provided permanent access for digital resources of back issues in the New York Public Library so as to detect the effect of cloud computing technology applied in permanent access of digital resources. This research played an important role in testing cloud computing technology applied in digital Library (Yingjun Lu et al., 2012).

2.2 Cloud Service Application Research Status of Domestic Libraries

China Academic Library & Information System (CALIS) is a literature guarantee system for colleges and universities launched by The Ministry of Education in 1998. CALIS joined nearly 800 university libraries to provide cloud service including E read, current contents of western journals(CCC), unified authentication system (UAS), interlibrary loan (ILL), distributed collaborative virtual reference system (CVRS) and the unified data exchange system (UES) (Wenqing Wang et al., 2009).

Jilin Province set up the union library which Included 50 public libraries, University libraries and scientific libraries. The union library started to plan and construct a cloud computing service platform in 2010. The platform provided IaaS for member libraries, PaaS for social users and integrated the resources of member libraries to provide unified services as SaaS.

National Science and Technology Digital Library (NSTL) is a virtual institution of science& technology literature information service. The institution involves in science, engineering, agriculture and medicine. And it includes 9 member libraries and 50 service stations all over the country. At present the co-construction and sharing among several member libraries is SaaS. And the standard open interface service oriented group users is DaaS. In 2012, the project was started to study the cloud service mode and build the cloud service platform (Xiaodong Qiao et al., 2010).

3 Cloud Service Demand Analysis of Digital Libraries

3.1 Service Situation and Problems of Digital Libraries

Due to the monopoly service of domestic companies such as CNKI, WANFANG and VIP in the Chinese literature resources, most of the libraries are not engaged in the electronic processing and network services of Chinese journals, dissertations, conference papers and so on. For readers, only buyers can access to these commercial database of Chinese literature resources such as CNKI. Many researchers have no effective way to obtain the required Chinese document information resources. In the processing and services of foreign literature resources, only CALIS and NSTL established the co-construction and sharing mechanism. It is common in provincial libraries and county libraries that resources are not complete and the service effects are unsatisfactory.

Besides, the functions of cataloguing system and data processing system are unsatisfactory in some libraries because of limited capital and technical strength. When processing literature data, data format conversion between different software results to inefficient data processing work.

3.2 Cloud Service Demand of Digital Libraries

Cloud service pattern will bring the library more flexible, rich and diversified service changes. For libraries, to innovating the service mode based on a cloud service alliance has become inevitable. Rich data resources are the basic guarantee of libraries' literature service. Because of their expensive price, foreign electronic resources of every library are not complete. Many libraries have their own distinctive Chinese resources.

With consciousness ascend of independent intellectual property rights, some organizations have begun to set up their own institutional repository (IR) not to sell their knowledge assets such as Ph.D. thesis to commercial data companies cheaply. As a representative, Chinese Academy of Sciences has established 77 institutional repository of subordinate institute. Island services are not only unfavorable for the readers but also not conducive to the development of libraries themselves. The service system of digital library based on cloud computing technology can provide efficient platform to process data and service for libraries which lack technical strength.

So the libraries at the same field can jointly build sharing alliance. Member libraries can carry out digital construction and service of Chinese-foreign resources to provide integrated retrieval and services for readers. The scale advantage of the library service will be helpful to ascend influence of the member libraries. Besides the interlibrary information sharing based on the same platform will be conducive to develop knowledge services and applications.

4 Cloud Service Mode of Digital Libraries Based on Sharing Alliance

In the future libraries services will focus on SaaS based on data resources. Several libraries establish the co-construction and sharing system to expand the scope and service mode of information resources. Cloud computing technology is adopted to establish SaaS cloud platform to which member libraries can submit the metadata of their Chinese-foreign resources. Large joint library can provide single sign-on and one-stop service with huge amounts of resources for readers.

4.1 Cloud Service Model of Digital Library

The cloud service hierarchy should be built based on the standards system and supported by the operation maintenance and security system. The hierarchy includes data layer, automated management system layer and service layer from bottom to top. The relationship and structure is shown in figure 1.

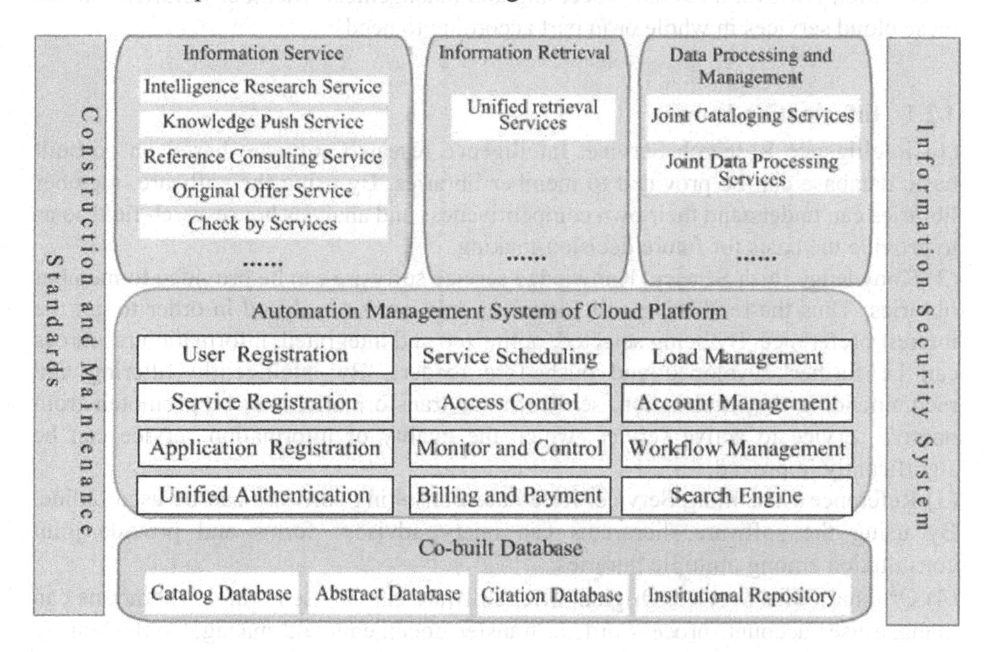

Fig. 1. Cloud Service Model of Digital Library

The data layer is the underlying structure of the three layer service model of cloud services. This layer consists of data resources built by member libraries of shared alliance such as catalog database, abstract database, citation database, institutional repository and so on.

The automated management system layer takes the system management based on data layer. The layer includes a set of middleware service such as user registration, service scheduling, unified authentication, search engine, load management, unified authorization, log statistics and service register.

The service layer is a set of services based on data layer and the automated management system layer. This layer is a software application model which provides cloud application service through the Internet transmission. The service can be divided into three categories: information service, information retrieval and data processing and management. The cloud services of data processing and management provide a cloud platform based on the unified metadata standard for member libraries. The information cloud service based on co-build data helps member libraries to jointly carry out intelligence research, knowledge push, the original offer, check by services and so on. The information retrieval cloud service provides readers with the retrieval tools to quickly find the resources needed (Chen Chen et al., 2012).

4.2 Cloud Service Mode of Digital Libraries Based on Sharing Alliance

The cloud services based on sharing alliance consist of information service, information retrieval and data processing and management. Member libraries can use these cloud services in whole or in part according to need.

4.2.1 Information Service

(1) Intelligence Research Service. Intelligence research software based on co-built basic database can be provided to member libraries. By using the software, member libraries can understand their own competitiveness and analysis hot research field so as to provide the basis for future decision-making.

(2) Knowledge Push Service. Knowledge service software can be provided to member libraries. Thus the readers' access historical data can be analyzed in order to get the interest preference. By being selected, optimized and integrated, information resources can be further developed and pushed to readers. By intelligently filtering and recommending the information, service mode transformation can be promoted from passive service to active service, which the quality of information service can be significantly improved.

(3) Reference Consulting Service. Reference consulting software can be used online. By using the software, librarians can query advisory forms and provide joint consultation among multiple libraries.

(4) Original Offer Service. Original offer software can be used online. Librarians can manage user account, process orders, transfer documents and manage settlement by using the software. Thus querying, acquisition and transmission services of all kinds of literature resources can be provided for researchers and graduate students.

(5) Check by Services. Check by services software can be used online. Librarians can receive and handle the entrusted orders. Query and delivery services facing all kinds of literature resources can be provided for domestic and abroad researchers (Xiaobo Xiao et al., 2012).

4.2.2 Information Retrieval

Unified retrieval function based on sharing alliance resources can be provided. By using the resources search engine, readers can retrieve all abstract information of books, journals, conference proceedings, reports, dissertations, standards and patent documents in cloud service platform to complete one-stop retrieval of printed and digital resources. The retrieval system can be used remotely or be embedded into the web systems of member libraries.

4.2.3 Data Processing and Management

(1) Joint Cataloging Services. Joint cataloging software can be provided for member libraries. By using the software, librarians can process metadata cataloging database of Chinese-foreign journals. Readers can log in the joint cataloging system of cloud services platform to search the resources of all member libraries.
(2) Joint Data Processing Services. Joint data processing software can be provided for member libraries. Librarians can adopt the unified standard to process data so as to provide data support for the retrieval service of Chinese-foreign information resources.

5 Conclusion

The wide application of cloud computing technology has brought a revolution to library information service mode. Cloud service has become a growth point of the digital library service innovation. As SaaS, cloud service based on library sharing alliance can significantly improve the efficiency of access to information and the utilization of information resources. In the future libraries should focus on the cloud service mode and play advantages guided by demand.

Acknowledgements. The work is supported by project of national science and technology library "Research on cloud service model and construction of cloud computing services platform for NSTL (2012XM02)", and the special fund project for Basic Science Research Business Fee, AII "Research on user behavior analysis of website of CAAS"(No. 2014-J-007).

References

1. OCLC News release. OCLC announces strategy to move library management services to Web scale (April 23, 2009). http://www.oclc.org/news/release/200927.htm (May 15, 2009)
2. Qiao, X., Liang, B., Li, Y.: NSTL strategic positioning,the recent advances and future development planning. NSTL Special Issue for 10 Years (Digital Library Forum, http://www.dlf.net.cn) (10), 11-17 (2010)
3. Wang, W., Chen, L.: The Model of CALIS Cloud Service Platform for Distributed Digital Libraries. Journal of Academic Libraries 4, 13–18 (2009)

4. Chen, C., Wu, W.: A Research on Cloud Services Pattern Architecture and Innovation for Digital Library Under Cloud Computing Environment. Research on Library Science 13, 70–74 (2012)
5. Xiao, X., Shao, J., Zhang, H.: The third Phase SaaS Platforms and Cloud Services of CALIS. Library and Information Service Online 3(52), 52–56 (2012)
6. Yingjun, L., Yiping, Z., Zhonghua, D.: Cloud Service in American Libraries. Library and Information (3), 16–21 (2012)
7. Wei, D., Xie, Q.: Research on the Grass-Roots Service Cloud of the National Digital Library. Journal of the National Library of China 4(82), 40–47 (2012)

Research on Detection Moisture of Intact Meat Based on Discrete LED Wavelengths

Li-Feng Fan[1,3], Jian-Xu Wang[1,2], Peng-Fei Zhao[1,2], Hao Li[1], Zhong-Yi Wang[1,2,3], and Lan Huang[1,2,3,*]

[1] College of Information and Electrical Engineering, China Agricultural University, Beijing 100083, China
[2] Modern Precision Agriculture System Integration Research Key Laboratory of Ministry of Education, Beijing 100083, China
[3] Key Laboratory of Agricultural Information Acquisition Technology (Beijing), Ministry of Agriculture, Beijing 100083, China
biomed_hl@263.net, hlan@cau.edu.cn

Abstract. The existing researches focused on using the commercial full wavelength spectrometer to determine the quality parameters of meat by detecting the meat emulsion, which were difficult to achieve online and non-destructive detection of the moisture content of intact meat. Moreover, the accuracy of pieces moisture detection is low, and people did not consider differences in the organizational structure of the pork meat itself. In this paper, we have developed a portable data acquisition system based on discrete wavelengths of spectral, and used it to detect the moisture content of fresh intact pork meat within a certain depth range. Based on the steady-state spatially resolved spectroscopy and considering the muscle fiber structure and direction of intact pork meat, we have designed a device with a symmetrical structure, which has a wavelength of 1300nm, 1450nm, 1550nm and 970nm LED light source for detecting the moisture content of the samples obtained from the Longissimus, within a certain depth range, and verified the stability and linearity of the system. The results show that the coefficient of determination is 0.49, and the detection range is 73.19%~77.654%. This study shows that scattering properties of meat is one of the main factors affecting the stability of detection.

Keywords: moisture, meat, discrete light, NIR, online.

1 Introduction

China is a great country of pork production and consumption. With the improvement of living standards, the requirement of the quality of fresh meat is increasing, and the moisture content of fresh meat is an important quality trait, which impacts on fresh meat processing, storage, transportation [1]. A high moisture content of fresh meat will accelerate the reproduction of bacteria and mold, which causes spoilage.

* Corresponding author.

© IFIP International Federation for Information Processing 2015
D. Li and Y. Chen (Eds.): CCTA 2014, IFIP AICT 452, pp. 411–418, 2015.
DOI: 10.1007/978-3-319-19620-6_46

Moisture content of fresh meat is not only an important attribute for consumers, but also affects the eating quality and nutritional quality of the meat [2]. Traditional detection methods such as sensory judgment and chemical testing, which are time consuming, more destructive, low efficiency, then it is necessary to develop the rapid detection technology and equipment research of pork quality, and realize the objective evaluation of pork quality.

As a fast, economic, non-destructive, real-time detection method, near infrared spectroscopy (NIR) has been used widely in the field of food in recent years. At present, many experts and scholars applied near-infrared spectral analysis to measure pork moisture detection research [3-6]. There have been many reports about water detection of mincemeat which have good results. Isaksson [3] et al has used an online detection system of mincemeat based on near-infrared reflectometer to detect the content of fat, protein and moisture in the minced beef. For pork meat moisture detection, with fresh pork as samples, Ji [7-9] et al used near infrared discrete illuminant based on the steady state spatial resolution method of diffuse reflection which can solve the problem of detecting depth, to test moisture content of fresh meat, and got better experimental results.

Thus, the aim of this paper is to investigate the arrangement of light source and detector in a probe, and develop a device for detecting intact pork meat and establish a model to predict the moisture of meat.

2 Principle and Measuring Device

2.1 Measuring Principle

The steady-state spatial resolution technology is based on the diffusion approximation theory, when the steady light source continuously penetrates a semi-infinite medium, the transmission of light within the medium is independent on the time, whose propagation path appears as a banana shape. See figure 1. The detecting depth is related to the space between light source and the detector. The greater the space, the deeper the detection. [10]

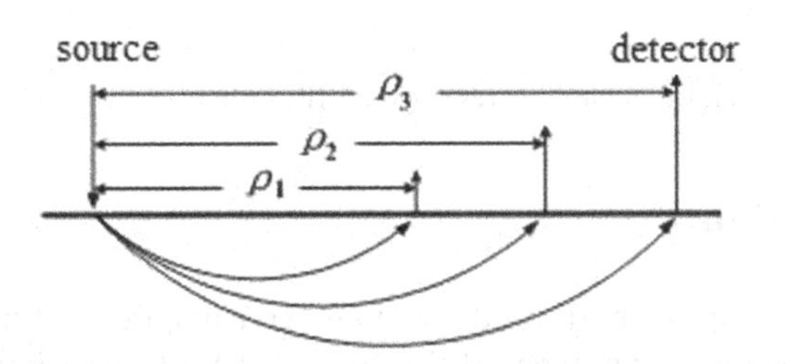

Fig. 1. The relationship of detecting depth and the space (ρ)

Meat mainly contains protein, moisture, fat and little of extractum. In the absorption characteristics of wavelength range, protein such as myoglobin and hemoglobin in the pork organization overlaps the moisture. Therefore, spectrum of pork is both a multicomponent absorption and scattering effect. As shown in figure 2, in the wavelength range of 600nm and 1700nm, the absorption peaks of moisture are 970nm and 1450nm [5]. By detecting the intensity of diffuse reflectance of each wavelength, we can associate the moisture of meat and the diffuse reflectance, and then build a prediction model related to the content of each component. To accurately express the original information, considering other background material in the pork organization, in this paper, we apply the design of multi-wavelength. Then we made the multiple linear relationships between spectral information parameters measured by the detector and pork moisture content value measured by the physical and chemical methods.

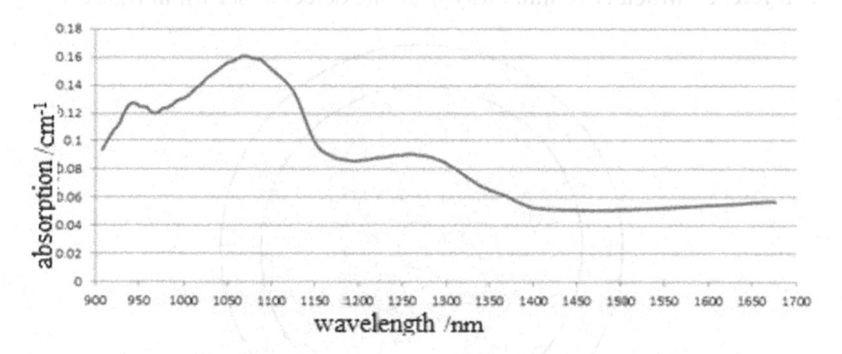

Fig. 2. The absorption in the wavelength range of 600nm and 1700nm

2.2 Measuring Device

We used infrared detection device based on discrete type illuminant, whose LED light source wavelengths are 970 nm, 1300 nm, 1450 nm and 1550 nm. The overall structure of the system is as follows:

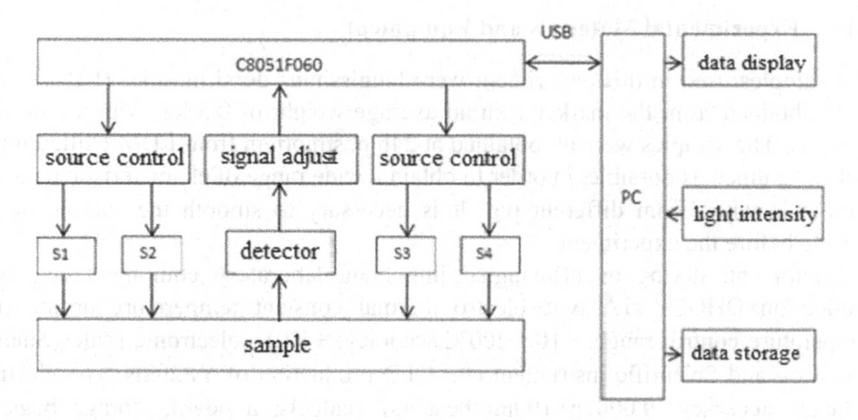

Fig. 3. The overall structure of the system

During the procedure of measurement, the main chip C8051F060 receives instructions from the PC. Then the four LED light source, i.e. S1, S2, S3, S4, at four wavelength 970nm,1350nm,1450nm,1550nmrespectivly, emitted light pulse signal by means of control circuit. The optical signal after scattering and absorption effect of pork samples was measured by the detector, converted to analog signals. The analog signals regulate steps such as isolation, amplification straight. When analog electrical signals go through the A/D conversion module of the master control chip, data is converted to spectrum. The data is transformed via USB to the computer. PC software can analysis data, display, storage, etc. In order to eliminate the difference between the sample and the influence of scattering, LED light sources uniformly distribute in the circumference which is 10 mm away from the detector, shown in figure 4.

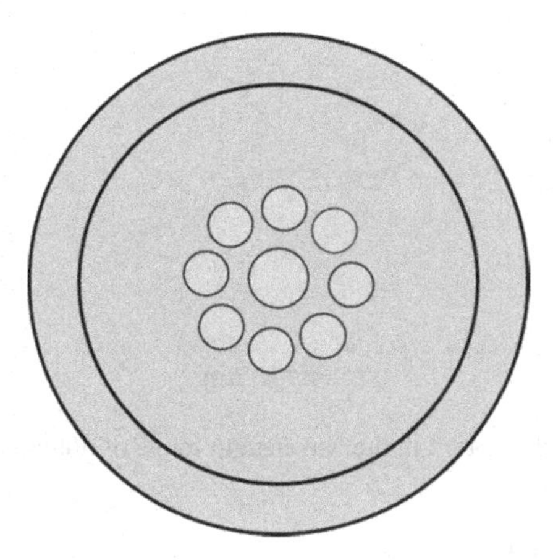

Fig. 4. The probe appearance

3 Experimental Materials and Method

3.1 Experimental Materials and Equipment

The samples used in this experiment were longissimus dorsi muscles (LDM), which were obtained from the market with an average weight of 0.5 kg, with a total of 60 samples. The samples were all obtained at 24h postmortem from LDM chilled for 24h (4°C). As much as possible, in order to obtain a wide range of characteristic data, each sample is taken from different pig. It is necessary to smooth the surface of pork muscle before the experiment.

Equipment: drying oven(Shanghai jing hong laboratory equipment co., LTD., production DHG29011A type electro thermal constant temperature drying oven, temperature control range: +10～200°C,accuracy: ±1°C)，electronic scales(Shanghai Precision and Scientific instrument co., LTD production of YP202N type electronic balance, accuracy: 0.0001g),100ml beakers，scalpels, a board, storage bags, our device.

3.2 Obtaining and Analyzing Data of Samples

Sample data includes the spectral data and moisture content. Spectral data was collected by the infrared detection device of discrete type illuminant. Experiments are conducted in an open environment and room temperature is 22 °C to 26 °C. The device is coupled directly to the surface of the pork sample. Because the luminous efficiency of LEDs with different wavelengths is not the same, and the diffuse attenuation degree through the pork organization is different, in order to avoid overflow, photoelectric diode detecting signal is adjusted by PC software automatically at the beginning of the experiment. It's necessary to make the light intensity be in a suitable range, and then interval sampling. The data is transformed via USB to the PC software. Do the linear regression with least square method to get Kn(each individual wavelengths in the channel). Each Kn is calculated by five groups of driving voltage DA and the matching output light intensity AD with linear regression, and then get the diffuse reflectance.

Use Chinese standard drying method (GB/T 9695.15 2008) to measure moisture content of pork samples. For a sample, three pieces each about 10g drying moisture measurement, finally take the average.

3.3 Results and Discussion

There are 51 samples of pork longissimus muscle. All samples are ordered by the moisture contents and renumbered. Samples whose number is 5 or multiples of 5 are classified to validation set, others into learning set. In this way, not only ensure validation set widely enough and avoid the validation set beyond learning set range. According to the classification method, learning set is a total of 41used to establish the model, accounting for about 80% of the total samples; the remaining 10 samples as a validation set are used to predict. The quantity of learning set can meet the demands of modeling. The table below is real value of the moisture content measured by Chinese standard method.

Table 1. Moisture content of the samples

Num	content	Num	content	Num	content	Num	content	Num	content
1	73.195	13	74.571	26	76.357	38	76.892	5	74.085
2	73.519	14	74.621	27	76.406	39	76.898	10	74.408
3	73.538	16	74.853	28	76.414	41	76.963	15	74.743
4	74.047	17	75.206	29	76.456	42	77.070	20	75.706
6	74.221	18	75.277	31	76.501	43	77.106	25	76.312
7	74.292	19	75.544	32	76.511	44	77.354	30	76.501
8	74.325	21	75.938	33	76.753	46	77.405	35	76.804
9	74.377	22	75.962	34	76.766	47	77.440	40	76.948
11	74.481	23	76.056	36	76.814	48	77.517	45	77.372
12	74.568	24	76.132	37	76.871	49	77.522	50	77.533
						51	77.654		

This device with four wavelength detector, can be measured with a total of four diffuse reflectance, the four diffuse reflectance used to determine moisture content of real value by using multiple linear regression model by the least squares method. We trained the learning set and validated test set respectively, learning set predicted results as shown in figure 5, the R^2 and the root mean square error are 0.492 and 0.914 respectively; Validation set predicted results as shown in figure 6, decision coefficient and the root mean square error are 0.778 and 0.645 respectively.

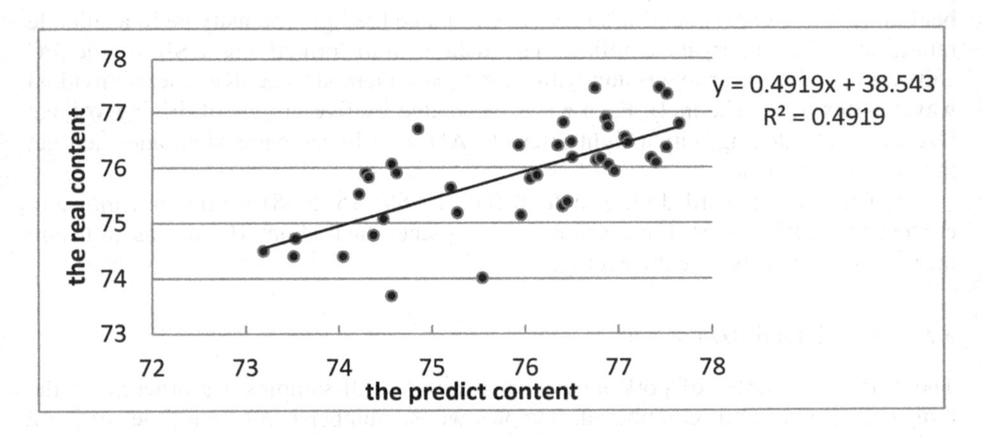

Fig. 5. Comparison predicted moisture content by near infrared spectroscopy using the proposed model consisting of fours wavelengths to measured moisture content (the learning set)

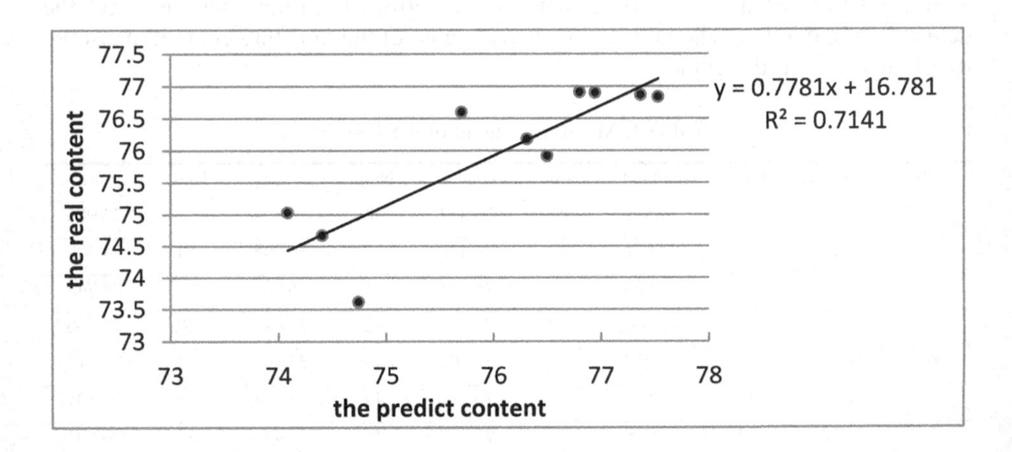

Fig. 6. Comparison predicted moisture content by near infrared spectroscopy using the proposed model consisting of fours wavelengths to measured moisture content (the validated test set)

As you can see from the above test, the R^2 needs to be improved. The main source of error is the following several aspects:

- That the different scattering between different samples of pork, is an important source of error. Pork tissue is the strong scattering medium. Scattering will occur within the organization on the boundary of the discontinuous refractive index, and is related to the structure, structure change, the number and distribution of tissue cells. Because of the experimental sample sources united hardly, and directions of fiber of different samples are different, those effect the scattering, namely detection under the influence of thickness of fibers and their distribution.
- Heterogeneity of the pork tissue leads to the difference of the scattering and absorption of the different parts [11]. The distribution of moisture is not uniform in pork tissues, which caused the moisture content in different areas of the same sample measured by the Chinese standard method are different.
- The breeds, growing environment and different slaughter time also have an important influence on the scattering and absorption of light. Tracking information of pork needs to be perfected.

The moisture content of samples which we bought from the market are mainly concentrated in between 74% and 77%, average from 75.5% to 76.5%.But the samples whose moisture content is below 74% or more than 77% are relatively little. This reduces the accuracy of predicting for the high moisture or low.

4 Conclusions

By the experimental results and data analysis, we can see the accuracy of moisture detection remains to be further strengthened. Reasons influencing the accuracy of the experiment are mainly that the tissue of fresh pork meat is not homogeneous, and the directions of fiber are different, and scattering. Fresh meat storage and unscattered sample moisture content data also have a certain influence on the accuracy of predicted results.

Acknowledgment. This research was supported by the National Key Technology R&D Program of China (Project No. 2012BAH04B02).

References

1. Liang, X., Lv, P., Zhang, X.: Research and development of the standards for moisture content of pork and beef meat (08), 38–43 (2001) (in Chinese) ISSN 1008-5467
2. Zhu, D., Wu, X., Liu, H., Xu, Y., Li, J.: Effect of fresh meat. Science and Technology of Food Industry (16), 363–366 (2013) (in Chinese)
3. Prieto, N., Roehe, R., Lavín, P., et al.: Application of near infrared reflectance spectroscopy to predict meat and meat products quality: A review. Meat Science 83(2), 175–186 (2009)

4. Barbin, D.F., ElMasry, G., Sun, D., et al.: Non-destructive determination of chemical composition in intact and minced pork using near-infrared hyperspectral imaging. Food Chemistry 138(2), 1162–1171 (2013)
5. Sun, T., Xu, H., Ying, Y.: Progress in Application of Near Infrared Spectroscopy to Nondestructive On-line Detection of Products/Food Quality. Spectroscopy and Spectral Analysis (01), 122–126 (2009) (in Chinese)
6. Xu, X., Cheng, F., Ying, Y.: Application and Recent Development of Research on Near Infrared Spectroscopy for Meat Quality Evaluation. Spectroscopy and Spectral Analysis (07), 1876–1880 (2009) (in Chinese)
7. Wen, X., Wang, Z., Huang, L.: Measurement of myoglobin in pork meat by using steady spatially-resolved spectroscopy. Transactions of the CSAE 26(supp. 2), 375–379 (2010) (in Chinese)
8. Zhang, G., Wen, X., Wang, Z., Zhao, D., Huang, L.: Measurement of Pork Tenderness by Using Steady Spatially-Resolved Spectroscopy. Spectroscopy and Spectral Analysis (10), 2793–2796 (2010) (in Chinese)
9. Ji, R., Hung, L., Liu, L., Wang, Z.: Method for Measuring Water Content in Fresh Meat Using Diffusion Reflectance Near Infrared Spectroscopy and Experiment. Spectroscopy and Spectral Analysis (08), 1767–1771 (2008)
10. Lai, J., Li, Z., Wang, Q., He, A.: System Model of Light Transporting in Biological Tissues and its Application. Acta Photonica Sinica (07), 1312–1317 (2007) (in Chinese)
11. Barbin, D., Elmasry, G., Sun, D.-W., et al.: Near-infrared hyperspectral imaging for grading and classification of pork. Meat Science 90(1), 259–268 (2012)

The Issues and Challenges in Copyright Protection for Agriculture Digital Publishing

Yuan-Yuan Tu[1,2]

[1] Agricultural Information Institute of Chinese Academy of Agricultural Sciences, Beijing, 100081, China
[2] Key Laboratory of Agricultural Information Service Technology (2006-2010), Ministry of Agriculture, The People's Republic of China, Beijing, 100081
tuyuanyuan@caas.cn

Abstract. With the rapid development of agriculture information technology and Internet, the agriculture digital publishing has become one of the most dynamic industries in China. Currently, agriculture digital publishing just begun and the related laws and regulations are being improved. However, agriculture copyright protection for digital publishing still faces many problems and challenges. It is very important to strengthen agriculture copyright protection in digital publishing industry. In this paper, the main concepts and challenges for agriculture digital publishing and its copyright protection are analyzed and some suggestions are also proposed.

Keywords: Agriculture digital publishing, Agriculture copyright protection.

1 The Concept of Agriculture Digital Publishing and Digital Copyright Protection

1.1 Agriculture Digital Publishing

The rapid development of computer, Internet and communication technology, for the news publishing industry has brought significance and profound changes. The concept of agriculture digital publishing has become increasingly prominent, and was concerned by agriculture academic world. At present, there is no unified definition of agriculture digital publishing. It is generally believed that agriculture digital publishing is a new way of publishing by using digital technology and disseminating the digital content throughout the network. Its main features include agriculture digital content production, agriculture editing of digital processing, agriculture digital printing, agriculture digital management process, agriculture digital product form, agriculture digital distribution and sales, agriculture digital communication channels and agriculture digital reading consumption (Shi Yongqin, Zhang Fengjie, 2012).

The agriculture digital publications includes agriculture electronic books, agriculture digital newspapers, numeral agriculture magazines, agriculture network literature, agriculture network educational publications, agriculture network map,

© IFIP International Federation for Information Processing 2015
D. Li and Y. Chen (Eds.): CCTA 2014, IFIP AICT 452, pp. 419–425, 2015.
DOI: 10.1007/978-3-319-19620-6_47

agriculture digital music, animation, agriculture network database publications, agriculture mobile phone publications (such as mobile phone MMS, customized ringing tone, mobile phone newspaper, mobile phone numeral magazines, mobile phone novels and mobile games) etc (see Table 1.). The speed of its development is amazing, and has begun to change the lifestyle and concept of consumption for the Chinese people, although the history of agriculture digital publishing in China is not for long.

Table 1. The character and life cycle of the agriculture digital publications

Category	Character	Life cycle
Agriculture electronic books	Authentic,	More than two years
Agriculture digital newspapers	Authentic	Update by days
Numeral agriculture magazines	Comprehensive	Update by month
Agriculture network literature	Fundamental	One year
Agriculture network educational publications	Authentic	More than 3 years
Agriculture network map	Universal, international	Real-time updates
Agriculture digital music	Popular	More than two years
Animation	New-emerging	One or two years
Agriculture network database publications	New-emerging	Long-term
Agriculture mobile phone publications	New-emerging	Update by days or month

1.2 Agriculture Digital Copyright Protection

Agriculture digital copyright protection is managing and protecting the agriculture copyright of digital content in the whole life cycle of creation, production, dissemination, sales and uses through the various technical methods, in order to ensure the legal possession of digital content, the use of digital content, the dissemination of digital content and the management of digital content. It is commonly known as agriculture digital rights management (ADRM) or agriculture information rights management (AIRM). From the technical point of view, agriculture digital copyright protection is a kind of software technology related with rights management and rights restriction. From the application perspective, it is a complete system engineering which consists of agriculture digital copyright protection technology, agriculture business model and the agriculture laws and regulations (Shi Yongqin, Zhang Fengjie, Ma Chang, 2012).

Agriculture digital copyright protection is produced in the situation of agriculture digital content production, the contents of the personalized demand and marketing communication network. The difference is that the traditional agriculture copyright protection only protects the interests of the copyright owner, but agriculture digital copyright protection manages and protects the interests of value chain of digital production in the entire life cycle, which relates to the agriculture content production, agriculture processing management, agriculture transmission and sales, etc. It is an indispensable digital tool which assists agriculture copyright management. Speaking from this meaning, agriculture digital copyright protection is not a single copyright protection. It includes the agriculture digital rights to network, agriculture communication rights of information network, agriculture multimedia rights and other unknown rights of new pattern of network communication (Zhang Li, Chen Hanzhang, 2006).

2 The Problems of Agriculture Digital Publishing and Copyright Protection

From now on, the first problem is that the base of agriculture digital publishing is still weak. The copyright protections are neglected. Strengthening the agriculture digital publishing has become the important thing.

In the meantime, technology provides space for agriculture digital publication, and enables consumer to become easy and simple on the copy of agriculture digital publishing works. Through the internet connection, the readers in the world are able to read agriculture digital publishing works quickly and easily, in order to meet their demands for fast reading. However, it increases the risk of agriculture digital publishing works piracy, and the legitimate rights of copyright owners get the great loss. In the era of agriculture digital publishing, the following main problems in agriculture digital copyright protection work are shown.

2.1 The Whole Agriculture Digital Publishing Recognize the Existence of a Misunderstanding

Generally speaking, the attention of publishing enterprises is insufficient, and the understanding of their own position is not clear, with certain blindness on the implementation of agriculture digital publishing work. For example, a lot of presses proposed the transformation, it means that they only consider the transmission and operation, and neglect the mechanisms for web content.

Both the traditional publishing enterprises and the new digital publishing enterprises have experienced difficulty, because of the agriculture digital publishing chain is disjointed and fragmented, and imbalance for the distribution of interests caused by the management difficulties. On the other hand, the agriculture digital copyright regulation has no legal basis for compensation, not enough punishment and the high cost of rights system. It defects the pain and suffered from piracy problem.

In the traditional publishing system, the publishers as content processors have the duration of protection for intellectual property rights at the same time. In the field of agriculture digital publishing, the publishing business which use the term of copyright protection regulations effectively will hinder the agriculture digital information resources, thus publishing enterprises need to coordinate the relationship between the traditional paper-based and agriculture digital resources for the coordinated unification. It can not only maintain the present status of traditional paper-based agriculture resources, but also can do the agriculture digital resource content provider (Yan Xiaohong, 2013).

In addition, many publishing enterprises which are professional in agriculture field only stay in a digital publishing platform of construction. Although spent a lot of manpower and material resources on agriculture field, the follow-up work is still in the attempt and exploration. It is not clear on the way of agriculture digital environment content delivery. The result is that the agriculture content is in small scale, and the high quality publishing copyright is nothing. It likes building the house, if only put up the reinforcing steel bars, and no pouring concrete, it is still regarded as useless at all.

2.2 Law of Agriculture Digital Copyright Protection is not Perfect

So far, it's no specialized in agriculture digital copyright protection laws in China. In the aspect of legislation, it lags behind the pace of digital publishing. Although some contents related with the promulgation of government regulations, most of them are for the technical measures, rather than the real content of work. It still can't be an agriculture system, and also can't solve all sorts of agriculture copyright disputes at present agriculture digital publishing era, even to mention the specialization and standardization of agriculture digital copyright protection.

For example, since March 1, 2013, the State Council revised "*Regulations for the protection of computer software*", "*People's Republic of China Copyright Law Implementation Regulations*" and "*The right of communication through information network*" has been officially implemented. For the three pieces of regulations, the administrative penalty fines were adjusted and modified. It only faces to all, and belongs to the technical measures. The adjustment and modification on the non technology level to the existing law is not even better. The new regulations on agriculture will be brought into being.

In a manner of speaking, the current law has appeared inadaptability gradually. It must be redesign the feasible mode of digital copyright protection. For example, a user sends a piece of original agriculture information on website, and then it will be copied or forwarded for tens of thousands of times in just a few minutes without notify. How to protect the copyright of original agriculture information? Is the "information" in the scope of agriculture copyright protection? This is a series of questions which bring the challenges to system and legal on agriculture digital copyright.

In addition, on the way of establishing digital copyright laws in China, it is very necessary to learn from the experience of western developed countries. USA made "*Copyright law in 1976*". It shows that the equipments or procedures in the field of entertainments and exhibitions contain the electronic delivery system and non-used or disclosed system, and provide the basis for the application of network copyright. After that, American digital publishing legal system has experienced the white paper "*1995 intellectual property and the national information infrastructure*", "*1997 online copyright liability limit method*", "*Clear digital copyright and technology education law*", "*1998 digital millennium copyright law*" (DMCA) and "*2009 digital consumer's right to know act*" and the development of other laws (Huang Xianrong, Li Weijuan, 2012).

USA is the one of most important countries which received attention to copyright legislation and enforcement in the world. In the continuous improvement of IPR legislation and enforcement of protection at the same time, protecting the legal interests of copyright owners effectively, and maintain the order of the copyright under the digital environment. It provides a model for establishing the legal system of digital copyright protection in China. It will effect on the development of agriculture digital copyright laws in the future.

2.3 Agriculture Digital Copyright Resources are Scattered and Unable to Form an Effective System of Agriculture Digital Copyright Protection

The scattered agriculture digital copyright resources refers that there is no clear boundaries among the publishing enterprises, the copyright owner and the third parties. Some for their respective camps, some two-two union, failed to form a set of effective specification of the agriculture digital publishing industry chain, and leading to the problem of agriculture copyright protection is more complicated and changeable.

If we suppose that a spurt of progress of technology of agriculture digital publishing third enterprises are growing, and the agriculture digital publishing activities is developing rapidly, the author of publishing enterprises or digital copyright will not get reasonable returns in the business, eventually they will be reluctant to digital copyright to cooperate with the third party of digital publishing enterprises. This is just one of these situations.

As it is not yet mature in the digital publishing business model, to digital publishing enterprises, the third party publishing enterprises and the author or copyright owners alike, no one can tell you how much really get from the agriculture digital publishing (Xu Xinghua, 2013).

A series of potential crisis also revealed, it confused the original intention of agriculture digital publishing. How to develop the agriculture copyright protection in the digital publishing industry? How to break through the bottleneck of agriculture digital copyright in order to form a healthy value chain? How to keep the maximize benefits on most aspects of the industry chain? These problems all need to be solved urgently.

2.4 Cloud Computing Brings New Opportunities for Agriculture Copyright Protection, also Brings New Problems

Cloud computing is a calculation based on internet (as shown in Fig. 1).

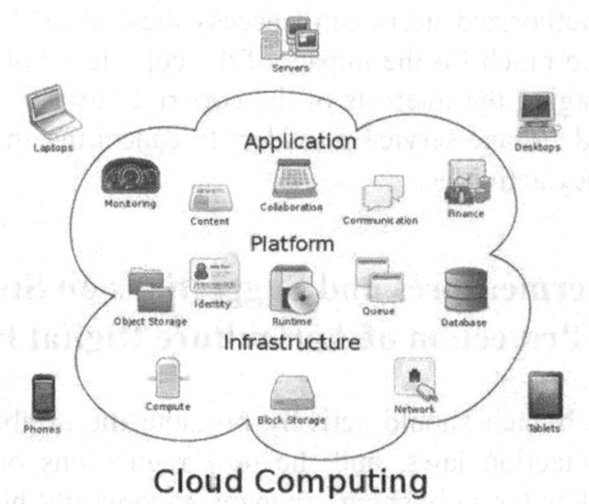

Fig. 1. What is cloud computing

In this way, the hardware and software resources and information can be provided to computers and other devices on demand. The technology applies after the agriculture digital publishing, its dissemination of information on the breadth and depth is unprecedented. It will be published in the traditional publishing enterprises, authors and the third-party of digital publishing enterprises was jointed on a chain. According to their interests, they plays different roles (as shown in Fig. 2.).

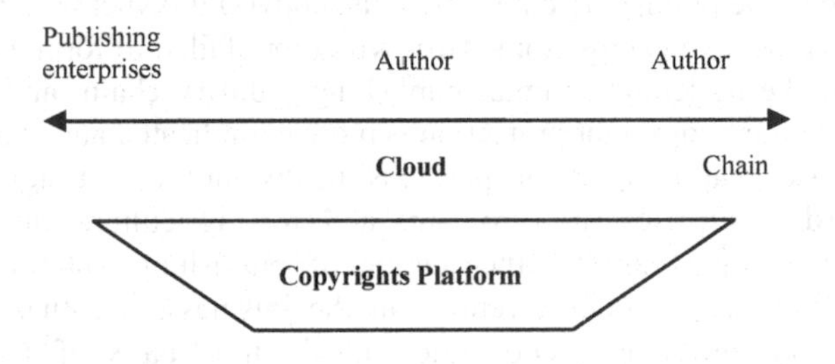

Fig. 2. The dissemination of roles

Meanwhile, the environment of agriculture digital publishing will be changed. The advantage of the improvement is that the distributing agriculture digital publishing resources on a unified and open platform for publishing services will be integrated on cloud platform. For example, the publishing companies can publish on the cloud platform by choosing different distribution channels which carried on the agriculture digital publishing technology to encrypt and regulate. A variety of channels can be integrated in the third-party of agriculture digital publishing for the terminal applications on the cloud platform. When enjoying the advantages of cloud computing, it has led to changes in transaction costs as a new technology, inevitably adds to challenge in the future (Zhou Liang, 2012). For example, the piracy illegal uses of agriculture works flood in market throughout a common cloud without the permission of the copyright holder. Now many third parties companies provide cloud storage service, but they have to review the data before storing. That is to ensure the data is safe and unauthorized users can't access these data. If the shared object is certain groups, it is so much for the impact of the copyright holder, but if it is for the individual, it is damaging the interests of the copyright owner. Even if the copyright holder requests cloud storage service providers to cancel the sharing links. They still can't control the piracy activities.

3 The Countermeasures and Suggestions on Strengthening Copyright Protection of Agriculture Digital Publishing

First, the legislative branch should actively promote the establishment of a special digital copyright protection laws, and the new regulations on agriculture will be brought into being. For the publishing companies, copyright holders and third-party digital publishing business, there are laws to go by. It ensures the sound development of agriculture digital publishing chain and the accomplishment of their interests.

Second, from a relatively weak awareness to a transformation process, the acceleration of the agriculture digital publishing business of copyright protection can avoid vicious competition among enterprises, thus driving to respond positively to the agriculture digital copyright protection, and ultimately form a good environment for agriculture digital rights protection.

Finally, the advanced digital technology (e.g. cloud computing) is a hardware protection of agriculture digital copyright protection. It can support digital copyright protection system, and promote the coordinated development of digital copyright protection laws and regulations in agriculture.

In the era of agriculture digital publishing, the developing of agriculture copyright protection is certainly full of hard troubles. The agriculture digital copyright protection is bound to meet new round of challenges in the spiraling process of continuous innovation and improvement.

Acknowledgements. The work is supported by the Academy of Science and Technology for Development fund project "The research about the content organizing and service pattern for website resources of Chinese Academy of Agricultural Sciences", and the special fund project for Basic Science Research Business Fee, AII (No. 2014-J-008).

References

1. Shi, Y., Zhang, F.: Beijing. Digital Rights Concept Analysis. China Publishing, pp. 61–64 (2012)
2. Shi, Y., Zhang, F., Ma, C.: The concept of digital copyright protection technology, types and its application in the field of publishing, Beijing, pp. 57–59 (2012)
3. Zhang, L., Chen, H.: The concept of digital publishing, Beijing (2006). http://tech.sina.com.cn/other/2006-09-29/16201166788.shtml
4. Yan, X.: Several problems about digital publishing and copyright. The academic lectures in publishing research center of Communication University of China, Beijing (2013)
5. Huang, X., Li, W.: The American legal system present situation and development trend of digital publishing. Chinese Publishing, Beijing, pp. 59–62 (2012)
6. Xu, X.: Digital copyright protection under the digital publishing perspective: the status quo, problems and countermeasures. Baoding College Journal, 116–119 (2013)
7. Zhou, L.: Cloud computing technology digital copyright protection. The master's degree thesis of Suzhou University, Suzhou (2012)
8. Li, X.: The analysis of copyright in digital publishing industry. Electronic Intellectual Property, 69–73 (2010)

Development of Glass Microelectrodes Pipette Puller Based on Monitoring and Controlling Heating Strength

Yuan Wang[1,2], Li-Feng Fan[1,3], Jian-Xu Wang[1,2], Yang Chen[1,2], Lan Huang[1,2,3], and Zhong-Yi Wang[1,2,3,*]

[1] College of Information and Electrical Engineering, China Agricultural University, Beijing 100083, China
[2] Modern Precision Agriculture System Integration Research Key Laboratory of Ministry of Education, Beijing 100083, China
[3] Key Laboratory of Agricultural information acquisition technology (Beijing), Ministry of Agriculture, Beijing 100083, China
{798711935,229670582,1012443804}@qq.com, yanchcy@163.com, biomed_hl@263.net, wzyhl@cau.edu.cn

Abstract. The pulling of the glass microelectrode is a necessary technology in the electrophysiological measurement techniques. In order to better meet the requirements of electrophysiological experiments, we designed a multi-functional microelectrode puller, and gave the design of functional modules, and discussed its controlling effect. In this paper, we analyze the impact on the glass microelectrodes pulling which is brought by the heating strength based on the correspondence between the heating temperature and the heating wire resistance. Experimental results showed that the direct measurement of the heating temperature during the glass microelectrodes pulling process provided more direct control basis for the intelligent control, and provided a subtle improvement ideas for achieving precise pulling of the pipette puller.

Keywords: heating measurement, microelectrode, pipette puller, intelligent control.

1 Introduction

A growing number of micro-electrode technologies show their superiority in the study of the electrochemical and electro analytical chemistry. Microelectrode usually refers to a very fine tip of the probe and is usually used in the measurement of both the intracellular and the extracellular domain[1]. The widely application of the microelectrode technology has a strong impetus to the research and development process of the electrophysiology[2-3]. The ion-selective microelectrode as the measurement of inside and outside biological cells and the new tools are widely used in biology, medicine and other disciplines[4]. SIET(self-referencing ion-selective microelectrode technique) is a new means of electrical physiology[5]. The micro-region ion flow dynamic change information in the plant cell and the tissue can be obtained by the non-destructive detection technology. For the ion-selective

[*] Corresponding author.

© IFIP International Federation for Information Processing 2015
D. Li and Y. Chen (Eds.): CCTA 2014, IFIP AICT 452, pp. 426–436, 2015.
DOI: 10.1007/978-3-319-19620-6_48

microelectrodes isthe basic measurement tool, the producing of the glass microelectrode is a crucial aspect in the plant ion flow study. Therefore, pipette puller is the necessary equipment used in the electrophysiology laboratory for ion flow detection.

Nowadays, the drawing instrument on the market is divided into horizontal and vertical. The P97 series produced by the United States Sutter is the mainstream horizontal product, while the WD-1 Vertical Pullers is the most universal vertical one. Compared with the vertical puller, the electrode drawn by the horizontal puller is more symmetrical and repeatability. Aa a concequence, this study sets multi-functional micro-electrode pullers as the design goal, and takes some mainstream pullers as reference. In order to complete a glass microelectrodes instrument with convenient interactive interface, stable performance, monitoring and controllable heating intensity, both the software simulation and the system experiment are carried out to validate the feasibility of the whole design scheme.

2 The Overall Design of the Control Circuits in the Microelectrode Puller

According to the theory of the glass processing technology, the time, the intensity of the heating, and the cooling time are important indicators which determin the final performance parameters of the microelectrode[6]. Microelectrode pullers control circuit includes a microcontroller, heating control module, pulling control module, keyboard input interface and several functional units. The overall hardware module is shown in Fig. 1.

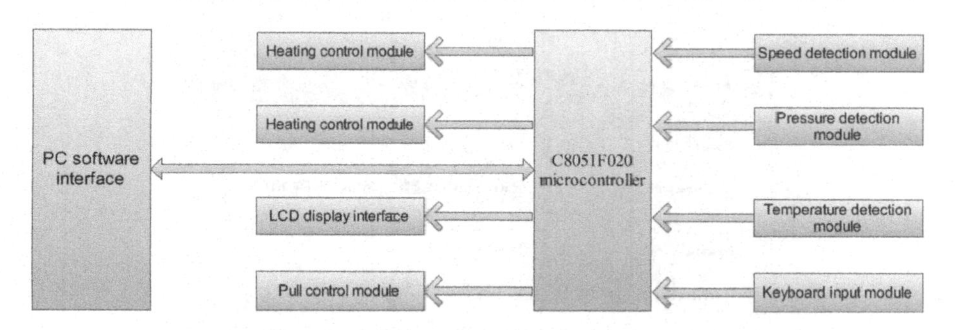

Fig. 1. The overall hardware design block diagram

2.1 Microcontrollers Selection

The core controller of the control circuit is C8051F020 microcontroller. And its ADC and DAC can make analog and digital signal processing more easily. Two UART serial ports can be used for serial communication between PC and microcontroller.

The large number of digital resources contained in the microcontroller can be allocated through priority crossbar compiler. Thus, it can ensure the flexibility of resource allocation.

2.2 Cooling Control Module

This module includes pump and solenoid, and the regulation of the air flow can be achieved by their coordination. The adjustment is based on the pressure measured by the pressure sensor, and the microcontroller will determine a feedback adjustment instruction according to the results compared with the pre-set value.

2.3 User Interface

The users can control the instrument in two ways, including the PC software and the keyboard. The drawing parameters can be set through the matrix keyboard directly in front of the puller. The serial port can also realize the parameter setting and saving. Serial communication is realized by the USB bus converter chip CH341A which is produced by Nanjing Heng Qin. And the chip is used to provide asynchronous serial bus, printer port, parallel port , the usual 2-wire and 4-wire and other synchronous serial interface via USB. In addition to the control parameters passed to the microprocessor, the serial communication module can also help upload the drawing parameters and save the document. In addition, the hardware circuit also includes a LCD screen display circuit, using YLF12864H-1 models with character screen. The entire interface display includes the welcome screen, the menu interface and parameter settings sub-menu interface.

PC software interface and LCD screen are shown in Fig. 2 and Fig. 3:

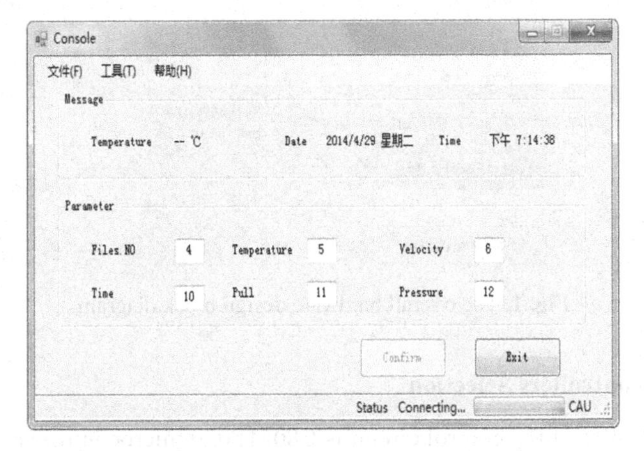

Fig. 2. PC software parameter setting interface

Fig. 3. LCD display interface

2.4 Heating Control Module and Pull Control Module

Heating control module and the pull control module is designed with FET which is a voltage controlled devices, namely, to achieve control of the drain current by the gate-source voltage. Heating control module uses AC-DC converter circuit, and its performance is also part of the circuit to do a more detailed experimental verification. Pull module selects the solenoid as the tension actuator. The size of the rally adjustment is realized through the current control in the solenoid. In the production process of glass, its melting, clarification, homogenizing, feeding, molding, and temperature system degradation process, are generally developed based on its corresponding viscosity. Therefore, glass viscosity variation has an important guiding significance on the control of the entire process of shaping glass. Speed detection value is mainly used to characterize the degree of softening of the glass tube blank. Therefore, it is an important indicator for the drawing process during the cooling, and heating control. Speed detection signals are mainly conditioned by amplifier circuits and filter circuits.

3 Signal Isolation Solutions and Experimental Verification

3.1 Digital Signal Isolation

Digital signal isolation methods are generally implemented by unidirectional digital signals, and the bidirectional digital signal isolation is achieved by employing two more elements. Common digital signal isolation methods are mainly (1) opt coupler, such as PC817, PS2501, 6N137, etc.; (2) ADI's magnetic isolation chip; (3) isolation [7-8] through the transformer. The design uses a si8422 single-chip and si8423 dual-channel digital isolators. Compared to traditional isolation techniques, ultra-low power digital isolation chip has advantages in the aspects of substantial data rate of CMOS devices, propagation delay, power, size, reliability. And it can remain stable under a wide range of temperatures and operating parameters during the life of the device so that it can be designed and widely used. What's more, the chip uses a narrow-body SOIC-8 package RoHS standard, which greatly saves installation space.

3.2 Analog Signal Isolation Method

In a scenario which accurate measurement is needed, the isolation cost of the analog signal will be an important consideration factor. More common analog signal isolation methods include: (1) isolation amplifier; (2) the frequency to voltage converter combined with digital isolation; (3) the use of DA / AD combined with the digital isolation solution for analog signals sample recovery; (4) linear opt coupler[9]. The design uses a linear opt coupler HCNR201 model[10], and the circuit in Fig 4 is used for the two DA signals isolation.

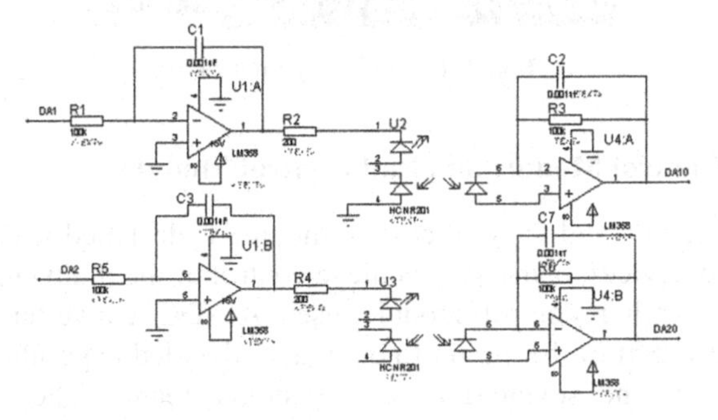

Fig. 4. Analog signal isolation circuit schematics

The relationship diagram which is shown in Fig. 5 is obtained by measuring the analog input and output signal isolation circuit. Fig. 5 shows a high correlation coefficient between the input signal and the isolated signals. That is to say, the circuit secures a true effect.

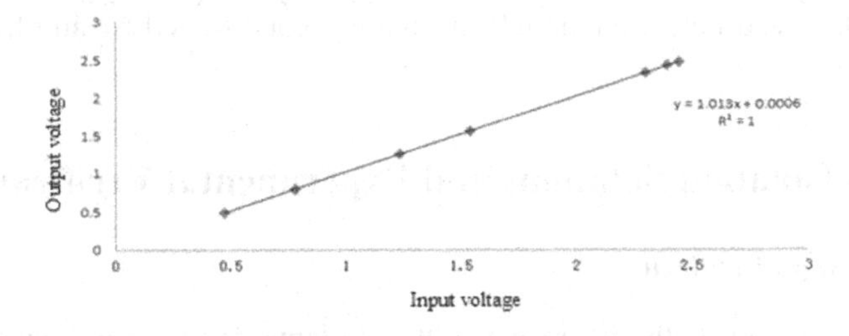

Fig. 5. Signal isolation between input and output voltage linear analog schematic

4 AC-DC Converter Circuit Solutions and Experimental Verification

4.1 AC-DC Converter Circuit Design

Common controllers which are applied to help small current control large current mainly include: SCR, FET and relay. Taking power, on-time, efficiency, cost and other factors into consideration, our design chose FET as the control device.

We designed a heating unit control module circuit with feedback regulation, and the circuit diagram is shown in Fig. 6.

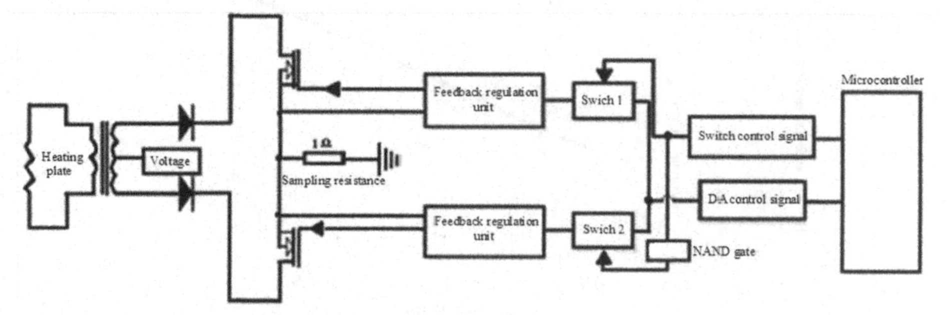

Fig. 6. DC-AC converter circuit diagram

The scheme used DA value as a feedback to adjust the parameters so as to achieve the control on the size of the drain current in the FET. At the same time, the symmetrical on both sides conducts in a certain adjustable frequency alternately. In the two-way symmetrical circuit, output signal terminal was connected to a transformer, which was used to provide heating. In the regulation of the DA, both the two output signal amplitude and the extent of FET conduction changed accordingly. The user can achieve primary side voltage by adjusting the heating transformer. After connecting the heating wire to the secondary side of the transformer, the heating intensity can be adjusted according to a certain adjustable frequency.

4.2 Circuit Experimental Verification

For the heating program, we used the unit circuit verification to test the design, and measured the corresponding amplitude relationship with the DA control parameter value between the primary side and the secondary side of the transformer. We concluded that the voltage varied with the DA changes according to exprimental analysis. And the obtained results are shown in Fig. 7.

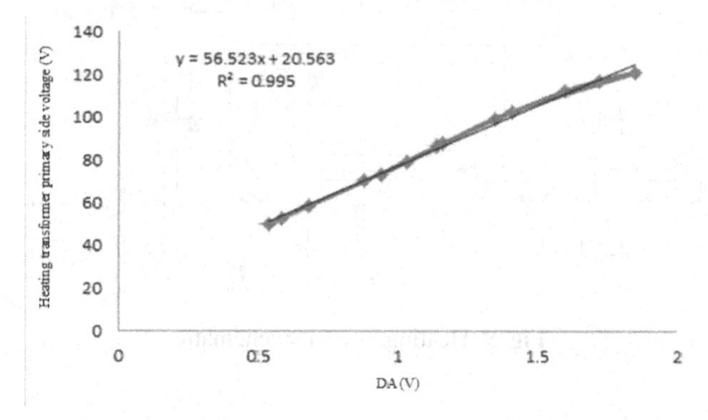

Fig. 7. Heating transformer primary side voltage (V) measurement curve under DA (V) regulation

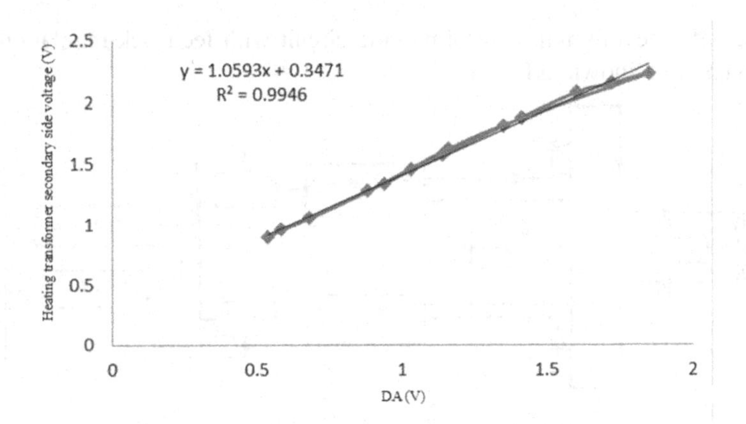

Fig. 8. Heating transformer secondary side voltage DA (V) measurement curve under DA regulation.

Fig. 7 shows that the primary side of the transformer can achieve a range between 45V to 125V as the DA value ranges from 0.4V to 2V. Besides, the secondary side Voltage control can be changed by adjusting the transformer turns ratio.

Fig. 8 shows the heating transformer secondary side voltage measurement curve under DA regulation. According to the transformer working theory, the secondary side varies according to the primary side. Thus, the secondary side voltage of the heating transformer will have a range as the DA regulates.

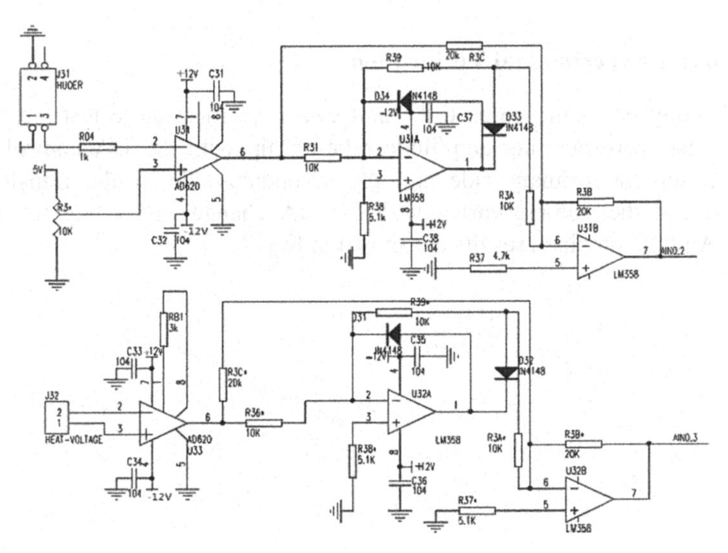

Fig. 9. Heating intensity schematic

4.3 Heating Intensity Detection Module

Heating intensity is the key factor during the entire drawing process. The constant resistance calculated by the voltage and current at both ends of the heating plate was used as the heating process indicator. And the heating voltage across the current measuring circuit is shown in Fig. 9.

By employing the nickel-chromium heating wire in the circuit, this experiment aimed to test the DA control status. The correspondence between the nickel-chromium heating wire resistivity values and the DA regulation is shown in Fig. 10.

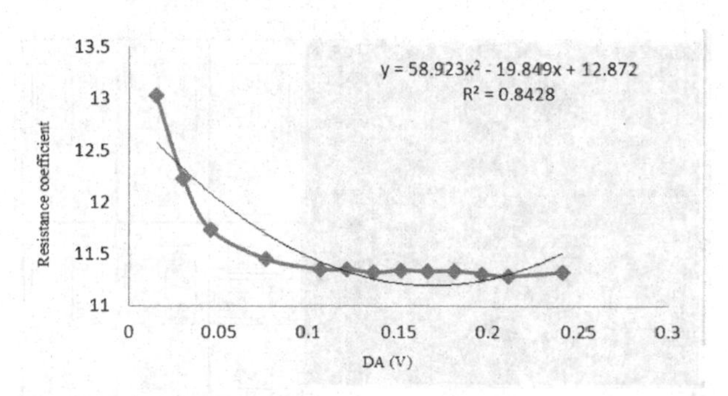

Fig. 10. The relationship between nickel-chromium alloy DA (V) and the DA

According to the Fig. 10, the resistivity value of nichrome DA regulation is a quadratic function, and the correlation coefficient is 0.8428. In the actual drawing process, the resistance is designed as the index of the heating strength.

5 Proteus Simulation

Proteus EDA tool simulation software is published by the British Lab center electronics. Not only does it has the function that other EDA simulation has, but also it can cooperate with the Keil simulation and the peripheral devices. The proteus is the best simulation tool in the industry and microcontroller peripheral device so far[11].

In order to validate the design effect of the module circuit, functional simulation for heating section and speed detection module by the proteus software are performed.

5.1 Simulation Results of Heating Module

In simulation, we selected the amps, resistors, capacitors, switches and other components from the library in accordance with the circuit diagram and the simulation schematic which are shown in Fig. 11. By real-time monitoring the waveforms of input and output signal from the virtual oscilloscope, the control signals obtained on the both sides of the symmetrical circuit are shown in Fig. 12.

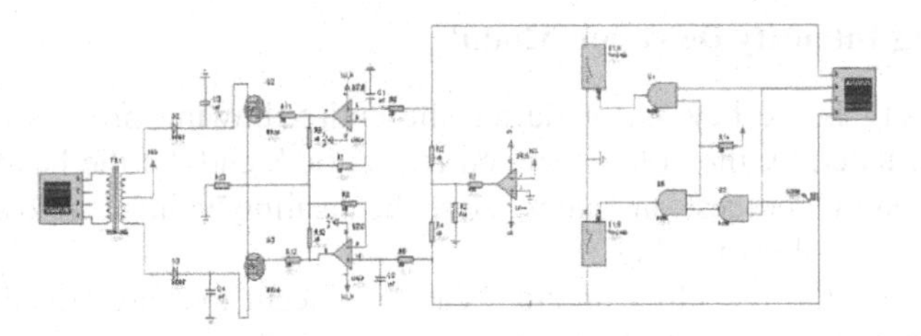

Fig. 11. Heating unit circuit simulation schematic

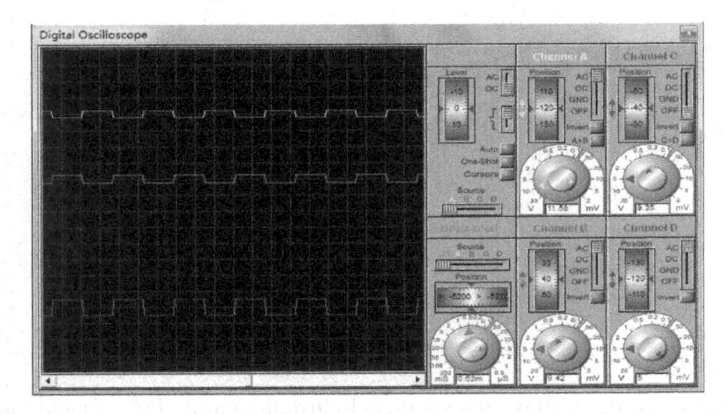

Fig. 12. Proteus simulation run chart

As shown in the virtual oscilloscope, the two sides are sequentially turned on in a certain frequency (the frequency is adjustable), which is equal to the frequency of the transformer's AC signal on the primary side. Thus, the DA control signal functions in the two sides in turn, and will control the size of the drain current of the FET. Therefore, the amplitude of the AC voltage in the heating transformer's primary side is controllable.

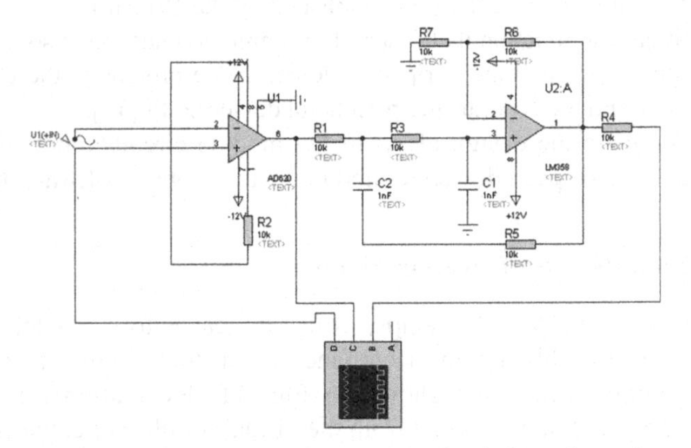

Fig. 13. Pressure monitoring module simulation schematic

5.2 Simulation on the Sensor Signal Conditioning Circuit

The speed sensor module and the pressure part used the similar signal conditioning circuit. Namely, the output voltage was turned into a certain range of the signal amplification by the AD620, and the active low-pass filter filtered out the invalid frequency. As the Fig. 13 shown, the three waveforms seen from the virtual oscilloscope are output voltage of the sensor, the amplified output signal and the waveform after AD620.

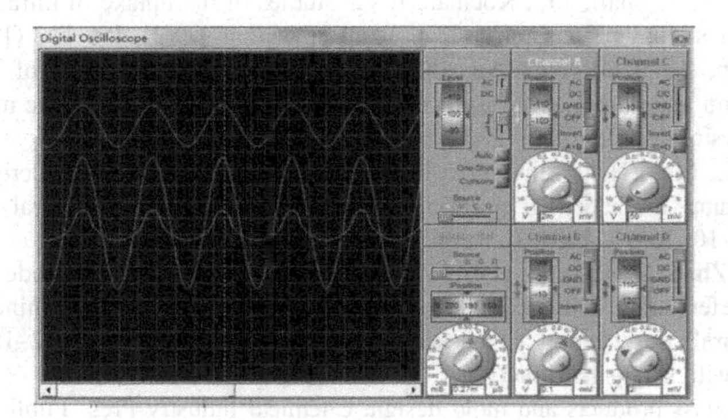

Fig. 14. Virtual oscilloscope schematic emulation

In Fig. 14, the output signal of the speed sensor used 1kHz sine wave to simulate. Analyzing the waveform seen in the virtual oscilloscope, the sensor output voltage signal is amplified through AD620 by 5.94 times. However, the signals above 1kHz was filtered out by the low-pass filter circuit to ensure the effective output signals. Simulation results consisted with the design requirements, so the circuit design can meet the desired effect.

6 Conclusion

This paper provided a design idea on the control unit circuit of the multi-functional micro-electrode pullers. The performance verification results were given by experiments and proteus unit circuit simulation. Analysis conclusion showed that the functional modules can play the desired control effect. Isolation of the analog signal can meet the drawing instrument control signal transmission accuracy. Meanwhile, heating control circuit under the control of DA values can range 45V-125V voltage regulator. The user can control the connected operating voltage and the current value performed on the load by changing the turns ratio between the heating transformer primary and secondary voltage.

The cooperation of every function module under the command of the microprocessor is the key factor. Future work will continue to make the control parts harmoniously work with the mechanical parts.

Acknowledgements. This is supported by National Key Scientific Instrument and Equipment Development Projects of China (2011YQ080052).

References

[1] Han, R., Gu, D.: Microelectrode production situation and its application. Hebei Normal University (Natural Science) 2, 025 (1991) (in Chinese with English abstract)
[2] Glass, A.D.M., Shaff, J.E., Kochian, L.V.: Studies of the uptake of nitrate in barley IV. Electrophysiology. Electrophysiology. Plant Physiology 99(2), 456–463 (1992)
[3] Henriksen, G.H., Bloom, A.J., Spanswick, R.M.: Measurement of net fluxes of ammonium and nitrate at the surface of barley roots using ion-selective microelectrodes. Plant Physiology 93(1), 271–280 (1990)
[4] Wang, X., Mao, H.P., Zhiyu, Z.: The application of ion-selective microelectrodes and patch clamp in detecting electrophysiology information. Agricultural Mechanization Research 10, 36–39 (2007) (in Chinese with English abstract)
[5] Xue, L., Zhao, D., Hou, P., et al.: Test and preparation of microelectrode in applications of self-referencing ion electrode technique. Transactions of the Chinese Society of Agricultural Engineering (Transactions of the CSAE) 29(16), 182–189 (2013) (in Chinese with English abstract)
[6] Yu, G.: Glass products and mold design. Chemical Industry Press Publishing Center of Materials Science and Engineering (2003) (in Chinese)
[7] Digital isolation techniques and technology. http://www.chinaecnet.com (in Chinese)
[8] Jin, H.: Selection and Application of Digital isolation devices. Electronic Design & Application World for Design and Application Engineers (2), 57–61 (2008) (in Chinese)
[9] Zhang, B., Wang, N.: Analog Isolation Board Based on High-linearity Analog Optocoupler HCNR200. Instrumentation Technology 5, 021 (2005)
[10] Li, H., Lin, H.: Application in current sampling with delicate linear optocoupler HCNR200. World of Electronic Components (11), 37–38 (2004) (in Chinese)
[11] Liu, H., Guo, F., Sun, Z.-X.: The application of proteus simulation technology in single-chip microcomputer teaching. Changchun Institute of Technology (Social Sciences Edition) 24(3), 96–98 (2007) (in Chinese with English abstract)

Effect of Low-Temperature Plasma on Forage Maize (*Zea mays Linn.*) Seeds Germination and Characters of the Seedlings

Changyong Shao[1,4], Decheng Wang[1,5,*], Xianfa Fang[2], Xin Tang[3], Lijing Zhao[4], Lili Zhang[4], Liangdong Liu[1], and Guanghui Wang[1]

[1] College of Engineering, China Agricultural University, Beijing, 10083, China
[2] Chinese Academy of Agriculture Mechanization Sciences, Beijing, 10083, China
[3] Shandong Agricultural And Engineering College, Jinan, 250100, China
[4] Shandong Province Seeds Group Co., Ltd, Jinan, 250100, China
[5] Key Lab of Soil-Machine-Plant System of Chinese Agriculture Ministry, Beijing, 100083, China
shaochangyong68@163.com

Abstract. Low-temperature plasma is a high-energy state of the material gathered. Plasma seed processing technology is the use of high energy aggregation on treating crop seeds within 20 seconds. Previous research elucidated that this technology could improve germination and seedling growth of welsh onion seeds. In this paper, the effects of different intensities of low-temperature plasma on forage maize seeds have been investigated. The vigor and rate of germination, length of root and shoot of seedlings, the green and dry weight, and the fiber root number of seedlings of the treated sample seeds were compared with those of untreated seeds. The results showed substantial changes in the vigor and rate of germination just like what we found in welsh onion seeds. Thus, the characters of seedlings change regularly with the intensity of low-temperature plasma, and the analysis indicated that 120W was the optimum treatment. Results showed that low-temperature plasma was effective in improving seedling growth, we believe it will be also effective in earlier flowering and maturity and higher yield. There is a great value of using and spreading on production.

Keywords: Low-temperature plasma, forage maize seeds, characters of seedlings.

1 Introduction

Besides the 'traditionally' known solid state, liquid and gas phase and the more recently found low-temperature states (BOSE-EINSTEIN condensate), high-temperature states, such as plasmas existing. Although the generation of a plasma from the gas phase

* Corresponding author.

© IFIP International Federation for Information Processing 2015
D. Li and Y. Chen (Eds.): CCTA 2014, IFIP AICT 452, pp. 437–443, 2015.
DOI: 10.1007/978-3-319-19620-6_49

(Figure 1) isn't strictly spoken to be a real phase transition, plasma was recognized as the 4th state of matter due to its distinct properties, which substantially discriminated it from the gas phase[1] .

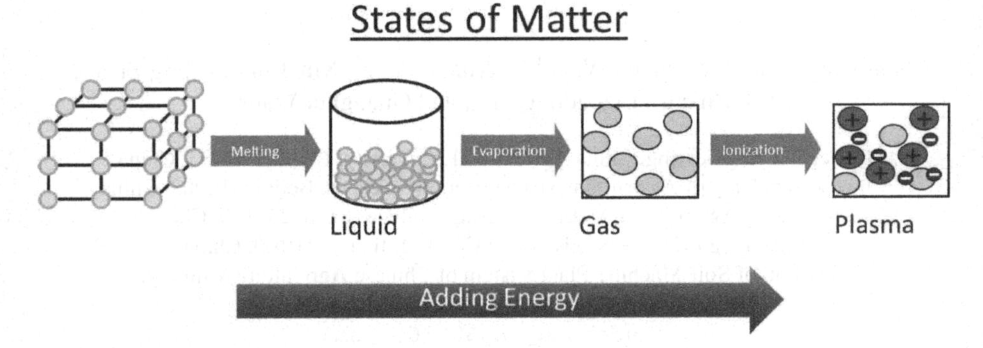

Fig. 1. Four states of matter. Plasma is characterized by a collective behavior of its free charge carriers. (Franziska Grzegorzewski, 2010)

Physical agriculture is the application of physical methods in agriculture and has been studied since the 1970s[2]. Low-temperature plasma (LTP) is an application of physical agriculture. Yin (2006) reported that LTP was utilized in biological applications in Russia[3]. There has been some research in LTP seed processor comprising a plasma generator was used to treat seeds in Russia and CIS countries (Li, 2010). Similar reports exist in America and Canada[4]. Stimulation of plants with a plasma field to promote seed germination and increase yield quantities and quality has attracted interest worldwide, with remarkable achievements reported in Japan[5]. At present, abroad study shows that both ionizing radiations, and even mass loading effect and exchange of charge are existing in LTP field. This field activated endogenous substances in the seeds, leading to improvements in the rate of seed germination, crop resistance and crop yield and an earlier maturity date[6-11].

Zea mays Linn., maize or corn, is one of the major cereals of the world and is the third largest field crop. Various kinds of technology have been used on maize seeds processing to promote germination and increase yield, such as magnetization and plasma[12-13].Our study focuses on the influence of LTP treatment on maize seeds in terms of the seed germination and biological characters of seedlings.

2 Materials and Methods

2.1 Experimental Set-Up

The equipment used in this study was provided by ChangZhou ZhongKe ChangTai Plasma Technology Co. and the device worked under the vacuum state. This machine can be carried out with large capacity in a continuous or batch process.

2.2 Sample Seeds

Tests were carried out both in laboratory and field conditions with LUZHONG 99118 maize seeds (selected by Shandong Province Seeds Group Co., LTD). This variety is food and fodder dual-purpose maize. There still have 12 to 13 green leaves when maturing, suit to be used as silage.

2.3 Experimental Design

2.3.1 Test Media

Laboratory plasmas can be generated by supplying energy to a kind of neutral gas. Electrical discharges are the most common for generating non-thermal plasmas. Neon is colorless and odorless inert gases. Neon conducts electricity 75 times better than air. The test result indicates a uniform stable glow discharge has been obtained between two planar electrodes when using neon as test media. Therefore, neon was used as media in our experiment.

2.3.2 LTP Treatment

Healthy and uniform maize seeds were treated with various plasma strengths (0, 60, 80, 100, 120, 140, 160, 180, 200W). 1 kilogram of maize seeds was used in each treatment.

2.3.3 Germination Tests and Analysis of the Germination Rate, Vigor

The germination tests were started on the 16th, 30th, 45th, 60th, 75th, 90th, 120th, 150th days after seeds treated by LTP. Each treatment was repeated four times to confirm the repeatability of the results, with 100 seeds for every repetition.

After treatment, the seeds were allowed to germinate in sand bed under laboratory. During the experiments, water was added to the sand bed to guarantee sufficient moisture for germination. The incubation temperature was 25°C.

A seed was considered germinated if both the radicle and coleoptile length were ≥2 mm. Seedlings were counted at 24-h intervals up to 7 days. The germination vigor is used to assess the quality of field emergence and is defined as the percentage of seeds that germinate within a short time (4 days after lined for maize according to GB/T 3543.4-1995, Rules for Agricultural Seed Testing-germination test). The germination rate is defined as the percentage of seeds that germinate in a specified time (7 days after lined for maize according to GB/T 3543.4-1995). These are calculated as follows:

Vigor of germination (%) = (Number of seeds germinated in 4 days/total number of seeds) × 100

Rate of germination (%) = (Number of seeds germinated in 10 days/total number of seeds) × 100

2.3.4 Investigation on Biological Characters of Maize Seedlings in Laboratory

On the 7th day of germination experiment, ten seedlings of each repetition were randomly selected and the length of shoot & root, green & dry weight of these seedlings were investigated. Length of shoot is defined as the length between the mesocotyl and tip of seedlings. Length of root is defined as the length between the mesocotyl and the root tip.

Samples used in this experiment were kept in a cool dry place for 90 days after treated by LTP. Also, both the green and dry weight of seedlings were measured.

2.3.5 Investigation on Biological Characters of Maize Seedlings in Field

In 90 days after treated by LTP, seeds were sowed in field. Ten days later. Ten seedlings of each repetition were randomly selected and the length of shoot & root, fiber root number were investigated.

3 Results and Discussion

This experiment was conducted to assess the influence of different intensities of LTP on seed germination and early growth, also the aging effect of this influence was investigated. The data of germination vigor for maize seeds are listed in Table 1, and the data of germination rate are listed in Table 2.

The optimum intensity of samples in different times after LTP treated is fugitive. But for each samples, we can both have the maximal vigor and rate in one certain intensity.

The effect of LTP on vigor and rate of maize seeds germination changed with time. LTP played an active role in seed germination within 16 days. And it dropped the seed germination rate and vigor during 30-120 days, 150 days after treated, and this inhibition had been weak.

Table 3 showed the shoot and root length of seedlings under lab condition, and the shoot & root length and fiber root number of seedlings in field were lined in table 4. The green and dry weight of seedlings in field were listed in table 5.

Table 1. Vigor of germination of forge maize seeds stimulated by low-temperature plasma (%)

Intensity of LTP (W) / Days after treatment	0 (ck)	60	80	100	120	140	160	180	200	
16	81b	88a	89a	91a	91a	91a	88a	88a	87a	87.83ab
30	82d	86bcd	86bc	87bc	90abc	90abc	92a	86cd	91ab	88.06ab
45	81b	91a	89a	86a	89a	88a	92a	90a	87a	88.17a
60	81b	91a	89a	89a	86a	86a	87a	89a	87a	87.22ab
75	81d	88abc	91ab	92a	86bc	86c	85cd	84cd	87bc	86.44b
90	80c	89a	86ab	86ab	91a	88a	88a	83bc	88a	86.44b
120	75d	80cd	81bc	84ab	86a	80bc	86a	81bc	78cd	80.33c
150	74cd	76abc	75bcd	79a	81a	80a	79ab	81a	71d	77.11d
	78.94e	85.5bcd	85.75abcd	87ab	87.38a	86.13abc	87.06ab	84.81cd	84.25d	/

Note: Columns with different lowercase letters: Low-temperature plasma treatment on Alexander Summer Squash bud potential, germination rate, significant differences in germination index ($P \leq 0.05$)

Table 2. Rate of germination of forge maize seeds stimulated by low-temperature plasma (%)

Intensity of LTP (W) / Days after treatment	0(ck)	60	80	100	120	140	160	180	200	
16	93c	94bc	94bc	96ab	96a	94bc	95ab	96a	93c	94.22a
30	93ab	92b	93ab	92b	94ab	92b	95a	94ab	92b	92.72bc
45	93bcd	93bcd	94abc	91d	95a	93bcd	94ab	92cd	92cd	92.79b
60	93ab	93ab	92abc	91bc	90c	93ab	93ab	94a	91bc	92.18cd
75	93ab	94a	94a	91b	93a	94a	93ab	92ab	92ab	92.83b
90	93abc	92bc	92abc	93ab	93abc	91c	92bc	93ab	93a	92.33bc
120	94a	92abc	93a	93ab	9a1bc	91abc	92ab	92abc	90bc	92.11cd
150	92abc	94ab	92abc	92bc	91c	91c	94ab	94a	91c	92.06cd
	92.56ab	92.68ab	92.87ab	92.75ab	92.69ab	92.13bc	93.25a	93.25a	91.63c	/

Note: Columns with different lowercase letters: Low-temperature plasma treatment on Alexander Summer Squash bud potential, germination rate, significant differences in germination index ($P \leq 0.05$)

Table 3. Length of shoot and root of seedlings under laboratory condition stimulated by low-temperature plasma (%)

Intensity of LTP (W)	0(CK)	60	80	100	120	140	160	180	200
Length of shoot (cm)	10.99cd	11.52bc	12.04ab	12.13ab	12.3a	11.50bc	11.65bc	11.36cd	10.74d
Length of root (cm)	19.11bc	19.55abc	20.82a	20.14ab	19.33bc	19.12bc	19.40bc	18.98bc	18.19c

Note: Columns with different lowercase letters: Low-temperature plasma treatment on Alexander Summer Squash bud potential, germination rate, significant differences in germination index ($P \leq 0.05$)

Table 4. Length of shoot and root and fiber root number of seedlings under field condition stimulated by low-temperature plasma (%)

Intensity of LTP (W)	0(CK)	60	80	100	120	140	160	180	200
Length of shoot (cm)	10.21cd	10.14d	10.33cd	11.42abc	12.04a	11.95ab	11.62abc	10.75bcd	11.03abcd
Length of root (cm)	31.04a	36.55a	35.53a	34.57a	33.05a	33.04a	36.07a	36.9a	35.01a
Fiber root number	4.4ab	4b	4.6ab	4.75ab	5ab	5.2a	4.6ab	4.6ab	4.5ab

Note: Columns with different lowercase letters: Low-temperature plasma treatment on Alexander Summer Squash bud potential, germination rate, significant differences in germination index ($P \leq 0.05$)

Table 5. Green and dry weight of seedlings stimulated by low-temperature plasma (%)

Intensity of LTP (W)	0	60	80	100	120	140	160	180	200
Green weight (g)	83.04cd	81.83d	85.77cd	91.62ab	95.10a	95.70a	87.96bc	83.47cd	83.46cd
Dry weight (g)	12.52bc	12.44bc	12.21cd	12.69bc	13.00a	12.75ab	12.55bc	12.76ab	12.45bc

Note: Columns with different lowercase letters: Low-temperature plasma treatment on Alexander Summer Squash bud potential, germination rate, significant differences in germination index ($P \leq 0.05$)

90 days after seeds treated by LTP, almost all the intensities (60 -200W) could increase the shoot & root length, green & dry weight and the fiber root number also increased.

It is clear that in the case of an ordinary plant leaf area will increase as growth proceeding, and with increasing leaf area the rate of production of material by assimilation will also increase, this will also lead to higher yield. LTP treatment has a great value of using and spreading on crop production, and the optimal treatment intensity for forage maize seeds was 120W.

Acknowledgment. Financial support from National High Technology Research and Development Program 863(2012AA10A505) and China Agriculture Research System (CARS-35) is gratefully acknowledged.

References

1. Hao, X., Qin, J.: Preliminary study on low temperature plasma treatment of seed. Journal of Shanxi Agricultural Sciences 26(2), 39–41 (1998) (in Chinese)
2. Li, R.: Plasma machine seed treatment technology. North Rice 4(4), 52–53 (2010) (in Chinese)
3. Yin, M.: Research of magnetized arc plasma on seeds biological effects. Dalian University of Technology, Dalian (2006) (in Chinese)
4. Zhang, Y., Zhang, J., Wang, Q.: The application of physical methods in sugar beet seed treatment. China Beet and Sugar (2), 20–22 (2005) (in Chinese)
5. Xu, Z., Chen, B., Wei, Z.: Various seed treatments on corn yield. Agricultural science & technology and Equipment (4), 15–16 (2011) (in Chinese)
6. Liu, S., Ouyang, X., Nie, R.: Application status and development trend of the physical methods in the crop seed treatment. Crop Research 5(2), 520–524 (2007) (in Chinese)
7. Shao, C., Wang, D., Tang, X., et al.: Arc plasma system and its application & development treads on pre-sowing seeds treatment. China Seed Industry (8), 1–3 (2012) (in Chinese)
8. Lianglong, H., Lijia, T., Zhichao, H., et al.: Application of physical agriculture techniques in cleaned seeds treatments. Journal of Anhui agricultural science 35(13), 3778- 3779 (2007) (in Chinese)
9. Yin, M., Huang, M.: Stimulatnig Effects of Seed Treament by Magnetized Plasma on Tomato Growth and Yield. Plasma Science & Technology 7(6), 3143–3147 (2005)
10. Filatova, I., Azharonok, V., Kadyrov, M.: Rf and Microvawe Plasma Application for Pre-Sowing Caryopsis Treatments. Publ. Astron. Obs. Belgrade (89), 289–292 (2010)
11. Shao, C., Fang, X., Tang, X., et al.: Stimulating effects of low-temperature plasma on seed germination characteristics of green Chinese onion. Transaction of the Chinese Society for Agriculture Machinery 06 (2013)
12. Fu, S., Zhang, F., Li, J., et al.: Several physical techniques in agriculture and Prospects. Agricultural Mechanization Research (1), 36–38 (2006) (in Chinese)
13. Shi, Y., Fang, X., Xu, D., et al. Effect of plasma treatment of soybean seed with different radiation intensity on biological traits, yield and economic output. Journal of Jilin Agricultural Sciences 35(6), 6-7 (2010) (in Chinese)
14. Luo, H., Ran, J., Wang, X., et al.: Comparision Study of Dielectric Barrier Discharge in Inert Gases at Atmospheric Pressure. High Voltage Engineering 38(5), 1070–1077 (2012)

15. Zeng, Q., Liang, C., Shen, D., et al.: Effects of Acid Rain on Seed Germination of Various Acid Fastness Plants. Journal of Agro-Environment Science 23(1), 39–42 (2004)
16. Niu, D., Sun, Z., Yu, F., et al.: Primary report on effects of plasma on maize. Journal of Farm and Cuiture 1, 43–44 (2011)
17. Yu, F., Zhang, R., Zhang, P., et al.: Effects of plasma on maize. Agricultural Technology &Equipment 148(4), 29–31 (2008)
18. Chai, S., Fang, X., Fu, X., et al.: Effects of plasma on maize. Modern agricultural science (23), 138–138 (2007)
19. Fa, X., Bian, S., Xu, K., et al.: Study on Maize Seeds Treated with Plasma to Influence Biological Properties and Yield of Maize. Journal of Maize Sciences 12(4), 60–61 (2004)
20. Du, H., Liu, S., Zhao, W., et al.: Effects of plasma on old maize. China Seed Industry (supplement), 22–23 (2011)

Comparison of Methods for Forecasting Yellow Rust in Winter Wheat at Regional Scale

Chenwei Nie, Lin Yuan, Xiaodong Yang, Liguang Wei, Guijun Yang, and Jingcheng Zhang*

Beijing Research Center for Information Technology in Agriculture, Beijing 100097, China
{nie_chenwei,byl6690,weiliguang}@126.com,
{yangxd,yanggj}@nercita.org.cn, zhangjc_rs@163.com

Abstract. Yellow rust (YR) is one of the most destructive diseases of wheat. To prevent the prevalence of the disease more effectively, it is important to forecast it at an early stage. To date, most disease forecasting models were developed based on meteorological data at a specific site with a long-term record. Such models allow only local disease prediction, yet have a problem to be extended to a broader region. However, given the YR usually occurs in a vast area, it is necessary to develop a large-scale disease forecasting model for prevention. To answer this call, in this study, based on several disease sensitive meteorological factors, we attempted to use Bayesian network (BNT), BP neural network (BP), support vector machine (SVM), and fisher liner discriminant analysis (FLDA) to develop YR forecasting models. Within Gansu Province, an important disease epidemic region in China, a time series field survey data that collected on multiple years (2010-2012) were used to conduct effective calibration and validation for the model. The results showed that most methods are able to produce reasonable estimations except FLDA. In addition, the temporal dispersal process of YR can be successfully delineated by BNT, BP and SVM. The three methods of BNT, BP and SVM are of great potential in development of disease forecasting model at a regional scale. In future, to further improve the model performance in disease forecasting, it is important to include additional biological and geographical information that are important for disease spread in the model development.

Keywords: Yellow rust, Disease forecast, Bayesian network (BNT), BP neural network (BP), support vector machine (SVM), fisher liner discriminant analysis (FLDA).

1 Introduction

Yellow rust (YR), caused by Puccnia striiformisWestend f.sp.tritici Eriks, is one of the most important epidemic diseases of wheat. It can cause significant loss on wheat at a global scale [1, 2]. It is of great importance to predict the YR effectively at an early stage, since it can provide critical information to agriculture plant protection

* Corresponding author.

© IFIP International Federation for Information Processing 2015
D. Li and Y. Chen (Eds.): CCTA 2014, IFIP AICT 452, pp. 444–451, 2015.
DOI: 10.1007/978-3-319-19620-6_50

departments to facilitate timely spray recommendation. So far, a series of studies have been conducted to forecast YR over a long time based on meteorological and agronomy data around the world. Hu et al (2000) constructed a BP model to predict YR in Hanzhong city, Shaanxi Province. The forecast results were highly consistent with actual situation of disease occurrence [3]. Chen et al (2006) predicted YR severities at a seasonal time step in both Maerkang county and Tianshui city using a discriminant analysis, with rewind accuracy and cross-validation accuracy greater than 78% [4]. Coakley et al (2006) developed an improved method to predict YR severity [5]. Wang et al (2012) conducted a study to develop a stable neutral network for predicting YR prevalence degree [6].

To date, it should be noted that there were few attempts made in regular YR forecasting (time step = 7 days) at a regional scale. Instead, efforts were made on forecasting seasonal severities of YR which rely on spores counting data and meteorological observations. Those models can achieve high accuracy at a local site, whereas it is difficult to apply those models in vast areas where the spores counting data are not available. In addition, given the distribution of the YR pathogen over large area is driven by oversummer and overwinter process at a regional scale which is closely associated with weather conditions, it is thereby necessary to develop a model that can be applied at a large spatial scale. However, such forecasting models are lacking recently.

Several critical weather factors associated with the occurrence of YR on winter wheat had been reported, they are air temperature, humidity, precipitation and sunshine duration [7]. It is important to relate YR occurrence with meteorological factors in the development of YR forecasting model. Several machine-learning techniques have been widely used for classification purpose since the intelligent learning feature allowing them to take advantages from large amount of data [8]. In this study, the Gansu province, which is an important YR prevalence region in China, was selected as our study area. Based on continuous YR field survey data over multiple years (2010-2012) and corresponding meteorological data, the potential of BNT, BP, SVM and FLDA in disease forecasting were examined and compared. The YR forecasting model was thus established to facilitate regular disease management at a regional scale.

2 Materials and Methods

2.1 Yellow Rust Survey Data

The YR survey data is collected by Gansu Provincial Protection Station. During 2010 to 2012, a weekly field surveys were conducted across southern area of Gansu province (Fig.1). In detail, the surveyed data include the initial date of disease occurrence, the prevalence status and the infected area. The climate of the study region is characterized by high humidity and rainfall, and YR disease occurs almost every year. A total of 22, 9, 30 counties were surveyed in 2010, 2011 and 2012, respectively. The distribution of surveyed counties is demonstrated in Fig.1. The investigation ranged from the beginning of March to the end of July in each year. There were 16 weeks field survey data totally in each year. For model calibration and validation, the surveyed data were randomly split into 60% versus 40% in each year.

In this study, prevalence status of YR disease in the week before forecasting day was chosen as one of the input variables and the occurrence status of YR disease during the week next the forecasting day was regard as the forecasting target. The occurrence status of YR disease was divided into four classes represented by D1, D2, D3 and D4, respectively. D1 meant there was no YR disease was found in the county, D2 indicate the YR disease was firstly found since the survey was conducted in the county, D3 indicate that the YR disease has been found in the county, but there was no development compared with last week, and D4 indicate the YR disease has been found before and it developed during the week after forecast day.

2.2 Meteorological Data

In this study, according to the research results of Cooke(2006) and Li & Zeng(2002)[2,9], four types of meteorological factors were chosen as the primary data, include air temperature(maximum air temperature, minimum air temperature and average air temperature), average humidity, precipitation and sunshine duration. The daily data of these meteorological factors from a total of 27 weather stations around the study area was acquired from Chinese Meteorological Data Sharing Service System. The time range of the data is from a week before YR occurrence (based on the investigation data) in spring to its mature stage in each year. There are 4 steps to process meteorological data, including removal of abnormal value, computing meteorological factors, choosing meteorological factors and interpolation of each factor to a resolution of 250m*250m. Considering some meteorological data have a strong relationship with altitude, the DEM (Digital Elevation Model) data was used the adjust the spatial maps of meteorological factors by interpolating the fitted residue across the region [10,11].

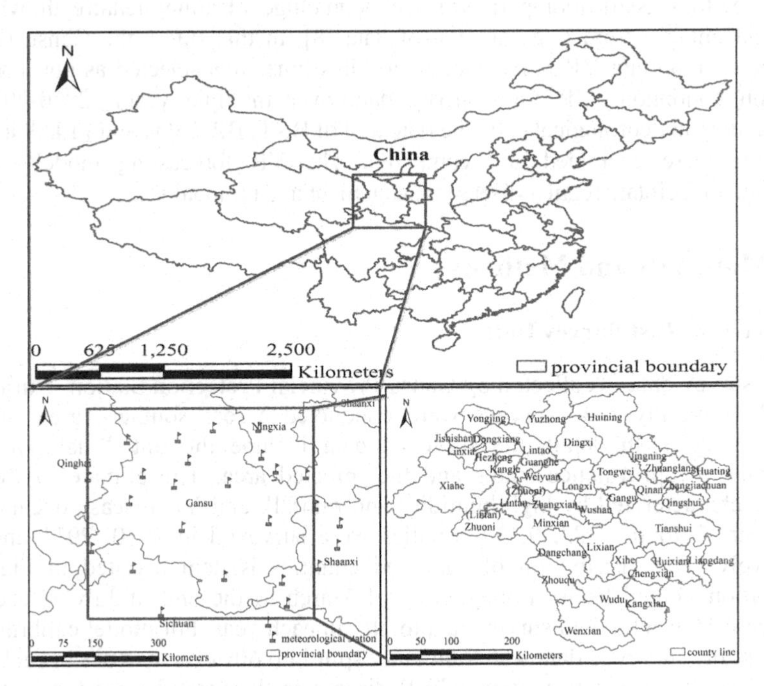

Fig. 1. Study area and the distribution of meteorological stations

Table 1. Meteorological factors used for predicting

Meteorological factors	Time range
Average minimum air temperature (aveminT)	From 7th day to 14th day before forecasting day
Average maximum air temperature(avemaxT)	From 7th day to 14th day before forecasting day
Days of Precipitation more than 0.25mm(Pdays)	From 7th day to 14th day before forecasting day
Average precipitation(aveP)	From march to the 7th day before forecasting day
Days of humidity lower than 50%(Hdays)	From 7th day to 14th day before forecasting day
Minimum of average humidity(minH)	From 7th day to 14th day before forecasting day
Average sunshine duration(aveS)	From march to the 7th day before forecasting day

As for select methods, the variance analysis between meteorological factors and disease prevalence classes and the correlation analysis among the same class meteorological factors were conducted by Tukey-Kramer method and Pearson method, respectively. For those meteorological factors have a *p-value*<0.05 were selected primarily, then if the R > 0.8 between two factors, the factor has a smaller *p-value* was chosen. And the meteorological factors were last chosen as shown in table 1. As for interpolation methods, the normality of the distribution of each meteorological factor was examined by Kolmogorov-Smirnov method. For those meteorological factors have a *p-value*<0.05, a kriging method is used to conduct interpolation. Otherwise, an inverse distance weighted method is adopted. Each meteorological factor was calculated on a county scale after the interpolation.

2.3 Methods

To find an appropriate method for predicting YR disease, in this study, four YR disease forecasting models were established with four classical methods, respectively, include BNT, BP, SVM and FLDA. The characters of all 4 methods are shown in table 2[12]:

Table 2. Characters of each method

Methods	Description
Bayesian network(BNT)	Based on traditional statistical theory. Has been used effectively to model those problems with characters uncertainty and non-linearity by incorporating prior knowledge extracted from selected sample datasets.
BP neural network(BP)	Based on traditional statistical theory. Has strongly adaptive and learning capability. Has been effectively to model those objects with characters uncertainty and non-linearity.
Support vector machine(SVM)	Based on statistical learning theory, has been widely introduced in disease recognition. Has been used effectively to model those problems with characters small sample, uncertainty and non-linearity.
Fisher linear discriminant analysis(FLDA)	An important branch of Statistical pattern recognition. Has been widely used in pattern recognition field, since it is a simple, rapid and efficient method.

BNT is a Directed Acyclic Graph(DAG), the nodes in the DAG structure representing domain variables, and the arcs between nodes represent probabilistic dependencies. In this research, the nodes were used to represent meteorological factors, prevalence status in last week before the forecasting day and the occurrence status during the week next the forecasting day. Each meteorological factor was graded according to the clinical characters of YR disease before the BNT model was established. Next, a BNT structure was constructed with the MCMC method and corresponding expertise. Then, the parameters were optimized using maximum likelihood estimation against the calibration data set. According to the chain rule of probability, the probability of an event occurs is calculated as:

$$p(x) = \prod_{i=1}^{9} p(x_i|x_1, \cdots, x_{i-1}) \tag{1}$$

One-hidden-layers BP forecasting model was constructed in this study with training function *trainlm* and adaption learning function *learngdm* under MATLAB environment. *Tansig* was used as the transfer function of both hidden layer and output layer. The structure of BP that is used in this study is shown in Fig.2. As a binary output, the y_i will be marked as '1' if it gets the maximum, which means the occurrence status of YR disease is D*i*. Otherwise, the y_i will be marked as '0', which means the occurrence status of YR disease is not D*i*.

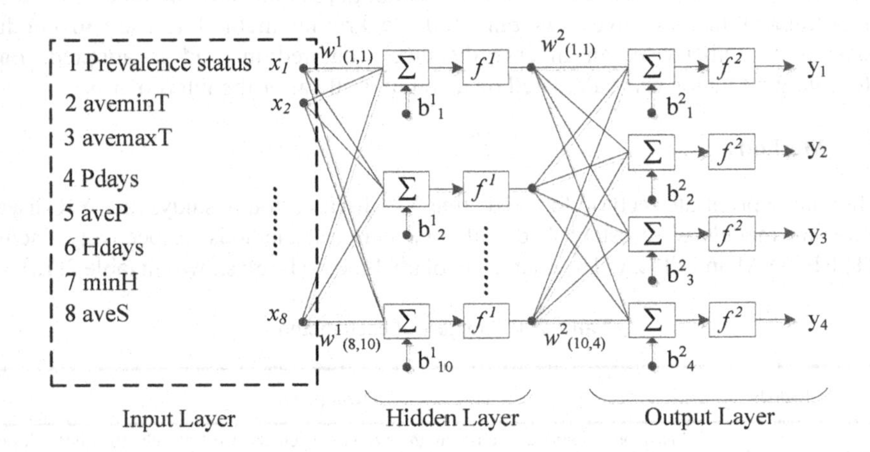

Fig. 2. Feed forward back propagation neural network

Note: **y**=Output vector, **X**=Input vector, f^1,f^2=Transfer functions on hidden layer and Output layer, b^1, b^2=bias on hidden layer and Output layer, w^1, w^2=weights in Input layer and hidden layer.

The output of the BP model is calculated as:

$$\mathbf{y} = f^2(w^2 f^1(w^1\mathbf{X} + b^1) + b^2) \tag{2}$$

In the third part of the experiment design, a SVM model was constructed under MATLAB environment with help of Libsvm tool and SVM_GUI tool. The RBF function was used as a kernel function. Grid search method was used as the parameters optimizing algorithm to search for the best penalty coefficient *c* and *gamma* with a step of 0.5. Both *c* and *gamma* have a range of [-8,8]. The structure of SVM model in this study is shown in Fig.3:

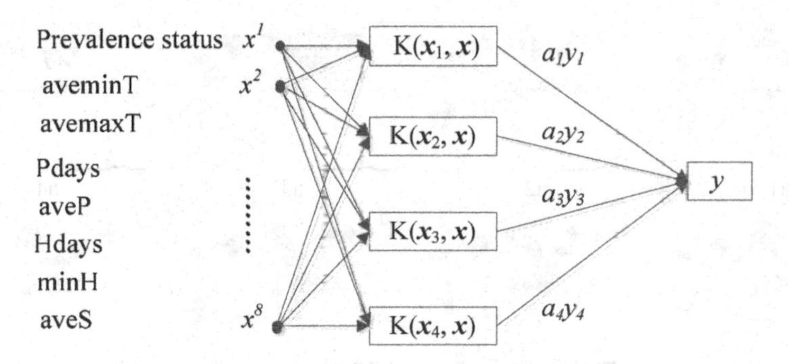

Fig. 3. Support vector machine design

Note: y is the output, $K(\boldsymbol{x}_i, \boldsymbol{x})$ is nonlinear transformation(inner product computation) function, $\boldsymbol{x}_i$ is support vector, $\boldsymbol{x} = (x^1, x^2, \ldots, x^8)$ is Input vector, a_i is Lagrangian coefficient.

The output of SVM model is calculated as:

$$y = \mathrm{sgn}(\textstyle\sum_{i=1}^{4} a_i y_i K(\boldsymbol{x}_i, \boldsymbol{x}) + b) \tag{3}$$

For FLDA forecasting model, likelihood ratio was used to assign observations to groups. It was run under MATLAB environment.

2.4 Evaluation of Disease Forecast Models

For the all 4 trained examples of the classifiers, a separate validation data set was used to evaluate the model accuracy. Different from the BP, SVM and FLDA, which classify a sample into several infection status directly, the BNT forecasting model produced result in probability, which would be converted to infection status by applying a certain threshold. A sample will be marked as D_i when the occurrence status D_i gets the maximal probability. The performances of four different forecasting models were evaluated by overall accuracy as comparing the forecasting results against validation data.

3 Results and Discussion

Table 3 summarized the forecasting results of BNT, BP, SVM and FLDA versus the validation data set. The results suggested that BNT, BP and SVM produced more accurate forecasts than FLDA in general. The SVM, BNT and BP produced approximately similar accuracies with SVM model having relatively high accuracy. The actual occurrence (Fig.4 (a)) and forecasting results (Fig.4 (b-e)) of disease distribution pattern of the 4 methods were demonstrated in Fig.4. As shown in Fig.4(a), the actual occurrence, there were only two counties infected by the disease in 9th week, however, the all 9 validation counties were infected in 13th week. It is obvious that the infection patterns estimated by BNT, BP and SVM are highly consistent with field survey records across multiple dates, with BNT outperformed BP and SVM.

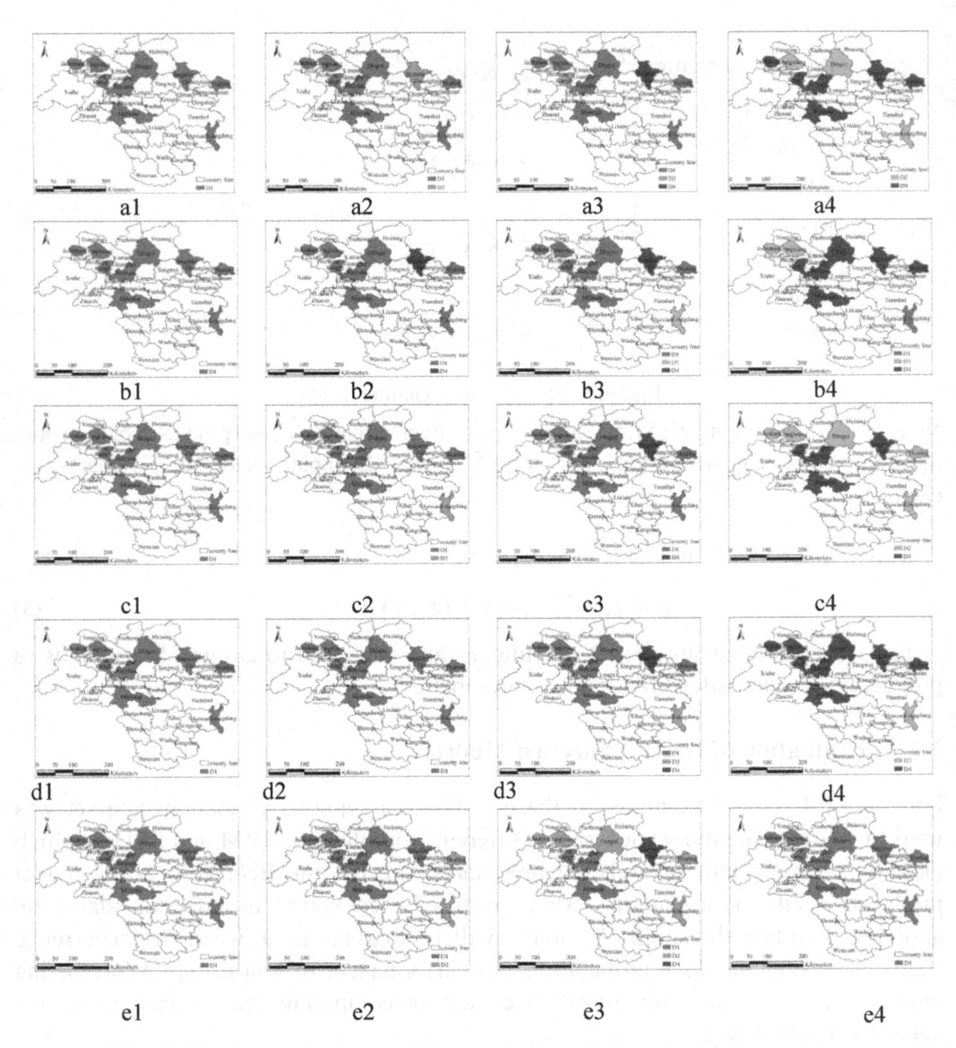

Fig. 4. The forecasting result of four methods and the true occurrence status

Note: a indicates the true occurrence status, b indicates the result of BNT, c indicates the result of BP network, d indicates the result of SVM, e indicates the result of FLDA. The number 1 indicates the first week, 2 indicates the fifth week, 3 indicates the ninth week and 4 indicates the thirteenth week.

Table 3. accuracy indices of tested methods

Methods	OAA	Kappa
BNT	82.29	0.73
BP	81.25	0.71
SVM	85.68	0.78
FLDA	68.75	0.56

To predict the YR disease more precisely, more information should be chosen as inputs of the forecasting model, since the YR disease is influenced by various factors, including meteorological conditions, growth vigor of wheat and number of fungus. As the remote sensing method can estimate the crop growth accurately and quickly in a large space scale, it would be used to forecast the YR disease integrating with meteorological information and fungus information in our future work.

4 Conclusions

A total of four methods including BNT, BP, SVM and FLDA were examined and compared in developing a forecasting model of yellow rust disease across vast area in this study. The performances of these models were evaluated against a weekly survey data during wheat's key growing stages from 2010 to 2012. The results confirmed that the disease forecasted results are able to reflect the spatio-temporal development and distribution pattern of YR except for FLDA. Further, a superior performance of BNT, BP and SVM also demonstrated that these nonlinearity methods are of great potential in forecasting yellow rust infection at a regional scale in weekly time step.

Acknowledgment. This work was supported in part by the Natural Science Foundations of China (Grant No. 41101395), Natural Science Foundations of Beijing (Grant No. 4122032) and Prior Sci-Tech Program for Scientific Activity of Overseas staff.

References

1. Shimai, Z.: Macro-phytopathology. Agriculture Press of China, Beijing (2005) (in Chinese)
2. Cooke, B.M., David, G.J., Bernard, K., et al.: The epidemiology of plant diseases. Springer, Dordrecht (2006)
3. Lei, Y., Shuqin, L.: Prediction of wheat stripe rust by wavelet neural network. Microcomputer Information 25(12-2), 42–43 (2009)
4. Chen, G., Wang, H., Ma, Z.: Forecasting wheat stripe rust by discrimination analysis. Plant Protection 32(4), 24–27 (2006)
5. Coakley, S.M., Line, R.F., McDaniel, L.R.: Predicting stripe rust severity on winter wheat using an improved method for analyzing meteorological and rust data. Phytopathology 78(5), 543–550 (1988)
6. Wang, H., Ma, Z.: Prediction of wheat stripe rust based on neural networks. In: Li, D., Chen, Y. (eds.) CCTA 2011, Part II. IFIP AICT, vol. 369, pp. 504–515. Springer, Heidelberg (2012)
7. Te Beest, D.E., Paveley, N.D., Shaw, M.W., et al.: Disease-weather relationships for powdery mildew and yellow rust on winter wheat. Phytopathology 98(5), 609–617 (2008)
8. Khan, M.N.A.: Performance analysis of Bayesian networks and neural networks in classification of file system activities. Computers & Security 31(4), 391–401 (2012)
9. Li, Z., Zeng, S.: Wheat rust in china. Agriculture Press of China, Beijing (2002)
10. Pan, Y.Z., Gong, D.Y., Deng, L., Li, J., et al.: Smart distance searching-based and DEM-informed interpolation of surface air temperature in China. Acta Geographica Sinaca 59(3), 366–374 (2004)
11. Wang, Z., Shi, Q.D., Chang, S.L., et al.: Study on spatial interpolation method of mean air temperature in Xinjiang. Plateau Meteorology 31(1), 201–208 (2012)
12. Bian, Z.Q.: Pattern Recognition. Tsinghua University Press, Beijing (2007)

The Standard of Data Quality Control Technology Based on the Share of Rural Science and Technology Data

Dan Wang, Xiaorong Yang, Jian Ma, and Yang Sun

Institute of Agricultural Information,
Chinese Academy of Agricultural Sciences, Beijing 100081, China
Key Laboratory of Agricultural Information Service Technology (2006-2010),
Ministry of Agriculture, The People's Republic of China
{wangdan01,yangxiaorong,majian,sunyang}@caas.cn

Abstract. The standard of data quality control technology based on the share of rural science and technology data is one of the important standards to share network information source. In the paper, data quality control system is presented. The technical specification of data quality control in the data collection, data input, subject indexing, data storage construction, data description and data unit was prescribed, too. At last, it produces workflow and estimate index of data quality control.

Keywords: Rural science and technology data, data quality control, technical specification, data sharing.

1 Introduction

The construction of rural science and technology data sharing platform (referred to as a Shared Platform) is an important part of agricultural informatization, which is an effective way to solve rural technology "islands of information" and "last mile" problem and the service to "agriculture, rural areas and farmers". "Shared Platform" has a variety of data types and a huge amount of data, creating a standard of data quality control technology based on the share of rural science and technology data to effectively control the quality of the data is very necessary in the process of data collection, processing and retrieval using.

2 "Shared Platform" Data Quality Control System

"Shared platform" data quality control system includes two parts of data quality control management specifications and technical specifications.

Data quality control management department is composed by the quality control system administration and quality control technology group. The former is mainly responsible for organization and management of quality control standards. The latter is mainly responsible for the technical specifications of quality control of data. "Quality control technical specifications" is the main indicator and basis for controlling rural science data quality, its main contents are as follows:

© IFIP International Federation for Information Processing 2015
D. Li and Y. Chen (Eds.): CCTA 2014, IFIP AICT 452, pp. 452–459, 2015.
DOI: 10.1007/978-3-319-19620-6_51

2.1 Quality Control of Data Collection

In the data collection process data quality control should be noticed from a few aspects about the scientificity and practicability and use effect of the data system structure and data content, and should follow the following technical specifications:

1) Numbers and text collected must be accurate, the accuracy of data must reach 99.5% ;
2) Document data must be complete, detailed and accurate ;
3) Numerical data are accurate, standardized, uniform units of measurement, and facilitate statistical analysis ;
4) Data quality of graphics and image includes a measurement error (error of the system, operator error, accidental error) Mapping errors, digital error, editing error of correction and analysis, data conversion error is less than 0.5%.

2.2 Quality Control of Data Entry

Comply with national standards of existing data entry. For non-standard input, professional department is responsible for data entry rules. For data sources involved by data of multi-sector should be unified using a sector's data. In the data entry process we should timely inspect the quality of data entry, at the same time, also should take the following quality assurance measures:

1) Entry personnel qualification: Input rate of 100 words per minute or so, bit error rate is less than five out of 10,000; Strictly obey the input rules and operational procedures;
2) In accordance with the requirements of entry indicators, if the program can be used to control data entry, entry control procedures must be used;
3) Strengthen proofreading, improve the quality of data entry (manual re-recorded proof method, machine proofreading method);
4) Select to ensure the accuracy of the input device and related information carrier;
5) There is a perfect data management and quality inspection system;
6) Data on the scan input should check and conform to the degree of the real data, requiring graphics and images are clear, identification accuracy is above 99.5%.

2.3 Data Classification and Indexing Quality Control

Classification indexing is a method of classification knowledge base on level enumerating method. It can more fully reflect the whole knowledge and its inner logical relationship. It is systematic, family of knowledge retrieval ability and enlarge/shrinkage function. "Shared platform" data classification indexing quality control to follow these guidelines:

1) Since rural science and technology data resource type is various, content is wide, there is a variety of classification indexing scheme, but must provide a classification scheme choice, that is as "rural science and technology data classification and code" technical specification as the basis of classification indexing;
2) Categories and classification codes of indexing object are mandatory fields, must be marked with the contents (values), allows duplicate content to appears (multi-value) with duplicate content between the":"character.

2.4 Subject Indexing of Data Quality Control

To ensure the effect of "Shared platform" data retrieval, data subject indexing is indispensable, the following subject indexing specification is adopted:

2.4.1 Subject Indexing Method

1) Subject indexing tool is selected. For example, the"Chinese Library Classification","Agricultural Professional Classification","Thesaurus of Agricultural Sciences"are as the word choice basis of subject indexing;
2) Strict enforcement word choice process of indexing work, word choice must be standard and accurate ;
3) Fully considering overall and specificity of theme analysis, the maximum to meet the requirements of precision and recall ;
4) Subject factors (the research object, material, method, process and conditions), common factor, location factor, time factor and data types constitute subject factors, we carry out the subject indexing from the above 5 factors;
5) Data of graphics, images and charts are indexed keywords which are connected in order to retrieving. Keyword choice must be standardized.

2.4.2 Subject Indexing Rules

1) Objectively reflect the information content, avoid introducing indexing personnel's personal views;
2) Keywords selected come from the thesaurus. Synonym keywords need to be converted into formal synonyms. Informal word cannot be used as indexing words;
3) When there is no specific corresponding keywords, you should choose the most directly related to a few keywords equipping indexing;
4) If the combination of indexing are still unable to meet the requirements, you should select the most direct hypernym Indexing;
5) If hypernym still not appropriate, you can use keywords indexing (free words);
6) Subject indexing depth is generally seven keywords (mean value), general indexing word are 5-10 keywords.

2.5 Subject Indexing of Data Quality Control

2.5.1 The Data Dictionary

To facilitate data sharing, improve the efficiency of data use and development, reduce development costs, “Sharing Platform” need to build the data dictionary of rural science and technology to regulate data storage structure. Data dictionary should include the following:

Data Item Name	Explanation
Data System Name	Name of data systems, such as rural science and technology data sharing system
The database name	The name of the database file
Data Name	The name of the data item
Data store name	The name of the data field
The data type	The type of data, such as digital type, character type, date type, etc.
The length of the data storage	Data stored in the computer space represented by byte
Unit	Measurement Units of data (unit of measurement)
Code description	the use of code system and coding rule
Precision	Effective number of minimum digit position
The lower limit of data	The reasonable lower limit of data
The upper limit of data	The reasonable upper limit of data
Access to the means of data	Data measurement methods or reference sources
Time and/or the environment	The time of get data and / or the environment
Remark	To add other instructions

2.5.2 Data Structure Specification

“Sharing platform” of agricultural science and technology data resource involves science and technology data of agriculture, forestry, water conservancy, meteorology and other fields. Data types have papers, books, journals, meetings, news, newspapers, patent information, policies and regulations, standards, non-book materials, etc. Data type is very complex, you should follow the following technical specifications in the design of the database’s data structure:

1) Follow the international and domestic existing metadata standards ; Or follow the Dublin core element set ; Or follow the MARC standard (USMARC, CNMARC) ; Or follow the "Geographic Information - Metadata" standard, (ISO 19115:2003, MOD, draft) ; Follow the “Metadata Specification of Rural Science Data Sharing Platform” based on the above criteria.

2) In the database of "Shared platform" the required data item is defined as: title, author, subject category, information source code, keywords, description, origin, date and ID number.
3) Data structure is designed to meet the requirements of rural science and technology data sharing system. Date structure has been recognized by experts in the relevant databases. Avoid arbitrariness of database structure definition.
4) Name of the data item must be standardized, meaning must be accurate.

2.6 The Technical Specification of Data Description

GB3793-83 in "Rural science and technology data resource sharing platform" is one of the cataloging rules must be obeyed.

2.6.1 Data Description of Graphics and Tables

There are two rules of the cataloging rules of graphics and bibliographic data:

1) For powerful database systems such as Oracle, you can put graphical information, form data directly into the field.
2) For less powerful database systems such as SQL Server, you can use the link technology, through the fields of subject (key words) associated with graphics and tables.

2.6.2 Data Record Consistency

1) Information source name, organization name should be consistent ;
2) Units of measurement should be consistent;
3) Description of special characters should be consistent;
4) Description of numbers and dates should be consistent

2.6.3 Data Record Consistency

"Sharing Platform" obey GB / T 17295-1998 international trade measurement unit code standards. Units of measurement appearing in Rural Science database must be strictly enforced this standard. Numeric database must have units of measurement instructions.

3 Data Quality Control as a Whole

3.1 Data Quality Control Process

Data quality control process is divided into self-examination, the quality control of a professional group and the overall group acceptance three stages. Each stage has its corresponding inspection standards, quality control flow chart below:

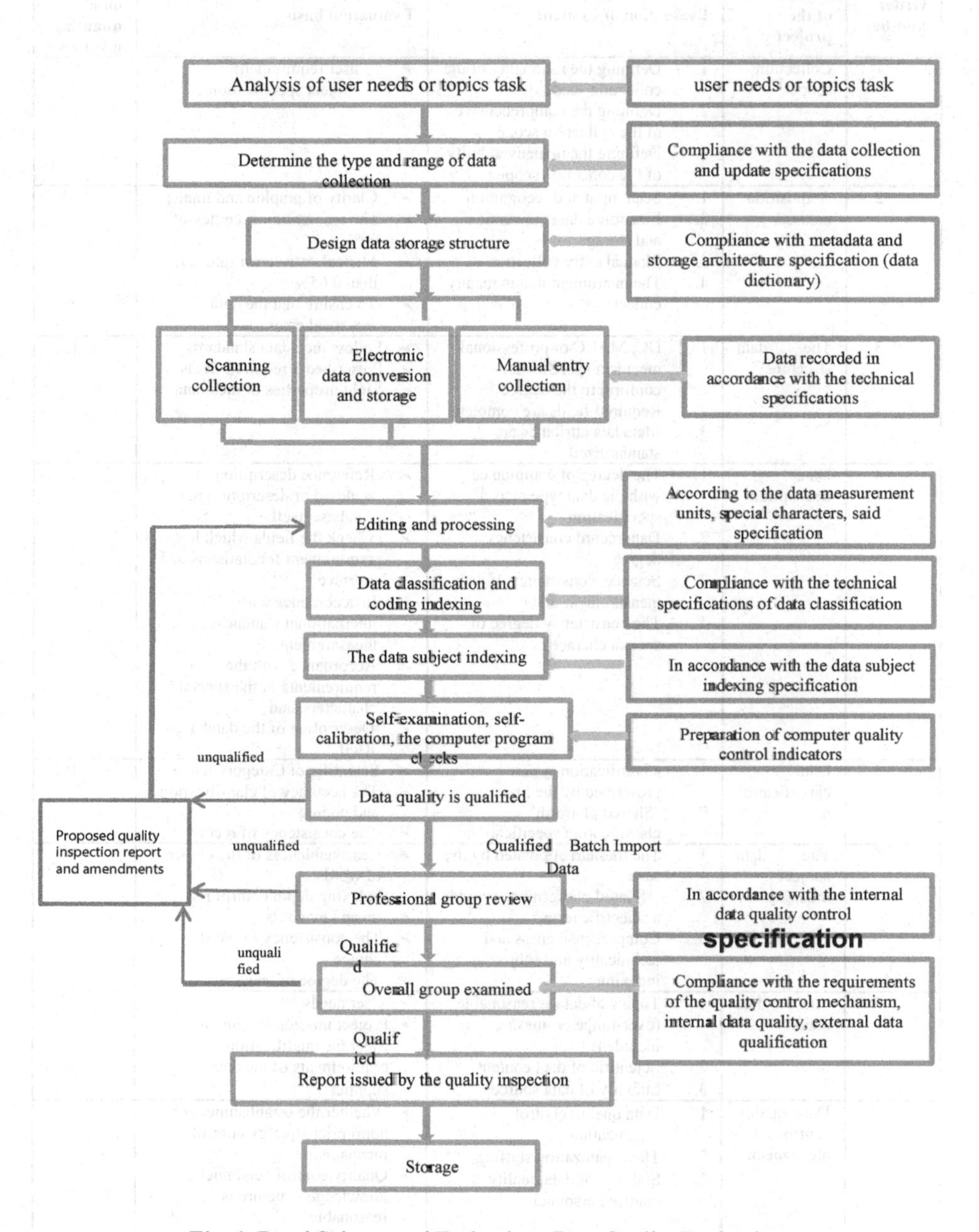

Fig. 1. Rural Science and Technology Data Quality Evaluation

Order number	Evaluation of the project	Evaluation of Content	Evaluation basis	Evaluation of quantized values
1	Collection scope	1. Defining the rationality of the collection scope 2. Defining the comprehensive of the collection scope 3. Defining the maneuverability of the collection scope	➢ user requirement ➢ Project specification	5
2	Acquisition methods	1. Scan input and recognition 2. Electronic data conversion and storage 3. Manual entry collection 4. The instrument automatically collect	➢ Clarity of graphic and image ➢ The recognition accuracy of 99% ➢ Manual entry error rate is less than 0.05% ➢ To ensure that the data statistical error	15
3	The data structure	1. DC, MARC or professional metadata specifications conform to the degree 2. Required fields are complete 3. Metadata attributes are standardized	➢ Follow metadata standards ➢ Prescribed 9 required fields ➢ Eight properties of metadata	15
4	Data description	1. The degree of compliance with the data type record specification 2. Data record consistency degree 3. Science, consistency of data measurement units 4. The consistency degree of special characters	➢ Reference description standard or description in the database itself ➢ Check the fields which has requirement for statistics and retrieve ➢ In accordance with international standards of measurement ➢ Accordance with the requirements of the special characters said ➢ Description of the database itself	20
5	Data classification	1. Classification system prescribed by the State 2. "Shared platform" classification specification	➢ Scientific of Category name ➢ The accuracy of classification and coding ➢ The consistency of record	10
6	The data subject indexing	1. The thesauri stipulated by the state 2. "Shared platform" to provide a scientific term 3. Comprehensiveness and technicality and subject indexing	➢ Reasonableness of the choice of words ➢ Indexing depth control for seven keywords ➢ The consistency of word choice ➢ The degree of leakage	10
7	The data content	1. Topics of data is reasonable (over-range or missing included) 2. Scientific of data content 3. Stability of data sources	➢ User needs ➢ Project mission statement Meet the qualification requirements of the data supplier	20
8	Data quality control mechanism	1. Data quality control organization 2. The organization staffing 3. Stability of data quality control personnel	➢ Whether the establishment of appropriate quality control mechanism ➢ Quality control personnel's knowledge structure is reasonable ➢ The stability of the data quality control personnel	5

4 Data Quality Assessment Report

After the overall quality control of the data, you shall issue the corresponding data quality assessment report. The content includes: the report name, data content review, the unit responsible for the data, data quality assessment, the person completing the report, the evaluation unit, reporting to fill time.

5 Application

This paper discusses the standard of data quality control technology has been used in China Agricultural Science and Technology Information Website and China Rural Science Information Website and other large national websites. These websites' data collection, data input, subject indexing, data storage construction, data description and data unit was prescribed. After these years of operation, these websites data quality control very well, these websites ranked among the best in the field of domestic agricultural websites.

Acknowledgment. The work is supported by the special fund project for Basic Science Research Business Fee "Website of CAAS content resources organizations and service models research", AII (No. 2014-J-008).

References

1. Xiaolin, Z.: Metadata Research and Application. Beijing Library Press, Beijing (2002)
2. Shao, Q.: Several Key Issues in GIS Database Development. Acta Geographica Sinica (Supplement), 34-42 (1995)
3. Huaji, P.: The Quality Control Method of Collecting Product Data. Applications of The Computer Systems (01), 77–79 (2003)
4. Liang, L., Wu, Y.: Verification and Quality monitoring of Information Network Data. Chinese Journal of Hospital Statistics (03), 192 (2003)
5. Fang, Y., Yang, D., Tang, S., Zhang, W., Yu, L., Fu, Q.: Data Quality Managements in Data Warehouse. Computer Engineering and Applications (13), 1–4 (2003)
6. Luo, M.: Quality Control of Open Databases. China Intelligence Information (02), 31 (1994)

The Impact of Climate Change on the Potential Suitable Distribution of Major Crops in Zambia and the Countermeasures

Yanqin Wang, Zhen Tan, and Guojun Sun

Institute of Arid Agroecology, School of Life Sciences, Lanzhou University, 730000 China
{wangyq2012,tanzh13,sungj}@lzu.edu.cn

Abstract. Climate change, as an inevitable process, will aggravate food shortage of the already vulnerable agriculture systems in Zambia. Rain-fed agriculture supports the livelihood of majority of smallholders in Zambia. To effectively adapt to foreseeable climate change, and to decrease risk of the food crisis, we analyzed potential suitable distribution of major crops (white maize (*Zea mays*), cassava (*Manihot esculenta*) and sorghum (*Sorghum bicolor*)) in Zambia under current and future (2080s) climates using the MCE-GIS (multi-criteria evaluation -geographical information system) Planting Ecological Adaptability model. The simulation results indicate that climate change will change the potential suitable area for maize from 66.8% to 48.6%; and that of cassava from 65% to 84%. The suitable regions of sorghum move northward although the total areas will not change. We conclude that future climate change will have different effects on various crops. Our modeling results can be used to make appropriate management decisions and to provide farmers with alternative options for their farming system in responding to climate change.

Keywords: climate change, agriculture crisis, Zambia, MCE-GIS Planting Ecological Adaptability model, potential suitability, adaptation countermeasures.

1 Introduction

About 80% of the variability of agriculture production is due to the variability in weather conditions, especially for rain-fed agriculture[1]. The lack of adaptive strategies to cope with this variability makes sub-Saharan Africa countries highly vulnerable to extreme weather events and climate change [2-5]. There is a growing number of evidence showing that in tropic and subtropical Africa, where crops have reached their maximum tolerance [1, 6], crops suitability and yields are likely to decrease due to the increased temperature [7, 8]. Thus, the declined crops yield eventually will cause substantial impact on food security of poor rain-fed rural communities in sub-Saharan Africa countries [1, 7].

Zambia is a landlocked country in central southern Africa. It has advantages of agricultural production, such as vast untapped agricultural lands, favorable climate conditions and adequate water resources [9, 10]. Despite all of these advantages,

© IFIP International Federation for Information Processing 2015
D. Li and Y. Chen (Eds.): CCTA 2014, IFIP AICT 452, pp. 460–472, 2015.
DOI: 10.1007/978-3-319-19620-6_52

Zambia has a very low food supply, and it is one of the 15 countries with the lowest food security worldwide [10]. Its farming system is depend on rainfall, and is characterized by low fertilizer inputs, low productivity and undeveloped markets [11]. Most of Zambia's farmers (75%) are smallholders who consume most of their own productions[10]. White maize is the leading crop in Zambia, accounting for about 70% of cereals planting area [12-14]. Cassava is the sixth world food crop for more than 500 million people in tropical and sub-tropical Africa, Asia and America [15]. Sorghum is an indigenous cereal from Africa that is well adapted to African semi-arid and sub-tropical agronomic conditions, and also a kind of main source of protein crop in Zambia. Thus it is important to determine the potential distribution of white maize and other crops cultivation zones under future climatic scenarios. Even though there have been many studies concerned the impacts of climate change on the agricultural system of Zambia [2, 10, 13], those studies have not focused on the response of individual crops to climate change.

In this study, we will use the MCE-GIS (multi-criteria evaluation -geographical information system) model to simulate suitable planting areas of three crops in Zambia. This model is particularly well suited to evaluate the potential suitability of the studied species in a given area according to its weather and geographical conditions, and to determine the crop potential suitable planting areas [16-18].

Our objectives are to: (1) rate the climatic suitability of white maize, cassava and sorghum cultivation distribution under current climate conditions in Zambia; (2) predict and simulate the white maize, cassava and sorghum suitable planting areas under future (2080s) climate conditions in Zambia; (3) analyze the variability of these three crops suitability and planting structure changes caused by climate change. It is expected that studies like this can provide a rational for better crop planning, improved land use and sustainable crop yields [3, 19], in the context of national scale adaptation to climate change [2, 13].

2 Material and Methods

2.1 Study Area

This study is conducted in Zambia, located in central southern Africa (15°S, 30°E) (Fig.1). Although the country has a tropical climate, temperatures remain relatively cool throughout the year due the high altitudes of the East Africa Plateau. There are two main seasons. The rainy season (November to April) corresponding to summer has rainfall 150mm to 300mm per month, and the daily highest seasonal temperature in this season are 22℃ to 27℃[20, 21]. Another is dry season (May to October), corresponding to winter (mean daily temperature is 15℃ to 20℃). The dry season is subdivided into cool dry season (May to August) and hot dry season (September to October). These months are very dry, receiving almost no rainfall between [21].

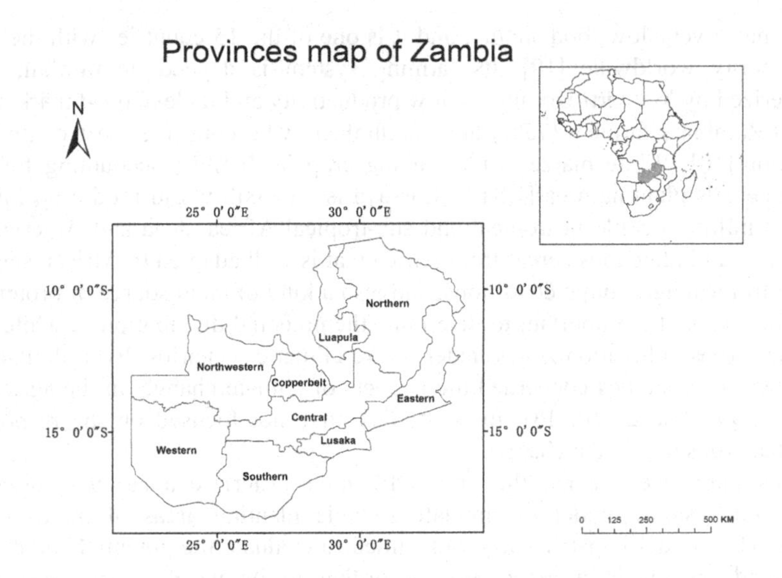

Fig. 1. Provincial boundary map of Zambia

The main crops in Zambia are white maize (*Zea mays*), sorghum (*Sorghum bicolor*), millet (*Pennisetum glaucum*), cassava (*Manihot esculenta*), wheat (*Triticum aestivum*), groundnut (*Arachis hypogaea*) and cotton (*Gossypium herbaceum*) [10]. The main crop growing season lasts from November to April, and almost entirely dependent on rains. There is a minor cropping season from June to September, in this dry season commercial farmers use irrigation to plant wheat. However no analysis is conducted for this minor season in this study.

2.2 Climate Indices and Meteorological Data

The climate indices affecting the distribution of cultivation zones of the three crops are listed below (all the meteorological data is for growth period from November to April): accumulative daily mean temperature above 10℃ (AT); monthly average maximum temperature (Max-T); monthly average minimum temperature (Min-T); and monthly precipitation (PRE)[15, 22-25].

Meteorological data (50 year normal of 1950-2000) is downloaded from the Website of WorldClim-Global Climate Data (http://www.worldclim.org) with 1 00square kilometer resolution [26]. It was interpolations of observed data.

Future conditions of the meteorological data (2080s) is the downscaled global climate model (GCM) data from Fourth Coupled Model Intercomparison Project (CMIP4) [19, 27]. We choose Coupled Global Climate Model (CGCM3) outputs under SRA1B emissions scenarios. The model was developed by Canadian Centre for Climate Modeling and Analysis (CCCma) [27]. This model produces a good simulation for future conditions meteorological especially for Sub-Saharan Africa [28]. These data sets are often used for environmental, agriculture and biological sciences [26]. The A1 scenario family describes a future world of very rapid economic growth, global population that peaks in mid-century and declines afterwards, and rapid introduction of new and more efficient technologies [19, 29]. The A1 scenario family divided into three groups(A1FI, A1B, A1T) describes an alternative directions of technological change in the energy system [19]. The three A1 groups are distinguished by their technological emphasis. The A1B is defined as not relying too heavily on one particular energy source and on the assumption that similar improvement rates apply to all energy supply and end-use technologies [19, 29]. The resolution of the data is 30 seconds.

2.3 Soil and Elevation Data

The soil data is from Harmonized World Soil Database, including soil pH (PH) data and soil texture data (Soil-T). (http://webarchive.iiasa.ac.at/Research/LUC/External-World-soil-database/HTML/).

We obtained the field moisture capacity data using equation 1 [30].

$$FMC = 0.003075 \times Sand + 0.005886 \times Silt + 0.008039 \times Clay + 0.002208 \times OM - 0.14340 \times Bulk\_Density. \quad (1)$$

Where:

FMC- field moisture capacity; Sand- soil sand content (%), the US system; Silt-soil silt content (%), the US system; Clay-soil clay content (%), the US system; OM-soil organic matter (SOM) content (%); Bulk_Density-soil bulk density(g/cm^3).

The elevation data is downloaded from the website of ASTER Global Digital Elevation Model (ASTER GDEM)[31] (http://www.jspacesystems.or.jp/ersdac/GDEM/E/4.html). The resolution of the data is 30m.The land-use data in this study is downloaded from Web-site Center for Earth System Science of Tsinghua University, (http://www.cess.tsinghua.edu.cn/publish/essen/index.html).

2.4 MCE-GIS Planting Ecological Adaptability Model

The MCE-GIS Planting Ecological Adaptability model has been effectively used to predict the crop distribution and evaluate land suitability [17, 18, 32, 33]. It is based on geographical information system (GIS) and the multi-criteria evaluation (MCE) [16, 34]. MCE can be defined as a set of systematic procedures that used to analyzing complex decision problems. This model transforms geographical data into resultant decision by using the regular criteria and an evaluation matrix. It also provides function to visualize the model outputs using GIS tool [35, 36]. All operations carried out in GIS and MATLAB software. A diagram of the methods used is shown in Fig.2. We run the MCE-GIS Planting Ecological Adaptability model for Zambia[37]. Some important procedures are further described below.

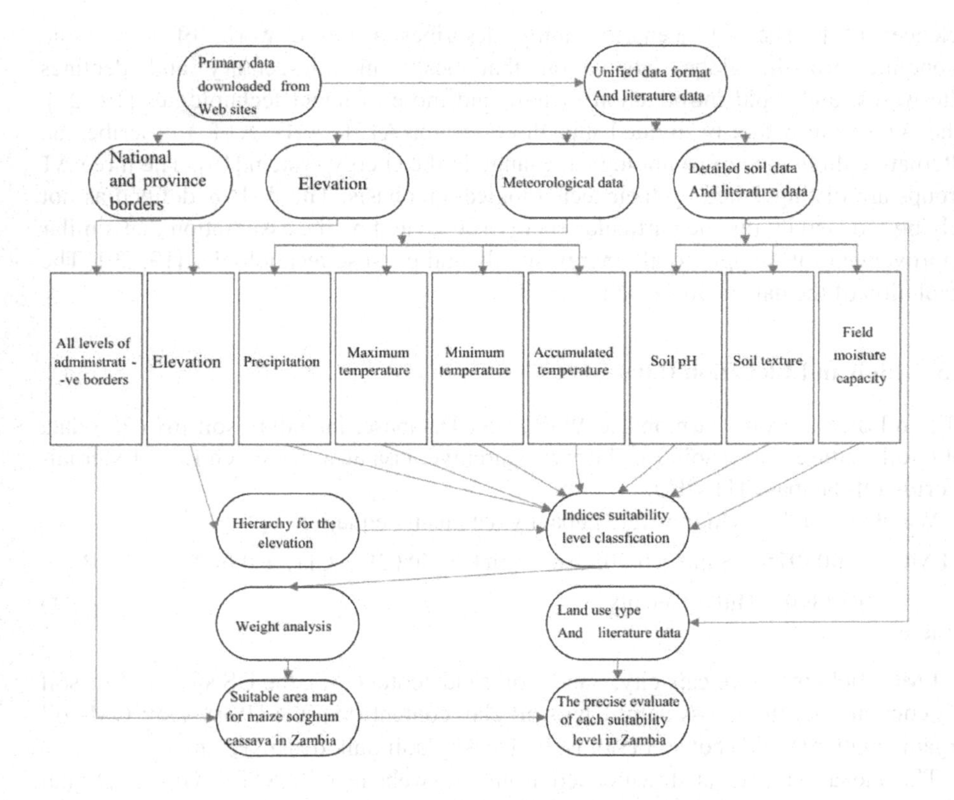

Fig. 2. The MCE-GIS Planting Ecological Adaptability model procedures

Firstly, we use eight indices data, (i.e. the accumulative of daily mean temperature above 10℃ (AT), monthly mean maximum temperature (Max-T), monthly mean minimum temperature (Min-T), monthly precipitation (PRE), elevation, soil texture (Soil-T), soil pH (PH) and field moisture capacity (FMC))to generate various thematic layers using GIS software. These indices were classified into five levels according to suitability (table.1; table.2; table.3). In this study, we adopted the FAO classification system for crops, which allows for land suitability rating based on soil and environmental characters. The FAO system has five classes of suitability rating for certain crops. The five classes are very high, high, medium, low and very low[38]. The specific suitability level of each index was defined based on experts' advice and literature reviews [15, 22-25, 39].

Secondly, we calculated weight of each of the indices using pair-wise comparison matrix known as analytical hierarchy process (AHP). The ratings were provided on a nine-point scale. The comparison concerns the relative importance of each two indices involved in determining the suitability of the stated objective, and the value ranges from 9 to 1/9. The weight coefficient is calculated with MATAB software (talbe.4). [15, 22-25, 39].

Finally, we simulated and derived the final potential suitability level maps of each crop using the indices and weight factors.

Table 1. Each index suitability level classification of white maize in Zambia

Index	Level of suitability				
	Very high	High	Medium	Low	Very low
Accumulative temperature≥10℃	3800-4300	4300-4800	3600-3800	4800-5000	<3600 or >5000
Precipitation (mm)	850-1050	1050-1200 or 750-850	1200-1300 or 650-750	1300-1500	<650 or >1500
Max Temp(℃)	28-31	31-32 or 26-28	32-33	33-34 or 24-26	<24 or >34
Min Temp(℃)	16-19	19-20 or 15-16	14-15	20-21	<14 or >21
Field moisture capacity (%)	0.30-0.35	0.25-0.30 or 0.35-0.4	0.23-0.25 or 0.4-0.45	0.18-0.23	<0.18 or >0.45
Soil pH	6.3-7.0	5.5-6.3 or 7.0-7.5	7.5-8.0	5.0-5.5	<5 or >8
Elevation (m)	<1200	1200-1300	1300-1500	1500-1800	>1800
Soil texture	Loam	Sand	Clay	___	Other

Table 2. Each index suitability level classification of cassava in Zambia

Index	Level of suitability				
	Very high	High	Medium	Low	Very low
Accumulative temperature≥10℃	4200-5000	3900-4200	5000-5300	3700-3900	<3700 or >5300
Precipitation (mm)	600-1000	500-600 or 1000-1200	1200-1400	1400-1500	<500 or >1500
Min Temp (℃)	27-31	25-27 or 31-33	33-34	24-25 34-35	<24 or >35
Max Temp (℃)	17-21	16-17 or 21-22	22-23	15-16	<15 or >23
Field moisture capacity (%)	0.3-0.38	0.25-0.3 or 0.38-0.42	0.42-0.45	0.18-0.25	<0.18 or >0.45
Soil pH	6-7	5-6	4.5-5	3.8-4.5 or 7-7.8	<3.8 or >7.8
Elevation	<1000	1000-1300	1300-1400	1400-1600	>1600
Soil texture	loam	sand	___	clay	other

Table 3. Each index suitability level classification of sorghum in Zambia

Index	Level of suitability				
	Very high	High	Medium	Low	Very low
Accumulative temperature≥10℃	4200-4900	3900-4200	4900-5200	3600-3900	<3600 or >5200
Precipitation (mm)	700-900	500-700 or 900-1050	1050-1200	1200-1400	<500 or >1400
Max Temp(℃)	28-31	25-28	31-33	33-35 24-25	<24 or >35
Min Temp(℃)	17-20	15-17 or 20-21	14-15	21-23	<14 or >23
Field moisture capacity	0.3-0.36	0.25-0.3 or 0.36-0.4	0.4-0.45	0.18-0.25	<0.18or >0.45
Soil pH	6.5-7.5	6-6.5	5.0-6.0 or 7.5-8.0	>8.0	<5.0
Elevation(m)	<1000	1000-1200	1200-1300	1300-1600	>1600
Soil texture	Loam	Sand	Clay	___	Other

Table 4. Each index weight coefficient of crops using AHP

Index / crop	AT	PRE	Max-T	Min-T	Eleva-tion	PH	FMC	Soil-T	total
White maize	0.1818	0.2045	0.1591	0.1364	0.0455	0.0682	0.1136	0.0909	1
Cassava	0.2045	0.1818	0.1591	0.1364	0.0455	0.0682	0.1136	0.0909	1
Sorghum	0.2222	0.1975	0.1728	0.1235	0.0494	0.0617	0.0988	0.0741	1

AHP- analytical hierarchy process; AT- Accumulative temperature≥10°C; PRE- Precipitation; Max-T- monthly mean maximum temperature; Min-T- monthly mean minimum temperature; PH- soil pH; FMC- field moisture capacity; Soil-T- soil texture

3 Results and Discussion

3.1 Simulation of White Maize, Sorghum and Cassava Suitability Under Current Climate Conditions

Under current climate conditions, maize is widely suitable in Zambia; the potential suitable provinces for white maize include Western, Eastern, Central, Southern and Lusaka provinces and part of Northwest province. The suitable area occupies more than 66.8% of the total country area and counts for 5.03×10^7 ha (Fig.3, maize). Cassava potential suitability area (65% of the total country area or 4.90×10^7 ha) is lower than maize, and covers Western, Eastern, Central, Southern and Lusaka provinces and part of Northwest province (Fig.3, cassava). Sorghum is the thirdly most suitable crop in Zambia (55.2% of total country area or 4.90×10^7 ha; Fig.3, sorghum). Most of potential suitable areas of sorghum are located in Western, Lusaka, Southern and Center provinces.

3.2 Simulation of White Maize, Sorghum and Cassava Suitability Under Future Climate Conditions

Under future climate conditions, the accumulated temperature will increase, the precipitation will increase particularly in the northeast of Zambia, monthly average maximum temperature and monthly average minimum temperature will increase about 3℃. In the southern and western regions of Zambia warmer rate is slightly more rapid than that in northern and eastern regions of Zambia.

With climate change, the maize suitable area decreases from 66.8% to 48.6% (3.66×10^7 ha) of the total country area (Fig.4), and the potential suitable region changes into most of the Central, Northwest and Northern provinces (Table.5). Cassava will benefit from increased temperature. It will become the most suitable one among these three crops in Zambia (84% of the total country area or 6.33×10^7 ha; Fig.4), covers almost the whole country except for part of the north border strip of

Zambia (Table.5). There is little change on potential suitability percentage in whole country of sorghum (55.5% of the total country areas or 4.18×10^{7} ha; Fig.4), however, potential suitable region moves northward. The southwest part of Western province becomes not suitable for sorghum, whereas, the Northern province from unsuitable becomes potential suitable area for sorghum planting (Table.5).

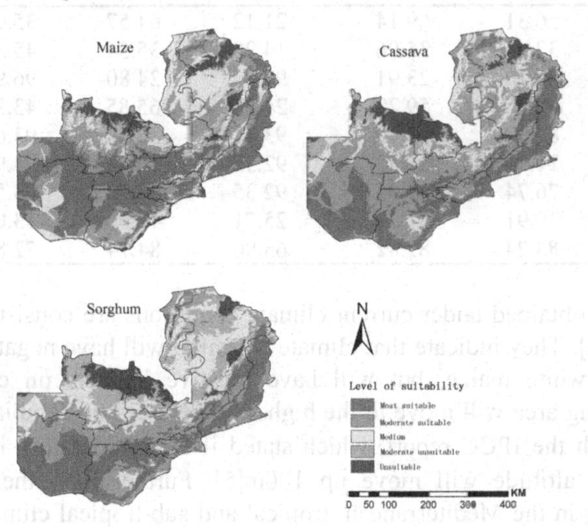

Fig. 3. White maize, cassava and sorghum potential suitable distribution area under current climate in Zambia

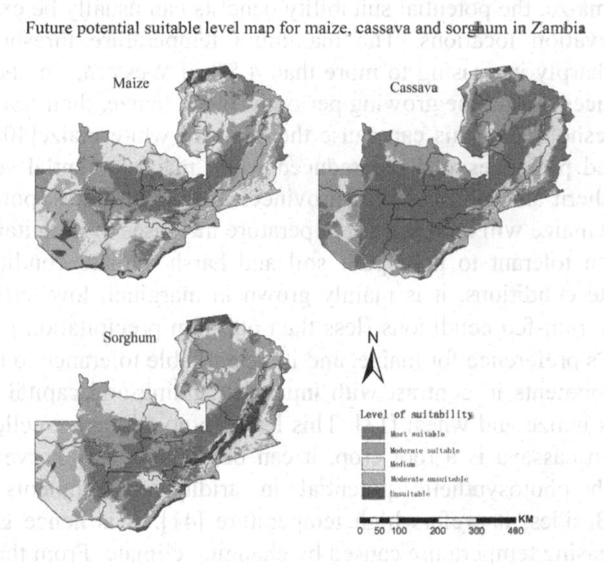

Fig. 4. White maize, cassava and sorghum potential suitable distribution area under future climate in Zambia

Table 5. Simulated planting area percentage for three crops in the current and the future climate conditions in each province of Zambia

Province Name	Potential planting area percent for maize (%)		Potential planting area percent for sorghum(%)		Potential planting area percent for cassava(%)	
	current	future	current	future	current	future
Northwester	56.31	69.14	21.12	64.57	35.07	95.76
Luapula	32.45	35.96	14.20	35.54	45.38	83.86
Eastern	87.16	23.91	96.04	24.80	96.98	59.40
Northern	40.95	59.28	24.23	65.85	43.29	83.80
Western	87.38	23.40	93.41	36.68	93.69	88.11
Lusaka	71.33	34.32	92.38	49.54	94.90	61.06
Southern	76.74	29.06	92.35	59.73	93.52	80.73
Copperbelt	39.91	51.79	25.71	47.66	13.04	81.16
Central	83.24	82.82	63.90	84.74	72.83	91.17

The results obtained under current climate conditions are consistent with previous studies [10, 13]. They indicate that climate warming will have negative impact on the suitability of white maize, but will have positive impacts on cassava. Sorghum suitable planting area will move to the high altitude areas of Zambia. This result is in agreement with the IPCC report, which stated if the temperature increases 1℃, the crops suitable altitude will move up 100m[5]. Furthermore, there is a generally agreement that in the Mediterranean, tropical and sub-tropical climatic zones, all the crop suitability will decrease with the warming trend of climate change, which contrasts with the increased suitability with global warming at higher latitudes[25]. In order to adapt to climate change, planting structure should be adjusted.

For white maize, the potential suitability benefits can usually be expected in cooler or higher elevation locations. The maximum temperature threshold is 34°C, as temperature sharply increasing to more than 4 °C in Western, Southern, Eastern and Lusaka provinces during the growing period of white maize, their temperature will be above the threshold, and this can cause the death of white maize[40]. Most areas of aforementioned provinces will see reduced white maize potential suitable areas. In contrast, Northern and Northwestern provinces at high altitudes, potential suitability areas of white maize will increase as temperature increases to its suitable limit.

Cassava can tolerant to both poor soil and harsh climate conditions[15]. Under current climate conditions, it is mainly grown in marginal, low-fertility acidic soils under variable rain-fed conditions (less than 600 mm precipitation per year). That is due to people's preference for maize, and its remarkable tolerance to nature stress and adverse environments in contrast with input-demanding and capital intensive cereal crops, such as maize and wheat [13]. This leads to no enough excellent farmland for cassava. Given cassava is a root crop, it can be continuously harvested. It can also maintain high photosynthetic potential in aridity environments with sporadic rainfall[15]. Besides, it prefers high temperature [41]. It is hence generally benefit from the increasing temperature caused by changing climate. From the result obtained by simulation model, we concluded that cassava's potential suitability areas expand into almost the whole country as temperature increases, except for the south edge of

Zambia where temperature sharply increases and exceeds the threshold. The expansion scenario is also supported by the sufficient rainfall around the country under future scenarios.

For sorghum, the potential suitability area is decreased as a result of warming in the south and west part of Zambia. In contrast the potential suitability of areas of sorghum in Northern, Northwestern and Central provinces are increased. If temperature raises less than 2°C, rainfall change can modulate the magnitude of the negative impact, but the warmer temperature exceeds 2°C in the whole Zambia, negative impacts caused by temperature rise cannot be mitigated by any rainfall increase [39]. Thus sorghum potential suitability areas northward migration is the result of temperature and precipitation increase caused by climate change.

The impact of climate change on crop planting and production may be a slow process, but the consequence is irreversible [7, 25]. In order to minimize the negative effects of climate, the stakeholders and society should formulate and implement adaptive measures.

There are two aspects of strategies we can take into consideration, one is national level strategies the other is farm level adaptations strategies. The former aspect including rural credit facility strategies, output products strategies, investment technological innovations and expansion of irrigation, and so on. Although relatively costly national strategies may effectively moderate negative impacts and result in biggest benefits, considering the limitations of subsistence farmers, it difficult to implement for smallholders, thus, suggest that small-scale and affordable solutions for subsistence farmers would be more benefit [1, 7, 42]. We focus on the second aspect, some inexpensive strategies that smallholders can implement based on this study. Firstly, switch crops from less potential suitable areas to higher potential suitable areas. Maize should be planted in most of the Northwest, Center and Northern provinces in the future. Sorghum should be planted move to northward, in Northwest, Center, Northern and Southern provinces. Cassava as a hot resistant crop, as temperature increase its suitable area scope will expand to the whole country. Secondly, cassava grows better than other staples including potato, maize, beans, sorghum, millet, banana from 24 climate different prediction models [43]. Cassava as the underutilized crop, following its high suitability in the future in Zambia, it is recommended to improve cassava availability. We have identified cassava as a key crop for the future from this study also from other former studies [43].

4 Conclusions

In this study, we simulated the distribution of white maize, cassava and sorghum under both current and future climate scenarios in Zambia. Our results indicate that, the suitability of these three crops will increase in high altitude regions in the northern Zambia; the growing temperature will have negative impacts on maize whereas cassava will benefit, and sorghum suitable areas will migration to the north of Zambia.

We outlined agricultural adaptation strategies in this study, that is, sorghum should be planted move to northward to the high altitude provinces; cassava as a hot resistant crop, its suitable area should be expanded to the whole country; reduce maize cultivation area.

Acknowledgment. The author is grateful for the funding from the Projects: 2012DFG31450 and 0S2014GO0481. Appreciate those constructive review and comments from Prof. Fernando T. Maestre, Mr. Xiaoyuan Geng, Dr. Jansheng Ye and Mr. Baocheng Jin.

References

1. Cooper, P.J.M., Dimes, J., Rao, K.P.C., et al.: Coping better with current climatic variability in the rain-fed farming systems of sub-Saharan Africa: an essential first step in adapting to future climate change? Agriculture, Ecosystems & Environment 126(1), 24–35 (2008)
2. Thurlow, J., Zhu, T., Diao, X.: Current climate variability and future climate change: estimated growth and poverty impacts for Zambia. Review of Development Economics 16(3), 394–411 (2012)
3. Howden, S.M., Soussana, J.-F., Tubiello, F.N., et al.: Adapting agriculture to climate change. Proceedings of the National Academy of Sciences 104(50), 19691–19696 (2007)
4. You, L., Wood, S., Wood-Sichra, U.: Generating plausible crop distribution maps for Sub-Saharan Africa using a spatially disaggregated data fusion and optimization approach. Agricultural Systems 99(2), 126–140 (2009)
5. Alexander, L.V., Allen, S.K., Bindoff, N.L., et al.: Climate Change 2013:Summary for Policymakers, pp. 19–21. Cambridge Univ. Press, Cambridge (2013)
6. FAO. Water for agriculture and energy in Africa: The challenges of climate change, pp. 129-131. The State of Food and Agriculture, Roma (2008)
7. Thompson, H.E., Berrang-Ford, L., Ford, J.D.: Climate change and food security in sub-Saharan Africa: a systematic literature review. Sustainability 2(8), 2719–2733 (2010)
8. Brown, M.E., Funk, C.: Food security under climate change. Science 319(5863), 580–581 (2008)
9. McSweeney, C., Lizcano, G., New, M., et al.: The UNDP Climate Change Country Profiles: Improving the accessibility of observed and projected climate information for studies of climate change in developing countries. Bulletin of the American Meteorological Society 91(2), 157–166 (2010)
10. Neubert, S., Kömm, M., Krumsiek, A., et al.: Agricultural Development in a Changing Climate in Zambia, pp. 1–9. German Development Institute (DIE), Bonn (2011)
11. Seshamani, V.: The impact of market liberalisation on food security in Zambia. Food Policy 23(6), 539–551 (1998)
12. Hamazakaza, P., Smale, M., Kasalu, H.: The Impact of Hybrid Maize on Smallholder Livelihoods in Zambia: Findings of a Household Survey in Katete, Mkushi, and Sinazongwe Districts. Indaba Agricultural Policy Research Institute Working Paper 73 1(1), 1–26 (2013)
13. Jain, S.: An empirical economic assessment of impacts of climate change on agriculture in Zambia. World Bank Policy Research Working Paper 1(4291), 2–24 (2007)

14. Byerlee, D., Eicher, C.: Africa's emerging maize revolution, pp. 3–23. Lynne Rienner Publishers, United State of America (1997)
15. El-Sharkawy, M.: Cassava biology and physiology. Plant Molecular Biology 56(4), 481–501 (2004)
16. Malczewski, J.G.: multicriteria decision analysis, pp. 3–260. John Wiley & Sons, Canada (1999)
17. Jia, C., Wang, L., Luo, X., et al.: Evaluation of suitability areas for maize in china based on GIS and its variation trend on the future climate Condition. In: Cao, B.-Y., Ma, S.-Q., Cao, H.-H. (eds.) Ecosystem Assessment and Fuzzy Systems Management, pp. 285–299. Springer International Publishing (2014)
18. He, W., Wang, L., Luo, X., et al.: The Trend of GIS-Based suitable planting areas for chinese soybean under the future climate scenario. In: Cao, B.-Y., Ma, S.-Q., Cao, H.-H. (eds.) Ecosystem Assessment and Fuzzy Systems Management. AISC, vol. 254, pp. 325–338. Springer, Heidelberg (2014)
19. Parry, M.: Climate Change 2007: impacts, adaptation and vulnerability: contribution of Working Group II to the fourth assessment report of the Intergovernmental Panel on Climate Change, pp. 80–248. Cambridge University Press (2007)
20. Reason, C., Hachigonta, S., Phaladi, R.: Interannual variability in rainy season characteristics over the Limpopo region of southern Africa. International Journal of Climatology 25(14), 1835–1853 (2005)
21. Usman, M.T., Archer, E., Johnston, P., et al.: A conceptual framework for enhancing the utility of rainfall hazard forecasts for agriculture in marginal environments. Natural Hazards 34(1), 111–129 (2005)
22. Hoogenboom, G.: Contribution of agrometeorology to the simulation of crop production and its applications. Agricultural and Forest Meteorology 103(1), 137–157 (2000)
23. Tadross, M., Suarez, P., Lotsch, A., et al.: Growing-season rainfall and scenarios of future change in southeast Africa: implications for cultivating maize. Climate Research 40(2-3), 147–161 (2009)
24. Masilionytė, L., Maikštėnienė, S.: The effect of agronomic and meteorological factors on the yield of main and catch crops. Zemdirbyste-Agriculture 98(3), 235–244 (2011)
25. Turner, N.C., Rao, K.: Simulation analysis of factors affecting sorghum yield at selected sites in eastern and southern Africa, with emphasis on increasing temperatures. Agricultural Systems 121, 53–62 (2013)
26. Hijmans, R.J., Cameron, S.E., Parra, J.L., et al.: Very high resolution interpolated climate surfaces for global land areas. International Journal of Climatology 25(15), 1965–1978 (2005)
27. Solomon, S.: Climate change 2007-the physical science basis: Working group I contribution to the fourth assessment report of the IPCC, pp. 11–18. Cambridge University Press (2007)
28. Washington, R., New, M., Hawcroft, M., et al.: Climate change in CCAFS regions: Recent trends, current projections, crop-climate suitability, and prospects for improved climate model information. Climate Change, Agriculture and Food Security, pp. 2–7 (2012)
29. Gaffin, S.R., Rosenzweig, C., Xing, X., et al.: Downscaling and geo-spatial gridding of socio-economic projections from the IPCC Special Report on Emissions Scenarios (SRES). Global Environmental Change 14(2), 105–123 (2004)
30. Gupta, S., Larson, W.: Estimating soil water retention characteristics from particle size distribution, organic matter percent, and bulk density. Water Resources Research 15(6), 1633–1635 (1979)

31. Jacobsen, K.: Comparison of ASTER GDEMs with SRTM height models. In: EARSeL Symposium, Dubrovnik Remote Sensing for Science, Education and Natural and Cultural Heritage, pp. 521-526 (2010)
32. Malczewski, J.: GIS-based land-use suitability analysis: a critical overview. Progress in Planning 62(1), 3–65 (2004)
33. Alejandro, C.S., Jorge, L.: Delineation of suitable areas for crops using a Multi-Criteria Evaluation approach and land use/cover mapping: a case study in Central Mexico. Agricultural Systems 77(2), 117–136 (2003)
34. Wander, M.M., Yang, X.: Influence of tillage on the dynamics of loose-and occluded-particulate and humified organic matter fractions. Soil Biology and Biochemistry 32(8), 1151–1160 (2000)
35. Eastman, J.: Multi-criteria evaluation and GIS. Geographical Information Systems 1, 493–502 (1999)
36. Drobne, S., Lisec, A.: Multi-attribute Decision Analysis in GIS: Weighted Linear Combination and Ordered Weighted Averaging. Informatica (03505596) 33(4), 459–474 (2009)
37. You, L., Wood, S.: An entropy approach to spatial disaggregation of agricultural production. Agricultural Systems 90(1), 329–347 (2006)
38. Kalogirou, S.: Expert systems and GIS: an application of land suitability evaluation. Computers, Environment and Urban Systems 26(2–3), 89–112 (2002)
39. Sultan, B., Roudier, P., Quirion, P.: Assessing climate change impacts on sorghum and millet yields in the Sudanian and Sahelian savannas of West Africa. Environmental Research Letters 8(1), 14–40 (2013)
40. Thornton, P.K., Jones, P.G., Alagarswamy, G., et al.: Spatial variation of crop yield response to climate change in East Africa. Global Environmental Change 19(1), 54–65 (2009)
41. Pellet, D., El-Sharkawy, M.: Cassava varietal response to phosphorus fertilization. I. Yield, biomass and gas exchange. Field Crops Research 35(1), 1–11 (1993)
42. Ziervogel, G., Bharwani, S., Downing, T.: Adapting to climate variability: pumpkins, people and policy. In: Natural Resources Forum, pp. 294–305. Wiley Online Library (2006)
43. Jarvis, A., Ramirez-Villegas, J., Campo, B.V.H., et al.: Is cassava the answer to African climate change adaptation? Tropical Plant Biology 5(1), 9–29 (2012)

A Model for Personalized Information Services of Agricultural Library Based on Multi-agent

Xie Meiling

Library of Agriculture University of Hebei, Baoding, China;
hebauxie@hotmail.com

Abstract. To realize the needs of the agricultural university's library information services, and to improve the situation that it is inconvenient for readers to access the resources in traditional libraries, On the basis of the traditional agent model, a multi-agent model was proposed and the personalized information service model was built. The information queried by the users was used as the sample, and the classification algorithm was introduced to classify the pages. The semantic-based user model was built according to preferences of the user's query and browsing. The main function modules of the model were designed, which included the update method of concepts and the update method of concept weights. Results of this paper are useful for further enhancing the level of intelligence service of agricultural libraries.

Keywords: Multi-agent mode, agricultural library, personalized service.

1 Introduction

With the development and application of information technology, the way of expression, information retrieval and transmission modes were changed, the library service model is also put forward new requirements. The service mode of the library should also change accordingly. Users' psychologies, behaviors, and interests were different. How to provide the individuality information service according to the characteristics and demands of the users was the hot research topic. Some research achievements have been obtained. These achievements include intelligent search engine based information retrieval mode [1] and mobile agent based information retrieval system [2] etc. More researches were focused on realize the personalized information service based on agent model [3-6]. Although Agent technology solves the problems of information retrieval and filtering, personalized information pushing and feedback were not targeted research; In addition, the accuracy of the expression of agent technology to the user's personality and demand was insufficient. To meet the demand of the agricultural universities library information service, this paper proposes a multi Agent model, to build personalized information service system, improve the level of agricultural library intelligence service.

© IFIP International Federation for Information Processing 2015
D. Li and Y. Chen (Eds.): CCTA 2014, IFIP AICT 452, pp. 473–479, 2015.
DOI: 10.1007/978-3-319-19620-6_53

2 Experiments and Methods

2.1 Design of the Information Service Model

Agricultural library users include researchers, teaching staffs, technical engineers and students, etc. Information in the information service has the characteristics of universality, professional, comprehensive, sociality etc. On this basis it is necessary to meet the precision rate and recall rate. The framework structure of the model of personalized service based on multiple Agents proposed in this paper was shown in figure 1:

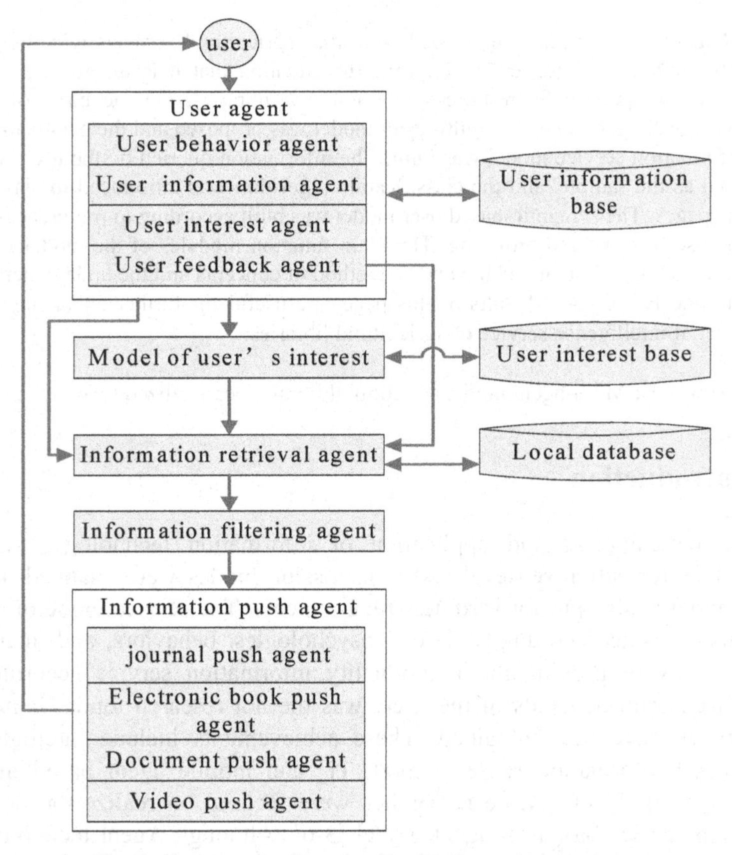

Fig. 1. Framework of personalized information service

The introductions of main modules in the framework introduction were as follows:

User Agent realized its function by child agents' calling each other. The user behavior Agent accessed to user's information through the interactive behavior of users and the computer. User's interest information was acquired by the analysis of the specific operation, pages browsed, residence time, etc. The above information were transmitted to the user Agent. User's interest was processed by classification algorithm of data mining. Processing result was transmitted to the user interest agent. According to the received information, user's specific interests and preferences were modeled by the user interest agent and the models were stored in user interest base. According to user feedback, user feedback agent updated user information base and improved the service quality.

The function of information retrieval Agent is to interact with all kinds of search engine. It retrieved the library of local data information and merged the retrieve information acquired. The information with low correlation degree to user demand was filtered by information filtering agent using certain filtering algorithm. There were two types of filtering algorithms. They were filtering algorithm based on the information content filtering and filtering algorithm based on collaborative filtering respectively. Another algorithm combined the advantages of the above two filter algorithms and realized the combination filtering.

According to the types of information and content, information push agent selected the corresponding agent for information push. These pushes included journal push, electronic book push, document push and video push etc.

The workflow of all agents and functional modules of the model was described below:

(1) Library users access information service platform. System calls the user's personal information registered. The system prompted the user to modify the information which has been changed. The information included the user's professional, educational background, interests, search history, etc and the information was stored in user information base.
(2) User login system and started operation. The user agent was triggered and tracked the user's operation. the user's specific behavior was stored in the user behavior base by the user behavior agent.
(3) User information agent classified user agent information. The classification results were sent to user interest agent.
(4) Interest model and interest base were established for the user by user interest agent according to the classification information. The user interest through analysis was transmitted to information retrieval agent.
(5) According to the user's specific interest, information retrieval queried agent relevant information and sent it to the information filtering agent.
(6) Information correlation was evaluated by information filtering agent. The information that closely related to the user interest was retained and sent to information push agent.
(7) The information was presented to user by information push agent. User's evaluation was transmitted to information retrieval agent. Repeated step (5) (6), until the user fully satisfied

2.2 User Ontology Model Based on Semantic

According to user's interest in digital library searching and browsing, this study aimed at the user interest model of personalized information service frame and built the user model based on semantic. The traditional information service mode was based on the query keywords. Due to the user's query demand diversification, and inaccuracy of the keywords submitted, it was difficult to accurately reflect the user's interest in the target domain. The query effect was unsatisfactory. In this study, the "ontology" improved was used in the user interest modeling and more accurate semantics was formulated.

Construction of the Model. Agricultural information query in the digital libraries can be regarded as a "conceptual" personalized view. In order to more fully express the user's interest in information query, in this paper, the traditional 4-tuple was redefined [7]. New element was added and the ontology model was constructed based on 5-tuple.

Definition 1: D represented the core message body. S indicated contacts between the core information and the following 5-tuple represented the ontology:

$$V = \{W, Q_D, X, R_D, U\} \tag{1}$$

Where W was the agricultural information concept set. Q_D was instance set. instance set was series of examples related to concept set. X was the concept weight set of the elements. R_D was concept instance declaration set.

W also can be further subdivided into W_D and W_S, representing the core information and relationships in the concept set respectively:

$$W = \{W_D, W_S\} \tag{2}$$

Definition 2: The concept weight set X can be subdivided into the core concept of weight X_D and relationship concept weight X_S, which can be represented by:

$$X = \{X_D, X_S\} = \{W_{dj}(t), W_{Sjk}(W_{dj}, W_{dk}, r)\} \tag{3}$$

Where dj was concept j and dk was concept k. S_{jk} was the connection between the concept j and k. W_{dj} was j-th element of the concept set and its weight was t. r was the weight of the relationship between W_{dj} and W_{dk}.

Definition 3: Define the concept instance declaration set R_D as follows:

$$R_D = \{W_{dj}(Q_{DI}, i)\} \tag{4}$$

Q_{DI} was instance I in the concept the j. i was its weight.

Definition 4: Define instance weights set as follows:

$$X_{QD} = \{Q_{DL}, Z\} \tag{5}$$

Weight of each example was expressed in Z.

Definition 5: Define time update set of the weight as follows:

$$U = \{U_t, U_r, U_z\} \tag{6}$$

Ontology Modeling Process. In this study, ontology with weights was constructed on the basis of improved "ontology". The basic idea was to guide the learner for different categories of data streams of continuous learning. The existing ontology has been corrected in the process of learning. The user's interest was processed by data mining of activating the diffusion process. The steps were as follows:

Step 1: The data stream of classification in use information was read to the data buffer by the user interest agent. According to the content of the data buffer, the learning device judged the scope of user requirements. Information needed was downloaded from the resource database and was stored in the local database.

Step 2: Users interest concept and the correlation between these concepts were obtained by learning device according to certain algorithm and were stored in user interest model. The ontology structure shown below was formed. (Here the entity concept "digital library" for example.)

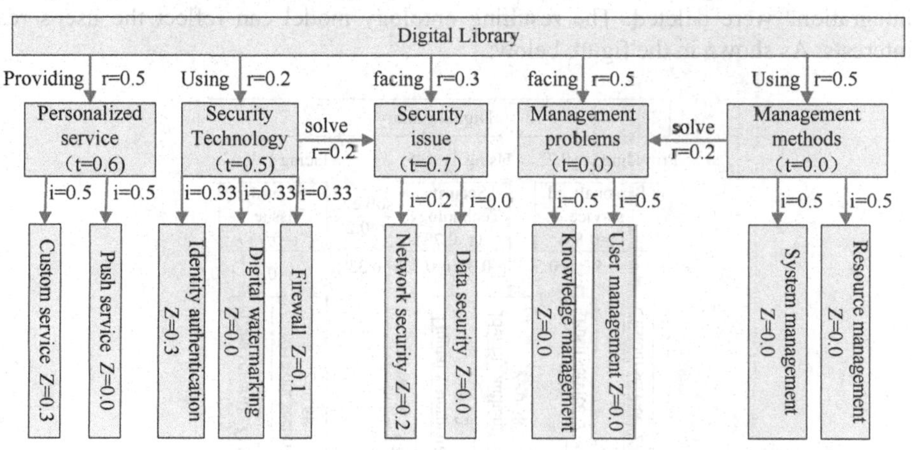

Fig. 2. Ontology Structure of preliminary study

The data streams were learned by learning device. Existing ontology was corrected. The weights of each concept and the weights between concepts were determined. The concepts with a weight of 0 were removed. User's scope of interest was more accurately determined. The response speed of system was improved.

Step 3: The user's interests in the ontology were searched based on spreading activation algorithm. The ontology was corrected by adjusting the weights of the concept of user interest and the interest of user was highlighted. The specific steps were as follows:

(1) The activation values were initialized. The initial node was set I_n and the weight was set Xn .
(2) According to the initial node weights of adjacent nodes, larger weights neighboring nodes are activated. Assuming a neighboring node I_m was activated and the initial weight was X_m , then the weight after activation was $I_m + I_{mn} * I_n(1-\alpha)$. I_{mn} is the weights between initial node and the adjacent node. α was the attenuation factor, said the loss in the process of diffusion, representing the loss in diffusion process.
(3) Repeat steps (2), until all adjacent nodes were activated.
(4) Repeat steps (1) - (3), until the concept whose weight was 0 were set to source node. And then activate the adjacent nodes.
(5) Set a certain threshold and remove the core concepts whose weight was under the threshold. Eventually the ontology model was formed.

According to the operation process, a preliminary study of ontology structure is optimized. "Management method" and its instance "system", "law" and "resource integration" were deleted. The resulting ontology model can reflect the user's real interests. As shown in the figure below:

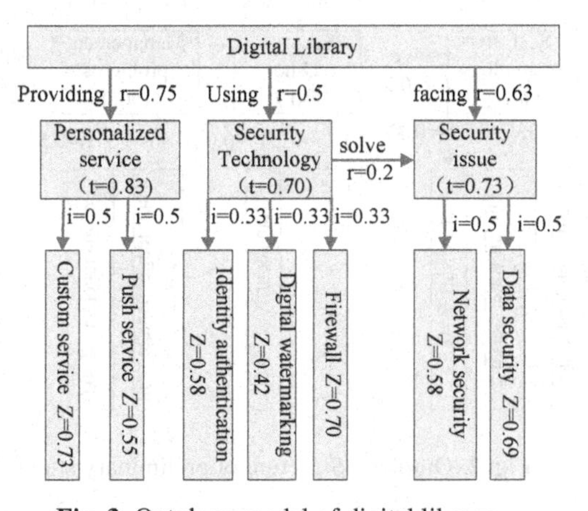

Fig. 3. Ontology model of digital library

Step 4: The user interests built was stored into user ontology base, which was used to retrieve the user requirements.

Ontology Model Update Strategy. User's interest and the demand changed constantly. So the ontology should be adjusted according to the changes. Time update set of weights was the time property of the ontology. Its role was to support the update of ontology. It included ontology concept update and weights update between ontology concepts.

For ontology concept updating, if the user's interest was found, user ontology base was queried by the system. If the concept interest point already existed, only its weight and time properties were updated. If the concept interest point did not exist, it

was inserted into the user ontology database and sorted according to weight. The no longer important points of interest were deleted.
About the weights updating between ontology concepts, the weights were determined according to the difference between the current moment U and the initial time t. As shown blow:

$$x'(U) = \frac{\partial}{\partial + (U - t)} * x(t) + x_n \tag{7}$$

Where t was the initial point. U was the current moment. x' was the updated weights. x_n was the initial weight. ∂ value depended on the specific circumstances. the resulting weights between the new concept of ontology. The new weight between the concepts of ontology was obtained.

3 Conclusions

On the basis of the traditional agent model, the semantic-based user model was built according to preferences of the user's query and browsing. The current intelligent agent was at the stage of the process of developing and perfecting constantly. In this study, more reasonable system modules were built according to the agricultural information users. The concept relations in the ontology were further expanded. The concept of implicit in the agricultural information was reflected, which was useful for providing more perfect personalized information service.

Acknowledgements. This research was supported by a grant from the National Natural Science Foundation of China (No. 31371532).

References

1. Chen, Y., Zhang, Y.: Application of intelligent technology in search engine. Journal of Information 2, 2–3 (2014) (in Chinese)
2. Song, Q., Kong, J.: Research on mobile agent applications in intelligent electronic commerce. China Management Information 11, 75–77 (2013)
3. Zhang, X.: Research on agent technology design personalized information service of digital library. Journal of Modern Information 1, 74–76 (2012)
4. Han, L.: Application of Agent technology in the personalized information service in Digital Library. Journal of Modern Information 4, 104–105 (2012)
5. Fu, H.: Intelligent Agent technology and the realization of the individualized information service. Journal of Information 1, 97–98 (2014)
6. Wang, R., Xu, X., Huang, H.: Intelligent Agent and its application in the information network. Beijing University of Posts and Telecommunications Press (2006)
7. Wang, H., Wu, J., Jiang, F.: A Study on Ontology Model Based on Description Logics. Systems Engineering 3, 101–106 (2013)

Interpolation Method of Soil Moisture Data Based on BMA

Wan Shu-jing, Zhang Cheng-ming*, Liu Ji-ping, Yu Ting, and Ma Jing

College of Information Science and Engineering,
Shandong Agricultural University, Tai'an, 271018, China
Research Center of Government Geographic Information System,
Chinese Academy of Surveying & Mapping, Beijing, 100830, China

Abstract. How to consider both the spatial distribution and the time series characters of the soil moisture data, which have an effect on the result of interpolation, is the key to improve the soil moisture data interpolation result. This paper proposes a new interpolation method : the BMA-I (BMA-Interpolation), which contains kriging method, and BMA(Bayse Model Average) method. Spatial forecasting model uses synergetic kring method. Time series forecasting model uses LS-SVM algorithm. The model average uses BMA method. Taking the project of Shandong Bohai granary as an example, through comparing with kriging method, the proposed approach can improve the problems existing in the kriging method effectively, and can give a reliable forecast uncertainty interval, the simulation results more accurate and reasonable.

Keywords: Soil moisture, the BMA -I method, Interpolation.

1 Introduction

Water is the most important resource in the earth, which is the basic element of life. About 97.2 percent water is in the sea 2.15 percent water is with ice, while the other 0.63 percent water is in the ground, and 0.005 percent water is in the soilall over the world[1]. Very small portion though the soil moisture occupies in the global water resource, it is a critical factor in the process of energy transition between the earth and the atmosphere[2]. Water moisture is an important part of land ecosystem water cycle, it has important significance for the study of water stress, drought monitoring and crop yield estimation. Soil moisture data of regionor even the world, is an essential parameter in the study of land process model, in improving the regional and global climate, in forecasting the area humidity. How to obtain high quality of the soil moisture data, has always been a focus of the study [3-5].

The development of sensor technology and communication technology have greatly improved the ability of the ground observation. People can use sensor networks to acquire soil moisture, land surface temperaturer, wind speed and other related parametersdirectly and real-timely. And the time resolution is very high, even it can up to second order of

---

* Corresponding author.

© IFIP International Federation for Information Processing 2015
D. Li and Y. Chen (Eds.): CCTA 2014, IFIP AICT 452, pp. 480–488, 2015.
DOI: 10.1007/978-3-319-19620-6_54

magnitude. But the observation data on behalf of the space is limited. Therefore, people often use the interpolation method which can transform limited observation data to surface data. Kriging method is a common interpolation method which is often adopted. Kriging method is demanded to satisfying the Second-order stationary hypothesis. Because the soil moisture is a surface variable, which is a kind of high spatial and temporal heterogeneity ,have an smooth effect on soil humidity interpolation, even ordinary kriging method, the kriging method, or collaborative kriging method. In otherwords, the smaller values are often overstated, and larger values are often underestimated. Estimate values can't reflect the real change characteristics of the real space. In order to solve this problem, researchers at home and abroad have made a wide range of research. Journel in Yao[7]' establishing a series of post-processing algorithms called kriging method, which is under the condition of sacrifice local estimation precision, this method processing the smoothing effect problem well. Olea [7], such as a combination of traditional kriging estimation and the characteristics of the random conditional simulation, puts forward the compensation kriging method, but the results of this method on the global optimality is better than random simulation, on the local accuracy is lower than the traditional kriging estimation. Yamamoto [8-10] in combination with the idea of Olea, puts forward a set of post-processing on the estimation of kriging method, Yang Yuting[11] uses his method of soil moisture interpolation and gets post-processing results.

As you can see, in the previous interpolation of soil moisture research , researchers mainly focuse on the influence of spatial variation characteristics of interpolation, and the time change characteristics of soil moisture is almost under no consideration. Related studies have shown that besides exclusion of precipitation and irrigation, the change of soil moisture have a relationship with temperature, illumination, wind speed, vegetation and other factors. And these factors can have some regularity changes on the time series. If considering both the space and time effects on the result of the interpolation, we can get a higher accuracy of the interpolated result.

Bayesian model averaging (Bayesian model averaging, BMA) is put forward by Rafery etc[11]. Bayesian model averaging uses a statistical processing method which uses multimodal set to make probability predictions. A particular variable probability density distribution function (PDF)predicted by BMA, is average weight which is a probability distribution of every single model forecasting. The PDF is possessed by a deviation correction. The average weight is a posteriori probability of corresponding model and it represents the relative prediction techniques of each model in the stage of model's training. Many researches have been carried out that the method of BMA shows great advantage [13-16].

This paper chooses the collaborative kriging method as spatial prediction model, and uses the time-series forecasting model as sequential forecasting model, based on the LS - SVM algorithm. This paper also applies the basic principle of the BMA, considering the soil moisture data of space and time changes at the same time. Establishing a kind of the interpolated method considering time and spatial variational characteristics simutaneously.

2 BMA-I Methods

Assuming that region Ω to be interpolated can be regarded as composed of M grids, each grid contains up to an observation point, namely$\Omega = \{pi\}, i = 1,2,\ldots, M$.Observation points set V = {v1, v2,..., vi,...}, vi∈ Ω.Collection of observation points were divided into .and the three V1, V2, V3mutually collections are disjoint.V1 is used as the input of kriging interpolation, V2 is used as the input of forecasting model, V3 is used as input validation.

The y is on behalf of forecast variables, namely soil moisture value of the grid .F = {fl, f2} present a set of predicting model, flis spatial interpolated model , f2 is time-series interpolated model.

According to the principle of the BMA, the probability density function of y is

$$\rho(y|V2) = \sum_{i=1}^{2} w_i g_i \tag{1}$$

For under the condition of given data V2, Fi is the optimal probability of the model, gi is the probability density function of y ,where V2at a given sample and model forecast variables Fi under the condition .Supposing that smaller regional of underlying surface is nearly the same, so the distribution of soil moisture in the area can be seen as a normal distribution. At this time, for every gi, the mean or expect meet:

$$y|gi, V2 \sim N(\alpha_i g_i + \beta i, \sigma_i^2) \tag{2}$$

α_i, βi are deviation corrections

The expectations for (1):

$$E(y|g1, g2) = \sum_{i=1}^{2} w_i \ (\alpha_i g_i + \beta i) \tag{3}$$

In order to have wi, α_i, βi,$\sigma_i^2$the variance is :

$$VAR(y|V2) = \sum_{i=1}^{n} w_i \ ((\alpha_i vi + \beta i) \ - \ \sum_{i=1}^{n} aivi + \beta i)^2 + \sum_{j=1}^{2} wj\sigma_i^2 \tag{4}$$

Taking the minimum value as constraint conditions, and using EM algorithm to find the solution .According to these formulas, we can obtain the parameters above, and then determine the value **y** of each grid.

3 Experiment

3.1 Study Region

Experimental area is in Shandong "Bohai granary demonstrative project of science and technology" in the project, the area contains Dezhou, Binzhou and Dongying district,115°45'-119°10' east longitude, 36°24'-38°10'north latitude, a total of 30 counties (city or area),and grain land is15 million mu, the cotton fields transformed to grain is 1 million mu, the land of sanlinization is 1 million mu. Area shown in figure 1 in the blue box. The layout of ground sensor acquisition node is shown in figure 2.

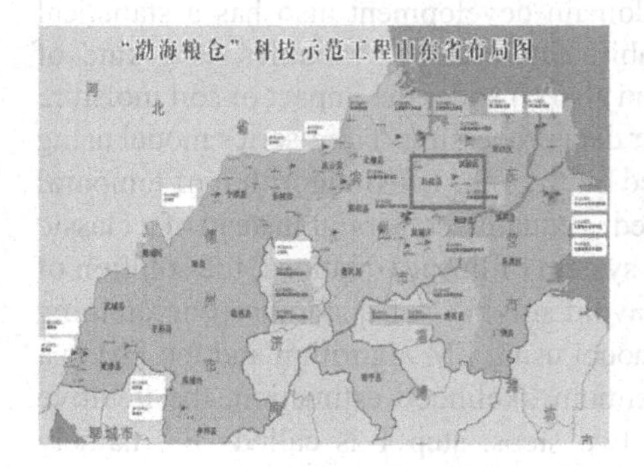

Fig. 1. Experimental area location

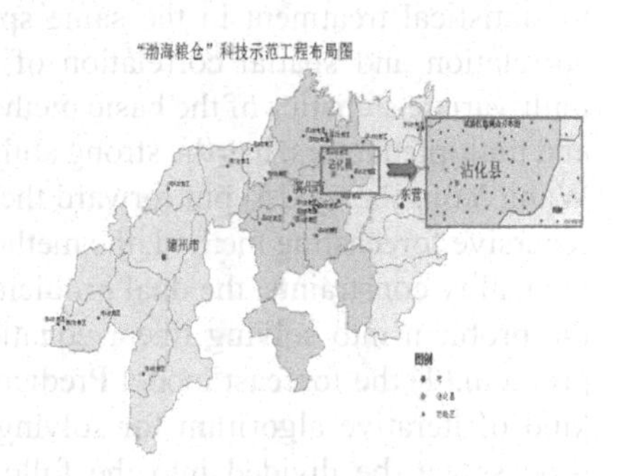

Fig. 2. Acquisition node distribution

3.2 Study Data

(1) The ground observation data

The ground observation data is obtained through the observation network, every node obtain soil moisture, wind speed, temperature and so on. We can obtain a data every 5 minutes. Data is on May 2, 2014 to May 28, 2014.

(2)The historical data of soil moisture

The historical data of soil moisture is being obtained by the writers' method developed by themselves. This method is a kind of is the inversion method. This method is using passive and positive remote sensing, based on genetic neural network algorithm. First of all, establish a BP neural network, and then use the genetic algorithm of BP network to optimize the node weight value; Lastly, the TM data (TM3, TM4, TM6), different polarized and polarized ratio (VV, VH, VH/VV) ASAR data as the input of neural network. At the same time ,the surface soil moisture content is the output of the network, with some measured data training and inversion on the network.The paperof this method have been received by Sensor Letter, details about this paper can be queried if you have interest.

3.3 Experiment Design

(1) observation point data preprocessing According to the need of the method, the observation point of every area can be divided into V1, V2, V3 three mutually disjoint collections, including V1 used as the input of kriging interpolation, V2 as input of forecast model, the V3 as input validation.(2) soil moisture history data preprocessing Soil moisture history data is preprocessed by ENVI software. Main work includes: registration of the vector data and raster data; To determine the grid position of each observation point ;Extraction with interpolation grid soil humidity value and and arrange in chronological order.(3) interpolated model and time-series forecasting model Interpolation model using collaborative kriging method, the method is improved to statistical treatment in the same space domain development also has a statistical correlation and spatial correlation of the ability of multiple variables, is a kind of multivariate statistics of the basic method, can also consider the impact of soil moisture and multiple factors, has the strong ability for data processing. Time-series model using Wang Yongsheng [15] put forward the based on the LS - SVM algorithm of temporal recursive forecasting method, the method used is equality constraint instead of a classic inequality constraints, the dual problem is a system of linear equations, the solution of the problem into solving linear equations, avoid solving of quadratic programming problem.(4) the forecast model Prediction model using EM algorithm and the EM is a kind of iterative algorithm for solving maximum likelihood estimation, the iterative process can be divided into the following two steps: step 1 is called "E", namely according to the result of the previous step to estimate the complete data set expectations of likelihood function; Step 2 is called a "M", namely the estimates made completely data set the parameters of the maximum likelihood function expectations; Then repeat these two steps until the iterative convergence, details refer to relevant literature.

3.4 Experiment and Results

In order to compare with the result of the experiment, We have used Ordinary Kriging method (Ordinary Kriging, OK) collaborating Kriging method (Cooperate Kriging, CK) and BMA_Interprose method(BMA - I) to interpolate, and then compared the results. We have chosen the data respectively on 9:00 am at May 3, 9:00 am at May 10, 9:00 am at May 21, 2014 to calculate. To get the interpolation result ,2400 grids were selected as the interpolation points. The results were shown in figure 3 (1) - (9).

Table 1 shows the statistical results of the three interpolation methods. By comparing the measured soil moisture and the soil moisture after two methods of kriging interpolation, we can see from the statistical results as follows:

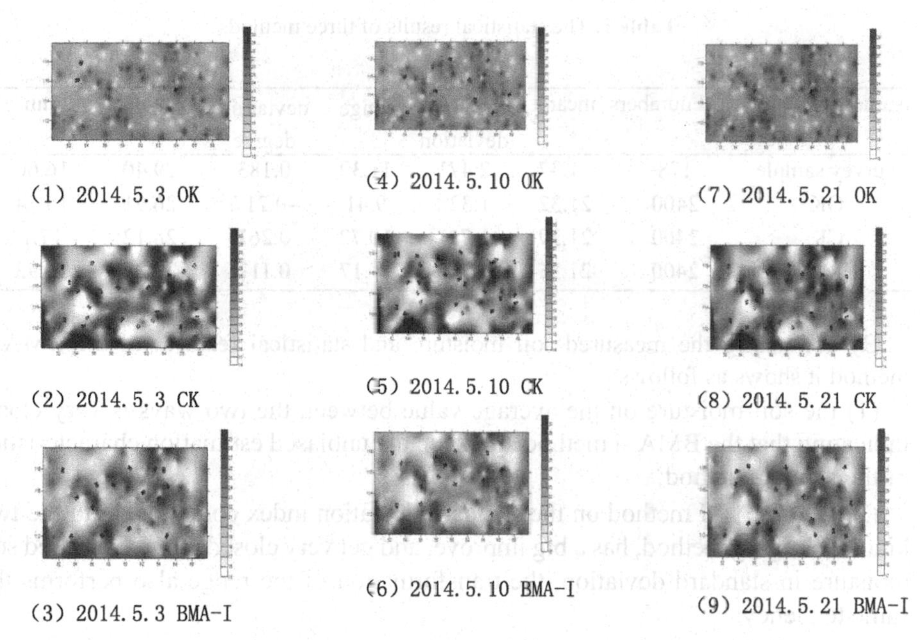

Fig. 3. The interpolation result

(1) The measured soil moisture and the mean of two kriging estimation of soil moisture are almost equal, It shows that the kriging estimation conform to the unbiasedness requirements;

(2) The standard deviation of the measured soil moisture obviously larger than the standard deviation of kriging estimation of soil moisture, this reflects that by using ordinary kriging method for spatial interpolation, produced obvious smooth effect, leading to the estimates of the variation degree is less than actual situation, the decline sharply of range can also explain the existence of the smoothing effect. Compared to the two kinds of kriging method we can found that both standard deviation indicators and The range index , collaborative kriging method is better than the ordinary kriging method. It shows that the result which only use the ordinary kriging method to interpolate can't reflect the real space distribution characteristic of soil moisture, but it will be improved clearly after Introduce correlated Variables.

(3) On the frequency distribution of soil moisture index, the performance of the two methods of kriging has a obvious difference. It shows us that using ordinary kriging method for spatial interpolation of soil moisture can not reflect the real soil moisture spatial distribution features, but after the introduction of relevant variables, it will be improved according to the result of the interpolation.

Table 1. The statistical results of three methods

Name	Grid numbers	means	Standard deviation	range	deviation degree	max	min
Survey sample	178	21.37	2.471	11.39	0.183	29.40	16.60
OK	2400	21.32	1.328	9.41	-0.217	26.91	15.34
CK	2400	21.39	1.711	10.72	0.261	28.12	17.1
BMAI	2400	21.35	2.317	11.17	0.117	29.27	16.93

By comparing the measured soil moisture and statistical features of the BMA-I method it shows as follows:

(1) the soil moisture on the average value between the two ways is very close, indicating that the BMA -i method still keep the unbiased estimation characteristics of the kriging method;

(2) the BMA - I method on the standard deviation index compared with the two kinds of kriging method, has a big improve, and get very closed to the measured soil moisture in standard deviation, the transformation of the range also performs the same tendency.

(3) in the skewness, both of them showed slight positive skewness. and other statistics also approximated.

In addition, using the OK method for interpolation,107 sample points scatter outside the scope of the measured sample, close to 5%, but CK method and BAMI method don't have this kind of situation. On the scope of data, the data range of CK method is obviously smaller than the measured sample data range, while the BMA -i method in getting data range and numerical range is closer to the sample. In order to illustrate better that the BMA -i method can reflect the soil moisture in the spatial distribution characteristics is improved, we have made the corresponding soil moisture frequency histogram, as shown in figure 4.

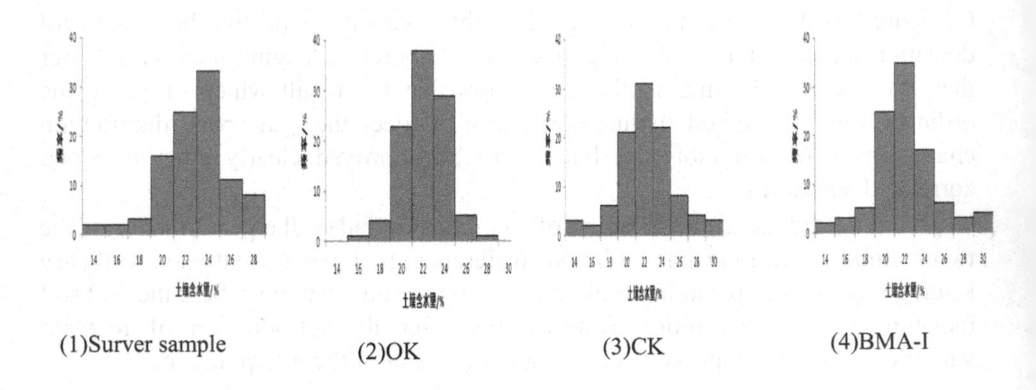

(1)Surver sample (2)OK (3)CK (4)BMA-I

Fig. 4. Soil humidity frequency distribution histogram

As can be seen from the figure 4, the ordinary kriging method to estimate the frequency distribution of the histogram is limited in a narrow range, the frequency distribution of the measured values and the histogram obviously has great discrepancy; Collaborative kriging method of frequency histogram has a significant improvement, but still with the measured values have some discrepancy; BMA-I estimate that the results with the measured frequency distribution histogram and are closer.

Through the above analysis, it shows that BMA-I method to obtain the estimate of basic can reflect the actual spatial distribution of the field soil moisture content.

4 Conclusion and Discussions

In this paper, we were using the measured soil moisture data, combined with the soil moisture remote sensing inversion historical time-series data, implements a space distribution and time distribution of give attention to two or more things interpolation method. From the point of processing result, due to the introduction of the time-series data, BMA-i method significantly improve the smooth effect brought by the kriging method, spatial variation characteristics of soil moisture also got reappear. Due to the expectations of the BMA -i method is members of the expected average Bayesian method, which can not only improve the prediction accuracy to a certain extent, and, more importantly, through further processing, this method can also provide reliable predicted uncertain interval.

In temporal changes, considering that irrigation and rainfall has a great influence to the soil moisture variation characteristics, we only use the data weeded out 5 days before and after irrigation and rainfall , the processing methods affect temporal feature extraction. In addition, this article only use small scale data validation and application. To the soil moisture data of different scale, we need to test the effectiveness of the proposed method more.

Acknowledgment. This study has been funded by National Natural Science Foundation of China (41101321), National High Technology Research and Development Program of China (863 Program) (2013AA122003), Foundation for Outstanding Young Scientist in Shandong Province (BS2011DX031), Science and Technology Develop Project in Shandong Province (2012GSF11713), National Special Fund(2012DFA60830) and Funded by Key Laboratory of Geo-informatics of State Bureau of Surveying and Mapping(201319).

References

[1] Yang, T., Gong, H., Li, X., et al.: Progress of soil moisture monitoring by remote sensing. Acta Ecologica Sinica 30(22), 6264-6277

[2] Han, N.: Retrieval of Bare Soil Moisture Using AMSR-E Data. Jilin University (2007)

[3] Wang, J.R., Engman, E.T., Shi, J.C., et al.: The SIR-B observations of microwave backscatter dependence on soil moisture, surface roughness and vegetation covers. IEEE Transactions on Geoscience and Remote Sensing 24, 510–516 (1996)

[4] Gao, T.: Study on Soil Moisture Inversion of Bare Random Surface Based on IEM model. Xinjiang University (2010)
[5] Li, J.: Soil Moisture Retrieval and Its Spatial Character Analysis in Bare Random Surface. Xinjiang University (2011)
[6] Olea, R., Pawlow Sky, V.: Compensating for estimation smoothing in Kriging. Mathematical Geology 28(4), 407–417 (1996)
[7] Goovaerts, P.: Accounting for estimation optimality criteria in simulated annealing. Mathematical Geology 30(5), 511–534 (1998)
[8] Yamamoto, J.K.: An alternative measure of reliability of ordinary kriging estimates. Mathematical Geology 32(4), 489–509 (2000)
[9] Yamamoto, J.K.: On unbiased back transform of lognormal kriging estimates. Computational Geoscience 11, 219–234 (2007)
[10] Yang, Y., Shang, S., Li, C.: Correcting the smoothing effect of ordinary Kriging estimation soil moisture interpolation. Advances in Water Science 21(2), 208–214 (2010)
[11] Wang, Y., Ouyang, Z., Wang, J., et al.: Predict the parametervarying chaotic time series based on LS-SVM. Computer Engineering and Applications 45(26), 114–117 (2009)
[12] Raftery, A.E., Gneiting, T., Balabdaoui, F., et al.: Using Bayesian model averaging to calibrate forecast ensembles. Mon. Weather Rev. 133, 1155–1174 (2005)
[13] Ajami, N.K., Duan, Q., Sorooshian, S.: An integrated hydrologic Bayesian multimode l Combination framework: Confronting input, parameter,and model structural uncertainty in hydrologic prediction. Water. Resour. Res. 43, W01403 (2007), doi:10.1029/2005WR004745
[14] Neuman, S.P.: Maximum likelihood Bayesian averaging of uncertain model predictions. Stoch. Environ. Res. Risk. Assess. 17, 291–305 (2003), doi:10.1007/800477-003-0151-7
[15] Raftery, A. E., Madigan, D., Hoeting, J.A.: Bayesian model averaging for linear regression models. J. Am. Stat. Assoc. 92, 179–191 (1997)
[16] Sloughter, J.M., Raftery, A.E., Gneiting, T.: Probabilistic quantitative precipitation forecasting using Bayesian model averaging. University of Washington, Department of Statistics, Technical Report 496 (2006)

Analysis of Soil Water Wetting and Dynamics in Trace Quantity Irrigation

Haobo Cui, Shumei Ren, Peiling Yang, Huang Lingmiao, Zixuan Ma, Xiaorui Zhang, Weishu Wang, and Zelin Li

College of Water Resources & Civil Engineering,
China Agricultural University, Beijing 100083, China
yinhexizhan@163.com, renshumei@126.com

Abstract. Compared with ordinary subsurface drip irrigation the trace quantity irrigation has some new features which will provide water according to the demand of the crops and continuous supply right and small amount water to the root of crops slowly, so the flow and moisture distribution is different from others and this Soil box experiments was used to study it. the result shows that when the trace quantity irrigation tape was buried into the soil the flow is approximately 53% of it in the air and the moisture distribution is divided into three kinds according to the depth.

Keywords: trace quantity irrigation, flow, moisture distribution.

1 Introduction

In order to improve the efficiency of water use in agricultural irrigation, we shall do some further research on the basis of existing water saving irrigation techniques to promote water conservation and the development of agriculture. Underground drip irrigation technology is a high efficient and water saving irrigation technique on the basis of the development of the drip irrigation technology[1].

Trace quantity irrigation technology is the perfection of the underground drip irrigation technology which is increasingly mature, making the quantity of irrigation close to the actual amount of water that the plants need as far as possible is the key word to minimize the invalid evaporation[2].The underground drip irrigation technology has greatly reduced the evaporation and the using of the high frequency of micro-irrigation technology will further reduce the quantity of it, on the premise of that high frequency will ensure the growth of crop , try to make the supply of water exactly equal to the quantity that absorbed by roots during the corresponding period of time. Theoretically, uninterrupted flow is demanded during whole life cycle of crop, if we can supply the flow to root zone along with the change of the requirements of different times that will has a certain extent reduce of evaporation. Based on the above ideas and trace quantity irrigation is developed based on soil capillary force principle and membrane filtration technology which will supply continuous, slowly amount of water to the crops[3,4].

© IFIP International Federation for Information Processing 2015
D. Li and Y. Chen (Eds.): CCTA 2014, IFIP AICT 452, pp. 489–494, 2015.
DOI: 10.1007/978-3-319-19620-6_55

2 Experiments and Methods

Soil box size is 1.2m × 0.8m × 0.6m, emitters spacing are 30cm, arranged four emitters, the two sides emitters are 15cm from the box edge.

Before buried trace quantity irrigation tape into the soil, conducted background value measurement, using soil auger which diameter is 3cm to take soil in the box, Randomly selected three points, one point take six layers are 0-10, 10-20, 20-30, 30-40, 40-50, 50-60cm, the oven drying method used for determining soil moisture content. Soil moisture background value shown in Table 1.

Table 1. Background values of soil moisture content

Depth/cm	0-10	10-20	20-30	30-40	40-50	50-60
Soil Moisture Content/%	5.77	6.17	8.09	9.17	6.96	9.39

Using Markov Bottles to supply water, Markov bottles will be placed on a tripod, adjust the height, maintain 1 meter constant head to supply water for trace quantity irrigation tape. The trace quantity irrigation tape was buried 15cm under the soil[5,6].

Collecting sample during irrigation 1d, 4d, 10d, 16d, respectively, using soil auger to get three points (three points are 5cm, 12.5cm and 25cm away from the emitters),each point take 0-10,10-20,20-30,30-40,40-50,50-60 cm layers of soil, the first sample we collect from the end of trace quantity irrigation tape.

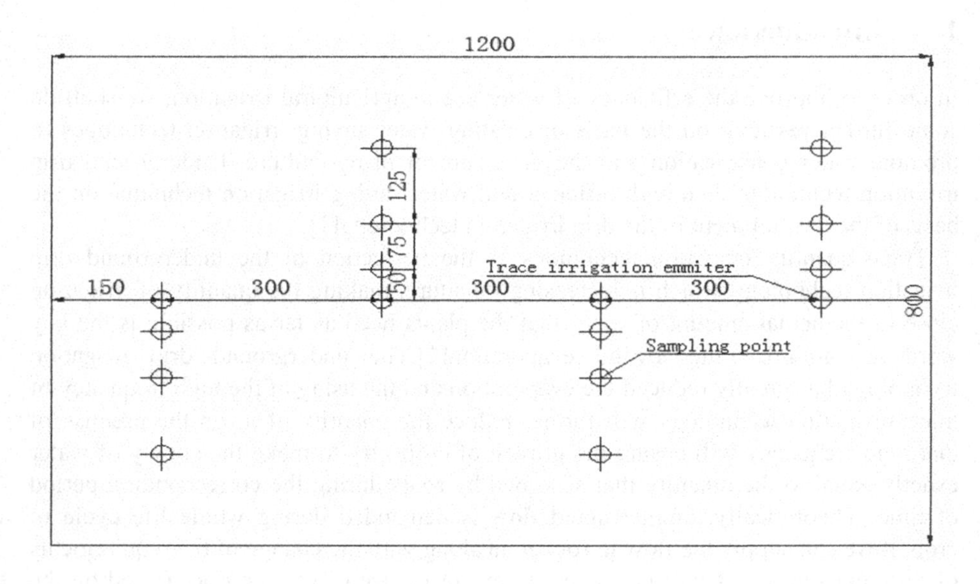

Fig. 1. Soil box floorplan and sampling points（unit：mm）

3 Data Analysis and Discussion

After continuous observation for 16 days, we regularly read Markov bottle readings, the single emitter flow amount is shown in the table below.

Before the experiment, we test the average flow of the four emitters in air under 1m head, the average flow was 0.36L/d.

Table 2. Single emitter flow amount

Date	Time	Cumulative flow amount /L	Average flow amount /L
13-Apr-14	4:50 PM	0.078	
13-Apr-14	5:39 PM	0.2756	
14-Apr-14	4:50 PM	0.462	0.1864
17-Apr-14	4:10 AM	1.053	0.197
19-Apr-14	5:00 PM	1.5574	0.2522
23-Apr-14	6:00 PM	2.4076	0.21255
24-Apr-14	3:13 PM	2.5766	0.169
27-Apr-14	3:40 PM	3.08	0.1678
29-Apr-14	4:00 PM	3.4138	0.1669
30-Apr-14	5:00 PM	3.5646	0.1508

During the experiment, We monitored the single emitter of trace quantity irrigation tape in the interval sampling time, the average daily flow amount showed a trend that first increase and then decrease, with soil gradually wetting. On April 14th, the average flow amount was 0.1864L/d, then on April 19th, peaked at 0.2522L/d, after that, rapidly decreased to 0.1508L/d. The mean flow amount was 0.1935L/d, approximately 53% flow amount in the air.

We took four times soil samples during the experiment, measured three points (three points are 5cm, 12.5cm and 25cm away from the emitters),each point take 0-10,10-20,20-30,30-40,40-50,50-60cm layers of soil, then draw line chart of soil moisture charge with layers by Excel, shown in Fig.2 (a, b, c, d, are 10-20, 20-30, 40-50, 50-60line chart, respectively).

Fig.2 indicated that the the trend of moisture content with the distant to the surface of the soil change. Analysis of the problem from the perspective of the wet area horizontal propagation to observe the change of moisture content, whether it is 5 cm, 12.5 cm or 25 cm away from the emitters,the soil moisture content is gradually increasing with time at 10-20cm,and when the area is nearer to the emitters the moisture content increased faster,after 10 days' experiment this trend begin to slow down and at the following 6 days' experiment the increased of moisture content is 15% of it increased at the beginning of the 10 days' experiment.To compared with

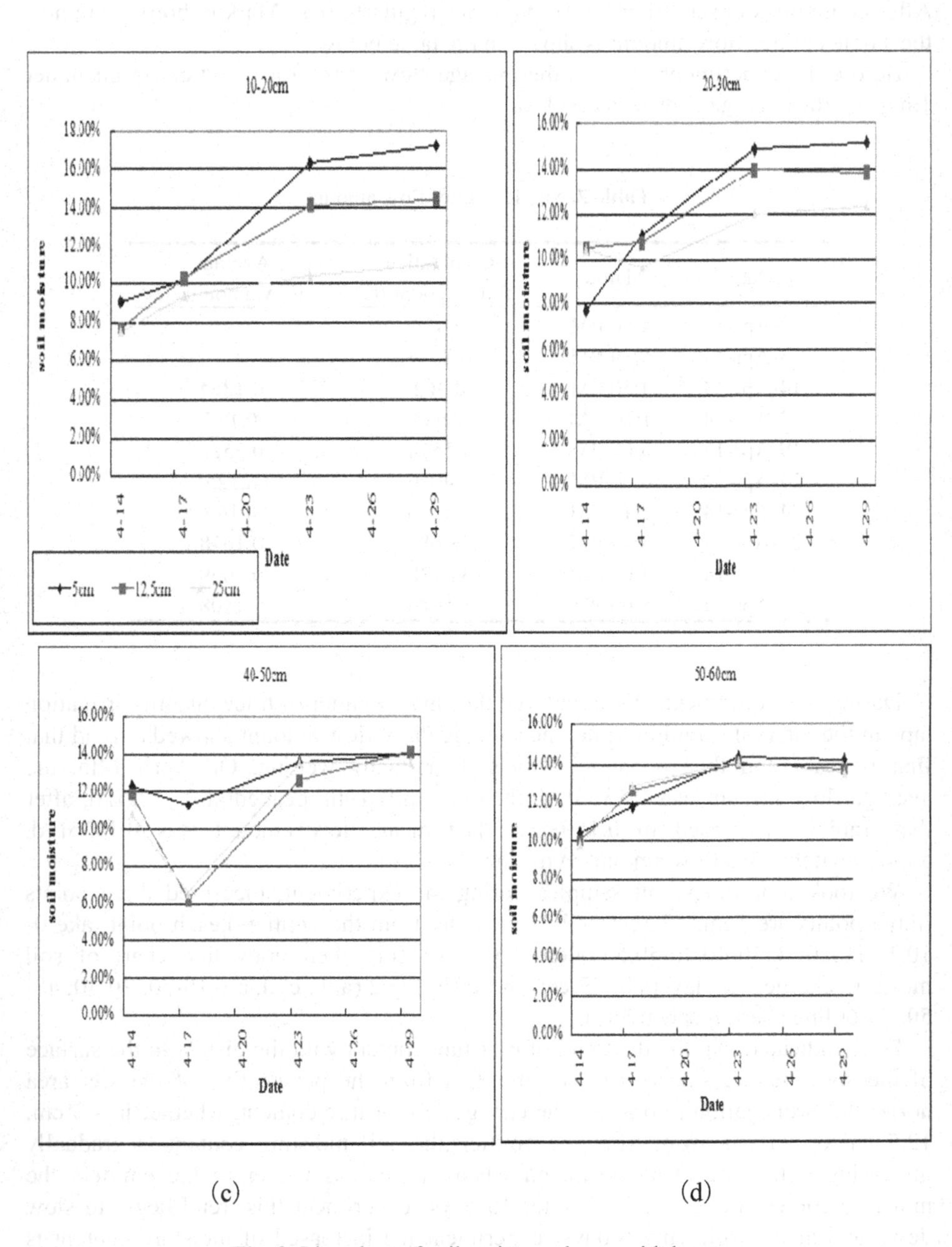

Fig. 2. Line chart of soil moisture charge with layers

10-20cm that the trend of increased at 20-30cm is slow down and even appeared to reduce,this phenomenon is sharped in the region of the 40 to 50 cm away from the surface of the soil, after 3 days' experiment at the beginning the moisture content fell down from 11% to 6%.This situation has changed at 50-60cm, soil moisture content in this region has a slowly increased at the fist and then tended to be stable , this process makes the soil moisture content rise 2% and this amount is 25% of the soil moisture content increased at 10-20cm only.

4 Results and Conclusion

Compared with ordinary subsurface drip irrigation the amounts of flow is small and when the emitters buried in the soil the flow amount is about 53% in the air, while the experiment demonstrated that under 1m head only, but the biggest characteristic of trace quantity irrigation is to use soil capillary force and supply water according to crop demand, if we increase the head to pursuit the flow only that is equivalent to give up this feature, but how is the crop growth under the condition of water absorbing and should we need appropriately add head still need further studies.

Due to the flow amount is very few and soil surface evaporation makes that even in the condition of trace quantity irrigation the soil moisture content growing slowly even has a reduce trend in the beginning of 3 to 5 days, but this is most important period to the germination of seeds, in order to make the seeds germinate well we shall let the trace quantity irrigation emitters closely to the surface of soil as far as possible , but the most suitable distance still needs further research.

Acknowledgements. We are grateful for the financial support from the National Science and Technology Ministry (ID: 2014BAD12B06), the Natural Science Fund of China (No. 51179190), Appreciation is also extended to Dr. Jin jishi for equipment installation.

References

1. Wu, F., Li, W., Li, J., Fan, Y., Feng, J.: Experimental study on hydraulic characteristics of subsurface drip irrigation emitter. Transactions of the CSAE 19(2), 85–88 (2003)
2. Yuan, Q., Zhu, D., Ai, P., et al.: Current Situation and Developing Trend of Subsurface Irrigation Technology in Protected Land Cultivation. Transactions of the Chinese Society of Agricultural Machinery 37(9), 199–203 (2006)
3. Zhu, J., Jin, J., Yang, C.: Effects of trance quantity irrigation on yield, dry matter portioning and water use efficiency of spherical fennel grown in greenhouse. Journal of Drainage and Irrigation Machinery Engineering 32(4), 338–342 (2014)
4. Yang, M., An, S., Zhou, J., et al.: Effects of Different Trace Irrigation Pipe Depths on Eggplant Growth, Yield and Water Utilization Efficiency in Solar Greenhouse. China Vegetables 20, 78–82 (2012)

5. An, S., Zhou, J., Liu, B., et al.: Effects of different fertilizing amount on yield and fruit quality of Pimento under trace irrigation. Zhongguo Yuanyi Wenzhai (4), 17–18 (2013)
6. Chen, X., Cai, H., Shan, Z., et al.: Yield and Quality of Tomato and Cucumber under Non-Pressure Subsurface Drip Irrigation at Crop Root Zone. Acta Pedologica Sinica 43(3), 486–492 (2006)

A Particle Swarm Optimization Algorithm for Neural Networks in Recognition of Maize Leaf Diseases

Jia Tao

College of Information Science and Technology,
Agricultural University of Hebei, Baoding, 071001
taojiand@126.com

Abstract. The neural networks have significance on recognition of crops disease diagnosis, but it has disadvantage of slow convergent speed and shortcoming of local optimum. In order to identify the maize leaf diseases by using machine vision more accurately, we propose an improved particle swarm optimization algorithm for neural networks. With the algorithm, the neural network property is improved. It reasonably confirms threshold and connection weight of neural network, and improves capability of solving problems in the image recognition.At last, an example of the emluatation shows that neural network model based on pso recognizes significantly better than without optimization. Model accuracy has been improved to a certain extent to meet the actual needs of maize leaf diseases recognition.

Keywords: neural network optimization, Particle swarm optimization, Opposition-Based Learning, maize leaf diseases.

1 Introduction

Grey speck disease, brown spot and leaf blight are important leaf diseases in the warm and humid maize areas in china, and they have become more and more serious for recent years[1-3]. The traditional identification method is observing with naked eyes by plant protection expert and seasoned farmers which is a time consuming and costly process [4-5]. In recent years, with the development of the image processing and pattern recognition technology, Automatic recognition by neural network has been an active topic in the field of disease recognition. But traditional neural network has the weaknesses such as slow convergent speed, easy getting into local minimum and low rate of correct motion pattern recognition. Particle swarm optimization (pso) is an evolutionary computation technique, it can be used to solve many kinds of optimization problem. In this paper, an improved particle swarm optimization (pso) algorithm is applied to solve the problem, and the improved methods is used in recognition of maize leaf diseases. Experiments show that the recognition accuracy and efficiency of this method are improved compared with other feature extraction methods. It builds a base for maize leaf diseases diagnosing and recognizing, and improve its research and application in this field.

© IFIP International Federation for Information Processing 2015
D. Li and Y. Chen (Eds.): CCTA 2014, IFIP AICT 452, pp. 495–505, 2015.
DOI: 10.1007/978-3-319-19620-6_56

2 Swarm Optimization Algorithm

Particle swarm optimization seeks and traversals the optimal particle in solution space. Here surpose the spatial dimension of situation is D and the particle swarm number is S.so particles i is expressed by the formula :

$$X_i(x_{i1}, x_{i2}, \cdots, x_{id})\,(i = 1, 2, \cdots, S; d = 1, 2, \cdots, D)$$

Running speed of i is expressed by Vi in the formula :

$$V_i(v_{i1}, v_{i2}, \cdots, v_{id})$$

The optimum point of i is expressed by the formula :

$$P_i(p_{i1}, p_{i2}, \cdots, p_{id})$$

Global advantage of the number of particles is expressed by the formula :

$$P_g(p_{g1}, p_{g2}, \cdots, p_{gd})$$

Position and speed of all the Particle swarm is shift by using iteration method.the iterative rule is expressed by the formula :

$$v_{id}^{k+1} = \omega_i v_{id}^k + c_1 r_1 (p_{id}^k - x_{id}^k) + c_2 r_2 (p_{gd}^k - x_{id}^k)$$

$$x_{id}^{k+1} = x_{id}^k + v_{id}^{k+1}$$

The value of k is related to the number of iterations and c1, c2 is related to the accelerated factor. Using the alterable parameter r1, r2 to adjust the speed and give expression to the randomicity of movement.The motion inertia is expressed by the parameter ωi. The following factors determine the real-time speed and position:

(1) The last speed and location;
(2) The trend to approach to optimal location;
(3) Members of particle swarm timely adjust the speed and location for information exchange.

Traditional neural network has low rate of correct motion pattern recognition, and slow convergence rule of the network [6-9].We introduce pso to improve the algorithm.

3 Neural Network Based on Improved PSO

3.1 Opposition-Based Learning

The principle is as the following [10]:

When searching the optimum solution x, the usual method is to begin from initial point χ which is determined randomly or according to the experience and obtains the global solution. To resolve the complex problem such as initialization of the weights of the neural network, the method adopted is determined by random samples. A problem to solve is when the random initial weights is apart from the optimal solution, calculating for optimization and searching is a CPU intensive work and convergence is difficult.So in theory, initial feasible solution can be investigated from at all positions and directions of the random point. Supposes that the comparative direction is beneficial for searching and the definition of point is given below firstly [11-14]:

Definition 1: Suppose $x \in [a,b]$ is a real numbers,and its opposite point $\chi$is definded as follows:

$$\chi = a + b - x$$

Similarly, opposite point in a multidimensional space is definded as follows:

Definition 2: Suppose that $P(x_1, x_2 ... x_n)$ is a point in n dimensional space, $x_1, x_2 ... x_n \in R$ and $x_i \in [a_i, b_i]$, $\forall i \in \{1,2......n\}$, the opposite point of P is definded as $P(\chi_1, \chi_2 \cdots \chi_n)$,in which $\chi_i = a_i + b_i - x_i, i \in \{1,2......n\}$.

The method of Opposition-Based Learning is provided below :

Suppose the function to be optimized is $f(x)$ and the fitness function is $g(.)$ which is used to evaluate the quality of candidate solution. $x \in [a,b]$ is a random initial point andχ is the opposite point of x. In the courses of iterative optimization,the values of x and$\chi$are first calculated. Then by comparing the fitness function of the two points to determine the larger one.If $g(f(x)) > g(f(\chi))$, values of x is regarded as the retention value, otherwise, χ.

For example, when optimizating a function of one variable defined on the interval $[a1, b1]$ as shown in fig. 1, the method is: through repeated iteration to evaluate the candidate solution and its opposite solution to find the optimal solution. Initial the x point firstly and obtain its opposite point x_0.then calculat the the distance of d and d_0 with the optimum solution respectively.if $d_0 < d$, binary search the space of x, otherwise of x_0. Iterate on till the distance with the optimum solution is less than the predefined thresholds.

Opposition-Based Learning is introduced in order to enhance the performance procedure is provided as following:

Step1: By using random generation method, the initial uniformly random distributed population is $X = \{Xi, Vi \mid i = 1,2,...N\}$.

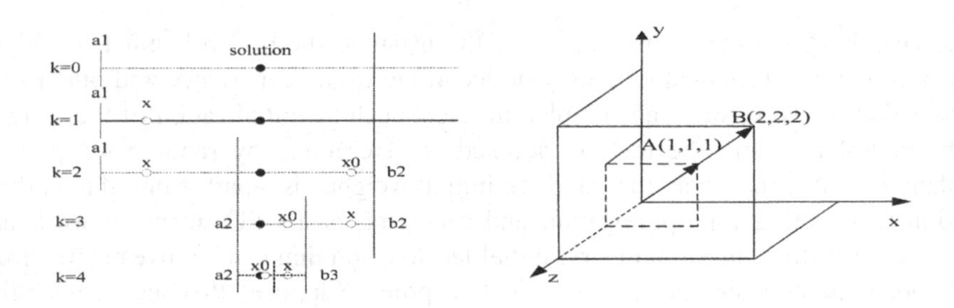

Fig. 1. Optimize single variable

Fig. 2. Simple definition of expand variatio

Step2: for each particle in populationX, constitute the opposite population from the calculation of the opposite particle $OX = \{OXi, OVi \mid i = 1, 2, \ldots N\}$ whose position and velocity are described as $ox_{id} = L_d + U_d - X_{id}$ and $ov_{id} = V_d^{\min} + V_d^{\max} - v_{id}$.

Step3: according to the fitness, choose n particles as initial population $X^0 = \{X_i^0, V_i^0 \mid i = 1, 2, \ldots N\}$ from X and OX.

In order to search out the global optimal solution in a higher dimensional space, we introduce the mutation model to obtain 2 particles by expansion and contraction. Fitness degree values among the 3 particles are compared and the best is retain for r the next iterative.

For example shown as the figure2, a particle expansion is from A (1, 1, 1) to B (2, 2, 2).

Its position changes as the formula :

$$x(i+1) = x(i) + a * x(i)$$

3.2 Particle Swarm Optimization

Based on particle swarm optimization algorithm, parameter t is used as the adaptability function. The formula is the expression:

$$t = \frac{1}{N} \sum_{i=1}^{N} \sum_{j=1}^{m} (y^d{}_{j.i} - y_{j.i})^2$$

The sum of example is denoted by N. The meaning of $y^d_{j,\ i}$ is the predict value from samples i and node j. Let $y_{j,\ i}$ denote the actual value related the predict one. The number of the nodes is denoted by m. the normal steps to establish neural network can be summarized as follow:

Step1: To design and train the structure of neural network by the training documents cluster; to determine the initial parameters;

Step2: To determine the particle swarm initial value based on the neural network which is constructed in Step1;
Step3: By initializing a random particle swarm, updating the velocity and position of particles in accordance with the fitness of particles, searches the optimal coordinates through iterative searching.
Step4: The scheme stops iterative computing when the iterative number is is correct. The global optimal solution and the network structure are obtained.If scheme condition is not satisfied, go to Step3.

4 Experimental Results

4.1 Acquisition and Processing

Diseased leaves of leaf blight, gray leaf spot, and brown spot are collected from experimental station of Hebei agricultural university in 2013. The collected images are saved as jpg file.

Image preprocessing schemes include gray processing, histogram equalization and Image segmentation.then from the open and close operations,Fig. 3 and Fig. 4 show the preprocessing effects:

Fig. 3. Median filtering before(left) and after(right)

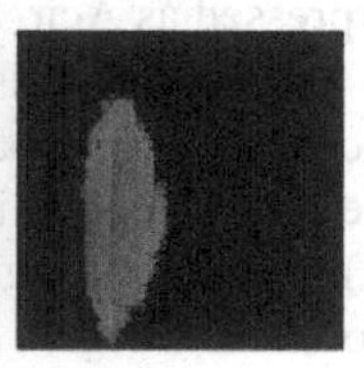

Fig. 4. Image segmentation before (left) and after (right)

4.2 Feature Extraction

Shape Features Extraction
(1) Lesion area
After image segmentation, the area parameter is the number of pixels of disease part, which is expressed as A_0:

$$A_0 = \sum_{i=1}^{N} f(x, y)$$

(2) Geometrical center
The centroid of 2d shape of leaf lesions is regarded as geometrical cente, in which R is the diseased spots:

$$\begin{cases} \overline{x} = \dfrac{1}{A_0} \sum\limits_{(x,y)\in R} x \\ \overline{y} = \dfrac{1}{A_0} \sum\limits_{(x,y)\in R} y \end{cases}$$

(3) Minimum exterior rectangle

Using the four vertex coordinates of encircle rectangle,this parameters is represented as

$(X_{\min}, Y_{\min})$ $(X_{\max}, Y_{\min})$ $(X_{\max}, Y_{\max})$ $(X_{\min}, Y_{\max})$, among which X_{min}, X_{max}, Y_{min}, Y_{max} denote respectively maximum and minimum of ordinates and abscissas.

(4) Rectangle degree

Rectangle degree means that the ratio of lesion area and minimum exterior rectangle which is expressed Rt :

$$Rt = A_0 / A_{cir}$$

The area of lesion is expressed as A0 and the area of minimum exterior rectangle is expressed as Acir. Thus it can be seen that Rt∈[0, 1].

If value of Rt is approaching to 1,the spot shape can be regarded more similar to rectangle, if approaching to$\pi/4$, the shape can be a circular. Other values indicate that the spot is mainly irregular in shape.

(5) Roundness degree

Roundness degree means that the similarity between the circular and the spot shape :

$$C = 4\pi A_0 / L^2$$

In the formula given above, the circumference of disease spot ferred to as L respectively. Thus it can be seen that C∈[0, 1]. If value of C is approaching to 1, the spot shape can be regarded more similar to circular.

(6) Figure complexity

This parametric reflects the discreteness of the spot shape which is expressed as S :

$$S = L^2 / A_0$$

L and A_0 denote respectively the circumference and area of disease spot. The larger the value of circumference is, the stronger the discreteness and image complexity will be.

4.3 Color Features Extraction

HSI color space is introduced to extract the color feature. Its main advantage is stable structure and less dimensions.So the image is converted from RGB model to HSI one by the formula [15-16] :

$$\begin{cases} \theta = \arccos\{\frac{\frac{1}{2}[(R-G)+(R-B)]}{[(R-G)^2+(R-G)(G-B)]^{\frac{1}{2}}}\} \\ S = 1 - \frac{3}{(R+G+B)}[\min(R,G,G)] \\ I = \frac{1}{3}(R+G+B) \\ H = \begin{cases} \theta, G \geq B \\ 360-\theta, G < B \end{cases} \end{cases}$$

The results of study show that abundant information of the source image is included in the lower order moments and middle order moments of color moment. The most remarkable characteristic of three components in RGB is the B component about maize leaf diseases.

4.4 Texture Features Extraction

We use a gray-primitive co-matrix to describe the feature more exactly. Let N be the the grey step in leaf diseases image. The gray level cooccurrence matrix is expressed by the formula :

$$M_{(\theta,d)}(i,j) = \frac{\{[(x_1,y_1),(x_2,y_2)] \in S | f(x_1,y_1)=i \wedge f(x_2,y_2)=j\}}{S}$$

In the formula, denote the gray level cooccurrence matrix by M(θd), in which θ represents the direction and d represents the spatial distances between pixels.So the meaning is the probability of coexisted pixels in one diseased spots which respectively has the gray level of i and j . Denote the coordinates of a pair of pixel by (x_1, y_1) and (x_2, y_2) . Thus, the pixel number is expressed as $f(x_1, y_1)$ and $f(x_2, y_2)$. The total number of coexisted pixels which satisfies the conditions is denoted by S. The following parameters are selected for texture features expression through a number of experiments.

(1) E(θ, d)

It represents the energy of the matrix which can be represented by the formula :

$$E(\theta,d) = \sum_{i}^{n}\sum_{j}^{n} M_{(\theta,d)}(i,j)^2$$

In the formula, M(θd) represents the gray level cooccurrence matrix, I and j represent the gray value of pixel pair in diseased spots. E(θ, d) has a higher value if there is most energy aronnd the diagonal.

(2)H(θ, d)

It represents the entropy of gray level cooccurrence matrix which can be represented by the formula :

$$H(\theta,d) = -\sum_{i}^{n}\sum_{j}^{n} M_{(\theta,d)}(i,j)\log_2 M_{(\theta,d)}(i,j)$$

The value of H(θ, d) directly proportional to image information quantum. H(θ, d) has a higher value if image texture distribution is equilibrium.

(3)I (θ, d)

It represents the moment of inertia of gray level cooccurrence matrix which can be represented by the formula :

$$I(\theta,d) = \sum_{i}^{n}\sum_{j}^{n} (i-j)^2 M_{(\theta,d)}(i,j)$$

The value of I (θ, d) is related to the image clarity of texture. I (θ, d) is very small if the center of matrix is near main diagonal which also reflects the image texture is roughness and obscure.

(4) C(θ, d)

It represents the correlation of inertia of gray level cooccurrence matrix which can be represented by the formula :

$$C(\theta,d) = \frac{\sum_{i}^{n}\sum_{j}^{n} i^* j^* M_{(\theta,d)}(i,j) - \mu_x\mu_y}{\sigma_x\sigma_y}$$

The value of I (θ, d) is related to the similarity of row and column elements. In the formula, μ_x represents the mean of each column sum in gray level cooccurrence matrix. μ_y represents the mean of each row sum. σ_x represents the variance of each column sum and σ_y the each row sum[17].

4.5 Recognition of Diseases

The design of a typical three-layer structure neural network is constructed by sigmod. The shape features of maize disease spots image consists of 6 attributes as follow: lesion area, geometrical center, minimum exterior rectangle, rectangle degree, roundness degree, figure complexity.

The color features consists of 6 attributes as follow: the first, second and third moment of B and H components.

The texture features consists of 8 attributes as follow: the respective mean value and standard deviation of E (θ, d), H (θ, d), I (θ, d), C (θ, d).

All of these attributes are used as the inputs of artificial neural network. It contains 20 neurons in total. The output neurons are composed of three major leaf disease of maize: leaf blight, maize brow spot and grey speck disease.

The number of the hidden layer nodes is 19 combining kolmogorov algorithms. After training the network based on the method of genetic algorithm the optimized weight and threshold are obtained.

The traditional neural network and the optimization methods are trained respectively and finally the suggestibility is compared.

Choose training samples and confirme learning speed as 0.01. A total of 400 effective samples are collected for neural network training which is composed of 175 grey speck disease images, 105 brown spot and 120 leaf blight. Experimental result show that the neural network model convergences at 22 and the optimization neural network at 10. The simulation results show that this algorithm can find the optimal solution more rapidly. A comparison is presented as figure5 and 6 below:

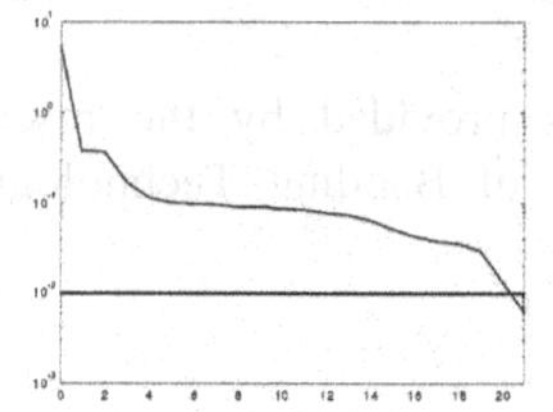

Fig. 5. Traditional neural network convergence

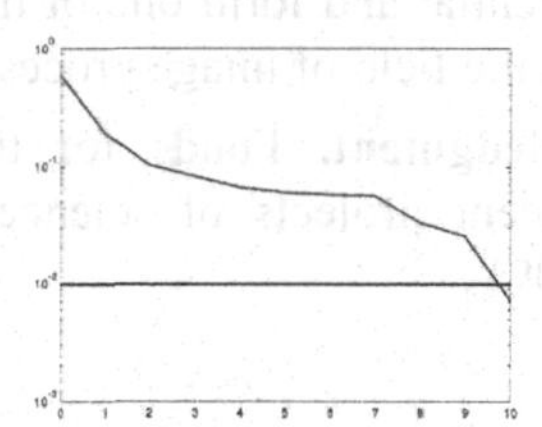

Fig. 6. Optimization neural network convergence

Test of the neural network is based on 60 images of maize disease respectively.the recognition result is compared below:

Table 1. I the maize leaf disease image recognition results

maize leaf diseases	grey speck disease	brown spot	leaf blight	recognition rate
total numbers of images	60	60	60	——
The recognition rate of traditional optimization network %	86.7	88.3	88.3	87.8
Optimization of optimization network identification rate %	91.7	93.3	95.0	93.3

The result showes that the identification rate of optimization network is 93.3%, As contrast, at the same time, the traditional neural network is 87.8%.conclusions were summarized that the recognition rate is more pronouncedly improved.

The predict value of optimization network is represented by the four spots. The predict value is expressed by abscissa and neural network output the ordinate.the actual value is expressed by the dashed line and the the fitting value is the solid line. Comparing the simulation results with the real data, we can find that they are similar whose fitting degree R is 0.997.

5 Conclusions

In this section, we propose an improved particle swarm optimization algorithm for neural networks and applied it for the recognition and diagnosis of main maize leaf diseases. The algorithm is based on Opposition-Based Learning and makes the pso high efficiency in searching for the best solution in the global area to improve neural network predictive model.

Research on neural network in image recognition continues at a rapid pace. This survey provides an introduction to the main concepts of an improved particle swarm optimization algorithm for neural networks and applied it to the diagnosis of maize disease. The simulation result shows the effectiveness of the method. However, the new optimal methods of neural network can obtain preferable purpose too, such as simulate anneal algorithm, Genetic algorithm et al. At present, the technique of disease detection based on image processing is still a new field of application.it show great potential and form one of the dominant research directions in both agricultural field and the field of image processing.

Acknowledgment. Funds for this research was provided by the research and development projects of science and technology of Baoding Technology bureau (13ZN009) .

References

1. Wang, Z.L., Li, Y.C., Shen, R.F.: Correction of soil parameters in calculation of embankment settlement using a BP network back-analysis model. Engineering Geology 91(2/3/4), 168–177 (2013)
2. Ji, X.D., Familoni, B.O.: A diagonal recurrent neural network-based hybrid direct adaptive SPSA control system. IEEE Transactions on Automatic Control 44(7), 1469–1473 (2012)
3. Kenned, J., Ebemart, R.C.: Partical swarm optimization. In: Proceeding of 1995 IEEE International Conference on Neural Networks, pp. 192–194. IEEE, New York (2013)
4. Cui, Y., Cheng, P., Dong, X., et al.: Image processing and extracting color features of greenhouse diseased leaf. Transactions of the CSAE 21(suppl. 2), 32–35 (2005)
5. Ma, X., Qi, G.: Investigation and recognition on diseased spots of soybean laminae based on neural network. Journal of Heilongjiang First Land Reclamation University 18(2), 84–87 (2006)
6. Lamedica, R., Prudenzi, A., Sforna, M., et al.: A Neural Net Work Based Technique for Short-term Forecasting of Anom-alous Load Periods. IEEE Trans. on Power Systems 11(4), 1749–1755 (2006)
7. Alfuhaid, A.S., El-Sayed, M.A., Mahmoud, M.S.: Cascaded Artificial Neural Networks for Short-term Load Forecasting. IEEE Trans. on Power Systems 12(4), 1524–1529 (2007)
8. Bakirtzis, A.G., Petridls, N.: A Neural Network Sort Term Load Forecasting Model for the Greek Power System. IEEE Transaction on Power Systems 11(2), 638–645 (2006)
9. Chow, T.W.S., Leung, C.T.: Neural Network Based Short-term Load Forecasting Using Weather Compensation. IEEE Trans. on Power Systems 11(4), 1736–1742 (2006)

10. Kim, J., Bentley, P.: Towards an Artificial Immune System for Network Intrusion Detection: AnInvestigation of Clonal Selection with a Negative Selection Operator. In: Proc. Congress on Evolutionary Computation, pp. 27–30 (2001)
11. Smith, D.J., Forrest, S., Perelson, A.S.: Immunological Memory Is Associative. In: Artificial Immune Systems and their Applications, pp. 105–112. Springer, Berlin (2008)
12. Endoh, S., Toma, N., Yamada, K.: Immune Algorithm for N-TSP. In: Proc. IEEE International Conferenceon Systems, Man, and Cybernetics, pp. 3844–3849 (2008)
13. de Castro, L.N., von Zuben, F.J.: Learning and Optimization Using the Clonal Selection Principle. IEEE Trans. Evolutionary Computation. 6(3), 239–251 (2002)
14. Varela, F., Coutinho, A.: Second Generation Immune Networks. Immunology Today 12, 159–167 (2001)
15. Xu, Y.: Wavelet transform domain filters: A spatially selectivenoise filtration technique. IEEE Trans. IP-3(6), 747–757 (2004)
16. Donoho, D.: L1De-noising by soft thresholding. IEEE Trans. Inform. Theory 41(3), 613–626 (2005)
17. Mallat, S.G.: Characterization of signals from multiscales edges. IEE Transaction on Pattern Analysis and Machine Intelligence 14(7), 710–732 (2002)

Applied Research of IOT and RFID Technology in Agricultural Product Traceability System

Guogang Zhao[1,2], Haiye Yu[1,2], Guowei Wang[1,2,3], Yuanyuan Sui[1,2], and Lei Zhang[1,2]

[1] College of Biological and Agricultural Engineering,
Jilin University, Changchun 130022, China
[2] Key Laboratory of Bionic Engineering, Ministry of Education, Changchun 130022, China
[3] School of Information Technology, Jilin Agricultural University, Changchun 130118, China
{zhaoguogang2000,41422306}@qq.com, {haiye,z_lei}@jlu.edu.cn,
suiyuan0115@126.com

Abstract. In recent years, more and more attention has been given to the safety of agricultural products, whose demands for the traceability system are increasingly urgent. This paper, in response to the demands, by using the C# programming language, database technology, IOT technology and RFID technology, realized the traceability system of agricultural products, and explained the implementation of network technology and RFID technology in detail. The traceability system can complete the functions of data collection, early warning, control and data automatic input & management during the production process. This system can well solve the problem of data entry in the farming operation, timely and effectively record the information involved in the production of agricultural products, which can ensure the authenticity of the data in the traceability system.

Keywords: agricultural products, the traceability system, IOT, RFID.

1 Introduction

With the improvement of people's quality of life, the agricultural products circulation are facing new situations esp.after China entered the WTO, and the security of agricultural products has caused more and more people's attention [1]. Because China is a large agricultural country, agricultural products' export plays an important role in the foreign trade of China. In recent years, vegetables, tea, mushrooms, meat, canned food and other agricultural products and processed food, which Chinese exported to America, Japan, the European Union and other countries, have had quality problems, which not only makes China suffer huge economic losses but also makes the exported agricultural products of China lose good reputation [2]. Traceability system is the important measures and guarantee to promote the production information transparency, improve the health and safety of agricultural products and increase the market competitiveness of agricultural products [3-4].

In recent years, some domestic experts, from the perspective of information technology, taking the primary products of some vegetables as the research objects,

© IFIP International Federation for Information Processing 2015
D. Li and Y. Chen (Eds.): CCTA 2014, IFIP AICT 452, pp. 506–514, 2015.
DOI: 10.1007/978-3-319-19620-6_57

has constructed the traceability system for safety production & management and quality of vegetables, has discussed the key technologies. Some of them has founded the traceability procedure taking the agricultural retrospective codes based on geographical coordinates and multiple encryption as the encoding, and some of them has founded the quality and safty traceability system of agricultural products, which depending on mobile phone 2D barcode recognition. At abroad, there also has some research on the product traceability system, such as the Australian animal identification and traceability system, food traceability system in Japan, the EU beef traceability system, American full traceability system for agricultural products and the Swedish agricultural products traceability management system [5-15]. Based on the above literature, it can be seen that the domestic and foreign experts and scholars have made some achievements in the aspects of vegetable traceability system,but since the complexity of the manufacturing process of agricultural products, the validity and objectivity of data acquisition would not be safeguarded. The IOT technology and RFID technology can solve this problem to a certain extent, which is the research direction of this paper.

2 The Structure of Vegetable Traceability System

The traceability system for the overall structure is shown in Fig.1. The system obtains the data from the process of vegetable production by the IOT and RFID system and record them truly and effectively to avoid manual intervention.

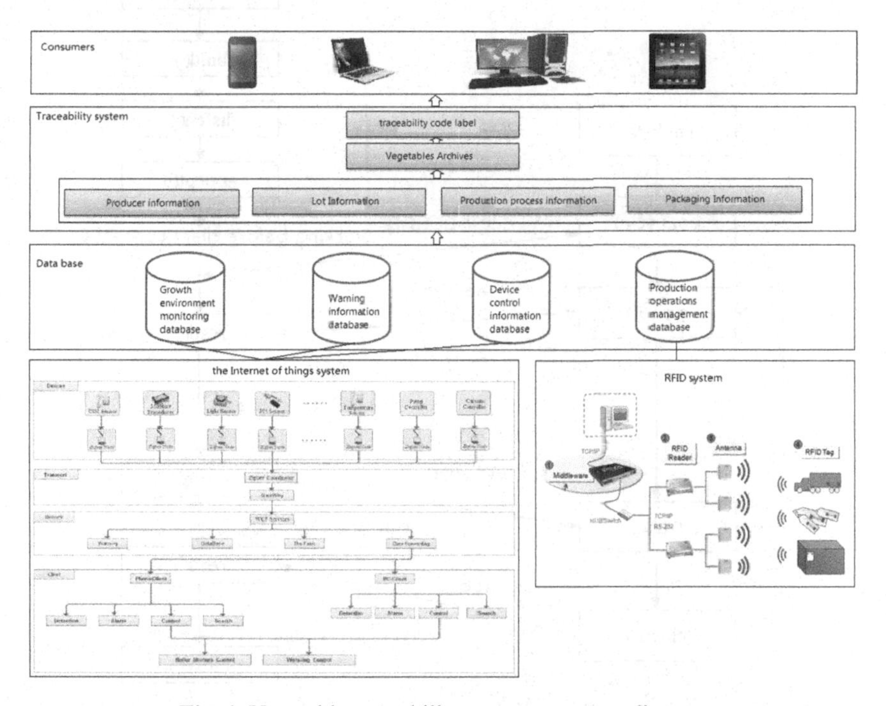

Fig. 1. Vegetable traceability system structure diagram

3 Technical Design of Internet of Things(IOT).

3.1 Data Acquisition of Vegetables Environment

One of the main process of data acquisition and storage: hardware gathered the data through the wireless network to the specifized gateway,finaly data which is collected will be transmitted the specified IP address (i.e., IP address of the server).The server opens a monitor to create a TCP connection with hardware gateway first and then listens to data sent from the hardware gateway (the process as shown in Fig. 2).Through data packets which is obtained from remote gateway, corresponding recognition and resolution are made to get a variety of distance parameter which we want.Because the node identifier of the corresponding hardware is exclusive, so the data acquisition of hardware packet is done by extracting the hardware identifier, then we can know an exact hardware belongs to which user when the data is compared one to one in the database. After various related parameters are gained, the collected data is stored in the database according to the original construction of the data table in order to inquire and analyze the data later. Meanwhile, the split of the data in chart form is reflected in the user interface to more directly show the changing various parameters in greenhouse condition.

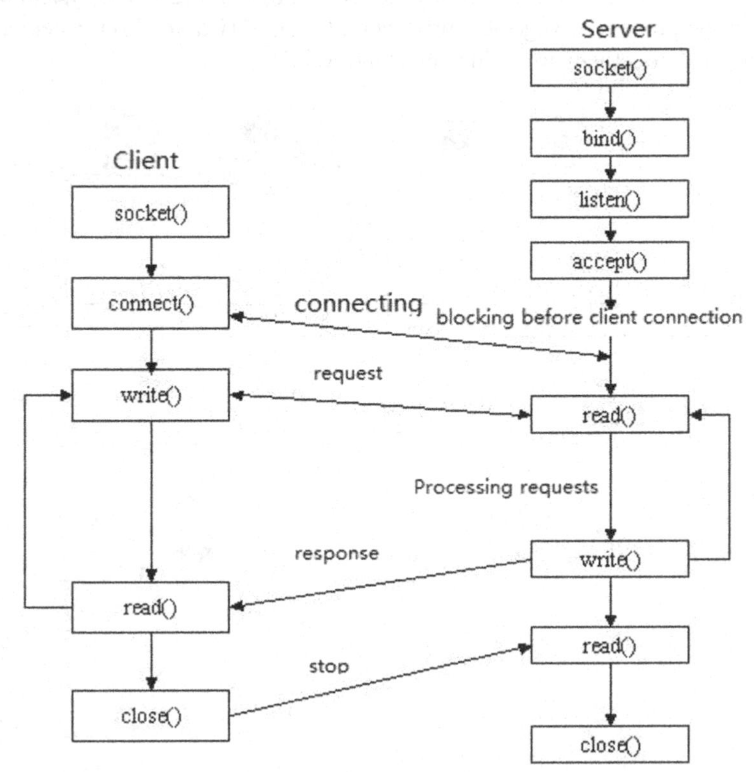

Fig. 2. Flow chart of data acquisition

3.2 Message Warning Model

When the server and remote hardware sensor connection is successful, through the analysis of relevant data transmission by remote hardware package, a rough picture of the remote environment can obtain in real-time. Each of the parameters of environment inside will be compared with the critical parameter values in the database.When either higher or lower than the critical value, we will send the corresponding message alerts to the users which the hardware of monitoring node belonged to. Surely, we have to be ready for this critical value given by experts, according to the data, we finally will be got the up-and-low range monitoring parameters by the variety of vegetables planted in a greenhouse with the parameter and the day (day and night), cloud or snow, crop planting stage. According to above, this domain system automatically decide whether to send the messages to the user to achieve intelligent early warning.

3.3 Equipment Control

When the early warning information is generated, the remote client through the establishment of communication can submit the instruction of opening or closing to the server and the server will transform the command into control information and then transmitted to the wireless sensor network. After the device receives the control information, the command will be taken into effect. But once a remote hardware changes, the server will send the implementation of results to the client. Meanwhile, for the sake of security, the server will also send the message to the exclusive user.

4 Technical Design of RFID

Bwhat RFID technology means lies in the control reader and writer software and acquire or input the magnetic card and electronic label with the correspoding vegetable traceability information whose function is realized by mi.dll.

4.1 Read the Information

Information of Integrated Circuit Card and Electronic label is read by read function.

```
private void btnRead_Click(object sender, EventArgs e)
        {
            byte[] byteuid = new byte[256];
            byte[] byteBuffer = new byte[256];
            string strErrorCode = "";
            string strFlag = "";
            int nFlag = 0;
            if (cbReadFlag.SelectedIndex == 0)
            {
                strFlag = "42";
```

```
                nFlag = 1;
            }
            else
            {
                strFlag = "02";
            }
            int nRet = MiReader.ISO15693_Read(Convert.ToByte(strFlag),
Convert.ToByte (textReadArea.Text), Convert.ToByte(textReadNum.Text), byteuid,
byteBuffer);
                string strText = "";
                int nCardNum = Convert.ToInt32(textReadNum.Text);
                strText += "Read Card:" + "\r\n";
                for (int nLoop = 0; nLoop < nCardNum; nLoop++)
                {
                    strText += ByteArrayToStr(byteBuffer, false, nLoop * (4 +
nFlag) + nFlag + 1, (nLoop + 1) * (4 + nFlag) + nFlag + 1) + "\r\n";
                }
                WriteLog(strText, nRet, strErrorCode);
            }
```

4.2 Write the Information

Information of Integrated Circuit Card and Electronic label is wrote by write function,

```
private void btnWrite_Click(object sender, EventArgs e)
        {
            byte[] byteuid = new byte[256];
            string strErrorCode = "";
            string strFlag = "";
            string[] reslut = strCutLength(textWriteData.Text.ToUpper(), 2);
            byte[] byteBuffer = StrToByetArray(reslut, 4);
            int nRet = MiReader.ISO15693_Write(Convert.ToByte(strFlag),
Convert.ToByte (textWriteArea.Text), Convert.ToByte(textWriteNum.Text),
byteuid,byteBuffer);
            string strText = "";
            for (int nLoop = 0; nLoop < 4; nLoop++)
              {
                if (reslut[nLoop].Length == 1)
                {
                    strText += "0" + reslut[nLoop] + " ";
                }
                else
                {
                    strText += reslut[nLoop] + " ";
                }
```

```
        }
        WriteLog("Write Card" + strText, nRet, strErrorCode);
    }
```

5 The Application of the IOT Technology and RFID Technology

This article, using the C# programming language and the database technology,data bank technology, has realized the data collection, early warning, control in the process of the growth and management of vegetable in the traceability system (e.g. Fig.3, shown in Fig.4), and, using the RFID technology, has realized the effective input of data from vegetables' packing, the processing and transportation to avoided the disturbance from artificial operation, to safeguard the data authenticity and to form the traceable vegetables file information as shown in Fig.5.

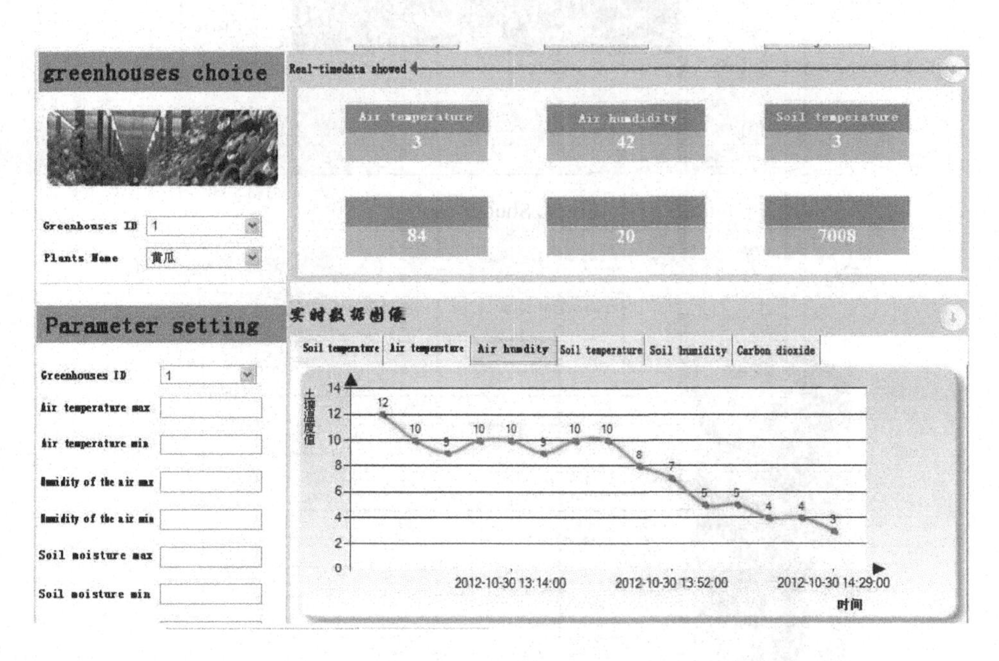

Fig. 3. Real time data acquisition

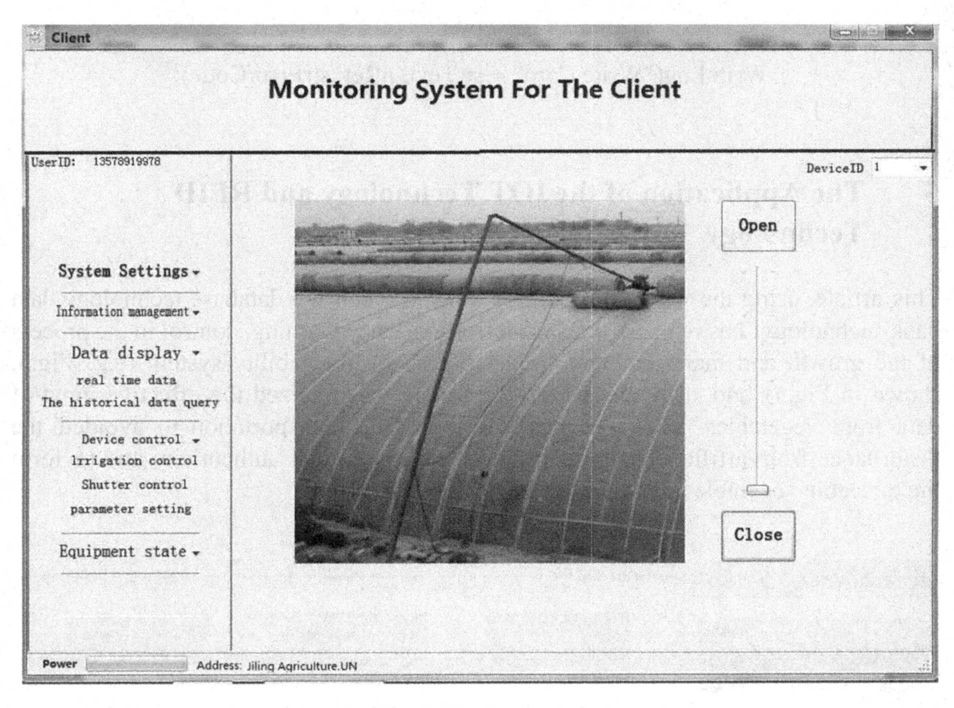

Fig. 4. Shutter control

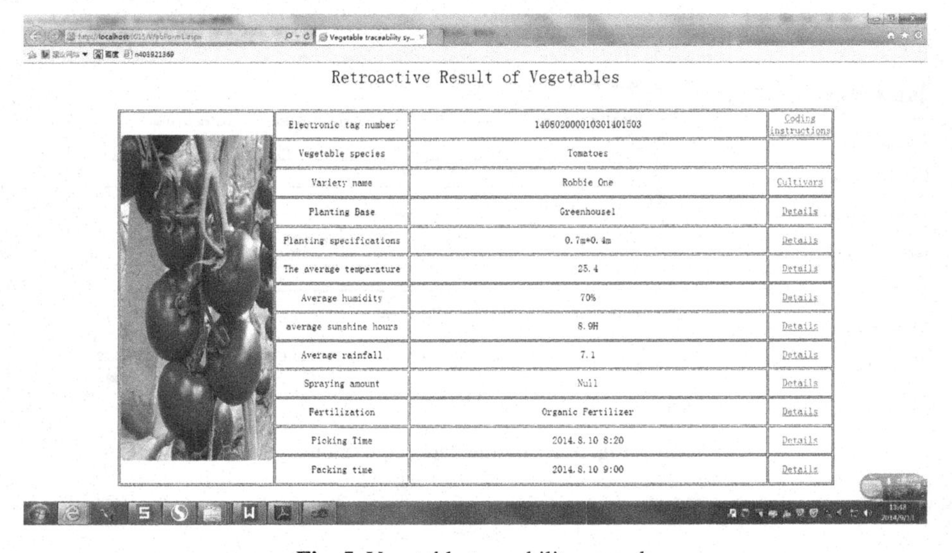

Fig. 5. Vegetable traceability records

6 Conclusions

The system, Using network technology, has realized the automatic acquisition of soil temperature and humidity, air temperature and humidity, illumination and CO2 concentration and so on. The system,depending on whether the collected data can produce early warning, can decide whether need control the corresponding devices to operate, such as rolling machine, irrigation etc. In the process the resulting data is automatically stored in the database, we can understand the process of vegetable production environment and production data for data security, from vegetable traceability system. Through the electronic tags and smart card access to get information and storage, using RFID technology for The farming operation (when or where to use pesticide, packaging, processing and transportation process), packaging, processing and transportation process of vegetable production. RFID technology, reduces artificial complex operation, reduces the information is not timely and not accurate, has solved problem of information reliability in traceability system.

Acknowledgment. Funds for this research was provided by National 863 subjects (2012AA10A506-4, 2013AA103005-04), Jilin province science and technology development projects(20110217).

References

1. Shen, G.-R., Zhao, X.-D., Huang, D.-F.: Consideration of Farm Product Safety System in China. Journal of Shanghai Jiaotong University (Agricultural Science) 23(1), 77–83 (2005)
2. Yang, T.-H., Chu, B.-J.: Study on Control System of Food Safety from Farm to Table. Food Science 26(3), 264–268 (2005)
3. Xie, J., Lu, C., Li, B., et al.: Implementation of pork traceability system based on .NET framwork. Transactions of the CSAE 22(5), 218–220 (2006) (in Chinese with English abstract)
4. Yang, X., Qian, J., Sun, C., et al.: Design and application of safe production and quality traceability system for vegetable. Transactions of the CSAE 24(3), 162–166 (2008)
5. Li, M., Jin, Z., Chen, C.: Application of RFID on products tracking and tracing system. Computer Integrated Manufacturing Systems 16(1), 202–208 (2010)
6. Pettitt, R.G.: Traceability in the food animal industry and supermarket chains. Revue Scientifique Technique-Office International 20(2), 584–597 (2001)
7. De Castro, N.M., Rodrigues, M.B.L., Pinto, P.A., et al.: Traceability on the Web: A Prototype for the Portuguese Beef Sector. In: EFITA 2003 Conference, vol. (2), pp. 5–9 (2003)
8. Mousavi, A., Sarhadi, M., Lenk, A., et al.: Tracking and traceability in the meat processing industry: A solution. British Food Journal 104(1), 7–19 (2002)
9. Zhao, L., Xing, B., Li, W., et al.: Agricultural products quality and safety traceability system based on two-dimension barcode recognition of mobile phones. Transactions of the Chinese Society for Agricultural Machinery 43(7), 124–129 (2012)
10. Meng, M.: Traceability system of agricultural products quality and safety based on B/S structure. Tropical Agricultural Engineering 34(3), 21–24 (2010)
11. Yang, X., Qian, J., Fan, B., et al.: Establishment of intelligent distribution system applying in logistics process traceability for agricultural product. Transactions of the Chinese Society for Agricultural Machinery 42(5), 125–130 (2011)
12. Wang, J., Li, M., Chen, M., et al.: Application of Dongguan agricultural products quality and safety traceability system. Guangdong Agricultural Sciences 39(14), 197–199 (2012)

13. Xu, J., Zheng, Y., Luo, W., et al.: Application of remote video monitoring in agricultural products quality traceability system. Agriculture Network Information (6), 5–8 (2012)
14. Liu, P., Tu, K., Hou, Y.: Traceability system of grain quality safety based on radio frequency identification middleware. Transactions of the Chinese Society of Agricultural Engineering (Transactions of the CSAE) (12), 145–150 (2009)
15. Qian, J., Yang, X., Zhang, B., et al.: RFID-based solution for improving vegetable producing area traceability precision and its application. Transactions of the Chinese Society of Agricultural Engineering (Transactions of the CSAE) 28(15), 234–239 (2012)

A Survey on Quality of Service Monitoring and Analysis of Network of Agricultural Science and Technology Resources

Ma Jian[1,2]

[1] Agricultural Information Institute of Chinese Academy of Agricultural Sciences, Beijing, 100081

[2] Key Laboratory of Agricultural Information Service Technology (2006-2010, Ministry of Agriculture, The People's Republic of China, Beijing, 100081

majian@caas.cn

Abstract. First, current situation on Network of agricultural science and technology resources is described. Then we pay much attention to the quality of service monitoring and analysis system of network resources. And finally, we come to the conclusion that the construction of Quality of service monitoring, analysis of network of agricultural science and technology resources is in great need.

Keywords: Network of agricultural science and technology resources, Quality of Service, Monitoring.

1 The Introduction

Now all kinds of agricultural science and technology resources website has walked into people's life, has gradually become an indispensable part of People's Daily life, people began to used to agricultural science and technology resource information from the Internet. With the speeding up of the society informatization process, people's concerns of the websites of the agricultural science and technology resources are changing, from the initial focus on the website of system quality and the quality of website information gradually shifted to focus on the website service quality, focus on the website and its management ability and willingness to meet user needs. For agricultural science and technology resources website service quality monitoring, analysis and assessment, and can make a website manager timely understanding and diagnosis of the problems existing in the service, in order to promote the service level of the website to provide theoretical basis and data support, has important practical significance.

Research and development "network of agricultural science and technology resources service quality monitoring, analysis and evaluation system" is the purpose of automatically to be monitored website service quality monitoring and analysis, so that you can image, intuitive, accurate and timely of monitoring objects, monitoring index and monitoring methods, analysis methods, assessment mechanism for

© IFIP International Federation for Information Processing 2015
D. Li and Y. Chen (Eds.): CCTA 2014, IFIP AICT 452, pp. 515–520, 2015.
DOI: 10.1007/978-3-319-19620-6_58

centralized control and management, such as security for performance of agricultural science and technology web site.

This article will systematically elaborate network of agricultural science and technology resources service quality related concepts, theories, methods and evaluation system. Subsequent content is organized as follows: section 2 is about the network current situation of the development of agricultural science and technology resources; Section 3 summarizes the current monitoring of the quality of service of technology; Finally, section 4 of this article are summarized.

2 Current Situation of the Development of Network of Agricultural Science and Technology Resources

At present, the agricultural websites have takes on an increase in number, rich content gradually, and huge potential for development of good posture. From 1994, the ministry of agriculture information center was the first to create "China's agricultural information website", the development of Chinese agricultural informatization construction has made great strides. According to the ministry of agriculture information center, agricultural websites in China is only 2200 at the end of 2002. In the year to December 2010 agricultural websites in China for more than 30 thousands, more than the developed countries such as France, Canada, China agricultural website has become the world's top ten.

This paper involved in agricultural science and technology resources is mainly refers to the government agriculture website. Government agriculture website refers to the competent department of agriculture to provide agricultural information to the public in the Internet, the Internet, interaction, and other personalized services platform. At present, the proportion of government web site is on the decline, this is caused by two reasons, one is because of government web sites around the increasingly standardized, integrating multiple websites together. The second is due to the websites of the agricultural enterprises, such as media site number increasing.

Although network of agricultural science and technology resources development situation is good, but there are many problems in developing:

(1) duplicate information, practicality is not strong

On the agricultural website in the content construction of our country exist a degree of repetition. One is due to a lack of unified planning of redundant construction problem, many sites have "regulations" column, its content is typically include China farm law, etc repeat information content. Another is due to the easy replication of network information, many websites will be other agricultural website copy and release of information.

Agricultural information services and the interests of the farmers lack of close contact and targeted, supply and demand of agricultural market information dissemination is not enough; The depth of the information resources development and utilization is not enough, directly related to the local agricultural information share is small, and the prospective predictive information is significant.

(2) the agricultural information standard not unified, difficult to Shared resources

Our country agriculture information resource distribution in different areas and departments, each management main body is according to their own work need to identify the information source, information collecting ways and presentation. Different sources of agricultural information due to a lack of specification, lost the foundation of communication and sharing. Between agricultural management departments, administrative departments of agriculture and frequent, complicated information between production and sales department can't realize the resource sharing, information exchange is often blocked, it is difficult to develop and integrate across departments of agricultural information resources.

(3) the site of utilization rate is not high, not attractive enough to farmers

In August 2009, the average agricultural class web site visitors online time is 5.81 seconds, It show that the majority of visitors for a short stay in agricultural sites, visitors is not interested in most of the content of the site .

3 Network Resources Service Quality Monitoring Technology

3.1 Monitoring Technology Overview

Abroad in the field of network information search, monitor, start early, developing very quickly. Among them, the United States in this area has been in a leading position, in addition to the established military "joint secret - computer network defense system" JTF – CND and the government of the federal intrusion detection network “FIDNet” , at the moment, is to build the global Internet for text, voice and image search analysis of comprehensive information monitoring system. Information world powers are analysis of network information search technology is adopted to establish the government monitoring infrastructure, finance and other key industry. Many foreign companies are actively research and develop this kind of technology and products. German Cobion company launched network image analysis techniques in 2003, such as: more information visual character recognition, face detection and recognition and so on, has established the Europe's largest network information gathering analysis center. In network quality of service technology research, a Web client and the HTTP request to provide a performance guarantee and service differentiation technology, Web service quality arises at the historic moment, and get more and more attention in scholars and businesses, become a new research field of quality of service technology and important academic branches. The research progress on the network quality of service, so far very rapidly, and has made many fundamental progress. Distributed network measurement infrastructure (DNMAI) is customized according to user demand survey and analysis solutions, provides customizable user interface functions, reached the international advanced level. IBM Tivoli Web Response Monitor can capture the end user experience on the performance of web-based applications. It measuring response time, URL information and so on.

In the field of network quality of service in China started late, but the government attaches great importance to it, and formulated a series of measures for the administration. Many scientific research institutions, army, enterprise, introduced a number of products, and good results have been achieved. Some companies in China in recent years have been introduced with the network information search technology related products, such as: Shanghai fudan guanghua company, guangdong Morgan company, Beijing capital industrial technology development company, Beijing sharp Ann company, launched network security monitoring products. But the product application scope is relatively single, difficult to adapt to the development of network information security domain at present.

The development of the domestic network monitoring technology mainly in the 1999 years later, and technology to focus on firewalls and network intrusion detection, for the contents of the information monitoring only simple seal on site, not to mention for the monitoring of network information service quality, in this aspect is almost blank. In this aspect of the monitoring can ensure the security of network and information, optimize the efficiency of information, strengthen the competitiveness of various aspects.

3.2 Monitoring Tool

According to the different of installing and measuring method of the monitoring software,the Internet monitoring method can be divided into the monitoring based on the server and client based monitoring two types.

At present, the vast majority of monitoring software are based on the server, the user buy the software, install it on the local server, for the performance of the local server, database and other hardware and software of the monitoring. The benefits of the monitoring software is that users can understand their local machine efficiency, also can understand the local network connection and some e-business processes at the local implementation. But the disadvantage is that can't reflect directly to the end user login web site and use of e-commerce services. Now offers the software company has many, like Freshwater , BMC, and CA.

At present domestic widely used monitoring method is based on the monitoring on the server. Domestic monitoring technology mainly oriented network performance test and speed of network monitoring, monitoring service way is mainly based on the monitoring agent, in this way, the monitoring agent is installed on a network server, monitoring agent specific monitoring tasks. Advantages of this approach is that powerful, able to monitor to the system of internal error; Stable operation, high reliability; The disadvantage is : due to the need to install to each machine running, so the maintenance cost is high; Run in the interior of the online services, rather than the client, so can't reflect all sorts of problems caused by network problems.

With the development of Internet technology and the transformation of the management concept, there is a new kind of monitoring service. The monitoring service is the software installed on the client, not installed on the network operators' local server. From the perspective of end users on the network monitoring , search a network bottlenecks. Using the principle of this monitoring service fundamentally embodies the user-centered management thought, because for the end user, when to log in, they don't care about network operators are using the HP servers or IBM servers, is NT operation platform, or Unix operating platform, also don't care about their use of database is Oracle database or Microsoft. Abroad have this kind of monitoring service have more in-depth research, domestic research on this aspect is less.

4 Concludes

Through the above research discussion, we can draw the following conclusion:

(1) network in our country agriculture science and technology resources have been gradually takes on an increase in number, content rich, the good situation of the development potential is huge, but many problems still exist in the process of development, such as information repeated, practicality is not strong, difficult to share resources, the use of the website's utilization rate is not high, not attractive enough for their users, etc.

(2) abroad in the field of network information search, monitor, start early, developing very quickly. In this field in China starts late, although the government attaches great importance to it, and a series of basic management measures had maked, many scientific research institutions, army, enterprises introduced a number of products, and good results have been achieved, but the application scope is relatively single, difficult to adapt to the development of network information security domain at present.

Therefore, the online resources of agricultural science and technology Service quality monitoring, analysis System construction has been on the agenda, Through monitoring and analysis , can pray for the important role for the healthy development of the websites of the agricultural science and technology .

Acknowledgements. The work is supported by the Academy of Science and Technology for Development fund project “intelligent searchbased Tibet science & technology information resource sharing technology”, the National Science and Technology Major Project of the Ministry of Science and Technology of China (Grant No. 2009ZX0300101901), and the special fund project for Basic Science Research Business Fee, AII (No. 2010-J-07).

References

[1] FIDNet, http://www.privacilla.org/government/fidnet.html
[2] IBM Tivoli Web Response Monitor and IBM Tivoli Web Segment Analyzer, http://publib.boulder.ibm.com/tividd/td/ITPathWRMon/fixpack120/en_US/PDF/wrm_wsa_20_fixpackIF0003.pdf
[3] Freshwater, http://www.freshwater-uk.com/
[4] BMC, http://www.bmc.com/
[5] CA, http://www.ca.com/us/default.aspx
[6] Richmond, B.: Ten C's For Evaluating Internet Sources[EB], http://www.montgomerycollege.edu/Departments/writegt/htmlhandouts/TenCinternetsources.Htm
[7] Atzeni, P., Merialdo, P., Sindoni, G.: Web Site Evaluation: Methodology and Case Study
[8] Harris, R.: Evaluating Internet Research Sources [EB], http://www.virtualsalt.com/evalu8it.htm
[9] Tillotson, J.: Web Site Evaluation: A Survey of Undergraduates. Online Information Review (6) (2002)
[10] Barjak, F., Li, X., Thelwall, M.: Which Factors Explain the Web Impact of Scientists Personal Homepages? Journal of the American Society for Information Science and Technology 58(2), 200–211 (2007)

An Automatic Counting Method of Maize Ear Grain Based on Image Processing

Mingming Zhao[1], Jian Qin[2], Shaoming Li[1], Zhe Liu[1], Jin Cao[1], Xiaochuang Yao[1], Sijing Ye[1], and Lin Li[1,*]

[1] China Agricultural University, Beijing 100083, China
lilincau@126.com
[2] IBM China Development Labs, CDL, Beijing, China

Abstract. Corn variety testing is a process to pick and cultivate a high yield, disease resistant and outstandingly adaptive variety from thousands of corn hybrid varieties. In this process, we have to do a large number of comparative tests, observation and measurement. The workload of this measurement is very huge, for the large number of varieties under test. The grain numbers of maize ear is an important parameter to the corn variety testing. At present, the grain counting is mostly done by manpower. In this way, both the deviation and workload is unacceptable. In this paper, an automatic counting method of maize ear grain is established basing on image processing. Image segmentation is the basis and classic difficult part of image processing. This paper presents an image pre-processing method, which is based on the characteristics of maize ear image. This method includes median filter to eliminate random noise, wallis filter to sharpen the image boundary and histogram enhancement. It also mainly introduces an in-depth study of Otsu algorithms. To overcome the problems of Otsu algorithm that background information being erroneously divided when object size is small. A new method based on traditional Otsu method is proposed, which combines the multi-threshold segmentation and RBGM gradient descent. The implementation of RBGM gradient descent leads to a remarkable improvement on the efficiency of multi-threshold segmentation which is generally an extremely time-consuming task. Our experimental evaluations on 25 sets of maize ear image datasets show that the proposed method can produce more competitive results on effectiveness and speed in comparison to the manpower. The grain counting accuracy of ear volume can reach to 96.8%.

Keywords: grain counting, Otsu, multi-threshold segmentation, RBGM gradient descent.

1 Introduction

During the corn variety testing, the grain numbers of maize ear is an important parameter to the corn variety testing[1]. At present, the grain counting is mostly done

* Corresponding author.

© IFIP International Federation for Information Processing 2015
D. Li and Y. Chen (Eds.): CCTA 2014, IFIP AICT 452, pp. 521–533, 2015.
DOI: 10.1007/978-3-319-19620-6_59

by manpower. In this way, both the deviation and workload is unacceptable. In addition, the subjective errors are not easy to be avoided and the efficiency of manual measurement is very low. So this paper proposes a fast and high-accuracy automatic counting method of maize ear grain based on image processing. For the special structure feature of maize ear grain, we need to find a new image segmentation algorithm and an automatic counting method of maize ear grain. It has high quality and high efficiency.

2 Review of Related Researches

Traditional image segmentation method mainly contains threshold segmentation, edge detection segmentation, region segmentation and segmentation method based on mathematical. At present some new image segmentation methods come up with the deep-research in image process.

In 1998 S.Beucher and C.Lantuéjoul proposed an image segmentation algorithm[2] [3], watersheds in digital spaces, based on immersion simulations. Roughly speaking, it was based on a sorting of the pixels in the increasing order of their gray values, and on fast breadth-first scannings of the plateaus enabled by a first-in-first-out type data structure. This algorithm turned out to be faster and behave well in image segmentation. However this algorithm often has the problem of over-segmentation due to the tiny noise on the image. As illustrated by Figure 1, the segmentation result of the maize ear grain based on watersheds has the problem of the over-segmentation, in this way, which is unacceptable.

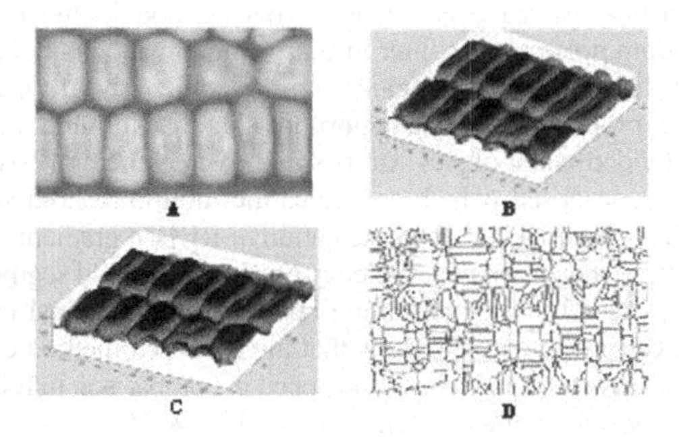

Fig. 1. Applying watersheds to the maize ear grain image. A: Original images from part of maize ear grain. B: Gray image of A. C: Gray image after filter. D: Result image of watersheds.

In these years, there are some popular image segmentation algorithms based on active contour model, such as Snakes and MS model[4]. Snakes are active contour models, MS models are level set models. Applications of this algorithm with regard to tracking the face activity, medicine CT image segmentation and cell image segmentation, but this algorithm has the problem of huge calculations and slow-speed in image segmentation. As shown by Figure 2, with the increase of iterations, it needs more time in image segmentation, which can't meet the need of corn variety testing.

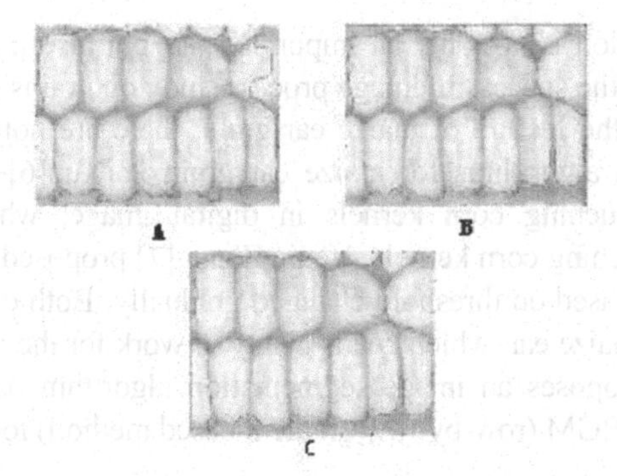

Fig. 2. Applying level set method to the maize ear grain image. A: Convergence result with iterations of 300 times. B: Convergence result with iterations of 1400 times. C: Convergence result with iterations of 4000 times.

The threshold segmentation is the most popular algorithm and is widely used in the image segmentation field[5]. The basic idea of threshold segmentation algorithm is to select an optimal or several optimal gray-level threshold values for separating objects of interest in an image from the background based on their gray-level distribution. The classical threshold segmentation algorithm include histogram shape-based methods, clustering-based methods (Otsu), mutual information methods, attribute similarity-based methods, local adaptive segmentation methods, etc. Among them, Otsu method has received more attention and frequently used in various fields.

As is illustrated in Figure 3, Otsu method behaves well in segmenting image of maize ear grain. But it doesn't give the satisfactory results because of the grains not separated completely. So to overcome this problem, this paper propose a new method that combines the multi-threshold segmentation and RBGM, based on Otsu method. It turns out to be high-accuracy and time-saving, which can meet the actually need of the corn variety testing.

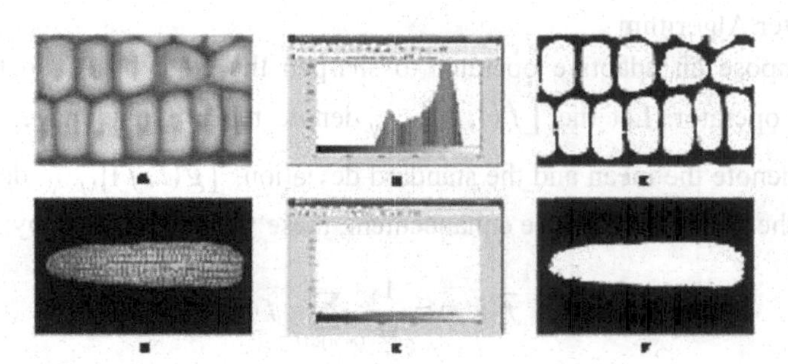

Fig. 3. Applying Otsu method to the maize ear grain image. A: gray image of maize ear grain. B: Gray histogram image of A. C: The segmentation image with Otsu of A. D: The gray image of complete maize ear grain image. E: Gray histogram image of D. F: The segmentation image with Otsu of D.

Image segmentation is not only an important part but also a challenge part in image process. At present, the successful image process study on grains is mostly about soybean and wheat. Due to the feature of maize ear grain, there are not particular and efficient image segmentation algorithms for maize ear grain. YiXun[6] proposed an automatic segmentation of touching corn kernels in digital image, which releases automatic segmentation of touching corn kernels. Yaqiu Zhang[7] proposed a method that separates corn seeds images based on threshold changed gradually. Both of these methods have to take grains off the maize ear, which needs plenty of work for the corn variety testing.

So this paper proposes an image segmentation algorithm based on multi-threshold segmentation and RBGM (row-by-row gradient based method) for maize ear grain image.

3 Image Preprocess

Image preprocess is an essential part to the image segmentation, which can enhance the visual appearance of images and improve the manipulation of datasets, including image resampling, greyscale contrast enhancement, noise removal, mathematical operations and manual correction. Enhancement techniques can emphasize image artefacts, or even lead to a loss of information if not correctly used. So this paper proposed a particular series of image preprocess methods based on the classical method for the feature of the maize ear grain, as follows:

Median Filter

Classical median filter algorithm only use the information of statistical in gray image, without considering the importance of other spatial information and different apex. So, we use the weighted median filtering method to remove noise, which is given by:

$$f'(x_0, y_0) = \left[\underset{(x_i, y_j) \in S}{Sort} \omega_{ij} f(x_i, y_j) \right]_{\frac{|S|+1}{2}} \quad (1)$$

Wallia Filter Algorithm

Wallia propose an adaptive operator to sharpen the edge based on the feature of Laplacian operator. Let the $[f(i,j)]_{M \times N}$ denote the original image, $\overline{f}(i,j)$ and $\sigma(i,j)$ denote the mean and the standard deviation, $[g(i,j)]_{M \times N}$ denote the pixel values of the image after image enhancement, these values are given by:

$$\overline{f}(i,j) = \frac{1}{M} \sum_{(m,n) \in D_{ij}} f(m,n) \quad (2)$$

$$\sigma(i,j) = \frac{1}{M} \sum_{(m,n) \in D_{ij}} [f(m,n) - \overline{f}(i,j)]^2 \quad (3)$$

$$g(i,j)=[\alpha m_d+(1-\alpha)\overline{f}(i,j)]+[f(i,j)-\overline{f}(i,j)]\frac{A\sigma_d}{A\sigma(i,j)+\sigma_d} \tag{4}$$

Histograms enhancement algorithm

Image enhancement is a mean as the improvement of an image appearance by increasing dominance of some features or by decreasing ambiguity between different regions of the image. Histogram processing is the act of altering an image by modifying its histogram, which is better suited for segmentation by multi-threshold algorithm.

4 Image Segmentation Algorithm

4.1 Multi-threshold Segmentation

In 1979, N. Otsu proposed the maximum class variance method (known as the Otsu method). For its simple calculation, stability and effectiveness, it has been widely used, was a well-behaved automatic threshold selection method, and its consumed time is significantly less than other threshold algorithms[8].

Set the pixels of segmentation image as N, there are L gray levels (0,1,…,L-1), n_i pixels whose gray level is i, then $N=\sum_{i=0}^{L-1}n_i$, and we express the probability density distribution with the form of histogram $p_i=\frac{n_i}{N}$, $\sum_{i=0}^{L-1}n_i=1$,$p_i\geq 0$. Let an image be divided into two classes C_0 and C_1 by threshold t. C_0 consists of pixels with levels $[1,\ldots,t]$ and C_1 consists of pixels with levels $[t+1,\ldots,L]$. Let u_0 and u_1 denote the mean levels, σ^2 denote the between-calss variances of the classes C_0 and C_1, respectively. These values are given by:

$$\mu_0=\frac{\sum_{i=0}^{t}ip_i}{p(t)}=\frac{\mu_0(t)}{p(t)} \tag{9}$$

$$\mu_1=\frac{\sum_{i=t+1}^{L-1}ip_i}{1-p(t)}=\frac{\mu_1(t)}{1-p(t)} \tag{10}$$

$$\sigma_t^2=p(t)(1-p(t))(\mu_1-\mu_0)^2 \tag{11}$$

The threshold t^*decided by maximizing the between-class variance proposed in Otsu is:

$$\sigma_{t^*}^2=Max(\sigma_t^2)\qquad t\in G \tag{12}$$

The shortage of Otsu algorithm is that Otsu algorithm is suitable on condition that there are two categories in the image; when there are more than two categories in the image, Otsu can't make the background and the target separate like Figure 3. So as to decide multi-threshold. The approach allows the largest between-class variance and the smallest in-class variance.

Based on Otsu, we can make out the multi-threshold as follows. Let an image be divided into n classes by threshold t. $t=\{t_k \mid k=1,2,...,n-1\}$. Let ω_k denote the probability of each class, μ_k denote the mean levels of each class, σ_k^2 denote variances of each class. These values are given by:

$$\omega_k = \sum_{i=t_k}^{t_{k+1}} P_i \quad k=0,1,....,n-1 \tag{13}$$

$$\mu_k = \sum_{i=t_k}^{t_{k+1}} iP_i / \omega_k \quad t_0 = 0, t_n = L \tag{14}$$

$$\sigma_k^2 = \sum_{i=t_k}^{t_{k+1}} (i-\mu_k)^2 P_i / \omega_k \quad 1 < t_k < L(k=1,2,...,n-1) \tag{15}$$

Let σ_W^2 means the within-class variance, σ_B^2 means the between-class variance, respectively:

$$\sigma_w^2 = \sum_{k=0}^{n-1} \omega_k \sigma_k^2 \tag{16}$$

Similar to the classical Otsu algorithm, we have multi-threshold constraint equation:

$$\sigma^2 = \sum_{i=1}^{L} (i-\mu)^2 P_i = \sum_{i=1}^{L} i^2 P_i - 2\mu \sum_{i=1}^{L} iP_i + \mu^2 \sum_{i=1}^{L} P_i = \sum_{i=1}^{L} i^2 P_i - \mu^2 \tag{17}$$

According to (15), we have:

$$\sigma_k^2 = \sum_{i=t_k}^{t_{k+1}} (i-\mu_k)^2 P_i / \omega_k = \frac{1}{\omega_k}\left(\sum_{i=t_k}^{t_{k+1}} i^2 P_i - 2\mu_k \sum_{i=t_k}^{t_{k+1}} iP_i + \mu_k^2 \sum_{i=t_k}^{t_{k+1}} P_i\right) = \frac{1}{\omega_k}\left(\sum_{i=t_k}^{t_{k+1}} i^2 P_i - 2\omega_k \mu_k^2 + \omega_k \mu_k^2\right) \tag{18}$$

Then

$$\sum_{i=t_k}^{t_{k+1}} i^2 P_i = \omega_k (\mu_k^2 + \sigma_k^2) \tag{19}$$

According to (14) (17) (19), we have:

$$\sigma^2 = \sum_{k=0}^{n-1} \omega_k (\mu_k^2 + \sigma_k^2) - (\sum_{k=0}^{n-1} \omega_k \mu_k)^2 \tag{20}$$

Then According to (16) (20), we have:

$$\sigma_B^2 = \sigma^2 - \sigma_w^2 = \sum_{k=0}^{n-1} \sum_{j=k+1}^{n-1} \omega_k \omega_j (\mu_k - \mu_j)^2 \tag{21}$$

4.2 RBGM Algorithm

When the threshold t^* decided by maximizing the between-class variance of multi-threshold proposed in $\sigma_B^2 = \sigma^2 - \sigma_w^2 = \sum_{k=0}^{n-1} \sum_{j=k+1}^{n-1} \omega_k \omega_j (\mu_k - \mu_j)^2$. However, with the increase of threshold from single to n, the resolution problem of σ_B^2 is changing from function of variable into multivariate function, which will needs plenty of time in selecting threshold. To overcome this problem, this paper take the method of the RBGM (row-by-row gradient based method). Given by:

$$x_i^{mN+j+1} = x_i^{mN+j} - \eta_m \Delta_i (x_i^{mN+j}), i = 1, 2, \ldots, N \tag{22}$$

$$\Delta_i (x^{mn+j}) = \begin{cases} \frac{\partial f(x^{mn+j})}{\partial x^i}, & j = i-1 \\ 0, & j \neq i-1 \end{cases}, \quad j = 0,1,\ldots, N-1; \; m = 0,1,\ldots \tag{23}$$

The RBGM method is described as follows:

Step 1: given an initial point $x^0 = (x_1^0, x_2^0, \cdots x_N^0) \in R^N$, error precision $\varepsilon > 0$, and Maximum Iterations $M_{\max}$. Set m : = 0 , j : = 0.

Step 2: complete a cycle to update x^{mN+j} as follows:

$$\begin{aligned}
&for \quad m = 0 : M_{\max} \\
&\quad for \quad j = 0 : N-1 \\
&\quad\quad x_i^{mn+j+1} = x_i^{mn+j} - \eta_m \Delta_i (x_i^{mn+j}) \\
&\quad\quad \Delta_i (x^{mn+j}) = \begin{cases} \frac{\partial f(x^{mn+j})}{\partial x^i} & j = i-1 \\ 0 & j \neq i-1 \end{cases} \\
&\quad j = j+1 \\
&m = m+1
\end{aligned}$$

Step 3: if $f(x^{mN+j})$ satisfy the ending rules, terminate the algorithm and output x^{mN+j} Otherwise go to step 2.

According to $\sigma_B^2 = \sigma^2 - \sigma_w^2 = \sum_{k=0} \sum_{j=k+1} \omega_k \omega_j (\mu_k - \mu_j)^2$ and the method RBGM, we have:

$$\begin{aligned} &\frac{\partial \sigma_B^2}{\partial t_i} \\ &= \frac{\partial (\sum_{k=0}^{n-1} \sum_{j=k+1}^{n-1} \omega_k \omega_j (\mu_k - \mu_j)^2)}{\partial t_i} \\ &= (p(t_i)\omega_{i+1} + p(t_{i+1})\omega_i)(\mu_i - \mu_{i+1})^2 + 2\omega_{i+1} p(t_i)(t_i - \mu_i) + 2\omega_i p(t_{i+1})(t_{i+1} - \mu_{i+1}) + \sum_{j=1, j \neq i, j \neq i+1}^{n} (\omega_i' \omega_j (\mu_i - \mu_j)^2 + 2\omega_i \omega_j \mu_i') \end{aligned} \tag{24}$$

Then we can get Multi-threshold segmentation iterative solution function:

$$x_i^{mN+j+1} = x_i^{mN+j} - \eta_m \frac{\partial \sigma_B^2}{\partial t_i}, i = 1, 2, \ldots, N \tag{25}$$

Given the initial point $\vec{t}_*$ and iteration step η_m, we can get the best threshold.

5 An Automatic Counting Model of Maize Ear Grain

Counting the number of grains in binary image after multi-threshold by connected component labeling method. Let $p(x, y)$ denote the value pixel of the point (x, y). The connected component in binary image are classified into two groups: the with value of one stands for maize grains, the $p(x, y)$ with value of zero stands for the background. Then, the binary image is followed by a line-by-line counter to find all the connected component with the value of one. Finally we can get the number of the grains in image, but we should find a method of counting the total number of grains.

According to the biological nature of maize ear grain, the number of maize ear's rows is always double. So we did plenty of experiment, we found that the number of maize ear's rows is always 12 or 14. In this way, we proposed a maize ear grain estimation model that the total number of the grains has a linear relationship with the number of grains in image we collect, on the basis the rows is straight and neat. The model is given by:

$$y = 1.9427x + 9.2498 \quad R^2 = 0.9664$$

y means the total number of grains, x means the number of grains in image.

6 Experiment and Result Analysis

6.1 Experiment

In this paper an automatic counting method of maize ear grain based in image process was proposed, and the detail experiment method was given as follows:

Step 1: We take pictures of 20 maize ear grain with digital with digital camera. After collect maize ear grain image, we count the grains number of each maize ear and that in image as training data to obtain the maize ear grain estimation model. The obtained original maize ear grain image is shown in Figure 4.

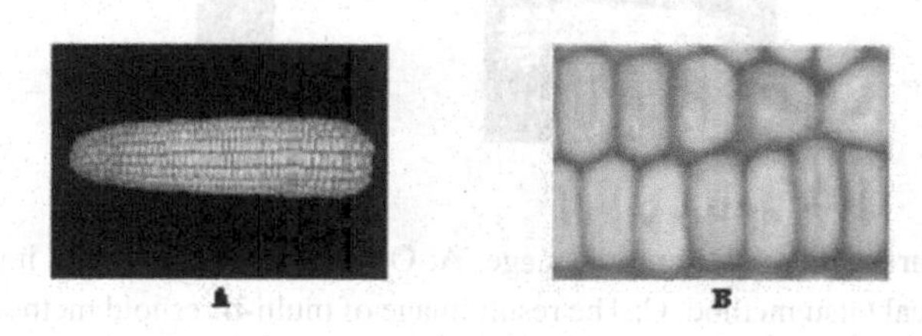

Fig. 4. Data collecting. A, B: Original maize ear grain image.

Step 2: A series of work with image preprocessing methods. First, the color image should be converted to the gray image. Then take median filter method with (1) to eliminate random noise of gray image, and wallis filter method with (4) to sharpen the image boundary. Finally, histogram enhancement method. The image preprocess image is shown in Fig. 5.

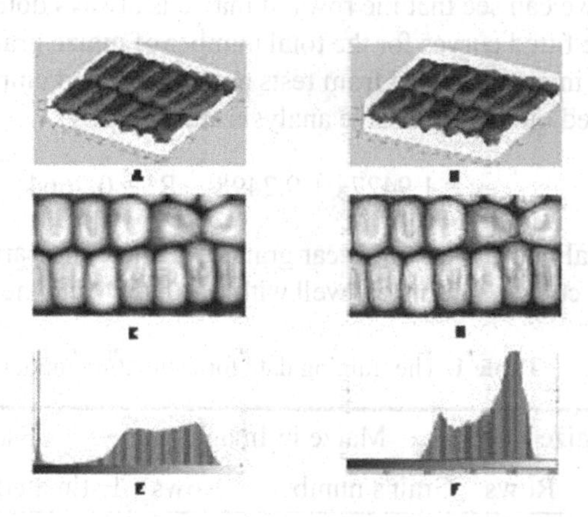

Fig. 5. Image preprocessing. A: Image before median filter method. B: Result image after median filter method. C: Image before wallis filter method. D: Result image after wallis filter method. E: Image before histogram enhancement method. F: Result image after histogram enhancement method.

Step 3: Image segmentation with multi-threshold algorithm by (25) and obtain binary image of maize ear grain. Automatic counting method with connected component labeling method. The obtained image is shown in Fig. 6.

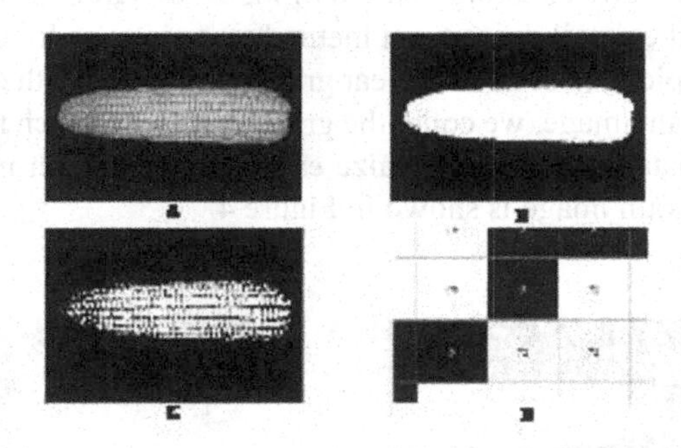

Fig. 6. Multi-threshold method result image. A: Original maize ear grain image. B: The result image of classical Otsu method. C: The result image of multi-threshold method. D: Result image with connected component labeling method.

Step 4: Maize ear grains counting with an automatic counting model y = 1.9427x + 9.2498 after obtaining the grains number in image, the results is shown in Table 1.

6.2 Analysis of Experiment Results

From the table 1, we can see that the rows of maize is always double and the number is most 12 or 14. The fitted curves for the total number of maize grains to the part number of maize grains in image obtained from tests are plotted, and empirical expressions for these curves worked out by regressive analysis are given by :

$$y = 1.9427x + 9.2498 \quad R^2 = 0.9664$$

y denote the total number of maize ear grains, x denote the part number of maize ear grains in image we collect. It turns out well with R^2=0.9664 and the average error is 1.2%.

Table 1. The training data for estimation model

ID	Total Maize		Maize in Image		Estimation Model	
	Grains number	Rows	Grains number	Rows	Estimated Value	Error
1	485	14	243	7	481	0.008
2	487	14	244	7	483	0.008
3	374	10	189	5	376	0.006

Table 1. (*continued*)

4	474	12	237	6	470	0.009
5	453	12	235	6	466	0.028
6	484	14	247	7	489	0.011
7	485	14	240	7	475	0.020
8	487	14	243	7	481	0.012
9	374	10	186	5	371	0.009
10	474	12	238	6	472	0.005
11	453	12	239	6	474	0.045
12	484	14	244	7	483	0.002
13	485	14	242	7	479	0.012
14	487	14	242	7	479	0.016
15	374	10	187	5	373	0.004
16	474	12	239	6	474	0.001
17	453	12	237	6	470	0.037
18	484	14	241	7	477	0.014
19	485	14	244	7	483	0.004
20	375	10	188	5	374	0.001

As is shown in table 2, the performance comparisons with multi-threshold segmentation algorithm and RBGM algorithm of this paper. The method of multi-threshold with RBGM performance better than the original algorithm and it turns out to be well in time-saving obviously. Because with the increase of threshold, the resolution problem is changing from function of variable into multivariate function, which will needs plenty of time in selecting threshold. While RBGM find the best value in the gradient direction, which saves much work under the same precision.

Table 2. Comparison original algorithm with RBGM algorithm

The threshold numbers	Original Method		RBGM Method	
	number of operation	operation time	number of operation	operation time
1	256	0.19s		
2	5536	0.78s	3000	0.51s
3	16777216	10s	50000	1s
4	4294967296	50min	10000000	9s

Table 3. The result of maize ear grain

	True Value		Measured Value		
ID	Grains number	Grains number in image	Grains number in image	Grains number	Accuracy
1	485	243	236	468	0.964
2	487	244	230	456	0.936
3	374	189	185	369	0.986
4	474	237	226	448	0.946
5	453	235	229	454	1.002
6	484	247	240	475	0.982
7	485	240	236	468	0.964
8	487	243	231	458	0.940
9	374	186	176	351	0.939
10	474	238	229	454	0.958
11	453	239	230	456	1.007
12	484	244	234	464	0.958
13	485	242	236	468	0.964
14	487	242	230	456	0.936
15	374	187	179	357	0.955
16	474	239	230	456	0.962
17	453	237	229	454	1.002
18	484	241	233	462	0.954
19	485	244	235	466	0.960
20	375	188	181	361	0.962
					0.964

From the Figure 3~Figure 6 and table 3, we can see: compared with traditional threshold algorithm, our method performance better in image segmentation of maize ear grain, which can make each grain separate obviously. Besides, multi-threshold algorithm also behaves well in image edge. As is shown in table 2, most measured value is closed to the true value, which have an accuracy over 96%. Some have a low precision, because the method may generate errors at places when two grains overlap in such a fashion which are very smooth or over-closed.

7 Conclusion

An automatic counting method of maize ear grain based on image process with the algorithm of multi-threshold and RBGM is proposed in this paper. Compared with classical threshold method, this method turns out that performances well in image segmentation of maize ear grain and have an advantage of time-saving. Especially, the accuracy of counting is 96.4%, which is acceptable for the corn variety testing.

Acknowledgements. This paper is supported by the "National Key Technology R&D Program under the Twelfth Five-Year plan of P.R. China" (Grant No. 2012BAK19B04-03). And National 863 Program (Grant No. 2011AA10A103-1).

References

1. Qin, J., Li, L., Li, S., Wang, L., Shi, Z.: New image segmentation method based on gradient. Journal of Computer Applications 29(8), 2071–2073 (2009)
2. Vincent, L.: Morphological Gray scale Reconstruction in Image Analysis: Applications and Efficient Algorithms. IEEE Transactions on Image Processing 2(2), 583–598 (1993)
3. Vincent, L., Soille, P.: Watersheds in digital spaces—an efficient algorithm based on immersion simulations. IEEE Transactions on Pattern Analysis and Machine Intelligence 13(6) (June 1991)
4. Kassm, W., Terzopoulos, D.: Snakes: active contour models. International Journal of Computer Vision 1(4), 321–331 (1987)
5. Chan, T., Vese, L.: An efficient variation multiphase motion for the Mumford- Shah segmentation model. In: Processing of Asiomar Conference Signals, Systems, and Computers, Pacific Grove, CA, USA, pp. 490–494. IEEE Press (2000)
6. Xun, Y., Bao, G., Yang, Q.: Automatic Segmentation of Touching Corn Kernels in Digital Image. Transactions of the Chinese Society for Agricultural Machinery 04, 033 (2010)
7. Zhang, Y., Wu, W., Wang, G.: Separation of corn seeds images based on threshold changed gradully. Transactions of the CSAE 27(7), 200–204 (2011)
8. Xu, X.: Characteristic analysis of Otsu threshold and its applications. Pattern Recognition Letters, 956–961 (2011)

Research of SF6 Pressure Gauge Automatic Reading Methods Based on Machine Vision

Song Yao[1,2], Liu Chunhong[1,3], Deng Qiao[1,2], and Wang Yixuan[4]

[1] College of Information and Electrical Engineering, China Agricultural University, Beijing, P.R. China, 100083
Sophia_liu@cau.edu.cn
[2] Modern Precision Agriculture System Integration Research Key Laboratory of Ministry of Education, Beijing, P.R. China, 100083
[3] Key Laboratory of Agricultural Information Acquisition Technology (Beijing), Ministry of Agriculture, Beijing, P.R. China, 100083
[4] State Grid Beijing Electric Power Company, Beijing, P.R. China, 100031

Abstract. With the rapid development of artificial intelligence and pattern recognition, digital image processing and recognition technologies become a popular research direction, especially, the using of it is quite extensive in power industry. Among various using, dashboard automatic reading is an important part of routing inspection of substation system by using robot. Automatic reading of SF6 pressure gauge pointer is based on image processing and automatic reading techniques, avoiding the influence of subjective factors of naked eye judgment. Designing and analyzing of identification algorithm ofSF6 meter pointer are shown in this paper. First, pre-processing operations were operated on the instrument image by using gray level transformation equalization and binarization to improve image quality, by using Hough line detection to realize pointer line extraction; determining the number by using the straight-line in mathematics. This traditional method of using morphological and Hough line detection method to determine reading have certain bias, so the using of Hough circle detection methods and centroid detection methods were proposed. The results showed that the improved method has greatly improved the accuracy of the readings, the method has better accuracy than traditional standard line Hough detection method.

Keywords: SF6 pressure gauges, Hough detection, centroid detection, automatic identification.

1 Introduction

With the rapid growth of economic, electricity is in demand sharply arise, the power supply appears tight trend, reliable operation of electrical equipment is very important [1]. Detection of Power Meter becomes more and more important.

Currently, many pointer instruments are in the configuration of the power system substation, such as: pressure gauge, oil temperature gauge, lightning protector, etc.

© IFIP International Federation for Information Processing 2015
D. Li and Y. Chen (Eds.): CCTA 2014, IFIP AICT 452, pp. 534–545, 2015.
DOI: 10.1007/978-3-319-19620-6_60

A large number of meter data depends mainly on the human eye observation, heavy labor need to be resolved. Assuming substation encountered harsh geographical conditions, high temperature, high altitude, arctic and other artificial outdoor work is very difficult. With the development and application of intelligent substation inspection robot [2], reading of meters have been improved to be automatic, using power equipment to detect and identify electrical safety hazards and removed promptly.

In order to be able to make better use of existing substations intelligent inspection system, based on the existing inspection system, using machine vision image processing and pattern recognition techniques to identify inspection equipment for automatic analysis whether be faulty or there is a security device hidden under the circumstances, notify the staff the cause of the alarm fault location and faulty equipment immediately, while providing video data to staff to be analyzed to help determine and judge [3].

SF6 (Sulfur Hexafluoride) gas has excellent insulating property and arc performance, so SF6is used in electrical devices extensively. Most current domestic power companies have adopted SF6 circuit breaker. Identification of SF6 pressure gauge can intuitively and accurately reflect the true state of the electrical equipment. In order to monitor and maintenance substation, gauge reading identification, monitoring, and electrical equipment of SF6 has great practical significance. Therefore, developing a suitable substation inspection robotSF6 pointer instrument quickly and accurately is important for automatic identification systems. Using machine vision technology for image recognition instrument automatically determines the location of the meter and the reading pointer above is the core technology of the system.

2 Pointer Recognition Algorithm

Pointer recognition algorithm is the core of the automatic identification system, including three parts: preprocessing of image, pointer recognition, reading calculate, as described below:

2.1 Image Preprocessing

Placed in the outdoor substation instrumentation, instrument inspection robot shooting angle image acquisition is extremely important. Shaded spot light changes and other equipment shelter, uneven brightness of image acquisition, if a straight line or circle detection, it may be included among the roughly circular parts when making circle detection error detection, or by light affect the test you want to get straight is not a pointer, so we only consider the positive SF6 gauge shot. Meanwhile, the instrument will be affected by the environment, the dial may be blurred, or even the naked eye can not see the pointer, the read pointer of the instrument dial when needed for image pre-processing, post-pointer to identify efficient, accurate, and its main steps shown in Figure 1 shows, image preprocessing shown in Figure 2.

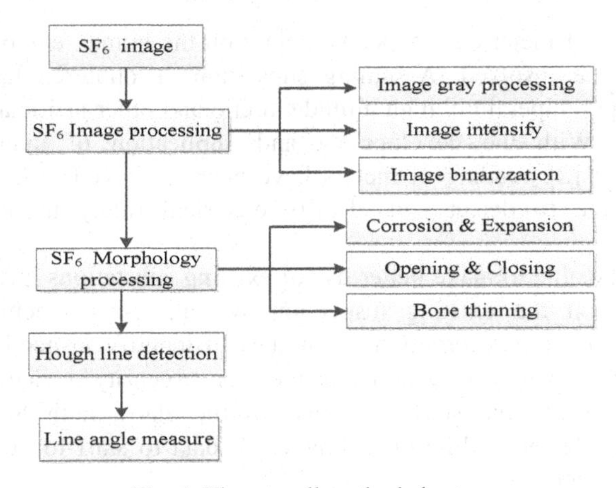

Fig. 1. The overall method chart

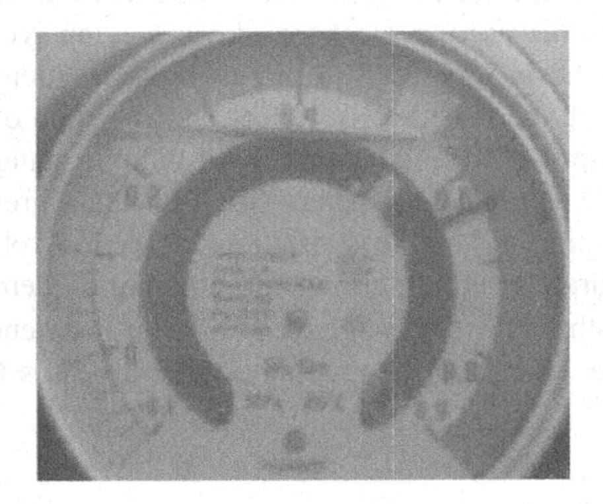

Fig. 2. Color image of SF6 pressure gauge

2.1.1 Image Gray Processing

In order to extract pointer needed for image segmentation, interference background must be removed. Transforming color image to gray-scale,using the average standard method [4], where g represents gray gradation value of formula (1) as follows:

$$g=0.30R+0.59G+0.11B \tag{1}$$

2.1.2 Image Enhancement

Acquisition and transmission for dash-board image are easy to be influent by noise and interference to various degrees. In the case of low SNR, this will causes decline in image quality and image blur. In order to eliminate the effects of noise, choose the appropriate method of image noise reduction processing is a reasonably important part of the reading pointer identification.

Histogram equalization [5] is acorrection of the original image by the transformation function to a uniform histogram equalization to the original image. After equalization process, the histogram of the image is flat, i.e., having the same frequency, since the uniform gray has even probability distribution, the image looks more sharp.

2.1.3 Binarization

In order to extract pointer from complex background after equalization, binarization [5] process will be carried on the gray image. Binary differentiate spixels into black and white in the image according to the given threshold. The Detail method: Define a template area whose size is (2w +1) × (2w +1), at the same time define (x, y) coordinates as the center of the template, define f (x, y) as the center of gray image, T (x, y) as the center binarization threshold, calculating threshold value for each pixel according to formula (2), binarized gray image according to equation (3) as follows:

$$T(x,y)=\frac{1}{(2w+1)^2}\sum_{k=-w}^{w}\sum_{l=-w}^{w}f(x+k)y+l \tag{2}$$

$$I(x,y)=\begin{cases}1, & if\ f(x,y)\geq T\\ 0, & if\ f(x,y)<T\end{cases} \tag{3}$$

Where I(x,y) is the image intensity of each pixel after binarization. After adaptive thresholding algorithm was used for image binarization, some noise was filtered, but the results are not satisfactory. Considering SF6pressure gauge dial has three large regions, ie, the red area, the white area and green area, while the pointer is black, and in the actual interpretation, reading of pointer will be affected by environment, so we use manual threshold, the lower the threshold, the extraction of the black pointer will be more obvious. Here we set the threshold value as 50 in a binarization process, so that if any changes in the external environment will not affect its accuracy.

2.1.4 Expansion and Corrosion

Binarized pointer may appear intermittently, so we use morphological dilation and erosion [5] methods to solve it. Corrosion is a process to eliminate boundary point so that boundary points will be shrink to internal. Expansion is the dual operation to corrosive, it is the expansion process to external boundary, so it can be used for the gap filler. After the expansion and corrosion, the pointer will be more complete and clear.

2.1.5 Thinning

Skeleton extraction method is thinning [5], prerequisite requirements is the topological of image remains unchanged. By thinning, the rough edge line from the outside in each pixel was stripped layer by layer, to obtain a final set of linking pixels, the set of points is called a skeleton image. Ensure to get the continuity and the main framework for skeleton topology are the biggest advantage of thinning algorithm.

2.2 Pointer Identification

2.2.1 Hough Line Detection

The basic idea of Hough transform [6] is to use the duality of lines and points. It uses the global characteristics, therefore the less noise, the more robust. Using polar coordinates to describe the linear equation Equation (4), with the following Hough transform functions:

$$\rho = x\cos(\theta) + y\sin(\theta) \tag{4}$$

If was limited to $[0, \pi]$ range, then the corresponding parameter of the line is unique. Each line in x-y plane corresponds to a point in $\theta - \rho$ plane. If point (x_i, y_i) is transformed to $\theta - \rho$ plane which defined in formula (4), then the problem of searching line in x-y plane is converted into the problem of finding the intersection of the curve in the $\theta - \rho$ plane, refer to equation (5) and Fig. 3.

$$\rho = x_i\cos(\theta) + y_i\sin(\theta) \tag{5}$$

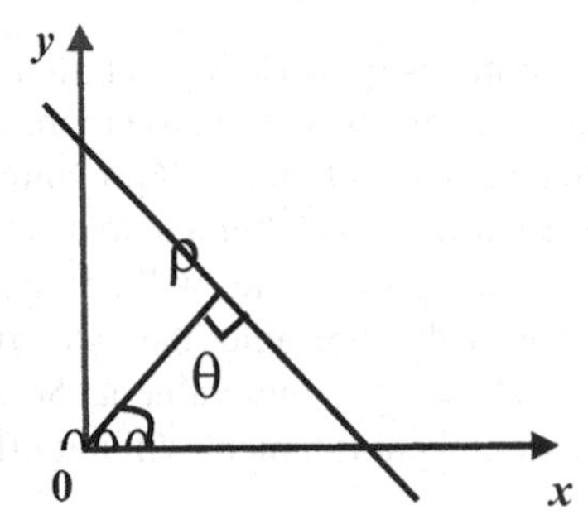

Fig. 3. Point (x, y) was transformed to $\theta - \rho$ parameter space

2.3 Readings Calculation

After a straight line was obtained, we get the coordinates of the two endpoints of the line, then angle can be got as follows:
Pointer is 270 degree between -0.1 to 0.9, calculating the angle from -1 to the straight line where you can get a pointer readings recorded as A, the formula is as follows:

$$1/270=(X+0.1)/A \tag{6}$$

3 Simulation Results

Adopting the proposed algorithm in MATLAB environment, take 220KV SF6 circuit breakers under pressure gauges for example.

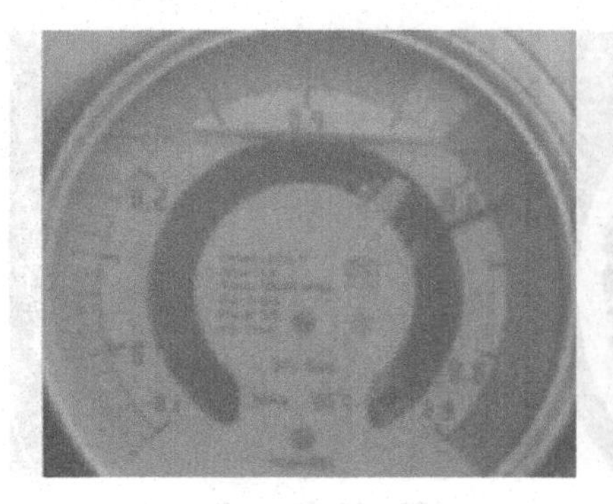

Fig. 4 Grayscale

Fig. 5. Equalization figure

Analysis: After histogram equalization was taken on gray image, the image was significantly enhanced the brightness and sharpness.

Fig. 6. Automatic threshold segmentation

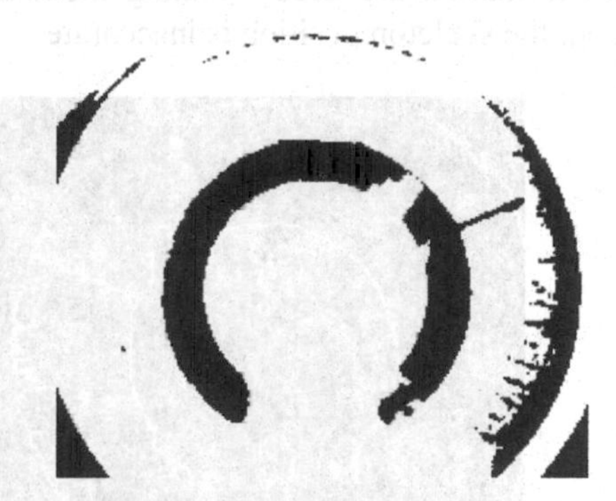

Fig. 7. Binarized figure with threshold set at 50

Analysis: After performing binarization, the segmented pointer is better by manual threshold at 50 than the value which got automatically, because image got by defining threshold manually screens reading seffectively and removed most noise.

Fig. 8. Binarization figure with threshold at 50

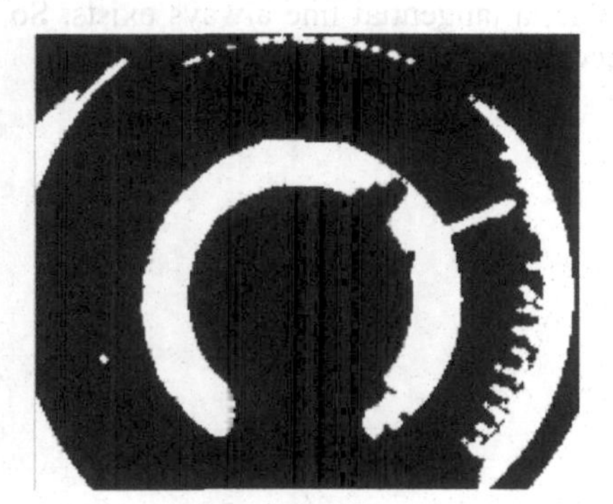

Fig. 9. Expansion figure

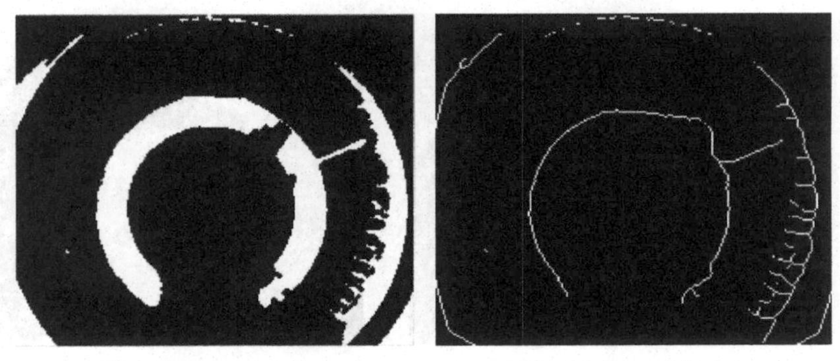

Fig. 10. Corrosion figure **Fig. 11.** Bone thinning

Analysis: The pointer in the binarized figure with conversion threshold at 50 is more obvious, in order for the pointer to be continuous, expansion and corrosion transformation are used, thinning method was used. The effect of thinning is not good, the skeleton position is inaccurate.

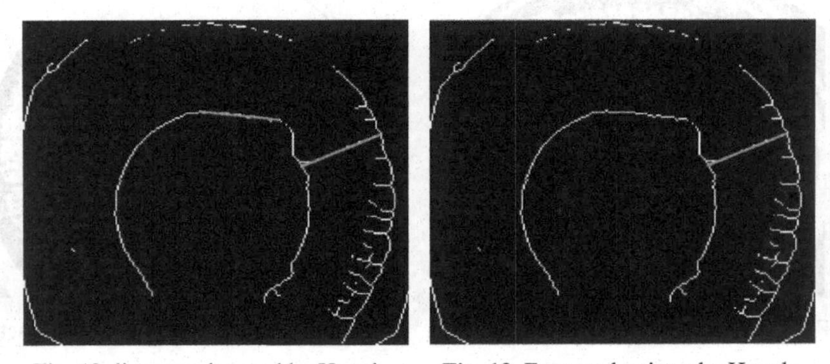

Fig. 12. lines are detected by Hough **Fig. 13.** Extracted pointer by Hough

Analysis: Initially, two straight lines are detected by Hough transform, the first line is inevitable, because in order to read SF6meter manually, the top part of shell SF6is bulge, a tangential line always exists. So we neglect the first line, only calculate the second detected line.

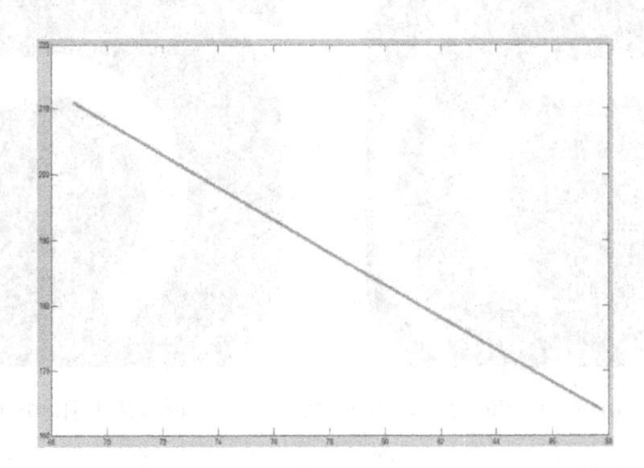

Fig. 14. A straight line obtained by Hough method

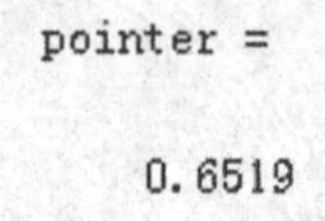

```
pointer =

    0.6519
```

Fig. 15. Pointer reading

Summary: After using traditional method of instrument readings, such as image gray, image enhancement, binary, morphological transformation, thinning and Hough detection, the resulting reading is erroneous, reading of 0.641 by naked eye. Thinning process produces a large deviation. Therefore, a new method is proposed to obtain meter readings. There must be a straight line when you use Hough circle detection, so detecting the image circle to get the coordinates of the center, another point can be obtained from centroid detection. Even dial projection is exists, there is a line after image processing, we determine a point through determining a round, another point is determined by centroid detection, which solved the above problem, no longer restricted by the dial, so meter reading can be more accurate, stable and reliable.

4 Improved Method

4.1 Hough Circle Detection

4.1.1 Edge Detection

Conducting round detection and edge detection at the same time. Edge is the junction between object and object or objects and background image, an edge exists in places where grayscale, color or texture changing rapidly. We can use Roberts operator, Sobel operator, Prewitt operator, Log operator and Canny operator to detect edge [5]. Various types of edge detection results are shown in Fig. 16 to Fig. 20.

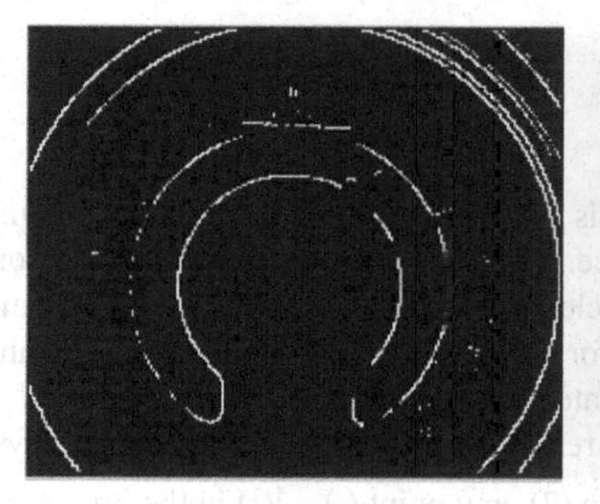

Fig. 16. Roberts edge detection result

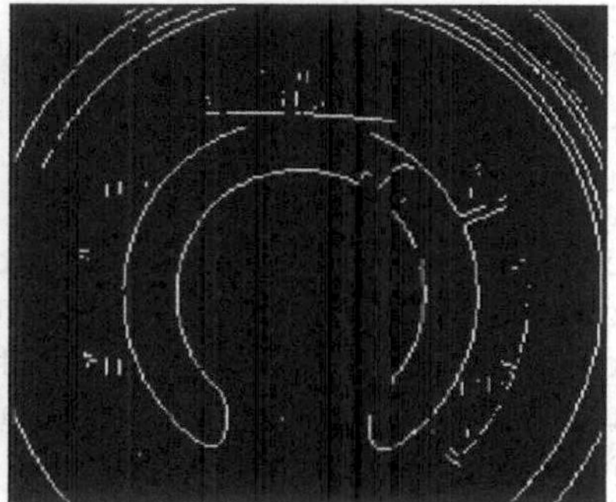

Fig. 17. Sobel edge detection result

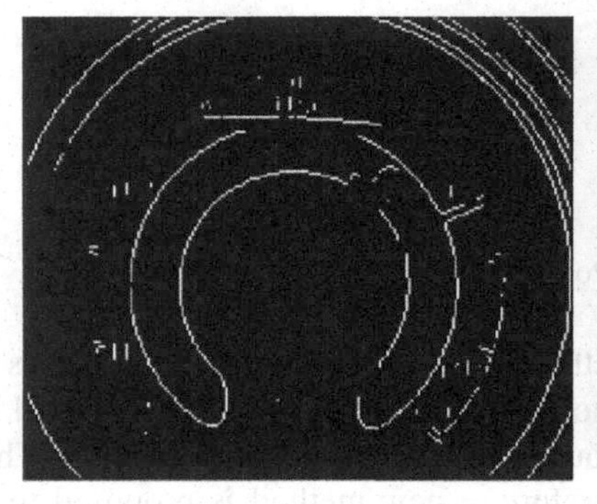

Fig. 18. Prewitt edge detection result

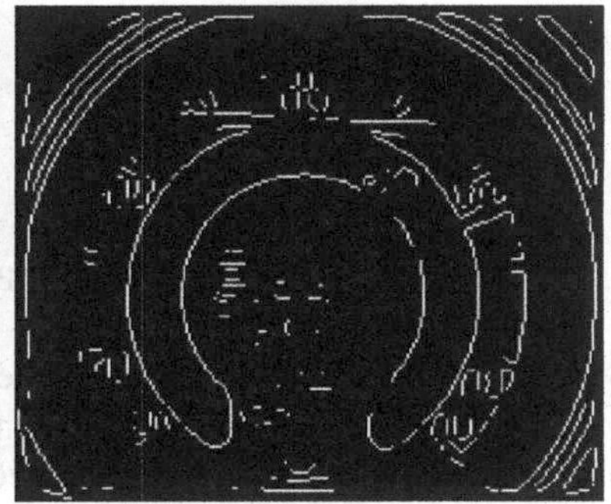

Fig. 19. Log edge detection result

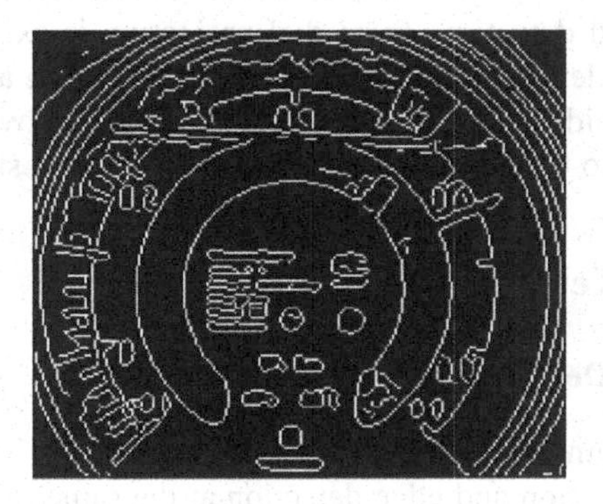

Fig. 20. Canny edge detection map

From the results of Fig. 16 to Fig. 20, different operators have significant difference, a significant portion of the edge is effectively extracted. Canny operator is more efficient to extract the weak edge, so we use Canny edge detection operator to detect circle.

4.1.2 Hough Circle Detection

Linear Hough transform parameter space is a two-parameter space, so other common curve shave corresponding parameter space. Three parameters defining a circle on the coordinate plane - the radius of the circle, the center of the x-coordinate and y-coordinate axis, so Hough transform [6] for a circle is a three-dimensional parameter space which use radius and center coordinates as parameter.

In an edge of the image which got by removing the background from a grayscale image, if a circle was described in equation (7), any point (x_i, y_i) in the image can be converted into the a-b-r parameter space 21 as shown in Fig. 21 and by Equation (8).

$$(x-a)^2 + (y-b)^2 = r^2 \tag{7}$$

$$(x_i - a)^2 + (y_i - b)^2 = r^2 \quad (8)$$

Each point in the edge of the image can be converted into a straight cone in the three-dimensional image as shown in Fig. 21. If the cones in parameter space which corresponds to the points in an edge image are intersect at one point, For example, all of these points is at the same circle which defined by the three parameters of the edge image. Results got by edge detection and round detection are shown in Fig. 22:

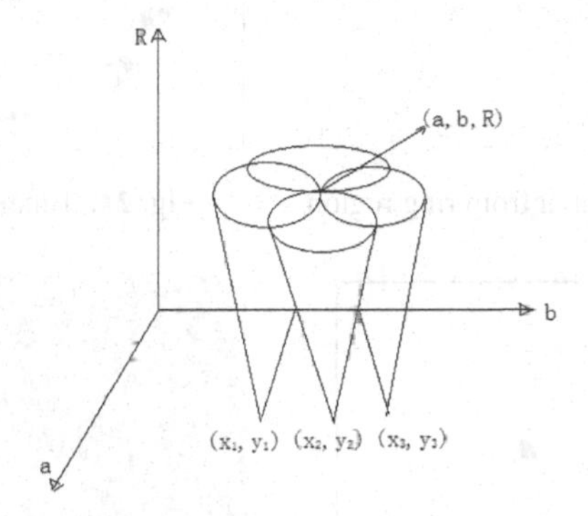

Fig. 21. Hough transformation parameters used during the image circle

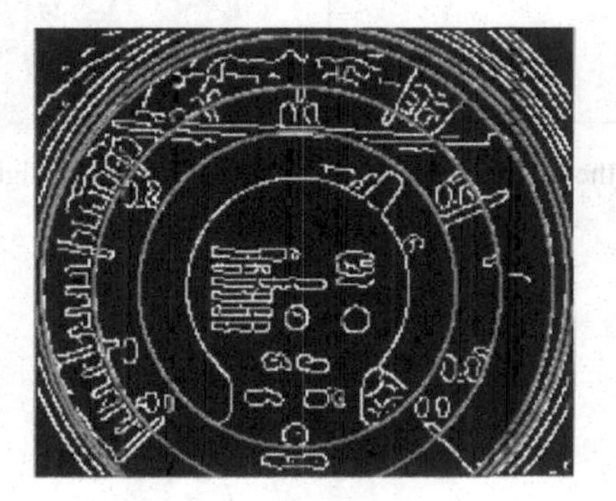

Fig. 22. Circle detection

4.2 Improved Methods - Centroid Detection

Centroid is the center of mass which means a material system is considered an imaginary point of the mass concentrating here. By Using centroid detection [7], which uses the centroid of a region to determine another point.

After circle detection on the original image, set a region of interest, which contains pointer. In this design, set (maximum radius -30) and (smallest circle +5) region, set other regions as 1which is white, refer to Fig. 23. Then binarized it, which is shown in Fig. 24. Select the largest area and get its centroid. Refer to Fig. 25. Connect the center of mass and the average centroid, then draw a straight line in the original image, which was shown in Fig. 26.

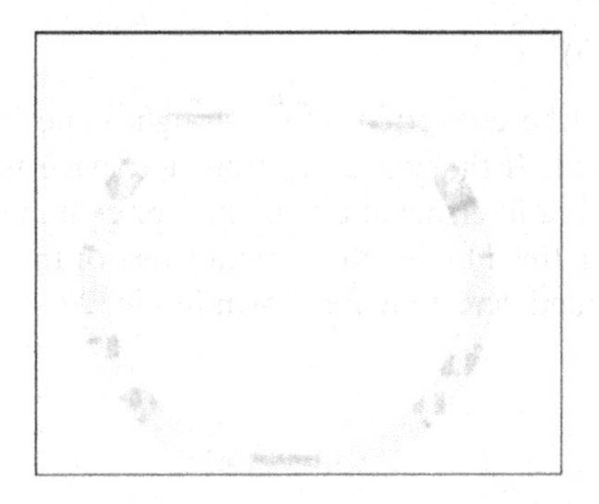

Fig. 23. Get interest area from ring region

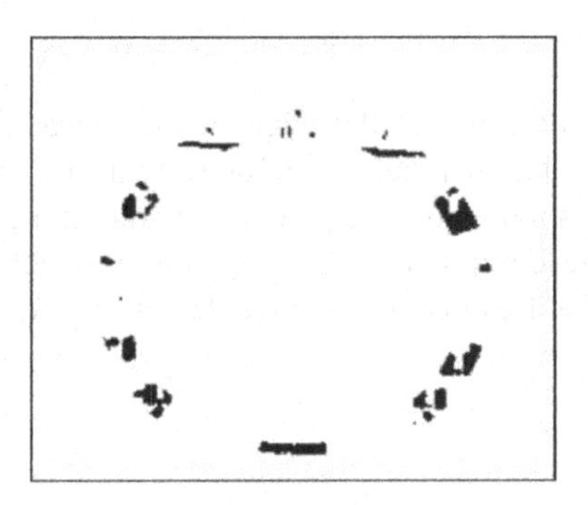

Fig. 24. .Binarize the ring area

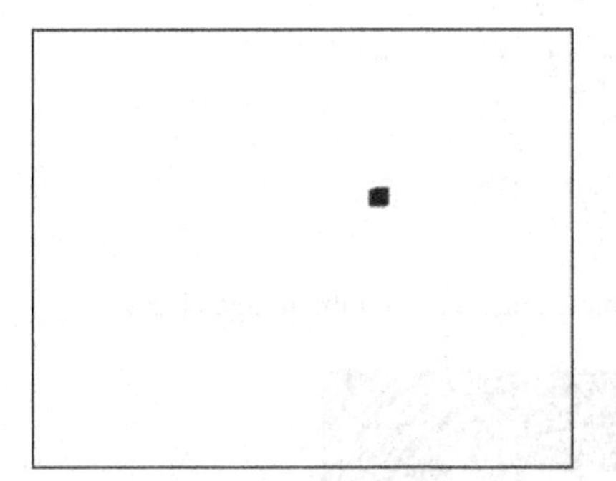

Fig. 25. Select the maximum area

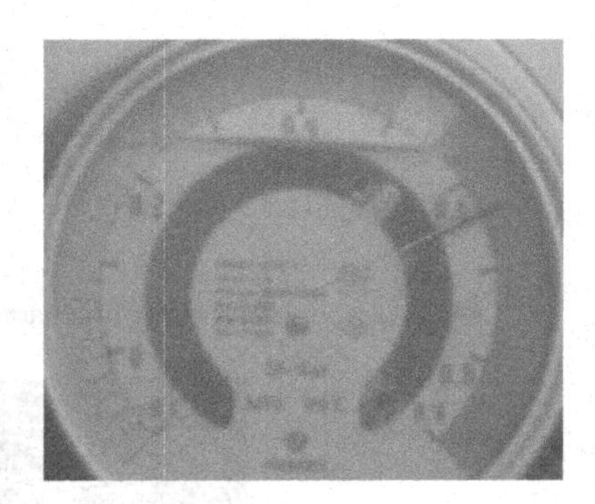

Fig. 26. Draw a straight line in the original image

```
pointer1 =

    0.6423
```

Fig. 27. Computer reading result

5 Conclusion

In order to solve the problem of the low reliability and poor reading accuracy in traditional instrument test process, based on machine vision theory and combined with image processing techniques,SF_6 instrument automatic identification method was improved. By HOUGH circle detection and regional centroid.

Acknowledgment. This paper was financially supported by the National International Cooperation Special Program(Grant No. 2013DFA11320).

References

1. Liu, Z.: Reader for Smart Grid. China Electric Power Press, Beijing (2010)
2. Zhou, L., Zhang, Y., Sun, Y., et al.: Intelligent substation inspection robot research and application. Automation of Electric Power Systems 35(19), 85–88, 96 (2011)
3. Qing, F.: switch position of image recognition and its application in power systems (Master thesis). China Electric Power University, Beijing (2005)
4. Zhu, X.-C.: Digital image processing and image communication. Beijing University of Posts and Telecommunications Publishing, Beijing (2002)
5. Pratt, W.K.: Digital Image Processing. Machinery Industry Press (2005); Hua, D., Zhang, Y. (transl.)
6. Hough, P.V.C.: Method and means of recognizing complex patterns. US: Patent 3069654 (1962)
7. Chen, J., Ding, Z.-L., Yuan, F.: Centroid detection method estimates of uncertainty. Optics 28(7), 1318–1322 (2008), doi:10.3321/j.2008.07.019, ISSN:0253-2239

Linking and Consuming Agricultural Big Data with Linked Data and KOS

Guojian Xian, Ruixue Zhao, Xianxue Meng, Yuantao Kou, and Liang Zhu

Agricultural Information Institution of CAAS, Beijing 100081, P.R. China
{xianguojian,zhaoruixue,mengxianxue,kouyuantan,zhuliang}@caas.cn

Abstract. This paper gives brief introduction about the big data, linked data and knowledge organization systems (KOS) and their relationships. As the authors mainly focus on the variety and value characteristics of big data, the linked data and KOS technologies are used to link and consume the large amounts of literature and scientific data in agricultural research community. The results show that it is a good way to describe, connect, organize, represent, visualize and access to big data effectively and semantically based on the linked data and KOS technologies.

Keywords: Big Data, Linked Data, Knowledge Organization System (KOS), Semantic Web, Scientific Data.

1 Introduction

Big data is now one of the hottest topics. Nowadays, we are generating huge amount of data every day, and the total volume of data would be doubled every 18 months, as for the rise of multimedia, social media, and the Internet of Things. It is true that of our activity, innovation, and growth are more and more based on the big data[1].

When talking about big data, people are likely to focus on technical issues, such as scalability, performance and how to deal with large quantities of heterogeneous data, but pay less attention to the connections, interoperability of data in disparate sources, and how to make sense of all the large data pools either. Actually, there are both latent and actual links, which are worth enriching, utilizing and publishing along with the raw dat. In most cases, it is the connections inside and outside of big data where the real value lies.

The formalized, structured and organized nature of linked data and its specific applications, such as the linked knowledge organization systems (KOS), have the potential to provide a solid semantic foundation for the classification, connection, representation, visualization of big data.

The reminder of this paper will firstly give a conceptual analysis of big data, linked data and KOS. And then we will illustrate how to create and consume semantic big data in agricultural research community, utilizing linked data technologies and knowledge organization systems as new tools for the describing, linking, organization, representation, visualization and access to big data.

© IFIP International Federation for Information Processing 2015
D. Li and Y. Chen (Eds.): CCTA 2014, IFIP AICT 452, pp. 546–555, 2015.
DOI: 10.1007/978-3-319-19620-6_61

2 Big Data, Linked Data and KOS

Big data always refers to large, diverse, complex, longitudinal, distributed data sets generated from instruments, sensors, internet transactions, email, video, click streams, and other digital sources. While there is no generally agreed understanding of what exactly is big data, an increasing number of V's has been used to characterize different dimensions and challenges of big data: volume, velocity, variety, value, and veracity[2, 10]. These terms means to the growing volume of different types of structured and unstructured data, the complex and heterogeneous nature and machine-processability. The organization, exploration, management, preservation, visualization and access to and use of these types of data pose technological and computational challenges.

Linked Data defines a set of guidelines, best practices or patterns for exposing, publishing, sharing, and connecting pieces of data, information, and knowledge on the Semantic Web using URIs and RDF[3, 11], which takes the WWW's ideas of global identifiers and links and applies them to (raw) data, not just documents[12-13]. Linked data can be effectively used as a broker, mapping and interconnecting, indexing and feeding real-time information from a variety of sources, and inferring relationships from big data analysis that might otherwise have been discarded, which is made all the more valuable by the connections (links) that tie it all together than the sum of its volume, velocity and variety.

Semantic text analysis, natural language processing, data mining and data visualization are the typical challenges in addressing the management and effective use of big data[4]. The knowledge organization systems, such as thesauri, classifications, subject headings, taxonomies, and folksonomies would increasingly important roles. As W3C's standard, the Simple Knowledge Organization System (SKOS) standard, aims to build a bridge between the world of KOS and the linked data community[17]. SKOS-based linked controlled vocabularies can provide a framework rich in semantics to effectively manage big data through combining, aligning and cross-linking multiple KOSs in order to automatic or semi-automatic analyzing, indexing and organizing text, and develop faceted, categorized or hierarchical views of big data[5].

A typical and successful example of combining the big data, linked data and KOS together is the new semantic web platform version of Agris: OpenAgris[6]. Based on nearly 4 million structured bibliographical records on agricultural science and technology, and AGROVOC vocabularies, alignment between KOS and ontologies, the OpenAgris also aggregates various open access data sources available on the Web, providing much data as possible about a topic or a bibliographical resource[7, 20].

3 Linking Agricultural Big Data

The 4th paradigm of science – data intensive scientific discovery has emerged within the last years, which means scientific innovations and breakthroughs will be powered by advanced computing capabilities that help researchers manipulate and explore massive datasets[8]. As for the several V's characteristics of big data, different disciplines highlight certain dimensions and neglect others. Fox example, people working on sensor and the internet of things may care more about the velocity, super-computing would be mostly interested in the volume dimension, and the research community pay more attentions to variety and value dimensions of big data.

The Chinese governments always pay high attentions to the agricultural science and technology innovation and the development of modern agriculture. With more investments are input to this sector, a large amount of agricultural big datasets are produced, such as the 3S data, scientific research data, and academic achievements (e.g. papers, books, proceedings, reports). These data have different formats, conceptualizations or data models, temporal and spatial dependencies[9]. Effective usage of these big data is very important to promote and advance the research work in a new round.

The following parts of this paper will explain how to describe, organize, integrate, and consume parts of these agricultural big data with linked data and KOS technologies, as we just focus on the variety and value dimensions of big data.

3.1 Linking Chinese Agricultural Thesaurus to Other KOS

As KOS can play important role to addressing some challenges of the big data, such as text analysis, natural language processing, data mining and data visualization of big data, the priority of our work is to convert the Chinese Agricultural Thesaurus (CAT) as linked data and link to other KOS.

CAT was developed by Agricultural Information Institution of CAAS in early 1990s and kept maintenance all the time. So far, CAT contains more than 60,000 Chinese descriptors and non-descriptors, most have corresponding English translations, and also include about 130,000 semantic relationships, such as UF, BT, NT and RT.

We describe CAT's concepts and their semantic relationships with SKOS and SKOS-XL standards. In addition, we map and link CAT to other well-known KOS such as AGROVOC, NALT, EUROVOC and LCSH, as shown in Fig.1.

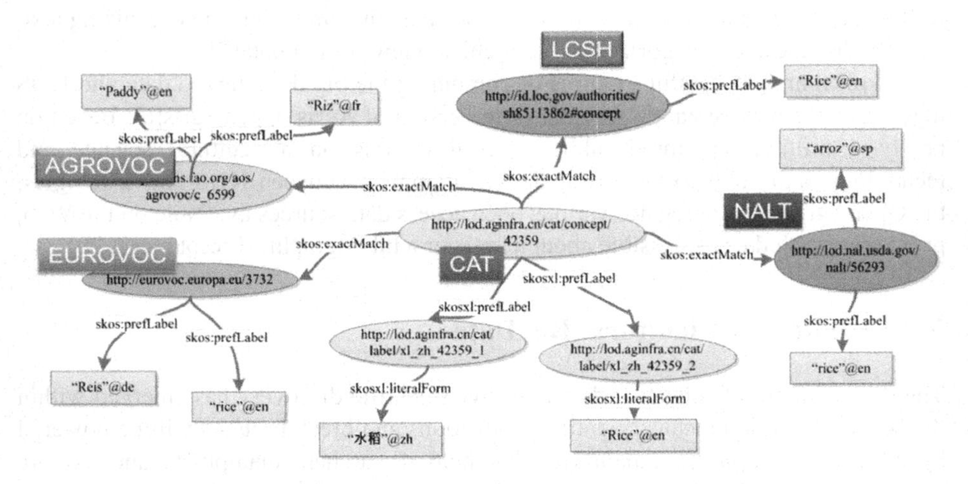

Fig. 1. The SKOS Model of CAT and Link to Other KOSs

We also develop a SKOS-based CAT linked data web system, providing services such as HTTP URI dereference, CAT concepts browsing and navigation, SPARQL query endpoint and RDF Triples Dumps, as shown in Fig.2. This work could greatly improve the CAT's visibility, accessibility and interoperability with other systems, and lay fundamental base to describe, organize and link other agricultural information resources semantically as well.

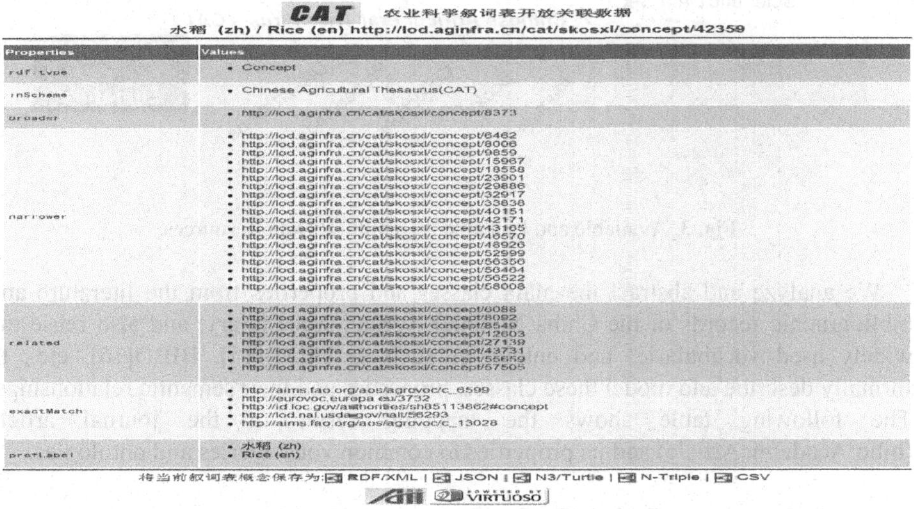

Fig. 2. The Linked Open data of SKOS-Based CAT

3.2 Linking Agricultural Literature and Scientific Research Data

As the current global research data is highly fragmented, by disciplines or by domains, from oceanography, life sciences and health, to agriculture, space and climate[21]. When it comes to cross-disciplinary activities, building specific "data bridges" are becoming accepted metaphors for approaching the data complexity and enable data sharing.

The millions of books, journals, proceedings and bibliographic records of the China National Agricultural Library, and also over 700 scientific datasets holds in the National Agricultural Scientific Data Sharing Platform, are the most valuable data materials for agricultural research communities. What we want to do most here is connect these literature and research data together in a light-weighted semantically way. Fig.3 shows the available and linkable data resources we could access to.

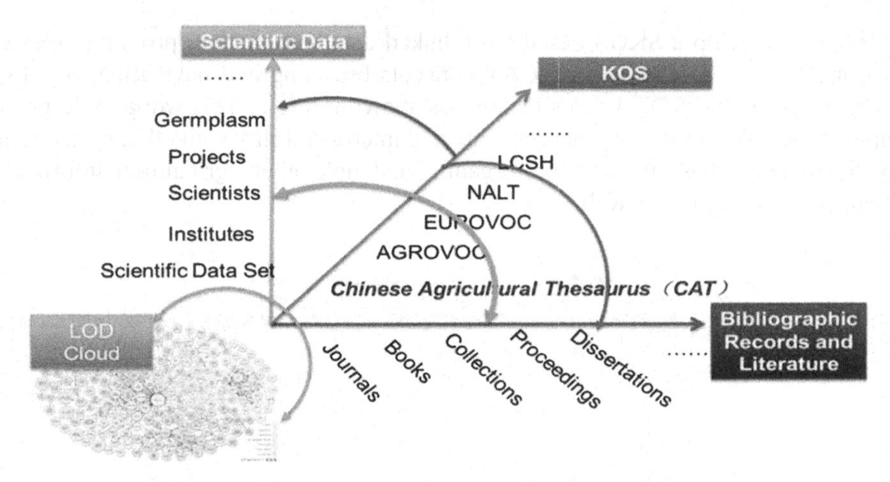

Fig. 3. Available and Linkable Agricultural Data Resources

We analyze and abstract the main classes and properties from the literature and bibliographic records of the China National Agricultural Library, and also reuse the widely used vocabularies and ontologies such as DCMI[15], BIBO[16], etc., to formally describe and model these classes, properties and their semantic relationships. The following table shows the mapping result of the journal article (bibo:AcademicArticle) and its properties to common vocabularies and ontologies.

Table 1. The Core Properties of Journal Article

Property	Type	Available Vacobulary
title	DataProperty	dc:title、swrc:title
alternateTitle	DataProperty	dcterms:alternative、prism:alternateTitle
author	ObjectProperty	dc:creator、foaf:maker、swrc:creator
keywords	ObjectProperty/ DataProperty	dc:subject、swrc:keywords、prism:keyword
abstract	DataProperty	bibo:abstract、dcterms:abstract、swrc:abstract
language	DataProperty	dc:language、swrc:language
startPage	DataProperty	bibo:pageStart、prism:startingPage
endPage	DataProperty	bibo:pageEnd、prism:endingPage
totalPage	DataProperty	bibo:numPages、prism:pageCount
DOI	ObjectProperty	bibo:doi、prism:doi

We also describe and connect the widely used literature, such as books, journals, collections, proceedings together, and the concepts and semantic relationships of CAT are taken into consideration while designing the describing and linking model (Fig. 4).

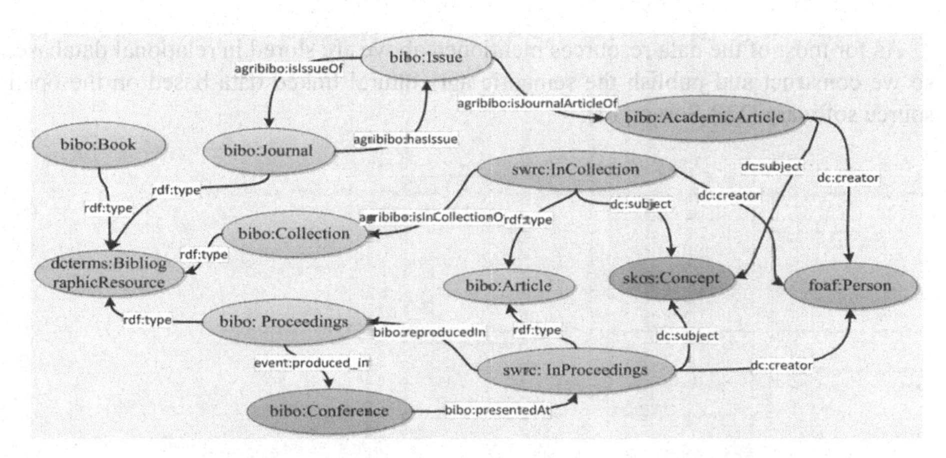

Fig. 4. Bibliographic Records and Literature Linking Model

The most interested and meaningful work we done is integrating and linking the SKOS-based CAT and several kinds of literature to the scientific research datasets. So far, we have modeled and linked about 700 core metadata of the scientific research datasets, and also some particular datasets hold the data about the institutes, researchers and projects of agricultural related domain, by reusing the well-known vocabularies or ontologies (e.g. VIVO, SWRC, FOAF)[19].The multidimensional semantic linking model covering the scientific data, literatures, and thesaurus we designed is as shown in Fig 5.

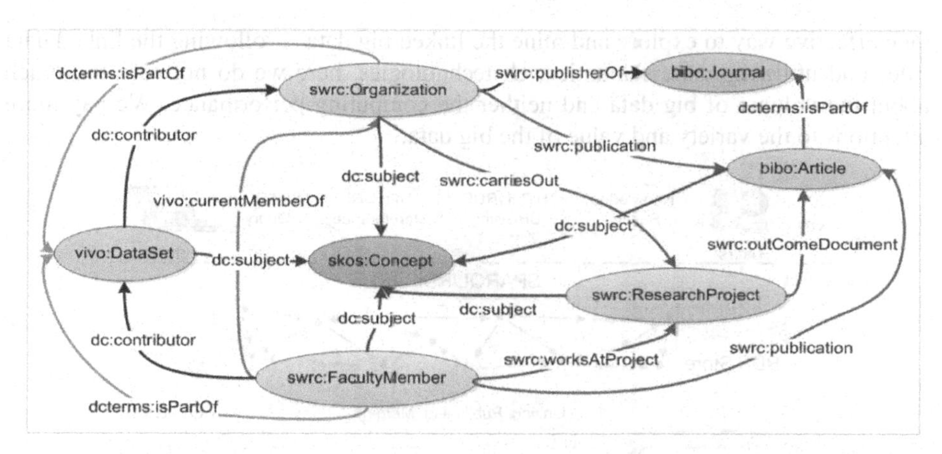

Fig. 5. Multidimensional Linking Model Covering the Scientific Data, Literatures and KOS

In order to process these resource based on the SKOS-based CAT, we developed an automatic text analyzing and indexing tool based on some open source tools, such as Lucene, IKAnalyzer, etc. The tool we realized could tag the concepts and semantic links of CAT into these literatures and scientific datasets. That is a very important step to make semantic connections between several kinds of data from different sources, with the professional knowledge of KOS.

As for most of the data resources mentioned above are stored in relational database, so we construct and publish the semantic agricultural linked data based on the open source software D2R Server[14].

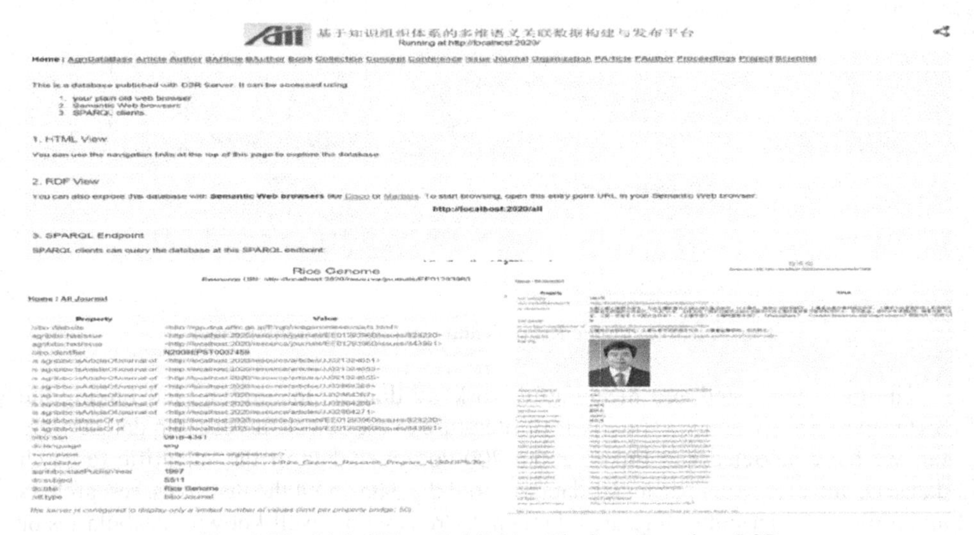

Fig. 6. The Linked Data Publishing Platform Based on D2R

4 Consuming Linked Agricultural Big Data

One effective way to explore and mine the linked big data is following the linked data rules and utilizing some semantic web technologies, here we do not care too much about the volume of big data and neither the computing performance. We pay more attentions to the variety and value of the big data.

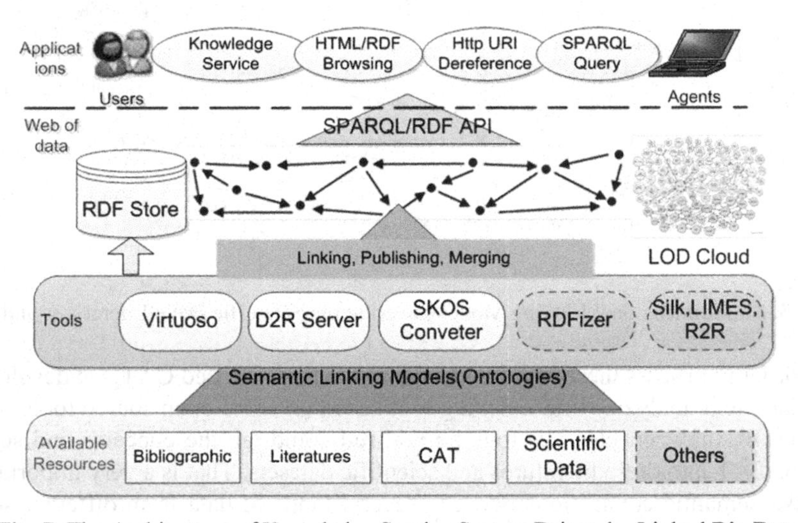

Fig. 7. The Architecture of Knowledge Service System Driven by Linked Big Data

The architecture and function models of domain knowledge service system have been designed, which totally driven by the linked data, as shown in Fig. 7. A prototype system was realized based on some key technologies such as SPARQL, Virtuoso[17] and so forth, see Fig.8. Some import service functions have been provided in this system, such as integrated browsing and discovery of domain knowledge, dynamic facet navigation and searching, SPARQL query endpoint, HTTP URI dereferencing, downloading RDF triples.

Fig. 8. The Domain Knowledge Service Prototype System Driven by Linked Data

5 Conclusions

The results of this study proves that it is one of the best practices to applying the ideas, principles and methodologies of linked data, to describe, organize and merge the huge amount of agricultural information resources in a more fine described, formally structured and semantically linked way. Linked data would play a great role to increase the popularity, visibility, accessibility and value of agricultural big data resources. The knowledge organization systems in SKOS formats would help us to align and match concepts to develop a broad and high level analytical framework for managing, representing and mining big data.

Further research work are needed because we at present just now focus on the variety and value dimensions of big data, we should actually address the volume, velocity issues in practice based on the cloud computing and other efficient big data infrastructure and technologies.

As massive amounts of data are available, linked and identifiable via URIs, big data, linked data and KOS would be an integral part of the future web infrastructure, the web of data, global data space and semantic web is beginning to take shape.

Acknowledgements. The related research work of this paper was support by the National Key Technology R&D Program "Knowledge Services Application and Demonstration Based on STKOS" (2011BAH10B06) during the Twelfth Five-year Plan Period.

References

1. Manyika, J., Chui, M., Brown, B., et al.: Big data: The next frontier for innovation, competition, and productivity[R/OL](May 01, 2011), http://www.mckinsey.com/insights/business_technology/big_data_the_next_frontier_for_innovation (June 08, 2014)
2. Hitzle, P., Janowicz, K.: Linked Data, Big Data, and the 4th Paradigm. Semantic Web 4(3), 233–235 (2013)
3. Heath, T., Bizer, C.: Linked Data: Evolving the Web into a Global Data Space, 1st edn. Synthesis Lectures on the Semantic Web: Theory and Technology, vol. 1(1), pp. 1–136. Morgan & Claypool (2011)
4. Suchanek, F., Weikum, G.: Knowledge harvesting in the big-data era. In: Proceedings of the 2013 International Conference on Management of Data, pp. 933–938. ACM (2013)
5. Shiri, A.: Linked Data Meets Big Data: A Knowledge Organization Systems Perspective. Advances In Classification Research Online 24(1) (2014), doi:10.7152/acro.v24i1.14672
6. Anibaldi, S., Jaques, Y., Celli, F., et al.: Migrating bibliographic datasets to the Semantic Web: The AGRIS case (2013)
7. Celli, F., Jaques, Y., Anibaldi, S., Keizer, J.: Pushing, Pulling, Harvesting, Linking: Rethinking Bibliographic Workflows for the Semantic Web. In: EFITA-WCCA-CIGR Conference "Sustainable Agriculture through ICT Innovation", Turin, Italy, June 24-27 (2013)
8. Hey, A.J., Tansley, S., Tolle, K.M., et al.: The fourth paradigm: data-intensive scientific discovery. Microsoft Research Redmond, WA (2009)
9. Janowicz, K., Hitzler, P.: The digital earth as knowledge engine. Semantic Web 3(3), 213–221 (2012)
10. Bizer, C., Boncz, P.A., Brodie, M.L., et al.: The Meaningful Use of Big Data: Four Perspectives-Four challenges. SIGMOD Record 40(4), 56–60 (2011)
11. Bizer, C., Heath, T., Lee, T.B.: Linked Data - The Story So Far. International Journal on Semantic Web & Information Systems 5(3), 1–22 (2009)
12. Linked Data - Connect Distributed Data across the Web [EB/OL] (June 18, 2012), http://linkeddata.org/ (May 28, 2014)
13. Mike. The Rise of the Data Web [EB/OL] (June 18, 2012), http://www.dataspora.com/2009/08/the-rise-of-the-data-web/ (May 28, 2014)
14. D2R Server-Publishing Relational Databases on the Semantic Web. [EB/OL] (February 16, 2010), http://www4.wiwiss.fu-berlin.de/bizer/d2r-server/ (June 15, 2014)

15. DCMI Metadata Terms [EB/OL] (June 14, 2012), `http://dublincore.org/documents/2012/06/14/dcmi-terms/` (February 03, 2013)
16. Darcus, B., Giasson, F.: Bibliographic ontology specification [EB/OL] (November 04, 2009), `http://purl.org/ontology/bibo/` (June 12, 2014)
17. Dolan-Gavitt, B., Leek, T., Zhivich, M., et al.: Virtuoso: Narrowing the Semantic Gap in Virtual Machine Introspection. In: 2011 IEEE Symposium on Security and Privacy (SP), pp. 297–312. IEEE, Berkeley (2011)
18. Isaac, A., Summers, E.: SKOS simple knowledge organization system primer [EB/OL] (February 21, 2008), `http://www.w3.org/TR/skos-primer/` (June 04, 2014)
19. Sure, Y., Bloehdorn, S., Haase, P., Hartmann, J., Oberle, D.: The SWRC ontology - Semantic Web for research communities. In: Bento, C., Cardoso, A., Dias, G. (eds.) EPIA 2005. LNCS (LNAI), vol. 3808, pp. 218–231. Springer, Heidelberg (2005)
20. AGROVOC Linked Open Data [EB/OL] (March 20, 2013), `http://aims.fao.org/standards/agrovoc/linked-open-data` (June 09, 2014)
21. Research Data Alliance [EB/OL] (June 20, 2014), `https://rd-alliance.org/` (June 22, 2014)

Research on Construction of Cloud Service Platform of Sci-tech Information for Agricultural Research System

Ruixue Zhao, Yuantao Kou, Ruopeng Du, Liangliang Gu, and Honglei Yang

Agricultural Information Institute, Chinese Academy of Agricultural Sciences,
Beijing 100081, China
{zhaoruixue,kouyuantao,duruopeng,guliangliang}@caas.cn,
363137894@qq.com

Abstract. In this paper, an overall architecture and service model of cloud service platform of sci-tech information based on cloud computing for agricultural research system were proposed, aiming at promoting the co-construction and sharing among information service institutions in agricultural research system. The platform will integrate resources and services of co-construction institutions into a shared body, and form a new science and technology information service system to provide network infrastructure, resources, platform and various application services for agricultural research system. And through dynamic management and distribution of a variety of services, it will meet the needs of users at different levels, such as institutions, academic teams and individuals and so on.

Keywords: agricultural research system, science and technology information, cloud service, cloud computing, digital library.

1 Introduction

Cloud computing is a new calculation model developed on the basis of distributed computing, parallel processing and grid computing and so on. Its core is to provide mass data storage and network computing services. Based on cloud computing, cloud service is a new IT service delivery model, aiming to combining technology and business delivery for the users [1]. Cloud is the center of the data storage and application service.

The emerging of cloud computing services has attracted more attention from various areas in the world. Google, IBM, Amazon, Microsoft and other IT giants have launched their own cloud initiative [1]. In the field of library and information, OCLC (Online Computer Library Center) launched the first cloud-based collaborative Web-level library management services [3][4]. In China, CALIS (China Academy Library and Information System) has carried out the development strategy of CALIS cloud computing at the earlier time, as well as research and exploration of cloud service platform of digital library [5].

Regarding scientific research information service system (library) as research object, this paper focused on the application of cloud computing in resources construction and services of agricultural research system.

© IFIP International Federation for Information Processing 2015
D. Li and Y. Chen (Eds.): CCTA 2014, IFIP AICT 452, pp. 556–564, 2015.
DOI: 10.1007/978-3-319-19620-6_62

2 Significance of Construction

Chinese agricultural research system contains three components, including independent agricultural research institutions of government, agricultural and forestry colleges and universities, and research institutes of private sector. In this paper, the scientific research system (ARS) consists of the Chinese Academy of Agricultural Sciences (CAAS), Chinese Academy of Fishery Sciences, Chinese Academy of Tropical Agriculture Sciences and the provincial academies of agricultural sciences. These institutions have a number of relatively independent professional institutes, and undertake tasks of agricultural science and technology innovation and industrial development. Meanwhile, these academies of sciences have specialized information service unit (library) which undertake research resources construction and information services of the internal system. For example, CAAS, a national comprehensive agricultural research institution, has 38 professional research institutes (centers) distributed throughout the country, one graduate school and a number of major national science and engineering projects, State Key Laboratory, Key Laboratory of the Ministry of Agriculture and Field Scientific Observation Station. For a long time, the Library of CAAS (National Agricultural Library) has undertaken important tasks of providing supporting scientific literature and information services for CAAS and the whole nation.

With the rapid development of internet and other modern information technology, and the increased pressure of agricultural science and technology self-dependent innovation in China, the requirements of information and services of agricultural research system are undergoing significant changes. Firstly, with the rapid development of the internet and a number of commercial information institutions, more pathways are provided to access information for researchers. The needs of researchers for information services provided by Library have gradually been weaken. Secondly, the demand for science and technology information of scientific research personnel is more urgent. In particular, they need to keep abreast of the latest research developments of disciplines and peers progress both at home and abroad. The needs of information content are also charactered by timeliness, specialty, comprehensiveness, dynamics and knowledge. The third change is that the way of access to information is featured with digitization, networking, and mobility. The network of scientific communication and collaboration environment become the mainstream of modern scientific research. The forth one is that researchers need more information finding ways and various information processing tools. They pay more attention to the preservation, sharing and utilization of the academic achievements.

Due to these changes, the library and other specialized information service organization have faced serious challenges. Because of lack of funding, manpower, knowledge, basic conditions and others, single information agency increasingly highlights limitations in conducting resource construction and providing services. Obviously, there need union, co-building, sharing and co-prosperity. What comes first is infrastructure construction. The infrastructure is weak in some Libraries of the academy system or information agency, and the capacity to update and maintain is inadequate. These affect the quality of the information service. But there are also some individual information services which have obvious advantages of the resources, technology, personnel, and equipment as well. The second issue is resource sharing. Although the academies of agricultural research system like CAAS have accumulated a certain amount of information resources for a long time via various channels, including the purchased electronic resources and self-built specialized databases, which played an important role in the service, it is still unable to meet the expanding needs of scientific users. And for a long time, the resource construction and service capacity are quite different and collaborative sharing is weaker among the various information institutions of agricultural research system. These lead to coexist of repeated construction and inadequate resources. That not only is very wasteful, but also affects the overall effect of service. Thirdly, there are also a lot of same subjects, although the academies of agricultural research system have their own unique fields of study. It is more prominent for the role of joint construction of information resources and providing service for researchers with similar research direction, although they are distributed in different research institutions.

Cloud computing technology provides an effective means to solve these problems. It opened up a new train of thought (library) of the co-construction and sharing for promoting information service organization (library) of agricultural research system in China. In this context, based on cloud computing, we proposed the construction of cloud service platform of agricultural scientific research system. Its significances lie in as follows:

(1) To promote the development of resource construction and service alliance of agricultural research system in China, to co-build a new system for agricultural research service system, achieve value adding and extension under limited conditions of resources, systems, services, technology etc.. It can provide broader and more convenient service for scientific users, and reduce a lot of problems such as scattered resources, self-limited service system, disorders of multi-service gateway when accessing system, and so on.

(2) To reduce duplication of construction through integration of high-quality resources of agricultural research system, and alleviating resources construction and service limitations of the individual information service institution especially due to lack of technology, talent, capital, ,to improve overall resources support ability and service level of agricultural research system.

(3) To adapt with the changing environment. With the popularization of internet services, it is supposed to take the advantages of professional information services, focusing on the construction of characteristic resources of agricultural subject, which are scarce on the Internet, expanding specialized knowledge services. And the private cloud or hybrid cloud service platform of sci-tech information for agricultural research system can be built when enjoying the services of internet public cloud, to improve competitiveness of agricultural specialized information service organization.

3 Construction Infrastructure

The construction of the proposed platform not only needs infrastructures such as good network environment, High-Performance Server, storage equipment, but also calls for rich agricultural professional data resources and service applications along with co-construction and sharing mechanism and strong desire in the system. As a result, we should fully consider the existing resources and conduct the optimized integration and promotion.

First of all, CAAS has provided a good network information environment, and established the first agricultural science and technology information network in China. The library of CAAS, also the agricultural Library of NSTL, takes responsibility of construction of the strategic guarantee system of national agricultural science and technology literature resource. It plays a crucial and pivotal role in the national agricultural specialized library and intelligence system. In 2005, the National Agricultural Library hosted and established the "National Agricultural Science and Technology Literature Information and Service System (NAIS)[6]" which has achieved effective integration and service integration of multi-source, distributed, and heterogeneous resources and provided "one-stop" information service of science and technology to the whole country.

Secondly, CAAS established the largest National Agricultural Scientific Data Sharing Center (Agridata)[7] in China with collaboration of Chinese Academy of Fishery Sciences, Chinese Academy of Tropical Agriculture Sciences, and some of the provincial academies of agricultural sciences, under the support of the National Science and Technology Basic Condition Platform Program. It is a national scientific data resource sharing and service system, which consists of a main center, seven sub-centers, and 15 provincial sharing services branches.

Thirdly, the information service institutions (library) of agricultural research system established the "Electronic Resources and Services Union of Agricultural Scientific Research system (Agricultural science union)" according to the initiative of library of CAAS, which collectively carries out group purchase and resource sharing. By now the Union has more than 30 member institutions (libraries).

Fourthly, in order to adapt to the new requirements of open scientific research information environment and scientific research innovation, the library of CAAS has built a technology system and a support platform for large-scale digital resource processing and distributed heterogeneous resources integrating based on the "cloud sever" open architecture since 2010. Meanwhile it has built a comprehensive agricultural scientific digital knowledge warehouse containing agricultural science and technology literature, scientific data, open access resources and other third-party Internet resources. CAAS Library also established a multi-level service system based on NAIS public and popular service and offers personalized services to areas, organizations, teams, subjects, and personal. A series of achievements has been achieved on resources construction and personalized technology, tools and platform service. These results have been preliminarily applied in CAAS institutes, academic teams and several provincial agricultural information service academies such as Beijing, Shanxi, Sichuan, Xinjiang and Tibet Academy of Agricultural Sciences. In this context, the prototype of cloud service platform of agricultural research system has formed.

4 Platform Overall Architecture

The proposed platform is a digital research information platform based on a variety of technologies and services, which provides diversified services for service institutions, research institutions, scientific researchers and even agricultural scientific and technological personnel all over the country. The overall architecture of the platform is shown in the Fig. 1. The platform adopts the open system architecture based on "cloud services", with support of the CAAS network infrastructure and the scientific research resources guarantee system. It integrates and uses other resources and services of agricultural research system, as well as external information environment of scientific research.

Cloud service center has a long-term, basic, stable resource and service support from NAIS, Agridata, NSTL, other agricultural research institutes, the Internet open access system, and third-party system. Among them, NAIS mainly focuses on literature construction and service, while Agridata aims at integration and sharing of scientific data. By integrating and reorganizing these sources, as well as encapsulating and clouding the core function modules of these service system, cloud service center builds a series of cloud service components with both versatility and reusability, including resources, tools, systems, Open APIs, and Web Services. It provides services for cloud users, such as professional information service agencies, research institutions, academic teams, individual users and third-party users, etc. As the resource and service center of agricultural research system, cloud service center can harvest resources shared by various users, thus forming a shared knowledge base of national agricultural

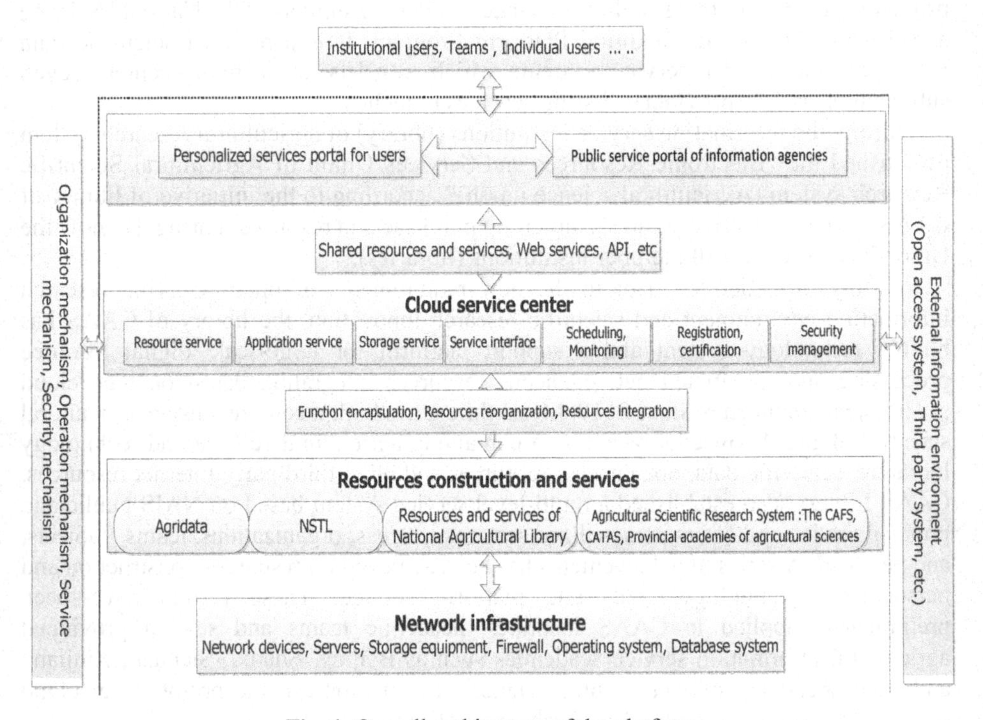

Fig. 1. Overall architecture of the platform

research. Meanwhile, it is responsible for global resource storage, organization, management, distribution and scheduling, conducting service monitoring and service statistics, maintaining login accounts and user rights, deploying various kinds of application service, and securing the safety of resource and service, etc. A resource sharing and collaborative service model can be formed through cloud service center, a public service portal of information institution and a personalized knowledge service platform for organizations, groups, individuals, disciplines and projects, which will enhance the overall service ability of agricultural research system.

5 Cloud Service Model

The platform is composed of a series of services, which can be divided into 5 layers as shown in the Fig. 2. The 5 layers are infrastructure layer, platform service layer, application service layer, cloud service layer and cloud clients, respectively [8-12].

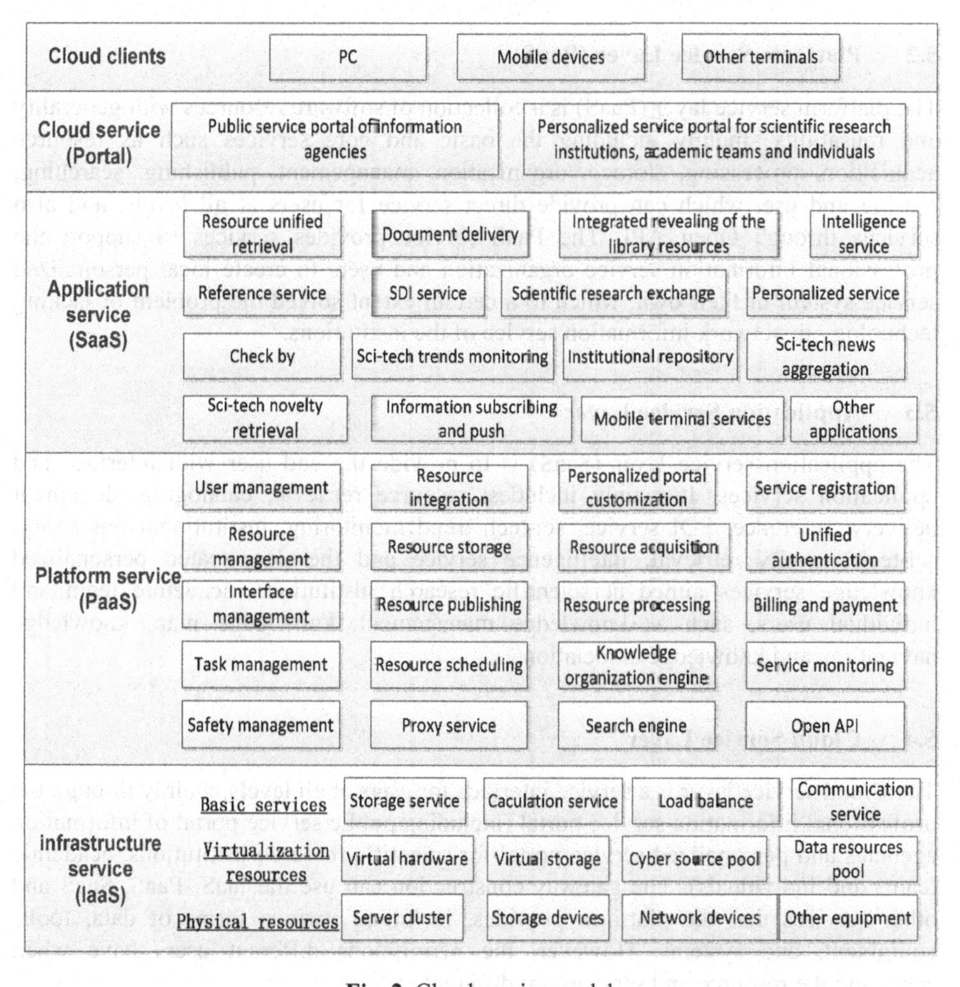

Fig. 2. Cloud service model

5.1 Infrastructure Layer (IaaS)

The infrastructure layer (IaaS) at the bottom is assembling of virtualized hardware resources and related management functions. It is the basis of the whole structure of cloud service platform, providing hardware, network and resource sharing service for the users of agricultural research system. It mainly includes three parts: 1) the physical resources which is composed of network devices, servers, storage devices and other equipment; 2) the virtualization resources which contains virtual server cluster, virtual storage, cyber source pool, data resources pool; 3) the basic management services based on virtual resources such as storage service, calculation service, communication service and load balance. The infrastructure cloud services can not only make full use of superior server resources, cyber source and storage resources within the system, but also decrease the idle resources, avoiding repeated purchase of hardware resources, and reducing the burden of weak institutions in the construction of hardware resources.

5.2 Platform Service Layer (PaaS)

The platform service layer (PaaS) is a collection of software resources with generality and reusability, mainly including the basic and core services such as resource acquisition, processing, storage, organization, management, publishing, searching, sharing and use, which can provide direct service for users at all levels, and also services through Open API. The PaaS service provides services to support the professional information service organization and users to create local personalized service system of their own, which to a certain extent solved the problem of lacking technology in network information service of the institutions.

5.3 Application Service Layer

The application service layer (SaaS) is to provide the end user with interface and application services. It mainly includes: resource retrieval, cataloging, document delivery, reference, SDI service, sci-tech trend monitoring, institutional repository, sci-tech novelty retrieval, intelligence service and the deep-seated personalized knowledge services aimed at scientific research institutions, academic team and individual users, such as knowledge management, knowledge map, knowledge navigation, and knowledge association.

5.4 Cloud Service Layer

The Cloud service layer is a service interface for users at all levels, mainly through the professional information service portal (including public service portal of information agencies and personalized service portal for scientific research institutions, academic teams and individuals). The gateway construction can use the IaaS, PaaS, SaaS and other services that the platform provides, including various forms of data, tools, equipment, and systems. However, the permissions different users have when accessing the resource and services are different.

5.5 Cloud Clients

The cloud clients are the main terminals with which users have access to the service portal and use the resources and services.

6 Service Mode

This platform can offer flexible, dynamic service configuration and service delivery according to users' needs. Considering permission to access, intellectual property and services and other factors, the Cloud service platform provides different services such as paid services, free services, and adjustment access by scenario sensitive.

The information service institutions of agricultural research system provide following services:

① The platform tools are provided which support the information service institutions to establish local personalized service portal. Those institutions will no longer need to build and develop their own information software. They only need access the needed services according to their business demands in accordance with the contract. This is the way to implement co-construction and sharing between application service systems and information services.

② The platform can offer service interface. Information service institutions can use directly according to the service authorization. Those service interfaces allow sub-sites access to the resource and service offered by cloud service platform. It enhances the convenience of access to resources by embedded services in the user local information environment. The general features include integrated retrieval, knowledge navigation, SDI service, reference, etc.

③ Cooperative service platform provides services such as group purchase of electronic resources, joint cataloging, joint processing, institutional repository alliance, collaborative research, document delivery, interlibrary loan etc. Co-construction units can formulate powerful resource and service superiority based on unified standards and specifications, as well as mutual coordination.

For professional research institutes, academic teams and individual users, the available services include:

① Users can build their own personalized service portal with cloud services. They can also embed required services into their local information environment through the embedded interface.

② Users can access to the resources and services of cloud service platform directly through the terminal access. The service in the agricultural research system is more comprehensive, individualized and integrated under the environment of cloud service system.

7 Conclusion

Undoubtedly, Cloud Computing brings unprecedented challenges and chances to the information service industry. The cloud service platform of agricultural research system presented in this paper will promote the resources construction and the development of sharing services union among information service institutions, and improve the overall information service of agricultural research system. However, there are still some problems to be solved. They include the issues of cloud computing security, resources copyright and ideas identification, how to reach an agreement about coalition building between agricultural research systems, and how to establish an efficient and mutually beneficial pattern effectively integrating the technologies, resources, services of member institutions. In short, in the global open information environment, the alliance of professional information institutions not only contribute to their development, but also facilitate more progress of agricultural sci-tech information service industry, with the continuous competition of the business information agencies, publishers, and under the condition of limited resources,. Moreover, it is beneficial to the broad masses of agricultural science and technology innovation and industrial development.

References

1. Yu, X., Wang, J.-Y.: Research on the Cloud Services Platform Architecture of Digital Library Based on Cloud Computing Technology. Information Science 30(12), 1854–1857 (2012) (in Chinese)
2. Guo, J.: Studying on the model of library cloud service under computing environment, vol. 5 (2012) (in Chinese)
3. Zhang, W., Feng, K., Hu, G., Hu, C.: Cloud Computing and Its Application and Obstacles in Library. Library and information Service 54(7), 42–45 (2010) (in Chinese)
4. Products & Services. [EB/OL] (January 23, 2014), http://www.oclc.org/en-asiapacific/services.html
5. Zhang, B.: Theory and Practice of Socialization of University Library Information Service-Library of Guangzhou University as an Example. Journal of Academic Libraries (4), 13–18 (2009) (in Chinese)
6. National Agricultural Science and Technology Literature Information and Service System (NAIS). [EB/OL] (January 23, 2014), http://www.nais.net.cn (in Chinese)
7. National Agricultural Scientific Data Sharing Center (Agridata). [EB/OL] (January 23, 2014), http://www.agidata.cn (in Chinese)
8. Chen, C., Han, J.: Improved Resource Management and Application Platform for Digital Library under Cloud Computing Environment. Journal of Modern Information 33(2), 18–20 (2013) (in Chinese)
9. Xie, Y., Zhang, J., Zhou, X.-X.: Design and Implementation of Cloud Computing Services Platform in Library Consortium. Information Science 30(12), 1854–1857 (2012) (in Chinese)
10. Yang, F., Dai, L.: Construction on Regional Digital Library Alliance Based on Cloud Services. Journal of Foshan University (Social Science Edition) 31(4), 87–91 (2013) (in Chinese)
11. Wang, X.: Research on Regional Digital Library Based on Cloud Computing. Digital Library Forum (11), 63–70 (2010) (in Chinese)
12. Wu, Z., Xu, G.: Discussion about the system of library consortia based on the cloud service. New Century Library (3), 8–10 (2011) (in Chinese)

Research on Construction of Agricultural Domain Knowledge Service Platform Based on Ontology

Yuantao Kou, Ruixue Zhao, and Guojian Xian

Agricultural Information Institute, Chinese Academy of Agricultural Sciences, Beijing 100081, China
{kouyuantao,zhaoruixue,xianguojian}@caas.cn

Abstract. Scientific researchers' increasing demand for knowledge service under the new situation, makes it urgent to embed information service into user research process, ad build an incorporate knowledge platform that integrates knowledge, skills, tools, and services of certain professional field. This paper put forward the technical solution of agricultural domain knowledge service platform based on ontology, including resource organization based on ontology, platform design and development. The construction progress of ontology base and service functions based on ontology are shown by application practice in rice domain.

Keywords: domain knowledge service, knowledge organization system, scientific research ontology, domain ontology.

1 Introduction

Scientific and technological innovation is the source of power that promotes social development and progress. In the era of knowledge economy, innovation and application of knowledge has become the most important part of scientific and technological innovation. Any research on scientific and technological innovation is inseparable from the support of science and technology information resources, and the construction of e-science environment.

Scientific users have an increasing needs [1,2] for service integration, knowledge management, knowledge services, communication and collaboration under the new situation. This makes it become urgent to build a research and knowledge service platform to integrate knowledge, technology, tools, services of certain domain, so as to embed information services into users' research process, achieve service innovation, and support the scientific and technological innovation.

Research on the construction of domain knowledge service platform has made significant progress internationally. Lots of countries are paying more and more attention to domain knowledge service platform, and some large, national level research projects and practical activities have been carried up. There are some systems or software tools that we can refer to, such as VIVO based on ontology [3,4], VRE based on SOA [5,6], Harvard Catalyst, Sciologer of Columbia University [7], SKE of

© IFIP International Federation for Information Processing 2015
D. Li and Y. Chen (Eds.): CCTA 2014, IFIP AICT 452, pp. 565–574, 2015.
DOI: 10.1007/978-3-319-19620-6_63

CAS [8], the open source software of virtual learning environment SaKai, [9-11], etc. These construction experiences provide a good foundation for this study.

Ontology is a good technical tool which shares common understanding of the structure of information among people or software agents and enables reuse of domain knowledge. This paper put forward the technical solution of agricultural domain knowledge service platform based on ontology, including resource organization based on ontology and platform design and development. Current construction and application situation in rice domain was introduced at the end.

2 Resource Organization Based on Ontology

Resources organization is the root of a knowledge service system. Knowledge organization system construction is foundation and core of realizing the transformation from scientific literature information services to knowledge service. Agriculture itself is a complex subject, so what needs to be solved in this system is the organization of core knowledge content. Using ontology which can represent the meaning and relevance of knowledge more accurately, laid the foundation for construction of the knowledge service application.

Regarding domain knowledge service application as the guidance, advanced international mainstream knowledge organization technology should be adopted in agricultural domain knowledge organization. Existing knowledge organization materials were used for reference in ontology design and ontology instance base construction. Ontology of agricultural domain knowledge service platform consists of scientific research ontology and domain ontology.

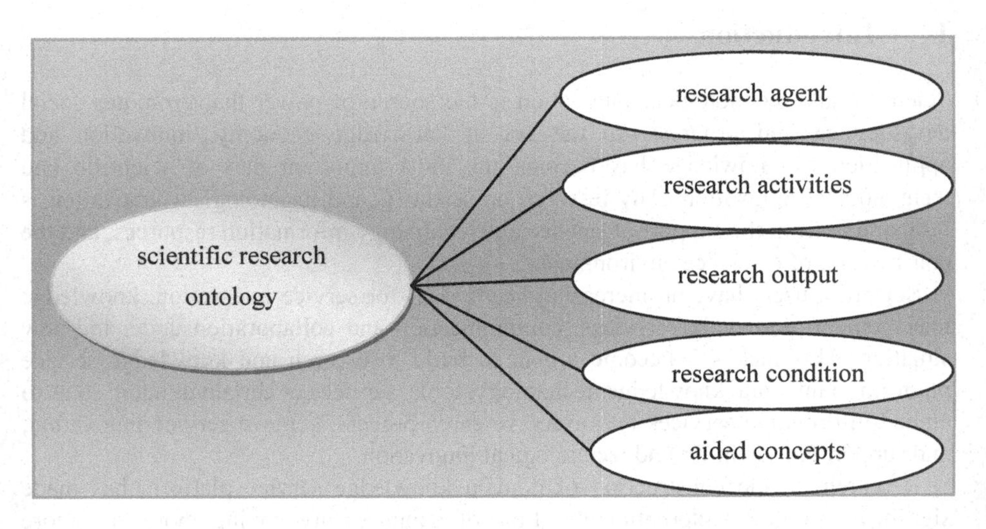

Fig. 1. Main class resources of scientific research ontology

2.1 Scientific Research Ontology

Scientific research ontology reveals and reflects research elements and their mutual connection, which is the foundation of knowledge discovery. VIVO, put forward by Connell University, is very useful reference. VIVO integrated with powerful ontology management tools, through construction of ontology around scientific experts ("people"), using Jena inference system to realize associated navigation and retrieval of research objects [12,13]. Considering scientific research agent, research condition, research activities and research output are core concepts during scientific research progress, in this paper, we proposed the scientific research ontology around research progress. Its main class resources are as Fig. 1.

Fig. 2 depicts the research output class. Its sub classes include journals, dissertations, proceedings, books, academic reports, scientific data, patents, and standards. The figure also shows the relationship between research output and other major classes through object properties.

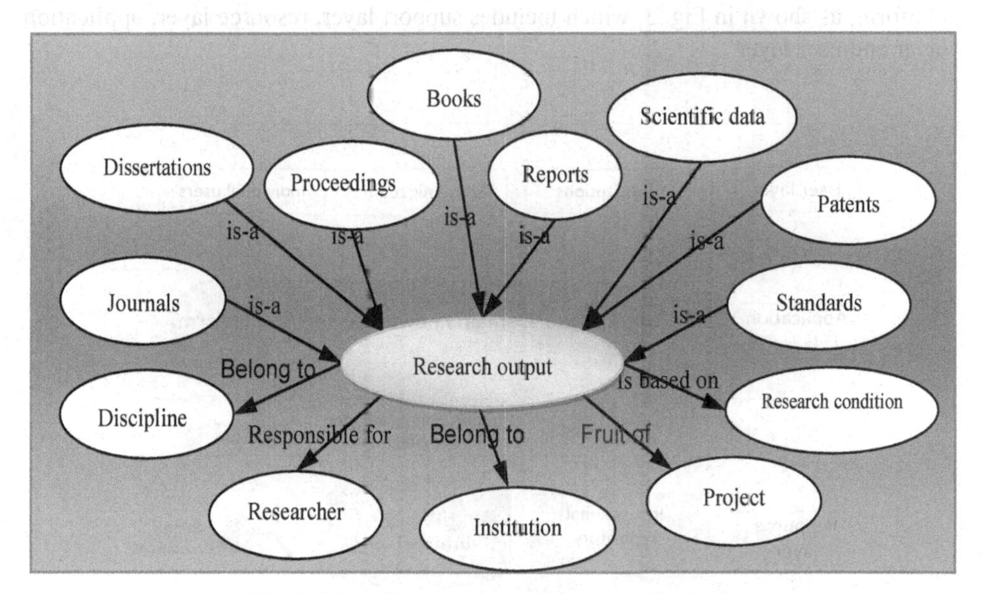

Fig. 2. Class diagram of scientific research output

2.2 Domain Ontology

Domain ontology is domain specific, and constructed on the basis of analysis of knowledge content and core concepts, their attributes and the relationship between them. And it's used for knowledge organization and knowledge association discovery of research objects and literature information, which are the base of domain knowledge navigation and knowledge structure exhibition.

Domain ontology can focus on the whole domain, or a small part of it, or even be a combination of several disciplines. According to the general ontology framework, extending ontology instances facing specific area is the key of content construction of agricultural domain knowledge service system. Therefore the domain ontology of this platform will be constructed after the target application domain is identified.

3 Platform Design and Development

Agricultural domain knowledge service platform is a personalized information platform for professional researchers, and also an integrated information system which is the organic integration of library services, users' research process and the internet service environment.

3.1 System Architecture

This paper presents the 4-layer architecture of agricultural domain knowledge service platform, as shown in Fig. 3, which includes support layer, resource layer, application layer and user layer.

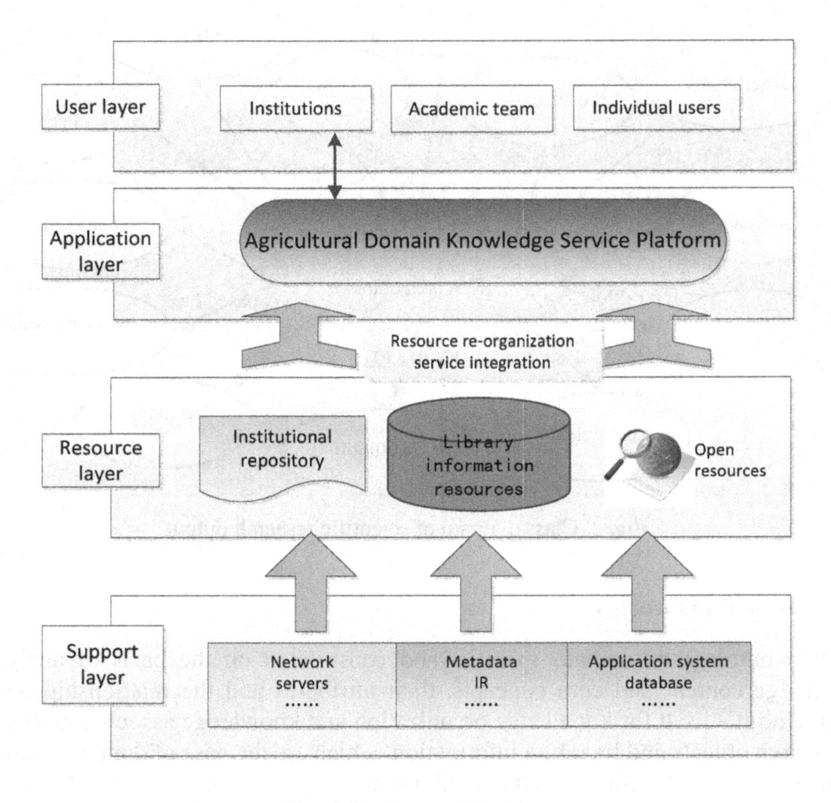

Fig. 3. System architecture

- The user layer is located at the top level of the whole structure, and gives out the primary users of agricultural domain knowledge service platform, including research users, manage users and third party users. Scientific users are the main service target which is made up with researchers, research teams and research institutions.
- The application layer, which mainly aims to solve the problem of platform's construction, realizes various services of agricultural domain knowledge service platform in the way of system functions and tools with seamless integration of third-party technology tools from existing service platform or open source systems.
- The resource layer, which is responsible for resource organization, clearly defines the resources elements of domain knowledge service platform, carries out the collection and procession of the resources, and builds the domain knowledge base using ontology.
- The underlying support layer aims at the building of hardware and software environment which is the technical foundation of the entire system.

3.2 Functional Structure

Fig. 4 shows the functional structure of agricultural domain knowledge service platform, and the green parts included are its main components:

- Knowledge service functions mainly provide the core services of agricultural domain knowledge service platform, including domain knowledge navigation, knowledge retrieval and acquisition, domain knowledge association, knowledge maps, knowledge sharing, knowledge management and other functional components.
- The domain knowledge organization and management tools mainly achieve content construction of agricultural domain knowledge service platform, maintenance of classes and attributes of scientific research ontology and domain ontology, as well as the processing and transformation of resource materials to database.
- Collaborative modules which mainly support researchers' tacit knowledge sharing and provide online communication and knowledge sharing environment, include collaborative work platform, scientific community, domain knowledge wiki, scholars' network and other functions.
- Data service interface mainly achieves the scalability and extensibility of agricultural domain knowledge service platform, supporting data exchange between the platform and third-party systems or services.
- The system basic functions such as system management, user and privilege management, and statistics support the system's normal operation.

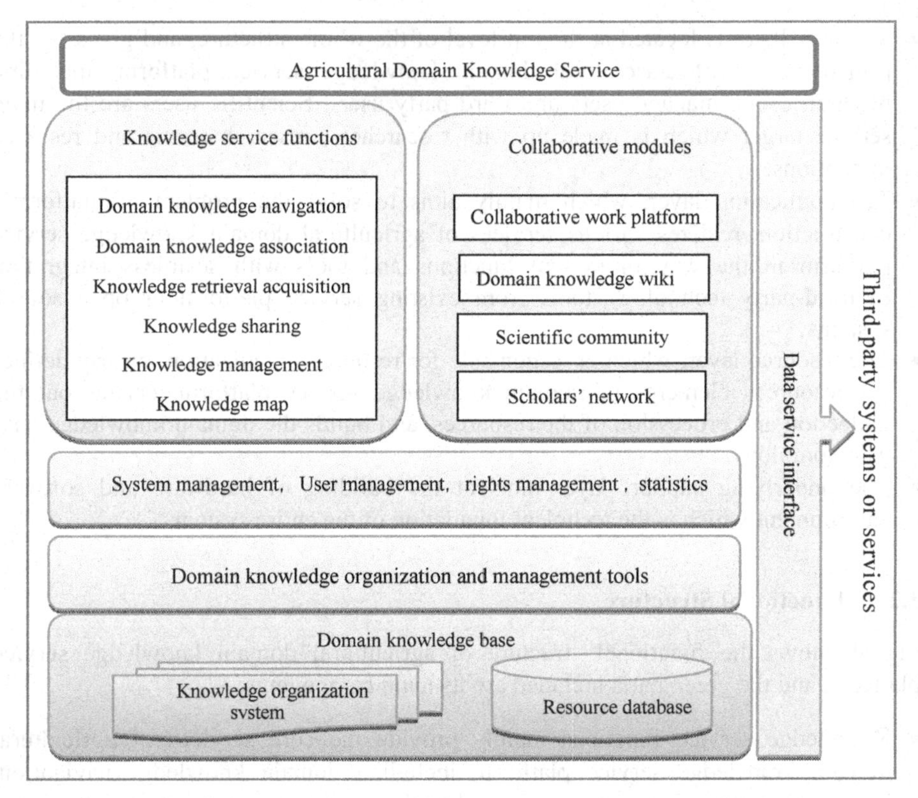

Fig. 4. Functional structure of Agricultural domain knowledge service platform

3.3 Platform Development

According to the characteristics of large volume data of the system, distributed architecture was used, so as to benefit sharing system computation and storage pressure. Java was the main development language, based on J2EE architecture, and MyEclipse was the development tool, with the environment of "Windows2008+ Tomcat6.0+Mysql5.5".

During construction of the knowledge base, O-R (ontology relation) mapping technique was applied to achieve dynamic storage of resources, and Hessian technology was used to obtain the cross language, cross platform and more openness of the knowledge base.

By adopting the SSH structure, knowledge application of the foreground system could dynamically adjust columns and information project from the knowledge structure, to support cross domain application. Knowledge service functions were realized by using AJAX, RSS, GIS, knowledge visualization and other related software technology. And knowledge organization and management functions in background system were implemented by "generic desktop" technology, which provides Windows operating experience under the Web system.

4 Application Practice in Rice Domain

Application to a specific domain is the "last mile" to realize construction target of agricultural domain knowledge service platform. Combined with country-level major projects, according to the discipline construction and development status of Chinese academy of Agricultural Sciences, we selected rice as the target domain and deployed the rice domain knowledge service system, which provides knowledge service to scientific workers engaged in rice research. Fig. 5 gives the rice knowledge service system homepage screenshots.

Fig. 5. Rice knowledge service system homepage

4.1 Ontology Base Construction

As to scientific research ontology base of rice domain, we carried out collecting, processing and arranging information resources about rice, including scientific research institutions, scientific research personnel, scientific literature, science data, knowledge organization system, etc.

Rice domain knowledge is extremely large and rich, covering rice varieties, rice cultivation, rice producing areas, rice biochemistry, rice diseases and insect pests, weeds and so on. So it's quite difficult to describe the knowledge content in detail for the whole rice domain with ontology. In view of this, this paper selected the sub field --plant diseases and insect pests of rice, as the research object of lightweight domain ontology construction.

Rice plant diseases and insect pests ontology mainly describes the concepts about symptom, pathogen, occurrence of damage, occurrence time, occurrence conditions, route of transmission, and prevention and control of rice disease and insect pests, as well as the relationships between these concepts. The construction method is, on the basis of professional books and database resources, analyze the knowledge content structure, disjunctive core concepts and relations, extract the pest discipline branches, complete the design of ontology framework, and finally complete the construction in protégé.

4.2 Service Functions Based on Ontology

In rice domain knowledge service system, besides the ordinary functions including domain knowledge retrieval, knowledge navigation, knowledge acquisition, and maintenance of knowledge organization system, there are two typical service functions based on ontology.

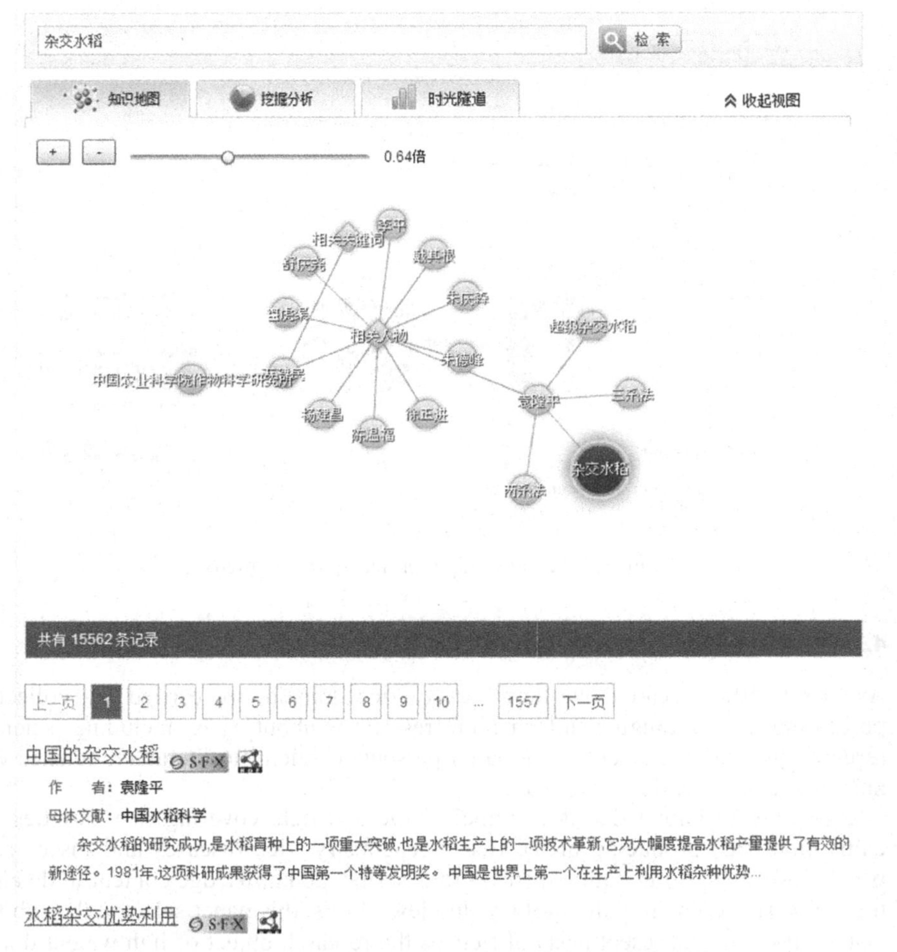

Fig. 6. Visualization of research entities and their relationships

First is visualization of research entities and their relationships. As shown in Fig. 6, scientific researchers, research topics, institutions, literature, and relationship between them are revealed in one "knowledge map".

Secondly, based on the semantic inference of domain ontology, rice plant diseases and insect pests diagnosis reasoning was realized. There are two forms of diagnosis in the platform: one is that users input the names of plant diseases and insect pests, and then the platform provides basic information of the disease or insect pest, as well as related symptoms, occurrence position, rice growth period, methods of prevention, drugs, and relevant literature; another is that, with some symptoms input by users, the platform presents suspected plant diseases and insect pests after reasoning based on ontology, so as to give treatment suggestion and relevant literature links (Fig. 7).

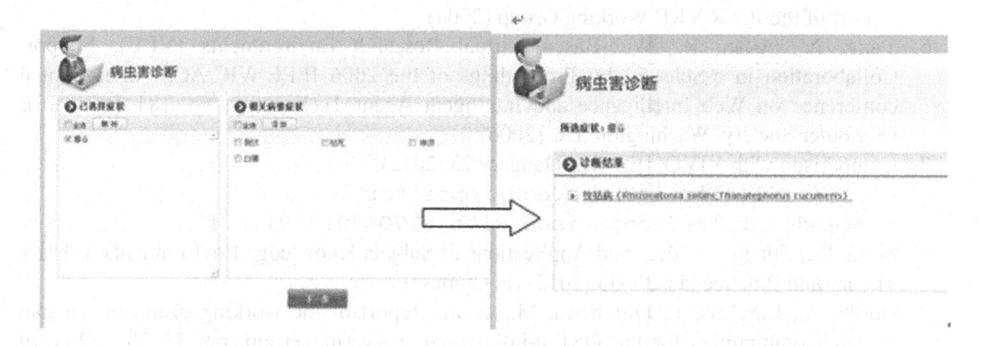

Fig. 7. Rice plant diseases and insect pests diagnosis

5 Conclusion

Domain knowledge service platform is not a new service model, but its construction is still in the exploratory. The practice we have tried using ontology in agriculture domain knowledge service is very meaningful. However, there are still a lot of issues that need further improvement and solvent in the future, such as core services, construction of domain ontology, data standards, users' research process and their work environment, non-technical factors, interdisciplinary knowledge exchange and so on. In short, building an integrated knowledge service platform for professional areas is a general trend, but carrying out the work in a large scale and showing the effectiveness still need a period of time. That needs constant exploration to new technologies, new methods of knowledge services to continuously promote technology innovation.

Acknowledgement. The related research work of this paper was support by the National Science & Technology Pillar Program "Knowledge Services Application and Demonstration Based on STKOS" (Grant No. 2011BAH10B06) during the Twelfth Five-year Plan Period of China.

References

1. Xiao, X., Wang, D.: Development trend of user information environment, information behavior and information demand. Library Theory and Practice (1), 40–42 (2010) (in Chinese)
2. Chen, C.: Users Information Requirements and Information Services under E-science Environment. Information Science 27(1), 108–112 (2009) (in Chinese)
3. Krafft, D.B., Cappadona, N.A., Caruso, B., et al.: VIVO: Enabling National Networking of Scientists. In: Web Science Conf. 2010, Raleigh, NC, USA, April 26-27 (2010)
4. Devare, M., Corson-Rikert, J., Caruso, B., et al.: VIVO Connecting People, Creating a Virtual Life Sciences Community. D-Lib Magazine 13(7/8) (2007)
5. Allan, R., Allden, A., Boyd, D., et al.: Roadmap for a UK Virtual Research Environment. Report of the JCSR VRE Working Group (2004)
6. Yang, X., Allan, R.: Web-Based Virtual Research Environments (VRE): Support Collaboration in e-Science. In: Proceedings of the 2006 IEEE/WIC/ACM international conference on Web Intelligence and Intelligent Agent Technology, pp. 184–187. IEEE Computer Society, Washington, DC (2006)
7. Comparison with VIVO. [EB/OL] (January 23, 2012), `http://swl.slis.indiana.edu/repository/systemdoc_files/comparison_with_VIVO%2012-01.pdf`
8. Song, W., Zhang, S.: Ideas and Applications of Subject Knowledge Environments. Library Theory and Practice (1), 30–33 (2012) (in Chinese)
9. Borda, A., Careless, J., Dimitrova, M., et al.: Report of the working group on vir-tual researchcommunities for the OST e-infrastructure steering group, pp. 33–36. Office of Science and Technology, London (2006)
10. Allan, R., Allden, A., Boyd, D., et al.: Roadmap for a UK virtual research environment. Report of the JCSR VRE working group (2004)
11. Yang, X.B., Allan, R.: Web-based virtual research environments (VRE): Support collaborationin e-Science. In: Butz, C., Nguyen, N.T., Takama, Y. (eds.) Proceedings of the 2006 IEEE/WIC/ACM International Conference on Web Intelligence and Intelligent Agent Technology, pp. 184–187. IEEE Computer Society, Washington, DC (2006)
12. Krafft, D.B., Cappadona, N.A., Caruso, B., et al.: VIVO: Enabling National Networking of Scientists. In: Web Science Conf. 2010, Raleigh,NC, USA, April 26-27 (2010)
13. Devare, M., Corson-Rikert, J., Caruso, B., et al.: VIVO: Connecting People, Creating a Virtual Life Sciences Community. D-Lib Magazine (13), 7–8 (2007)

Analysis on Snow Distribution on Sunlight Greenhouse and Its Distribution Coefficient

Chunhui Dai[1], Xiugen Jiang[1], Min Ding[1,*], Dongxin Lv[1], Guilin Jia[1], and Peng Zhang[2]

[1] College of Water Resources & Civil Engineering, China Agricultural University, Beijing 100083, China

[2] China Aerospace Construction Group Co., Ltd, Beijing 100071, China

{daichunhui,dingmin}@cau.edu.cn, jiangxiugen@tsinghua.org.cn, {1056026017,1251613521}@qq.com, zhangpengcau@sohu.com

Abstract. There is no corresponding snow distribution coefficient for snow distribution on widely used sunlight greenhouse in the current Greenhouse structure design load code (GB/T 18622-2002). To present snow distribution and its distribution coefficient on sunlight greenhouse, the collapse reason of sunlight greenhouse under snow load in recent years was discussed by investigation and analysis. Failure modes of sunlight greenhouse under extreme snowstorm were achieved. According to the current greenhouse structure design load code in our country, 7 snow distribution types of sunlight greenhouse were put forward and its distribution rules were analyzed; Take basic snow pressure in the Beijing for example, the mechanical behavior of sunlight greenhouse under 7 snow distribution types were analyzed by using the finite element analysis software ANSYS considering the greenhouse skeleton initial defect. The calculation model of snow distribution coefficient for sunlight greenhouse was proposed. The fruits are useful to the perfection of snow load calculation method for sunlight greenhouse.

Keywords: sunlight greenhouse, snow load, distribution rule, snow distribution coefficient.

1 Introduction

Because of little material, low construction cost, strong adaptability to the climate and plant efficiency is relatively high, sunlight greenhouse skeleton has been widely used in our country in recent years. Greenhouse in addition to providing basic space of animal and plant growth and production and the suitable temperature and humidity environment, also need to undertake include abnormal operating condition and various loads produced by extreme natural disasters (blizzard, wind, hail) [1-3].

From the perspective of the structure, sunlight greenhouse is a typical light structure, its load-bearing skeleton is generally consists of prefabricated steel pipe or steel welded together. Design load is the basic foundation of structure design, also is

* Corresponding author.

© IFIP International Federation for Information Processing 2015
D. Li and Y. Chen (Eds.): CCTA 2014, IFIP AICT 452, pp. 575–588, 2015.
DOI: 10.1007/978-3-319-19620-6_64

the primary factor for structure reliability and economy. Snow load is one of the main live load for sunlight greenhouse, its definition, value calculation directly affect the safety and economy of the greenhouse structure. In our country, building structures in the specification for industrial and civil buildings have been made the snow load definition and calculation values clear. But now, the industrial and civil building code and relevant standards abroad are still used for the greenhouse structure snow load calculation. There is a big difference in the degree of safety, importance, roofing and other aspects of snow distribution coefficient, what is worse, it lack of unity and pertinence.

Now the greenhouse has been widely research by scholars at home and abroad. Liang Zongmin[4]、Gong Wanting[5] 、Morcous G[6]、Castellano S[7-8] et al studied the wind and snow load in the greenhouse structure design and its effect on the bearing capacity of greenhouse. Zhang zhaoqiang[9] pointed out that, base on the contrast analysis of loads provided in foreign and China standards, under the present conditions, It is feasible in calculation snow load when considering the importance factor, the exposure factor and the heating factor according to China architectural structure load standards (GB50009 – 2001). Through the study of Wang Dongxia[10], probabilistic distribution model of wind and snow load in arbitrary design reference period is established by adopting Bernoulli trials, calculation of characteristic value of wind and snow load is put forward on the basis of "mode method". Yu YongHua[11] apply nonlinear finite element and the method of artificial stiffness, establish a calculation model of greenhouse film load effect analysis, and combined with the relevant specification, get the greenhouse film load calculation formula and the value of safety factor.

Snow distribution rule is one of the important factors affecting the snow distribution coefficient, and influence on the mechanical performance of the greenhouse structure. The greenhouse structure design load standard of GB/T 18622-2002 [12] (hereinafter referred to as the "Chinese standard") only provisions snow load distribution coefficient for the slope roof, single span arch roof, double slope multi span roof and multi span arch roof, But the snow distribution coefficient for the solar greenhouse, which is widely used in China, is not given. The distribution of the current standard stipulated by the snow load distribution coefficient can't fully meet the needs of its structural design in our country.

This paper study the collapse reason of sunlight greenhouse under snow load in recent years in China; According to the current greenhouse structure design load code of GB 50009-2012 [13] in our country, put forward its snow distribution rules; The mechanical behavior of sunlight greenhouse under different snow distribution types were analyzed by using the finite element analysis software ANSYS.

2 The Collapse of Sunlight Greenhouse under Snow Load

2.1 Disaster Instances

In north China, frequent extreme weather to a snow disaster, every year the heavy snow caused collapse accidents occur frequently, which brought huge losses to agricultural production. Through Xinhua network to collecting the collapse of sunlight greenhouse under snow load in recent years in China, analysis is as follows:

(1) On November 1, 2009, the town of LiXian, Daxing district, Beijing, were affected by the cold air, the snow lasting since more than 2 o 'clock in the morning to the afternoon 16 o 'clock, snow thickness was up to 12 cm, which caused great damage on the agricultural production development in LiXian. After detailed statistics, from the department of agriculture, Liyi , Hebei and Hebeitou , etc, total 42 village of greenhouses suffered from snow storms. Including 1216 mu of greenhouse facilities, 40 greenhouse structures collapsed. As shown in Fig. 1 (a) and (b).

(a) Store in the front of the roof

(b) Store in the front of the roof

Fig. 1. Da'xing, Beijing

(2) On November 10 to 12, 2009, big blizzard has fallen in He'nan province, a lot of anti-season vegetables were damaged by the collapse of sunlight greenhouse under snow load in Hua county. As shown in Fig. 2 (a) and (b).

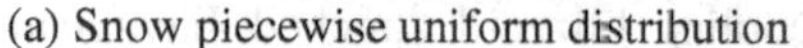

(a) Snow piecewise uniform distribution

(b) Snow piecewise uniform distribution

Fig. 2. Hua county,He'nan province

(3) On November 11, 2009, A severe blizzard struck the Dali county, Shanxi province, which seriously caused great losses to the local agriculture. The number of villages and towns, whose agricultural facilities had been affected, was up to 16. The affected area was 16675 mu, of which there are 473 sunlight greenhouses collapsed, 95% of greenhouses overwhelmed by heavy snow, 70% of bamboo poles were broken by snow pressure, 50% of the film was damaged, more than 1000 mu of cucumber

seedling suffered from cold, more than 10000 mu of autumn capsicums were total crop failure, direct economic loss was up to 54.52 million yuan. As shown in Fig. 3 (a) and (b).

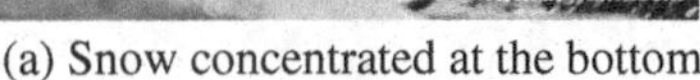

(a) Snow concentrated at the bottom (b) Snow concentrated on the top

Fig. 3. Dali, Shanxi province

(4) From February 28, 2010, 16 o 'clock to March 1, 3 o 'clock, Shouguang, a city of Shandong province, was suffered the biggest snowfall in 20 years, rainfall was up to 25.6 mm, The thickness of the snow on the ground was more than 20 cm. About 30% of the sunlight greenhouse deformation was very large, 5% of them collapsed, which seriously caused great losses to the local agriculture. As shown in Fig. 4 (a) and (b).

(a) Snow concentrated on the top

(b) Snow concentrated on the top

Fig. 4. Shouguang, Shandong province

(5) The influence of snow disaster on greenhouse in Gansu province in March 2010. As shown in Fig. 5 (a) and (b).

On March 5 to 8, a sustained strong cooling blizzard attacked Jiayuguan city, rainfall was up to 9.8 mm, from March 6 20 o 'clock to March 7 20 o 'clock, when rainfall was 6.3 mm. It can be a blizzard.

On March 7, Qingquan town of Yumen city suffered once every 50 years of blizzard. Thereby it cause loss to the local crops and greenhouse, such as the direct economic loss was up to 1.3748 million yuan. According to the statistics, 49 sunlight greenhouses were crushed and 121 mu of onion seedling suffered from cold until March 10 in Qingquan town.

(a) Snow piecewise uniform distribution (b) Snow concentrated on the top

Fig. 5. Yumen,Gansu province

(6) On March 10 to 11, 2010, March 13, 14, after a heavy snowfall in Hailun city, Heilongjiang province, local area can reach blizzard degree, part of the crops were affected, especially Helen town, Lunhe town, Gonghe town and Haibei town. A lot of tent poles were broken, a great many seedling suffered from cold. According to statistics, The number of sheds and greenhouses was more than 170, the affected area was 5.2 hectares, 1 hectare was total crop failure, direct economic loss was up to 2.55 million yuan. As shown in Fig. 6 (a) and (b).

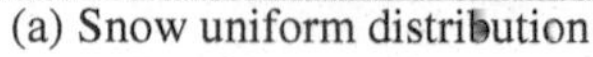

(a) Snow uniform distribution (b) Snow uniform distribution

Fig. 6. Hailun, Heilongjiang province

(7) On 18-20 January, 2011, after three days of consecutive heavy snow, a lot of greenhouse vegetables were under different degree of loss in Wanzai county, Yichun city, Jiangxi province. According to preliminary statistics, more than 1500 greenhouses were affected, including more than 1100 bamboo greenhouses and more than 400 steel greenhouses, the direct economic loss was up to more than 3 million yuan. As shown in Fig. 7.

Fig. 7. Wanzai, Jiangxi province

2.2 Collapse Reason and Analysis

Through collecting a large number of sunlight greenhouse under snow load in recent years, at the same time analysis the pictures listed above, you can get the following information:

(1) The snow distributions on the roof were not evenly distributed, it can be influenced by wind speed and temperature, snow on front of the roof may produce a concentrated load, then, generate adverse impact on greenhouse structure.
(2) The less damage was caused by covered film strength, more for roof snow when sunlight greenhouse under the snow. Greenhouse skeleton instability caused the collapse of the whole structure, the insufficient of skeleton strength lead to fracture, then, the whole structure was destroyed.
(3) The back of the greenhouse roof width is narrower, and cant for right angle; thin film surface is smooth and easy to slide, so snow is not easy to be destroyed. In the pictures above, the back of the greenhouse roof is not damaged.

3 Snow Distribution Rule of Sunlight Greenhouse

3.1 Snow Distribution Type

The collapse reason of sunlight greenhouse under snow load in recent years was discussed by investigation and analysis. According to the current greenhouse structure design load code in our country, 7 snow distribution types of sunlight greenhouse were put forward and its distribution rules were analyzed. As shown in Fig. 8.

3.2 Snow Distribution Rule

(1) Distribution form 1: snow was uniform distribution, because the roof slope was slow, what was more, the front and back of roof slope were similar, and there was no wind or wind was small when snowed. Fig. 6 (a) and (b) both belong to this situation.
(2) Distribution form 2: snow was uniform distribution, just because the front and back of roof slope were large different, so the thickness of snow. The back of the roof slope was steep, so snow was easy to slide down, snow thickness is very small. The front of the roof slope was slow, so snow thickness is very big. Fig. 7 belongs to this situation.
(3) Distribution form 3: the snow thickness was biggest in the middle parts of the roof, backwards gradually reduced. The reason was that the indoor temperature of greenhouse was higher. So snow melted down and slid down, finally stored in the front of the roof; otherwise, some deformation was due to the front without pillar, greenhouse structure was weak and skeleton was instability, local sag, snow is piling up. Fig. 1 (a) and (b) belong to this situation.

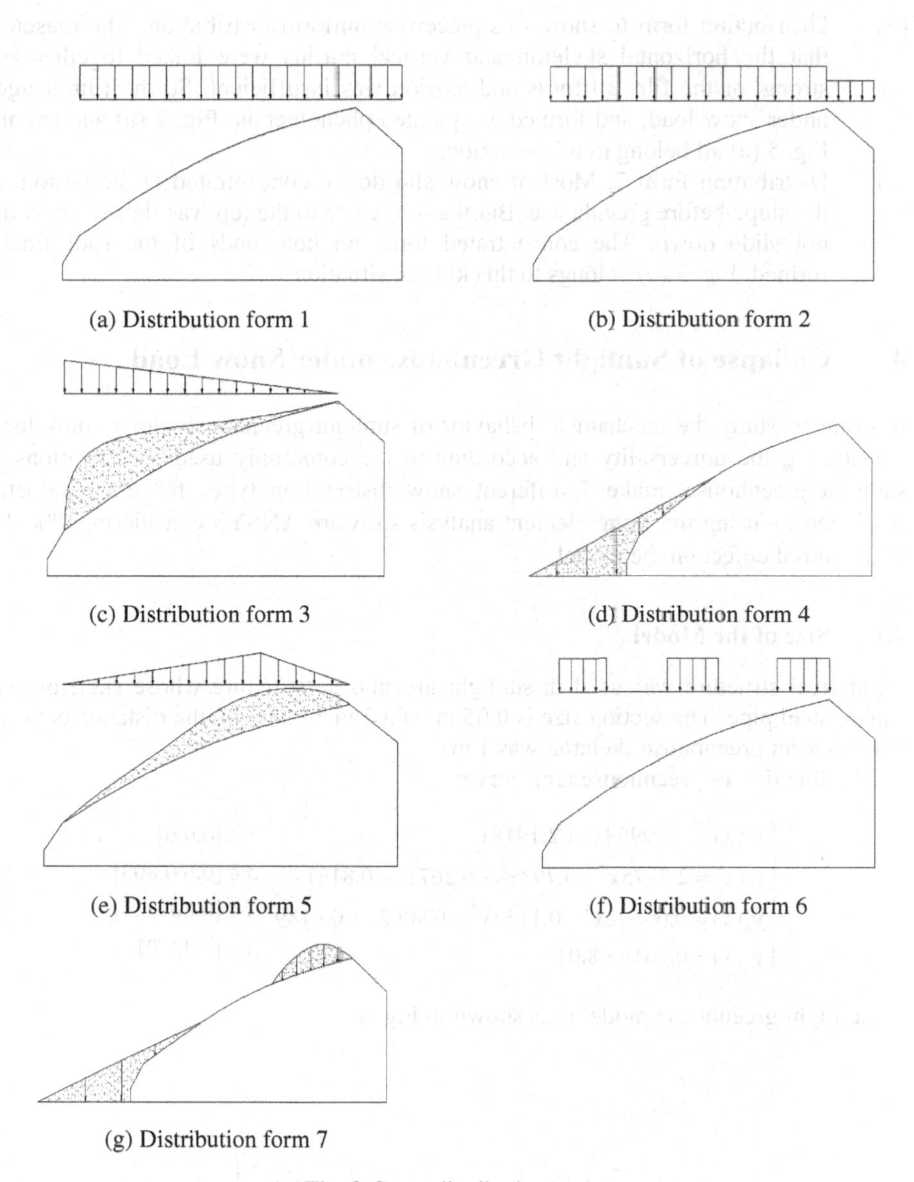

(a) Distribution form 1

(b) Distribution form 2

(c) Distribution form 3

(d) Distribution form 4

(e) Distribution form 5

(f) Distribution form 6

(g) Distribution form 7

Fig. 8. Snow distribution type

(4) Distribution form 4: Most of snow slid down, concentrated at the bottom of the slope before greenhouse. Fig. 3 (a) belongs to this situation.

(5) Distribution form 5: the snow thickness was biggest in the middle parts of the roof, forwards and backwards gradually reduced. The reason is that the front of the roof was the windward side, so snow moved back in under the action of wind, finally snow stored in relatively flat parts, and made local pressure very big. Fig. 4 (a) and (b) and Fig. 5 (b) all belong to this situation.

(6) Distribution form 6: snow was piecewise uniform distribution. The reason is that the horizontal skeleton and vertical purlins were linked together and strong, or the film stiffness and tension was insufficient. So the film sagged under snow load, and formed a separated phenomenon. Fig. 2 (a) and (b) and Fig. 5 (a) all belong to this situation.

(7) Distribution form 7: Most of snow slid down, concentrated at the bottom of the slope before greenhouse. But the roof close to the top was flat, so snow did not slide down. The concentrated force on both ends of the roof finally formed. Fig. 3 (b) belongs to this kind of situation.

4 Collapse of Sunlight Greenhouse under Snow Load

In order to study the mechanical behavior of sunlight greenhouse under snow load, considering the universality and according to the commonly used specifications of sunlight greenhouse, make 7 different snow distribution types for the parameter. Analyzed by using the finite element analysis software ANSYS considering 2% [11] as the initial defect on the model.

4.1 Size of the Model

Light steel structure was used in sunlight greenhouse structure whose skeleton was square steel pipe. The section size is 0.05 m * 0.05 m * 0.003 m, the distance between two adjacent greenhouse skeleton was 1 m.

The function of greenhouse arch curve:

$$\begin{cases} y_0(x) = -0.9958x^3 + 2.1918x & x \in [0,0.6] \\ y_1(x) = 2.7775x^3 - 6.792x^2 + 6.2671x - 0.8151 & x \in [0.6,0.803] \\ y_2(x) = 0.00602x^3 - 0.1156x^2 + 0.9062x + 0.6189 & x \in [0.803,6.4] \\ y_3(x) = 0.761x + 8.011 & x \in [6.4,8.0] \end{cases} \tag{1}$$

Sunlight greenhouse model is as shown in Fig. 9.

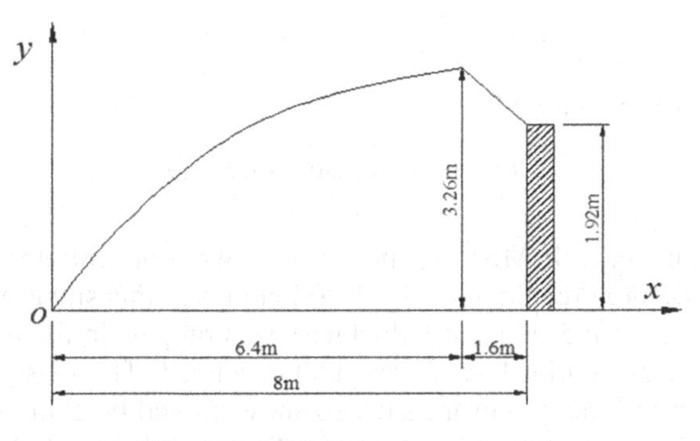

Fig. 9. Sunlight greenhouse model

4.2 Material Performance

Chinese Q235 steel was used in sunlight greenhouse skeleton, material parameters were shown in Table 1.

Table 1. Parameters of material

Category	Density/kg·m^{-3}	Elasticity modulus /GPa	Poisson ratiol	Yield strength / MPa
Skeleton	7850	206	0.3	235

4.3 Unit and Meshing

The mechanical behavior of sunlight greenhouse under 7 snow distribution types were analyzed by using the finite element analysis software ANSYS. Choose Beam188 unit, Structure was divided into 40 nodes. Number them from 1 to 40. It would be divided into a single unit when between two adjacent nodes. Structure was divided into 40 units.

4.4 Load and Constraints

The research content was one single skeleton in this paper. Snow load was surface load. Firstly, transform the surface load into line load; secondly, the mechanical behavior of sunlight greenhouse skeleton was analyzed with curved beam line load decomposition method [15]. The ends of the greenhouse skeleton were fixed. In this paper, we mainly considered the mechanical behavior of sunlight greenhouse skeleton with different snow distribution forms

For snow distribution form 1, take the same snow distribution coefficient for the front and back slope, According to the construction load standard GB 50009-2012 [13] (hereinafter referred to as the "specification") regulations, take the arched roof uniform distribution coefficient values as snow distribution coefficient.

For snow distribution form 2, according to the specification, take the arched roof uniform distribution coefficient values as front slope snow distribution coefficient, and the back slope was half of the front slope.

For snow distribution form 3, the front slope of snow was not evenly distributed. According to the specification, take the maximal arched roof snow uneven distribution coefficient as maximum value. The snow distribution coefficient of the back slope was 0.

For snow distribution form 4, considering the snow slide to the bottom, part of the snow scattered on ground, take snow distribution coefficient approximate to 1.

For snow distribution form 5, the maximal snow distribution coefficient of the front slope was the same as the snow distribution form 3, but the snow load was triangular distribution on the front slope.

For snow distribution form 6, according to the specification, take the arched roof uniform distribution coefficient values as front slope snow distribution coefficient, the whole structure was divided into three range, loaded in turn.

For snow distribution form 7, considering the snow slide to the bottom, take snow distribution coefficient of the front slope approximate to 1; part of snow was on the roof, take the arched roof uniform distribution coefficient values as snow distribution coefficient.

4.5 Calculation Model

According to the specification, the start loading position was on that the angle between tangent of the curve and the horizontal was 60 °. The calculation model was as shown in Table 2.

Table 2. Parameters of every computation model

Category	Snow distribution form	Start loading position α/°	snow distribution coefficient μ_1	snow distribution coefficient μ_2	The length of the load /m
Standard condition	1	60	0.49	0.49	7.76
Loading angle condition 1	1	50	0.49	0.49	7.52
Loading angle condition 2	1	65	0.49	0.49	7.92
Snow distribution condition 1	2	60	0.49	0.49/2	7.76
Snow distribution condition 2	3	60	2.0	0	6.0
Snow distribution condition 3	4	—	1.0	0	2.4
Snow distribution condition 4	5	60	2.0	0	6.0
Snow distribution condition 5	6	—	1.0	0	1.2×3
Snow distribution condition 6	7	—	1.0(0.49)	0	2.4+1.2

Annotation: ① Loading angle conditions 1 and loading angle conditions 2 considering the snow loading angle distribution form 1 changing in standard condition.
② The basic snow pressure of the Beijing was 0.4 kN/m^2 [13].

4.6 The Results and Analysis

According to the load model, calculate the results by using ANSYS finite element software. For example, in standard condition, the start loading position was on that the angle between tangent of the curve and the horizontal was 60 °, in other words, load from number 4 node, The length of the load was 7.76 m. Results of mechanical analysis were shown in Fig. 10.

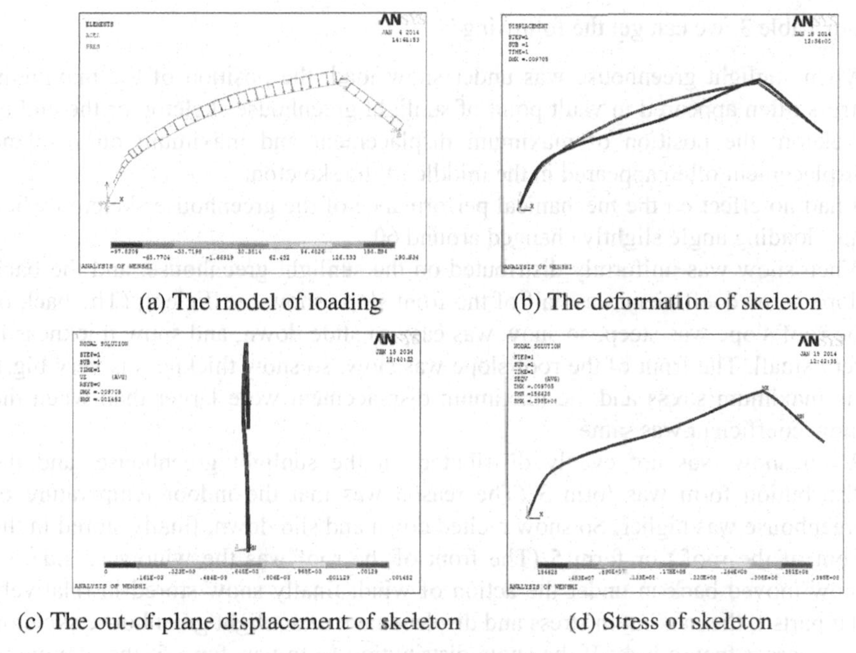

(a) The model of loading (b) The deformation of skeleton

(c) The out-of-plane displacement of skeleton (d) Stress of skeleton

Fig. 10. Results of mechanical analysis

As the stress nephogram and deformation pattern above shown, the maximum stress was 39.5MPa, the position was on the outer edge of the skeleton near vault point (position: near node 33); the maximum displacement of whole structure was 9.7 mm (position: near node 22), the maximum out-of-plane displacement of whole structure (UZ) was 1.5 mm (position: near node 22).

Summary of sunlight greenhouse mechanical analysis is shown in Table 3.

Table 3. Summary of sunlight greenhouse mechanical analysis

Category	Maximum stress		Maximum displacement		maximum out-of-plane displacement	
	σ_{max} /MPa	Position	d_{max} /mm	Position	UZ_{max} /mm	Position
Standard condition	39.5	Vault point	9.7	No. 22 node	1.5	No. 22 node
Loading angle condition 1	39.4	Vault point	9.7	No. 22 node	1.5	No. 22 node
Loading angle condition 2	39.5	Vault point	9.7	No. 22 node	1.5	No. 22 node
Snow distribution condition 1	41.6	Vault point	10.3	No. 21 node	1.4	No. 22 node
Snow distribution condition 2	74.4	Vault point	21.9	No. 19 node	1.9	No. 22 node
Snow distribution condition 3	30.9	The end of back slope	10.3	No. 16 node	0.7	No. 22 node
Snow distribution condition 4	114	Vault point	30.7	No. 21 node	3.2	No. 22 node
Snow distribution condition 5	24.2	Vault point	5.8	No. 21 node	0.9	No. 22 node
Snow distribution condition 6	28.2	Vault point	8.3	No. 17 node	0.9	No. 22 node

From Table 3, we can get the following:

(1) When sunlight greenhouse was under snow load, the position of the maximum stress often appeared in vault point of sunlight greenhouse skeleton or the end of skeleton; the position of maximum displacement and maximum out-of-plane displacement often appeared in the middle of the skeleton.
(2) It had no effect on the mechanical performance of the greenhouse skeleton when start loading angle slightly changed around 60°.
(3) When snow was uniformly distributed on the sunlight greenhouse, and the back slope snow coefficient was half of the front slope snow coefficient, (The back of the roof slope was steep, so snow was easy to slide down, and snow thickness is very small. The front of the roof slope was slow, so snow thickness is very big.), the maximum stress and the maximum displacement were larger than when the snow coefficient was same.
(4) When snow was not evenly distributed on the sunlight greenhouse, and the distribution form was form 3 (The reason was that the indoor temperature of greenhouse was higher. So snow melted down and slid down, finally stored in the front of the roof.) or form 5 (The front of the roof was the windward side, so snow moved back in under the action of wind, finally snow stored in relatively flat parts.), the maximum stress and displacement of sunlight greenhouse skeleton were larger than others; If the snow distribution form was form 5, the maximum stress and displacement of greenhouse structure were the biggest of all.
(5) When sunlight greenhouse was under snow load, snow was uniformly distributed on the roof or not, so this paper suggested that the snow distribution coefficient can be calculated as shown in Fig. 11.

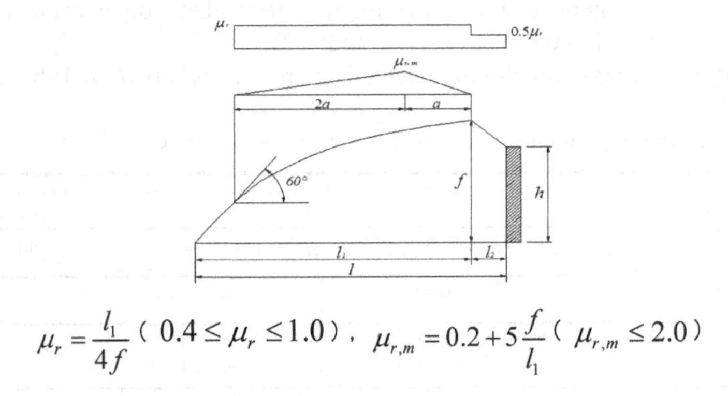

$$\mu_r = \frac{l_1}{4f}\ (0.4 \le \mu_r \le 1.0),\quad \mu_{r,m} = 0.2 + 5\frac{f}{l_1}\ (\mu_{r,m} \le 2.0)$$

Fig. 11. Proposed calculation model of snow distribution coefficient for sunlight greenhouse

5 Conclusions

To study the mechanical behavior of sunlight greenhouse under snow load, the method of investigation and numerical simulation were used, and the different snow distribution forms of sunlight greenhouse, which were made the snow as variables, were summarized. The analysis results reflected effect on the mechanical behavior of sunlight greenhouse when sunlight greenhouse was under different snow distribution forms.

(1) When sunlight greenhouse was under snow load, the position of the maximum stress often appeared in vault point of sunlight greenhouse skeleton or the end of skeleton; the position of maximum displacement and maximum out-of-plane displacement often appeared in the middle of the skeleton. So strengthen two ends of the greenhouse skeleton and vault point were necessary, as well as in making effective vertical and lateral support in the middle of skeleton.
(2) When sunlight greenhouse was under snow load, the snow distribution coefficient can be calculated as shown in Fig. 11.

Acknowledgment. Support for this research by Beijing Natural Science Foundation No. 3144029, Chinese Universities Scientific Fund No. 2011JS126, and the Specialized Research Fund for the Doctoral Program of Higher Education of China No. 20110008120017.

References

1. Zhou, C.: Modern greenhouse project, pp. 1–2. Chemical Industry Press, Beijing (2003) (in Chinese)
2. Zhou, C.: The development of China's modern greenhouse and standardization. Transactions of the Chinese Society of Agricultural Engineering 19(suppl.), 88–91 (2003)
3. Zhou, C.: The enlightenment of the greenhouse collapse caused by snowstorm. Agricultural Engineering Technology (12), 13–15 (2007)
4. Liang, Z.: Design theory on wind-resistance reliability of structure for multi-span greenhouse. China Agricultural University, Beijing (2004) (in Chinese with English abstract)
5. Gong, W., Liang, Z.: Numerical simulation of wind pressure on the surface of the new greenhouse, pp. 73–74. Beijing modality of the 17th academic essays, Beijing (2011)
6. Morcous, G.: Performance of conservatories under wind and snow loads. Journal of Architectural Engineering 15(3), 102–109 (2009)
7. Castellano, S., Candura, A., Scarascia-Mugnozza, G.: Greenhouse Structures SlS Analysis: Experimental Results And Normative Aspects. In: International Conference on Sustainable Greenhouse Systems – Greensys (2004)
8. Castellano, S., Mugnozza, G.S., Vox, G.: Collapse Test on a Pitched Roof Greenhouse Structure. Colture Protette 35(1), 55–61 (2006)
9. Zhang, Z.: A preliminary study on calculating of snow load for greenhouse structural design. Construction Technology of Low Temperature (3), 36–37 (2003)
10. Wang, D.: Study of nominal value of wind and snow load of the greenhouse structures with unified warranty probability. Academic Essays of the Second Session of the National Civil Engineering Graduate Student Academic BBS, Shanghai (2004)

11. Yu, Y., Wang, J., Ying, Y.: Nonlinear finite element analysis of the bearing capacity of arch structure in plastic greenhouse on snow load working condition. Transactions of the Chinese Society of Agricultural Engineering 23(3), 158–162 (2007)
12. General Administration of Quality Supervision, Inspection and Quarantine of the People's Republic of China, GB/T18622-2002. The greenhouse structure design load. China Building Industry Press, Beijing (2002)
13. General Administration of Quality Supervision, Inspection and Quarantine of the People's Republic of China, GB50009-2012. Load code for the design of building structures. China Building Industry Press, Beijing (2002)
14. Wang, X.: ANSYS numerical analysis of engineering structures, pp. 353–356. China Communications Press, Beijing (2007) (in Chinese)

Agricultural Library Information Retrieval Based on Improved Semantic Algorithm

Xie Meiling

Library of Agriculture University of Hebei, Baoding, China
hebauxie@hotmail.com

Abstract. To support users to quickly access information they need from the agricultural library's vast information and to improve the low intelligence query service, a model for intelligent library information retrieval was constructed. The semantic web mode was introduced and the information retrieval framework was designed. The model structure consisted of three parts: Information data integration, user interface and information retrieval match. The key method supporting retrieval was designed. The traditional semantic similarity algorithm was improved according to its shortages. An algorithm based on semantic distance was designed and tested. The results can improve the recall ratio and precision of information retrieval, improving information retrieval performance.

Keywords: Improved semantic algorithm, Agricultural library, Information retrieval.

1 Introduction

Agricultural digital library was the carrier of agricultural information resources. The current digital library retrieval query service there are some disadvantages, including low query intelligence, independent results and low degree of shared information data. The traditional pattern of information query was based on the keyword query. This kind of query can't reflect the deep meaning of user query demand[1]. Semantic web query mode was a hot research topic in recent years. In semantic query mode, user's semantics of natural language can be understood by the machine to a certain extent and the semantic level of retrieval was implemented finally. How to build the information query model based on semantic technology was a subject to be solved in which the most key problem was the semantic similarity computation. Traditional semantic similarity algorithms include algorithm based on semantic distance, algorithm based on the concept features and algorithm based on the amount of information. The disadvantage of the above algorithms was easy to lead to errors[2]. Some other researchers studied the information retrieval based on semantic retrieval[3-6]. In this paper, aiming at the shortcomings of the similarity calculation in traditional semantic algorithm, the traditional semantic algorithm was improved. Using the similarity algorithm based on semantic distance, the model of information query based on semantic technology was built and information retrieval precision was increased.

© IFIP International Federation for Information Processing 2015
D. Li and Y. Chen (Eds.): CCTA 2014, IFIP AICT 452, pp. 589–594, 2015.
DOI: 10.1007/978-3-319-19620-6_65

2 Information Query Architecture and Improved Similarity Algorithm

2.1 Information Query Architecture of Digital Library

In the human-computer interaction environment, the information resources in the digital library can be combined with the user semantic body through the semantic web technology. The specific meaning of the language user used in specific environment can be accurately expressed. Using standardized semantics, understanding between the user and the system was implemented and user information demand was accurately obtained from the perspective of semantics. And finally the information the user needed was obtained through information retrieval. Digital library retrieval model based on the semantic technology was constructed, as shown in the figure below:

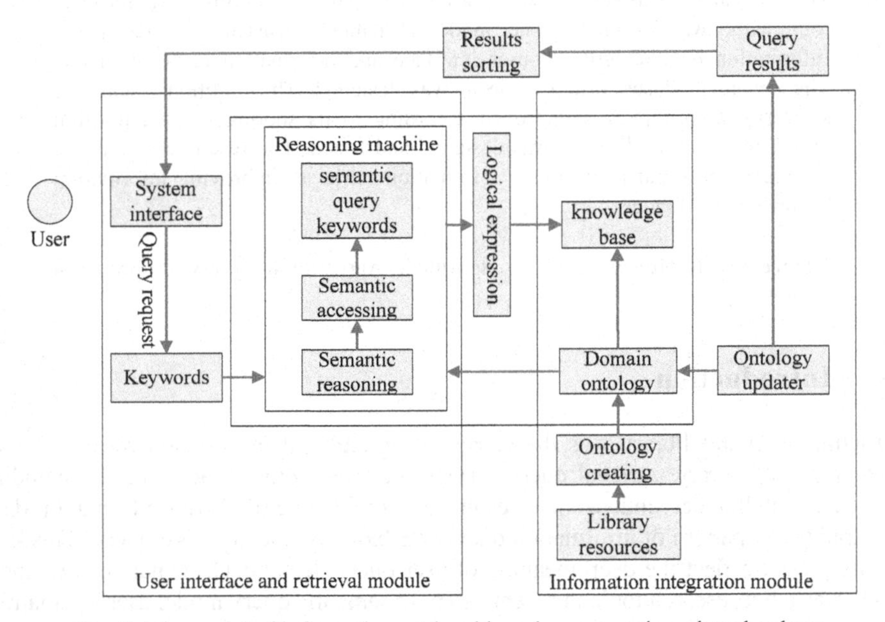

Fig. 1. The model of information retrieval based on semantic web technology

The figure 1 showed that in the semantic web technology, the retrieval model was divided into three modules.

(1) User Interface and Retrieval Processing Module
Using semantic web technology, the module processed query conditions submitted by the user in the interface based on human natural language. Query keywords was pretreated and converted to ontology query. According to the domain knowledge ontology, natural language query submitted by the user was resolved by the reasoning machine. Semantic reasoning based on semantic similarity calculation was executed.

(2) Information Integration Module
According to the resources in the semantic web annotation, this module created the ontology model for the information of the digital library. And domain ontology was formed According to the evolution of the information, ontology updater was used to expand and refresh the domain ontology periodically. Using XML technology, the information be standardized processed. Metadata information with high relevance to user requirements was acquired and stored in knowledge base.

(3) Match and Output Module
According to the search keywords pretreated, this module searched results in the domain ontology knowledge base. Semantic reasoning was conducted based on ontology by reasoning machine unit according to the search keywords. Logical expression of user query and retrieval was constructed and submitted. The result set complied with the expression was searched out by the system. The result was sorted using semantic similarity algorithm. Then the system returns the user interface.

With the support of these modules, the user retrieval process can be described as:

(1) key information in the sources of was acquired. Using the semantic web technology such as XML (Extensible Markup Language), RDF (Resource Description Framework), these key information was standardized processed according the metadata such as MARC(Machine Readable Catalog) ,DC (Dublin Core Element Set). The key information was converted into metadata and was stored in metadata database.
(2) According to the data in the metadata database, the relationship between the concepts were derived using logical reasoning. Domain ontology was constructed and stored in the knowledge base.
(3) Search conditions entered by the user were read. Based on the domain ontology constructed in (2), user's search conditions were converted into standard formats. Semantic extraction operations were conducted according to semantic similarity. Preliminary treatment of user query needs was finished.
(4) Through user query needs to build query expressions, system knowledge base was traversed. Matching information collection was searched out. All elements in the set were screened using domain Ontology-based semantic similarity. Query results are sorted and displayed according to semantic similarity.

3 Improved Semantic Similarity Algorithm

Key of library information retrieval was to determine the semantic similarity between concepts. The higher the degree of similarity, indicating the query results returned to the user and the user query needs more match. Semantic similarity algorithm based on distance described the connections between the concepts using hierarchical network. The geometric distance of the concept in the hierarchical network was as semantic distance. The traditional method to measure the similarity between the concepts was:

$$similar(v_1, v_2) = \frac{2(len-1) - dis(v_1, v_2)}{2(len-1)} \tag{1}$$

Where len was the depth of the hierarchical network possessed. $dis(v_1, v_2)$ was the directed edge number contained in the minimum value of paths between the two concepts. The traditional similarity algorithm based on semantic distance had some disadvantages. Only the lengths of the paths between the nodes were paid attention. Semantic distance effects on similarity were ignored. Node hierarchy of semantic effect was not considered. These led to the difference between the calculated results and the actual situation[7]. Ontology hierarchy in the following as an example:

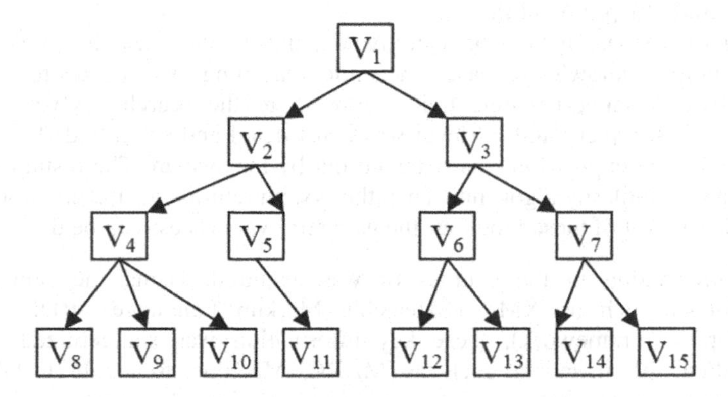

Fig. 2. Ontology hierarchy chart example

Assuming the instance belonged to the complete structure of domain ontology, the semantic similarities between nodes v6 and v12, v6 and v3 were computed respectively using traditional semantic similarity algorithm. The results were both 0.83. In the complete structure of domain ontology, the closer it got to the ontology layer, the lower the semantic distance between the elements values should be. So the truth should be the semantic similarity between the node v6 and node v12 was greater than the semantic similarity between the node v6 and node v3. The traditional method did not reflect the actual situation and needed to be improved.

In order to overcome the shortage of the traditional methods, the improved semantic distance similarity algorithm was designed in this paper. Semantic depth as an important factor was introduced to the semantic distance similarity calculation. If the number of connected path between two nodes were equal, the semantic similarity between nodes and ontology hierarchy were related. The higher the sum of two conceptual levels, the greater the similarity was. The greater the difference between the hierarchies of the two concepts, the smaller the similarity was. This can be represented by the following expression:

$$\frac{|depth(v_1) - depth(v_2)|}{depth(v_1) + depth(v_2)} \tag{2}$$

This expression was a reduction function. Therefore, the depth of the semantic ontology nodes was expressed by α:

$$J \tag{3}$$

Where $depth(v_1)$ and $depth(v_2)$ were the numbers of layers (depth) of node v1 and node v2 respectively. The depth of the root node of the ontology hierarchy network can be regarded as the value 1. Along with the increase of the layer, the depth was also increasing. According to (3), the semantic similarity between nodes v6 and v12 was 0.86 and the semantic similarity between nodes v6 and v3 was 0.80.

The position of the node in the structure should be considered in semantic similarity calculation based on the semantic distance. Generally, if the two nodes belong to parent-child relationship, the parent node contained all the properties of the child nodes and the child nodes contained some properties of the parent node. Similarity between high-level and low-level nodes should be lower than the similarity between low-level and high-level nodes. J was defined as a status factors of the node in the network hierarchy, expressed as:

$$j = \frac{1}{2}\left(1 + \frac{depth(v_1) - depth(v_2)}{len_{\max}}\right) \tag{4}$$

Where $len_{\max}$ was the max depth of the hierarchical network. The semantic similarities between nodes v6 and v3, v3 and v6 were computed according to (4) and the results were 0.63 and 0.38 respectively. Based on the above analysis shows that the semantic similarity between the nodes affected by the 3 kinds of properties: path length, semantic depth and node hierarchy. Namely:

$$similar(v_1, v_2) = \left(\frac{2(len-1) - dis(v_1, v_2)}{2(len-1)} * i * j\right)^{\frac{1}{2}} \tag{5}$$

4 Results and Discussion

Using the improved algorithm and traditional algorithm to calculate the similarity between different nodes respectively, the results were shown in table 1.

Table 1. Composition of the experimental samples

	Traditional algorithm	i	j	Improved algorithm
semantic similarity between v12 and v6	0.83	0.86	0.63	0.67
semantic similarity between v6 and v3	0.83	0.79	0.63	0.65
semantic similarity between v3 and v6	0.83	0.79	0.38	0.49

The chart showed that using the traditional semantic similarity calculation method to calculate the similarity between different nodes, the results were the same and did not reflect the actual situation. Table 1 showed that using the algorithm presented in this paper, the semantic similarity(0.67) between node v12 and v6 was higher than the semantic similarity(0.65) between the node v6 and node v3. The semantic similarity(0.65) between node v6 and v3 was higher than the semantic similarity(0.49) between the node v3 and node v6.

The semantic similarity between node v12 and v6 was the largest, followed by the semantic similarity between node v6 and v3. The semantic similarity between node v3 and v6 was the smallest. This was consistent with the actual situation. This showed that the optimization algorithm designed in this paper can more accurately reflect the actual situation.

5 Conclusions

With the improvement of the level of agricultural information, digital libraries should also improve their level of query to provide users with information services in this trend. .A model for intelligent library information retrieval was constructed. The semantic web mode was introduced and the information retrieval framework was designed. The model structure consisted of three parts: Information data integration, user interface and information retrieval match. The key method supporting retrieval was designed. The traditional semantic similarity algorithm was improved according to its shortages. An algorithm based on semantic distance was designed and tested. This showed that the optimization algorithm designed in this paper can more accurately reflect the actual situation. There were no algorithm can fully realize the semantic similarity search. Therefore, this study also needs further improvement.

References

1. Sheng, X.: Knowledge organization of digital library. Library and Information Service (3), 26–29 (2011) (in Chinese)
2. Qin, J.: Semantic Web and Ontologies. In: 2nd Joint Adcanced Workshop in Digital Libraries, Beijing (May 2012)
3. Shi, X., Niu, Z., Song, H., et al.: Intelligent agent-based system for digital library information retrieval. Journal of Beijing Institute of Technology 12(4), 450–454 (2003)
4. Wang, J., Yang, X.: A distributed cooperative approach to Web information retrieval using metadata and Z38.50. Journal of Software (4), 620–627 (2001)
5. Li, H., Wu, W., Zhang, C.: Implementation research on information retrieval based on semantic. Journal of Gansu Lianhe University (Natural Sciences) 22(2), 86–89 (2008)
6. Klyuev, V., Oleshchuk, V.: Semantic Retrieval of Text Documents. In: Proceedings of the 7th IEEE International Conference on Computer and Information Technology, pp. 189–193 (2007)
7. Li, S.: Research of Relevancy between Sentences Based on Semantic Computation. Computer Engineering and Applications 38(7), 75–76 (2012)

Assessment of Agricultural Information Service Based on Improved BP Network

Xie Meiling

Library of Agriculture University of Hebei, Baoding, China
hebauxie@hotmail.com

Abstract. According to the specific needs of the agricultural information service assessment, ant colony algorithm was adopted to optimize the traditional neural network to avoid its disadvantages of low convergence speed and being prone to fall into the minimum. Evaluation Index system of agricultural information service was built and the neural network model was designed. Learning and training were carried out using the sample data of agricultural information service system. The final evaluation result was obtained through learning of 56 sample data and training of 24 sample data. The result showed that iterations of the BP model optimized was significantly reduced, learning rate and stability were also improved. The model was able to evaluate scientifically and objectively the service of agriculture intelligence agencies.

Keywords: Agriculture Information Service, BP Network, ant colony algorithm.

1 Introduction

With the development of modern agriculture, scientific management methods are widely used in library. How to conduct scientific, quantitative assessment of the agriculture intelligence service of the library, so that it can better meet the agricultural research and engineering staff requirements for intelligence and information has become a hot research focus. Some researchers have made some achievements: Tan(2009) analyzed the evaluation index of the library service and discussed the five evaluation index system on university library personalized service performance. Qiu(2013) built a evaluation index system of library information service quality based a constructed model. Ran(2013) designed the main models of knowledge service in university libraries using fuzzy synthetic appraising method. Some other researchers(Walters 2003, Liangzhi 2005, Nimsomboon 2003) studied the assessment method of service of libraries. However, current researches do not give a method based on uniform standards and intelligence service evaluation system. Based on existing researches, this paper constructed a reasonable evaluation index using APH method, and combined with neural network algorithm to get an overall assessment result. Traditional neural network has slow learning speed and local optimum over-fitting and other defects, affecting the accuracy of the assessment. In this paper, the

© IFIP International Federation for Information Processing 2015
D. Li and Y. Chen (Eds.): CCTA 2014, IFIP AICT 452, pp. 595–602, 2015.
DOI: 10.1007/978-3-319-19620-6_66

neural network was optimized by using genetic algorithms(GA). The right values and the structure of the network were obtained through training and learning. The algorithm was reliable and the low convergence speed, easy falling into local minima and other defects can be avoided. The algorithm can automatically find the inherent nature of property law of the samples. And finally objective and accurate assessments of the agricultural information service quality of library were calculated.

2 Construction of Evaluation Index System

2.1 Construction of Index System

By combining the features of agricultural information services and users' requirements for information, the evaluation system was divided into three categories: (1) personnel assessment of Information Service (2) Resource assessment of Information Service (3) achievements assessment of Information Service. Indicators were further refined and the index system was got as shown below:

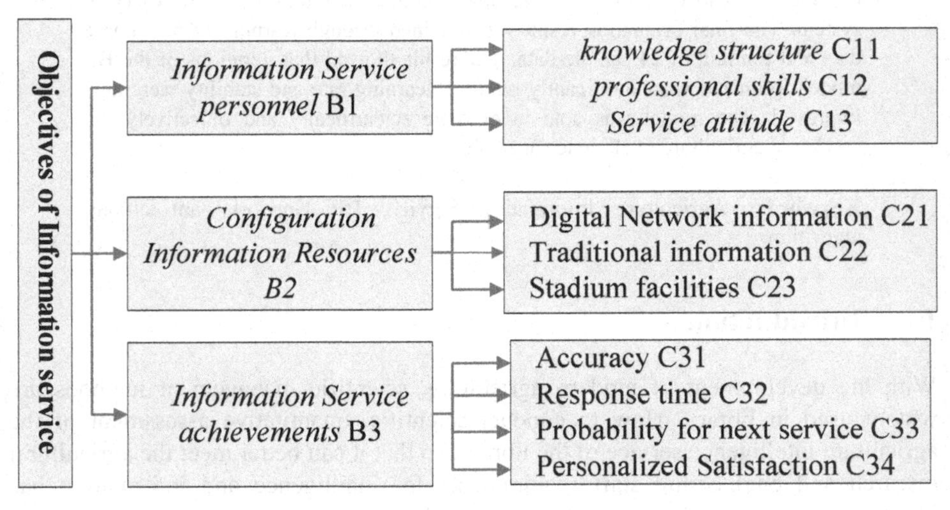

Fig. 1. Index system

From the above, the secondary indicators to assess the information services of the library can be expressed as {B1, B2, …, Bn}(n=3 in this research).

The function of information retrieval Agent is to interact with all kinds of search engine. It retrieved the library of local data information and merged the retrieve information acquired. The information with low correlation degree to user demand was filtered by information filtering agent using certain filtering algorithm. There were two types of filtering algorithms. They were filtering algorithm based on the information content filtering and filtering algorithm based on collaborative filtering respectively. Another algorithm combined the advantages of the above two filter algorithms and realized the combination filtering.

2.2 Consistency Test

The weights of the Evaluation indexes were given by the agriculture information specialists. The weights matrices A-B, B1-C, B2-C, B3-C were created after the statistical analysis. Column vectors of the matrices were normalized. The matrices rows were summed and finally the weights of evaluation indexes of the specialist. Consistency test of the evaluation was conducted and the result was shown in the following table:

Table 1. Result of consistency test of evaluation indexes

	A-B	B1-C	B2-C	B3-C
Max characteristic root λmax	3.0020	3.0146	3.0187	4.0533
Consistency index CI	0.0011	0.0013	0.0127	0.0159
Random Consistency RI	0.4400	0.4400	0.4400	0.7800
Consistency ratio CR	0.0025	0.0030	0.0289	0.0204

The above table showed that the consistency ratio CR were all under 0.1. This showed that the consistency of evaluation was within the acceptable range(Liu, 2008). The weight values of the evaluation indexes were shown in the following table:

Table 2. Weights of the evaluation indexes

Evaluation index	Weight W	Weight in percentage
C11	0. 0492	4%
C12	0. 1054	11%
C13	0. 0833	8%
C21	0. 1396	14%
C22	0. 0642	7%
C23	0. 1461	15%
C31	0. 1228	12%
C32	0. 0902	9%
C33	0. 0640	7%
C34	0. 1343	13%

The consistency test of weights was conducted and the result was $CR = CI/RI < 0.1$. Therefore, the consistency of the evaluation could be determined in the acceptable range.

3 Evaluation of Information Service

3.1 Optimization of Neural Networks

The traditional neural network was optimized using genetic Algorithms, so that it can adapt to the assessment of the information services. This optimization algorithm improved the initial weights of the neural network by genetic variation. Essence of the algorithm is designed to train the network by two steps. The first step is the introduction of genetic algorithm and optimization of the BP network weights by GA. The second step is to achieve the goal of network training using the BP network. By encoding the neurons that may be present in neural networks, the initial population was randomly generated. The neural network was optimized by selection, crossover, mutation and other genetic operators. The specific steps are as follows:

(1) Initial population, the crossover scale and a weight threshold
(2) Calculate the evaluation function of each individual by the following formula:

$$P_i = f_i / \sum_{j=1}^{N} f_j \tag{1}$$

Where f_i was the fitness value of individual i. f_i was measured by error sum squares E, that is

$$f_i = 1 / \frac{1}{E(i)} \tag{2}$$

Where $i = 1, 2, \ldots N$, N is the number of the chromosomes.

(3) Conduct crossover and mutation among the group, thus making it possible to build the next generation of groups.
(4) Place the new population of individuals generated by crossover and mutation in group P and then recalculate the new individual evaluation function. Repeat step 2 to step 4 and a set of thresholds were obtained. And finally training targets complied with the requirements of this study, the algorithm ended.

3.2 Determination of Neural Network Parameters

(1) Selection of the Network Hidden Layers

Increasing the number of hidden layer can improve the nonlinear mapping ability of BP network But if the number of hidden layer exceeded a certain value, it caused performance degradation. of the network. According to the characteristics of neural network and the research needs, this network in this study was designed a three-layer network. The 10 evaluation index values were the input vector of BP neural network. The number of neurons in neural networks was 10. There was only one user evaluation result. Therefore, there was one node in output layer of the neural network.

(2) Determination of the Number of Neurons in the Hidden Layer

The number of neurons in the hidden layer had a relatively large impact on nonlinear of the neural networks. It was also related to the complexity of the problems to be solved. The complexity of solving a problem is difficult to quantify. In current time there is no an algorithm for the accurate number of neurons in the hidden layer. If relatively few nodes in the hidden layer were set, the neural network data is difficult to obtain enough information, network training can not be achieved. If more hidden layer nodes were set, the neural network would become over-fitting and the performance would be affected. Assume herein that the number of nodes in the input layer of BP network was I and J in the output layer. The number of hidden layer nodes was determined by the following formula:

$$a = \frac{I+J}{2} < L < (I+J)+10 = b \tag{3}$$

10 factors were selected in the input layer. The output layer was the evaluation result of the user's. The output layer was in the range of [0,100], 100-90 for excellent, 89-80 for good, 79-70 for medium, 69-60 for acceptable and under 59 for unacceptable. Therefore, in this paper, I=10 and J=1. It can be seen hidden layer node range was between 5 and 21. After several tests, when the hidden layer neural network structure located in the 8-node structure, minimum mean square error (MSE) is obtained as 0.0025113. Therefore, the 10-8-1 three layer neural network was constructed.

(3) Determination of the Learning Rate

In order to ensure the neural networks without oscillation, a smaller value should be selected as the learning rate. Under normal circumstances the learning rate interval is [0.01,0.9]. To reduce the cost of training, rate based on the adaptive adjustment was selected in this study.

(4) Processing the Sample Data

Due to the different units between each sample collected, sample sets were normalized to [-1,1] range. Thus BP network training speed was increased.

3.3 Neural Network Training

Five University Libraries have been selected, namely library of Agricultural University of Hebei, Shanxi Agricultural University Library, Baoding Agricultural Higher Academy Library, Beijing Agricultural College Library, Inner Mongolia Agricultural University Library. 100 users in each University were selected for evaluation, there were 500 evaluation results. 100 samples were randomly selected as the study sample, among which the number of the excellent, the good, the medium, the acceptable, and the unacceptable are 20 respectively. The other 400 samples were for testing. After the training of BP network, the weights obtained when the learning process finished were used as the weights of BP neural network. Neural network model was finally showing convergence. The simulation results showed that the neural network model optimized by genetic algorithm run 184 times when it met the requirement of error(<0.001) . The Training stopped and Network error change leveled off. The neural network optimized by genetic algorithm can achieve rapid optimization. The measured values and the curve fitting values are as follows:

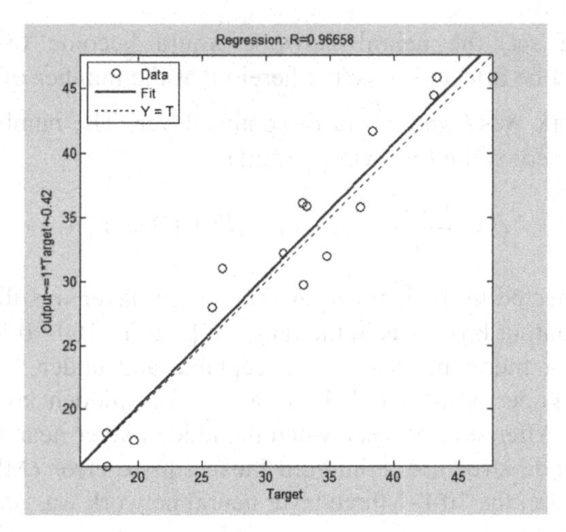

Fig. 2. Curves of measured values and the fitting values

The correlation coefficient was close to 1. The algorithm Optimized constructed in this study achieved an ideal result. Learning training error performance curve of basic BP network was shown below:

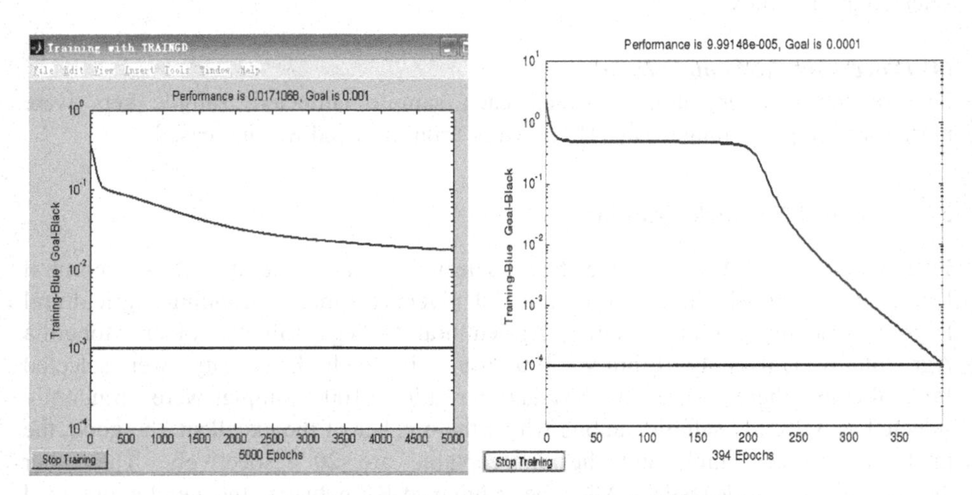

Fig. 3. The error performance curve of basic BP network

The horizontal axis represents the number of training and the vertical axis represents the performance of the error, the black line is the training goal line. The blue line is the training error curve. As can be seen from the graph, the basic BP neural network did not meet the target error 0.001 after 5000 times of training, indicating that the basic training of BP network convergence speed was very slow. Learning and training error performance curve of improved adaptive neural network was shown in the following figure:

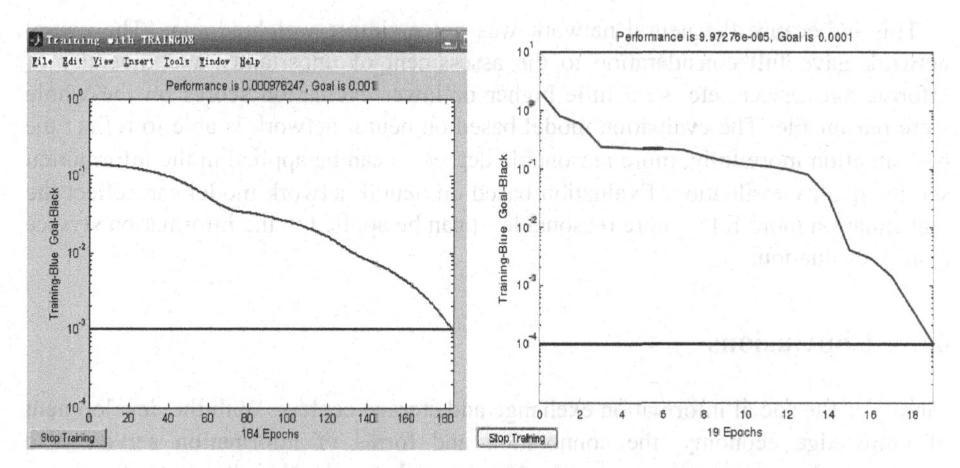

Fig. 4. Learning and training error performance curve of improved adaptive neural network

The neural network model optimized by genetic algorithm met the requirement of error (<0.001) after 184 times operation. From the results, the fitting curve was smooth and good simulation performance was got. The remaining data samples 400 were tested. And the test results were compared with the results using by the conventional method. The comparison result was shown below:

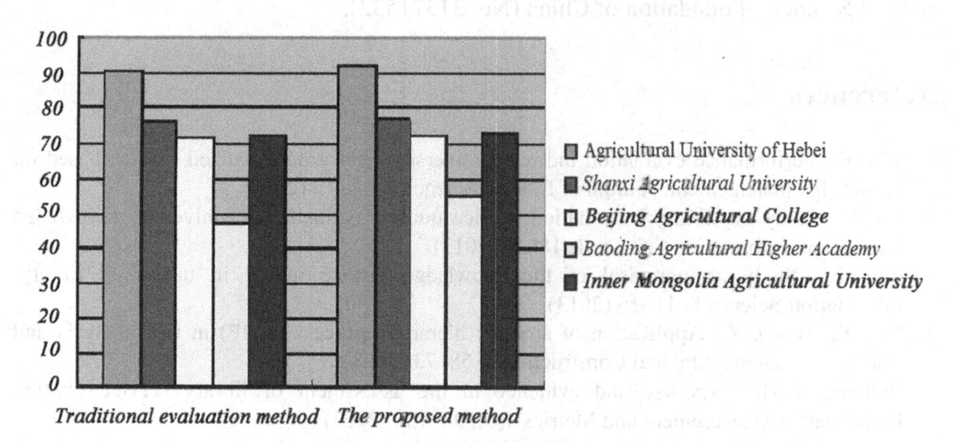

Fig. 5. Comparison of proposed method and the traditional method

Two sets of values are as follows, the traditional method test results were 90.5, 76.2, 71.4, 47.9, 71.4. Neural network method test results were 92.5, 77.2, 72.7, 55.6, 73.5. The university library information service quality sorting results were consistent, both are Baoding Agricultural Higher Academy Library, Beijing Agricultural College Library, Inner Mongolia Agricultural University Library, Shanxi Agricultural University Library, library of Agricultural University of Hebei. Figure 5 illustrates the evaluation scores based on neural network are higher than the original evaluation results.

This is because the neural network was not a simple weighted sum. The neural network gave full consideration to the assessment of uncertainty and diversity of information service, etc. so a little higher or lower evaluation scores on the whole were reasonable. The evaluation model based on neural network is able to reflect the real situation more fully, more reasonable degree, can be applied in the information service quality evaluation. Evaluation based on neural network model can reflect the real situation more fully, more reasonable. It can be applied in the information service quality evaluation.

4 Conclusions

Library is the social information exchange and storage centers. With the development of knowledge economy, the connotation and forms of information service also changed with users' requirements. The library needs to raise their level of information service to meet the user's information needs. This paper constructed the evaluation index system based on AHP and assessment method based on the optimized neural network. This study was helpful to libraries for finding problems and constantly improving their services.

Acknowledgements. This research was supported by a grant from the National Natural Science Foundation of China (No. 31371532).

References

1. Tan, F.: Performance evaluation index of university library personalized service based on knowledge management. Journal of Library Science 3, 11–15 (2009)
2. Qu, C.: Study on library information service quality evaluation in university knowledge innovation. Information Science 1, 18–23 (2013)
3. Ran, X.: Study on appraisal of the knowledge service model in university library. Information Science 8, 11–16 (2013)
4. Liu, Y., Wang, Z.: Application of analytic hierarchy process (AHP) in risk analysis and evaluation. Engineering and Construction 1, 68–73 (2008)
5. Walters, W.H.: Expertise and evidence in the assessment of library service quality. Performance Measurement and Metrics 4(3), 98–102 (2003)
6. Liangzhi, Y., Song, G., Zheng, Z.: SERVQUAL and Quality Assessment of Library Services: A Review of Research in the Past Decade. Journal of Academic Libraries 1, 014 (2005)
7. Nimsomboon, N., Nagata, H.: Assessment of library service quality at Thammasat University Library System. University of Library and Information Science, Japan (2003) (retrieved February 6, 2006)

Measurement System of Reducing Temperature Fluctuation of Thermostat Bath for Calibrating Thermocouple

Min Zhang, Feixia Liang, Yue Xie, Ruguo Huang, Haitao Yuan, and Jiahua Lu

College of Food Science and Technology, Shanghai Ocean University,
Shanghai 201306, China
zhangm@shou.edu.cn,
{1375176575,313929785,895251475,784991342}@qq.com,
lujiahua.home@yahoo.com.cn

Abstract. Based on the periodic unsteady state heat conduction theory, a new measurement system of reducing temperature fluctuation of thermostat bath was developed in order to obtain a liquid environment with uniform and constant temperature controlled for the measurement requirements of calibrating thermocouple. The experimental results show that the temperature stability in this measurement system is superior to that in traditional system. The measurement system had the advantage of calibrating multipoint thermocouple at the same time and completing the data acquisition and control automatically.

Keywords: Temperature fluctuation, Thermostat bath, Thermocouple, Calibrate.

1 Introduction

Low-temperature thermostat bath is essential laboratory equipment in food, biological, chemical scientific research and industrial applications[1]. It is mainly used to provide liquid controlled environment with uniform and constant temperature. It can be used as heating or cooling source or secondary heating or cooling source [2-3]. It is often used as thermocouple calibration equipment in the measurement of food thermal properties. So the performance of low-temperature thermostat bath have a directly impact on the reliability and accuracy of thermocouple calibrating [4]. However, there is a defect of bigger temperature fluctuations and consistent unsteady temperature in moderate low-temperature thermostat bath because the temperature of liquid in thermostat bath is affected by compression engine running and fluid convection heat transfer. As a result, it is difficult to get high precision of temperature calibration of thermocouple which is put in the liquid[5-8]. Especially in the process of calibrating multiple thermocouples at the same time, multipoint stable temperature environment is essential. To solve these problems, on the basis of the traditional

© IFIP International Federation for Information Processing 2015
D. Li and Y. Chen (Eds.): CCTA 2014, IFIP AICT 452, pp. 603–609, 2015.
DOI: 10.1007/978-3-319-19620-6_67

thermostat bath, a new measurement system of reducing temperature fluctuation of thermostat bath was developed and T-type thermocouple temperature calibration value had been analyzed.

2 Principle

Based on the Seebeck effect, temperature was measured by thermocouples[9]. To reduce errors of temperature measurement system, the temperature dividing should be calibrated due to different thermocouples materials which would cause difference of thermoelectric properties and allowable deviation [7]. The temperature deviation would be revised according to the measured temperature value of standard thermometer and thermocouple in the same liquid environment with uniform and constant temperature controlled which is provided by low-temperature thermostat bath. The thermocouples and standard thermometer are directly put into a low-temperature thermostat bath. But the inner liquid would produce cyclical fluctuations of heating and cooling and then it would pass to measured surface. The reason is that temperature control is achieved by means of hot and cold offset.
According to the theory of heat transfer[10] , the measured object can be considered as a homogeneous semi-infinite body. The temperature field in boundary conditions of periodic change for semi-infinite homogeneous object, can be descripted by non-steady-state heat conduction differential equation [11] :

$$\frac{\partial t}{\partial \tau} = \alpha \frac{\partial^2 t}{\partial x^2} \tag{1}$$

The boundary conditions can be considered to be a harmonic. Then the temperature changes in the surface of object can be obtained as cosine function:

$$\theta = A_w \cos \frac{2\pi}{T} \tau \tag{2}$$

By separation of variables method, equation (1) (2) can be integrated and the internal temperature distribution of object is as follow:

$$\theta(x,\tau) = A_w \exp(-x\sqrt{\frac{\pi}{\alpha T}}) \cos(\frac{2\pi}{T}\tau - x\sqrt{\frac{\pi}{\alpha T}}) \tag{3}$$

Where t is object temperature (K), τ is heating time (s), α is thermal diffusion coefficient (m^2/s), x is distance from the surface(m), θ is instantaneous surplus temperature on the surface of the object(K), $\theta(x,\tau)$ is instantaneous surplus temperature inside the object(K), A_w is temperature fluctuation on the surface(K), T is the cycle of wave period (s).

By equation (2) and (3), the surface temperature and the internal temperature will be periodic changes over time. But the fluctuation amplitudes of internal temperature is less than that of surface temperature. Meanwhile, the amplitude decreases greatly with increase of x. The extent of the amplitude attenuation can be expressed:

$$\upsilon = \exp(-x\sqrt{\frac{\pi}{\alpha T}}) \tag{4}$$

It shows that internal wave amplitude decreases with the deepening of the temperature wave. It is one of the important characteristics of periodic heat conduction. That is to say, the temperature wave amplitude increases gradually as the surface distance attenuation. From the theory, reducing the test ambient temperature fluctuation degree can ensure a liquid environment with uniform and constant temperature controlled.

3 Test System and Method

3.1 Test System

As shown in Figure 1, thermocouple calibration measurement system is based on the above principles. The test device includes low-temperature thermostat bath, temperature balance outer pipe, temperature balance inner pipe, blender, insulation cover, insulated container, standard platinum resistance, thermocouple, data acquisition device and computer.

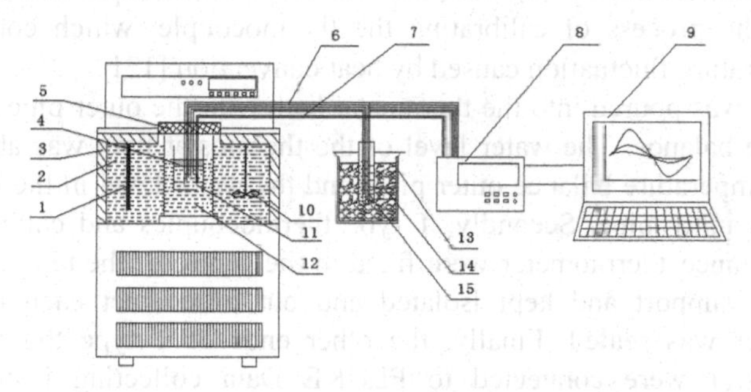

Fig. 1. Schematic diagram of measurement system

1. Heating tuber 2. Thermostat bath 3. Insulation layer 4. insulation cover 5. Outer pipe insulation cover 6. Thermocouple 7. Standard platinum resistance 8. Data acquisition device 9. Computer 10. Blender 11. Balance outer pipe 12. balance inner pipe 13. Insulated container 14. Ice-water mixture 15. Heat transfer oil

The outer pipe and inner pipe of temperature balance are located in the thermostat bath. Flowing medium level within thermostat bath is above temperature balance outer pipe. Flowing medium level within temperature balance outer pipe is higher than temperature balance inner pipe. An anti-mutual contact device is set between two pipes and make inner pipe fix in the middle of outer pipe. The anti- mutual contact device is not touched the bottom and sides wall of temperature balance outside tube. A supporting insulation cover is used on the outer pipe and inner pipe of temperature balance. A calibrated standard platinum resistance thermometer and several groups of non- touching thermocouples were put in the temperature balance inner pipe at the same time. Platinum resistance thermometer and thermocouples fixed by stents located in the inner pipe nozzle, and not touched bottom and sides wall of temperature balance pipe. Then temperature data of standard platinum resistance thermometer and thermocouples were collected and transferred to computer by FLUKE data acquisition instrument.

3.2 Standard Test Method

DC2006 low-temperature thermostat bath was used for this work. The flume dimension was length 250mm, width 200 and depth 200mm. The temperature range was 20-95°C. With water as a working fluid, the heating system is made of L Type stainless steel heating pipe. The heating end was at the bottom of thermostat bath. The evaporator in refrigeration system was made of copper pipe, which was placed directly in the sink in order to increase the cooling rate. T copper-constantan thermocouples with diameter 0.5mm were selected. The reference terminals were immersed in transformer oil surrounded by ice-water mixture. The temperature balance pipes were made of high borosilicate glass material. The outer pipe was external diameter 100mm, depth 100mm and wall thickness 3mm, while the inner pipe external diameter 75mm, depth 100mm and wall thickness 3mm. Due to the presence of glass containers, the distance between thermocouple and fluid walls was increased in the process of calibrating the thermocouple, which could effectively prevent temperature fluctuation caused by heat convection [12] .

First, water was poured into the thermostat bath with the outer pipe and inner pipe of temperature balance. The water level of the thermostat bath was above the water level of the temperature balance outer pipe, and the water level in the outer pipe was higher than in inner pipe. Secondly, T type thermocouples and calibrated standard platinum resistance thermometer were fixed in the centre of the temperature balance inner pipe by support and kept isolated and out of contact each other. Thirdly, insulated cover was sealed. Finally, the other ends of T type thermocouples and platinum resistor were connected to FLUKE Data collecting instrument into a computer.

Temperature data were collected every 30 seconds during testing. The changes of temperature were observed in real-time detection. The temperature stability data could be collected when adjusting to the required temperature [13]. T type thermocouple and calibrated standard platinum resistance thermometer could be compared and corrected at the same temperature. Every degree of temperature value of T type thermocouple was calibrated in turn and was contrasted directly place the thermocouple in water and only place the thermocouple in temperature balance inner tube.

4 Results and Discussion

4.1 Temperature Fluctuations

Fig. 2, Fig. 3 and Fig. 4 showed T type thermocouple temperature test case when liquid temperature in the thermostat bath was set to 20°C, respectively.

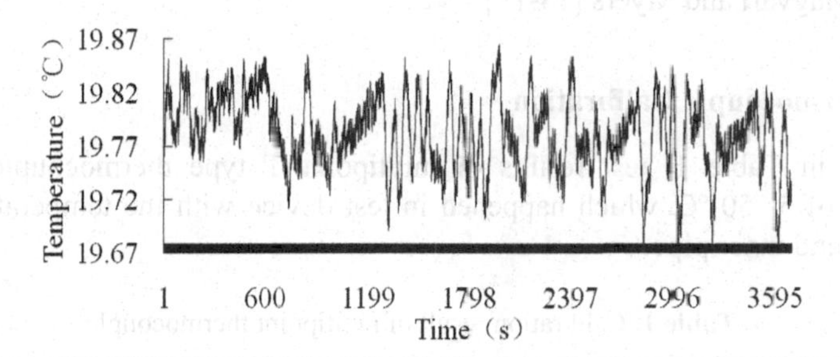

Fig. 2. Temperature test curve of thermocouple put in water

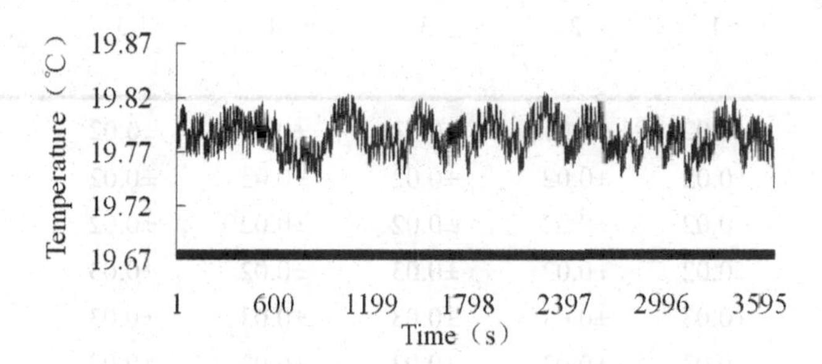

Fig. 3. Temperature test curve of thermocouple put in temperature balance tube Water

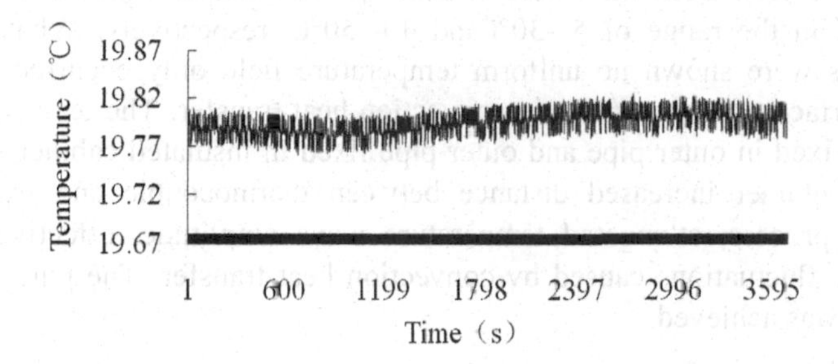

Fig. 4. Temperature test curve of thermocouple put in measurement system

As can be seen from the diagram, temperature wave amplitude of T Thermocouple in contact with fluid directly was greater than other two conditions. Calibration temperature wave amplitude of the T type thermocouple in contact with fluid directly was ±0.075°C. Calibration temperature wave amplitude of only increasing outer pipe is ±0.03°C, Calibration temperature wave amplitude of simultaneously increasing outer pipe and inner pipe was ±0.02°C. The temperature stability raised 73.3% and 33.3% than the previous two conditions, respectively.

It was shown that the thermal resistance of the glass could prevent temperature fluctuation causing by convection. Test results were in accord with the analytical solutions for natural convection boundary layer equations by Pohlhausen and research results of Magyari and Myers [14-15] .

4.2 Thermocouple Calibration

As shown in Table 1, test results of multipoint T type thermocouple under the conditions of 5- 50 °C, which happened in test device with the temperature balance outer pipe and inner pipe.

Table 1. Calibration result of multipoint thermocouple

Thermostat setting temperature (°C)	Temperature fluctuations of thermocouple test point (°C)					Temperature fluctuations in average (°C)
	1	2	3	4	5	
5	±0.02	±0.01	±0.02	±0.01	±0.02	±0.02
10	±0.02	±0.02	±0.02	±0.02	±0.02	±0.02
20	±0.02	±0.02	±0.02	±0.02	±0.02	±0.02
30	±0.02	±0.02	±0.03	±0.02	±0.03	±0.02
40	±0.03	±0.03	±0.03	±0.03	±0.03	±0.03
50	±0.03	±0.03	±0.03	±0.03	±0.03	±0.03

As can be seen from the Table 1, average temperature fluctuation was ±0.02°C and±0.03°C in the range of 5 -30°Cand 40- 50°C, respectively. What is more, the experiments were shown no uniform temperature field only occurred in closing to the wall surface of heat transfer in convection heat transfer. The temperature balance inner pipe fixed in outer pipe and outer pipe fixed in insulated cabinet. The presence of glass container increased distance between thermocouple and wall surface in calibration process, attenuated temperature wave amplitude, effectively prevented temperature fluctuations caused by convection heat transfer. The purpose of precise calibration was achieved.

5 Conclusions

The designed measurement system device included low-temperature thermostat bath, temperature balance outer pipe, temperature balance inner pipe, blender, insulation cover, insulated container, standard platinum resistance, thermocouple, data acquisition device, computer and so on. T type thermocouple under the conditions of 5-50 °C was calibration with average temperature fluctuation ±0.02°C and±0.03°C in the range of 5-30°Cand 40-50°C, respectively. It is simple, convenient and reliable. There will be a bright prospect for determining temperature in scientific research and engineering application.

Acknowledgment. Funds for this research was provided by the Natural Science Foundation of China, Grant No 31371526.

References

1. Merlone, A., Iacomini, L., Tiziani, A., et al.: A liquid bath for accurate temperature measurements. Measurement 40(4), 422–427 (2007)
2. Wu, J., Liu, Z., Wang, F., et al.: A new research of the high-precision thermostatic bath at low temperature for fluid thermal physical measurement. Journal of Xi'an Jiao Tong University 38(5), 504–507 (2004)
3. Merlone, A., Marcarino, P., Iacomini, L., et al.: A liquid bath for accurate platinum resistance thermometers calibration at IMGC. In: The 9th International Symposium on Temperature and Thermal Measurements in Industry and Science, pp. 929–933 (2004)
4. Zhang, P., Xu, Y.X., Wang, R.Z., et al.: Fractal study of the fluctuation characteristic in the calibration of the cryogenic thermocouples. Cryogenics 43(1), 53–58 (2003)
5. Hu, F., Chen, Z., Luo, D., et al.: A research of the precise low-temperature thermostat bath for refrigerants PVT Development for experiment. Chinese Journal of Scientific Instrument 23(4), 414–416 (2002)
6. Abdelaziz, Y.A., Megahed, F.M., Halawa, M.M.: Stability and calibration of platinum/palladium thermocouples following heat treatment. Measurement 35(4), 413–420 (2004)
7. Villafañe, L., Paniagua, G.: Aero-thermal analysis of shielded fine wire thermocouple probes. International Journal of Thermal Sciences 65(3), 214–223 (2013)
8. Sarma, U., Boruah, P.K.: Design and development of a high precision thermocouple based smart industrial thermometer with on line linearisation and data logging feature. Measurement 43(10), 1589–1594 (2010)
9. Danisman, K., Dalkiran, I., Celebi, F.V.: Design of a high precision temperature measurement system based on artificial neural network for different thermocouple types. Measurement 39(8), 697–700 (2006)
10. Holman, J.P.: Heat Transfer, 10th edn. McGraw-Hill, New York (2002)
11. Yang, S., Tao, W.: Heat transfer, 4th edn. Higher Education Press, Beijing (2007)
12. Nouanegue, H., Muftuoglu, A., Bilgen, E.: Conjugate heat transfer by natural convection, conduction and radiation in open cavities. International Journal of Heat and Mass Transfer 51(25-26), 6054–6062 (2008)
13. Yu, W., Zhao, W.: The computer measurement and control system based on agilent 34970A Data acquisition instrument. Automation Instrumentation 10, 46–48 (2004)
14. Magyari, E.: Backward boundary layer heat transfer in a converging channel. Fluid Dynamics Research 39(6), 493–504 (2007)
15. Myers, T.G.: An approximate solution method for boundary layer flow of a power law fluid over a flat plate. International Journal of Heat and Mass Transfer 53(11-12), 2337–2346 (2010)

Design and Implementation of Monitoring and Early Warning System for Urban Roads Waterlogging

Yang Liu[1,2,*], Mingyi Du[1,2], Changfeng Jing[1,2], and Guoyin Cai[1,2]

[1] School of Geomatics and Urban Information,
Beijing University of Civil Engineering and Architecture, Beijing 100044, China
[2] The Key Laboratory for Urban Geomatics of National Administration of Surveying, Mapping and Geoinformation, Beijing university of Civil Engineering and Architecture, Beijing 100044, China
{liuyang,dumingyi,jingcf,cgyin}@bucea.edu.cn

Abstract. Waterlogging happens almost in every rainstorm in city roads and the concave type of overpasses in Beijing. It causes to serious effect for the security of people's daily travel. For resolving this kind of problem, this paper designed a monitoring network by water level sensors and other technologies of internet of things, and developed a monitoring and early warning system for detecting the real-time information of waterlogging conditions in urban roads. This system can capture the real time information of waterlogging depth, make a fast query statistics for certain road or area, and also provide the short message service for the public. It is very convenient for the managers working in department of flood control and command center to collect the real-time information of waterlogging status around urban roads. This is helpful for the local government to perform the correspondingly emergency plan and issue waterlogging information for the public, so as to reduce the impact on the security of people's travel caused by the suddenly bad weather.

Keywords: urban roads waterlogging, monitoring and early warning system, design and implementation, sensors.

1 Introduction

The speed of city construction has become more and more quickly with the rapid development of urbanization in recent years. As side consequences, the disaster of regional flood in urban areas has increased significantly, and some other weaknesses in city planning, construction and management have been also emerged at the same time [1]. A large number of concave type of overpasses and tunnels have been constructed to alleviate traffic pressure and to ensure travel unimpeded in some mega-cities such as Beijing and Shanghai[2-4]. A big challenge should be faced by the people working in the department of city flood control because some sudden extreme rainfall weather happens now and then in recent years. For example, the 7 roads were

* Corresponding author.

© IFIP International Federation for Information Processing 2015
D. Li and Y. Chen (Eds.): CCTA 2014, IFIP AICT 452, pp. 610–615, 2015.
DOI: 10.1007/978-3-319-19620-6_68

in traffic disruption, the 27 bridges were covered with water, and the traffic was resumed until 6 hours later during the severe natural disaster happened at July, 21, 2012 in Beijing. The main reason for this phenomenon is due to the unreasonable planning of the city drainage system, additionally, there was no systematic monitoring for the status of the road waterlogging which caused the delay of the early warning and forecasting because of no real time monitoring and transmission of the data with real time water condition[5-6].

For resolving the above mentioned problems, this paper designed a road waterlogging monitoring network covering the whole Beijing urban areas by means of sensor technology and wireless transmission technology. Based on this network, a road waterlogging monitoring and early warning system was developed. The system can provide real-time information for managers performing decision-making strategy, and offer technical support for the organization of municipal drainage management. This system can be also used to release water monitoring information to the public through the micro-blog and other rapidly updated media, so as to guide the public travel. The monitoring method, system design and implementation were discussed in this paper.

2 Methods

As shown in Fig. 1, sensors were installed near the concave type of overpass or low-lying courtyard road where often suffer from waterlogging. Sensors in the whole study areas constitute a waterlogging monitoring network. The mobile GMS/GPRS communication network was used to transmit water information to the monitoring center at certain preset intervals. The road waterlogging monitoring system was developed through the information collected by sensors. The waterlogging locations are displayed in this system. It can be used to query the real-time water information and make an early warning for urban flood. This monitoring system software is installed on the WEB server, which can provide service of real time water information to the public. The workers in department of flood control only needs a computer with internet access to operate the software, realizing the real-time query of road waterlogging, water retention, reminding of early warning, water site inquiries with exceed the standard water, and 12 hours of water values etc..

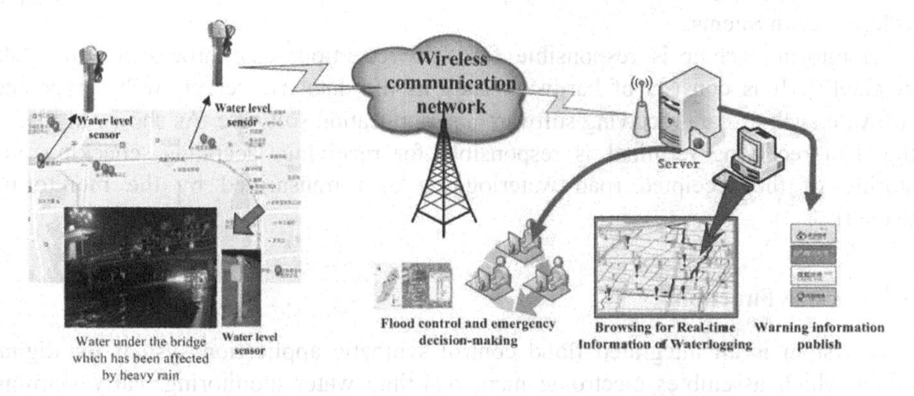

Fig. 1. Schematic diagram of monitoring method for urban roads waterlogging

3 System Design

3.1 System Architecture

This system includes two parts, that is, road waterlogging monitoring stations and monitoring center, as shown in Fig. 2.

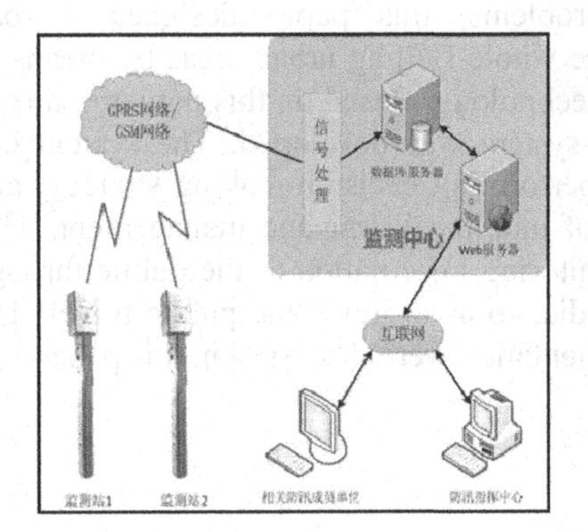

Fig. 2. The system structure diagram

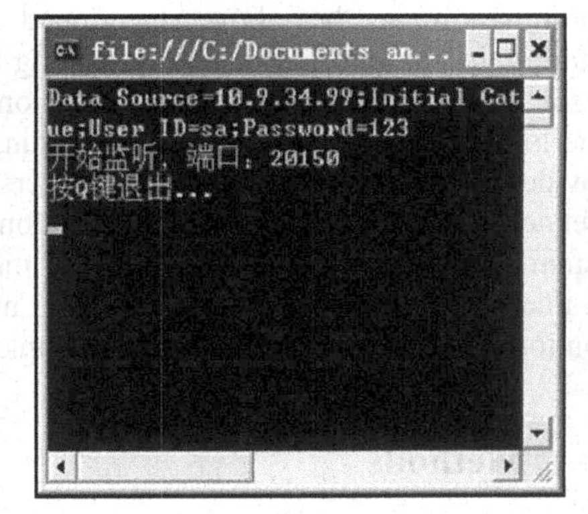

Fig. 3. Data receiving Service

Road waterlogging monitoring station is consists of electronic gauge and wireless terminal, which are responsible for data acquisition and data transmission of real time road waterlogging. Electronic gauge and wireless client machine are the important components of road waterlogging monitoring station. Ectronic gauge is used to collect data of water level. It is equipped with advanced microprocessor chips for digital water level controller, and built-in communication circuit of sensor. It has the characteristics of high reliability and anti-interference performance with measurement accuracy of 1cm.The wireless client machine: it is used to data transmission. It is a whole new generation of intelligent telemetering terminal through assembling RTU, GPRS/GSM communication module, protection module and the charge controller. It can directly access the 5-10W solar panels or other accords with the power supply voltage requirements.

Monitoring center is responsible for data reception, data processing and data display[7]. It is consists of hardware such as the database server, web server and software such as data receiving software and application software. As shown in Fig. 3, the data receiving terminal is responsible for receiving, decoding, checking and storing of the receipted road waterlogging data transmitted by the monitoring network.

3.2 Core Functions

This system is an integrated flood control synthetic application system for digital urban which assembles electronic map, real-time water monitoring, early warning analysis. As shown in Fig. 4, the core functions of this system can be divided into 3 major parts, also known as a subsystem. They are water information receiving module, water monitoring and early warning module, and a warning short message service module, each module has its own corresponding function.

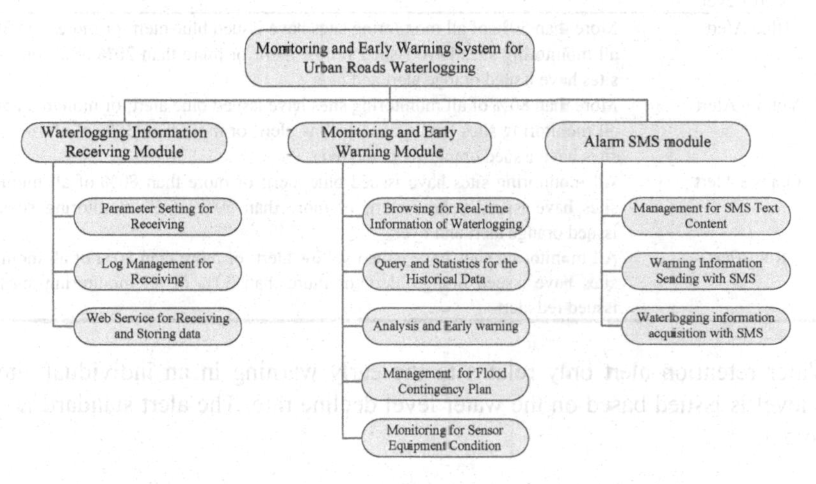

Fig. 4. The core functions of monitoring and early warning system for urban roads waterlogging

3.3 Road Waterlogging Alert Levels

There are four alert levels designed in this system. They are blue, yellow, orange and red alert in an ascending order level. The alert level in a single monitoring station is determined by the waterlogging time and water depth. The alert standard for different warning levels is shown in table 1.

Table 1. The alert level of real-time monitoring data for urban roads waterloggin (single monitoring site)

The Depth / Duration	5 cm - 10cm(Ex)	10 cm - 15cm(Ex)	15 cm - 20cm(Ex)	>= 20cm
0-15 min(In)	Yellow alert	Orange alert	Red alert	Red alert
15-30 min(In)	Yellow alert	Orange alert	Red alert	Red alert
30-60 min(In)	Blue alert	Yellow alert	Orange alert	Red alert
> 60 min	Blue alert	Blue alert	Yellow alert	Red alert

The synthetic alert in the whole urban region is based on the comprehensive analysis of all the monitoring sites. The alert standard is shown in table 2.

Table 2. The alert Level of real-time monitoring data for urban roads waterlogging (taking all the monitoring sites into account)

The Comprehensive Alert Level	The Specific Standards
Blue Alert	More than 50% of all monitoring sites have issued blue alert; or more than 30% of all monitoring sites have issued yellow alert; or more than 20% of all monitoring sites have issued orange alert and over.
Yellow Alert	More than 80% of all monitoring sites have issued blue alert; or more than 50% of all monitoring sites have issued yellow alert; or more than 30% of all monitoring sites have issued orange alert and over.
Orange Alert	All monitoring sites have issued blue alert; or more than 80% of all monitoring sites have issued yellow alert; or more than 50% of all monitoring sites have issued orange alert and over.
Red Alert	All monitoring sites have issued yellow alert; or more than 80% of all monitoring sites have issued orange alert; or more than 50% of all monitoring sites have issued red alert.

Water retention alert only relates to the early warning in an individual site. The alert level is issued based on the water level decline rate. The alert standard is shown in table 3.

Table 3. The alert Level of water retention in urban roads

Withdrawal Rates (unit: cm / hour)	2.0(In) - 5.0	2.0(In) - 3.0	1.0(In) - 2.0	< 1.0
Alert Level	Blue alert	Yellow Alert	Orange Alert	Red Alert

4 System Implementation

City Road Waterlogging Monitoring and warning system is designed as a browser / server (B / S) software system architecture. Flex framework and C# were used as the programming language, and SQL Server 2008 as the database server. ArcSDE is adopted to store the spatial data, and ArcServer is employed to provide the map service. The interface is shown below, as shown in Fig. 5.

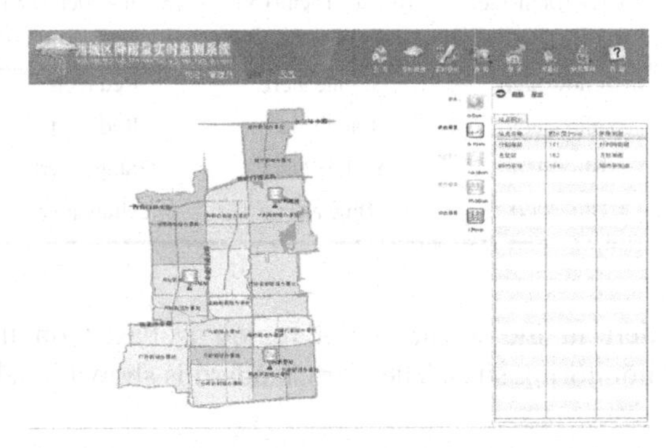

Fig. 5. The Interface of Monitoring and Early Warning System for Urban Roads Waterlogging.

5 Conclusions

With the rapid development of social economy in urban region, the disaster of waterlogging or lost caused by sudden rainstorm or other bad weather might become more and more serious. It makes the managers in department of flood control and command center to collect the correct and real time information of waterlogging much more convenient that the establishment of the monitoring and early warning system [8-9]. This system built by means of sensors not only provide reliable information for the local government to perform some decision-making in flood control and command, but make the working be more of efficiency and the decision-making policy be more of scientificalness, so as to guarantee the security of people's daily travel much more better.

Acknowledgment. The research performed is supported by the National Natural Science Foundation of China (No. 41101444), Beijing Nova Program (No. Z121106002512025), and is partially funded by Beijing Municipal Organization Department talents project(No.2012D005017000001). The work was performed under the Key Laboratory for Urban Geomatics of National Administration of Surveying, Mapping and Geoinformation (No. 20111210N). We wish to thank Guoyin Cai for his careful proofreading of the manuscript.

References

1. Xia, Z.C., Zhong, X.J., Ruan, F., et al.: Design of monitoring and alarm system for city stagnant water. Electronic Test 61(5), 56–57 (2013)
2. Wan, S.L., Zuo, L.Y., Wang, S.J.: The design of urban road in Laixi water monitoring system. Shandong Water Resources 65(1), 61–62 (2011)
3. Pan, C.L.: The design of city overpass water automatic monitoring system in Shanghai. Water Resources Informatization (3), 42–44 (2010)
4. Xu, H.J., Liu, L.H.: Application of electronic gauge in the city road water monitoring system. Water Resources Informatization 57(3), 45–47 (2010)
5. Zhou, R.Y., Bai, F., Sui, H.L., et al.: Urban waterlogging remote monitoring system based on embedded. Water Conservancy Science and Technology and Economy (4), 109–112 (2013)
6. Ramos, M.H., Leblois, E., Creutin, J.D., et al.: Form point to areal rainfall: linking the different approaches for the frequency characterisation of rainfalls in urban areas. Water Science and Technology 54(6/7), 33–40 (2006)
7. Caudle, L.: Flood waring. Roads & Bridges 48(6), 16–17 (2010)
8. Zhao, R.: Safety evaluation of Urban Flood Control System Based on Variable fuzzy pattern recognition. In: Micro Nano Devices, Structure and Computing Systems, pp. 264–269 (2011)
9. Akter, S.T., Islam, M.Z., Islam, M.N., et al.: An investigation on maintenance problems of some flood embankments in and around Dhaka city. Indian Journal of Power and River Valley Development 59(3/4), 38–49 (2009)

Development of Early-Warning Model for Intensive Pig Breeding

Nanxin Chen, Qingling Duan, Jianqin Wang, and Ruizhi Sun

China Agricultural University, College of Information and Electrical Engineering, Beijing 100083, China

Abstract. Following the rapid development of intensive pig breeding in China, the impact of environmental factors on the production and health of pigs has become increasingly apparent, and the monitoring of these environmental factors recognized as critical for improved breeding productivity. Based on the effects of environmental factors on pig growth, this paper established an early-warning model of the piggery environment. Using the model and the environmental factors, which were obtained in real time from a piggery, it was possible to obtain timely warning information, conducive to both creating an appropriate breeding environment for pigs and reducing the incidence of disease. In this article, we established the environmental early-warning indicators relating to pig breeding and then demonstrated the method based on single-factor and fuzzy comprehensive multi-factor models of the piggery environment. Finally, the two models were analyzed based on the experimental results, which showed that the fuzzy comprehensive early-warning model performed better than the single-factor model, and that it could be applied in an intensive farming environment to provide timely warning of environmental deterioration, to maintain the safety of the pig-breeding environment.

Keywords: environmental factor, single-factor early-warning model, fuzzy comprehensive early-warning model.

1 Introduction

In China environmental factors are becoming increasingly important to the health and productivity of pigs because of the rapid development of pig breeding. As a result, control of the piggery environment has become a critical factor in improving the efficiency of pig breeding. The environmental factors and other elements that directly or indirectly affect the growth, development, reproduction, and health of livestock include temperature, humidity, noise, illumination, and harmful gases [1-3]. For example, high temperature can cause a pig to reduce its nourishment intake and thus affect its growth. In this article, we analyzed the effect of each selected environmental factor and based on the results, built an environmental early-warning model. According to environmental data acquired in real time, warning information could be delivered by the early-warning model, making it conducive to both creating an appropriate breeding environment for pigs and improving the efficiency of pig breeding.

© IFIP International Federation for Information Processing 2015
D. Li and Y. Chen (Eds.): CCTA 2014, IFIP AICT 452, pp. 616–626, 2015.
DOI: 10.1007/978-3-319-19620-6_69

At present, research on the breeding of livestock and poultry is concentrated mainly on the influence of different environmental factors on the growth of the animal [2-3] and on monitoring the breeding environment [4-7]. Hansen et al. adopted a model-based control design method to develop a controller to regulate the airflow of the breeding environment to control its inner temperature and humidity [8]. Soldato et al. used robust nonlinear feedback control and feed forward control of the inner temperature and humidity to maximize livestock productivity [9]. However, in the field of pig breeding, little research has been reported on early-warning models, especially studies of early-warning models combining multiple environmental factors. At present, common early-warning models can be divided into single and multi-factor models. Single-factor models always incorporate the theory of "certainty" in the index model, and establish the worst factor for the warning of the piggery environment by comparing various factors with early-warning standards [10]. The common feature of multi-factor early-warning models is the inclusion of techniques such as artificial neural networks, the N.L. Nemerow Index method, and fuzzy comprehensive evaluation methods. Warnings achieved using artificial neural networks are better, but require additional training data [11]. The N.L. Nemerow Index method, despite considering the most significant factors, cannot reflect the relative importance of the various factors [11]. A fuzzy comprehensive early-warning model is a relatively mature method used widely in water-quality prediction that can produce realistic results. It describes the warning level by membership degree in fuzzy mathematics, and then determines the weight of each factor, evaluates it, and generates the warning [12-14].

In this article, we selected five environmental factors as early-warning indicators for the pig-breeding environment, and designed four grades of environmental warning: suitable, mild warning, moderate warning, and severe warning. Then, we demonstrated the method based on single- and fuzzy comprehensive multi-factor models for the piggery environment, and an experiment was designed to validate the model.

2 Construction of Early-Warning Indicators

The factors that affect the pig-breeding environment are complex and they can act collectively or individually to affect the animals in many ways. The proper growth environment differs during every stage of development of the pigs and therefore the environmental indicators should be assessed according to the animals' growth characteristics and the corresponding environmental standards. In this article, we only consider environmental factors of standardized scale pig farms, and although fattening pigs serve as our source material, we do not consider the effects of feed.

At present, the national standards of environmental parameters regarding pigs are principally the "Standardization Construction Standard of Scale Pig Farms", and "Scale Farms Environmental Parameters and Environmental Management" [15, 16]. According to the national standards and by consulting breeding experts, the principal environmental factors affecting pig breeding were determined to include temperature, humidity, harmful gases (NH_3, H_2S, and CO_2), dust, sunlight, noise, and airflow. Because intensive breeding occurs in buildings equipped with relatively complete facilities and relatively stable conditions of lighting, noise, dust, and airflow, we selected significant changes of temperature, humidity, and air pollutants as the environmental early-warning indicators. These factors could be divided into two categories: single threshold factors (STFs) and double threshold factors (DTFs). STFs include NH_3, H_2S, and CO_2 because they are harmful to the pigs when their concentrations exceed certain standards. DTFs include temperature and humidity because values of these factors that are either too high or too low will affect the health of the pigs.

According to the environmental standards of pig breeding [15, 16], the grading standards of environmental warnings are as shown in Tables 1 and 2.

Table 1. STF early-warning grading standard

	Suitable	Mild warning		Moderate warning		Severe warning	
		low	high	low	high	low	high
Temperature (°C)	15-23	10-15	23-30	1-10	30-35	<1	>35
Humidity(%RH)	65-75	45-65	75-80	40-45	80-95	<40	>95

Table 2. DTF early-warning grading standard (mg/m3)

	Suitable	Mild warning	Moderate warning	Severe warning
CO_2	<1500	1500-7857	7857-39285	>39285
NH_3	<25	25-30	30-35	>35
H_2S	<10	10-30	30-75	>75

3 Construction of the Early-Warning Model

3.1 Single-Factor Early-Warning Model

The single-factor early-warning model (SFM) is evaluated on single environmental factors of the piggery, and the worst factor taken as the indicator of the piggery environment.

The definition of the STF's early-warning index IS_i is as follows:

$$IS_i = \frac{C_i}{S_i} \tag{1}$$

where C_i is the measured value of environmental factor i and S_i is the standard value of environmental factor i; the value is exceeded when $IS_i > 1$.

For DTFs, the early-warning index ID_i is defined as follows:

$$ID_i = \frac{s_{li} - c_i}{s_{ui} - s_{li}} \quad c_i < s_{li} \tag{2}$$

$$ID_i = 0 \quad s_{li} \le c_i < s_{ui} \tag{3}$$

$$ID_i = \frac{c_i - s_{ui}}{s_{ui} - s_{li}} \quad c_i > s_{ui} \tag{4}$$

where C_i is the measured value of environmental factor *i*, S_{li} is the standard low threshold of factor *i*, and S_{ui} is the standard high threshold of factor *i*; the value is exceeded when $ID_i \neq 0$.

3.2 Fuzzy Comprehensive Early-Warning Model

The result of the SFM for the piggery reflects the effect of a single environmental factor only. However, the interaction of multiple environmental factors influences the growth of the pigs, and environmental quality is a fuzzy concept without clear boundaries for the effects of the factors. Therefore, this paper presents a fuzzy comprehensive early-warning model (FCM) for a piggery environment.

In the FCM, a subjection degree matrix is initially established according to a set of subjection degrees for each environmental factor. Then, a comprehensive evaluation vector, which is the product of the subjection degree and weight of each environmental factor, is obtained. This vector reflects the subjection degree of the current environment belonging to each environmental early-warning grade [8].

(1) The subjection degree calculation of STF

This paper defined four warning grade standards: suitable, mild warning, moderate warning, and severe warning, which correspond to values of '*j*' = 0, 1, 2, and 3, respectively, and designate the subjection degree function for each grade for each factor.

In this paper, rs_{ij} denotes the subjection degree of the STF '*i*' on the environmental standard grade '*j*'. The STF includes NH_3, H_2S, and CO_2, which correspond to values of '*i*' = 0, 1, and 2, respectively. The subjection degree function is constructed using the following trapezoid formula.

The subjection degree function of grade 0 for the STF is defined as follows:

$$rs_{ij} = \begin{cases} 1 & C_i \le S_{i0} \\ \dfrac{S_{i1} - C_i}{S_{i1} - S_{i0}} & S_{i0} < C_i \le S_{i1} \\ 0 & C_i > S_{i1} \end{cases} \tag{5}$$

The subjection degree function of grade j (j = 1, 2) is defined as follows:

$$\mathrm{rs}_{\mathrm{ij}}=\begin{cases}0 & \mathrm{C_i} \le S_{i(\mathrm{j}-1)} \\ \dfrac{C_i - S_{i(\mathrm{j}-1)}}{S_{i\mathrm{j}} - S_{i(\mathrm{j}-1)}} & S_{i(\mathrm{j}-1)} < C_i \le S_{ij} \\ \dfrac{S_{i(\mathrm{j}+1)} - C_i}{S_{i(\mathrm{j}+1)} - S_{ij}} & S_{i\mathrm{j}} < C_i \le S_{i(\mathrm{j}+1)} \\ 0 & C_i > S_{i(\mathrm{j}+1)}\end{cases} \tag{6}$$

The subjection degree function of grade 3 is defined as follows:

$$\mathrm{rs}_{\mathrm{ij}}=\begin{cases}0 & \mathrm{C_i} \le S_{i2} \\ \dfrac{C_i - S_{i2}}{S_{i3} - S_{i2}} & S_{i2} < C_i \le S_{i3} \\ 1 & C_i > S_{i3}\end{cases} \tag{7}$$

where C_i is the measured value of the STF I and S_{ij} is the maximal value of the warning grade j in the early-warning grading standard for the STF.

For instance, S_{1j} = (0, 10, 30, 75) are maximal values of each warning grade for H_2S. We assume C_1 = 17 mg/m^3; then, $RS_{1j} = (0,\ 0.65,\ 0.35,\ 0)$.

(2) The subjection degree calculation of DTF

For the DTF, this paper also defined four warning grade standards: suitable, mild warning, moderate warning, and severe warning. The warning grade of the environment is denoted by $|j|$ = {0, 1, 2, 3} and j ∈ {-3, -2, -1, 0, 1, 2, 3}. If the measured value is bigger than the maximal value of grade 0, then j = {0, 1, 2, 3}; otherwise, j = {-3, -2, -1, 0}.

In this paper, rd_{ij} denotes the subjection degree of the DTF 'i' on the environmental warning grade 'j'. The DTF includes temperature and humidity, which correspond to values of 'i' = 0 and 1, respectively. The subjection degree function is constructed using the following trapezoid formula.

The subjection degree function of grade -3 for the DTF is defined as follows:

$$\mathrm{rd}_{\mathrm{ij}}=\begin{cases}1 & \mathrm{C_i} \le S_{i(-3)} \\ \dfrac{S_{i(-2)} - C_i}{S_{i(-2)} - S_{i(-3)}} & S_{i(-3)} < C_i \le S_{i(-2)} \\ 0 & C_i > S_{i(-2)}\end{cases} \tag{8}$$

The subjection degree function of grade j (j = -2, -1, 0, 1, 2) is defined as follows:

$$\mathrm{rd}_{\mathrm{ij}}=\begin{cases}0 & \mathrm{C_i} \le S_{i(\mathrm{j}-1)} \\ \dfrac{C_i - S_{i(\mathrm{j}-1)}}{S_{i\mathrm{j}} - S_{i(\mathrm{j}-1)}} & S_{i(\mathrm{j}-1)} < C_i \le S_{ij} \\ \dfrac{S_{i(\mathrm{j}+1)} - C_i}{S_{i(\mathrm{j}+1)} - S_{ij}} & S_{i\mathrm{j}} < C_i \le S_{i(\mathrm{j}+1)} \\ 0 & C_i > S_{i(\mathrm{j}+1)}\end{cases} \tag{9}$$

The subjection degree function of grade 3 is defined as follows:

$$\mathrm{rd}_{\mathrm{ij}}=\begin{cases}0 & \mathrm{C_i} \le S_{i2} \\ \dfrac{C_i - S_{i2}}{S_{i3} - S_{i2}} & S_{i2} < C_i \le S_{i3} \\ 1 & C_i > S_{i3}\end{cases} \tag{10}$$

where C_i is the measured value of the DTF i and S_{ij} is the maximal value of the warning grade j in the early-warning grading standard for the DTF.

For instance, S_{0j} = (1, 10, 15, 19, 23, 30, 35) are maximal values of each warning grade for temperature. We assume C_0 = 27 °C; then, RD_{0j} = (0, 0.43, 0.57, 0).

(3) The calculation of weight

In this paper, $W = [w_0, w_1, w_2, w_3, w_4]$ denotes the weight of each factor. Considering that environmental factors that have higher warning grades or that have more serious effects on the pigs should have greater weight, the method of calculating the weight was designed as follows.

(i) According to the influence of environmental factors on the pigs, this paper designed the initial weights as follows:

$$w_b = \{0.07, 0.06, 0.07, 0.11, 0.09\} \tag{11}$$

(ii) The remaining 0.6 of the weight is distributed according to the severity of each factor.

In this paper, z_i denotes the severity of the environmental factor.

For STFs:

$$\mathrm{z_i}=\frac{c_\mathrm{i} - b_i}{s_i - b_i} \tag{12}$$

For DTFs:
when the monitoring value is higher than the suitable value:

$$z_i = \frac{c_i - s_{ui}}{s_i - b_i} \tag{13}$$

and when the monitoring value is lower than the suitable value:

$$z_i = \frac{s_{li} - c_i}{s_i - b_i} \tag{14}$$

In the formula, z_i means the *i*-th factor's weight, c_i is the measured value of the *i*-th factor, s_i is the maximum of grade *j* of the warning grading standard, and b_i is the minimum of grade *j* of the warning grading standard.

The weight of each factor is calculated as follows:

$$w_i = \frac{z_i}{\sum_j z_j} \times 0.6 + w_{bj} \tag{15}$$

where w_{bi} is the base weight of the *i*-th factor.

(4) The calculation of the early-warning grade
Using the methods stated earlier, we can obtain a subjection degree matrix *R* that consists of a subjection degree for each grade of each factor, and the fuzzy comprehensive warning vector is the product of the weight vector *W* and the subjection degree matrix *R*.

The subjection matrix *R* is calculated as follows:

$$R = \begin{bmatrix} rs_{00} & rs_{01} & rs_{02} & rs_{03} \\ rs_{10} & rs_{11} & rs_{12} & rs_{13} \\ rs_{20} & rs_{21} & rs_{22} & rs_{23} \\ rd_{00} & rd_{0x_1} & rd_{0x_2} & rd_{0x_3} \\ rd_{10} & rd_{1x_1} & rd_{1x_2} & rd_{1x_3} \end{bmatrix} \tag{16}$$

In the formula, $(x1, x2, x3)$ could be (1, 2, 3) or (-1, -2, -3).

Then, the fuzzy comprehensive warning vector *B* is calculated as follows:

$$B = W \times R \tag{17}$$

Finally, we calculated the grade of fuzzy comprehensive evaluation as follows:

$$I = b_1 \times 1 + b_2 \times 2 + b_3 \times 3 + b_4 \times 4 \quad (18)$$

When the result matches:

$0 \leq I < 1$, the warning grade is 1;
$1 \leq I < 2$, the warning grade is 2;
$2 \leq I < 3$, the warning grade is 3;
$3 \leq I$, the warning grade is 4.

4 Warning Process

Based on the early-warning model stated earlier, this paper designed an environmental warning process, as shown in Figure 1.

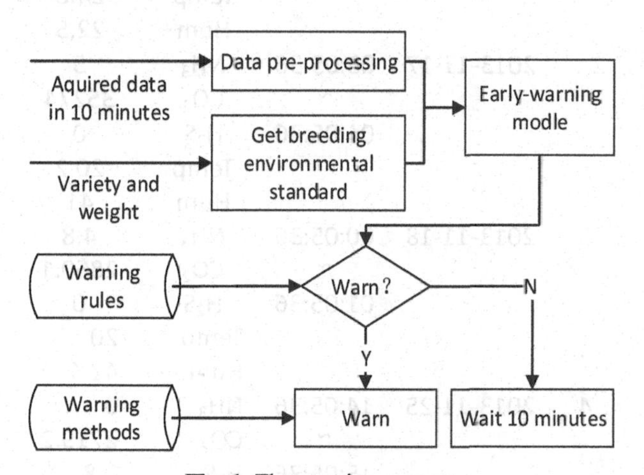

Fig. 1. The warning process

The detail of the warning process is as follows:

(1) Search the breeding environmental standard from the breeding information library according to the pigs' breed and weight;
(2) Every 10 minutes, data from the sensors are filtered using the Distribution method [17], and the average of these data calculated to evaluate the environment during that 10-minute interval using the early-warning model;
(3) Check whether a warning is required according to the evaluation results. If a warning is not needed, wait another 10 minutes and return to step (2); otherwise, choose the appropriate warning from the warning method database according to the evaluated result.

5 Experiment and Analysis of Results

We installed a set of sensors of environmental parameters at an experimental station in Zhuozhou. The parameters monitored included NH_3, CO_2, H_2S, temperature, and humidity. We chose four periods for taking the experimental measurements, as shown in Table 3 (Temp is temperature, Hum is humidity, and the units of NH_3, CO_2, H_2S, Temp, and Hum are mg/m^3, mg/m^3, mg/m^3, °C and %RH, respectively), and the warning results of the SFM and FCM are shown in Table 4.

Table 3. Measured values

No.	Date	Time	Factor	Value
1	2013-11-15	22:05:36	NH_3	0
		~	CO_2	1288.5
		23:05:36	H_2S	0
			Temp	24.8
			Hum	22.5
2	2013-11-17	03:05:36	NH_3	5
		~	CO_2	35274
		04:05:36	H_2S	0
			Temp	20.2
			Hum	41
3	2013-11-18	00:05:36	NH_3	4.8
		~	CO_2	2860.1
		01:05:36	H_2S	0
			Temp	20
			Hum	42.4
4	2013-11-25	14:05:36	NH_3	0
		~	CO_2	2719.2
		15:05:36	H_2S	0.8
			Temp	20.8
			Hum	39.2

Table 4. Warning results of SFM and FCM

No.	SFM	FCM
1	4	4
2	3	4
3	3	3
4	4	3

We can draw several conclusions from the experimental results.

(1) In No. 1, the result of the FCM is grade 4, which is the same as the result of SFM. As can be seen from Table 3, in No. 1, the humidity is very low and the temperature is outside the optimum range; therefore, the warning results are appropriate to the actual situation.

(2) In No. 2, the result of the FCM is grade 4, but in No. 3, it is grade 3, whereas the result of the SFM is grade 3 in both. As can be seen from Table 3, the concentration of CO_2 in No. 2 is obviously higher than in No. 3; therefore, the results obtained by the FCM are most reasonable.

(3) In No. 4, the result of the FCM is grade 3, while the result of the SFM is grade 4. As shown in Tables 3 and 1, in No. 4, humidity is 39.2%, which is a little lower than the threshold (40%), but the other environmental factors are very good. Considering all factors, a grade 3 warning is deemed most suitable based on the actual situation.

6 Conclusions

The main contributions of this paper are as follows:

(1) Many environmental factors affect the breeding and growth of pigs, but not all have a significant impact. This paper consulted authoritative environmental standards and considered the environmental factors of temperature, humidity, CO_2, NH_3, and H_2S as the principal factors for an early-warning model. The experimental results show that these factors satisfactorily reflect the actual situation of a pig-breeding environment.

(2) This paper compared a single-factor early-warning model with a fuzzy comprehensive early-warning model. The result of the experiment shows that when one or more factors exhibit significant deterioration, both models are able to describe the degree of deterioration of the environment accurately. However, when no single factor is seriously deteriorated, the performance of the fuzzy comprehensive early-warning model is superior to the single-factor early-warning model.

(3) The warning model presented in this paper is shown to perform well and therefore it could act as a timely warning of environmental deterioration.

Acknowledgements. This paper was supported by the National High Technology Research and Development Program (863 Plan) (2013AA102306), and the National Science and Technology Supporting Plan Project (2012BAD35B06), and the Program of International S&T Cooperation(2013DFA11320).

References

1. Li, F.: The impact of light on the pig and the lighting control technology. Veterinary Journals of China Animal Husbandry (12) (2012)
2. Yang, C., Xu, C.: Effect of environmental condition on reproductive performance of swine. Heilongjiang Journal of Animal Reproduction (03) (2012)

3. Qu, J., Niu, J.: Study on the effect of different environmental factors on the growth of piglets. Feed Review (03) (2005)
4. Sun, L.: Design of livestock environment monitoring system. Nanjing Agricultural University (2011)
5. Yu, X.: Study on multi-environment factor monitoring system of the livestock breeding. Sensors and Transducers (11) (2013)
6. Gao, L., Li, D.: Study on the construction and management of the Internet of things application system of aquaculture supervisory. Shandong Agricultural Sciences (08) (2013)
7. Sun, Z.: Design of a telemonitoring system for data acquisition of livestock environment. In: Livestock Environment VIII - Proceedings of the 8th International Symposium (2008)
8. Hansen, M.: Temperature and humidity control in livestock stables. In: 11th International Conference on Control, Automation, Robotics and Vision, ICARCV 2010 (2010)
9. Soldatos, A.: Nonlinear robust temperature-humidity control in livestock buildings. Computers and Electronics in Agriculture (03) (2005)
10. Ma, Z.: Study ofwater quality early warning model for Litopenaeus vannamei intensive aquaculture. Ocean University of China (2010)
11. Yang, L., Lu, W.: The application of improved N.L.Nemerow index method and fuzzy comprehensive method in water quality assessment. Water Resources and Power (2012)
12. Li, X., Zhou, Q., Zhang, L.: The application of fuzzy comprehensive assessment of heavy metal pollution in the soil on farmland which irrigation with swine breeding wastewater. Journal of Agro-Environment Science (03) (2009)
13. Li, M.: Heavy metal pollution in soil and the evaluation of farmland. Shandong Agricultural University (2009)
14. Wang, R., He, Y., Fu, Z.: Quality early warning model of pond water for freshwater aquaculture. Journal of Jilin Agricultural University (01) (2011)
15. GBT 17824.3-2008, Scale farms environmental parameters and environmental management
16. NYT 1568-2007, Standardization construction standardized of scale pig farms
17. Xia, Z.: Application of distribution method in the treatment of mistake errors. Practical Measurement Technology (2002)

An On-Line Monitoring System of Crop Growth in Greenhouse

Wu Lixuan[1], Sun Hong[1], Li Minzan[1], Zhang Meng[2], and Zhao Yi[2]

[1] Key Laboratory of Modern Precision Agriculture System Integration Research, China Agricultural University, Beijing 100083, China
[2] Key Laboratory of Agricultural information acquisition technology, Ministry of Agriculture, Beijing 100083, China
zhwlx22@163.com, {sunhong,pact,caupac_zm,YiZhao}@cau.edu.cn

Abstract. It is significant to estimate the growth of crops during their growth stages rapidly and non-destructively, the spectral analysis technology could satisfy the requirement and it is one of the important techniques to support in the precision agriculture management. In the article, a system was developed for estimating the growth of crops in greenhouse, which was based on spectrum and WSN (wireless sensor network) technology. The system consisted of three parts: the intelligent sensor nodes, intelligent gateway and the software in remote server. The ZigBee wireless communication modules were embedded in the intelligent sensor nodes and gateway to create WSN. The intelligent sensor nodes contained four optical channels, which allowed these instruments work at the wavelengths of 550, 650, 766 and 850 nm. They were used to collect spectral data and transmit to the intelligent gateway by ZigBee wireless network. The intelligent gateway worked as a coordinator of the whole wireless network waiting for sensor nodes to join in and receiving all the spectral data from sensor nodes, and transmitting the data to remote server by GPRS. The software in remote server analyzed the spectral data, calculated the NDVI (Normalized Difference Vegetation Index), by which, the growth of the plants can be detected effectively. In order to test the effectiveness and stability of the system, the strawberry of greenhouse was detected. Several experiments were carried out in the greenhouse of ZhuoZhou Experiment Station in Hebei. Results showed that the canopy spectral data of strawberries could be acquired by the intelligent sensor nodes stably, and the intelligent gateway could connect the sensor nodes and the remote server effectively. The spectral data can be received by the remote server normally as well. The system provides a kind of technical support and theoretical basis for crop growth estimation in greenhouse.

Keywords: spectral analysis, WSN, crop growth, greenhouse, ZigBee.

1 Introduction

Evaluating crop growth status is one of important steps for the plant monitoring and predicting yield [1]. Nowadays, the chlorophyll and nitrogen content are generally selected as nutrient indicators to monitor crop growth status. At present, the most

© IFIP International Federation for Information Processing 2015
D. Li and Y. Chen (Eds.): CCTA 2014, IFIP AICT 452, pp. 627–637, 2015.
DOI: 10.1007/978-3-319-19620-6_70

common way to get the chlorophyll and nitrogen content is by chemical analysis method in a laboratory, which is expensive, complicated and time-consuming [2-3]. Previous research revealed that nitrogen had a great effect on the chlorophyll content of the crop leaves, and could further cause a change of the spectral reflectance of the crop canopy [4]. And it was the theoretical basis for obtaining biochemical parameters of crops by spectral methods, which made it possible to use spectroscopy to estimate nitrogen content.

In recent years, researches were carried out on variation of crop canopy reflectance spectra and nitrogen sensitive band selection [5]. Simultaneously, a variety of portable crop canopy detectors based on vegetation index had been developed. The Green Seeker, developed by Marvin et al [6], collecting the crop canopy reflectance in the red and near infrared, by which, they can calculate the NDVI (Normalized Difference Vegetation Index) to analyze crop growth. Ni et al [7] (2013) developed a portable crop growth estimator by detecting the crop canopy reflectance at the wavelength of 710nm and 820nm, then calculating vegetation index. Zhang et al [8] (2006) developed a handheld spectral instrument to estimate the growth status of the crop in a greenhouse using optical fibers. Li et al [9-10] (2009, 2012) developed two generations of crop growth detector, working at red and near infrared band, and based on this, they developed a 4-waveband crop growth detection sensor. Sui et al [11] (2005) developed a device for detecting nitrogen status in cotton plants by measuring the spectral reflectance of the cotton canopy at four wavebands (blue, green, red, and NIR). However, these instruments mentioned above cannot achieve the real-time detection of crop growth.

Therefore, in this article, a system for crop growth estimation in greenhouse was developed, which consisted of the intelligent sensor nodes, intelligent gateway and the software in remote server. The system can detect the crop canopy spectral data in greenhouse at 4-waveband, and then calculate the NDVI, by which, the growth of the plants can be detected effectively. The system provides a kind of technical support and theoretical basis for crop growth estimation in real time.

2 Development for Crop Growth Estimation System

2.1 Structure of Crop Growth Estimation System

The system was made up of intelligent sensor nodes, intelligent gateway and software in remote server. It was shown in Fig.1. Several intelligent sensor nodes connected with the intelligent gateway based on ZigBee wireless sensor network (WSN). On one hand, as the coordinator of the WSN, the intelligent gateway set up WSN, waiting for sensor nodes to join in, and receiving spectral data detected by the sensor nodes, on the other hand, the gateway connected with the remote server, transmitting spectral data to the remote server by GPRS. Meanwhile, the software in the remote server analyzed and processed the spectral data, and then calculated NDVI. The intelligent sensor nodes consisted of an optical part and a circuit part. The optical part contained four optical channels at the wavelengths of 550, 650, 766 and 850 nm respectively. Since the system used sunlight as light source, the sunlight intensity should be measured as well besides measuring the crop canopy reflectance spectra.

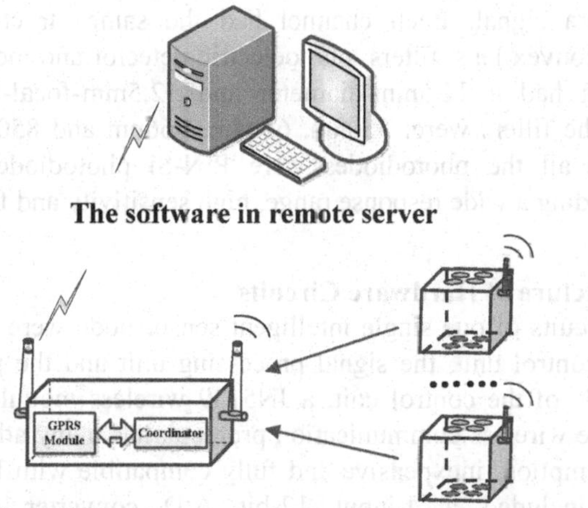

Fig. 1. Structure of the system of crop growth estimation

2.2 Development of the Intelligent Sensor Nodes

The intelligent sensor nodes were designed to collect optical signal, realize photoelectric conversion, process and amplify electrical signal, send data etc. As an example shown in Fig.1, Fig.2 illustrated the block diagram of an intelligent sensor node. The whole sensor node mainly included an optical system, hardware circuits and software design.

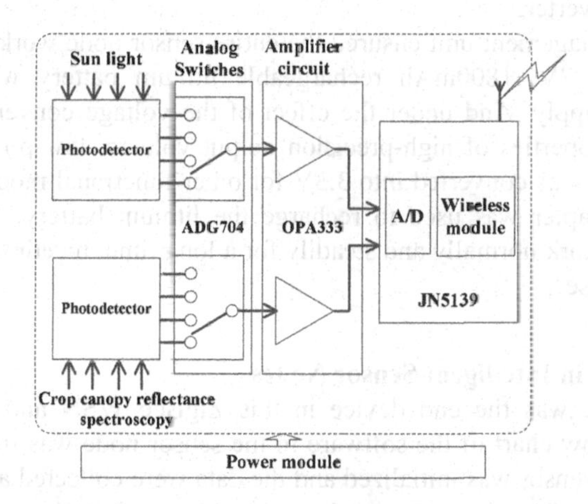

Fig. 2. Block diagram of intelligent sensor node

2.2.1 Development of Optical System

The whole optical system was designed with two parts: four optical channels to collect sunlight signal, and the other four optical channels to collect the crop canopy reflectance spectra signal. Each channel had the same structure, which mainly consisted of the convex lens, filters, photoelectric detector and mechanical enclosure. The convex lens had a 12.5mm-diameter and 12.5mm-focal-length. The center wavelengths of the filters were: 550nm, 650nm, 766nm and 850nm, the bandwidth was 20nm. And all the photodiodes were PIN-Si photodiode which has many advantages, including a wide response range, high sensitivity and fast response.

2.2.2 The Structure of Hardware Circuits

The hardware circuits of one single intelligent sensor node were mainly made up of three parts: the control unit, the signal processing unit and the power management unit. As the MCU of the control unit, a JN5139 wireless module (Jennic Co. UK), embedded ZigBee wireless communication protocol, has many advantages, including low power consumption, inexpensive and fully compatible with IEEE802.15.4. This microcontroller included a 4-input 12-bit A/D converter unit, meeting the requirements of the signal acquisition. And JN5139 Integrated 16MHz 32bit RISC MCU core, high-performance 2.4GHz IEEE 802.15.4 transceiver, 192KB ROM and 96kB RAM. Thus, the MCU provided a versatile low cost solution for wireless sensor network applications.

The signal processing unit mainly processed the photoelectric signal detected by the photoelectric detector. Firstly, a 4:1 time sharing analog multiplex chip ADG704 was applied to select four optical signals. Secondly, an OPA333 amplifier, which had the properties of high-precision, low quiescent current and low power consumption, was chosen to amplify the optical signal selected by ADG704. The output of the OPA333 was sent to the MCU, where the analog signal can be converted to digital signal in A/D converter.

The power management unit ensured the entire sensor node work normally. In this development, a 3.7V, 1800mAh rechargeable lithium battery was chosen as an external power supply. And under the effect of the voltage converter chip SP6201, which had the properties of high-precision output voltage, low-power-consumption. The input voltage was converted into 3.3V for other functional modules. Meanwhile, a 4.2V power adapter was used to recharge the lithium battery. So the intelligent sensor node can work normally and steadily for a long time, meeting the requirements using in greenhouse.

2.2.3 Software in Intelligent Sensor Nodes

Each sensor node was the end device in this ZigBee WSN and shared the same workflow. The flow chart of the software in the sensor node was illustrated in Fig.3. Once started, the sensor was initialized and the data were collected automatically with a certain sampling frequency. By setting the address of analog switch, the sensor selected the appropriate channel and collected data. Data acquisition of each channel was repeated for 10 times and averaged. When data collections of all the channels were completed, the data were sent to the coordinator via wireless module embedded in JN5139. Once finished, another five minutes was started for the next round of measurement and acquisition.

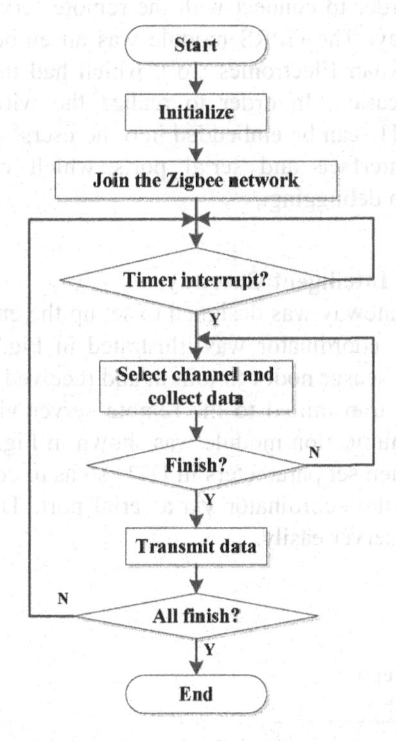

Fig. 3. Flow chart of software in sensor node

2.3 Development of Intelligent Gateway

The intelligent gateway was mainly consisted of two parts: the coordinator and the GPRS module. The coordinator was designed to set up wireless sensor network, waiting for intelligent sensor nodes to join in, receiving optical data, and then transmitting data to the remote server via the GPRS module. The structure of the intelligent gateway is illustrated in Fig. 4.

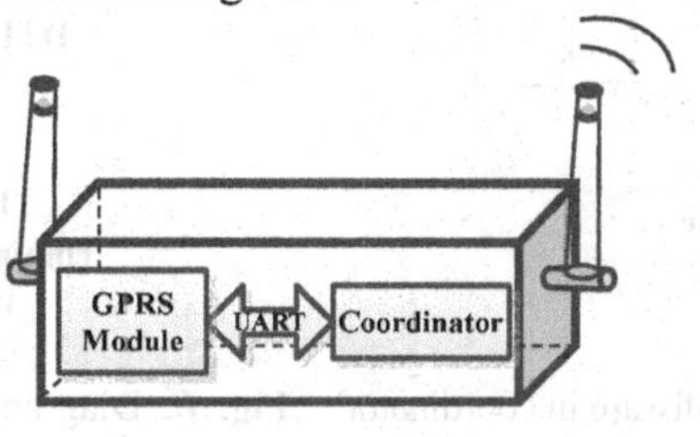

Fig. 4. Diagram of intelligent gateway

2.3.1 Hardware Circuit in the Intelligent Gateway

On one hand, as the coordinator of the WSN, the gateway connected with the sensor nodes, therefore, a wireless module JN5139 was required as the MCU in the gateway.

On the other hand, in order to connect with the remote server, a GPRS module was embedded in the gateway. The GPRS module was an embedded DTU ZWG-28DP GPRS (Guangzhou Zhiyuan Electronics Co.), which had the properties of compact size and flexible application. In order to realize the wireless communication in different devices, the DTU can be embedded into the users' devices easily. The DTU had a configuration interface and serial ports which can be easily used for configuration and system debugging.

2.3.2 Software in the Intelligent Gateway

The coordinator in the gateway was designed to set up the entire WSN. And the flow chart of the software in coordinator was illustrated in Fig.5. Once initializing, the coordinator searched for sensor nodes to join in and received the data sent from them. The received data were transmitted to the remote server via a GPRS module. The diagram of GPRS communication module was shown in Fig.6. A static IP was set in the remote server, and then set parameters in DTU so as to connect the remote server. The DTU connected to the coordinator via a serial port. Therefore, the coordinator connected to the remote server easily.

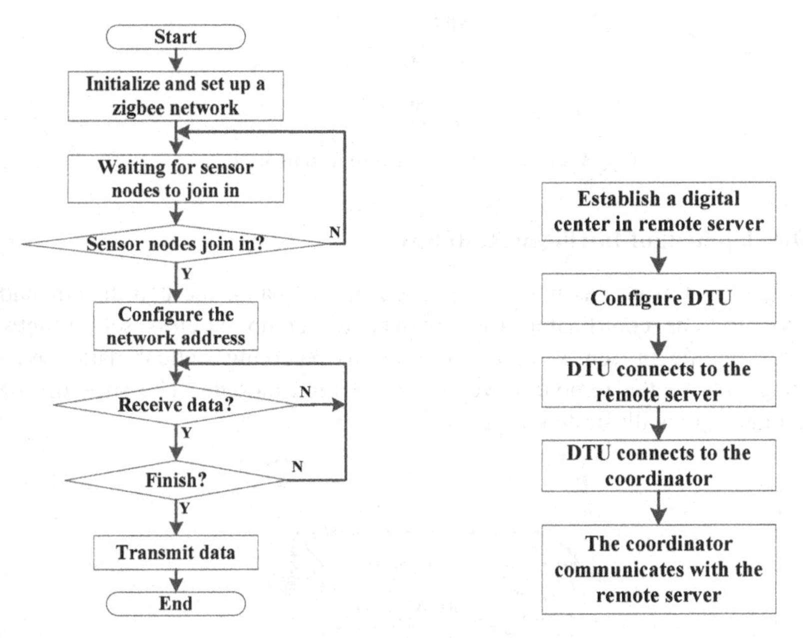

Fig. 5. Flow chart of the software in coordinator

Fig. 6. Diagram of GPRS communication module

2.4 Software in Remote Server

According to the principle of the GPRS communication module shown in Fig. 6, parameters of remote server and the DTU in the gateway were set. A vegetation index such as NDVI could be immediately calculated in the software once the data was transmitted to the remote server. Fig. 7 showed the interface of the software, it mainly included a log-in page and a data acquisition system. And the flow chart of the software in remote server was illustrated in Fig. 8.

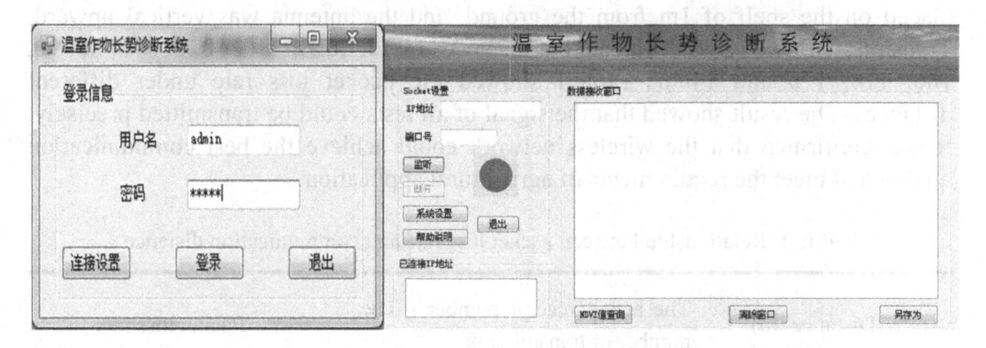

Fig. 7. Interface of the software in remote server

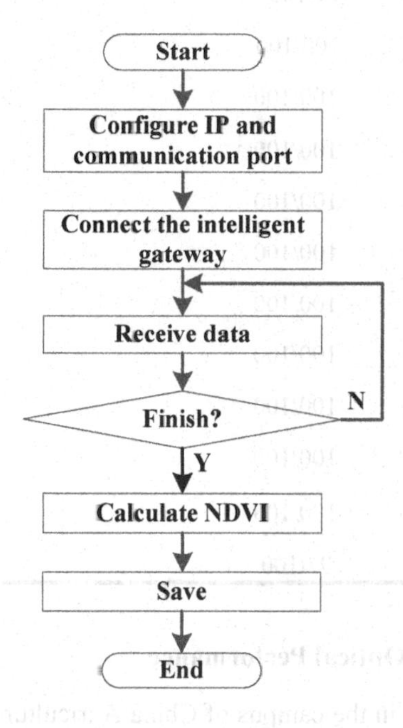

Fig. 8. Flow chart of the software in remote server

3 Experiments and Analysis

3.1 Test of the Wireless Performance

This experiment analyzed wireless performance by testing LQI (link quality indicator) in the different communication distance. The wireless performance was tested at an experimental field located in Changping District, Beijing. There were no obstacles between sensors and gateway in the open field. Sensor nodes and gateway were placed on the shelf of 1m from the ground, and the antenna was vertical upward. Transmission quality was evaluated at distances of 10, 20, 30, 40, 50, 60, 70, 80, 90, 100, 110, 120 and 140m. Table.1 showed the packet loss rate under different distances. The result showed that the signal of all tests could be transmitted precisely. It was confirmed that the wireless network could achieve the best communication quality and meet the requirements of agricultural application.

Table 1. Relationship between packet loss rate and communication distance

Distance (m)	The actual receipt number / The number of transmission	Packet loss rate
10	100/100	0
20	99/100	0.01
30	100/100	0
40	100/100	0
50	100/100	0
60	100/100	0
70	100/100	0
80	100/100	0
90	100/100	0
100	100/100	0
110	100/100	0
120	100/100	0
140	97/100	0.03

3.2 Calibration of the Optical Performance

Calibration was carried out in the campus of China Agricultural University. The sensor nodes were used to measure the incident sunlight and the reflected light of a standard white panel while an illuminometer was used to measure the sunlight. This panel was made of polytetrafluoroethylene (Anhui Institute of Optics and Fine Mechanics, China) and assumed to have 100% of relative reflectivity. Both illuminometer and standard white panel were set in a horizontal plane. The tests were carried out every 10 minutes from 9:00 am to 2:00 pm. Data output from illuminometer and from the signal of each optical channel were compared respectively. The result was shown in Table.2. The minimum R^2 between illuminometer and each optical channel of the sensors was 0.919. It was showed that the developed sensor was sensitive to measure the sunlight (Upward) and reflected light from objects (Downward).

Table 2. Correlation coefficients between each channel of the detector and the illuminometer

	550nm	650nm	766nm	850nm
Incident channels	0.936	0.952	0.956	0.919
Reflecting channels	0.975	0.940	0.946	0.962

3.3 Experiment in Greenhouse

In order to test the effectiveness and stability of the system, the strawberry planted in greenhouse was detected. Several experiments were conducted in the greenhouse of ZhuoZhou Experiment Station in Hebei. Two sensor nodes were placed in the greenhouse to acquire the canopy spectral data of strawberries, an intelligent gateway was placed in the greenhouse as well to connect the sensor nodes and the remote server, the intelligent gateway was placed in the laboratory in China Agricultural University to receive the spectral data transmitting from the greenhouse.

According to the different combination of visible light and near infrared, software in the remote server calculated the Normalized Difference Vegetation Index (NDVI) as a detection parameter of strawberry nutrition content. The calculation formula of NDVI [12] was as follows:

$$NDVI = \frac{R_{\mathrm{nir}} - R_{\mathrm{r}}}{R_{\mathrm{nir}} + R_{\mathrm{r}}} \quad (1)$$

In the formula, R_{nir} means near-infrared spectral reflectance, and R_{r} means visible light reflectance.

The four optical channel of each sensor node can measure the waveband of 550nm, 650nm, 766nm and 850nm. NDVI can be calculated according to the different combinations of different wavebands. Results showed that from November 2013 to May 2014, during the growth stages of strawberry, the system was able to collect canopy spectral data of strawberries stably, and based on the combination of different wavebands, NDVI were calculated easily, which was used to judge the growth of strawberries. Therefore, the system provides a kind of technical support and theoretical basis for crop estimation in greenhouse.

4 Conclusions

Based on spectral analysis technology, electronic technology, and WSN technology, a system was developed for the crop growth estimation in greenhouse. The system was mainly made up of several intelligent sensor nodes, an intelligent gateway and software in remote server. After performance test, calibration and greenhouse experiments, the following conclusions were obtained.

(1) The sensor nodes were compact and small sized. Transmission quality of the sensor nodes was evaluated at different distances and the signal could be transmitted precisely with low packet loss rate in the tests. It was confirmed that the wireless network could achieve the best communication quality and meet the requirements of agricultural application.
(2) Calibration experiments showed that the accuracy of the optical components was high enough for application. The measured values between the monitor and illuminometer had a good correlation, and the minimum R2 between illuminometer and each optical channel of the sensors was 0.919.
(3) The result of the greenhouse experiment showed that the system was able to collect canopy spectral data of strawberries stably during the entire growth stages of strawberry, and based on the combination of different wavebands, NDVI were calculated easily to reflect the growth of strawberry. Therefore, the system provides a kind of technical support and theoretical basis for crop estimation in greenhouse.

Acknowledgment. This research was supported by NSFC Program (31271619), the Doctoral Program of Higher Education of China (Grant No.20110008130006) and Chinese National Science and Technology Support Program (2012BAH29B04).

References

1. Mistele, B., Schmidhalter, U.: Estimating the nitrogen nutrition index using spectral canopy reflectance measurements. European Journal of Agronomy 29(4), 184–190 (2008)
2. Guo, J., Zhao, C., Wang, X.: Research advancement and status on crop nitrogen nutrition diagnosis. Soils and Fertilizers Sciences in China (4), 10–14 (2008)
3. Li, M., Han, D., Wang, X.: Spectral Analyzing technique and Applications, pp. 177–194. Science Press, Beijing (2006)
4. Filella, I., Serrano, L., Serra, J., et al.: Evaluating wheat nitrogen status with canopy reflectance indices and discriminant analysis. Crop Science 35(5), 1400–1405 (1995)
5. Xue, L., Cao, W., Luo, W., et al.: Monitoring leaf nitrogen status in rice with canopy spectral reflectance. Agronomy Journal 96(1), 135–142 (2004)
6. Stone, M.L., Needham, D., Solie, J.B., et al.: Optical spectral reflectance sensor and controller: U.S. Patent 6,596,996[P] (July 22, 2003)
7. Ni, J., Yao, X., Tian, Y., et al.: Design and experiments of portable apparatus for plant growth monitoring and diagnosis. Transactions of the Chinese Society of Agricultural Engineering 29(6), 150–156 (2013)

8. Zhang, X., Li, M., Cui, D., et al.: New method and instrument to diagnose crop growth status in greenhouse based on spectroscopy. Spectroscopy and Spectral Analysis 26(5), 887–890 (2006)
9. Li, X., Li, M., Cui, D.: Non-destructive crop canopy analyzer based on spectral principle. Transactions of the Chinese Society for Agriculture Machinery 40(suppl.), 252–255 (2009)
10. Li, X., Zhang, F., Li, M., et al.: Design of a Four-waveband Crop Canopy Analyzer. Transactions of the Chinese Society for Agricultural Machinery 42(11), 169–173 (2012)
11. Sui, R., Wilkerson, J., Hart, W., Wilhelm, L., Howard, D.: Multi–spectral sensor for detection of nitrogen status in cotton. Applied Engineering in Agriculture 21(2), 167–172 (2005)
12. Yang, S.: A measuring technology for normalized difference vegetation index. Journal of Basic Science and Engineering 12(3), 328–332 (2004)

Evaluation Model of Winter Wheat Yield Based on Soil Properties

Wei Yang, Minzan Li, Lihua Zheng, and Hong Sun

Key Laboratory of Modern Precision Agriculture System Integration Research, Ministry of Education, China Agricultural University, Beijing, China
{cauyw,gpac,zhenglh,sunhong}@cau.edu.cn

Abstract. In order to realize precision management of winter wheat, two prediction models of winter wheat yield based on soil parameters were proposed and compared. The field tests were carried out in two years. The variety of the experimental winter wheat was Jingdong 12, and the test area was divided into 60 zones with 5m×5m grids. The sampling point was put in the center of the zone, and the depth of the sampling point was 5cm. Soil EC was measured by a DDB-307 EC meter, and the winter wheat yield data were provided by a CASE2366 grain harvester with GPS receiver. Gray theory were used to analyze the gray relation between soil EC value and each of other soil parameters, such as total nitrogen content, K^+、NO_3^- and pH of soil. Results showed that there were high gray relation between soil EC and three other indexes, total nitrogen content, K^+, pH of soil. Since soil organic horizons had high correlation with soil negative charge capacity, when soil had more organic horizons, there were more soil negative ions, and the soil EC was higher. Hence, the gray relation between K^+ and EC was high. Using nitrogen fertilizer could removal caution from soil, and increase the content of K^+, Na^+, Ca^{2+} and Mg^{2+}, so that there was also high correlation between total nitrogen content and EC. The reason of high correlation between EC and soil pH was attributed to that the change of pH had influence on negative charge. After analyzing the correlation between winter wheat yield and soil EC, total nitrogen content, K^+, NO_3^-, pH of soil in different growth period respectively, two prediction algorithms of yield were proposed, Least Square-Support Vector Machine (LSSVM) and Fuzzy Least Square-Support Vector Machine (FLSSVM). LSSVM prediction model took soil EC, total nitrogen content and K^+ as the input factors and winter wheat yield as output. While FLSSVM prediction model took soil EC, nitrogen content, K^+ and gray relation as input factors and also winter wheat yield as output. Results showed that the prediction and validation R^2 of LSSVM model were 0.772 and 0.685 respectively. Prediction R^2 of FLSSVM was 0.8625, and validation R^2 of FLSSVM model was 0.8003. FLSSVM used Fuzz Similar Extent to fuzz input samples so that it could avoid over-training. Also because it was based on membership function, it had several advantages such as simple structure, efficient convergence, precise forecasting, and etc. FLSSVM had high accuracy prediction result and could be used in estimating yield and providing theory and technical support for precision management of crops.

Keywords: near infrared spectroscopy, waste cooking oil, support vector machine, parameters optimization.

© IFIP International Federation for Information Processing 2015
D. Li and Y. Chen (Eds.): CCTA 2014, IFIP AICT 452, pp. 638–645, 2015.
DOI: 10.1007/978-3-319-19620-6_71

1 Introduction

Wheat is one of China's three major food crops, and most of wheat is winter wheat. Water and fertilizer management is very necessary to winter wheat. Soil electrical conductivity (EC) is a basic index to reflect soil electrochemical properties and fertility characteristics. Soil fertility, as the basic property of soil and the foundation of land productivity, is defined as the soil ability to supply the essential plant nutrients, as well as related to the nutrient supply capacity of various soil properties and state by Soil Science Society of USA. In recent years, combined with soil physical, chemical, biological and environmental conditions, some researches on comprehensive evaluation of soil fertility showed that, soil EC was closely related with the soil properties and could be used as a quantitative index of comprehensive evaluation of soil fertility. Since the measurement of soil EC, can effectively control the salt concentration, soil water status and other properties, and also can timely diagnose agricultural production problems, the soil EC in guiding agricultural production, precision cultivation, has a special status and important role [1-3]. Zhao Yong stated that the relationship between soil EC and yield of winter wheat was linear correlation, so that we can use soil EC value of winter wheat in heading period as evaluation index of winter wheat yield [4]. Min thought that high EC value could improve tomato's lycopene, glucose, fructose and soluble solids content [5]. Jiang Dongyan studied the effects of irrigation amount under field condition and nitrogen fertilizer on wheat yield and soil nitrate nitrogen content. The results showed that increasing the amount of nitrogen fertilizer, the grain yield, protein content and protein yield were significantly increased [6]. However, the above research mostly only analyzed the influence of some factors on the yield, the variety of the soil parameters were not considered. In this paper, winter wheat was chosen as the object of study, and the influences of soil parameters on the yield of winter wheat were analyzed. A forecasting model of winter wheat yield was established to realize the wheat yield prediction by monitoring the soil parameters in the growth process, and the theory and technology were provided to support for the realization of fine management in the field of winter wheat.

2 Materials and Methods

2.1 Experimental Material

The test was conducted in two years in National Precision Agriculture Demonstration Base. The experimental area was 30m×60m and divided into 60 cells, each of which was a 5m×5m zone. Sampling point was located in the intersection position of the cell and sampling depth was 5cm.

2.2 Experimental Methods

A soil EC meter DDB-307 (Shanghai Precision Scientific Instrument Co. LTD, Shanghai, China) was used to measure standard soil EC value. Each sample was cleaned by deionizer water before each measurement, so as not to affect the measurement result. Winter wheat yield information was provided by a combine harvester CASE2366 (CASE IH, USA) with global positioning system (GPS). When the GPS receiver provided accurate position information, the whole plot of yield can be obtained based on the flow data of CASE2366. And then the relationship between soil EC and yield of wheat were analyzed [7].

3 Results and Discussion

3.1 Analysis of Influence Factors of Soil EC

As the amount of charge in soil can affect soil EC, the influence of nitrogen content, K^+, NO_3^- and pH value of soil was analyzed. Gary relation analysis method was used, which was used to compare the geometric relationship of comparative sequence data. If two comparative sequences in each moment overlapped together, the association degree is 1. Otherwise, the association degree is smaller than 1. Specific methods are introduced as follows.

X_0 is reference sequence $X_0=\{x_0(k)|k=1,2,...,n\}$, X_i is comparative sequence $X_i=\{x_i(k)|k=1,2,...,n\}$, where n denotes the number of data of reference sequence and comparative sequence. Then correlation coefficient is calculated as:

$$\zeta_{i,0}(k)=\frac{\min\limits_i \min\limits_k |x_0(k)-x_i(k)|+\rho \max\limits_i \max\limits_k |x_0(k)-x_i(k)|}{|x_0(k)-x_i(k)|+\rho \max\limits_i \max\limits_k |x_0(k)-x_i(k)|} \tag{1}$$

ρ is called resolution coefficient, is a pre-fixed constants between zero to one, generally taking as 0.5. $\zeta_{i,0}(k)$ represents the relative difference of comparative sequence X_i and the reference sequence X_0 at the moment k, and reflects similarity between the different sequence and the similarity of the reference sequence in the same point. Because the correlation coefficient is comparative sequence and reference sequence at each time point (i.e. curve points) degree value, the values of the efficient are bigger than 1, and the information is too scattered to facilitate an overall comparison. This study using average correlation coefficient of each times represents the correlation relationship between comparative sequence and the reference sequence correlation formula is as follows:

$$r(x_i,x_0)=\frac{1}{n}\sum_{i=1}^{n}\zeta_{i,0}(k) \quad ; \quad r(x_i,x_0)\in(0,1] \tag{2}$$

According to formula (2), the gray relation between soil EC and one of the other soil parameters, nitrogen content, K^+, NO_3^- , and pH, were calculated, and all results are shown in Table1[8-10]. Form Table1 we can see that the gray relation results were K^+>nitrogen content >pH>water content >weight > NO_3^-. Soil organic matter and soil cation exchange capacity (CEC) had a high significant positive correlation, which meant that the soil had more organic matter and more soil cation. Anion adsorption on soil colloids is more, and the greater the soil EC is. The grey correlation degree between K^+ and soil EC was high. Application of nitrogen fertilizer can promote the dissociation of cation adsorption on soil colloids, increase K^+, Na^+, Ca^{2+}, Mg^{2+} content, so that the grey correlation degree of nitrogen content in soil and soil EC was relatively high[11-14]. Soil EC and pH had high correlation; it might be because the pH change affected the amount of charge soil colloid and soil EC. The gray relation between winter wheat yield and one of the soil parameters, soil EC , nitrogen content, K^+, NO_3^-, pH, were calculated. Form Table1 we can see that the gray relation results were soil EC>nitrogen content >pH>soil water content > NO_3> K^+>soil weight.

Table 1. Gray relation between wheat yield and all indexes

Name	Correlation with EC	Gray relation with yield
Weight of soil	0.6891	0.2376
Water content of soil	0.7352	0.4783
Nitrogen content	0.8215	0.6421
K^+	0.8358	0.4206
NO_3^-	0.6548	0.4529
pH	0.7361	0.5632
Soil EC	1.0000	0.5839

In order to better understand the relationship between soil EC and nitrogen content, K^+, NO_3^- , pH and yield, the coefficients of correlation were calculated and results are shown in Table 2.

Table 2. Correlation between wheat yield and all indexes in different growth time

Name	Correlation coefficient (heading stage)	Correlation coefficient (filling stage)	Correlation coefficient (milky stage)
Soil EC	-0.3887	-0.5385	-0.6274
Nitrogen content	0.5360	0.5842	0.5925
K^+	0.4863	0.4247	0.4014
NO_3^-	-0.2992	-0.2674	-0.2258
pH	-0.1462	-0.0979	-0.1036

From Table 2 we can find that soil EC, nitrogen content and K^+ had notable correlation with yield in the heading stage, grain filling stage and milk stage. NO_3^- and pH had low correlation with yield; it was because although the ammonium nitrogen and nitrate had the same value as nitrogen source of plants, if the two kinds of nitrogen source could be chosen, the relative absorption rate of different plants still had obvious difference. Generally, tobacco, cotton crop response to nitrogen is better, while wheat is more sensitive to ammonium nitrogen. The correlation analysis between soil EC and yield showed there was a linear correlation between them. Since there were more soil nutrients in the early stage of growth of winter wheat, soil contribution to the winter wheat was larger, and there were more obvious changes in the value of soil EC. Since winter wheat photosynthesis was strong in the late stage of the growth, winter wheat got less nutrition from soil, and the soil EC was stable, and yield significantly associated with soil EC.

3.2 Prediction Model of Winter Wheat Yield

After analyzing the correlation between soil EC, nitrogen content, K^+ and yield, soil EC, nitrogen content and K^+ were chosen as input, and yield was used as output, LSSVM and FLSSVM were used to build winter wheat yield prediction model.

(1) LS-SVM

The core of LS-SVM (least square-support vector machine) is using the kernel function to change the nonlinear problem into a linear problem, and get better data classification or regression results in latent feature space operations. At present, the common kernel function of SVM is linear kernel, polynomial kernel function (ploy), radial basis function (RBF) kernel function and sigmoid kernel function. Among them, the RBF kernel function with fewer parameters and high regression accuracy is adopted in this study. For RBF kernel function, the optimal parameters include the penalty parameter "C", "γ" parameters and loss parameters "ε". "C" is the regularization parameter, and is used to handle the complexity and the training error of the model and to obtain good generalization; Parameter "γ" reflects the data distribution or range property, which determines the width of the local neighborhood. Because "C" and "γ" have larger effect on the model, the two parameters are the key point to establish the high precision model based on SVM. As shown in Figure 2, using SVM model to predict yield, and predict R^2 is 0.772, verification R^2 is 0.685.

(2) Wheat yield forecast based on fuzzy least squares support vector machines

Least squares support vector machine (LSSVM) is sensitive to outliers, and it will bring over-fitting problems. Thus, this paper tried to use the concept of fuzzy membership LSSVM. A fuzzy least squares support vector machine support vector domain description (FLSSVM) was proposed in the paper. Data samples were mapped into a high dimensional space and the smallest enclosing ball was built, and then the membership values based on the sample from the heart to the ball were determined, while the isolated points outside the centralized ball was given a small positive number to reduce the impact. In this way it can improve noise immunity and fitting results of FLSSVM. FLSSVM used fuzzy membership as fuzzy input samples, and gave samples different degree of membership values based on their importance, and then trained samples. In this paper, input vectors of sample were soil EC, nitrogen content and K^+. Furthermore, introducing the concept of fuzzy membership degree of the input vector, which was called gray correlation weights, the new input vectors now were soil EC, nitrogen content, K^+ and gray correlation. The gray correlation degree represented the weight vector between the variables and yield. The gray correlation between soil EC, nitrogen content, K^+ and yield are shown in Table 2. Different samples corresponding gray correlation as the membership value would make the forecast result more accurate. Specific algorithm of FLSSVM is shown below:

The LSSVM optimization problem can be expressed as:

$$y_i = \omega\phi(x_i) + b + \xi_i$$
$$Q = \min\left\{\frac{1}{2}\|\omega\|^2 + \frac{c}{2}\sum_{i=1}^{n}\xi_i^2\right\} \tag{3}$$

ω is weight vector, b is offset, ξ_i is error term, c is penalty parameter. The index ψ_i was used to distinguish the importance of different data in the training process in FLSSVM, which was called fuzzy membership and the formula above can be changed as follow:

$$Q = \min\left\{\frac{1}{2}\|\omega\|^2 + \frac{c}{2}\sum_{i=1}^{n}\psi_i\xi_i^2\right\} \tag{4}$$

The steps of FLSSVM are shown below :

(a) Normalizing the data; (b) Forming the training samples and the testing samples; (c) Establishing objective function of Q with the training sample; (d) Solving the objective function; (e) Building prediction model for wheat yield using samples. The prediction results are shown in Figure 2, R_v^2 and R_c^2 are 0.8625 and 0.8003. The result shows that, soil EC, nitrogen content and K^+ can be used to predict winter wheat yield. Based on the three soil parameters, increasing the degree of grey incidence as a weight vector to be fourth input vector and then building the FLSSVM winter wheat yield prediction model has a high forecasting precision.

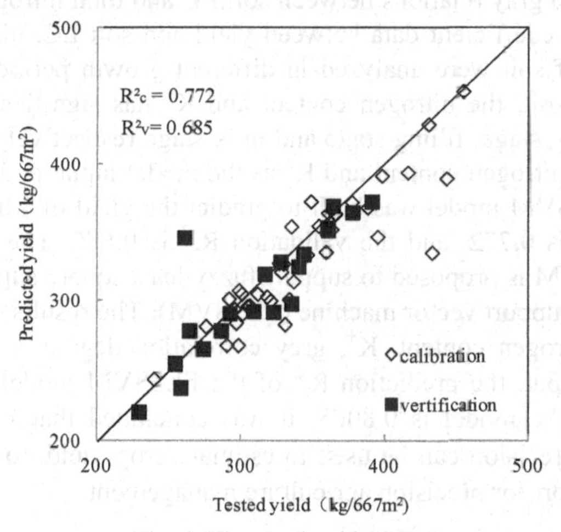

Fig. 1. The result of LSSVM

FLSSVM used fuzzy membership to fuzzy input samples, and gave different samples with different degrees of membership values, and then trained model. This method can solve the problem of small sample learning, and avoid the defects of over learning. AS using the membership data for the weight, the method has high precision and is easy to realize. It is obtained that using FLSSVM to predict winter wheat yield has higher prediction accuracy and is practical.

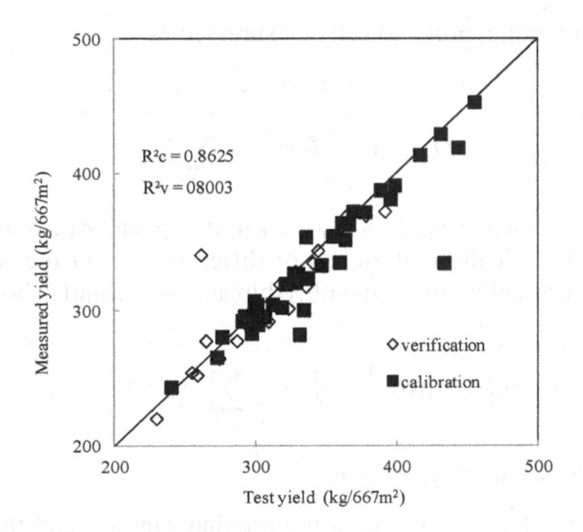

Fig. 2. The result of FLSSVM

4 Conclusions

Gray theory was used to analyze the gray relation between soil EC value and each of other soil parameters, total nitrogen content, K^+, NO_3^- and pH of soil. Results showed that there were high gray relations between soil EC and total nitrogen content, K^+, pH of soil Correlation coefficient data between yield and soil EC, nitrogen content, K^+, NO_3^- , pH value of soil were analyzed in different growth periods of winter wheat. The EC value of soil, the nitrogen content and K^+ has significant correlation with yield in the heading stage, filling stage and milk stage respectively. Therefore, taking the soil EC value, nitrogen content and K^+ as the model input, and yield as the output of the model, LS-SVM model was built to predict the yield of wheat. The prediction R_c^2 of the model is 0.772, and the validation R_v^2 is 0.685. The fuzzy membership concept into LSSVM is proposed to support fuzzy least square support vector domain description of the support vector machine (FLSSVM). The results show that, we using soil EC value, nitrogen content, K^+, grey correlation degree as the input, and the wheat yield as output, the prediction R_c^2 of the FLSSVM model is 0.8625, and the validation R_v^2 of the model is 0.8003. It was concluded that FLSSVM prediction model with high precision can be used to estimate crop yield, to provide theoretical and technical support for precision agriculture management.

Acknowledgment. This research was supported by National Science, Technology Support Plan (2013AA102303), Chinese Universities Scientific Fund (Grant No. 2014JD140) and Chinese National Science and Technology Support Program (2012BAH29B04).

References

1. Liu Guangming, Yang Jinsong, Ju Maosen, et al. Technology of chronometry using electromagnetic induction and its application in agriculture. Soil, 2003, (3): 27-29.
2. Wang Qi, Li Minzan, Wang Maohua. Development of a portable detector for soil electrical conductivity. Journal of China Agriculture University, 2003, 8(4): 20-23.
3. Naiqian Zhang, Kyeong-Hwan Lee, et al. Simultaneous Measurement of Soil Water content and Method Salinity Using a Frequency-Response, 2004 CIGR International Conference, 11~14 October 2004, Beijing, China.
4. Zhao Yong, Li Minzan, Zhang Junning. Correlation between soil electrical conductivity and winter yield [J]. Transactions of the CSAE, 2009, 25(Sl): 34~37. (In Chinese with English abstract)
5. Min Wu, Chieri K. Effects of high electrical conductivity of nutrient solution and its application timing on lycopene, chlorophyll and sugar concentrations of hydroponic tomatoes during ripening[J]. Scientia Horticulturae, 2008, 116(2): 122-129.
6. Lin Yicheng, Ding Nengfei, Fu Qinglin, et al. The measurement of electric conductivity in soil solution and analysis of its correlative factors[J]. Acta Agriculturae Zhejiangensis, 2005, 17(2): 83-86.
7. Shi Yongchen, Sui Jidong, He Chuanqin. Fast determination of soil electrical conductivity [J]. Journal of Heilongjiang August First Land Reclamation University, 2000, 12(4): 15-18.
8. Xie Naiming, Li Sifeng. Discrete GM(1,1) and mechanism of grey forecasting model [J]. System engineering-Theory&practice, 2005, 25 (1): 93-99.
9. Chen Zhujun, Wang Yijun, Xu Anmin, et al. Effects of the application of different nitrogen fertilizers on the ion compositions in solution of the greenhouse soil. Plant nutrition and fertilizer science, 2008, 14(5): 907-913.
10. Ritter C, Dicke D, Weis M, et al. An on-farm approach to quantify yield variation and to derive decision rules for site-specific weed management [J]. Precision Agric., 2008, 9(3): 133-146.
11. Nadler A. Effect of soil structure on bulk soil electrical conductivity (Eca) using the TDR and 4P techniques [J]. Soil Sci., 1991, 152: 199-203.
12. Kitchen N R, Sudduth K A, Drummond S T. Soil electrical conductivity as a crop productivity measure for clay pan soils [J]. Prod Agric., 1999, 12(4): 607-617.
13. Rhoades J. D., Corwin D. L. Determining Soil Electrical Conductivity-depth Relations Using an Inductive Electromagnetic Conductivity Meter. Soil Sci. Soc. Am., 1992, 45: 255-260.
14. McBride R. A., Gordon A. M., and Shrive S. C. Estimating Forest Soil Quality from Terrain Measurements of Apparent Electrical Conductivity. Soil Science Society America Journal, 1990, 54: 290-293.

Design of a Measurement and Control System for Delinting Machine

Liming Zhou, Yanwei Yuan, Junning Zhang, and Xin Dong

Chinese Academy of Agricultural Mechanization Sciences, Beijing 100083, China
{haibo1129,yyw215,zjn990210}@163.com, 191942384@qq.com

Abstract. To improve the working efficiency of cottonseed delinting machine, an automated measurement and control system for the machine is designed. The system hardware is comprised of three torque sensor, master controller, data acquisition and regulation module, displacement transducer, linear actuator and variable frequency device. The operating parameters such as the rotation torque and speed of every delinting cylinder and outlet opening are obtained on line by the apparatus and then sent to the master controller. With the help of LabWindow/CVI, the master controller program is developed to get and deal with the information. Since the feed rate and discharge rate directly affect the working load of the cylinders, the speed of feeding motor and outlet opening can be manually regulated by means of the program to facilitate the working load test experiment. In order to achieve the optimization load allocation of the delinting cylinder, the optimal decision is made on the basis of the third cylinder's working load to regulate the outlet opening. The experiment results show that the outlet opening regulation accuracy reaches ±1mm.This system is also easy to use and provide a good graphical user interface and has higher accuracy, higher efficiency for the real-time control.

Keywords: delinting machine, measurement and control system, LabWindows/CVI.

1 Introduction

Cottonseed is not only a major by-product of cotton processing, but also an important resource that we need to improve the utilization. At present, there are two important delinting methods: chemical acid delinting [1] and mechanical delinting [2-3]. The foamed sulfuric acid is often used in the acid delinting. Due to churning action, cotton fuzz is uniformly subjected to the acid reaction. After a moment, the acid treats seeds and the slurry is washed with the water. This method is a simple technology, but there are some disadvantages such as high production costs, dangerous operation, and environment pollution. Mechanical delinting relies on physical friction to remove the fuzz forcibly. Since no acid is used in the process, there are many advantages for this method. For example, the environment pollution is prevented and the production cost is reduced and the fuzz can be recycled. Steel brush organization is the main form of structural mechanical delinter. With the help of centrifugal force, the rotating cylinder

© IFIP International Federation for Information Processing 2015
D. Li and Y. Chen (Eds.): CCTA 2014, IFIP AICT 452, pp. 646–652, 2015.
DOI: 10.1007/978-3-319-19620-6_72

with opposite spinning rotary brush can remove fuzz from cottonseed easily. A number of rotating cylinders are often used at the same time to improve the efficiency of delinting machine. Due to the abrasive action of the brush, care must be taken to prevent overheating the seed and damaging the seed coat. During the mechanical delinting process, there is an obvious relationship between working load of cylinder and delinting quality. So far, many efforts have been made to improve the quality of cotton seed delinting. Wang [4] used the automatic control technology and information processing technology to the parameters such as concentration, temperature and so on during the acid delinting process. Zuo [5] developed an automatic feed rate control system for the mechanical delinter by comparing the desire value with real value of feeding motor current. Chen [6] designed a frequency automatic setting system of seed feeding mechanism to keep the feed rate stable. These improvements are based on the simple open loop control to regulate the feed rate and do not take into account the working load of cylinder. So, it is difficult to improve the delinting quality effectively.

Therefore, in order to improve the delinting efficiency and reduce the damage rate of cottonseed, the measurement and control system for third level delinting machine is developed based on virtual instrument in this paper. According to the working load of the third rotating cylinder, the system can adjust the outlet opening automatically so as to keep the working load stable. When the outlet is in good condition, the rotating cylinder will work at its best. On this condition, the delinting efficiency is high, and the damage rate of cottonseed is low. Meanwhile, for the convenience of delinting test experiments, the system also integrates the manual adjustment of feeding motor speed and outlet opening function.

2 Materials and Methods

2.1 Cottonseed Delinting Machine

The delinting machine studied in this paper includes main frame, feed hopper, three rotating cylinder and fuzz recycling unit. The cylinders are comprised of screen mesh and steel brush. The simple structure of delinter machine is shown in Fig.1. To ensure the cottonseed pass through the delinting machine smoothly, the rotation direction of first cylinder is opposite to that of the second cylinder, and is the same to that of the third cylinder. These cylinders are driven by an AC motor, which power is 45kw.The power of feeding motor is 0.75kw.

2.2 Hardware of Measurement Control System

The main function of the control system is to acquire the rotating cylinders torque and speed information to get the working status of the cylinders and carry out the torque control automatically. The system also can regulate the speed of feeding motor and outlet opening easily by means of manual mode to facilitate the experiment. The system is comprised of three torque sensor, master controller, data acquisition and regulation module, displacement transducer, linear actuator and variable frequency device. Its structure is shown in figure 2.

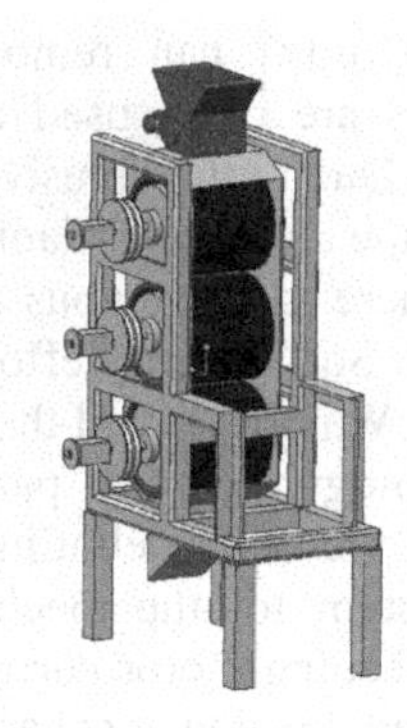

Fig. 1. Structure of cottonseed delinting machine

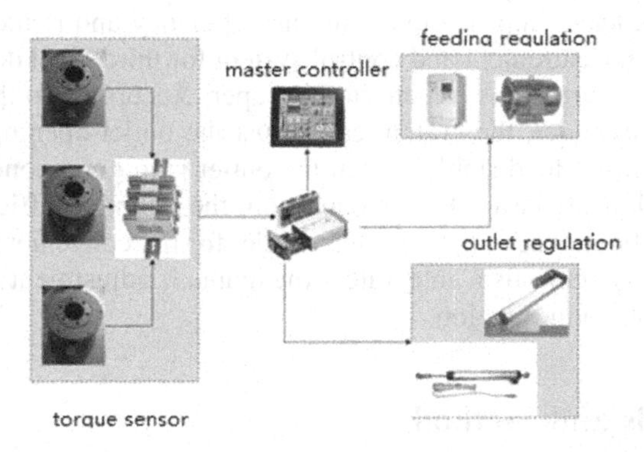

Fig. 2. Schematic of the measurement and control system for delinting machine

2.2.1 Torque Sensor

In order to obtain working load of the cylinder accurately, the belt pulley torque sensor is selected in this paper. The sensor can provide extremely accurate torque measurement and speed measurement over a broad range. Unlike the tradition torque sensor, the pulley torque sensor can measure belt or chain torque on pulley system directly [7]. It is applicable to a mechanical rig without set screw and with the rotary cylinder driven by a chain/belt case. The selected torque sensors are mounted directly on the outside of pulley, without having to disconnect the drive shaft. According to the cylinders motor drive power, the working torque of every cylinder is estimated, then the measurement range of 0-300 Nm can be determined. The output of torque sensor is a digital pulse signal to avoid the electromagnetic disturb.

2.2.2 Data Acquisition and Regulation Module

Data acquisition and regulation module is mainly used to obtain signals of torque sensors, displacement transducer and feeding motor speed sensor and carry out the outlet opening regulation. STC12C5A60S2 microcontroller and the corresponding peripheral module are combined with the data acquisition circuit. Acquisition circuit includes a microcontroller, input pulse optocoupler isolation, 82C54 pulse counter, analog to digital conversion circuit, RS232 communication interface. Input pulse optocoupler isolation circuit is mainly used as the amplitude converter of torque sensors signals. It can convert rotation torque and speed signals and feeding motor speed signal into TTL-level signals. Then, the signals are sent to the programmable interval timer 82C54 to count. Analog to digital conversion circuit is mainly used to convert the displacement sensor signal to digital codes. Finally, the information on rotation torque and speed of cylinder and outlet opening is sent to the master controller by STC12C5A60S2 via RS232.

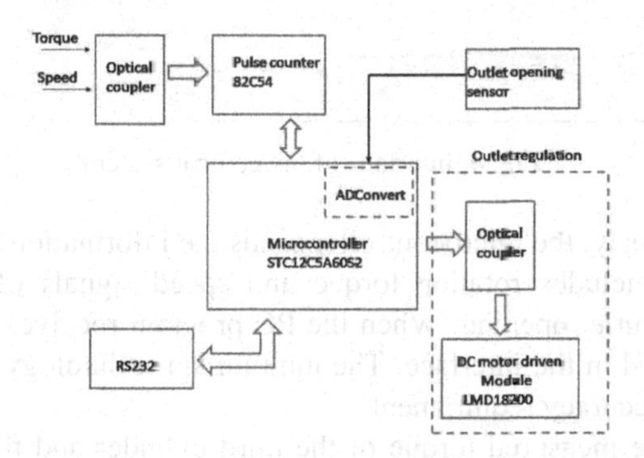

Fig. 3. Schedule model of the data acquisition and regulation module

Linear actuator is used to regulate the outlet opening of delinting machine. For the high accuracy of outlet opening regulation, the displacement transducer is arranged to provide the out electrical signal proportional to the mechanical movement of linear actuator. An H-Bridge LMD18200 is selected to drive the linear actuator. LMD18200 has a perfect logic control and protection circuit [8]. It can provide good support for PWM control. The PWM and direction signal of LMD18200 are provided by the microcontroller and isolated by optocoupler.

2.3 Software Design of the System

Virtual Instrument tool LabWindows/CVI is chosen as the software design platform. LabWindows/CVI is an ANSI C programming environment for test measurement developed by Nation Instruments Corporation. It provides powerful function libraries and a comprehensive set of software tools for data acquisition, analysis and presentation [9-10]. In the program development process, we can compiled C object modules, dynamic library link, C libraries, and instrument drivers in conjunction with ANSI C source files. The system developed in this paper has some key functions as follows: auto-acquiring parameters of delinting machine such as rotation torque and speed of cylinder, automatic torque control of third cylinder, adjusting feed rate and outlet opening by manual. The software can also display, save and print these signals in graphic and numeric form. The program interface is shown in Fig.4.

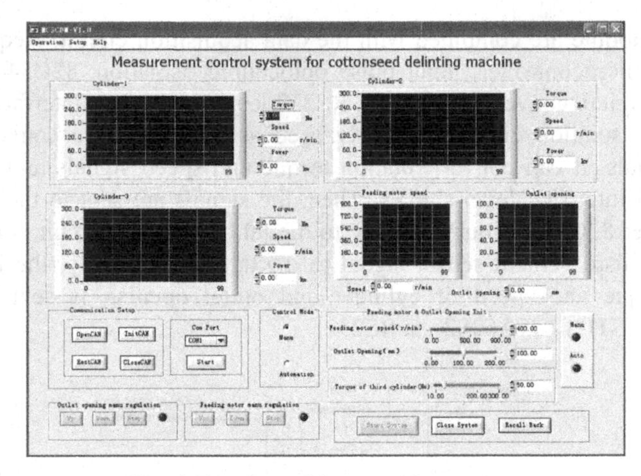

Fig. 4. Interface of the control system

As noted previously, the microcontroller sends the information to master controller. The information includes rotation torque and speed signals of cylinders, feeding motor speed and outlet opening. When the PC program receives these data, they are saved and displayed in the interface. The multithread technology is applied to satisfy the real time and accuracy requirement.

According to the measured torque of the third cylinder and the set value, we can use incremental PID control algorithm to regulate the outlet opening in order to keep the torque stable. In addition, the feeding motor speed is also adjusted manually by the program through the transmission frequency commands from the PC to the variable frequency device via RS485.The outlet opening is regulated manually by the program through the transmission commands to the microcontroller via RS232.

3 Results and Discussion

3.1 Outlet Opening Regulation Accuracy

To qualify the outlet opening regulation accuracy, the experiment about outlet opening control was practiced. First, we set some desire value of outlet opening at the program interface. Then, the linear actuator moved until the outlet opening reached the desire value. At the same time, the actual value could be obtained by the displacement sensor. The comparison results of desire value and actual value are shown in table 1.

Table 1. Comparision results of desire value and actual value of outlet opening

Desire value ,mm	Forward actual value ,mm	Absolute error, mm	reverse actual value , mm	Absolute error, mm
30	29.4	-0.6	30.7	0.7
63	62.5	-0.5	64.0	1.0
96	95.2	-0.8	96.7	0.7
129	128.3	-0.7	129.6	0.6
162	161.4	-0.6	162.9	0.9

From the table1, it is clear that the regulation of outlet opening is very precise, and the maximum error of it is 1mm.

3.2 Relationship between Outlet Opening and Torque of the Third Cylinder

In order to realize the torque control, we should study the relationship between outlet opening and torque of the third cylinder firstly. The result is shown in Fig. 5. From the figure, we know that the relationship between the torque and outlet opening is very strong. With the outlet opening becoming larger, the torque of third cylinder gradually decreases. Therefore, the torque can be adjusted by changing the outlet opening.

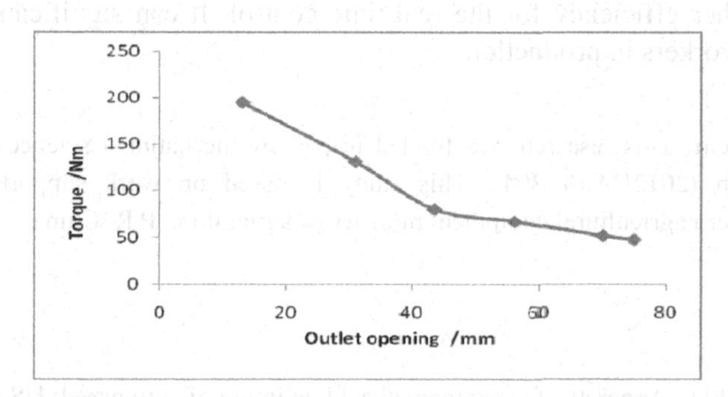

Fig. 5. Relationship between outlet opening and torque of third cylinder

3.3 Torque Automatic Control

The torque of the third cylinder is closely related to the delinting efficiency and quality. If the torque is in a good range, the delinting efficiency and quality will also be very good. The experiment started with an initially torque of 76 Nm and the reference was settled to 140 Nm. After the transitory regime, the PID controller reached the set point. It is shown in Fig.6. From the figure, we know that the controller is able to reject the load disturbance effect and keep the cylinder in good condition.

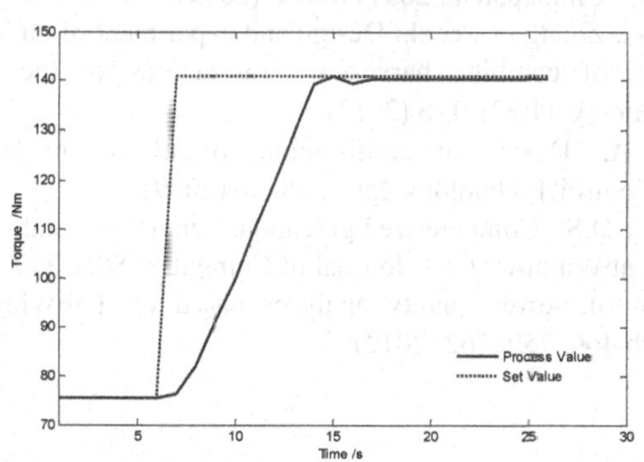

Fig. 6. Step response of torque of the third cylinder

4 Conclusions

In this study, a design of delinting mechanical measurement control system based on LabWindows/CVI is presented. According to the experiments, the system is proved that it has many advantages such as high testing precision, accuracy, and convenience for delinting experiments. The system can monitor automatically the data of torque of delinting cylinders, feeding motor speed and outlet opening. Because of the strong relationship between the outlet opening and torque of the third cylinder, the system can regulate the outlet opening automatically in order to keep the torque stable. According to the result, the precision of outlet regulation can reach 1mm. This system is also easy to use and provides a good graphical user interface and has higher accuracy, higher efficiency for the real-time control. It can significantly reduce the workload of workers in production.

Acknowledgment. This research was funded in part by the national science and technology supporting plan (2012BAF07B04). This study is based on work supported by the key laboratory modern agricultural equipment ministry of agriculture, P.R. China.

References

1. Downing, J.D.: Apparatus for the foamed acid delinting of cottonseed: US Patent, 4259764 (April 7, 1981)
2. Kincer, D.R., Kincer, D.L., Kincer, L.T.: Mechanical cottonseed delinter: US Patent, 4942643 (July 24, 1990)
3. Jones, C.: Method of delinting cotton seed: US patent, 5249335 (October 5, 1993)
4. Wang, G.F., Guan, H.J.: The application research on the production control of dilute acid delinting. In: Industrial Mechatronics and Automation International Conference 2009, Chengdu, May 15-16, pp. 37–40 (2009)
5. Zhen, Z., Jingdong, Z.: Automatic control system for delinter: China patent, 201220150875 (December 5, 2012)
6. Chen, X., Han, S., Sun, Z., et al.: Frequency automatic setting system of a delinter seed feeding mechanism: China patent, 203191861U (2013)
7. Liang, X., Chen, Z., Zhang, X., et al.: Design and experiment of on line monitoring system for feed quantity of combine harvester. Transactions of the Chinese Society for Agricultural Machinery 44(z2), 1–6 (2013)
8. Wei, X., Song, A.: Design of multi-motor control system based on C8051F340. Measurement & Control Technology 28(1), 38–40 (2009)
9. Yang, S., Simbeye, D.S.: Computerized greenhouse environmental monitoring and control system based on LabWindows/CVI. Journal of Computers 8(2), 399–408 (2013)
10. Yang, F.: Design of power quality analyzer based on LabWindows/CVI. Advanced Materials Research 466, 759–762 (2012)

Automatic Navigation System Research for PZ60 Rice Planter

Liguo Wei[1,2], Xiaochao Zhang[2], Quan Jia[2], and Yangchun Liu[3]

[1] Key Laboratory of Modern Agricultural Equipment, Ministry Agriculture, P.R. China, Nanjing 210014, China
[2] Chinese Academy of Agriculture Mechanization Sciences, Beijing 100083, China;
[3] National Key Laboratories in areas of Soil-Plant-Machine System Technology, Beijing 100083, China
weilg78@126.com, zxchao2584@163.com, jiaquan301@163.com, lyc@caams.org.cn

Abstract. In order to satisfy the agricultural demond of rice transplanting, the transplanter work should guarantee the escapement even in straight line transplanting that is convenient for field management and harvest later. Because of variable soil conditions and hard work environment, the driver driving level and the long boring driving to follow line is a big influence to the accuracy of the rice transplant. It is easy to produce overline, leak line and cause losses in yield. Aimed at above problem, this text introduces an automatic navigation system developed on PZ60 rice transplanter based on global navigation satellite system (GNSS). The steering, transmission and transplanting control system of the rice transplanter were modified from manual control system to electro-hydraulic control system using electro-hydraulic proportional valve. According to the position information of the rice transplanter acquired from GNSS receiver and vehicle sensors, the close-loop feedback control system of steering was builded. The system can accurately control rice transplanter to follow row navigating and turn around at the end of field by the self-adaptive fuzzy control method. The road and field experiment results indicated that the lateral tracking error could be kept within 100 mm when the speed of the rice transplant is not greater than 1.0m/s. The control system can satisfy the requirment of rice transplanting.

Keywords: rice transplanter, global navigation satellite system, automatic navigation.

1 Introduction

Rice is one of the most important food crop in China, the national rice planting area accounts for about 30% of the grain crop area, close to half of the total output of grain yield. Rice suitable transplanting operation time is usually 7-10 days, the rural labor resources are insufficient, the labor cost is increasing year by year. Paddy field working environment is poor, and transplanting machine take the personnel, not only energy consumption, there are also operating personnel safety hidden danger.

Automatic navigation technology is widly applied in modern agriculture and gradually becoming an important part of the agriculture engineering. At present,

© IFIP International Federation for Information Processing 2015
D. Li and Y. Chen (Eds.): CCTA 2014, IFIP AICT 452, pp. 653–661, 2015.
DOI: 10.1007/978-3-319-19620-6_73

GNSS, machine vision and multi-sensors data fusion are the most widely applied automatic navigation technologies in the agricultural project. The technology of machine vision has the price advantage, but its application on the rice planter is restricted in the weak light intensity. As GNSS accuracy is continually rising and GNSS cost is continually falling, GNSS technology has a wide application on precision irrigation, fertilizer and farm robots automatic navigation[1-3]. The implementation of GNSS automatic navigation on the rice transplanter can relieve the labour intensity of the driver, extend the operating time and be conducive to the improvement of productivity and the yield of rice. The objective of this study is to develop an automatic navigation system that was developed on PZ60 rice transplanter based on RTK-GNSS navigation technology in the paper. The steering, transmission and transplanting control system of the rice transplanter were modified from manual control system to electronic control system[4-7]. The automatic navigation system integrating GNSS technology and fuzzy control strategies can control the rice transplanter trace the presetted path and achieve automatic navigation transplanting.

2 Experiments and Methods

It is key that the manual driving mode of farm machinery is modified to electronic control mode in order to achieve mechanical automatic navigation control and variable assignments. The PZ60 high-speed rice transplanter has four-wheel-drive,six rows,engine power 8.3 kilowatt, the escapement 300 mm, the maximum speed of over 1.6 m/s. In order to realize automatic navigation working of the rice transplanter, the steering, HST no level shift system, braking system of the PZ60 rice transplanter were modified as shown in figure 1.

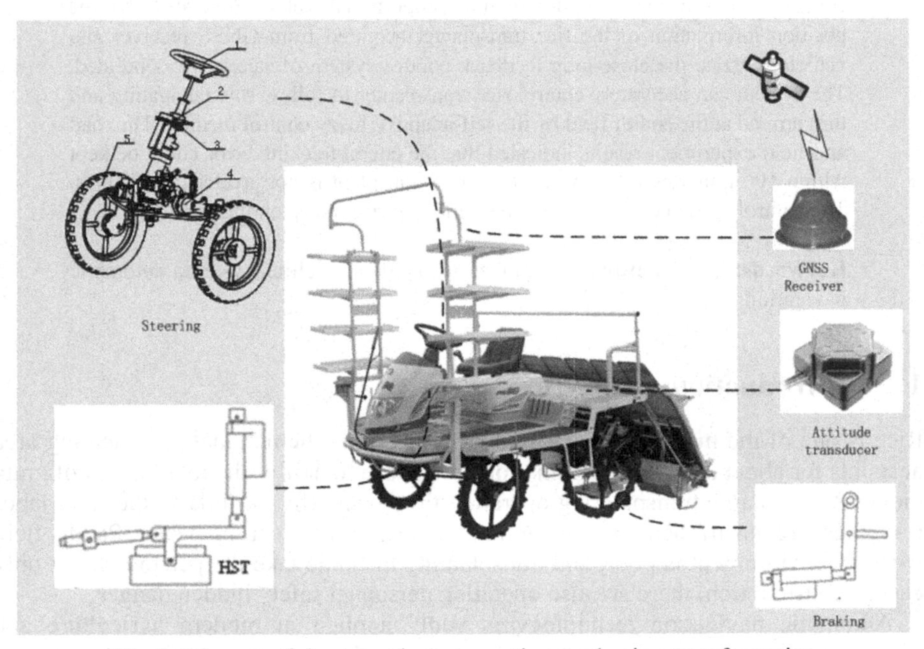

Fig. 1. Diagram of rice transplanter operating mechanism transformation

2.1 HST No Level Shift System

The rice transplanter is carried out going forward and backward infinitely variable speeds through adjusting HST hydrovalve flux and flow direction. The manual operation of HST action bar was modified to driving by electric handspike. The distance of the electric handspike is 80 mm, rating torque $75N \cdot m$, the range of work voltage 12~24V. The electric handspike control the hydrovalve through driving connecting plate rotation. The shift sensor fixed at the connecting plate feeds back the control position of the hydrovalve. According to testing, the electric control mechanism costs 2.3 s to complete HST shift system from stop to most gear forward and costs 2.1 s to complete HST shift system from stop to most high gear backward. It can satify the demand of the field working.

2.2 Steering Mechanism

The steering mechanism of PZ60 rice transplanter is composed of steering wheel, hydraulic booster and clutch gear box. The manual steering wheel was modified to driving by turning motor. Turning motor is SGMJV-04AAA61 AC servo motor made in Japan, rating power 400 W, rating torque $M_0 = 1.27N \cdot m$, rating speed $n_0 = 3000r/min$. The motor decelerated by speed reducer pass drive to hydraulic booster by clutch gear box and then turn the steering wheel by the main transmission case. The angle sensor fixed at the end of steering axle gathers steering angle. The clutch gear box was designed for automatic and manual operation switch. as shown figure2. The ininative gear joggles the tooth inlay gear. The tooth inlay can be shifted up and down through the control stick drives shifting fork. When the tooth inlay links the tooth inlay gear together, the steering mechanism is in automatic mode through motor driving. Whereas, the steering mechanism is in manual mode through the steering wheel.

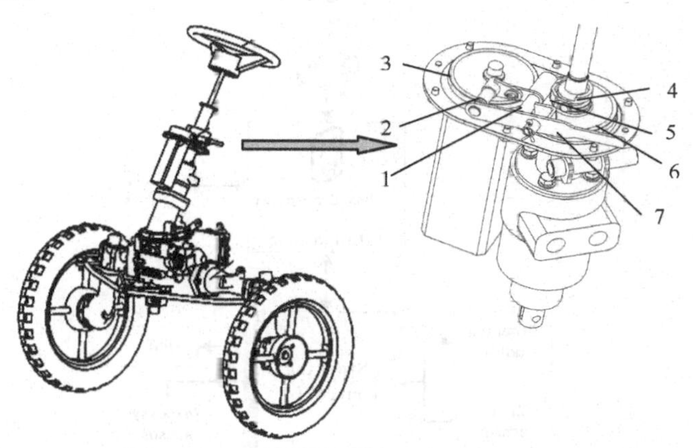

Fig. 2. Structure of clutch gear box 1.Fork rotor 2.Control stick rotor 3.Ininative gear 4.Tooth inlay 5.Fork 6.Tooth inlay gear 7.Clutch control stick

According to mensuration, driving the steering wheel needs torque $M_z \geq 10N \cdot m$ in normal. The retardment rate of motor retarder i_0 is $30:1$ in the design. The transmission rate of the clutch gear box i_1 is $1:1$. When the motor driving is by way of the clutch gear box, the output torque is:

$$M_1 = M_0 \cdot i_0 \cdot i_1 = 38.1N \cdot m > M_z \tag{1}$$

The retardment rate of the main gear box i_2 is $i_2 = 10.5:1$. The rating speed of the steering output axis is:

$$n_1 = \frac{n_0}{i_0 \cdot i_1 \cdot i_2} \approx 9.5r/min \tag{2}$$

The limit steering angle of the inside wheel is 45 degree. When the steering wheel turns from the left limit angle to the right limit angle, the turning angle θ of the steering output axis is 72 degree. It costs time:

$$t = \frac{\theta}{n_1 \times 360/60} \approx 1.3s \tag{3}$$

Therefore, The power and performing time of the automatic steering mechanism satisfy the need of operation.

2.3 Automatic Navigation System

The automatic navigation system of the rice transplanter is composed of GNSS, on-board computer, steering controller, sensors and actuators as shown in figure3.

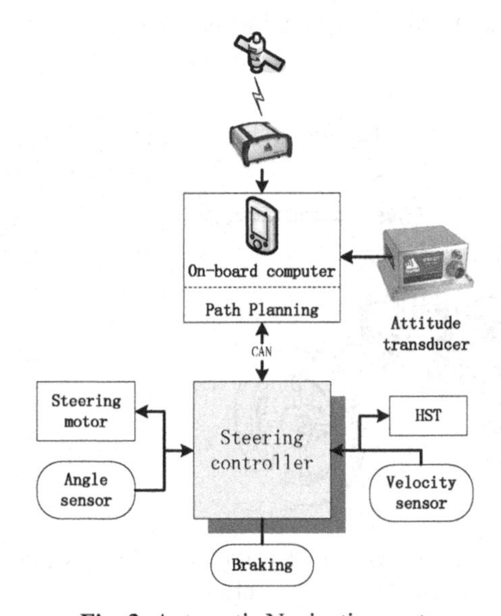

Fig. 3. Automatic Navigation system

GNSS adopts RTK locate mode that takes FlexPak6™ receiver as base station, ProPak-V3™ receiver as remotion station. Accuracy of static positioning is less than 20 mm. The output frequency of positioning information is set for 10Hz. The transplanter navigation control center is a high-powered on-board computer which can receive the informations acquired by GNSS receiver and angle sensors, make intelligent behavioral decision making and output navigation information instructions. The steering controller(SC) used the AD module to acquire the attitude information of transplanter in the navigation control process which measured by angle sensor and velocity sensor, and received the navigation information instructions from on-board cumputer at the same time. SC controlled the steering motor to move accurately, adjusted the HST speed control system to make the vehicle speed within limit, made the transplanting mechanism up or down and controlled the ground-contour-following device. Data transmission between SC and on-board computer was through the serial port. Steering motor thansmitted the driving force to make the steering wheel deflexion and then change the direction timely and accurately.

Steering control is the major part of automatic navigation control to the rice transplanter. The steering motor controller used PID position control mode. Figure 4 is its control block diagram. Microcontroller used angle sensor to gather the current angle information of steering axis and send it to on-board computer as feedback information. According the horizonal deviation, yaw angle, the current angle information of steering axis and the preestablished fuzzy control rules, on-board computer made the steering control strategy and ordered microcontroller to output control signal which drove the steering motor. All of these constitute a steering closed-loop control system[8-10].

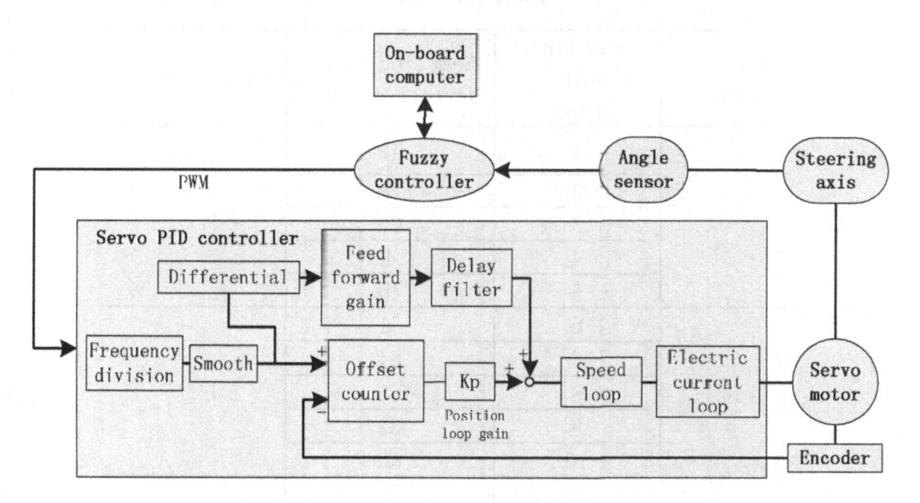

Fig. 4. Diagram of steering control

3 Results and Discussion

3.1 Blacktop Experimentation

In order to measure the presion of rice transplanter automatic navigation sysytem, the experimentations were done at a blank blacktop nearby beijing xiao wang village experiment station of chinese academy of agricultural mechanization sciences.
First, Two points A and B was fixed on blacktop groud, and the distance between point A and point B was longer than 80 m. The latitude and longitude information of two points A and B was achieved using GNSS receivers, and through on-board computer path planning software to prearrange a straight path of AB. A dropping style scriber fixed on rice transplanter was used to record the tracking of the rice transplanter on the ground. The scriber and GNSS receiver were fixed on the same centre axis of rice transplanter. After starting up the navigation control system, the rice transplanter begined automatic navigation in a definite velocity, and the scriber recorded the tracking on the ground at the same time.

After the rice transplanter completed the process of automatic navigation, a white fishing line as the navigation datum line was fixed between point A and point B on the ground. Tracking route of the rice transplanter recorded by scriber compared with the navigation datum line, and used meter ruler to measure the deviation as shown in figure 5. The experiment data is shown in table 1. According to the experiment testing, the rice transplanter automatic navigation error is not greater than 100 mm when the rice transplanter is not greater than 3.6 km/h.

Table 1. Experiment data

Segment number	Sampling point number	Sampling point deviation (mm)	Speed (km/h)
S1	1	42	2.0
	2	28	
	3	20	
	4	4	
	5	45	
S2	6	51	2.5
	7	64	
	8	60	
	9	82	
	10	80	
S3	11	75	3.2
	12	87	
	13	13	
	14	40	
	15	21	

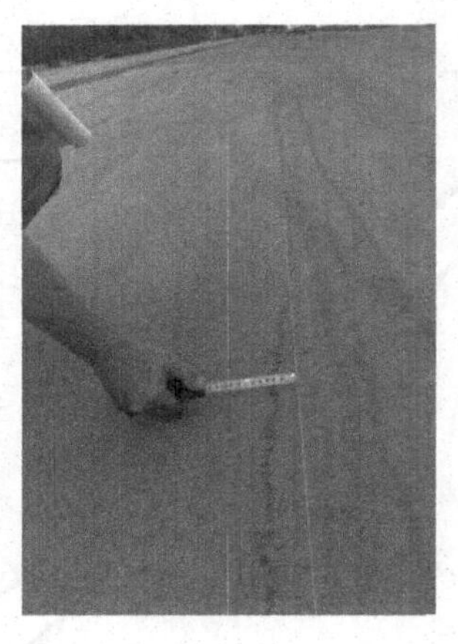

Fig. 5. Deviation measurement

3.2 Paddy Field Experimentation

The paddy field experimentations were promoted in MiYun agricultural mechanization demonstration centre. The range of mud depth in the paddy field was 200~300 mm. The paddy field expermentations were as shown in figure 6. The velocity of the rice transplanter was 0.7 m/s. The tracking of the rice transplanter was recorded by GNSS receiver as shown in figure 7, real line for prearranged route, dashed line for following route. It can be seen from figure 5, following route essentially coincides with prearranged route in addition to following error at the end of the field. A large tracking errors appears in the joint of line following and turning at the end of the field, except with the maxiam mechanism steering angle correlation and control strategy also needs to be further improved.

According to measurement, the following line error of the rice transplanter is not greater than 100 mm and the automatic navigation system has good controlling precision and stability. It can satisfy the agricultural demand of rice transplanting

Fig. 6. Paddy field experimentations

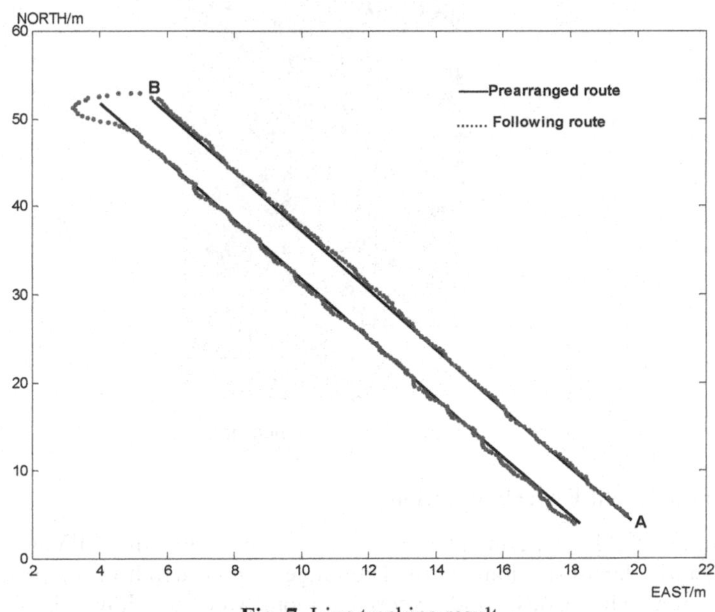

Fig. 7. Line tracking result

4 Conclusions

(1) In order to study agricultural machinery navigation and variable working, the steering, transmission and transplanting control system of the rice transplanter were modified from manual control system to electronic control system drived by servo motors. The navigation controller designed based on GPS technology realizes the rice transplanter automatic navigation working.
(2)The tracking error of automatic navigation system is not greater than 10 cm when the rice transplanter velocity is not greater than 1m/s in paddy field experimentations. It can satisfy the precision demand of rice transplanter automatic following line.
(3)Rice transplanter structure is complex, more joysticks need operation, the full realization of the unmanned is difficult. The autopilot system is transformation on the basis of the original, the steering system reform is relatively easy, and the transformation of the shift, clutch, throttle lever is more difficult. Paddy field working environment is poor, odds mud depth and wheel slipping impact on the effect of automatic navigation control. So, how to advance the performance of automatic navigation system from mechanical and control strategies that need more experimentation research.

Acknowledgment. Funds for this research was provided by the National Science and Technology Plan Projects (2013BAD28Q01). This study is based on work supported by the key laboratory modern agricultural equipment ministry of agriculture, P.R. China.

References

1. Reid, J.F., Zhang, Q., Noguchi, N., et al.: Agricultural automatic guidance research in North America. Computers and Electronics in Agriculture 25, 155–167 (2000)
2. Li, J., Lin, M.: Research progress of antomatic guidance technologies applied in agricultural engineering. Transaction of the Chinese Society of Agricultural Engineering 22(9), 232–236 (2006)
3. Zhang, Z., Luo, X., Zhou, Z., et al.: Design of GPS Navigation control system for rice transplanter. Transactions of the Chinese Society for Agricultural Machinery 37(7), 95–97 (2006)
4. Nagasaka, Y., Umeda, N., Kanetai, Y., et al.: Autonomous guidance for rice transplanting using global positioning and gyroscopes. Computers and Electronics in Agriculture 43, 223–234 (2004)
5. Zhang, Q., Cetinkunt, S., Hwang, T., et al.: Use of adaptive control algorithms for automatic calibration of electrohydraulic actuator control. Applied Engineering in Agriculture 17(3), 259–265
6. Hu, L., Luo, X., Zhao, Z., et al.: Design of electronic control device and control algorithm for rice transplanter. Transaction of the Chinese Society of Agricultural Engineering 25(4), 118–122 (2009)
7. Zhang, Q.: A generic fuzzy electrohydraulic steering controller for off-road vehicles. Automobile Engineering 217, 791–799 (2003)
8. Sun, H., Chen, Z.: Machines and mechanisms theory, 6th edn. Higher Education Press, Beijing (2001)
9. Hu, S.: Automatic control theory, 4th edn. Science Press, Beijing (2000)
10. Chen, W.C., Gao, G.W., Wang, J., et al.: The Study of the MEMS Gyro Zero Drift Signal Based on the Adaptive Kalman Filter. Key Engineering Materials 500, 635–639 (2012)

Plant Image Analysis Machine Vision System in Greenhouse*

Jianlun Wang**, Xiaoying Cui***, Dongbo Xu***, Shuangshaung Zhao***, Hao Liu***, Shuting Wang***, and Jianshu Chen***

College of Information and Electrical Engineering, China Agricultural University, Beijing, 100083, China

wangjianlun@cau.edu.cn, cxyfly617@163.com,
{404635349,1127381060,sky-sun1030,578477613,283922224}@qq.com

Abstract. In this paper, we use greenhouse field plants as studying objects to build a plant image analysis agricultural intelligent machine vision system based on web control. The system can provide data support for intelligent decision when manage a producing process by acquiring agronomic parameters of plant development and plant nutrition in real-time and remote through image analysis algorithm and relevant hardware and software platforms. For the software part of the system, we use an assembled installing environment based on Windows, Apache, PHP, MySql as the web application platform and establish the data structure of B/S network model based on Web. We build an image data base and use the agricultural parameters of plant development and plant nutrition to analyze the image and dynamically publish it on Web. At present, we have measurement algorithm modules include segmentation algorithm, shape recognition algorithm, 3D reconstruction algorithm, ranging algorithm, chlorophyll and nitrogen contents measure algorithm. We use these algorithms to obtain agricultural parameters such as nutrition, developing size, quality, diseases and pests and put thoroughly monitors and alerts into practice. For the hardware part of the system, it is consists of remote Web server, machine vision control equipments in field and sensors. The equipments on the platform can practice close-cycle control and condition management. These functions make it possible for the assemblage of internet controlled hardware system.

Keywords: agronomic parameters, plant image processing, Machine vision system, greenhouse.

---

* This work was supported by the National scientific and technological support projects and the Urban Agricultural Subject Clots Project. It was undertaken at China Agricultural University.

** Corresponding author.

*** These authors contributed equally to the article.

© IFIP International Federation for Information Processing 2015

D. Li and Y. Chen (Eds.): CCTA 2014, IFIP AICT 452, pp. 662–690, 2015.
DOI: 10.1007/978-3-319-19620-6_74

1 Introduction

The study and application of machine vision system in agriculture date from the late 1970s. The study was focusing on using machine vision system to conduct quality detect and classification of agricultural products [1]-[4].

In 1995, Xiaoguang Chen [5] used computer image processing technique to identify characteristics such as contour line and coordinate position of vegetable seeding, analyze and judge their growth, provide the necessary information for transplant and thinning in the whole growth process. In 1996, P.Ling [6] has established an automated plant monitoring system by using machine vision system, the turntable can realized all-around monitoring, the machine vision system can determine the change of nutrition in 24 hours, it can calculate the reflectance projection area of lettuce seedlings, and reflectance projection area can reflect lack of fertilizer. In 2002, Guili Xu[7] measured the plant leaf area by using machine vision system and Reference substance method, designed sample light box, and optimized light box parameter, the method improved the calculation efficiency and accuracy of leaf area, made average error reduced to 2.86%.Qiaoxue Dong[8] designed greenhouse computer distributed automatic control system, it can realize real time read and store environment parameter values and alarm information of the greenhouse, then monitor the operation of the greenhouse. In 2003, Changying Li [9] used computer vision system to monitor the condition of cucumber seedling growth. By using image processing technology to extract reflectance projection area and plant height, and it is concluded that the average height measured by image processing is 0.927 relevant to artificial measurement results, the system can save measure consumption, quick and easy. In 2007, Libin Zhang [10] used machine vision system to detect greenhouse cucumbers with near-infrared spectral imaging. Zhiyu Ma [11] monitored the growth information of plants using technology of machine vision and image processing in 2010, the CCD camera of its monitoring system was installed in the precious rotation PTZ which was surrounded by ten plants to be researched. Near infrared filter was installed on the camera, and near infrared light was installed on the top of the camera. The PTZ rotated automatically in a certain time interval, and the CCD camera would take pictures of the tested plants.

With the development of information technology, network controlled agriculture intelligent control system, which is based on Web and Internet, has becomes the combination of multidisciplinary theory and technologies. It raises the management of agricultural production to a new technology platform.

In this article, we use greenhouse field plants as studying objects to build a plant image analysis agricultural intelligent machine vision system based on web control. The system can provide data support for intelligent decision when manage a producing process by acquiring agronomic parameters of plant development and plant nutrition in real-time and remote through image analysis algorithm and relevant hardware and software platforms [20].

Combining industrial network control technology, Internet technology, Web technology, Database technology, SOC technology, model technique, GIS technology, machine vision system and agronomy technology into a powerful extra management platform, not only can we solve problems in agricultural production, but also establish a precise operation control of the whole real-time online production process of machine vision, establish the acquisition and backtracking of agricultural field real-time data, publish and share resources.

Thus, the agricultural production can meet the standards of industrialize, precise and intelligent. It greatly improves the management efficiency and the quality of agricultural production.

2 Machine Vision System Demand Analysis

According to the requirements of the greenhouse production and the controlling problems of the machine vision system, we need to firstly consider the climatic conditions of agricultural production, the production layout of equipments and network conditions. On this basis, we can analyze the image characteristics of plants and by combining the plant physiology cycle with the production management process and the occurrence of plant diseases and insect pests, we can fully understand the condition for getting clear images. Then we can analyze control problems and set up reasonable control targets as well as their logical relationship according to the functions of the machine vision system. At last, we can analyze the relationship between control activities and controlled object in detail during the control process, derive the relationship between input and output, the order of the various movements, or rule of time of the movement, develop system control scheme and the control system structure, according to the actual conditions.

2.1 The Network Stability, Reliability and Security Requirement Analysis

The control network which responsible for monitoring and controlling the agricultural production site, the information transmission direct to production process, it need to meet the requirement such as real-time, high reliability, harsh environment adaptability. At the same time, for the characteristics of openness, decentralization and low cost, agricultural control network requires to add office automation system, upper middle class network communication such as control management layer and control layer, as well as communication between the field devices.

Due to its simple protocol, open, stable and reliable characteristics, industrial Ethernet can be used as the reference of agricultural production network. This network has advantages such as good compatibility, easy connection to Internet, low cost, high development potential, high communication speed.

There are some problems correspond to the agricultural operating environment of industrial control. With the continuous improvement of Ethernet technology, there has development corresponding key technology to adapt the requirements such as stability and reliability or certainty and real-time, as well as the standardization of interconnection and security and so on.

(1) The technique which is used to ensure certainty and instantaneity of communication.

① Fast Ethernet and switched Ethernet technology reduce the network transmission delay and the collision probability by increasing communication rate.
② The Ethernet interchanger with the star topology structure divides network into segments, not only does it has the function of data storage and forwarding, makes the data frame buffering in the port between input and output and has no collision, but also filters the data .Data between the nodes in the segments was limited to the local network, without occupying other network or bandwidth of backbone segments, in order to reduce the load and reduce the data frame collision probability.
③ The full duplex technology makes two twisted pair or fibers between ports can also receive and transmit message frame at the same time, they can avoid conflict, greatly improving the confirmation and real time communication of the industrial Ethernet.

(2) The technique which is used to ensure the stability of the system

① According to the adverse industry site environment such as vibration, dust, high temperature, low temperature, high humidity and so on, we put forward higher request to the tolerance and the installation of the equipment, also develop corresponding products.
② Redundant Ethernet can improve the system recovery, and guarantee the reliability of the industrial control system.
③ The technology which is to ensure the network security.
Control network for production process, especially the structure of B/S network, has a high requirement of real-time, reliability, security and data integrity. We can take strict privilege management, key information encryption, and security technology which based on interchangers such as flow control, access control list, and secure connection layer.

As a result, the existing Ethernet control technology based on TCP/IP protocol stack can ensure the security, stable and reliable operation of the greenhouse machine vision system.

2.2 The Demand Analysis of Equipments and on Site Factors of Greenhouse System

① Environmental factors

Generally, the normal working condition of a greenhouse machine vision system is with natural sunlight, high humidity, high temperature and wide differences in temperature. These factors require the equipment has characteristics of moisture resistance, radiation resistance, and an extreme endurance of high temperature. In the process of designing and installing the equipment, there are a lot of factors for us to concern. Such as working conditions, install space, whether the positions of the shooting target are suitable, whether the circuits are safe enough, and whether the controlling methods are proper.

② Shoot object factors

The shooting objects of greenhouse machine vision system are fruits, leaves and stems of plants. There are relatively taller fruit trees (e.g. cherry trees and flat peach trees), liana plants which people need to build stands (e.g. tomatoes and cucumbers), ground cover plants (e.g. strawberries), medium height plants (e.g. lily), etc. It is necessary to adjust object numbers and shooting distance according to the different needs of image analysis.

The growth areas of leaves and fruits are different. We need to adjust shooting height, distance and angle according to the size and position of leaves and fruits. Controlling the angle, motion and speed of camera platforms is also needed.

According to the characteristics of agricultural production, we need platforms to make proper progresses for the camera to be adjustable in an effort to thoroughly monitor the trends of plants growing within the whole producing process. To ensure that the cameras scanning the whole producing area without influencing the production, we require the devices abilities to responsibly carry the cameras, at the same time move within the greenhouse. Using this kind of device, we no longer need multiple camera sets in order to capture clear pictures. Thus, we may progressively simplify the system structure, cut the cost, and maintain the cost.

③ Image analysis factors

To meet the demand of making clear pictures, we need to estimate the cost performance of cameras and lenses. And control of focal distance, zooming and aperture is required for the chosen camera.

④ Benefit cost factors

Cost effective of production management decide the economic system of greenhouse machine vision system. There is a simple equipment design requirements. The system should be convenient for operation and maintenance at same time. The camera shooting requires that the movement of the motor can be remote automatic controlled in orbit.

⑤ Infrastructure needs

Infrastructures are needed for running mechanical visual systems. There are 3 factors to be considered. They are Internet connection, the safety of circuit, and the quality of the greenhouse construction.

2.3 The Analysis of the Needs of Software Modular Integration

The concept, which we used for this article is IMS (Integrated Modular System), contains the characteristics of integration and sharing of information resources, distributed functional areas and parallel processing, and function module.

IMS requires a compositing use of both software and hardware. In other words, it needs to be independent, transplantable and replaceable between the operating system and first floor software.

Establishing distributed control and management is the synthetically miniaturization of function modular.

To ensure the system health monitor, we need the software to be openly built to ensure its reliability and safety.

3 The Overall Structure of the Field Machine Vision System

3.1 Frame Structure System

In this paper, according to the requirement of greenhouse production and control problem of machine vision system, as well as design thoughts of software system [27], combine with the algorithms, we have designed the following structure as shown in Figure 1.

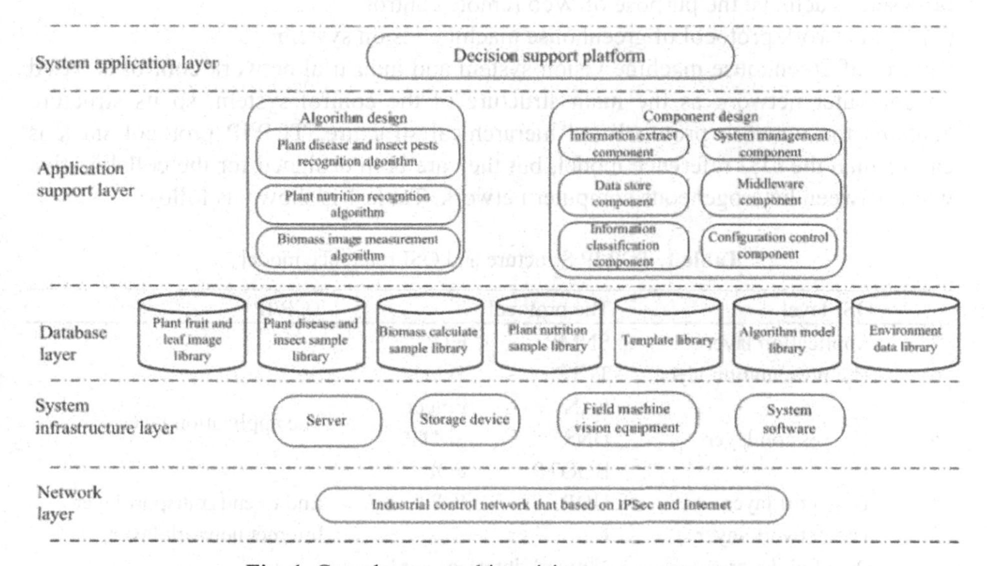

Fig. 1. Greenhouse machine vision system structure

3.1.1 Network Architecture

Network interconnection equipment is the key to industrial network, according to the physical structure, from low to high is divided into the network access layer, network distribution layer, core layer network remote device interconnection layer. Network access layer devices mainly include Internet access serial interface, embedded Web server module, remote server network interface card and interchangers. Distribution layer devices are mainly consist of routers and responsible for the wireless LAN access lines [28]. Core layer is mainly responsible for high-speed connection and its access equipment is mainly interchanger which is used to connect the data exchange between different greenhouses. Remote interconnection equipment which mainly refers to 3G router, can communicate with a remote server through the establishment of VPN channel.

This system with computer network as the main body has designed industrial control network which based on Internet and IPSec, use sensors to measure monitoring data and do real-time transmission, such as plant image, soil temperature and humidity, CO2 concentration and radiation and so on. Using the wireless network to send the data which is measured by the sensors or the machine vision system to the server, then it is published by the server in real time.

The system also uses B/S mode to exchange data, and uses the embedded web to access device control points to network. The network structure uses flat network, routers directly connect to switches, and switches drive workstation work, and then connect to the next switch. At the same time, combine with the Ethernet technology to achieve remote control devices, and it can publish data in real time on the Web, use B/S mode for remote monitoring. Ordinary users can use this system through a browser, to achieve the purpose of Web remote control.

(1) The network protocol of greenhouse machine vision system

Control of greenhouse machine vision system and industrial network control is based on computer network as the main structure of the control system, so its structure contains the network protocol and hierarchical structure. TCP/IP protocol stack is earlier than the OSI reference model, but they are both designed for the collaborative work between heterogeneous computer network. Table 1 is shown as follow.

Table 1. TCP/IP Structure and OSI reference model

	OSI level	The protocol		TCP/IP
7	Application layer	SNMP	FTP,	The application layer
6	The presentation layer	TFTP	Telnet,	
		FAN	Finger,	
5	The session layer	DNS	SMTP,	
		BOOTP	POP	
4	Transport layer	UDP	ICP	End to end transport layer
3	The network layer	IP		Internet network layer
2	Data link layer	Network interface card		The network interface layer
1	The physical layer	Transmission media		

Machine vision network control system is composed of computer, routers, switches, camera, PTZ, the logic control module and the controller hardware composition, and they link through different levels of agreement rules.

Repeaters of the machine vision system network work in the physical layer of the OSI reference model, is responsible for the physical layer of two nodes on a transmission of information and signal amplification of Ethernet, and the hub can be used as a multiport repeater, is used to gather the cable, it transmit data by sharing.

In this paper, the bridge in the equipment system work on the data layer of the OSI reference model and it can be connected to the same network in order to achieve the interconnection between remote LAN.

Switch operates at the data link layer of the OSI reference model, which is transmitted by way of exchanging data, can connect the workstations, hubs, and servers in Ethernet.

Routers which work at the network layer of the OSI reference model, is responsible for transferring data between two Ethernet network layer according to the data packet, change its physical address when forwarding packets, in order to be interconnected heterogeneous networks with multiple subnets or the Internet.

Gateway belongs in the application layer of the OSI reference model and it is mainly used to connect two different network architectures. In TCP / IP protocol, the gateway is an IP address leads to another network.

The network adapter on the machine vision system remote control program is the interface between the computer and the network physical transmission medium. With the driver you can achieve a variety of functional data link layer on the computer, but also as an integral part of the physical layer.

(2) Internet-based industrial control network structure of the machine vision system

Network structure of the machine vision system grew out of industrial control network based on Internet, with the development of Ethernet technology and Internet applications in the control area, making network-based remote monitoring system based on Web technology has become an important part of the industrial Ethernet technology. With the development of applications of embedded Web server in PLC, inverter and other control devices, B / S model has become an important way of industrial data exchange. In this paper, a machine vision system built network also uses the B / S mode of data exchange, with using of embedded Web access network control points of each device.

With the development of microelectronics, integrated circuits and embedded technology [30], 32-bit processing chip allows embedded systems into TCP / IP protocol, which has Internet access performance. Embedded operating system will regard mini Web server as part of the operating system to support its Web features that make embedded devices become simplified Web server.

Devices with embedded Web server can communicate via HTML pages and allow Web server on the Internet to visit the device. Networked Intelligent IP sensors or actuators can put sensing, signal processing, control, Ethernet, TCP / IP protocols, real-time operating system and simplified Web server and other hardware and software together, by B / S structure of the system can remotely maintain it.

In this technical background, machine vision control system adopts this mode of Internet-based network structure (see Figure 11), the devices which need remotely control [29]are all adopt the equipments with embedded interface or RJ45,or that can be connect to Internet via serial server.

In machine vision systems, interfaces between devices and configuration interface use application layer protocols, as industrial Ethernet allows different devices application layer protocol on the same network running .Configuration software uses Microsoft's ActiveX, a set of standard (device drivers) application software development technologies including COM technology, methods and properties for the communication between different devices on site to provide public interface that provides data exchange standards for different applications , unified incompatible industrial Ethernet protocols. This industrial control network based on Internet has a lot of features.

① Information network and the control network have no difference, and we simplify the control architecture of the network, so that the control network and the information network further “flat” (see Figure 11).

② Field devices with embedded WEB server on the configuration can be directly connected to the corporate information network and become an Internet node, realize plug and play.

③ Remote control, maintenance and management are true sense, also local control and remote control together.

④ Scheduling for networked control systems, security and stability issues need to be further analyzed.

3.1.2 System Facility

Based on WAMP installation environment, including integrated Windows, Apache, MySql, PHP as a Web application platform on which the software and hardware platform Web-based data exchange B / S network structure model. Through the establishment of a database of images, using plant growth and nutrition agronomic trait parameters measurement module for image analysis algorithms and dynamic Web publishing. In this paper, the PHP syntax mixed C, Java, Perl and PHP syntax itself, to meet the needs of all levels of application software designed in this paper.

System's servers include Web server, database server, map server. System software includes the Windows operating system, Apache, MySql, PHP, C++ and so on. Data storage devices include remote databases, site caching devices. Field devices include embedded Web equipment, VPN, 3G router, switches, NPORT, PLC, PTZ cameras, data acquisition instruments, etc.

3.1.3 The Database

The system according to system requirement analysis and data type designs database as below: plant fruit and leaf image library, plant disease and insect sample library, biomass calculates sample library, plant nutrition sample library, template library and algorithm model library.

Plants fruit leaf image feature library collects plant fruit and leaf color, texture and geometric features of the data. They are used to get agronomy trait parameters for plant growth [20].

Plant disease and insect sample library contains types and symptoms of common plant diseases and insect pests. Combine with corresponding algorithm [18]-[19], we can diagnose plant diseases and give prevention measures.

Biomass calculate sample library is used to store all kinds of image analysis to calculate the status information for plant growth, leaf area, stem diameter, plant height, fruit size, color, shape, and so on, they are used to determine the growing plants.

Plant nutrition sample library is used to store the chlorophyll and nitrogen contents in the process of plant growth and related characteristic data, it is used to determine the nutritional status of plants, and we can combine with the plant disease part, to determine whether it is suffering from disease such as nutrient deficiency in the plant growth cycle.

Template library contains template and characteristic data of plant leaves and fruits used in the process of algorithm design.

Algorithm model library contains all image processing algorithms: plant growing biomass algorithm module, plant disease and insect pest module, plant nutrition module.

3.1.4 Application Support Layer Design

Application support layer mainly contains algorithm design and component design. Algorithm design further comprises: plant disease and insect pest recognition algorithm, plant nutrition recognition algorithm and biomass image measurement algorithm. Plant disease and insect pest recognition algorithm aims at identifying the disease and making early warning for plants in the whole plant growth process. Plant nutrition recognition algorithm helps get the nutritional status of the process of plant growth, like chlorophyll content, get the nutritional status of different stages, and assisting users to make corresponding measurement. Biomass image measurement aims at getting agronomic traits of plant growth, like leaf area, and judging plant growing conditions of different stages.

Component module further comprises: information extract component, information classification component, data store component, system management component, middleware component, configuration control component. Through information extract component we can obtain the needed information from acquired images, and use classification component to classify the large amount of information, then store in the data store component. Middleware component includes PLC and PC data transfer drive, PTZ control drive, camera control drive, to realize the network control of trolley track, PTZ and camera. Configuration control component can realize online access or modify parameters of camera and PTZ remotely, control the movement of the orbit.

3.1.5 System Application Platform

There is a decision making platform in system's server, the decision making platform mainly used in fertilizer [13], irrigation [23], diagnosis and treatment of diseases[12] and other agricultural production projects, which provide decision making support for users in different plant growth stages. The decision making platform has its own algorithm module, it contains fertilization decision making algorithm [15], irrigation fuzzy control algorithm and diseases and pests warning algorithms. Decision support platform supports remote data collection and analytical decision-making and services to provide decision support for plant production management of fertilization, irrigation, pest and disease management.

3.2 Algorithm Software Platform

Machine vision algorithm layer contains image feature extraction module and image parsing module. They can obtain plant growing biomass (leaf area, stem diameter, plant height) and fruit quality (fruit size, fruit shape and fruit color), plant diseases and insect pests and plant nutrition.

Image feature extraction module uses a mathematical model to extract the image edge of leaf or fruit, skeleton, the three-dimensional shape information and optical properties [16- 17, 21-22, 24]. Image feature extraction used to extract plant leaves or fruit edge and skeleton and other information, image parsing module can obtain plant nutrition, disease and other conditions by extracting plant image information. The algorithm module is shown as Figure 2

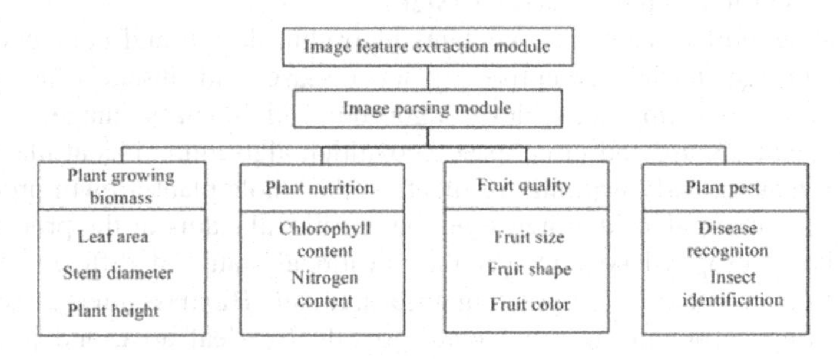

Fig. 2. Image analysis algorithm module

3.2.1 Image Feature Extraction Module and Image Feature Library

1 The method of image feature extraction

(1) The geometric feature extraction

Geometry model consist of point, line, surface, cube, sphere, ellipsoid, which are the basic unit of analytic geometry. The point set of the feature space, which is constructed by these basic unit, map to high dimensional space. The point with similar attributes can be spatial clustering.

Brief introduction of these basic models are as follow.

① Point model can be expressed as follow.

$$\mathrm{D}_i = x_i \tag{3-1}$$

D_i is on behalf of a hardware primitive of classification i. $x_i = \{x_1, x_2 \cdots\cdots x_n\}$ represents a n dimensional sample point. Point set model record each sample point, and the point set model is nearest neighbor classifier.

② Line model is the change trajectory of two original images. The change trajectories express by line and the trajectories construct a feature sub-space. The linear model is show as follow.

$$X^i_{mn} = span(x^i_m, x^i_n) \tag{3-2}$$

X^i_{mn} is on behalf of a hardware primitive of classification i. x^i_m and x^i_n are sample points of classification i. $span$ represents extension. Line is the extension of point. If there is a distance threshold, the feature lines of line model can construct cylinders. The scope of sample space is the superposition of each cylinder.

③ Feature surface is a promotion of feature line. The expression is as follow.

$$M^i_{mnk} = span(x^i_m, x^i_n, x^i_k) \tag{3-3}$$

M_{mnk}^{i} is on behalf of a hardware primitive of classification i. x_m^i, x_n^i, x_k^i are sample points of classification $i(m \neq n \neq k)$. *span* represents extension. The extension of three points is a plane.

④ Hypercube model simulates the learning process of the human brain. The basic idea of it is to select some points as memory point of learning. Then we increase sample constantly, and expand the hypercube, and increase the weight of successful learning. The model is shown as follow.

$$C_{hj}^{i} = span\_G(x_{j_1}^i, x_{j_2}^i, \cdots, x_{jn}^i) \quad (3\text{-}4)$$

Hypercube can be considered as the extension of i sample points of $x_{j_1}^i, x_{j_2}^i, \cdots, x_{jn}^i$. The Hypercube contains the points between the i sample points.

⑤ The hyper sphere model is on behalf of a cluster of points. A sample data can be regarded as a high dimensional space point. A classification of points is a point set. It is shown as follow.

$$F = \|x - a\|_2 \leq r \quad (3\text{-}5)$$

a is on behalf of super ball center. r represents the hyper sphere radius. This can be similar to expression of hypercube, also.

⑥ Ellipsoid model is extension of sphere model. Ellipsoid expresses better than sphere model. Ellipsoid can have different length in each feature dimension, and its spindle can be arbitrarily rotated. Ellipsoid can be recognized directly, and can be combined with many kinds of classifications to improve the precision of recognition, also. The concrete representation is as follow.

$$T_i = (x - u)^T \Sigma^{-1} (x - u) \quad (3\text{-}6)$$

T_i is on behalf of a hardware primitive of classification i. u represents the center of ellipsoid and $\Sigma^{-1}(x - u)$ represent covariance of matrix. The center of ellipsoid and the covariance of ellipsoid are shown as follow.

$$\overline{u_i} = \frac{1}{N_i} \sum_{\bar{x} \in P_i} \bar{x} \quad (3\text{-}7)$$

$$\Sigma_i = \frac{1}{N_i} \sum_{\bar{x} \in P_i} (\bar{x} - \overline{u_i})(\bar{x} - \overline{u_i})^T \quad (3\text{-}8)$$

Ellipsoid model is one of the most complex models. For its complexity, the length and direction of the spindle can be different. Its expression ability can be stronger. It can express data by fewer units.

(2) Color feature extraction

The common used color models are RGB model and CMY model of hardware,YUV model and YIQ model of display device, and HSI model,HSB model, HCV model, HVB model for perception, etc.

For the usage of the extracted information, we extract mean value, variance, energy, twist, kurtosis, entropy of the image color, extract them directly in every quadrant of RGB color space or change to other color space (for example HSI). We extract feature data of image gray, image shape, image texture from varies color space quadrants, or multiple color space quadrants.

(3) Texture feature extraction

Texture is a pattern of the change of gray or color space. Texture closely relate to high frequency components of the image spectrum. The texture analysis methods commonly used are statistical method, structure method, and spectrum method. Gray level co-occurrence matrix is established by pixel relative position in gray area space, and is used to define and calculate the texture descriptor. It is shown as follow.

$$p_{\bar{d}} = \begin{bmatrix} p_{\bar{d}}(0,0) & \cdots & p_{\bar{d}}(0,j) & \cdots & p_{\bar{d}}(0,L-1) \\ \vdots & \vdots & \vdots & \vdots & \vdots \\ p_{\bar{d}}(i,0) & \cdots & p_{\bar{d}}(i,j) & \cdots & p_{\bar{d}}(i,L-1) \\ \vdots & \vdots & \vdots & \vdots & \vdots \\ p_{\bar{d}}(L-1,0) & \cdots & p_{\bar{d}}(L-1,j) & \cdots & p_{\bar{d}}(L-1,L-1) \end{bmatrix} \tag{3-9}$$

Let S be a spatial associated pixel pair set of target area R, then the elements of the normalized co-occurrence matrix P can be defined as follow.

$$p(g_1, g_2) = \frac{\#\{[(x_1,y_1),(x_2,y_2)] \in S \mid f(x_1,y_1)=g_1 \& f(x_2,y_2)=g_2\}}{\#S} \tag{3-10}$$

The numerator of above formula is the number of pixel pairs with a kind of spatial relationship, and the gray value of the pixel pairs are g_1 and g_2. The denominator of the formula is the total sum of the pixel pairs. # represents number.

The area texture descriptors which are defined on the basis of co-occurrence matrix, for example, angular second moment, contrast, relevance, differential moments, inverse difference moment, sum average, sum variance, sum entropy, entropy, differential variance, differential entropy, related information measure, the maximum correlation coefficient, etc.

2. Feature Databases

Feature databases can be divided into the following three categories according to the feature extraction method.

① Shape feature database contain all kinds of leaf shape features, all kinds of fruit shape features, and all kinds of plant type features.
② Color feature database contain fruit color features, leaf chlorophyll level color features, and leaf nitrogen level color features.
③ Texture feature database contains all kinds of plant disease textures and all kinds of plant insect attack textures.

We set up the database according to the following requirements.

① Classify the data according to plant species, part and growth stages.
② Collect the feature data in accordance with the requirements of image parsing process.
③ Establish algorithms that corresponding to the search and match of database.

3.2.2 Pattern Recognition and Algorithm Model Database

(1) Image pattern recognition method

There are three directions of image pattern recognition. One is statistical pattern recognition which is based on the classical decision theory. The second is structure pattern recognition which is based on formal language. The third is fuzzy pattern recognition which is based on fuzzy mathematics theory.

In general, the pattern of images consists of characteristics. Pattern category is composed of a set of patterns that have similar characteristics. Pattern recognition process is the process to analyze and describe the pattern categories. Pattern usually quantitatively and structured shown as vectors, strings and tree structure. Pattern vector, string, and tree structure is commonly used in quantitative and structured representation.

For a pattern class set $m_1, m_2, \cdots, m_n$, if its corresponding discriminant function set is $P_1(X), P_2(X), \cdots, P_n(X)$, and unknown pattern X has $P_i(X)_{max}$ in the solution set, then X is belong to pattern m_i.

If there is $P_i(X) = P_j(X)$, then X is the boundary condition of m_i and m_j.When $P_i(X) > P_j(X)$, X belongs to P_i. When $P_i(X) < P_j(X)$, X belongs to P_j.

The pattern class of nearest neighbor classifier is each pattern itself, but the discriminant function is Euclidean distance. It means that for unknown sample X, if there is a pattern m_i closet to X, then X belongs to m_i. For instance, the K neighbor rule classifies X to its nearest class.

(2) Image parsing module

Image analysis can base on pattern recognition, or according to characteristics of other images. According to the change rule of color, texture and shape, we build characteristic patterns in order to analyze or recognize images. Greenhouse machine vision system, which is used to analyze plant field images, has the following parse modules.

① Image segmentation module

Plant image segmentation module mainly analyzes plant leaves, stem and fruits. We extract the feature at the edges of these parts, for the subsequent local feature extraction of 3D reconstruction, texture analysis, etc.

② 3D reconstruction module

The 3D reconstruction module recovery the images of leaves, stems and fruits, to get the exact leaf area, stem diameter, fruit size and shape.

③ Distance measure module

Distance measure module combines with field equipment and three-dimensional reconstruction module to measure leaf area, stem diameter, plant height, fruit size and shape.

④ Growing biomass measure module

Through field image parsing, we can get the plant growing data, which mainly contains the correlation model of growth. And build plant growing model in order to combine with production manage process such as water and fertilizer management, and to further build the virtual model in the whole process of plant growth.

⑤ Plant disease and insect recognition module
We can get features like texture, color and shape through image parsing [31], and build the analysis models of diseases. We can effectively monitor the status of plant disease and insects through real-time image data, provide data support for disease and pest control.
⑥ Plant nutrition monitor module
Through the analysis of plant leaf images, we can measure the chlorophyll and nitrogen content, and build image analysis measurement model. We can effectively monitor the nutritional status of plants through real-time data, and cooperate with water and fertilizer management.
(3) Algorithm model library
Algorithm model library contains growing biomass module, plant disease and insect module, nutrition module.

Growing biomass module includes segmentation algorithm [14], three-dimensional reconstruction algorithm and ranging algorithm.

① Segmentation algorithm includes fitness function algorithm module [21, 25], preprocessing module, partial differential parsing algorithm module, morphological algorithm module, edge extraction module, split operation module, shape discrimination module, character description module and clustering module, we can extract the key parts from plant images such as plant leaves and fruits, to do particular area study.
② Three-dimensional reconstruction [20] algorithm contains Lambert reflector model, normal vector and vector model, and leaf surface model, they are used to transfer the leaf image or other part images from two-dimensional coordinate to three-dimensional coordinate, in order to take further measurement.
③ Ranging algorithm includes difference registration algorithm model, motion single visual distance algorithm model, zoom single visual distance algorithm model and real measure algorithm model, they are used to measure leaf area, plant height, stem diameter, fruit color, fruit shape and quality.

Plant disease and insect module contains feature description operator, clustering operator or template classifier and template database. Feature description operator is used to describe all kinds of plant diseases and insects, determine the corresponding disease category through the clustering algorithm [26], in preparation for diagnosis and treatment.

Nutrition module includes feature description operator, clustering algorithm module, information fusion algorithm module, chlorophyll recognition module and nitrogen identity module. Through feature description operator we can get three-dimensional shape, texture and color of plant key parts. Through the combination with clustering fusion algorithm, we can extract the chlorophyll and nitrogen content of plant growth and different stages, judge the plant growth situation.The design and content of algorithm model library as shown in Figure 3.

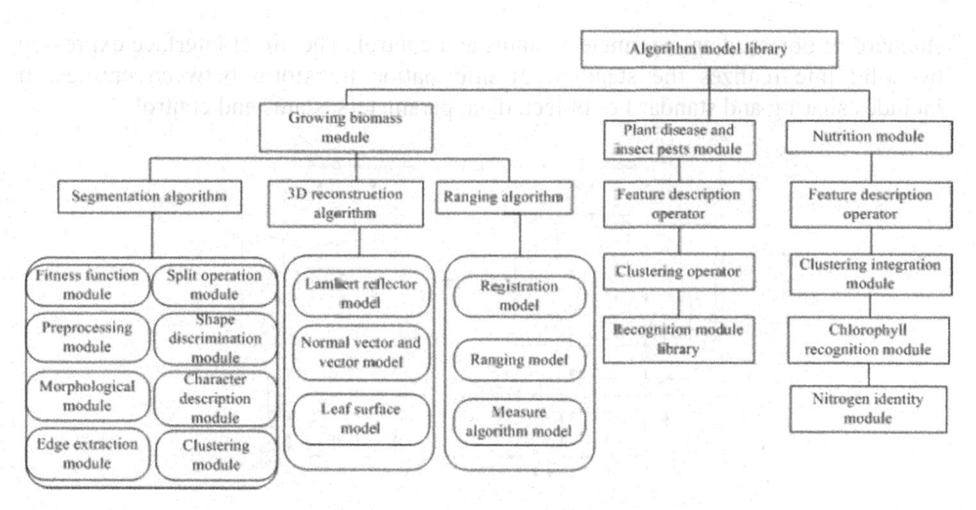

Fig. 3. Algorithm model libraries

3.3 The Software Platform of Machine Vision System

The GOA model defines the structure of software, hardware and interface. The frame of hardware and software contains four logical layers, and nine logical interfaces.

The software platform of greenhouse machine vision system in this paper designs based on universal "General open architecture (GOA) framework" [28], this standard is an important standard of open system structure. It is shown in Figure 4.

GOA model includes application layer, system service layer, resource access service layer and physical resource layer. The application layer contains 4L logical interface among software application components. System service layer provides public service of application software, and it contains the operating system components and 4D as well as 3L interfaces. Resource access service layer provides direct access to the software component of hardware, such as device drivers and storage that have IO definition, 3D interface, 2D interface between software and hardware, peer 2L interface. Physical resource layer provides the direct interface 1D and 1L between different physical components, it contains the physical definition and data link definition of the bus, 1D defines the electrical/mechanical requirements and 1L defines the decoding.

Among 4 layer structures, simplify defines the common interface point, realize the interaction of the logical independent layers and that of equivalence relation layer, through the interaction and isolation of logical/physical interface.

This 4 layer 9 interface mode realizes the functions of the machine vision software and hardware, and facilitates the expansion of the ready-to-use components, provides the unified basis platform for the application of system components. The system platform can perform operation and constraint instructions, and realize the interoperability between devices of every level. As shown in Figure 4.

The logical interface which is expressed by the dotted line realizes standard of support information sharing between point to point entities, it includes sharing and standard of object, data, parameters, status and control. The direct interface expressed by solid line realizes the standard of information transform between entities. It includes sharing and standard of object, data, parameters, status and control.

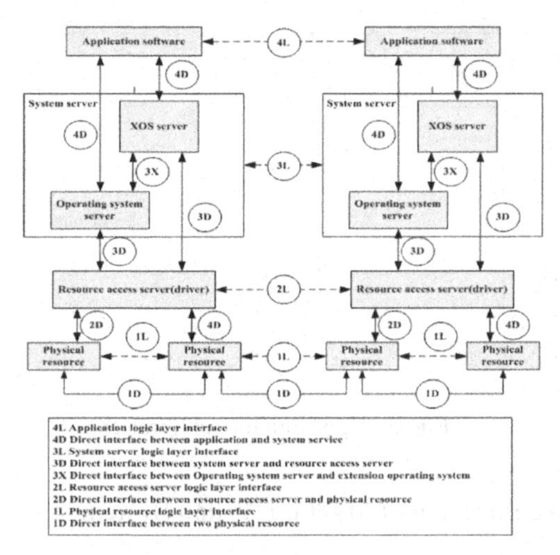

Fig. 4. Algorithm model libraries

3.3.1 The System Software Architecture

In this paper, the machine vision system software builds on the open software architecture, it can realize following functions. ① Apply and test software running normally under the operating system. ② Construct and test of drivers and complex routing under multiple network environments. ③ Access and test between file system and network file system server. ④ Manage the resource of the processor and CPU. ⑤ Manage the application node. ⑥ System time configure. ⑦ Announce. ⑧ Data configure and use. ⑨ System monitor and receive agent. The architecture is shown as Figure 5.

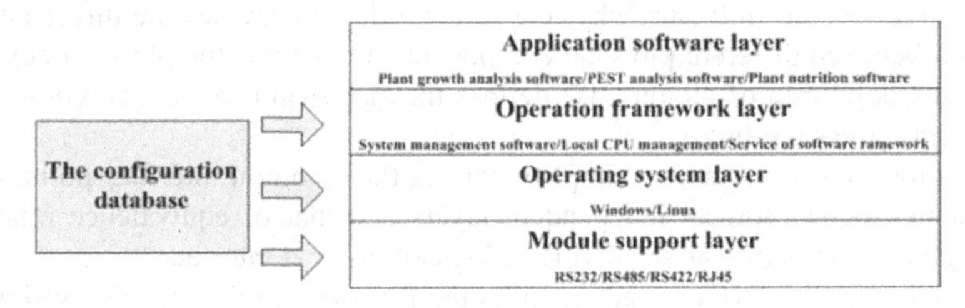

Fig. 5. Software framework

Under the system software architecture, the operation management corresponding to functions contains the following parts.

① The service of operating system. It includes task management, time management, memory management and external events management.

② Node administrate. Mainly supervise and control the state of application software. It includes node online or offline query and control service, application software initialization and restart, access control and maintain control.

③ Message manage. It provides exchange type for the data of application software and control information. The objects are system information, application information, general information, urgent information and broadcast message. System information is the universal message that defined by software running framework, including node online or offline information, test information, etc. Application information contains the data and control information among all application software and internal communication. General information is that of source and destination address in the message queue. Urgent information doesn't wait in the message queue. Broadcast message is multiple receiving. The main function of information management is to provide interface, routing, broadcast news and urgent information support of application message sending and receiving.

④ System communicate control. It manages all the network communication interface of the system. Including TCP/IP network manager and wireless broadband management, Rapid IO interface management, RS-422 interface management and 1394 interface management. Wireless broadband management contains initialization of the control network switch, network routing load, network configuration management, the maintenance or monitor of switch and the backup of network controllers. Rapid IO interface management contains communication interface management within system module or between system modules.

⑤ System time serve manage. The system software platform provides unified time service for all users, including the system clock calendar and real-time services.

⑥ System reconstruct services. Use the configuration items in database to construct the whole system, including network reconfiguration service, application software refactoring service, migration service of task software or add software to specified module, power control service.

⑦ The management of data access. It provides unified interface for data storage, including removing access, query, encryption and other functions. Manage object includes large capacity storage module, etc.

⑧ Integrated monitoring management. Including monitoring agent, news monitoring, node status monitoring.

⑨ Health manage. Including the test result and error report service of hardware layer, receiving and reading the health data, fault processing services of equipments of all levels.

3.3.2 Data Structure

In this paper, we describe the data structure of greenhouse machine vision system take Java custom universal data structure or Java class set for example as shown in Figure 6.

As shown in figure 6 is a class structure, the four kinds of structure are Field, Field Map, Field Map Set and Field Map Node, the system allows for the business scenario select data structure.

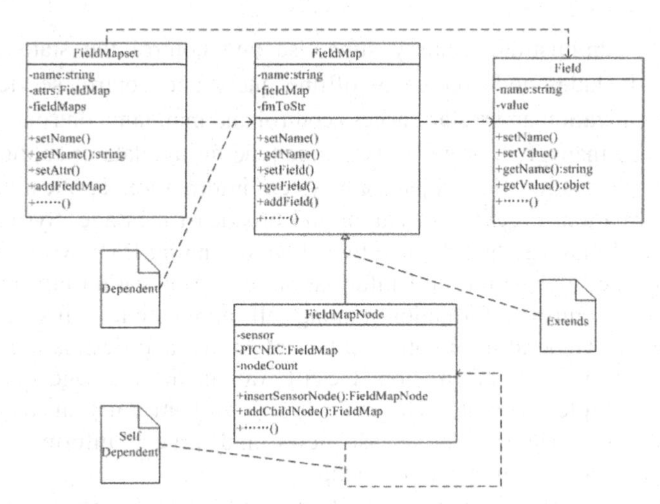

Fig. 6. Field, Field Map, Field Map Node, and Field Map Set classes map

① Field is the smallest data structure. Field contain name and value, the typical XML structure corresponding to Field is as follow.

<Sensor Name>Temperature</Sensor Name>
<Value>29</Value>

② Field Map is used to assembly or classify different Field and show a specific business. It mainly contains member variable such as name, Field Map. Name stores the name of current Field Map. The type of Field Map is Map<String, Field>, it can assemble different Field. XML structure corresponding to Field Map is as follow.

<Sensor>
<Sensor Name> Temperature </Sensor Name>
<Value>29</Value>
</Sensor>

③ Field Map Set is the set of Field Map, it can assemble complex business data, also can contain Field directly. This Field is the Attribute of Field Map Set. XML structure corresponding structure is as follow.

<Sensor Infos>
<Fid>PICNIC01</Fid>
<Sensor>
<Sensor Name > Temperature </Sensor Name>
<Value >29</Value>
</Sensor>
</Sensor Infos>

④ Field Map Node use nested recursive fashion to load all nodes, similar to directly use DOM analysis.

⑤ CommonMsg depends on the above four kinds of data structure, it is the external unified interface of application. Its construction mode is as follow.

```
CommonMsg ;
    String strMsg= "..." ;
  CommonMsg cm=New CommonMsg() ;
cm.set MsgString (strMsg)
```

CommonMsg is used to hand all the switch control to the XML protocol, let the application which is based on protocol specification to analyze the data.

3.3.3 Data Analysis

In the field of industrial control, in the interaction process of the data which belong to the communication of upper and lower machine, there exists a variety of network communication protocol such as OPC, UDP.

The transmission data format is simple text data, like TXT, XML, etc. Data specification is used to explain data format and construct data relationship, then acquire the initial data and conduct data provide constraints.

Through the data acquisition interface, we encapsulate the logical operation module of specific data access, transfer and integrate process, analyze the data source into a unified format of metadata 'name-value' structure.

The main function of the logical processor is to operate the metadata and return the result according to the configuration transformation rules of data source.

The main function of custom visual control is intelligently manage the call of data providing interface, abstract the data attribute by defining, transfer parameters required for logical processor and return the result to caller.

The generic data acquisition and monitoring software, through dynamic script, DLL call and execute logical control, realizes the real-time acquisition, display and storage. Take OPC for example, the data analysis process is shown as figure 7.

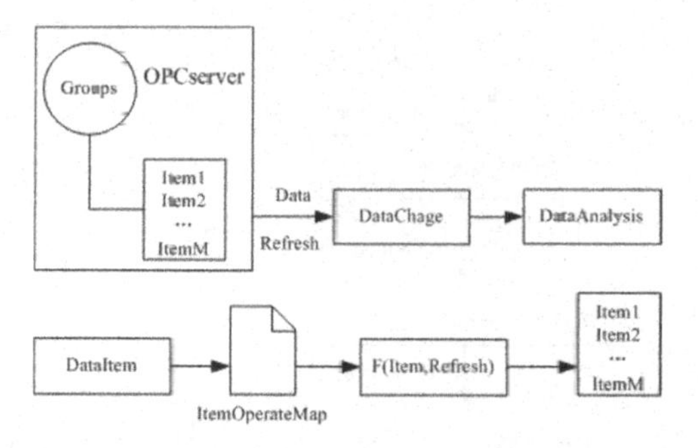

Fig. 7. Data analysis and logic processor matching process

On the database side, file management interact operation such as upload, download, delete and modify controlled by the MySql file control module. Define the interactive interface and metadata extract mode in MySql. Specific steps are as follows.

① Establish the corresponding fields and field types of metadata in the database table.

② Customize a template to achieve a batch conduct of data files, and establish the template library.
③ Parse the Data-XML of XML schema, and store the data in database.

3.3.4 System Data Flow Chart

The data in this system is from camera real-time data, environmental sensor real-time data and knowledge library experimental data, these data is submitted to the terminal processing algorithm on the server through all kinds of middleware, process and store, and through the dynamic Web page call and process data and display result.
The algorithm database can extract and analyze image characteristics, and submit the result to decision support module. The data which has been parsed by decision support algorithm can be used in production decision and management.

Data is maintained and copied by administrator, after filtering and screening it can be released on the Web platform. The data flow diagram of the system shown as Figure 8.

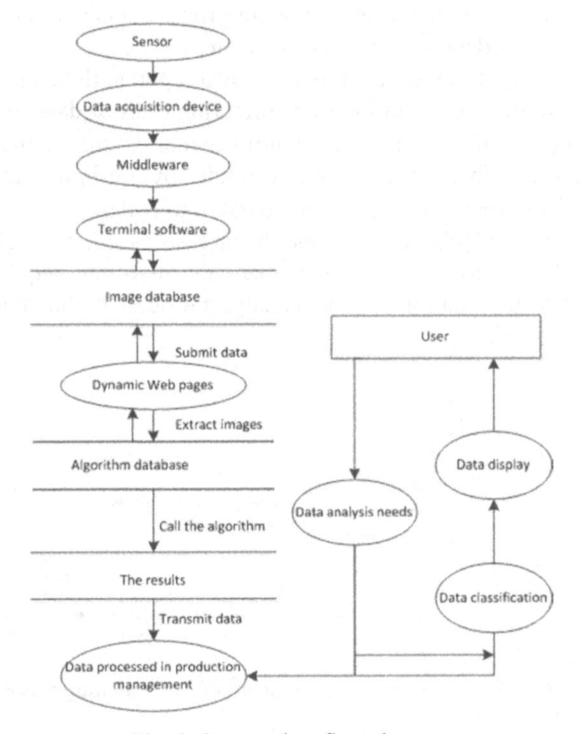

Fig. 8. System data flew chart

3.4 The Hardware Platform of Machine Vision System

The hardware platform of greenhouse networked machine vision system consists of remote Web server, Web database server, field machine vision control system and sensors. Devices in the platform can do closed-loop control and state management based on Web, and realizes the integration of network controlled hardware system. The hardware system structure is shown as Figure 9.

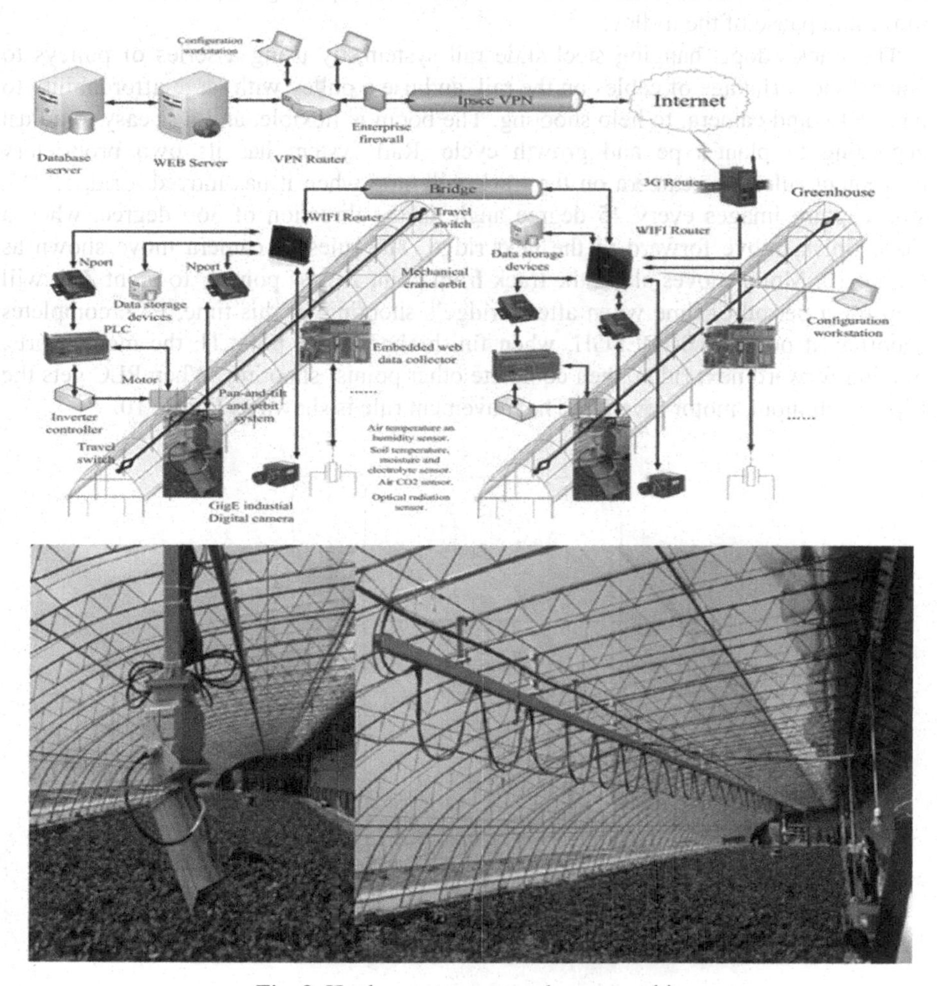

Fig. 9. Hardware structure and camera orbit

Hardware system contains motion control part, machine vision part, environment data acquisition part, configuration software, middleware and industrial controlled network.

3.4.1 Motion Control Part

Motion control part contains: PLC, trolley and rail, frequency converter.

PLC uses a series of programmable memory, for its internal storage procedures, perform user-oriented instructions such as logical calculate, sequential control, timing, counting and arithmetic operations, and through digital or analog input/output control various types of machinery or production process. It can modify online or query the parameters of camera, PTZ, lens and rail, control the motion state of devices, realize the intelligent control of the valve opening and closing as well as move and pause of the trolley.

The track adopts hanging steel slide rail system, by using a series of pulleys to control the shrinkage of cables on the rail, and use a pulley with large affordability to hang PTZ and camera, to help shooting. The boom is flexible, and it is easy to adjust according to plant type and growth cycle. Rail system has its own proprietary movement rules: the camera on the track will stop when it has moved a ridge, PTZ stop capture images every 45 degree angle in the direction of 360 degree, when a ridge finish, move forward to the next ridge. The rules of camera move shown as Figure 10. Motor moves along the track from point X pass point Y to point Z, it will stop for a period of time when after a ridge's shooting, at this time, PTZ completes shooting at points ABCDEFGH, when finish shooting at point H, the motor starts, moving forward next ridge, then complete other points' shooting. When PLC gets the trip switch input, motor reversal. The movement rule is shown as Figure 10.

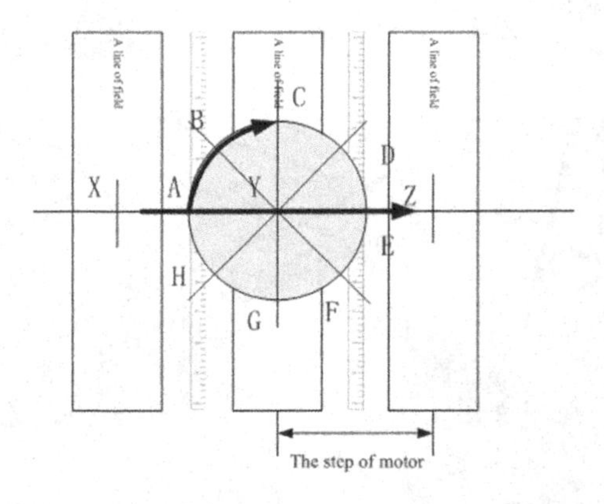

Fig. 10. The movement rules of motor

Frequency converter is used to control the movement of the trolley speed

3.4.2 Sensors and Data Acquisition Equipment

① Vision sensors and related equipment

Vision sensors and related equipment contains camera, intelligent PTZ, lens.

The system uses a GigE industrial network digital video camera, adopts the frame exposure CCD progressive scan sensor, has high image quality and high cost effective, color reproduction is good, it is suitable for the needs of plant image analysis. GigE gigabit network digital camera can trigger an acquisition or continuous acquisition by external signals. It can control brightness, gain, frame rate, exposure time and asynchronous reset by programming. It is also suitable for greenhouse orchard remote monitoring requirements.

The PTZ can change speed intelligently and rotate 360 degree endless, they have a wide perspective, and they can get the greenhouse plant images in a wide range.
The lens is industrial lens which is megapixels, the aperture, focus and zoom can be controlled by PTZ.

② Environmental data acquisition part
Environment data contains soil humidity, specific conductance, temperature, PH value, the air temperature and humidity, CO_2 concentration and irradiance. They are acquired by corresponding sensors.

3.4.3 Configuration Software and Middleware

The system configuration software is King View. Monitoring functions of configuration software can control the state of devices and running status of controller as well as travel switch via Internet, the data statistic function of configuration software can obtain data by directly reading the register of related equipment, it supports SQL Server, Access, able to get the data processing, the required data is automatically stored in the database, facilitate later query analysis.

The function of record state is characteristic of the configuration software, it provides a convenient tool for reporting statistics, according to the summary finally generate different statements, it can also visually dynamic display the status of various devices. The system configuration software and PLC connected into a remote monitoring system via the network, it can remotely modify and maintain equipment parameters and achieve the purpose of remote control. The device interface component of the system mainly the device driver provided by equipment manufacturers, and is installed on the PC side by configuration control software. The middleware consists of PLC and PC data transmission driver, PTZ control driver, camera control driver.

The PC act as upper computer in this system, the middleware can conduct configure, monitor, modify parameters online, real-time alarm and record data to the PLC system, PC can also provide a development environment and download the code to PLC via Ethernet or serial port, this can shorten the development cycle and simplify the programming process. The visual control configuration software on PC can control simulation modeling and generate code for motor, the code can be downloaded to PLC via Internet and consist a control loop, as well as remote monitoring and management.

PTZ control driver and camera control driver can connect the PTZ and camera with PC, modify the parameters of the camera and PTZ on the PC, and change their motion state according to need. The configuration controlling is shown as Figure 11.

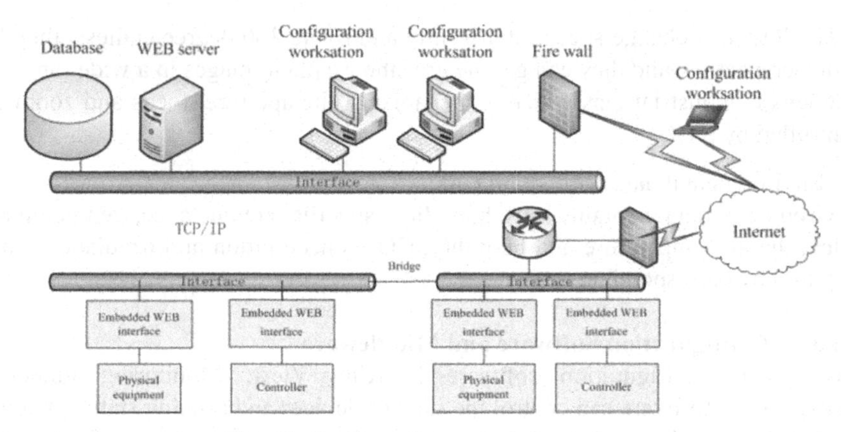

Fig. 11. Configuration control sketch

3.4.4 Industrial Controlled Network

The control nodes of Web server has been embedded in the system, interactive information services can be provided through a network server, provides a remote monitoring user interface that comply the Http protocol and conduct information exchange. The remote high-definition cameras with large amount of information use remote monitoring system of independent Web server. Remote monitoring security measures by using the system security management function is provided by King View.

The remote loop control system is implemented by variable frequency inverter, motor, PLC, travel switch and PC. PC can conduct functions as configure, monitor, modify parameters and record data, PC also can provide a development environment and download the code to PLC via Ethernet or serial port, this can simplify the programming process.

PLC communicates with PC via TCP/IP protocol, signals of the remote controller interact with the controller feedback signals via Ethernet, form remotely close-loop control, and design special motor movement rules, to ensure that all plants in greenhouse images can be collected. By programming in remote and upload to the PLC to control the camera rail. When they are online, we can go directly to PLC, by modifying the output port to control the movement of the camera track.

Through the simple system that composed by configuration and PLC, to realize real-time control of Machine vision system. King View has monitor function, it can control the status of devices and the running state of controller as well as travel switch via Internet, it can obtain data from registers of related devices directly, the record function can provide report and summary into different statements.

4 Functions of Plant Image Analysis Machine Vision System

The system can realize image acquisition, image processing and greenhouse production intelligent management. The image processing module through the segmentation algorithm to get accurate leaf edge, the addition of leaf shape recognition algorithm, makes the segment result more accurate. Through the three-dimensional reconstruction algorithm can obtain accurate leaf area. Through image analysis, building the relationship between image and parameters that needed to calculate the parameters, use the software to replace hardware or manual measurement.

(1). Image acquisition module: including camera equipment, track motion device, network closed loop control system. They are used to acquire field leaf images under the condition of natural light.

(2). Image processing module: including image feature extraction module and image analysis module. By using the image segmentation algorithm which based on adaptive threshold, image recognition algorithm that based on leaf shape discrimination, image three-dimensional algorithm that based on single leaf image and plant leaf nutrition analysis algorithm by means of image parsing to establish the calculate model from leaf image enhancement, image segmentation, edge extraction, leaf shape feature extraction, shape discrimination, leaf position information acquisition to leaf three-dimensional surface reconstruction, ranging, leaf area calculate model, chlorophyll and nitrogen measurement model, in order to achieve the machine visual measurement of leaf area, and get the plant growing biomass, nutrition, disease and insect pests condition.

(3). Greenhouse orchard production and management intelligent decision support module includes the digital management intelligent decision system that based on machine vision, plant leaves and fruit parameters obtained by image processing module can realize remote management specific to the digital block and plants, provide managers of decision information and support tools in the whole agricultural production process (including fertilization[15], irrigation, pest and disease diagnosis and treatment and early warning). The system functions are shown in Figure 12.

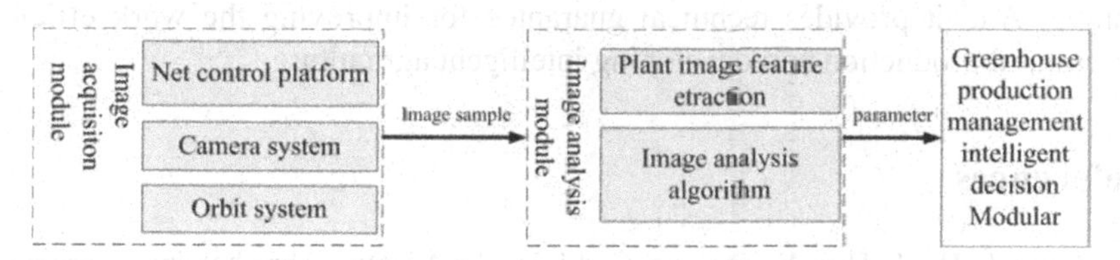

Fig. 12. System function sketch

5 Summary

The greenhouse machine vision system use the greenhouse field plant as research object, it is a kind of network controlled agricultural intelligent visual control system that based on Web and Internet. The system selects devices and networking according to demand analysis such as the geographical location of greenhouse, network conditions, equipment conditions and greenhouse environmental conditions, and integrates the agricultural production machine vision equipment which based on TCP/IP network control. The system combines the Ethernet control technology, can make the machine vision system realize real-time remotely acquire parameters of plant growth or nutrition and online monitor the growth condition of the plant, through image parsing algorithm and corresponding software and hardware platform. It provides data support for the control of the whole production process and intelligent decision.

The system software sets up the WAMP integrated installation environment as the web application platform, the platform realizes the data structure of B/S network model based on Web. Through the image, parameters and model database, establish agronomy trait parameters measurement algorithm module, to measure the agronomic parameters. Publish the results as dynamic Web page, the published result contain image data, image parsing result, decision support data, etc.

The hardware of the greenhouse machine vision system consists of remote Web server, Web database server, the machine vision control equipment in field and sensors. Equipments of the platform can realize close-cycle control that based on Web as well as condition management, making the integration of network controlled hardware system come true.

This system organic combines computer technology, automatic control technology, image parsing technology, machine vision technology with the agronomic trait parameters acquire demand in the agricultural production process. The process object is the important agronomic variables that difficult to measure or need large amount of data. Through the logical process such as data collection and analysis of the plant key part, the selection and extraction of characteristic variables, image parsing, construction of mathematical model relationship between image model and parameters which need to be measured, we can calculate or statistic the needed parameters. It provides important fundamental methods and basis for application work such as plant growth condition survey, prevention and control of plant disease and insects, plant nutrition monitoring, water and fertilizer management or decision making. And it provides technical guarantee for improving the work efficiency of agricultural production and developing intelligent agriculture.

References

1. Wang, J., He, J., Han, Y., Ouyang, C., Li, D.: An Adaptive Thresholding algorithm of field leaf image. Computers and Electronics in Agriculture 96, 23–39 (2013)
2. Wang, J., Han, Y., Fu, Z., Li, D., Chen, J., Wang, S.: Edge geometric measurement based principal component analysis in strawberry leaf images. In: Li, D., Chen, Y. (eds.) CCTA 2012, Part I. IFIP AICT, vol. 392, pp. 58–68. Springer, Heidelberg (2013)
3. Wang, J., Dong, J., Wang, Y., He, J., Changqi, O.: The design of an optimal decision-making algorithm for fertilization. Mathematical and Computer Modelling 54(3-4), 1100–1106 (2011)
4. Wang, J., Dong, J., Li, L., Wang, Y.: Design and implementation of an integrated office automation/geographic information system rural E-government system. In: 2010 World Automation Congress, WAC 2010, pp. 377–384 (2010)
5. Cheng, H., Chen, X., Yu, H., Zhou, Y.: The identification of characteristics of vegetable seeding by the using of image processing technology. Transactions of the Chinese Society of Agricultural Engineering 04, 23–26 (1995)

6. Giacomelli, G.A., Ling, P.P., Morden, R.E.: An automated plant monitoring system using machine vision. Acta Horticulturae 440 (1996)
7. Xu, G., Mao, H., Hu, Y.: Measuring Area of Leaves Based on Computer Vision Technology by Reference Object. Transactions of the Chinese Society of Agricultural Engineering 01, 154–157+3 (2002)
8. Dong, Q., Wang, Y.: Research and Development of Greenhouse Computer Distributed Auto-Control System. Transactions of the Chinese Society of Agricultural Engineering 01, 154–157+3 (2002)
9. Li, C., Teng, G., Zhao, C., Qiao, X., Wu, C.: Development of non-contact measurement on plant growth in greenhouse using computer vision. Transactions of the Chinese Society of Agricultural Engineering 03, 140–143 (2003)
10. Zhang, L., Yang, Q., Xun, Y., Chen, X., Ren, Y., Yuan, T., Tan, Y., Li, W.: Recognition of greenhouse cucumber fruit using computer vision. New Zealand Journal of Agricultural Research 50, 1293–1298 (2007)
11. Ma, Z., Qing, S., Gu, S.: Non-destructive measurement system for plant growth information based on machine vision. Transactions of the Chinese Society of Agricultural Engineering 09, 203–209 (2010)
12. Ouyang, C., Li, D., Wang, J., Wang, S., Han, Y.: The research of the strawberry disease identification based on image processing and pattern recognition. In: Li, D., Chen, Y. (eds.) CCTA 2012, Part I. IFIP AICT, vol. 392, pp. 69–77. Springer, Heidelberg (2013)
13. He, J., Wang, J., He, D., Dong, J., Wang, Y.: The design and implementation of an integrated optimal fertilization decision support system. Mathematical and Computer Modelling 54(3-4), 1167–1174 (2011)
14. Dong, J., Wang, J., Li, D., He, J., Wang, Y.: Complex target image of field jujube leaf segmentation based on integrated technology. Nongye Jixie Xuebao/Transactions of the Chinese Society of Agricultural Machinery 42(1), 165–170 (2011)
15. Wang, J., Li, D.: Fertilization processing method and system, CN101578936 (A) CN101578936 (B), CN101901477 (B) (January 05, 2011)
16. Wang, J., Dong, J.: Method and system for extracting field image edges of plant leaves, CN101901477 (A) (March 06, 2013)
17. Wang, J., He, J., Han, Y., Ouyang, C.: Real-time on-line system-based field leaf image edge extraction method and system, CN101964108 (A), CN101964108 (B) (January 23, 2013)
18. Chen, J.: The research on diseases of straw berry image features matching search. China Agricultural University, Beijing (2013)
19. Ouyang, C.: The study of the diseases recognition of strawberry based on image processing and pattern recognition. China Agricultural University, Beijing (2012)
20. He, J.: The research on the field jujube leaves 3D reconstruction based on single image. China Agricultural University, Beijing (2011)
21. Wang, Y.: The study of jujube leaf feature extracting and matching. China Agricultural University, Beijing (2011)
22. Han, Y.: The research on extracting image edge of strawberry based on multi scale analysis. China Agricultural University, Beijing (2012)
23. Wang, S.: The study of remote automatic irrigation fuzzy control algorithm based on PLC in strawberry greenhouse. China Agricultural University, Beijing (2013)
24. Dong, J.: The Edge extraction of jujube leaf image. China Agricultural University, Beijing (2010)

25. Nam, Y., Hwang, E., Kim, D.: A similarity-based leaf image retrieval scheme: Joining shape and venation features. Computer Vision and Image Understanding 110(2), 245–259 (2008)
26. Mao, H., Zhang, Y., Hu, B.: Segmentation of crop disease leaf images using fuzzy C-means clustering algorithm. Transactions of the Chinese Society of Agricultural Engineering 24(9), 136–140 (2008)
27. Steger, C., Ulrich, M., Wiedemann, C.: Machine Vision Algorithms and Applications. Wiley-VCH (2008)
28. Song, X., Zhao, M.: Key technology of machine vision system. Computer World, Sect. B11 (2007)
29. Xu, Z., (director Yang, X.): Research on remote monitoring technology in industrial production. Nanjing Technology University (2006)
30. Gong, A.: Research on information acquisition and processing technology based on Embedded Machine Vision. Zhejiang University (2013)
31. Zhang, Y.: Image engineering. Tsinghua University Press (2005)

A Review on Spectrometer of Pb(II) in Water

D. Zhang[1], M. Sun[2], and L. Zou[3]

[1]Key Laboratory of Modern Precision Agriculture System Integration Research, Ministry of Education
adalyme@hotmail.com

[2] Key Laboratory of Agricultural Information Acquisition Technology, Ministry of Agriculture
sunming@cau.edu.cn

[3] College of Information and Electrical Engineering, China Agricultural University, Beijing 100083, China
1475126006@qq.com

Abstract. Heavy metals have many characteristics such as susceptible to bio-accumulation, ecological amplification effect, and high toxicity and so on. Heavy metal pollution in water not only destroys the ecological environment, but also threat to human health and life seriously. Pb (II) enters the body through the enrichment of animals and plants in water, and damage to the reproductive ability, nervous system and body function of human. Source controlling is the key to the prevention of heavy metal pollution, so we urgently need to apply a plenty of heavy metal fast-detection instrument. we know that the research and development of heavy metal fast-detection instrument is becoming more and more important. However, the existing methods of detection of heavy metals have many problems like its procedure is too complicated, its equipment is expensive and operation process is not suitable for site test. The detection instruments are also numerous and varied, but the classified papers about heavy metal rapid-detecting instrument based on different principle seldom appear. In this paper, the current situation of existing heavy metals rapid-detection instrument based on different principles was reviewed, and the development direction of rapid-detection instrument was pointed out.

Keywords: Rapid, detection equipment, Pb (II), portable.

1 Introduction

There are some heavy metals entered our body through food disrupting the body's normal physiological function and harming our health, and they are known as toxic heavy metals mainly including Hg, Cd, Cr, As, Pb, etc [1].Water environment mainly includes rivers, lakes, reservoirs, marine and industrial water by human processing, discharge water and drinking water etc [2].Heavy metal pollution of water environment refers to the heavy metals discharged into water exceeding the self-purification ability of water. It makes the composition and properties of the water

© IFIP International Federation for Information Processing 2015
D. Li and Y. Chen (Eds.): CCTA 2014, IFIP AICT 452, pp. 691–701, 2015.
DOI: 10.1007/978-3-319-19620-6_75

change, the biological growth conditions in water environment deterioration, and bad influence to human life and health. Since the heavy metals in the water environment have some characteristics of multi-source, strong concealment, slight transport property, high toxicity, complex chemical behavior and ecological effect, non-biodegradable, biological accumulation effect, its low concentration have larger toxicity and so on, they become important pollutants in the water environment [3-5]. If they are absorbed by the food chain of aquatic plants, they will harm humans and animals.

Pb (II) pollution is defined as the most harmful environmental pollution to humans by WHO, and Pb (II) in food comes mostly from the enrichment of animals and plants in the water or soil [2]. Pb (II) to the nervous system especially the child's nervous system is extremely sensitive toxins. Infant exposed to high concentrations of Pb (II) is easy to cause poison, and then lead to toxic encephalitic. The main symptoms are excitement or sleepy, gastrointestinal discomfort, movement disorder, headache suffering, and it even can lead to coma when serious. If the children exposure to the low concentration of lead by long-term, it can cause changes in behavior functions, the analog learning ability decreased, movement disorders, inattention, impulsiveness, hyperactivity, and IQ decline [5-9]. In addition, lead pollution can affect reproductive ability, sexual function and nervous system, and cause high blood pressure .

To protect the water environment, control the water pollution, maintain ecological balance and protect human health, china has formulated a series of standards and methods corresponding to the evaluation of water quality [7]. The Indicators of Heavy Metal Pollution of Water Environment: "GB 3838-2002 Surface Water Environmental Quality Standard", "GB/T 14848-93 Groundwater Quality Standards", "GB 5749-2006 Life Drinking Water Health Standards "," GB 3097-1997 Water Quality Standards "," GB/T 11607-1989 Fishery Water Quality Standards ".

2 Methods for Heavy Metal Pb (II) Detection

The most usually used traditional indeterminacy of heavy metals is chemical analysis methods, such as Atomic Fluorescence Spectroscopy (AFS), Atomic Absorption Spectroscopy (AAS), Plasma Emission Spectroscopy (ICP), Reagent Colorimetric etc. With the advances of material science, information technology and sensing detection, the progress of a new indeterminacy based on mufti-disciplinary promoted the development of rapid-detection technology of heavy metals. Common methods for fast-detection include Electrochemical methods, UV-visible Spectrophotometry (UV), X-ray Fluorescence Spectrometry (XRF), Near Infrared Spectroscopy, Immunology Detection, Enzyme Inhibition and Bio-sensor Technology [10-26].

2.1 Traditional Indeterminacy of Heavy Metals

Atomic fluorescence spectrometry (AFS) is a analytical method of measuring the target element atomic vapor fluorescence emission intensity excited by the radiation to measure the target element [27]. The limit of detection of Atomic fluorescence spectrometry is less than that of atomic absorption spectrometry, and Atomic fluorescence spectrometry has simple lines and less interference, wide linear range, but the application of finite elements.

Atomic absorption spectrometry (AAS).In the method of ample vapor phase, the ground state atoms of the measured element create resonance absorption which is response for the characteristic of the atomic narrowband radiation emitted by the light source or hollow cathode lamps , outer electrons from the ground state to an excited state to produce atomic absorption spectrometry. Within certain limits, the absorbance is proportional to the concentration of the measured elements of the ground state atoms in a vapor phase, the quantitative relationship can be expressed Lambert-Beer's law [28]. The downside is that it requires a specific element corresponding hollow cathode lamp as a light source, cannot be achieved at the same time multi-element determination.

Inductively coupled plasma mass spectrometry (ICP-MS). The inductively coupled plasma mass spectrometry high temperature ionization combined with the advantages of sensitive fast scanning, mainly used for inorganic elemental and isotopic analysis test [29]. Excitation light source is inductively coupled plasma torch, and the test substance is ionized and then is separated by charge-to-mass ratio. Measure the spectral peak intensity to determine the ion concentration. The analysis method is highly sensitive, fast, little interference and can be completed within a few minutes while quantitatively determine dozens of elements, but easy to appear some of the ion background interference in ICP high temperature environment in the process of some elements determination.

2.2 Methods for Fast-Detection of Heavy Metals

Electrochemical analysis method based on the electrochemical properties established by the substances in solution and on the pole [31]. The electrical signals of power, potential, current and conductivity can be measured directly in the electrochemical analysis method, and do not need to signal conversion. The electrochemical analysis method is recognized as the trace and ultra trace analysis method of accurate, rapid and sensitive in the chemical composition analysis, and its concentration can be measured as low as 10~12g/L (metal ion). The electrochemical detection of heavy metals mainly includes potentiometry, polarography, conductometry, voltammetry etc.

UV visible spectrophotometry is a qualitative, quantitative and structural analysis method according to the absorption characteristics of the electromagnetic wave in the range of 200~760nm of the material molecular [32]. Its detection principle is some heavy metals make complexion reaction with one chromogenic agent in solution, and react with a colored complex, and its color depth is proportional to the concentration of the solution. Then make the spectral scanning in certain wavelength of ultraviolet visible light wave band to realize the detection of heavy metal content.

X-ray fluorescence spectrometry is also a qualitative, quantitative analysis method using the sample absorption characteristics to X-ray, which changes with the ingredients and concentration of the samples [33]. The method is simple, rapid, less spectral interference, small destructive and its elemental analysis range is wide, spectral line is simple. This method has low detection limit and high accuracy, but the X-ray fluorescence spectrometry need to train the operation staff before the experiment, so operation is more trouble and not suitable for on-site detection.

Near infrared spectroscopy can simultaneity determine a variety of data only through one time near infrared spectrum scanning of the samples in a short time, and this method need a small amount of sample, without damage and pollution, with the advantages of high efficiency, fast, low cost and environmental protection. [34]Near infrared light is mainly for the multiplier frequency and sum frequency absorption of the vibration of hydrogen groups. Different organic contains different groups, and different groups have different energy levels, so the absorption wavelength of the near infrared to all groups in different physical and chemical environment has significant difference. While the composition of aquaculture water is complicated, we cannot realize the direct detection of heavy metals using near infrared light needing the combination with other methods.

Biosensor method's principle is fixing specific protein on the electrode material, making the metal ion combined with the specific protein, and then the protein structure will change and transfer the signal to the capacitance sensor, and make the quantitative detection of heavy metals by using the change of capacitance signal sensitivity [35]. In recent years, such as the specific protein biosensor is developed for the determination of toxic compounds in aqueous solution. But the lifetime of the biological sensor mainly depends on the biological activity; the organisms generally live for a short time, which restrict the application and development of biological sensors. The principle of enzyme inhibition method is the heavy metal ions combining with the active center of the enzyme of methyl sulfhydryl or sulfhydryl, and then the enzyme changes its structure and properties and its activity decreased, making the color, conductivity and absorbance of the chromogenic agent changed [36]. We build the relationship between heavy metal concentration and enzyme system change through the photoelectric signal obtained. This method is used for qualitative measurement of heavy metal in environmental, food, water and vegetables. The enzyme inhibition method has advantages of fast, convenient and economic, which can be used in field and rapid detection, but compared with the traditional detection technology, its sensitivity and accuracy is low. Immunological analysis method has a high specificity, but its sensitivity is also high. We classified this method according to the immunoassay method of the type of antibody to heavy metal ions [37].We must finish the work in two aspects before analyzing the heavy metal ions in the use of immunoassay, the first step is to select a suitable complexes connecting with metal ion, guaranteed a certain spatial structure, so as to produce original reaction. The second step is to the combined metal ion compounds attached to a carrier protein, producing an immunogenic. The choice of compounds is the key to prepare specific antibody.

3 Instruments of Heavy Metal Pb (II) Detection

The research and development level of scientific instruments represents the scientific prowess of countries and overall national strength to some extent. The research and development of corresponding instrument based on the above traditional or rapid heavy metal detection method has become the focus of current research. The traditional instruments of heavy metal detection have high accuracy and good precision, but the body of that is too large and the price is expensive, the pre-processing and detection process is complicated, the detection time is too longer. The concrete function and price of those is shown in table 1:

Table 1. Traditional instrumental analysis methods of function and price

Analysis methods of function and price detection limit(g/L)	Atomic fluorescence spectrometry (AFS)10^{-9}	Atomic absorption spectrometry		Inductively coupled plasmamass spectrum	
		Flame 10^{-7}	Grahite funace10^{-9}	ICPP-AES10^{-9}	ICP-MS10^{-10}
Range of linearity	3-5order of magnitude	narrow		wide	
accuracy	high	high		high	
Multielement analysis at the same time	yes	no		yes	
Element valence state analysis	yes	no		yes	
Number of metal elements	More than twenty	More than seventy		More than eighty	
Instrument price/ten thousand yuan	10-99	1-99		10-999	

As for the rapid detection technology of heavy metal, the corresponding instruments have the characteristics of high sensitivity, good accuracy and timeliness strong. The research focus at home and abroad is mainly reflected in the development of the products of rapid detector of heavy metals. The key research direction is to implement the instrument which is low cost, small and portable, whose detection is suitable for field detection combining equalization with quantitation [10-12]. According to the characteristics of existing portable rapid-detector of heavy metal, we generally divide them into two classes of physicochemical instruments and biological instruments.

3.1 Portable Instruments Based on Physicochemical

Physicochemical instrument refers to which are designed and developed on the basis of spectrophotometry and electrochemical technology, through the material innovation and the small simple design including the instruments based on electrochemical method, UV Vis spectrophotometry (UV), near infrared spectroscopy and X-ray fluorescence spectrometry (XRF). These kinds of product have a lower price and wide detection range, its pre-treatment and operation process has improved than the precision instruments, but they still complicated.

3.1.1 Portable Instruments Based on Physicochemical

The electrochemical methods work with electrochemical sensor, and its performance is mainly limited by the range of potential scanning. The potential of the anode direction is determined by the materials of working electrode of electrochemical sensor. The oxidation potential in the electrolyte is different with electrode material. The potential of the cathode direction is determined by the over potential of hydrogen which depends on the influence of support electrolyte solution [11-13]. The electrochemical detection of heavy metals mainly includes ion-selective electrodes, polarography and voltammetry. The instrument using Anodic Stripping Voltammetry has developed rapidly in recent years, and Anodic Stripping Voltammetry is an electrochemical analysis method based on the theory of the classical polarography and combined with constant potential electrolysis enrichment and voltammetry determination. China has promulgated a national standard of Anodic Stripping Voltammetry which is suitable for the determination of metal impurities in chemical reagents. The Anodic Stripping Voltammetry method can determine varieties of metal ions in one time continuously, and it has the common advantages of electrochemical method that is low detection limit, high sensitivity, simple instrument and convenient operations [11-14].

The detecting instrument of electrochemical in foreign develop fast. As early 2007, Japan University of Tsukuba and Chemical Industrial Co. Ltd. have developed a model of the THMA-101 analyzer jointly. This instrument weights 4.2kg working with the theory of Anodic Stripping Voltammetry, which can detect Pb, Cd, Hg, Sb, Sn and some other heavy metals continuously. Its detection limit is l01-10mol/l and the analysis time is less than 5min [15] .The Company of wagtech in UK has developed a series of upgrades named HM1000-HM5000 based on the theory of Anodic Stripping Voltammetry. One of the latest products is HM5000 portable detectors of heavy metal, whose power supplied by rechargeable battery. Its weight reduces to 3.5kg, and the fastest sample detection time is 3min, and the detection limit is less than 1ppb [16]. Domestic electrochemical detection instrument is also quite mature. Skyray Instrument Company in JiangSu province of China combined the method of Anodic Stripping Voltammetry which have high sensitivity and recognized authority with National Standard Methods "Colorimetric method" which have fast detection rapid and good immunity, and developed a series of HM products. The resolution of the latest product of HM5000P is 0.01ppb, the detection time is less than 5 minutes, and the fastest detection time is less than 30 seconds [17-18]. The analysis of heavy metals based on the theory of Differential Pulse Stripping Voltammetry was developed by Cai Wei, and measured four kinds of heavy metal ions using the methods of pre-plating with mercury film. He puts forward the test method with different background of pure water solution and sea solution. The detection accuracy of background with pure water solutions is high, the precision of which can reach about 5%. It can also effectively determine with the seawater background solution, the precision of that is controlled within 15%, and its detection limit is 0.1-0.3ng/L. The above electrochemical detector all have the characters of low detection limit, high sensitivity and convenient operation, but the cost is relatively high.

3.1.2 Instruments Based on Spectrophotometry

The theories of spectrum instruments include ultraviolet visible spectrophotometry (UV), near infrared spectroscopy and X-ray fluorescence spectrometry (XRF). The three methods are a kind of analysis about qualitative, quantitative and structural based on the absorption properties of substance molecule on different wavelength range of the electromagnetic wave [20-22]. The detection theory in aspects of heavy metals is the organic compounds of chromogenic agent making complexion reaction with heavy metal, generating colored molecular group or producing fluorescence. The depth of color or the intensity of fluorescence is proportional to concentration. So we can detect the heavy metals under the specific wavelengths using the corresponding method. The amount of the sample to be detected of the spectral method is little. There is no damage and no pollution in the process of detection. So these methods have the advantages of efficient, fast, low cost and environmental protection etc. [10-22]. The above portable instruments at home and abroad based on the theory of UV-visible spectrophotometry (UV), near infrared spectroscopy are rarely reported, but the instruments based on X-ray fluorescence spectroscopy (XRF) technology have developed more mature.

The Checkboy Soil analyzer developed jointly by central Science Japan and Research Institute of resources and Environmental System. There is a built-in and small portable spectrophotometer and it powered by battery. It is suitable for on-site analysis the content of heavy metals in soil and water. It can also be simultaneous to determine nitrogen, phosphorus, COD and other indicators. Ten samples can be detected per hour. This instrument has been used for the rapid screening of environmental pollution source in Tokyo [25].

America Thermo Fisher Scientific Company launched the NITON-XL3t-600 portable instrument whose accuracy and precision is high, and the testing time of the sample is only 120s-200s. Seiko yingsi Electronic Technology Co., Ltd. specifically launched the SEA1100 of X-ray fluorescence spectrometer for the detection of heavy metals in soil; it also can be used in soil laboratory rapid for the detection for heavy metals. The instruments named X-MET 5000 analyzer developed by British Oxford using the Oxford high-energy, long-life and miniature X-ray tube. It can rapid analyze the content of heavy metals of Cd, Cr, Hg, Pb, Cu, Zn, Sn, Sb in the soil and environment in a few seconds [11]. Wang Jihua published a portable soil heavy metal analyzer specifically for rapid detection of heavy metals in soil, and developed a portable detector of XRF-7 which can test heavy metals in the soil. But the drawbacks of this kind of instrument are the high cost, and the detection limit of some heavy metals is not enough low [24-25].

3.2 Portable Instruments Based on Biological and Others

Biological instrument primarily based on the theories of heavy metals immunological or enzyme inhibition, using the development of biological specificity and sensor technology to achieve. Including the instruments based on immunological test, enzyme inhibition and biosensor technology. As early 1985, Reardan through a antigens of metal chelating agent isolated single antibody for the first time, making the immunological detection technology began to be used in heavy metal detection [26] .Subsequently, the complex antibody of metal chelates agent of the heavy metal ions of Hg2+, Cd2+, Pb2+ had successfully developed [27-29]. Enzyme inhibition technology is determining the content of heavy metals using the inhibitory effect of enzyme to metals. Biological Sensor Technology is fixing the specific protein, enzyme and its composite system on the electrode or biofilm and realizing the corresponding rapid detection of the heavy metals. The detection accuracy and sensitivity of such instruments used biological specificity principle is affected by antibody or enzyme specificity, which belongs to the indirect detection instruments [20-22].

Blake [39] initially developed an instrument named KinExA based on the theory of heavy metals immunological of ELISA. This instrument can make a rapid detection to heavy metal ions, which is expected to realize the field detection. There are few reports about this kind of instrument at domestic. This kind of instrument although can overcome the shortcomings of traditional instruments such as time-consuming and laborious, expensive, needing for lots of samples pretreatment, and the detection must be carried out in a large indoor laboratory. At the worse, the screening of heavy metals for small molecule specific monoclonal antibody is more difficult, it needs a long cycle time, its development cost is expensive, and its pH detection range is narrow.

Beijing Agricultural Product Quality Detection and Farmland Environment Monitoring Technology Research Center and South Korea Makar Heath Co. Ltd. jointly developed the "fruit and vegetable pollutant three in one portable detecting instrument", in which the effective range of detection based on enzyme inhibition technique of heavy metal lead is between0.5mg/mL and 4.0mg/mL [40]. Lehmann [41] using fermented glutinous rice recombinant yeast strains as biological components, developed a current type microbial biosensor for special measuring copper ion, the biosensor can detect the CuSO4 solution in the concentration range of 0.5 m/mol to 2m/mol. Karlen [42] the use of basophilic bacteria into the biological sensor, and used for the measurement of the concentration and bioavailability of Zn2+ in sewage , the concentration in the range of 0mg/l to 10mg/l, its shortcoming is the determination is single.

4 Portable Instruments Based on Colorimetry

Near infrared spectrum technology has been developing rapidly, It has many advantages, such as fast analysis, no needing for sample preparation and no environment pollution, so it is getting more and more attention. Since the 1980s, china agricultural university introduced the Fourier transform infrared spectrometer to carry out the research on agricultural products quality analysis, combined with the chemometrics methods, and established data analysis model. In the early years, the key technology of portable near infrared spectrometer was almost mastered by foreign, and our country can only rely on imports using a similar system. In 2006, Chongqing University began to develop a portable near infrared spectrometer, breaking through the key technologies of the structure design and products can be applied on the pesticide composition and content of vegetable base on-site rapid detection. Beijing Ying Xian has now successfully commercialized portable near infrared spectrometer. The reduction of volume and quality of series of products, fast analysis, high precision measurement, give the way to implement of low cost, low detection limit, detection a variety of metal detector.

5 Summaries

In conclusion, at this stage, portable rapid detection of heavy metals is lack of detector which is low cost, low detection limit, early simple treatment and suitable for a variety of heavy metals. The near-infrared spectroscopy and reagent colorimetric method, can be simplify the complex pretreatment of colorimetry. The tested object converses to a measured compound to build detection model, and the model is applied to a portable near-infrared spectrometer to achieve the detection of heavy metals. The portable detector of heavy metals will become low cost, low detection limit, early simple treatment and be suitable for many heavy metals detection equipment.

Acknowledgment. Funds for this research were provided by the twelfth five year science and technology support program vegetable and aquatic production safety key technology research and application of the Internet of things (NO.2012BAD35B03).

References

1. Cai, W.: Researches on Micro Electrochemical Sensors and Automatic Analysis Instruments for Heavy Metal Detection in Aqueous Environment. Doctor Paper of Zhejiang University, Hangzhou (2012)
2. Xu, J.G., Wang, L., Xiao, H.Y., Gao, J., Li, J.: Status of Water Pollution by Heavy Metal and Advance in Determination Methods on Heavy Metal in China. Chinese Journal of Environmental Science 29(5), 104–108 (2010)
3. Dong, D.M., Hua, X.Y., Li, Y., et al.: The adsorption in the natural water environment and its biofilm on heavy metals. Published by Science Press (2010)
4. Cooperative, R.G.: Research Report Series of Investigation and Prevention of Environmental Pollution in Beijing eastern suburbs, pp. 1976–1979 (2010)
5. Xu, H.C.: Design of the Rapid Detection Instrument of Heavy Metal Ions. Hebei University of Science and Technology, Shijiazhuang (2013)
6. Lalit, K., Pandey, D.K., Arpana, Y., Jyoti, R., Gaur, J.P.: Morphological abnormalities in periphytic diatoms as a tool for bio-monitoring of heavy metal pollution in a river. Ecological Indicators 36(1), 272–279 (2014)
7. Zhai, S.J., Xiao, H., Shu, Y., Zhao, Z.J.: Countermeasures of heavy metal pollution. Chinese Journal of Geochemistry 32(4), 446–450 (2013)
8. Liu, Y., Kelly, D.J.A., Ymg, H.Q., Lin, C.C.H., Kuznicki, S.M., Xu, Z.: Environ. Sci. Technol. 42, 6205 (2008)
9. Levinton, J.S., Poehron, S.T., Kane, M.W.: Environ. Sci. Technol. 40, 7597 (2006)

10. Wang, J.H., Han, P., Lu, A.X., Pan, L.G.: Rapid-indetermination of Heavy Metals and Research about the Applications of instruments. Quality and Safety of Agricultural Products 1(1), 48–52 (2012)
11. Su, Q.M., Qin, W.: Advances indetermination of lead in sea water. Marine Sciences 33(6), 105–111 (2009)
12. Zhu, G.Y.: Review of Electrochemical Detection Methods on Trace Heavy Metals. Modern Scientific Instrument 04, 21–29 (2013)
13. Tania, L., Eleni, B., Maxim, B.J., Julie, V.M.: In Situ Control of Local pH using a Boron Doped Diamond Ring Disk Electrode: Optimizing Heavy Metal (Mercury) Detection. Anal. Chem. 86(1), 367–371 (2014)
14. Yan: The high sensitivity, high precision and rapid heavy metal device successfully developed [EB/OL] (August 28, 2007), `http://www.jst.go.Jp/pr/info/inf0418/index.html` (November 15, 2011)
15. Wagtech Projects CO. LTD. Metalyser HM5000 [EB/OL] (February 2012), `http://www.wagtechprojects.com/products/Metalyser-Field-Pro-HM5000.html`
16. China Jiangsu Skyray Instrument Co., Ltd.HM-3000P, Portable detector for Heavy metal in water (April 24, 2012)
17. China Jiangsu Skyray Instrument Co., Ltd. HM-5000P, Portable detector for Heavy metal in water (November 22, 2012)
18. Du, Y.P., Huang, Z.X., Tao, W., Wei, X.M.: Determination of Trace Lead by Near Infrared Spectroscopy (NIR) with Enrichment Technique. In: Proceedings of Chinese Chemical Society of 26th Annual Meeting of Chemical Informatics and Chemometric Venue (2008)
19. Cang, J.S., Wu, S.B., Zhu, X.S., Liu, D.J.: Determination of Trace Cobalt in Samples by UV-Vis after Cloud Point Extraction. Chemistry World 04, 209–212 (2009)
20. Wu, J.J., Liu, T., Zhao, S.H., Li, Y., Xie, W.B., Ruan, Y., Sun, G.J., Chen, H.X.: Determination of Heavy Metal Plumbum in Textiles by XRF. Journal of Silk 51(5), 21–25 (2014)
21. The central science, environment and Resource System Research Institute Series, Heavy metals in soil simple and rapid analysis method, 1-9 [EB/OL] (March 30, 2011), `http://www.kankyo.metro.tokyo.jp/chemical/soil/information/analysis/heavymetals.html` (November 15, 2012)
22. Wang, J.H., Huang, W.J., Zhao, C.J., et al.: A portable heavy metal soil instrument: China. 200710175770.X (July 14, 2010)
23. Lu, A.X., Wang, J.H., Pan, L.G., Han, P., Han, Y.: Determination of Cr, Cu, Zn, Pb and As in Soil by Field Portable X-Ray Fluorescence Spectrometry. Spectroscopy and Spectral Analysis 30(10) (2010)
24. Reardan, D.T., Meares, C.F., Goodwin, D.A., et al.: Antibodies against metal chelate. Nature 316, 265–268 (1985)
25. Wylie, D.E., Lu, D., Carlson, R.: Monoclonal antibodies specific for metal ions. Proceedings of the National Academy of Sciences of the United States of America 89, 4104–4108 (1992)
26. Johnson, D.K., Combs, S.M., Parsen, J.D.: Lead analysis by anti-chelate florescence polarization immunoassay. Environmental Science and Technology 36, 1042–1047 (2002)
27. Lajunen, L.H.J., Perämäki, P.: Atomic fluorescence spectrometry (2007)
28. Welz, B., Sperling, M.: Atomic absorption spectrometry. John Wiley & Sons (2008)
29. Beauchemin, D.: Inductively coupled plasma mass spectrometry. Analytical Chemistry 82(12), 4786–4810 (2010)

30. Thayer, J.R., McCormick, R.M.: High-performance liquid chromatography (1996)
31. Cleven, R.F.M.J., Van Leeuwen, H.P.: Electrochemical analysis of the heavy metal/humic acid interaction. International Journal of Environmental Analytical Chemistry 27(1-2), 11–28 (1986)
32. Pinheiro, H.M., Touraud, E., Thomas, O.: Aromatic amines from azo dye reduction: status review with emphasis on direct UV spectrophotometric detection in textile industry wastewaters. Dyes and Pigments 61(2), 121–139 (2004)
33. Potts, P.J., Ellis, A.T., Holmes, M., et al.: X-ray fluorescence spectrometry. Journal of Analytical Atomic Spectrometry 15(10), 1417–1442 (2000)
34. Cogdill, R.P., Drennen, J.K.: Near-infrared spectroscopy. Drugs and the Pharmaceutical Sciences 160, 313 (2006)
35. Bontidean, I., Ahlqvist, J., Mulchandani, A., et al.: Novel synthetic phytochelatin-based capacitive biosensor for heavy metal ion detection. Biosensors and Bioelectronics 18(5), 547–553 (2003)
36. Amine, A., Mohammadi, H., Bourais, I., et al.: Enzyme inhibition-based biosensors for food safety and environmental monitoring. Biosensors and Bioelectronics 21(8), 1405–1423 (2006)
37. Raps, A., Kehr, J., Gugerli, P., et al.: Immunological analysis of phloem sap of Bacillus thuringiensis corn and of the nontarget herbivore Rhopalosiphum padi (Homoptera: Aphididae) for the presence of Cry1Ab. Molecular Ecology 10(2), 525–533 (2001)
38. Khosraviani, M., Pavlov, A.R., Flowers, C.G., et al.: Detection of heavy metals by immunoassay: optimization and validation of a rapid, portable assay for ionic cadmium. Environmental Science and Technology 32, 137–142 (1998)
39. Blake, D.A., Jones, R.M., Blake, R.C.: Antibody based sensors for heavy metal ions. Biosensors and Bioelectronics 16, 799–809 (2001)
40. Luan, Y.X., Han, P., Lu, A.X., et al.: Application of fruit and vegetable pollutants three in one portable tester. Journal of Agricultural Machinery 40(Suppl.), 146–149 (2009)
41. Lehmann, M., Riedel, K., Adler, K.: Aerometric measurement of copper ions with a deputy substrate using a novel Saccharomyces cerevisiae sensor. Biosensors and Bioelectronics 15(3-4), 211–219 (2000)
42. Karlen, C., Walinder, I.O., Heerick, D., et al.: Run of rates and eco-toxicity of zinc induced by atmospheric corrosion. Science of the Total Environment 277(1-3), 169–180 (2001)

System Design of Online Monitoring and Controlling System Based on Zigbee in Greenhouse

Fengmei Li, Yaoguang Wei*, Yingyi Chen, and Xu Zhang

College of Information and Electrical Engineering,
China Agricultural University, Beijing 100083, China
{497840887,1585891028}@qq.com, weiyaoguang@gmail.com,
chyingyi@126.com

Abstract. To solve the problem that online monitoring and controlling system in greenhouse, a wireless monitoring and controlling system which based on ZigBee has been proposed in this paper. The system consist data acquisition module and data gathering module. The data acquisition module adapted DHT11 as the temperature and humidity sensor, it monitor the information of temperature and humidity and transmit the information to the data gathering module by wireless RF Module CC2530, the data acquisition module is powered by solar battery. It realized wireless monitoring and easy to maintain. When received the monitoring results, the data gathering module will take logical judgment and realize automatic controlling of the temperature and humidity. The wireless monitoring and controlling network was connected by star topology, it adapted the ZigBee protocol. The network has the function of self-organize and self-adaptive. The result has shown that the system is more suitable for greenhouse monitoring and controlling.

The paper has constructed star self-organization network based on ZigBee, the developed system can solve the technical problems of transmission of remote data, and also avoid the drawbacks of installing a large number of sensors and terminal equipments. The system meets the requirements of multi-point, multi-factor, mobility and convenience in environmental monitoring. Finally, the experimental results show that the network can achieve good self-organization, scalability, stability, performance systems built for wireless monitoring and control of temperature and humidity in greenhouse-site needs. The working performance of the system built is stable, suitable for wireless monitoring and control of temperature and humidity in greenhouse-site.

Keywords: Temperature and humidity, measurement and control, ZigBee.

1 Introduction

In China, the greenhouse occupies an important position in vegetable production, quality and yield of vegetable is closely related to the temperature and humidity [1]. With the development of computer network and wireless sensor technology, wireless sensor network technology has played a revolutionary role in the data acquisition process [2]. It's difficult to lay the wires which have interference when we transmit information in long-distance. This traditional way brings a lot of inconvenience to the

---

* Corresponding author.

© IFIP International Federation for Information Processing 2015
D. Li and Y. Chen (Eds.): CCTA 2014, IFIP AICT 452, pp. 702–713, 2015.
DOI: 10.1007/978-3-319-19620-6_76

practical application [1]. In addition, the traditional measurement ways commonly adopt thermocouples, thermistors, temperature sensors and humidity sensors, which are susceptible to the sites of measurement and environmental restrictions. However, it will degrade performance of measurement system in long-term, and the inspecting and replacing the cable transmission regularly is indispensable in traditional ways. To sum up, it's inconvenient to update and maintain the system with the traditional methods on account of many shortcomings, for instance, the difficult mobility, the weak anti-interference, the inaccuracy of measurement and so on [3].

In view of the current situation lacking of effective means of online measurement and control in greenhouse, difficulties of monitoring and wiring and other issues, respectively from the collection of environmental information, wireless transmission and intelligent control, this paper designed and developed temperature and humidity acquisition module combining DHT11 and 8051, and developed wireless transceiver module based CC2530 to achieve a temperature and humidity information. In addition, the LCDs adopted in system display data in real-time and the actuators achieve the control of environment in greenhouse. All of them are for purpose of satisfing the environmental requirements in the greenhouse. The ZigBee technology can transmit the remote data and read each greenhouse's data to be unattended in a real application. The monitoring system has lower power, lower cost and larger network capacity compared with traditional wireless technology.

2 The Overall Structure of System

The ZigBee network supports star, tree and mesh, this paper adopts star including a coordinator node and three temperature and humidity acquisition terminal nodes. The overall structure is simple and the cost is very low, the system is easy to transplant and maintain. In a star network [4], all nodes can only communicate with the coordinator, the communication between each other is prohibitive.

This system includes a coordinator node and three temperature and humidity acquisition terminal nodes, the hardware components of coordinator node includes microcontroller AT89S51, LCD 1602, ZigBee module (coordinator), 5V DC power supply. The acquisition terminal node includes temperature and humidity sensor, microcontroller AT89S51, 1602 LCD display, ZigBee module (terminal node), 5V DC power supply. Figure 1 is a star network.

(1) Acquisition node: It collects the real-time temperature and humidity data, display on the LCD screen and transmit wirelessly to the coordinator node through the ZigBee.
(2) Coordinator node: It receives temperature and humidity data from the acquisition nodes through ZigBee, displays which on the LCD 1602 every 30 seconds depending on the needs in the different season and real-time parameters through man-machine interface [5]. You can set upper and lower parameters of temperature and humidity via touching screen. If data exceeds the predesigned threshold, four different alarm lamps will be lit up by micro-controller, and the relays open related device which can change the temperature and humidity, such as the humidifiers, fans, sprayers and heating.

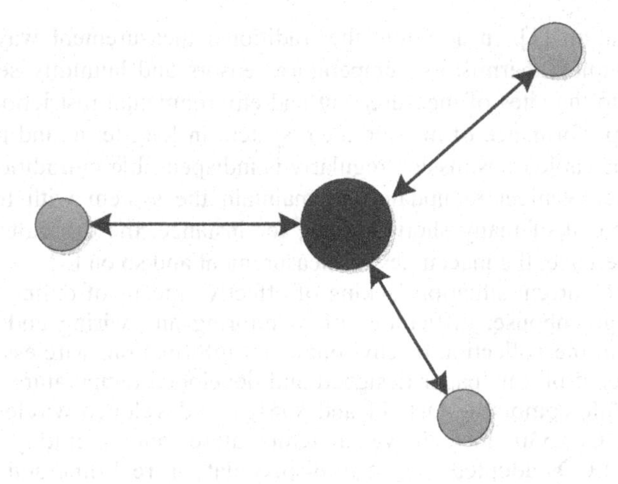

Fig. 1. Star topology

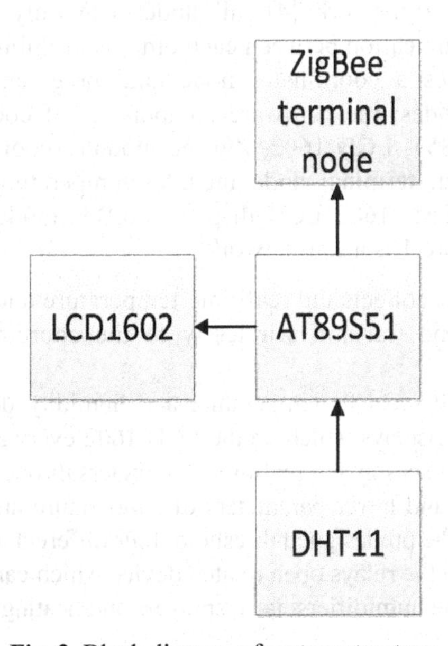

Fig. 2. Block diagram of system structure.

3 Materials and Methods

3.1 The Overall Structure of the System

The ZigBee[6-8] wireless sensor network consists of three terminal nodes and a coordinator node. The coordinator is the core of the sensor network because it is responsible for establishing, connecting, joinning and exiting network. In addition, it can assign network address for terminal nodes. The terminal nodes deployed in the monitoring area collect the information which is sended to the collected temperature and humidity data regularly to coordinator [3]. If finished, the terminal equipments enter into dormancy after receiving the coordinator's response. This method can ensure power consumption is ultra-low. The coordinator node is Full Function Device (FFD), and the terminal equipments are Reduced Function Device (RFD) [9].

A wireless sensor node contains the data acquisition module, data processing module, wireless transceiver module and power supply module, the node structure is shown in Figure3. Data acquisition module is responsible for collecting temperature and humidity data adopting a digital sensor DHT11, the data signal is amplified and transmitted to the processing module via the port P1_1. The data processing module uses CC2530 chip as micro-controller converting analog signals into digital signals. The wireless transceiver module is responsible for communicating with other sensor nodes, the power supply module provides energy [1] [9] for the other three modules.

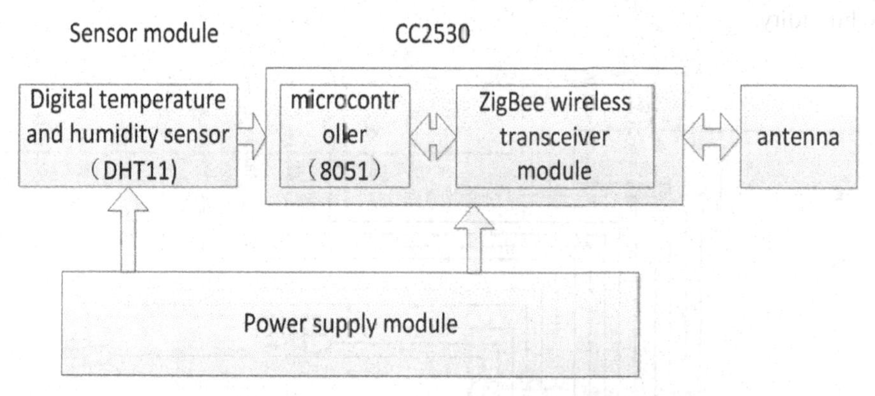

Fig. 3. Structure chart of acquisition node

3.2 Data Acquisition Node of Temperature and Humidity

(1) Temperature and Humidity Sensor Node

The DHT11 is a composite temperature and humidity sensor containing digital output calibration table. The sensor uses these calibration coefficients stored in OTP (One Time Programmable) memory in the process of detecting signals, with this method, it has high reliability and excellent long-term stability. Because the sensor consists of a NTC (Negative Temperature Coefficient) temperature sensor and a resistance element sensitive to humidity, it measures the temperature and humidity accurately at the same time. More importantly, the transmission distance is up to 20 meters with ultra-small size and low power consumption. To sum up, the sensor has many excellent qualities, such as the power consumption, the quick responsiveness and strong anti-interference. It's the best option for applications in greenhouse.

The DHT11 is connected to an 8 bit micro-controller with high performance. It is a four-wire package transmitting temperature and humidity signals to CC2530's P1_1 port via data line DATA.

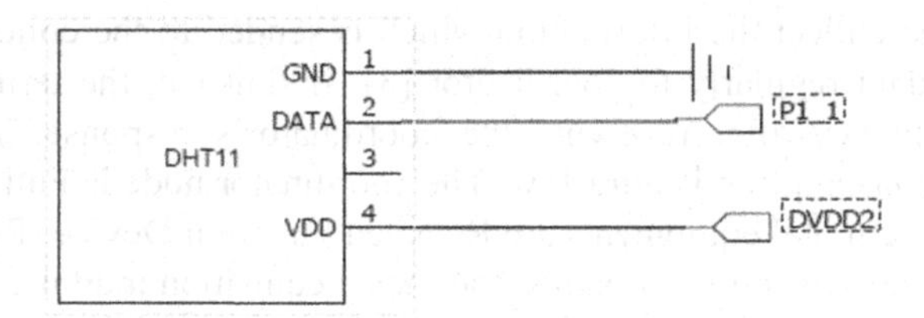

Fig. 4. The Wiring diagram of DHT11with single chip microcomputer.

(2) LCD Screen

The LCD 1602 displays letters, numbers, symbols and other dot matrix. LCD 1602 means the displayed content is 16*2, that's to say, it can display two lines and 16 characters each line. The first line shows the temperature and the second line shows the humidity.

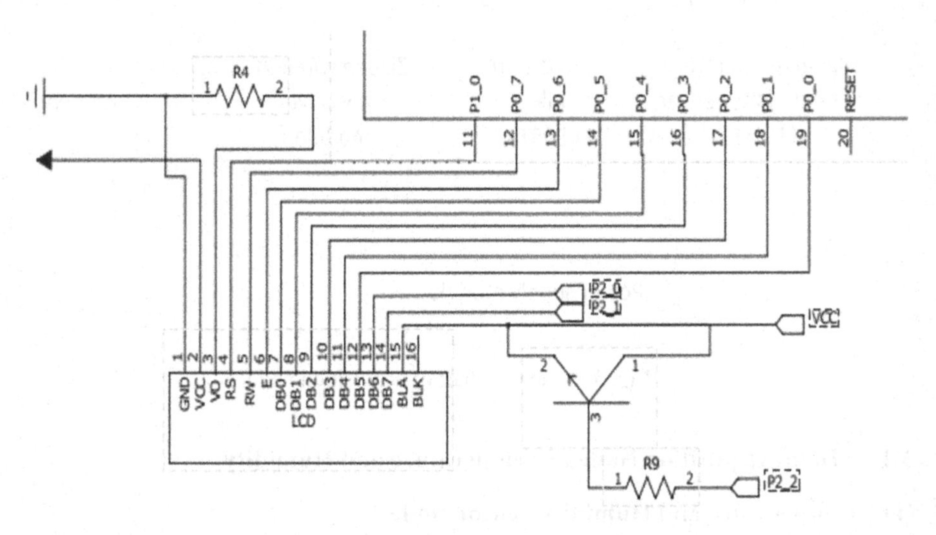

Fig. 5. Interface diagram of LCD screen with single chip microcomputer

The LCD displays the data post-processed by the microcontroller, the CC2530's port P0, P2_0, P2_1, P2_2 connect with data ports of LCD 1602, port P1_0, P0_7, P0_6 are respectively connect with RS, R/W, E of the 1602 modules.

(3) CC2530 Processing Module

The CC2530 is an IEEE 802.15.4 on-chip system that integrates an enhanced 8051 kernel, so it's easy to set up their wireless communications network combined with TI Z-STACK protocol stack. The internal configuration includes: enhanced 8051 CPU, programmable 256 kb flash memory, 8 kb RAM, a 16-bit timer, an 8-bit timer, an IEEE 802.5.4 MAC timer, a watch dog timer, a AES security coprocessor, a 5-channel DMA and 2 USART. The CC2530 support a variety of serial communication protocols with high stability, low power consumption, less peripheral devices and so on[10]. In addition, The CC2530 has different operating modes, and the transition time between modes is very short, so it can ensure low energy consumption [11]. The application of CC2530 circuit is shown in figure 6.

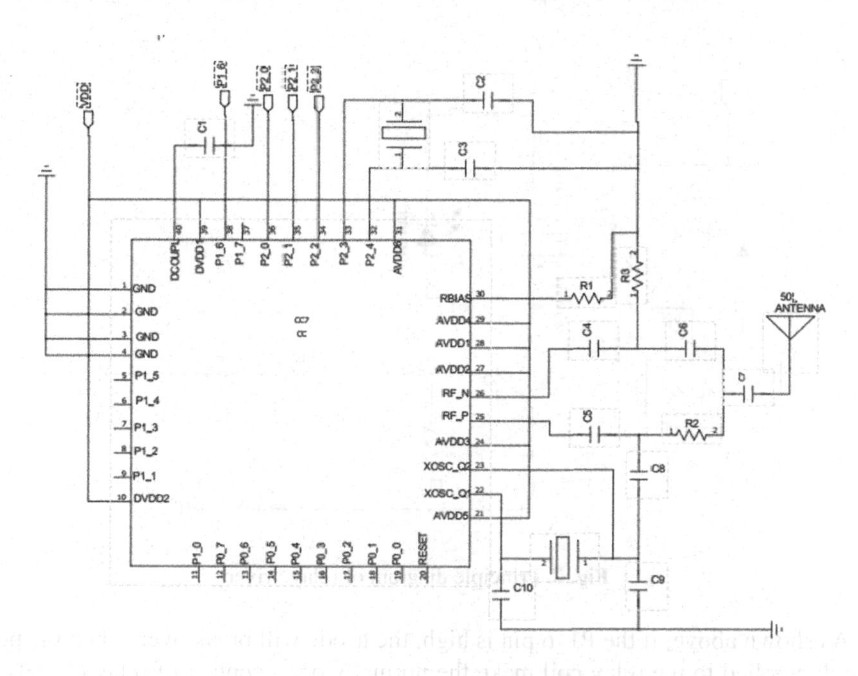

Fig. 6. Application of CC2530 circuit

(4) Energy Supply Module

The temperature and humidity monitoring nodes are used in agricultural greenhouse where environment is relatively poor, but battery-powered nodes can be changed frequently, integrating the above two points, the greenhouse requires the batteries with high-performance, large-capacity, long life and maintenance-free. Without any low-power measures, 2 ordinary battery can work for acquisition node over one day[12], and ZigBee wireless transceiver can control data sampling interval and take other low-power measures. Judging from the experimental results, 2 batteries can insure acquisition node working for more than one week, so the energy module reduces energy consumption significantly and farmers' production costs.

3.3 Hardware Design of Coordinator Node

Compared with the sensor node, the coordinator does not have a sensor module but increases the alarm module and relay module, The LCD displays temperature and humidity data of the three sensor nodes alternately at intervals of 10 s, the SCM (Single Chip Micyoco) triggers the corresponding relay when the alarm module lit the different LEDs. According to the actual situation of the greenhouse, the actuators connected with the relays can adjust the temperature and humidity in the greenhouse effectively.

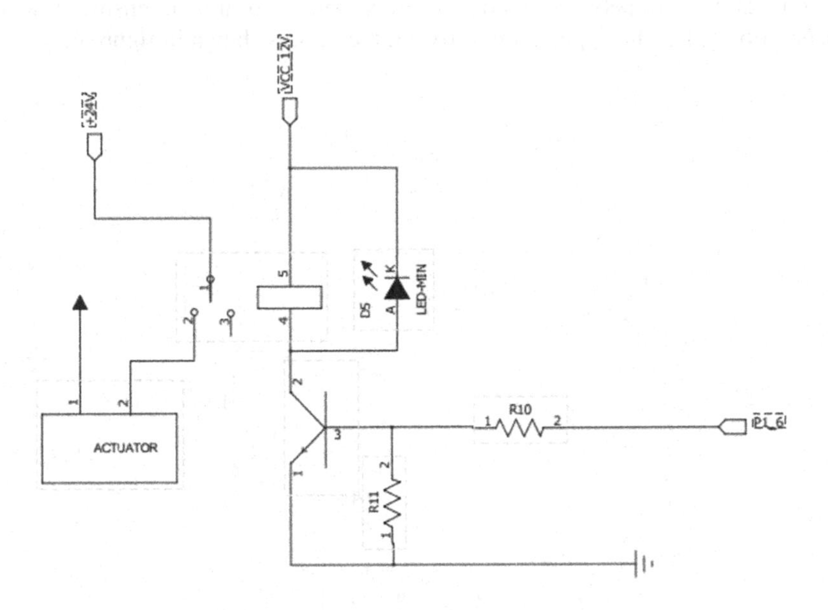

Fig. 7. Principle diagram of relay driver

As shown above, if the P1_6 pin is high, the triode will break over. Then the power supply applied to the relay coil make the normally open contact of relay closed. This equate to closing switch and triggering the related motors. On the contrary, if the output of P1_6 pin is low, the triode will cut off and there is no potential difference between the relay coils. It means the normally open contact of relay is released, it equate to switch off.

Alarm module use SCM to drive four LED lights L1, L2, L3 and L4. When the greenhouse temperature exceeds a preset maximum temperature lit L1, when the minimum temperature is lower than a preset minimum temperature lit L2, when the humidity is higher than the preset maximum humidity lit L3, when the humidity is lower than the preset minimum humidity lit L4.

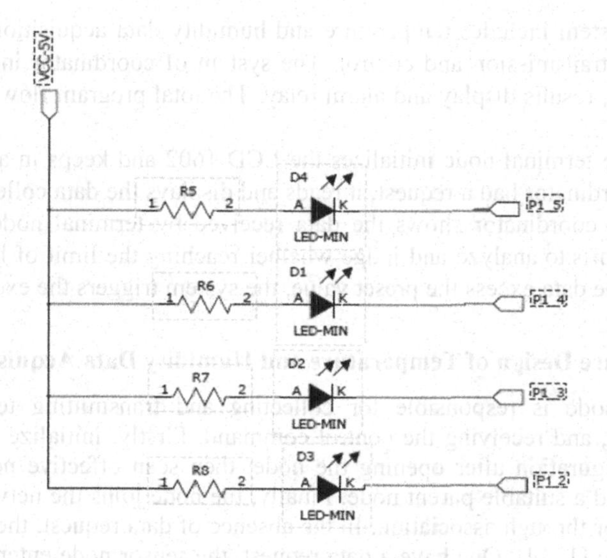

Fig. 8. Principle diagram of LED alarm

4 Design of System Software

The CC2530 chip is the core of ZigBee and built-in ZigBee protocol stack, we can develop software on the application layer reducing the difficulty of programming, so we can shorten the development period of the system. The programming environment of CC2530 is the IAR embedded workbench [13]. The temperature and humidity

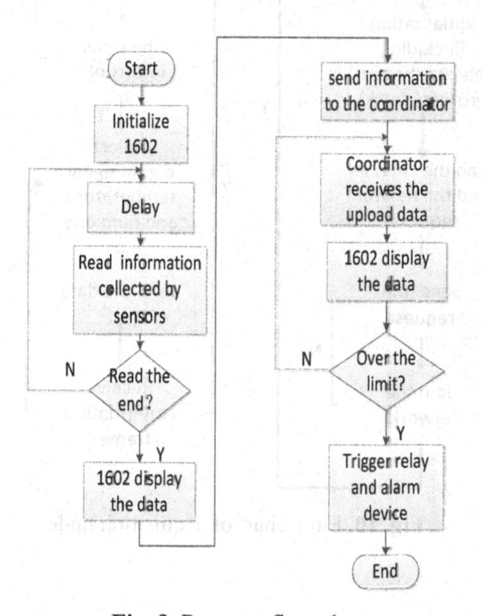

Fig. 9. Program flow chart

acquisition system includes temperature and humidity data acquisition, data display, wireless data transmission and control. The system of coordinator includes wireless data receiving, results display and alarm relay. The total program flow chart is shown in figure 9.

At first, the terminal node initializes the LCD 1602 and keeps in a dormant state. When the coordinator had a request, it reads and displays the data collected by sensor. Secondly, the coordinator shows the data received by terminal nodes. Finally, the coordinator starts to analyze and judge whether reaching the limit of temperature and humidity, if the data excess the preset value, the system triggers the executor.

4.1 Software Design of Temperature and Humidity Data Acquisition System

The sensor node is responsible for collecting and transmitting temperature and humidity data, and receiving the control command. Firstly, initialize the system and network configuration after opening the node, then scan effective network channel actively to find a suitable parent node. Finally, the node joins the network created by the coordinator through association. In the absence of data request, the sensor node is in a sleep state [1, 14]. Onc have a data request, the sensor node enters working state immediately to parse and response to the data request. That's to say, when during the period of dormancy, if there is an external interrupt or timer interrupt, the node will be restored to working condition and begin to perform the task.

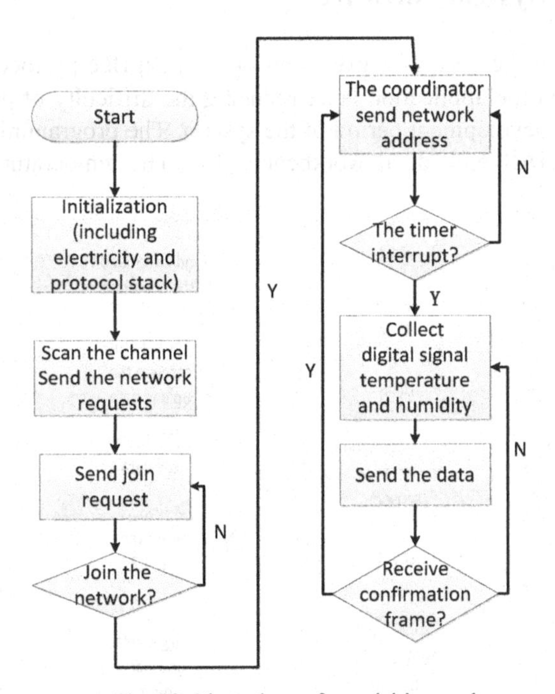

Fig. 10. Flow chart of acquisition node

Because the DHT11 is equipped with a standard interface without an external A/D converter, it is easy to control and implement. The flow chart of acquisition node is shown in figure 10.

4.2 Software Design of Coordinator Node

The purpose of network coordinator is to establish and manage a wireless network, transfer the data to LCD through the serial ports. Firstly, the coordinator establishes a wireless LAN. Secondly, scan a channel for receiving a new node to join, which is assigned to a specific address, it's called the binding between the nodes. Finally, the coordinator initializes the LCD and trigger, if acquisition node sends data successfully the coordinator will reset the serial port.The coordinator program flow chart is shown in figure 11.

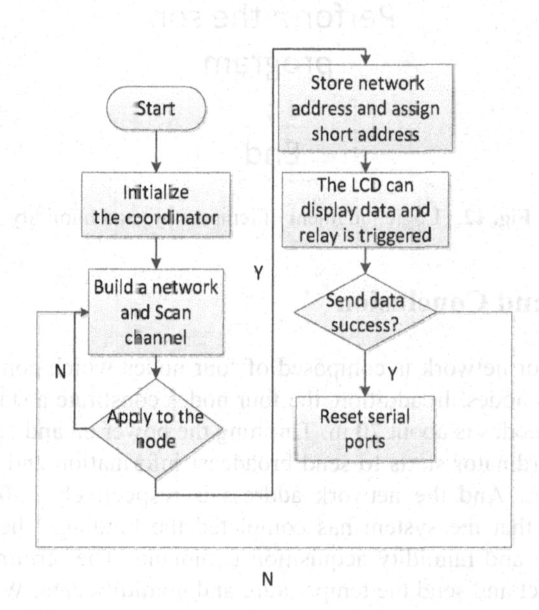

Fig. 11. Flow chart of coordinator program

After the coordinator receives temperature and humidity data from acquisition sensor nodes, the information takes turns to show data every 10 s in LCD1602. The coordinator analyzes and judges whether the number is over the acquisition nodes, and then executes a subroutine to judge whether reaching the limit of temperature and humidity, if reaching the limit, the system will trigger corresponding relay. The logic judgment flow chart is shown in figure 12.

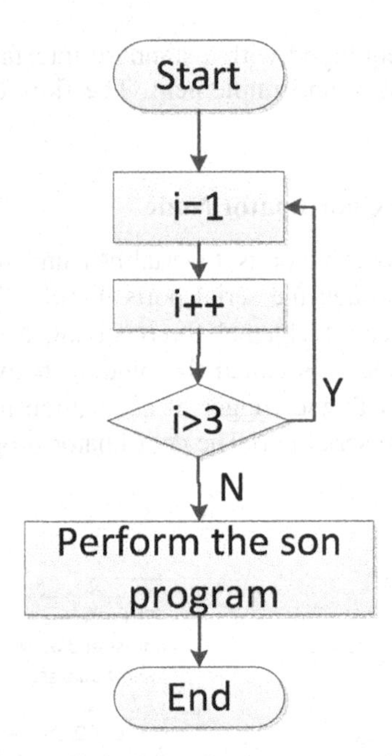

Fig. 12. Logic judgment of temperature and humidity

5 Results and Conclusion

The wireless sensor network is composed of four nodes which contain a coordinator and three terminal nodes. In addition, the four nodes constitute a star network and the distance between nodes is about 70 m. Turnning the power on and seting the baud rate to 38400, the coordinator starts to send broadcast information and distribute address for terminal nodes. And the network address is respectively 0x0035, 0x0036 and 0x0037, showing that the system has completed the binding. Then the coordinator sends temperature and humidity acquisition command. The terminal nodes respond command to collect and send the temperature and humidity data. When you remove a terminal node, the LCD corresponding to this terminal node displays 0. And if you join it again, the LCD starts to display the real temperature and humidity. So the system has a self-healing capability. During the experiment, we programed aim at ZigBee protocol to realize the three to one wireless communication between nodes, so we achieved the collection and control of temperature and humidity in greenhouse.

In this paper, by using the technology of forming the network automatically between terminals and coordinator, we can achieve the data transmission of many to one and monitor the temperature and humidity of several greenhouses at the same time. Except for gathering environmental information, the coordinator is devoted to triggering alarm lamp and striking the corresponding relay when the parameter value exceeds the predetermined value. Therefore the coordinator can control temperature and humidity by controlling the motor drive. This system has several features: Firstly, terminal node adopts digital temperature and humidity sensor which has the features of high accuracy and good stability. Secondly, the SCM adopts CC2530, integrating RF-chip and microcontroller unit, which has the different operating modes to ensure low energy consumption with short conversion time. Thirdly, the wireless transceiver module adopts ZigBee wireless network technology to reduce the problems caused by wiring. Fourthly, the system has advantages of simple structure, easy to transplant and high accuracy to meet the modern intelligent management in agriculture.

Acknowledgment. This work was supported by "Special Fund for Agro-scientific Research in the Public Interest" (201203017).

References

1. Zhuang, L., Lu, Q., Wang, X.: The design and implementation of temperature and humidity wireless acquisition node based on CC2530 in greenhouse. Hubei Agricultural Science (3), 585–582 (2014)
2. Li, S., Duan, Y.: The design of environment information collection system based on CC2430 in greenhouses. Micro Computer and Applications (19), 31–34 (2012)
3. Li, Z., Zhang, X., Liu, H.: The design and implementation of a temperature and humidity monitoring system based on CC2530. Measurement and Control Technology (05), 25–28+ page 39 (2013)
4. Liu, Q., Song, L.: The research of ZigBee wireless sensor network (WSN). Development and Application of Computer (6), 44–45 + 48 pages (2008)
5. Pang, D.: The design of agricultural greenhouse temperature and humidity control system based on single chip microcomputer and PLC. Hubei Agricultural Science (02), 450–448 (2013)
6. Ren, Z., Huang, Y.: The network monitoring platform of wireless sensors based on CC2530. Electronic Technology Applications (10), 125–122 (2012)
7. Lu, N.: The monitoring system of wireless greenhouse temperature and humidity based on ZigBee technology. Modern Electronic Technology 31(15), 98–100 (2008)
8. Li, J., Hu, Y.: The application design of ZigBee communication network based on CC2530. Electronic Design Engineering (16), 111–108 (2011)
9. Wang, Q., Chen, Z., Chen, X.: The design of the warehouse temperature and humidity acquisition system based on ZigBee. Computer and Digital Engineering (9), 211–207 (2009)
10. Yuan, Z.: The design of greenhouses wireless monitoring and control system based on ZigBee technology. Jiangsu Agricultural Science (11), 397–396 (2012)
11. Tan, Z., Zhang, Z.: The ZigBee wireless street lamp energy-saving intelligent monitoring system based on CC2530. Micro Computer and Applications (19), 83–81 (2011)
12. Huang, J., Yang, F., Zhang, Y.: The design and application of greenhouse wireless sensor node based on CC2430. Journal of Huazhong Agricultural University (5), 123–119 (2013)
13. Jin, H.: The design of wireless sensor network node and research of communication based on Zigbee, p. 76. Hefei University of Technology (2007)
14. Han, Y.: The design of wireless sensor network node in aquatic product refrigeration truck based on CC2530. Agricultural Mechanization Research (4), 178–174 (2013)

Study on the Temporal and Spatial Variability of Maize Yield in Precision Operation Area

Yueling Zhao[1], Haiyan Han[2], Liying Cao[1], Li Ma[1], and Guifen Chen[1,*]

[1] Jilin Agricultural University , Jilin 130118
[2] Changchun University, Jilin 130000

Abstract. Jilin province is an important agricultural production resource in our country. Study on the temporal and spatial variability of maize yield are significant to understand the potential nutrients in soil, and guide agricultural practice for the recent years in precision operate area. The every year yield distribution belonged to the normal distribution by testing K-Sand the coefficient of variation of them were 16.33%, 11.85%, 12.90% respectively. That precise operation can reduce the yield difference in every operate cell. As a result, the temporal and spatial variability of maize yield can be successfully applied to yield prediction in Jilin Province. Partition results can not only guide soil nutrient evaluation, and can be used to implement variable input and precise fertilization recommendation.

Keywords: Temporal and spatial variability, maize yield, precision agriculture.

1 Introduction

It is directly affect agricultural production regenerative potential, that yield prediction status are understood, so the maize yield level of the comprehensive analysis and evaluation is very necessary, it is the use of agricultural resources and technology in an effective way. In precision fertilization can be employed in management zones instead of grid sampling technology, and the extensive application [1-3]. In recent years, with the precision production technology system improvement and application in our country, some researcher used crop yield data of many years in comprehensive analysis in agricultural management partition extraction research.

With the precise operation technology in agricultural production in China continues to expand and application of crop production, continuously improved, the maize problem is getting more and more attention under the background, how to predict the future of food production for a period of time has become a research hotspot[4]. To investigate the relationship between the precision maize yield and time series, and the results of the comprehensive analysis and study is very necessary, it is a way of effective use of agricultural resources and technology. Through the analysis of precision for variation characteristics of maize yield, on the one hand can be scientific,

* Corresponding author.

© IFIP International Federation for Information Processing 2015
D. Li and Y. Chen (Eds.): CCTA 2014, IFIP AICT 452, pp. 714–719, 2015.
DOI: 10.1007/978-3-319-19620-6_77

reasonable, effective evaluation on it, on the other hand, the workers engaged in agricultural production not only can improve the use efficiency of soil nutrients, but also save the resources, obtain better economic benefits, to protect the ecological agriculture resources and the quality of environment. The temporal and spatial variability of maize yield were studied for some years in precision area in the paper.

2 Materials and Methods

2.1 The Situation of the Study Area

The study area is located in yushu City of Jilin Province. The researcher area was located at latitude 44.999369 -45.002761, longitude 126.3139 -126.31671. The genus sub-humid temperate continental monsoon climate, annual average temperature is 4.6 °C -5.6 °C, annual precipitation is the middle of in 500 -600mm, and the accounting for 90% of annual rainfall is concentrated in the warm season, the soil is more fertile, soil types are suitable for planting maize, bean, rice and other various crops. The distribution map of soil samples can be shown from the Fig.1.

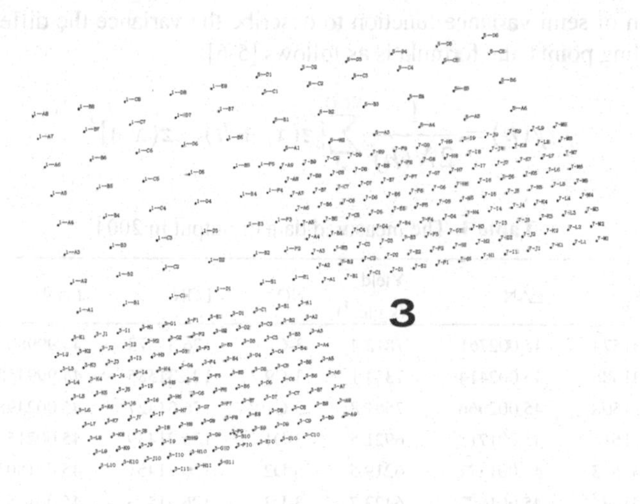

Fig. 1. Distribution map of soil samples

2.2 Process and Methods

We used DGPS(Differential Global Satellite Positioning System) devices in USA company Trimble for accurate positioning, set point, draw the boundary line and on the acquisition of land area. Then the researchers are back to the lab with the data by using ArcGIS software to the plots of the data collected by the spacing is according to 40m*40m as a unit area of the anchor point, classification, determination sampling points and sampling points map making. The actual measured area of a sample, the number of samples of corn plants, rough calculation can count per plant grain number per unit area yield of maize, the calculation according to the following formula: 1000 grain weight (grams) * Mu * panicle number /1000000 = kg / mu, and the yield of unit conversion into kg.ha^{-1}. The results can be shown from Tab.1

2.3 Data Analysis

All the experimental data were analysed using SPSS 18.0 and ArcGIS as tools in the processing. At first, the samples data were standardized, then in order to achieve data sharing, the maize yield results are be saved to the excel table and input to the precision agricultural database.

A quantitative tool of geostatistics as a soil variable space analysis, can be analysed by variation function, discrimination, contrast observation data, it is used to describe the spatial difference between the variables of the form of function, it is a kind of distance measure function H. It describes the similarity between the two sampling points, the similarity decreases as the distance between sampling points. When the distance between them to a certain extent, they often have no correlation, this time they will become two independent. It is a tool to describe the function of the spatial variability of soil properties in different distance between observation and reflect the changes of values, so the half of the definition of semi variance function to describe the variance the difference between the two sampling points, the formula is as follows [5-6]:

$$\gamma(h) = \frac{1}{2N(h)} \sum_{i=1}^{N(h)} [z(x_i + h) - z(x_i)]^2$$

Table 1. The measured data of output in 2004

NO.	LON	LAN	Yield (kg.ha^{-1})	NO.	LON	LAN	Yield (kg.ha^{-1})
3-A1	126.31574	45.002761	7812.4	3-C8	126.31573	44.999805	7869.4
3-A2	126.31586	45.002414	7354.1	3-C9	126.31585	44.999458	8077.2
3-A3	126.31598	45.002066	7561.8	3-C10	126.31427	45.002498	8102.7
3-A4	126.3161	45.001718	6921.5	3-D1	126.31439	45.00215	7589.6
3-A5	126.31623	45.001371	6219.5	3-D2	126.31451	45.001803	7687.5
3-A6	126.31635	45.001023	6423.7	3-D3	126.31463	45.001455	6986.4
3-A7	126.31647	45.000676	6635.4	3-D4	126.31475	45.001107	7152.8
3-A8	126.31659	45.000328	7324.1	3-D5	126.31488	45.00076	7354.6
3-A9	126.31671	44.99998	6523.7	…	…	….	….
3-A10	126.31525	45.002674	6952.2	3-K3	126.31242	44.998843	9482.7
3-A1	126.31574	45.002761	7812.4	3-K4	126.31083	45.001884	8897.7
3-A2	126.31586	45.002414	7354.1	3-K5	126.31095	45.001536	8046.2
…	…	….	….	3-K6	126.31107	45.001188	8157.9
3-C7	126.31561	45.000152	8260.6	3-K7	126.31119	45.000841	7982.5
3-C8	126.31573	44.999805	7869.4	3-K8	126.31132	45.000493	8534.6

In the formula, h is the space distance of the sample points, also known as step or potential difference (Lag); N (h) represents the spatial isolation distance apart as the number of samples h. The value of Z (X_i) and Z (x_i+h) were measured in two position. γ(h) is the value of the semi-variance function. Variation as a function of distance is the semi variance of H, and increases with the increase of H in the semi variance function within a certain range, but when the maximum distance between the observation point is greater than the distance between the regional variables, the values of the function tends to a constant, and contains steady state.

3 Results and Analysis

3.1 The Descriptive Statistics of Soil Nutrient Data

In the study of soil science, the magnitude coefficient of variation (CV) usually represented spatial variability strength of soil nutrient characteristics . It is usually weak variability if variation coefficient of $CV \leq 0.1$, $0.1 < CV < 1$ as moderate variability, $CV \geq 1$ strong variability [7-8]. Study on the variation characteristics of maize yield is each are not identical, soil of different yield properties have bigger difference.

Table 2. Attribute characteristics in different years of production

Case	Mean	SD	Variance	Kurtosis	CV%	Min	Max	K－S
2004	8422.668	1375.178	0.605	0.31	16.33	6005.8	12328.2	0.702
2007	9069.098	1170.657	-0.3	-1.02	12.9	6606.4	10986.4	0.139
2009	7303.635	865.239	0.105	0.33	11.85	5182.23	9765.11	0.966

The cell yield of these three years by K-S test, its distribution accords with the normal distribution, The average maize yield is the mean 8422.67kg.ha-1 in 2004, the average maize yield is average 9069.1 kg.ha-1 in 2007, the average maize yield is 7303.63 kg.ha-1in 2009 from the table 2.The coefficient of variation of them were 16.33%, 11.85%, 12.90%, coefficient of variation, as can be shown in Figure 2.The yield decreased gradually, and engaged in precision during the operation, the coefficients of variation were less than 20%, belongs to the medium variation.

3.2 The Temporal and Spatial Variability of Maize Yield

Observation of 2004 to yield spatial distribution map of 5-14 in 2009 2004, can be seen in the cell changes in output frequency is larger, and with the precision of the year by year, change of output frequency in cells decreased obviously, the first two years of the analysis of production trends can be seen, the east-west direction on the production line is steeper, after several wheel precise operation, each cell analysis of the yield curve changes in the east-west direction, they yield trend line close to the line is smooth, it also shows that the spatial variation of yield decreased, difference in. That precise operation can reduce the yield difference within a cell.

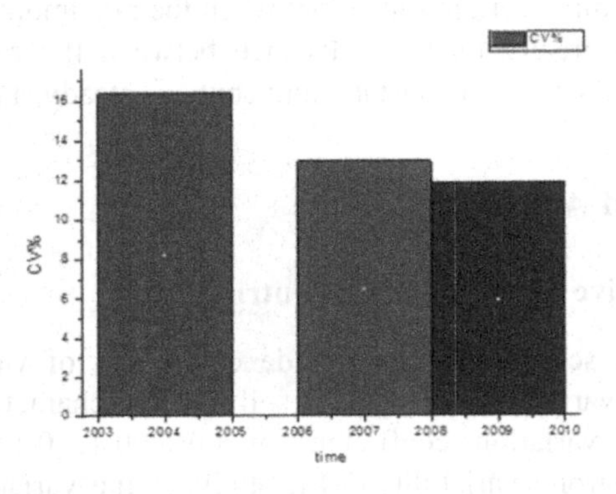

Fig. 2. Three years of variation coefficient of yield comparison chart

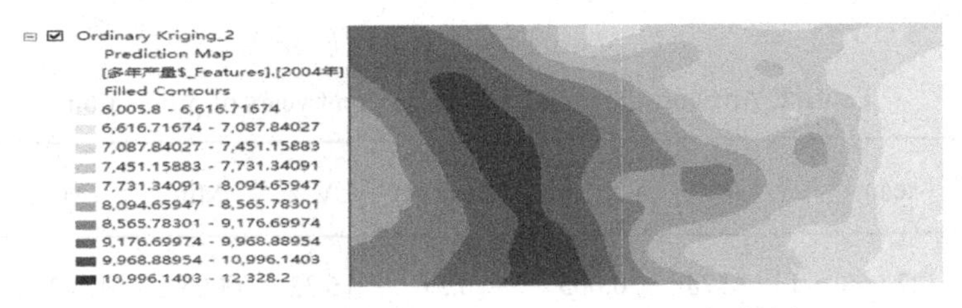

Fig. 3. The spatial attribute graph of output in 2004

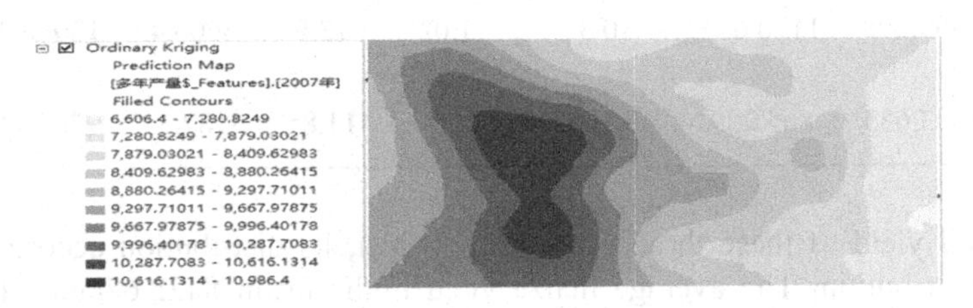

Fig. 4. The spatial attribute graph of output in 2007

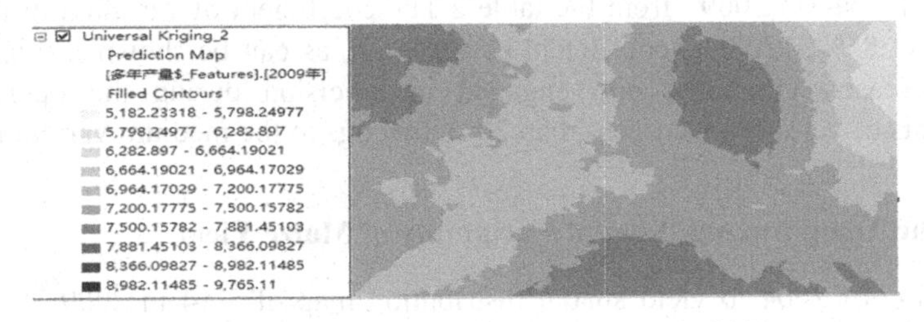

Fig. 5. The spatial attribute map of output in 2009

4 Results and Discussion

In the actual process of agricultural production, researchers know Soil Nutrient timing variability exists, don't also aware of the crop yield in the study area also exists obvious spatio-temporal variation. The application of traditional time the total output of grain in Shandong province of the prediction model and the ARIMA model predicts that the regression equation by Zhang Xiaojie et al. The prediction effect is relatively good [9]. Of course, the variability of maize yield is affected by many factors, such as variety, environment, climate, plant diseases and insect pests, cultivation mode and management mode, through the analysis of precision operation area in recent years, soil nutrient and yield data, mining and grasp the changes of crop yield in space and time, the characteristic of temporal variation of maize yield was given in recent years.

Acknowledgments. This work is supported by the Jilin province education science planning projects (Grant No. GH13180) ; Jilin Agricultural University general topic (Grant No. 2012xjyb048); department of education of Jilin Province (Grant No. 2013(68)); Science and TechnologyBureau of Changchun city (Grant No. 13KG71). The authors also gratefully acknowledge the helpful comments and suggestions of the reviewers, which have improved the presentation.

References

1. Zhang, H., Zhang, G.L.: Farm scale spatial variability of soil quality indicators. Chinese Journal of Soil Science 34(4), 241–245 (2003)
2. Schepers, A.R., Shanahan, J.F., Liebig, M.K., Schepers, J.S., Johnson, S.H., Luchiari Jr., A.: Appropriateness of management zones for characterzing spatial variability of soil properties and irrigated corn yields across years. Agronomy Journal 96, 195–203 (2004)
3. Wu, X.-L., Wang, D.-Q., Xu, B., et al.: Spatial Variability of Black soil Nutrients in Hilly Land. Chinese Journal of Soil Science 41(4), 825–829 (2010)
4. Lizarraga, H.S., Lai, C.G.: Effects of spatial variability of soil properties on the seismic response of an embankment dam. Soil Dynamics and Earthquake Engineering 64(9), 113–128 (2014)
5. Wang, T., Franz, T.E., Zlotnik, V.A., et al.: Investigating soil controls on soil moisture spatial variability: numerical simulations and field observations. Journal of Hydrology (March 16, 2015) (accepted manuscript)
6. Glendell, M., Granger, S.J., Bol, R., et al.: Quantifying the spatial variability of soil physical and chemical properties in relation to mitigation of diffuse water pollution. Geoderma 214–215(2), 25–41 (2014)
7. Narendra Babu, C., Eswara Reddy, B.: A moving-average filter based hybrid ARIMA–ANN model for forecasting time series data. Original Research Article Applied Soft Computing 23(10), 27–38 (2014)
8. Tiao, G.C.: Time Series: ARIMA Methods International Encyclopedia of the Social & Behavioral Sciences, 2nd edn., pp. 316–321 (2015)
9. Zhan, X.J., Zhang, Z.: Application of time series analysis model on total corn yield of Shandong province. Research of Water Conservation 14(3), 309–311 (2007)

Soil Water Sensor Based on Standing Wave Ratio Method of Design and Development

Yinli Xu[1,2,3], Weizhong Yang[1], and Zhenbo Li[1,2,3]

[1] College of Information and Electrical Engineering, China Agricultural University, Beijing 100083, China
[2] Key Laboratory of Agricultural Information Acquisition Technology, Ministry of Agriculture, Beijing 100083, China
[3] Beijing Engineering Center for Advanced Sensors in Agriculture, Beijing 100083, China
jjkeyxu@163.com, yamantaka.yang@gmail.com, zhenboli@126.com

Abstract. Soil moisture is one of the most important parameters in the soil properties, the traditional methods of soil moisture measurement is time-consuming. And it can't meet the soil moisture measurement's real-time and fast requirements. This paper analyzes the relationship between soil moisture and its dielectric properties, research the standing wave ratio method for measuring soil water content. Firstly, design hardware circuit, plate making, and welding circuit board, then insert the probe into the soil, by measuring the reflection coefficient, we can know the probe impedance which depend on the soil dielectric constant. At last we can find the internal relation between output voltage and soil water content, by measuring the multiple sets of different water content of soil.

Keywords: Soil moisture, Standing-wave ratio, Dielectric property.

1 Introduction

Soil moisture is an important part of the soil, and plays an important role in the growth of crops. Our country is lack of water resources, agricultural water accounts for about 70% of the total national water, of which more than 90% for irrigation, the current national average irrigation water utilization rate is only 43%, Unit water output of agricultural products is far lower than developed countries, and so, the water saving potential is very big also. The current water conservation and efficient water use by governments around the world increasingly widespread attention, promote water-saving technologies and equipment is to save water resources, improve the utilization of water resources in an effective way.

Water is not the only one important factor in the growth of crops, and it is an important prerequisite for effective utilization of fertilizer can be crops, particularly effective use of nitrogen fertilizer, if the soil is too dry will result in fertilizer use cannot be fully absorbed by crops, Cause soil salinization. If too much soil moisture, will cause water leakage, Fertilizer with water will penetrate into the groundwater has become an important reason for groundwater contamination, Due to soil structure and

© IFIP International Federation for Information Processing 2015
D. Li and Y. Chen (Eds.): CCTA 2014, IFIP AICT 452, pp. 720–730, 2015.
DOI: 10.1007/978-3-319-19620-6_78

soil moisture spatial variability, create the different soil water content in the same field, which requires promoting soil moisture monitoring and variable irrigation technology.

Our country is a large agricultural country in the world, seeking fast and efficient soil moisture measurement technology is particularly important. China accounted for 22% of the world population, while only 8% of the world's freshwater resources total. Per capita water resources only 2300m3/y, equivalent to only a quarter of the world average. Is one of the world's water resources per capita the poorest 13 countries, water resources in our country is very poor, on the other hand wasteful and very large, at present, China's annual irrigation water to about 380,000,000,000 m3, The effective utilization rate of it is only 30% ~ 40%, while developed countries reached 70% ~ 80%,China's per cubic centimeter of water less than 1 kg of grain production capacity, Advanced countries has amounted to 2.35 kg, one of the reasons for this situation is that there is no an effective soil moisture measurement technology to ensure the implementation of a system of water-saving agricultural technology, So with an urgent need to study in our country to develop a suitable for application in production of soil moisture rapid measurement technology.

The study and implementation of precision agriculture support to solve problems faced in the agricultural production provides the theory and technology. As the main part of the information technology, sensor technology is the important subject in the process of precision agriculture research, is also an important guarantee of realization of soil moisture monitoring and variable irrigation technology. Electronic information experts, soil scientist, agronomists, water conservancy, and some engineers are trying to seek a kind of high performance soil moisture quickly measuring sensor technology. Currently, a variety of soil moisture measurement technology has matured, based on TDR, FDR, and SWR principle dielectric sensor has a more widely used in agriculture, water conservancy, meteorology and other departments and university research. This paper describes a method based on the measurement of soil moisture standing principle.

2 Methods Introduction

2.1 Common Measurement Methods

(1) Drying Method

This is the only way can direct measuring soil moisture, is currently on the international standard method. It is generally believed that traditional drying measured soil moisture value is credible, can be used for other kinds of soil moisture measurement calibration standards. But its drawback is obvious, drying method is time-consuming, laborious, deep sampling is difficult, will destroy the soil sampling.

(2) Tension Meter Method

Tension meter method is a widely used method to measure soil moisture, and it is measuring the soil matric potential. Due to the relationship between energy and soil moisture is very complicated, non-linear, and easily affected by many physical and chemical properties of soil, this method lagged and loopback, affect the measurement speed. The existence of the above defects greatly limits the popularization and application of the method.

(3) Near Infrared Method
Near infrared reflection method uses the water in the soil resonance absorption characteristics to measure soil water content. Its advantage is that can realize the non-contact measurement of the soil, but only can measure the soil water content, when measuring deep soil moisture we need to slot soil.

(4) Dielectric Property Law
Soil is composed of soil particles, water and air, a special dielectric, within a certain frequency electromagnetic wave, the dielectric constant of soil particles is about 4, and dielectric constant of soil is about 1, the dielectric constant of water is about 77, far more than the air and soil particle dielectric constant, so we can get the water content of soil by measuring the dielectric constant when the soil is under a certain frequency.

There are many other methods to measure the soil moisture, according to the purposes of the research on soil moisture and the need of agricultural production and ecological environment construction, corresponding measurement method can be used.

2.2 Standing Wave Method Measuring Principle

Standing wave ratio method, also known as the impedance method or amplitude domain reflection method, it is based on radio frequency technology in soil moisture measuring method of standing wave ratio principle, Experiment shows that the change of the three states of mixture dielectric constant Ka can be cause significant change of standing wave ratio on a transmission line. It measured the dielectric constant of soil by measuring the impedance of the probe in soil media. Signal source sends Electromagnetic wave, which spread along the transmission line, in the joint of transmission lines and soil probe due to the reflection impedance mismatches occur, the reflected wave and the incident wave to form a standing wave in the transmission line.

Therefore, the reflection coefficient is determined by measuring the voltage at both ends of the transmission line, and then we can know the impedance of soil probe. While the geometry of the probe is determined, probe impedance depends on the dielectric constant of the soil, and dielectric constant of the soil is mainly affected by soil moisture, so we can achieve the purpose of measuring soil water content by measuring the impedance of the probe.

This method has the advantage of high precision measurement, Fast response, and it is not sensitive to electromagnetic interference, Can be used long-term embedment fixed monitoring or instant insert using. All circuit integration within the sensor probe, using flexible, sensor output is the standard analog signals, can be connected to any general analog acquisition equipment. But precision is less than TDR, due to the low working frequency, the general is 100 MHZ, it is greatly influenced by soil types, this is also a measure of this article will focus on the study.

The structure principle block diagram is as follows

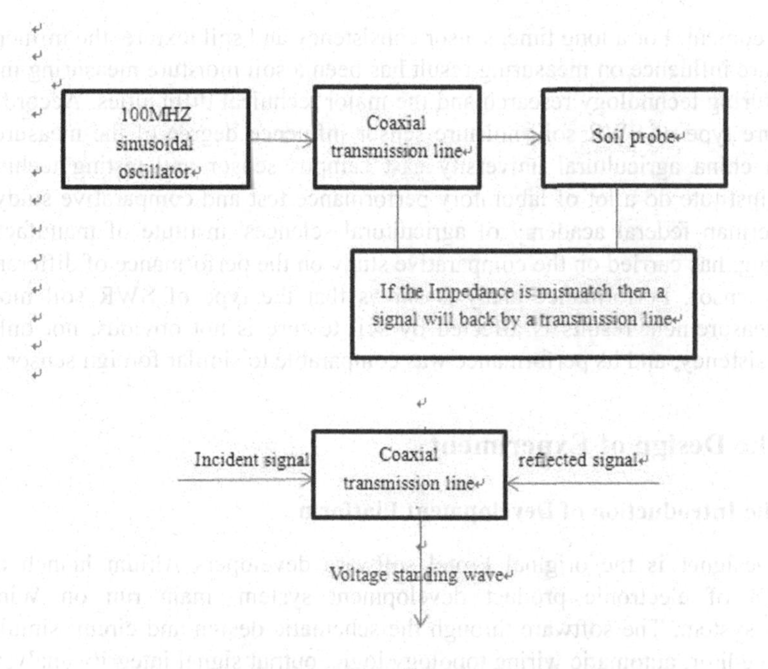

Fig. 1. Structure principle diagram

2.3 The Soil Dielectric Constant Relationship with the Soil Moisture Content

Soil dielectric constant has close relation with soil moisture content, because the dielectric constant of soil particles, water and air difference is very big. Dielectric constant of air is 1, dry soil dielectric constant is 3 ~ 7, free water in the temperature of 20°C the dielectric constant is 80.36. Therefore the different proportion of a certain volume of water in the soil, There will be significant changes in the dielectric constant. In 1980 found that soil volumetric water content and dielectric constant relations expressed by a 3 a number of empirical formula, but as a result of various types of soil in our country, and different basic physical and chemical properties of soil, therefore, the dielectric constant is affected by many factors, to reduce the error of the general formula, china Agricultural University, through the study of several typical Chinese soil permittivity and moisture relations, fitting a new $\boldsymbol{\theta_V}$-$\boldsymbol{\varepsilon_r}$concise model:

$$\theta_{\mathrm{v}} = 0.1219\varepsilon_{\mathrm{r}}^{0.5} - 0.1846 \tag{1}$$

Soil is composed of water and air, solid porous medium, where the water dielectric constant of about 80, solid dielectric constant is about 4, while air dielectric constant is about 1.So that, the dielectric constant of aqueous soil should be mainly decided by the water. By measuring the dielectric constant of the soil can achieve the goal of measuring soil volumetric moisture content.

Standing wave method of measuring soil moisture by measurement of standing wave ratio on a transmission line can achieve the goal of measuring soil volumetric moisture content. For a long time, sensor consistency and soil texture, the influence of temperature influence on measuring result has been a soil moisture measuring method and measuring technology research and the major technical difficulties. According to soil texture type of SWR soil moisture sensor influence degree of the measurement result, in china agricultural university east campus sensor and testing technology research institute do a lot of laboratory performance test and comparative study, and in the German federal academy of agricultural sciences' institute of manufacturing engineering, has carried on the comparative study on the performance of different soil moisture sensor. Performance analysis shows that the type of SWR soil moisture sensor measurement results is affected by soil texture is not obvious, not only has good consistency, and its performance was comparable to similar foreign sensor.

3 The Design of Experiment

3.1 The Introduction of Development Platform

Altium Designer is the original Protel software developers Altium launch of the integration of electronic product development system, main run on Windows operating system. The software through the schematic design and circuit simulation, PCB map editor, automatic wiring topology logic, output signal integrity analysis and design techniques such as perfect fusion. Provides designers with a new design solution, the designer can easily design, familiar with the software will make the quality and efficiency of circuit design is greatly increased.

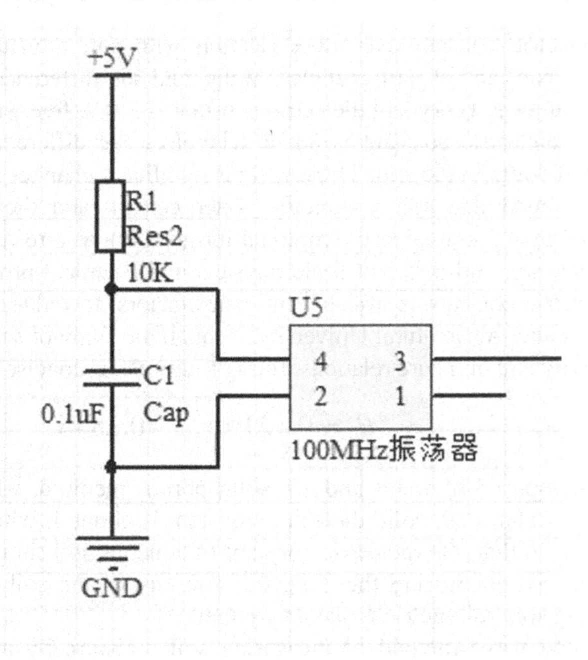

Fig. 2. Signal Source

3.2 The Describes of Hardware Circuit

The use of Altium Designer is to design of circuit principle diagram, mainly divided into the signal source, transmission line and probe part.

3.2.1 Signal Source Part

The system uses the signal source 100 MHZ oscillating sine wave generator.

3.2.2 Coaxial Transmission Line

When the signal source produces an incident signal V_0

$$V_0 = a \sin 2\pi \mathrm{ft} \tag{2}$$

Where a is the amplitude, f is the frequency of oscillation and t is the time since some arbitrary instant, then the reflected component at the same point, then the reflected component at the same point will be

$$V_0 = a \sin 2\pi ft + a\rho \sin \pi f (t - \frac{2l}{V_p}) \tag{3}$$

Where l is the length of the transmission line, V_p is the velocity of propagation of signals along the line and ρis called reflection coefficient .If the transmission line is made $V_p/4f$ in length, then this reduces to

$$V_0 = a \sin 2\pi ft - a\rho \sin 2\pi ft = a(1-\rho)\sin 2\pi f \tag{4}$$

At that point can be concluded that peak voltage as the biggest, at the same time on the transmission line, the superposition of incident wave and reflected wave form a standing wave, the standing wave of peaks and troughs in the ends of the coaxial cable.

Standing wave makes voltage amplitude of each point on a transmission line is different

$$\Delta U = 2a\rho = 2a\frac{(Z_l - Z_0)}{(Z_l + Z_0)} \tag{5}$$

Where z_l is the probe impedance, and z_o is that of the transmission line.
And the standing wave on a transmission line rate can be expressed as follows

$$\Gamma = \frac{1-\rho}{1+\rho} \tag{6}$$

In this way, through the measurement of standing wave ratio on a transmission line can achieve the goal of measuring soil volumetric moisture content.

3.2.3 The Probe

By China agricultural university, institute of information and electrical engineering professor Wang Yiming leading group of the successful development of science and technology based on stainless steel type standing wave ratio of the soil moisture sensor probe has three kinds of different specifications, probe diameter is 6 mm, probe spacing is 4 cm, around the probe length is 4 cm, intermediate probe lengths ranging 10,15,20cm to meet the requirements for measuring the soil of different depth.

This study was conducted through three probes collected, when the probe geometry must probe impedance depends on the soil dielectric constant, and the dielectric constant of the soil mainly affected by soil water content, so by measuring the dielectric constant of the soil can achieve the goal of measuring soil volumetric moisture content.

3.3 The Soil Sample Making

Take enough for calibration of soil spread in a cool, ventilated place dry, until can sieve, will be the first ground soil, with soil sieve pore diameter 3 mm, then use soil sieve pore diameter of 1 mm, for soil samples. The soil sample was flat out on the stainless steel tray, in 105 ~ 110°C oven drying to constant weight 24 hours a day, then remove them in the dryer to cool to room temperature(20°C).

From the dry soil to nearly saturated 10 kinds of quality water content of soil sample, use electronic weigh and drying quality for Gs soil samples respectively and water quality for Gw, the soil mixed with water and stir well, in a sealed plastic bag in the balance after 24 hours, stir again, then placed in sealed plastic bags rebalancing 24 hours, You can get quality water content for the θm series of soil sample.

3.4 Weighing Method to Measure Soil Column the Actual Moisture Content

From each soil column of about 50 g of soil samples, respectively in the aluminum box, with a sense of 0.1 g electronic weighing soil total quality w_o, drying in an oven at 105 ~ 110 °C for 24 hours to constant weight, cooling to room temperature(20°C) in the dryer, weighing dry soil quality w_s. The experimental data fill in the table 1.

3.5 Measuring the Sensor Output Voltage

Insert the SWR soil moisture sensor into each standard soil samples in the soil column, the output voltage V is measured with a digital voltmeter, and record in table 2, 3 ~ 5 in each soil column can be repeated measurement points.

Table 1. Soil moisture measurement data

NO.	Soil quality (g)	dry soil quality (g)	water quality unit (g/g)	weight volumetric (g/cm3)	water content (cm3/cm3)
1	13.63	13,56	0.005	1.244	0.006
2	14.12	13.65	0.034	1.254	0.043
3	16.81	15.71	0.070	1.257	0.088
4	18.35	16.18	0.134	1.249	0.125
5	20.34	17.58	0.157	1.251	0.197
6	18.20	15.04	0.210	1.451	0.305
7	26.34	21.93	0.201	1.511	0.302
8	24.26	19.36	0.253	1.399	0.354
9	29.35	22.51	0,304	1.402	0.426
10	27.15	20.23	0.342	1.337	0.458

Table 2. Sensor measurement data

NO.	The sensor output voltage					The average (v)
	1	2	3	4	5	
1	0.030	0.027	0.024	0.026	0.028	0.027
2	0.325	0.322	0.328	0.326	0.325	0.325
3	0.524	0.533	0.532	0.535	0.525	0.530
4	0.719	0.714	0.718	0.722	0.724	0.719
5	1.214	1.219	1.215	1.213	1.214	1.215
6	1.579	1.543	1.538	1.543	1.563	1.533
7	1.907	1.903	1.893	1.91	1.903	1.903
8	2.211	2.202	2.198	2.189	2.203	2.201
9	1.850	1.855	1.846	1.855	1.845	1.850
10	1.916	1.950	1.909	1.909	1.911	1.910

4 Results and Discussion

4.1 The Experiment Results Analysis

In Table 2 obtained from the average value of the sensor output voltage V as independent variables, and based on the actual volumetric water content of the obtained in table 1θm as the dependent variable, the calculated results in table 3, using least squares regression equation.

Table 3. Sensor experimental data

NO.	1	2	3	4	5
Θv (cm3/cm3)	0.006	0.043	0.088	0.125	0.197
V(v)	0.027	0.325	0.530	0.719	1.215
NO.	6	7	8	9	10
Θv (cm3/cm3)	0.305	0.302	0.354	0.426	0.458
V(v)	1.533	1.903	2.201	1.850	1.910

Elected to take linear function relation is the function of regression curve type, get regression curve shown in figure 3 as follows:

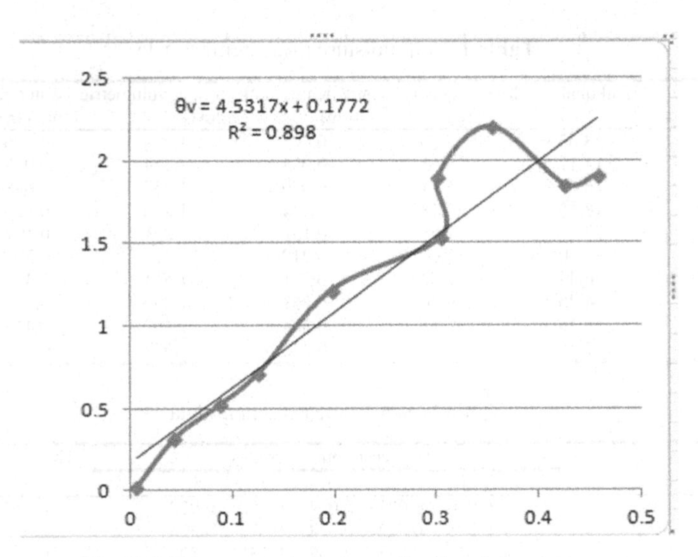

Fig. 3. Linear regression curve of soil moisture sensor

When choosing a cubic polynomial for regression function type, the regression curve is shown in figure 4:

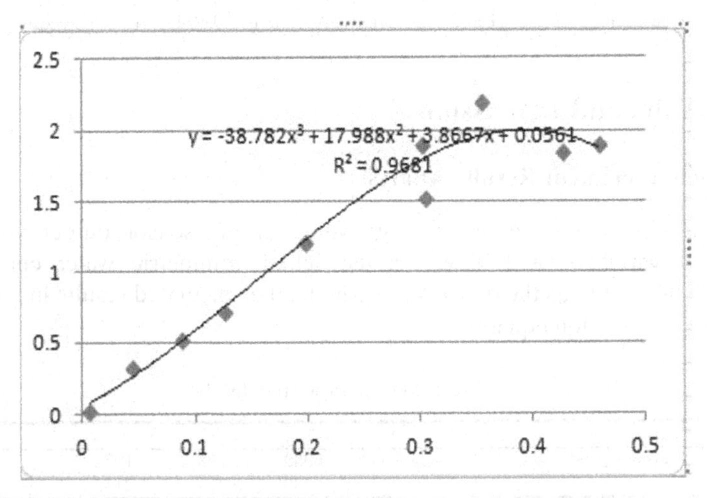

Fig. 4. Three regression curve of soil moisture sensor

Compare two curve, curve fitting of three times better than linear, linear regression equation of relative coefficient R2 = 0.9681, greater than the correlation coefficient of linear equation, so the fit of the regression equations of three times to far better than the linear regression equation, and has a high significance and accuracy.

5 Conclusions

By analyzing the data from experiment, proving that a standing wave method of the principle of water content of soil moisture sensor is a good relationship with the voltage value curve, is a kind of high performance soil volumetric water content measurement method.

1. This paper starting from the basic principle of electromagnetic wave interaction with the media, through the theoretical analysis of the characteristics of the soil dielectric demonstrates the dielectric method to quickly measure the soil moisture content is a kind of high sensitive wide applicable affected by spatial soil physical and chemical characteristics of spatial variation of small fast measurement method.
2. Design scheme of the method of soil column experiment, through the experiment and regression analysis, we establish the static mathematical model of SWR type moisture sensor. In the pattern measurement, the correlation between the measured value well.

Acknowledgment. This research is financially supported by the National Science and Technology Support (2012BAD35B07).

References

1. Bausch, W.C., Duke, H.R., Heemann, D.F.: assessing and mapping the plant nitrogen status in irrigation corn. Precision Agriculture 97
2. Knight, J.H.: Sensitivity of time domain reflectometry measurements to lateral variations in soil water content. Water Resources Research 28(9) (1992)
3. Roth, K., Schulin, R., Fluhler, H., et al.: Calibration of time domain reflectometry for water content measurement using a composite dielectric approach. Water Resour. Res. 26(10), 2267–2273 (1990)
4. Topp, G., Davis, J.: Measurement of soil water content using time-domain reflectrometry (TDR): a field evaluation. Soil Science Society of America Journal 49(1), 19 (1985)
5. Kwok, B.P., Nelson, S.O., Bahar, E.: Time-Domain Measurements for Determination of Dielectric Properties of Agricultural Materials. IEEE transaction on Instrumentation and Measurement IM-28(2), 6 (1979)
6. Mullins, C.E., Mandiringana, O.T., Nisbet, T.R., Aitken, M.N.: The design limitations, and use of a portable tensiometer. Journal of Soil Science 37, 691–700 (1986)
7. Wheater, H.S., Langan, S.J., Miller, J.D., Ferrier, R.C.: The determination of hydrological flow paths and associated hydrochemistry in forested catchments in central scotland. In: Forest Hydrology and Watershed Management Symposium, Vancouver (August 1987)
8. Bell, J.P.: Neutron probe practice. Institute of Hydrology Report No. 19 (1973)
9. Wellings, S.R., Bell, J.P., Raynor, R.J.: The use of gypsum resistance blocks for measuring soil water potential in the field. NERC Rep. No. 92 (1985)
10. Dean, T.J., Bell, J.P., Baty, A.J.B.: Soil moisture measurement by an improved capacitance technique. 1-sensor design and performance. Journal of Hydrology 93, 67–78 (1987)

11. Yanuka, M., Topp, G.C., Zegelin, S., Zebchuk, W.D.: Multiple reflection and attenuation of time domain reflectometry pulses: theoretical considerations for applications to soil and water. Water Resources Research 24(7), 939–944 (1988)
12. Curtis, L.F., Trudgill, S.: The measurement of soil moisture. Technical Bulletin of the British Geomorphological Research Group N. 13 (1974)

A New Method for Rapid Detection of the Volume and Quality of Watermelon Based on Processing of X-Ray Images

Ling Zou[1], Sun Ming[2], and Di Zhang[3]

[1] Key Laboratory of Modern Precision Agriculture System Integration Research, Ministry of Education,

[2] Key Laboratory of Agricultural Information Acquisition Technology, Ministry of Agriculture,

[3] College of Information and Electrical Engineering, China Agricultural University, Beijing 100083, China

1475126006@qq.com, sunming@cau.edu.cn
adalyme@hotmail.com

Abstract. Real-time online detection of fruit quality system has been applied to production practice because online testing and grading of fruits screening technology has matured. However, fruit size and quality online testing have always been difficult. Many detection methods of fruit size and quality are very complicated and time consuming, which cannot meet the needs of real-time detection. In this paper, a new method for rapid detecting small watermelon of volume and quality was based on the X-ray image processing. The volume and quality of the relevant model established the predicting outcome of coefficient of determination R^2 obtained were 0.9858 and 0.9922 respectively, showing that these two models have high correlation and predicting the volume and quality of watermelon accurately as well as laying the foundation for the watermelon online detection and classification.

Keywords: X-ray image, small watermelon, quality, rapid detection.

1 Introduction

High fruit production and export capacity need large online testing and grading equipment to replace manual work to save costs and increase efficiency. There are many researches about nondestructive testing technology based on the spectrum [1-4]. The fruit size will affect light permeability resulting in error detections, particularly such a large watermelon. The volume and quality greatly affects its quality testing, so we need to measure volume and quality to correct its model.

Depending on the detection principle, the main agricultural product quality nondestructive testing methods are acoustic detection technology, machine vision inspection technology, X-ray detection technology, electronic nose detection technology, the physical characteristics of the dielectric detection techniques, magnetic resonance imaging detection technology and near-infrared spectroscopy technique [5]. Currently,

© IFIP International Federation for Information Processing 2015
D. Li and Y. Chen (Eds.): CCTA 2014, IFIP AICT 452, pp. 731–738, 2015.
DOI: 10.1007/978-3-319-19620-6_79

these nondestructive testing technologies have applications and researches in different degrees pachyderm fruits internal quality detection. For example, Cooke [6] proposed the Young's modulus to evaluate grade of watermelon maturity, and Chuma [7] studied the relationship between watermelon fruit resonance characteristics and quality with the sound plus vibration mode. Although many scholars at home and abroad studied the internal quality of watermelon based on the acoustic characteristics and some detection effect is ideal, these results largely confined to laboratory testing since there are so many factors that affect the actual testing process. Iwamota [8] invented a method and apparatus for detection of fruit sugar content based on near infrared transmission spectroscopy NDT. As for Near-infrared spectroscopy, there are some challenges needing to solve in terms of detection of thick-skinned fruits. Since Roentgen discovered X-rays in Germany at 1895, X-ray has made considerable development in many areas after long_term efforts, such as medical CT [9], environmental monitoring, forestry and food testing. Compared with other detection techniques, X-ray detection technology has more advantages in agricultural products on the internal quality testing and evaluation. In recent years, X-rays detection of internal defects in agricultural products and foreign substance examination has become a hot research.

The traditional methods of measuring the volume of the fruit are vertical and horizontal diameter law, surface area and drainage method [10], in which the method of drainage is more commonly used, but the method of its operation is not suitable for real-time online testing. Currently, non-destructive testing techniques have been widely used in agricultural product quality testing, such as getting fruit volume by processing the image of the visible region [11]. Generally, those techniques conclude more complex algorithms and impact the online testing speed, meanwhile, it is difficult to obtain internal information accurately because of its large volume of watermelon. The researches of watermelon nondestructive testing methods of quality are different with other ripe fruit, and the related reports are also uncommon in non-destructive testing of its volume and quality indicators. This paper presents a method based on X-ray images which can detect the volume and quality of watermelon quickly and effectively.

2 Materials and Methods

2.1 Materials

The materials were picked from a watermelon shed Yanqing County Beijing after pollination. Ten samples were taken every two days during growing period between 21 days to 37 days and a total of 94 samples of the different growing season were obtained, which the samples growth (growth is the difference between picking and pollination period) distribution was shown in Table 1. Volume was calculated with water displacement method and quality was measured by JJ500 Precision electronic balance, and collecting results of box diagram distribution were shown in Figure 1. Maximum volume of 94 samples was 1614 cm^3, the minimum was 398 cm^3, with an average of 765.0969 cm^3. With a big difference between maximum and minimum sample and its large range of variation, the experiment had a strong representation.

Table 1. Sample growing distribution table

After pollination, the growing season (days)	Number of samples (a)	After pollination, the growing season (days)	Number of samples (a)
21	12	31	11
23	10	33	10
25	11	35	10
27	11	37	9
29	10		

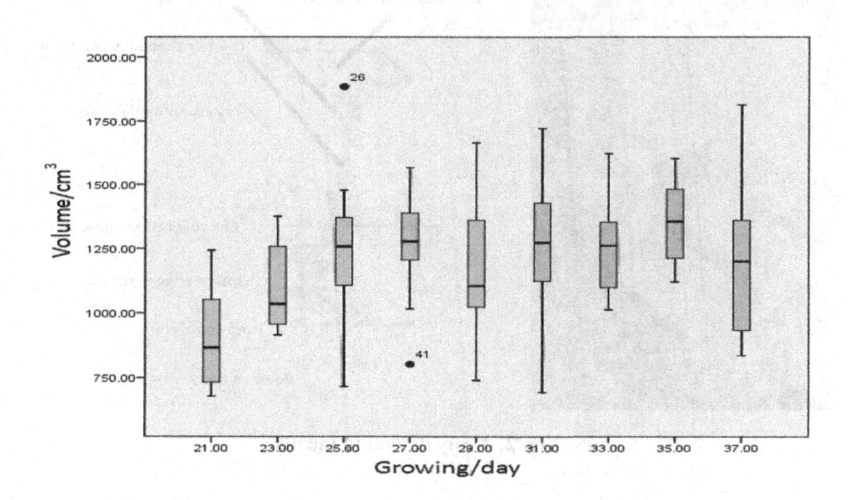

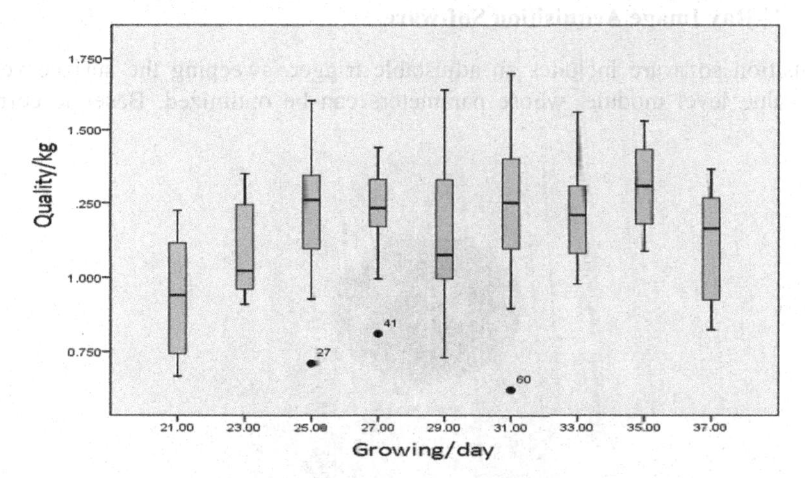

Fig. 1. Volume and mass distribution

2.2 Experiment Device

Experimental devices were designed and manufactured, including the X-ray generator, a linear array detector, conveyor system, image acquisition card, host computer, image processing software and etc al (as shown in Figure 2) . The conveyor belt transmitted test samples, to meet test requirements for obtaining a sharp image, after several transformation of voltage and current experiments, the information could be obtained relatively rich images by adjusting the X-ray source voltage to 80 kV and current to 1 mA. Only the detector integration time matched conveyor speed, did we get the real picture. By adjusting the detector integration time to 2ms, conveyor belt speed to 17.8cm / s, undistorted image with vertical and horizontal proportion could be obtained accurately.

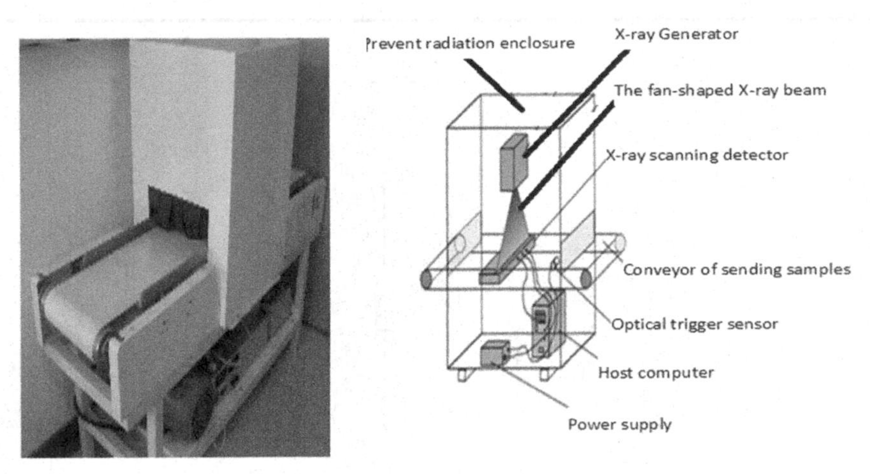

Fig. 2. X-ray detecting means

2.3 X-Ray Image Acquisition Software

Acquisition software includes an adjustable trigger, sweeping the surface velocity, gray value level module, whose parameters can be optimized. Baseline correction

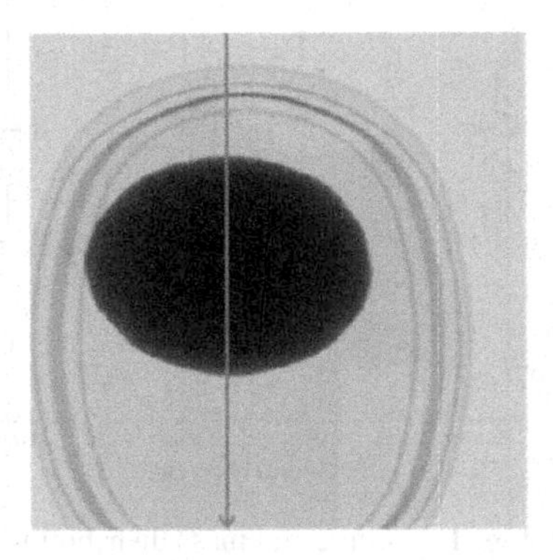

Fig. 3. X-ray image

X-ray image was collected as shown in figure 3, where in the black part belonged to watermelon. Image captured was saved as two formats, and one was an image file format which gray scale range was 0-255, the other document format was text which gray scale range was 0-16383. It was convenient to make intuition analysis with image and text facilitated to accurate calculation. Data processing in this article used text document format for its greater range of gray and the result was accurate.

2.4 X-Ray Image Preprocessing

For watermelon's round shape, if it was placed directly on the conveyor, belt would shake transport, resulting in the acquisition of image distortion. In order to ensure access to get clear, uniform image, homemade base would be fixed and all samples would be collected images using the same fixed base experiment. All samples were placed upright, collecting vertical images. To remove the background noise on the base of image processing, it was necessary to determine the threshold value, which distinguishes watermelon itself and background in the image.

To determine threshold value, we needed to find the centroid of watermelon first, gray value curve was extracted through vertical line of the watermelon centroid image which was shown in Figure 4. The abscissa represented the number of pixels from top to bottom through image centroid vertical lines of the image, and the ordinate was the vertical line gray value of the pixel. Difference was much apparent by comparing the difference of the curve between the sample and background and the threshold value was 1600 at the knee, which could effectively remove background interference.

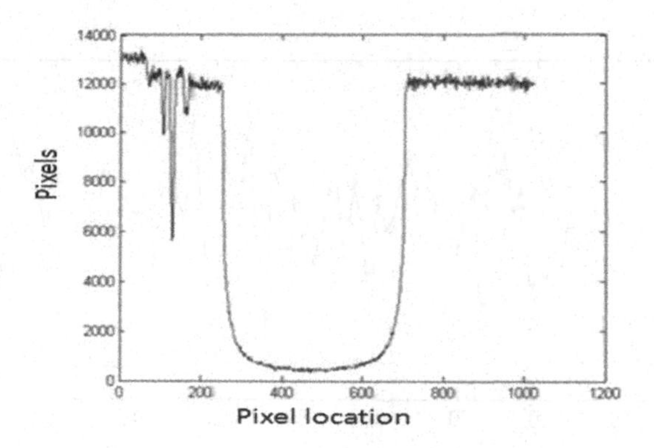

Fig. 4. Determining a threshold value

3 Results and Discussion

Using text document format of 0-214 gray scale to calculate the data, the number of pixels whose gradation value summed was below 1600 of each sample. Indicators statistical results of 94 sample's mass, volume, and the number of calculated pixels were shown in Table 2.

Table 2. Indicators statistical results

Index	Distribution	Mean	Standard deviation
Volume (cm^3)	678-1884	1190.3351	263.32813
Weight (kg)	0.621-1.697	1.16091	0.234394
Pixels	74198-207392	147067.8191	30312.43563

The three sets of index data were for standardization (raw data minus the average of the data and then divide the standard deviation), then the data were displayed on the same graph (as shown in Figure 5). The dotted line represented pixels change trend of 94 samples, and chain line represented volume change trend of the true value, and the linear represented change trend of the true mass. The discovery was that the number of pixels calculated with the method in this article had a good linear relationship with the true value of watermelon volume and quality. When penetrating the material, X-rays were in line with the index variation. While the index trend lines within a certain small range could be approximated as a straight line, the number of pixels calculated by the X-rays image had a good correlation with quality and volume. It could be established linear correlation model for predicting the value of the volume of watermelon.

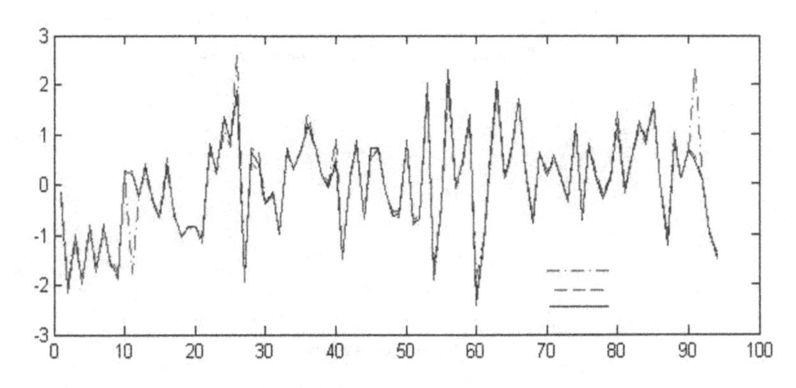

Fig. 5. Index change trend

As the using of the drainage volume method for measuring volume in this measurement, manual operation resulted in abnormal samples. In which 21-11, 25-4, 37-6, that was No.11 sample whose growing season was 21 days, the No. 4 sample whose growing season was 25 days, the No. 6 sample whose growing season was 37 days. The remaining 91 samples after removing outliers were for a linear regression, and quality was measured using an electronic balance with high accuracy so there are were no obvious abnormalities samples. After linear regression of 94 samples, modeling results of the volume and quality were shown in Figure 6, where volume forecasting model $R^2 = 0.9858$, quality prediction model $R^2 = 0.9922$, showing that this two prediction models had been relatively high correlation respectively.

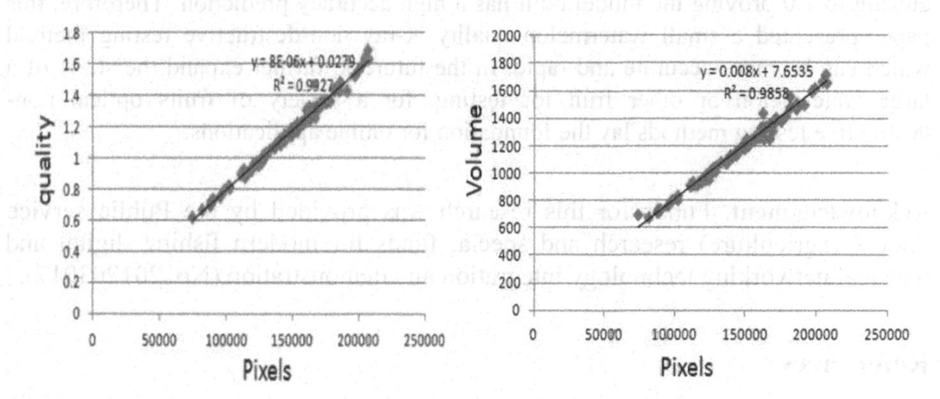

Fig. 6. Linear regression models

Small watermelon mass and volume values were calculated by linear regression equation, where the calculated value and the true value had a linear relationship shown in Figure 7. From the linear correlation equation y = k x + b ,k coefficients were 1.0308 and 0.9826 respectively closing to 1, and the accurate prediction results proved that it was feasible to use the number of pixels by the X-ray image to estimate the volume of watermelon.

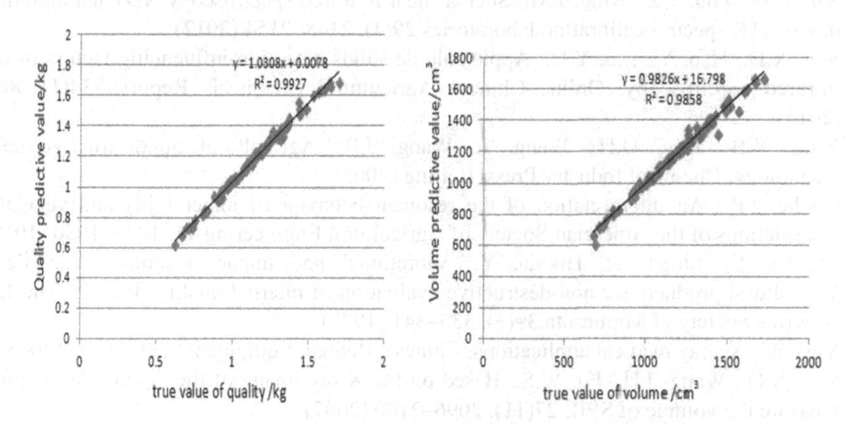

Fig. 7. The linear regression model of predictive value and the true value

4 Conclusions

A method of measuring the volume and quality of small watermelon based on X-ray image processing was proposed, which built the best predictive model. Since the watermelon volume and quality determination coefficient of the linear regression model created reached 0.9927 and 0.9858, respectively, indicating that the model has a high correlation. In the predicting the value and the true value of the linear correlation equation y = k x + b k coefficients are 1.0308 and 0.9826, respectively, closing to 1.0 proving the model built has a high accuracy prediction. Therefore, this paper presented a small watermelon quality X-ray non-destructive testing method which can be more accurate and rapid. In the future to further expand the study of a large watermelon or other fruit for testing, for a variety of fruits optical non-destructive testing methods lay the foundation for online applications.

Acknowledgment. Funds for this research was provided by the Public service sectors (agriculture) research and special funds for modern fishing digital and physical networking technology integration and demonstration (No. 201203017).

References

1. Bobelyna, E., Serbanc, A.S., Nic, M.: Postharvest quality of apple predicted by NIR spectroscopy. Study of the effect of biological variability on spectra and model performance. Postharvest Biology and Technology 55, 133–143 (2010)
2. Chew, L., Ma, X.G.: Compare several commonly used methods for measuring the volume of the fruit of Chinese fruit (2), 44-45 (1995)
3. Liu, Y.D., Chen, X.M., Ou, Y., Ai, G.: Non-destructive measurement of soluble solid content in Gannan navel or anges by visible/ near-Infrared spectros copy. Acta Optica Sinica 28(3), 478–481
4. Xin, R.G., Zhu, Z.Z., Ning, J.M.: Such as near in frared spectroscopy NDT tea ingredients based ATR spectral calibration laboratories 29(4), 2148–2154 (2012)
5. Sun, X.D., Hao, Y., Liu, Y.D.: Apple soluble solids detection influencing factors of near-infrared spectroscopy. Online Chinese Agricultural Chemicals Reported 34(1), 86–90 (2013)
6. Ying, Y.B., Han, D.H., Wang, J., Wang, J.P.: Agricultural nondestructive testing technology. Chemical Industry Press, Beijing (2005)
7. Cooke, J.R.: An interpretation of the resonant behavior of intact fhjits and vegetables. Transactions of the Amercian Society of Agricultural Engineering 15, 1075–1080 (1972)
8. Chuina, Y., Shiga, T., Hikida, Y.: Vibrational and impact response properties of agricultural products for non-destructive evaluation of internal quality (Part 1). The Japan America Society of Minnesota 39(3), 335–341 (1977)
9. Yan, W.: X - ray medical applications. Chinese Medical Equipment 2(3), 31–32 (2005)
10. Sun, X.D., Wang, J.H., Fu, W.S.: Based on the X-ray image of the Apple online quickly measure the volume of SPIE 27(11), 2096–2100 (2007)
11. Ye, A.O., Zhang, X.W.: SiPLS algorithm based on near-infrared spectroscopy pear soluble solids content. Spectroscopy Laboratory 30(1), 68–72 (2013)

An Improved Method for Image Retrieval Based on Color and Texture Features

Jun Yue[1], Chen Li[2], and Zhenbo Li[1,*]

[1] College of Information and Electrical Engineering,
LUDONG University, YanTai, 264025, China
[2] College of Information and Electrical Engineering,
China Agricultural University, BeiJing, 100083, China
yuejun509@gmail.com, yuejuncn@sina.com

Abstract. With the development of technology and Internet,people need more and more information.The carrier of information is also changed from the original text into the current image, video, etc. Image retrieval plays an increasingly important role in the field of study. This paper is a study of image retrieval based on color and texture features, the main algorithms include color histogram,color moment,gray level co-occurrence matrix(GLCM) and Tamura texture features,etc.Then it made further improvements that combined color histogram and Tamura texture features method (we made Tamura texture retriveal again based on color histogram firstly) based on search results.Experimental results demonstrate that the rates of retriveal by comprehensive approach is higher than using a single method.

Keywords: Color histogram, color moment, gray level co-occurrence matrix, gray-scale diffience statistics, Tamura texture features.

1 Introduction

With the development of multimedia network technology,we can get information from many kinds of methods. Besides text,currently digital imaging is widely used in various areas including medical science,industrial manufacturing,aerospace engineering,remote sensing,etc. In order to effectively manage and retrieve image, the traditional methods are to mark the image with text and create relation between them. But sometimes it is difficult to describe the complete meaning of the images using only a few key words, and it is subjectivive(people have different descriptions of the same image).Using the traditional methods lead to differences between input keywords and the keywords in database,so users won't find the target image. To overcome this disadvantage, people proposed a content-based image retrieval [1] which includes extracting the features of the image, matching the target image, and the images in the database, and then we can search the

* Corresponding author.

© IFIP International Federation for Information Processing 2015
D. Li and Y. Chen (Eds.): CCTA 2014, IFIP AICT 452, pp. 739–752, 2015.
DOI: 10.1007/978-3-319-19620-6_80

target image. Image features, including text features and visual features , visual features can be described by color, texture and shape.

Color is the most straightforward descriptor of the images, Swain [2] first proposed a method using the color histogram feature to search images. Stricker,took the first three-moment matrix of colors to characterize the content of the image. Texture is an important feature of the image, and extracting the texture feature is also an important way to retrieve image. In 1970s, Haralick proposed a method gray level co-occurrence matrix; Tamura [3] proposed a form of expression texture including coarseness, contrast, directionality, linelikeness, regularity, and roughness.

In this paper, we retrieve images from both color and texture. The main algorithms include color histogram,color moment,GLCM,gray-scale diffidence statistics and Tamura texture features.The recall ratio and precision were used as the performance of different algorithms. Based on the experiments,we also proposed a method combining the color histogram and Tamura texture features. The results demonstrated that the rates of retrieval by comprehensive approach is higher than only using a single method.

This retrieval process is divided into five modules, and the procedure is as follow:

1 Create an image database
2 Extract the feature of the image, form a feature library
3 Input search image and extract the feature
4 Feature matching
5 Output query results and compare the accuracy

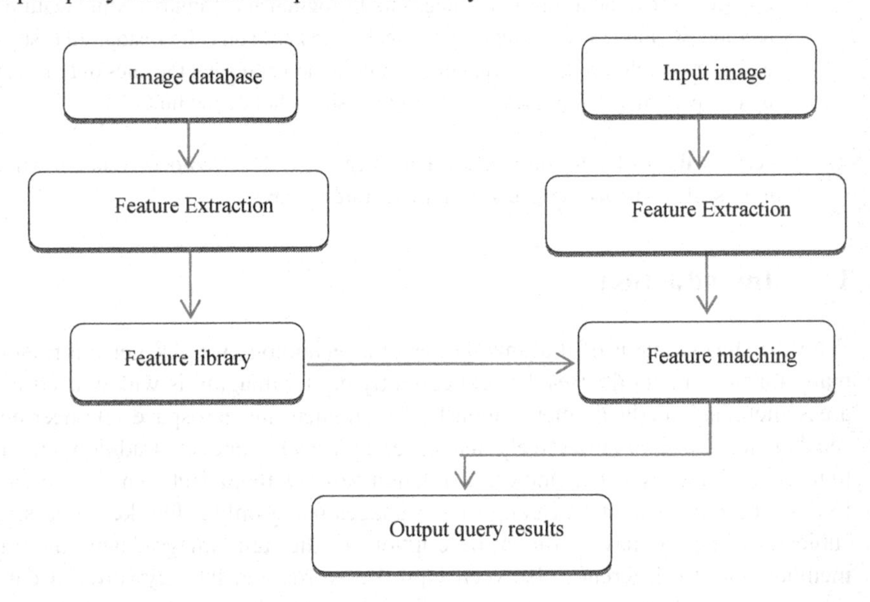

Fig. 1. Image retrieval flowchart

2 Feature Extraction

2.1 Color Feature

Color is the most widely used visual feature of image retrieval, because color is highly related to objects and scenes of image. The common methods that we used to describe-the color feature contain color moment, color correlogram, color histogram [4], etc.

(1) Color Moment
Color moment is based on numerical methods and it can describe the distribution of color by calculating the moment.Since the color distribution information mainly concentrated on the lower order moments, so first moment (mean), second moment (variance) and third moment (skewness) are commonly used methods to represent the color distribution of image.The mathematical formula is as follow:

First moment:

$$\mu_i = \frac{1}{N}\sum_{j=1}^{N} p_{ij}$$

Second moment:

$$\sigma_i = \left[\frac{1}{N}\sum_{j=1}^{N}\left(p_{ij} - \mu_i\right)^2\right]^{1/2}$$

Third moment:

$$S_i = \left[\frac{1}{N}\sum_{j=1}^{N}\left(p_{ij} - \mu_i\right)^3\right]^{1/3}$$

Where p_{ij} represents the probability of the color image of the i-th color component of the channel appears as the gray pixel j, N represents the number of pixels of the image.

(2) Color Correlation
The color correlogram is another expression of the distribution of image color.It reflects the spatial correlation between each two neighboring pixels, as well as the correlation between local pixel distribution and overall pixel distribution.It is easy to calculate and has good effect.

2.2 Texture

Texture is an attribute of an image area, it needs to contact the context because the texture of one pixel is meaningless. Texture reflects the relationship between the gray value of image pixels. Some common texture descriptors are introduced as follow.

(1) GLCM [5]
Before calculating GLCM,we should preprocess images. The grayscale of images is generally large,so it can be time-saving and reducing the amount of computation greatly if we compress grayscale first.

GLCM describes texture by doing a study of the spatial correlation properties of gray, and it is based on second-order gray level co-occurrence matrix statistical features. It describes as follow:

Entropy:

$$\mathrm{H} = -\sum_i \sum_j p(i,j) \log\left(p(i,j)\right)$$

Entropy is a measurement of the amount of information of images.It shows the texture complexity and non-uniform of images. When the values of GLCM are uniformly distributed, the entropy is large.

Energy:

$$\mathrm{E} = \sum_i \sum_j p(i,j)^2$$

Energy is the sum of squares of the element values of GLCM.It reflects the degree of gray uniformity and texture coarseness. If the values of GLCM concentrated in a piece, the energy value is great; if the values distribute evenly, the energy value is small.

Moment of inertia:

$$\mathrm{I} = \sum_i \sum_j (i-j)^2 p(i,j)$$

The moment of inertia is contrast that reflects the image clarity and the degree of the groove depth. The more pixels in large contrast, the greater moment of inertia.

Correlation:

$$U = \frac{\sum_i \sum_j ijp(i,j) - \mu_x \mu_y}{\sigma_x \sigma_y}$$

Correlation measures the degree of similarity on GLCM elements in a row or column direction.So it reflects the relationship between the local gray of the image. When the matrix element values are average, the correlation is large, whereas the matrix element values vary greatly, the correlation value is smaller.

(2) Gray-Scale Diffience Statistics
Assumed that the gray value of any point (x, y) of the image is g(x, y).A point deviating from (x,y) is (x+Δx,y+Δy), the gray value of it is g(x+Δx,y+Δy). The difference between gray values is:

$$g_{\Delta}(x,y) = g(x,y) - g(x+\Delta x, y+\Delta y)$$

We can get the histograms of $g_{\Delta}(x,y)$ and texture features by the formula. In this paper, texture features are the mean, contrast and entropy. Gray value reflects the smooth degree of texture, the smaller the mean, the more smooth the texture. Contrast reflects the clarity of image, the greater the contrast, the more easily distinguish the image.Entropy reflects image texture roughness,the higher the value, the more coarse the texture.

2.3 Comprehensive Color Histogram and Tamura Texture Feature

Color histogram is the simplest and most common image features, which describes the proportion of different colors in the whole image and has the invariance of rotation, translation and scale.But it is not concerned with the spatial position of each color.Therefore,diffierent objects with same color histogram might be determined as one target when using color histogram based methods.Ultimately,the error of retrieval results increases significantly. It is suitable to use color histogram based methods when images are difficult to segmented or the spatial distribution of color is not important.

Based on the texture of visual perception, Tamura et al.proposed an expression of texture features. Calculating the six components of texture which is corresponding to the texture features of the six attributes in psychological [6]: coarseness, contrast, directionality, regularity, linelikeness, and roughness. Typically, coarseness, contrast and directionality is sufficient to express the texture features of images. Therefore,we get the roughness, contrast and directionality of the image in the database as eigenvectors, calculating the Euclidean distance, matching the image and searching the target image. The following describes the mathematical expression of these three attributes[7]:

(1) Coarseness

First, taking a sliding window(the size is $2^k \times 2^k$ per pixel) scan image, calculating the mean of gray value of all the pixels falls in the window at each slide.Calculating the average intensity difference of pixel in horizontal direction and vertical direction which falls on overlap window, the formula is:

$$E_{k,h}(x,y) = |A_k(x+2^{k-1}, y) - A_k(x-2^{k-1}, y)|$$

$$E_{k,v}(x,y) = |A_k(x, y+2^{k-1}) - A_k(x, y-2^{k-1})|$$

We can set the optimum size of the window by K value which is the maximal of E.The size is $S(x,y) = 2^k$.Then the coarseness is the average of all $S(x,y)$ that we calculate, the formula is:

$$F_{crs} = \frac{1}{m \times n} \sum_{i=1}^{m} \sum_{j=1}^{n} S(i,j)$$

(2) Contrast

$$F_{con} = \frac{\sigma}{\partial_4^{1/4}}$$

Contrast is the distribution of gray values of all the pixels by statistics. Defined by the formula $\partial_4 = \mu_4 / \sigma^4$,and μ_4 means fourth moment.

(3) Directionality

Directionality is determined by calculating the gradient of the neighboring pixels. The formula of mode and direction of the gradient is:

$$|\Delta G| = (|\Delta_H| + |\Delta_V|)/2$$

$$\theta = \tan^{-1}(\Delta_H / \Delta_V) + \pi / 2$$

Δ_H represents the change in horizontal direction and Δ_V is vertical.

As we all know,color [8] describes the surface properties of scene in the image or block as a global features.Because of color is not sensitive to the change of direction and size of the image,it can not capture the local features greatly. Texture reflects the relationship between the gray value of image pixel. The global regularities and local irregularities of the image may be certain or random.From this experiment we can get that single image retrieval method has advantages,but there are also various shortcomings, resulting in retrieval efficiency is not very high.Therefore,in order to remedy these shortcomings, this paper presents a comprehensive method of color histogram and Tamura texture feature.

The method that integrates color histogram and Tamura texture feature can search the image fully. The main principle is:based on the color histogram,we make secondary use of retrieval by using Tamura texture feature.We compare the similarity of texture after color,judging by the two indexes so that the experimental result is more accurate and has higher retrieval efficiency. Experiment shows that the retrieval rates of comprehensive color histogram and Tamura texture feature is higher than single retrieval method.

3 Image Similarity Measure

The measurement of image similarity plays an important role in image retrieval,and it affects the accuracy of image matching. When matching the image features, the common method calculating the distance between two points. The methods of calculating the distance include Absolute distance, Secondary Type distance and Euclidean distance [9]. The Euclidean distance is a commonly used method which has the advantage of low complexity, so this paper measure the similarity of the image by calculating the Euclidean distance. The formula is as follows:

$$d = \left[\sum_{i=1}^{n} (a_i - b_i)^2 \right]^{1/2}$$

4 The Experimental Results

First, create an image database.We choose ten kinds of flowers and each of them are ten. It's pink Chinese Rose,orange Chinese Rose, Narcissus, Lotus, Magnolia,Blue Enchantress, Sunflower,green Rose,Peach Blossom and Winter Jasmine. Sequentially numbered from 1 to 100, the size is 128 * 128pix and the resolution ratio is 72 * 72.The same kinds of flowers have similarities but is not higer,so the search process may has large deviations and needs to be improved.

Each image has been searched in five ways in the experiment, the results as the following examples:

Selecting the retrieved picture:

Fig. 2. The retrieved image

(1) Color Histogram

Fig. 3. The results of color histogram

(2) Color Moment

Fig. 4. The results of color moment

(3) Gray-Scale Diffience Statistics

Fig. 5. The results of gray-scale diffience statistics

(4) GLCM

Fig. 6. The results of GLCM

(5) Tamura Texture Feature

Fig. 7. The results of Tamura texture feature

As can be seen from the above results, the number of images that retrieved by five ways is:7,3,3,4,8. So for figure 2, the best way of retrieval is Tamura texture, followed by color histogram, the worst way are color moments and gray-scale difference statistics.But this is only one picture in database and is not representative,then we search the remaining 99 images and add up data.

5 Performance Evaluation

5.1 The First Method of Evaluation

We need to have some indexes to measure if retrieval is accurate and how much the degree of accuracy is. The common indexes are recall ratio and precision ratio[10].The definition is as follows:

Recall ratio: n/N
Precision ratio: n/T

n refers to the number of images in the query results are associated with the key figure [11]. N refers to the number of images in the test set are associated with the key figure.And T refers to the number of images returned by the query. According to the choice and the number of image database,we know N is 10, T is 20.

Experiments were conducted on each image(total 100) with five kinds of retrieval methods(Color Histogram, Color Moment, GLCM, Gray-scale diffience statistics, Tamura texture features). We get a total of five sets of data and calculate the average recall and precision ratio, the following is a table:

Table 1. The experimental results

Retrieval method	Average recall ratio	Average precision ratio
Color histogram	66%	33%
Color moment	37.6%	18.8%
Gray-scale difference statistics	31.3%	15.7%
GLCM	36.2%	18.1%
Tamura texture feature	37.8%	18.9%

The analysis of the results:from the table 1 we can get the best retrieval method is color histogram with average precision ratio of 33 and the worst is gray-scale difference statistics with 15.7%.

In addition, the system response time is also a index of retrieval results. From the experimental we can know the fastest retrieval method are color histogram and color moments with only one second. The gray differential statistics with a time of about 10 seconds followed by GLCM with a time of 20 seconds and the longest is Tamura texture with more than one minute. Therefore, for the purposes of retrieval efficiency, color histogram and color moments have the highest efficiency.

In summary, using color histogram for image retrieval can achieve better results in this image database of experiment.

5.2 The Second Method of Evaluation

In addition to calculating recall and precision ratio, we can collate data and get a line chart so that we have seen what is the better retrieval method. In the 100 samples we selected 50 pictures with equal intervals(taking five images of each kind of flower),a line chart is as follow:

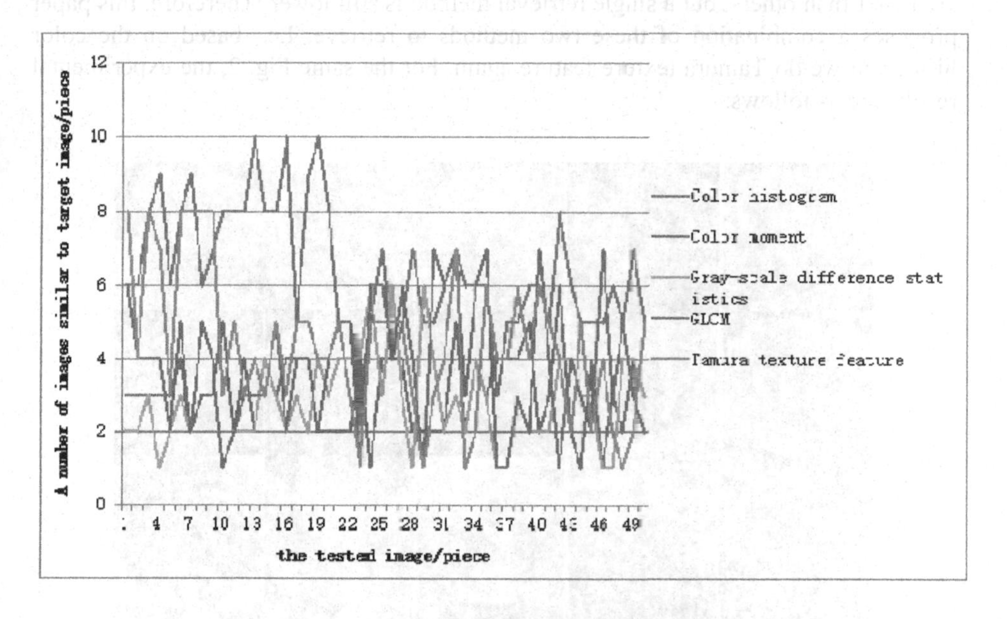

Fig. 8. A line chart of search results

Calculating the average of this 50 pictures with five method, we get the following table:

Table 2. The average number of retrieval with different method

Retrieval method	Average numbers
Color Histogram	6.82
Color Moment	3.64
Gray-scale difference statistics	3.02
GLCM	3.32
Tamura Texture Feature	3.72

From Fig. 8 and Table 2 we can see that color histogram is better than others especially for the former 20 pictures with the average number of retrieval 6.82. Tamura texture feature is the next one. The overall line trend of gray-scale difference statistics located below the other four retrieval method with a poor result, the average number of sheets to retrieve is only 3.02. It is also consistent with the experimental results in Table 1.

6 Improved Methods

As can be seen from the above experiment Tamura texture feature and color histogram are better than others, but a single retrieval method is still lower. Therefore, this paper proposes a combination of these two methods to retrieve, I.e., based on the color histogram we do Tamura texture feature again. For the same Fig. 2, the experimental results are as follows:

Fig. 9. The result of a combination of color histogram and Tamura texture feature

Analyzing by contrast:for the same Fig. 2, it has 7 images of former 10 are the same kinds of flowers with targeted image only by color histogram,just show as Fig. 3. It has 5 images of former 10 are the same kinds of flowers only by Tamura texture feature, just show as Fig. 7.

But the combination of color histogram and Tamura texture feature is better, it has secrched 8 images that are similar with targeted image showing as Fig. 9.

In order to make it representative,we selected ten images (numbered 11,21, ... 81, 91)from database to test and get the number of similar images of the former 10,the Table is as follows:

Table 3. The comparison of single retrieval method and integrated method

The image number	Clolr histogram	Tamura texture feature	Integrated method
1	7	5	8
11	7	2	6
21	4	5	6
31	3	1	7
41	3	1	5
51	8	2	4
61	7	2	5
71	6	2	7
81	5	2	7
91	5	2	6
The average number	5.5	2.4	6.1

To make the results more clear, we draw a line chart by the above table:

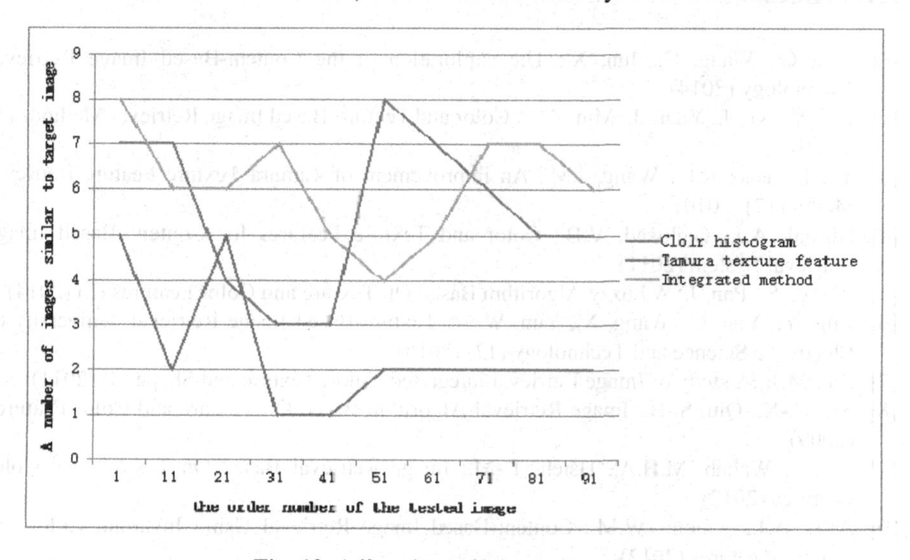

Fig. 10. A line chart of improved method

From Table 3 and Fig. 10, for the numbered images 51 and 61,we can see that the integrated method is worse than color histogram, but is better than Tamura texture feature. Overall, the efficiency of a single retrieval method is relatively poor, and integrated color histogram method and Tamura texture feature enables retrieval rate increased.

7 Summary and Expectation

This paper is a study of image retrieval based on color and texture features, the five methods we used include color histogram,color moment,GLCM, gray-scale difference statistics,and Tamura texture features.Subsequently we do further comparison after calculating the recall ratio and precision ratio. From this experiment we can see that color histogram is better than other methods whether on time or the accuracy. Tamura texture feature provides better results than other methods,but it is not suitable for real-time applications because it is time-consuming. The preliminary results showed that all methods based on single feature cannot provide acceptable performance,and thus a retrieve model was proposed using both color-histogram and texture.The experimental results showed that the retrieval accuracy of the improved model are increased.

In future, the study will be focusing on three aspects. In order to avoid effects caused by irregular images, all will be automatically cropped and saved into the database. The retrieval efficiency on large-scale database needs to be further studied and optimized. Additionally, the selection of similarity measurement is crucial to improve the retrieval accuracy, and various similarity calculations should be tested.

References

[1] Wei, G., Wang, C., Jun, X.: The exploration of the Content-Based Image Retrieval Technology (2014)
[2] Hui, Y., Xu, J., Yuan, J., Min, Z.: A Color and Texture-Based Image Retrieval Methods (5) (2009)
[3] Yu, B., Hao, R.L., Wang, J.M.: An improvement of Tamura Texture Feature Retrieval Method (7) (2010)
[4] Nirmal, A.J., Gaikwad, V.B.: Color and Texture Features for Content Based Image Retrieval System (2011)
[5] Wang., S., Pan, J.: A Fuzzy Algorithm Based On Texture and Color Features (2) (2014)
[6] Jing, Y., Yan, C., Wang, X., Yun, W.: A Texture-Based Image Retrieval. University of Electronic Science and Technology (12) (2010)
[7] Zhi, M.J.: A study of Image Retrieval integrated Color. Texture and Shape (5) (2011)
[8] Yu, C.-X., Qiu, S.-B.: Image Retrieval Algorithm Based On Texture and Color Features (2009)
[9] Su, C., Wahab, M.H.A., Hsieh, T.-M.: Image Retrieval Based On Texture and Color Features (2012)
[10] Afifi, A.J., Ashour, W.M.: Content-Based Image Retrieval Using Invariant Color and Texture Features (2012)
[11] Wang, Y.: The Research and Implementation Based on Color and Texture Image Retrieval (2011)
[12] Fang, Y.: Image Retrieval Research Based On Texture and Color Features (2012)

Author Index

Printed in the United States
By Bookmasters